Modern processing plant facility

MINERAL PROCESSING PLANT DESIGN

Co-Editors

Andrew L. Mular
Professor of Mineral Engineering
University of British Columbia
Vancouver, B.C.
Canada

Roshan B. Bhappu
Vice-President & General Manager
Mountain States Research and
Development, Inc.
Tucson, Arizona

This volume was originated by the Mineral Processing Division of the Society of Mining Engineers of AIME to serve as a practical textbook of current plant design and practice.

Published by
Society of Mining Engineers
of the
American Institute of Mining, Metallurgical, and Petroleum Engineers, Inc.
New York, New York • 1978

Printed in the United States of America
by Port City Press, Inc., Baltimore, Maryland

Library of Congress Catalog Card Number 77-26531

ISBN 0-89520-251-4

MINERAL PROCESSING PLANT DESIGN SYMPOSIUM

Cochairmen

Andrew L. Mular
University of British Columbia
Vancouver, B.C.
Canada

Roshan B. Bhappu
Mountain States Research and
Development, Inc.
Tucson, Arizona

Executive Editorial Committee

Dennis K. Mortensen
Anamax Mining Company
Sahuarita, Arizona

Gerald V. Jergensen
Barber-Greene Company
Tucson, Arizona

F. Milton Lewis
Mountain States Research & Development
Tucson, Arizona

John H. Bassarear
Cyprus Pima Mining Company
Tucson, Arizona

SESSION CHAIRMEN

Financial and Feasibility Studies

J. T. Reagan
United Bank of Denver, N.A.
Denver, Colorado

W. P. Dyrenforth
Ortloff Minerals Services Corp.
Golden, Colorado

Sampling and Testing

D. K. Mortensen
Anamax Mining Company
Sahuarita, Arizona

G. V. Jergensen
Barber-Greene Company
Tucson, Arizona

Crushing and Grinding

L. E. Lipphardt
Allis-Chalmers Corp.
Tucson, Arizona

C. B. Dugger, Jr.
Minnesota Electric
Tucson, Arizona

Screening-Classification-Gravity Separation

G. M. Meisel
Consultant
El Cerrito, California

C. F. Lagergren
Applications Corp.
Tucson, Arizona

Flotation-Magnetic-Electrostatic Separation

H. B. Treweek
Wemco Division, Envirotech Corp.
Sacramento, California

J. H. Bassarear
Cyprus Pima Mining Company
Tucson, Arizona

Solid-Liquid Separation and Instrumentation

S. M. Moos
Pincock, Allen and Holt, Inc.
Tucson, Arizona

C. A. Richardson
Joy Manufacturing Company
Tucson, Arizona

Auxiliary Operations

I. L. Sandberg
Galigher Division, Baker Oil Tools, Inc.
Tucson, Arizona

J. Miller
Miller Sales
Tucson, Arizona

Engineering-Construction and Future Plant Design

T. Henderson
Asarco, Inc.
Tucson, Arizona

J. M. Keane
Mountain States Mineral Enterprises, Inc.
Tucson, Arizona

FOREWORD

This volume has been published by the Mineral Processing Division of the Society of Mining Engineers of AIME. As the reader progresses through each chapter, one important factor will tend to stand out, namely, the requirement to decrease costs in all phases of plant design and practice. In this regard, progress has been real—but only because of gradual improvements in design philosophy and plant practice over the past 40 years. In all, these improvements have been substantial. Primary crushers now handle 1.2-m (4-ft) diam ore; cone crushers draw up to 522 kW (700 hp); ball mills with wraparound motors that draw 5968 kW (8000 hp) can be designed; flotation cells are 16 980 L (600 cu ft) and more in volume; semiautogenous mills are in use; computers are being used for design purposes. These, included with many innovative design features and concepts, along with radically new operating philosophies associated with instrumentation and automation, have (in terms of relative dollars) been responsible for cost reductions that permit economical treatment of extremely low-grade ores. And this, despite stiff environmental policies and taxation laws.

The primary objective of this text is to document the current state of the art in mineral processing plant design and practice. Operators, designers, suppliers, contractors, consultants, researchers, educators, and students will find chapters to satisfy corresponding interests. In fact, this book will serve as a practical structure for mineral processing plant design courses offered by colleges and universities; participants in the Mineral Processing Plant Design Symposium, who have contributed the chapters herein, deserve complete credit. Their work will stand for years to come.

The efforts by Gerry Jergensen, Dennis Mortensen, Milton Lewis, and John Bassarear to ensure proper content and appropriate structure are truly appreciated, and not to be forgotten. Skillful support and executive assistance by Marianne Snedeker, Manager of Publications, Society of Mining Engineers, is gratefully acknowledged. The typing assistance of Melba Weber, Department of Mineral Engineering, University of British Columbia, and Susan Valentine, Mountain States Research and Development, deserves mention.

A. L. Mular
R. B. Bhappu

Co-Editors

TABLE OF CONTENTS

INTRODUCTORY REMARKS

by

Norman Weiss
Consultant
Tucson, Arizona

We may presume that each new mill—and indeed each modernization or enlargement—is an embodiment of the advanced state of the art of mill design, and this volume has for its purpose the assessment of the many elements of progress that have brought our industry up to the present. Indeed, just as "the past is prologue," and we are cautioned by Santayana that we forget it at our peril, we are told by Longfellow not to trust the future, however pleasant.

In man's progress to put to use the metals accessible in the earth's crust, concentration of the raw materials was understandably one of the earliest and most important steps, and we usually date this by the writings of Georgius Agricola in Saxony over 430 years ago as translated in the years preceding 1950 by Herbert Clark Hoover and his wife Lou Henry Hoover. This translation is a gift of inestimable worth to our industry. After Agricola, nearly 200 years passed before Schluter's great contribution to metallurgy about 1750, but the modern beginnings of mineral dressing were in 1864 (for convenience and at least a fair approximation), at which time this reference point was firmly established by the French engineers Huet and Geyler, who through a careful and detailed survey drew a firm line between past and (then) modern milling practices.

The integration and control of milling processes and the effective use of continuous operation were the keystones of this new era. The subsequent explosion in the mining industry coincided with the beginnings of MIT and the Columbia School of Mines, and the establishment of AIME, all of which had a large impact in modern industrialized nations. The technology of ore milling and concentrate reduction grew apace in physical and chemical processes. The usefulness of the wettability of minerals by water and oil (Haynes, 1860) started a half-century of progress which culminated in a commercial froth flotation mill at Basin, Montana (Butte and Superior) in October 1911. Only a dozen years later we had alkaline circuits, Sheridan and Griswold's patents on cyanides as depressants, Cornelius Keller's patent on xanthates, and the use of soaps for flotation of nonmetallics. The five years from 1921-1925 were without equal in the four centuries after Agricola. The effect on flotation of large additions of acids or alkalis was appreciated by this time, but the effect of minor changes in hydrogen-ion concentrations (pH) was not recognized until the work by Weiss in 1926 and the studies by Gaudin in 1929.

From the early '20s until after World War II, the need for metals increased so fast that literally we were too busy producing them to devote much attention to improvements in design for increases in efficiency. As a personal note, it was late in this period, early 1946 to mid-1947, that I terminated my work in Mexico for Asarco, and traveled extensively for Dow Chemical Company in the Western

Hemisphere. I examined and reported on over 150 mills in this period, and it was very clear that a revolution in milling methods was overdue, and that advances in mill design were to be the key. My return to Asarco in 1947 was timely, and together with the 10 years from retirement in 1967 to the end of consulting work in 1977, I was privileged to witness — and play a small part in — 30 years of progress.

In effect, this volume covers these 30 years, certainly a most fertile period, yielding great growth in ion exchange, autogenous grinding, control instrumentation, computers, and entirely new technologies in the metallurgy of ores of uranium, iron, and molybdenum (associated with copper). Froth flotation, which started in North America about 1911, was a mature process by 1947, and most of its advances in the 30-year period we are now contemplating have been in the increase in size of operations and need to cope with coarse particles as a result of diminishing grades of ore and increasing cost of energy. In-situ technology has begun to play an important part in this advance as hydrometallurgy has grown.

To summarize the accomplishments of that period in a few words is impossible, but the following gains would be prominent: larger unit operations, greater plant flexibility, more automatic control and the application of information theory, new process applications, and a growing awareness of technology's responsibility to society and nature. This volume, which commemorates the Symposium on Mill Design, Denver, February 1978, creditably recounts the gains of these 30 years, but we should look upon it also as an introduction to a future that we can see only "as through a glass, darkly." In the apprehension that is generally felt today over shortfalls in petroleum products and very great increases in price, the industrial nations are also concerned over the exhaustion of other irreplaceable raw materials in the earth's crust.

Whether or not North Americans are running out of many metals and some nonmetallic ores can hardly be argued, but the question of exhaustion of these in the earth's crust — which averages only 35 km in average thickness (Kanamori) on land masses, and closer to 5 on ocean bottoms — involves the very important matters of rates and cost of production. Common sense should assure us that only an infinitesimal part of the earth's riches has been extracted, but it also tells us that economic considerations will drastically limit our prospects in future centuries.

Nevertheless, for as many years as any of us will live, and our children and grandchildren as well, the mineral processing industry will meet the challenges with an ever-expanding technology. This volume is encouraging evidence of that.

Tucson, Arizona Norman Weiss
November 28, 1977

1

Financial and Feasibility Studies

Chairmen: J. T. Reagan
W. P. Dyrenforth

Chapter 1

FEASIBILITY STUDY

By Edward S. Frohling and F. Milton Lewis

President
Mountain States Mineral Enterprises, Inc.

Metallurgical Consultant
Mountain States Mineral Enterprises, Inc.

ABSTRACT

The information which is required for a preliminary feasibility study and the additional information which is required for the final feasibility study on a new mineral deposit are discussed. The preliminary feasibility study is made from very little information and at a minimum cost to the owner. However, the outline for the preliminary report is similar to the one used in the final report. In this manuscript, it is assumed that the results from the preliminary study are favorable and the owner authorizes a final feasibility study by an outside engineering company. The type of organization for completing this final feasibility study is described. The final reports includes all information and data relative to this new mineral deposit. This information is presented so that it may be understood by financial people as well as by technical people.

INTRODUCTION

This manuscript describes procedures that are commonly used for developing a newly discovered mineral deposit into a promising viable project through feasibility studies. These are planned so that the owner will spend minimum funds by progressing by steps from an initial financial appraisal of the deposit to the final feasibility study. The results from each step will permit the owner to decide whether or not to continue the study. It is assumed that the results from each step are favorable and that the owner proceeds with the study until the final feasibility report is completed.

INITIAL FINANCIAL APPRAISAL

Shortly after discovering a mineral deposit and analyzing the results of initial sampling, an economic evaluation should be made to determine if this deposit has possibilities for a viable project. As practically no information is available, this evaluation must be made entirely on assumptions, which should be on the optimistic side. If the economics are unsatisfactory after using these optimistic assumptions, then it is prudent not to spend any additional money on exploration on this deposit.

PRELIMINARY FEASIBILITY STUDY

It is assumed that the results from the initial financial appraisal are favorable, and the owner will request a preliminary feasibility study. This phase will again be based on very little information and at a minimum cost to the owner, by either his staff or by an outside engineering firm. In either case, the following will be included in the preliminary report:

- Location
- Description of the Project
- Summary and Conclusions
- Geologic Information
- Possible Ore Reserves
- Metallurgical Results from Preliminary Laboratory Test Work
- Proposed Mining Plan
- Production Schedule
- Sales Value of Products
- Map of Mine Area and Plot Plan of Surface Plant and Ancillary Facilities
- Flowsheet of Processing Plants with Material Balances
- Process Equipment Lists and Budget Quotations
- Capital Cost Estimate
- Operating Cost Estimate
- Environmental Considerations
- Economic Evaluation of the Proposed Project

FINAL FEASIBILITY STUDY

Although more substantial information has been developed, it is to be understood that this preliminary study is still a rough one and will not be wholly accurate. It is assumed that the results from this study are favorable and that the owner will want to continue the study and will provide funds for preparing a final feasibility report. It will require several months to complete this study and the cost will be in a range from a half to one and a half percent of the total estimated capital cost. Generally, an engineering company will be employed to make this final feasibility study, as the engineering department in most mining companies is oriented toward production rather than designing and evaluating new mining projects.

The owner should be rather careful in selecting the engineering company for this final feasibility study. The engineering company's records and performances on similar studies, and the qualifications and experiences of the key personnel, who will be responsible for the study, should be investigated thoroughly. From the conclusions of this investigation, the owner should have confidence in the engineering company and the key personnel who will be reponsible for making this feasibility study. When this confidence prevails and the owner assigns at least one liaison representative on a full-time basis, the study should progress smoothly and efficiently.

The size and skills of the engineering company's organization which is to be responsible for the study depends on the work that will be involved in completing the study and preparing the final report. However, in all typical mining feasibility studies for major projects, the engineering company's organization will contain the following key personnel, with varying numbers of subordinates:

Project Manager - Responsible for the complete feasibility study.

Project Engineer - Responsible for all plant design engineering.

Senior Mining Engineer - Responsible for the complete mining study and mine design.

Chief Geologist - Available for consultation and approval of detailed geological study.

Senior Metallurgical Engineer - Responsible for the metallurgical processes.

Process Engineer - Responsible for the flowsheet, material balance, equipment selection and operating cost estimate.

Design Engineers - Responsible for the design criteria of all surface plants and ancillary facilities in all of the engineering disciplines.

Cost Estimating Engineer - Responsible for all capital cost estimates for the project.

Financial Analyst - Responsible for the financial evaluation of the project.

A smaller staff may be adequate for minor projects where one person may act in more than one of the foregoing positions.

Immediately after the project manager and his staff are designated and approved, there should be a meeting between them and representatives of the client to define in detail the scope of the feasibility study. At this meeting, the Project Manager should present for approval an outline of contents that will be included in the final feasibility report, also the procedures that will be discussed. It should be mutually agreed that the Project Manager and a designated representative of the client will approve and finalize promptly all criteria connected with the study. A schedule for completing the feasibility study should be prepared by the Project Manager. An understanding should be established on interim reports which will be required to keep the client informed on the progress of the study. The Project Manager should summarize the decisions that were made in this initial meeting in a report to the client for his approval.

Shortly afterwards, the Project Manager with certain members of his staff and representatives of the client should visit the site for visual observation of the topography of the area, the approximate boundaries of the mine, possible locations for the surface plants, tailings disposal area, miscellaneous buildings and access roads. While at the site, the source of water, power, facilities required to receive freight at the site, type and availability of local labor, taxes and other miscellaneous items should be checked.

After the Project Manager and his staff have returned to the home office, work should be started concurrently on several parts of the study and this work can be expanded as the study progresses. When the final feasibility report is finished, the following work will have been completed and will be included in the report:

Exhaustive surface geological study and mapping of immediate area.

Extensive drilling to establish grade and geometry of deposit, retaining all core rejects for metallurgical process development.

Development of the economic pit limit and mining sequence and establishment of a preliminary cut-off grade.

Calculation of ore reserves.

Establishment of most likely extraction rate.

Selection of mine equipment.

Evaluation of the economics of outside contracting for the removal of overburden versus removal by the owner.

Determination of metallurgical, chemical, and physical properties of the ore.

Testing of ore hardness to size comminution equipment.

Test work to establish concentrator flowsheet, recovery, and grade of concentrate.

Production schedule for proposed facilities which may include concentrator, leaching plant, smelter and waste gas treatment.

Equipment lists for proposed facilities.

General description of site and facilities.

Subsoil investigation.

Location and dimension of site facilities.

Plot plan.

Preliminary general arrangement drawings.

Preliminary structure design.

Foundation sketches.

Architectural criteria.

Utilities - steam, water, and power requirements.

Preliminary motor lists.

Substation designs and specifications.

Capital cost estimates for mines and surface plants which are accurate within 20 percent.

Operating costs for mining, concentrating, leaching, smelting and general expenses.

Financial analysis of the project.

The report will be lengthy. It should be adequately indexed, easy to read, and presented in a clear and logical manner so that it will be readable by financial people as well as technical people. The report will be divided into sections and include the following:

<u>Introduction</u> - States briefly all of the information that has been gathered relative to the project.

Scope of the Study - Describes the activities of the personnel who contributed to the study and what is included in the study.

Summary - Presents the results and observations from all of the succeeding sections. Preferably, these are shown in tabular forms and are complete so that a reader may be informed about the property, the proposed facilities, the production schedule, capital and operating costs and profitability of the project without reading the remaining sections of the report.

Ore Reserves - Includes the size and geometrical shape of the deposit, a geological analysis, mineralization, types of ores, sampling by drilling or by tunnelling. Calculations are included for establishing the cut-off grade, based on the forecast metal prices and operating expenses, the tonnage and the grade of these reserves.

Mine Preproduction Schedules - Describes in detail all work which is required for developing the mine, a practical timing schedule for completing this work, manpower requirements, equipment list, and the estimated annual or semi-annual cash requirements for all mine preproduction work.

Mining Plan - Describes the mine layout and method for extracting the ore, ratio of waste rock to ore, percent of the total ore which will be recovered, mine production schedule, transportation of waste rock to dumps and ore to the crushing plant, equipment list and supplies, work schedule and manpower requirements.

Metallurgical Testing - Review of laboratory and pilot plant testing, proposed further test program.

Production Schedule - Tabulates the operating schedule and production rates for mining and ore processing units, the recoveries, the product grades and product weights.

Process Description and Design Criteria - Includes the ore processing flowsheets, the primary crushing, secondary crushing and milling rate, material balances throughout the ore processing facilities, and the design criteria for:

- Primary crushing plant
- Coarse ore storage
- Secondary and tertiary crushing plant(or fine crushing plant)
- Fine ore storage
- Concentrator
- Filter plant and load-out facilities
- Reagent facilities
- Tailing disposal
- Industrial and potable water facilities
- Reclaim water facilities
- Considerations for future leaching of dumps

Civil Work and Utilities - Includes a project area map showing:

- Electrical system
- Water supply
- General shops
- Changehouse
- Offices
- Warehouses
- Compressed air
- Roads and parking
- Fences and lighting
- Sanitary sewers

List of Major Equipment - Specifications and cost quotations for all major equipment are tabulated within the following areas:

- Mine
- Primary crushing plant
- Coarse ore storage and secondary crushing plant
- Fine ore bin and concentrator
- Thickening, filtering and concentrate load-out facilities
- Tailing disposal and reclaim water system
- Reagent facilities
- Civil works and utilities

Capital Cost Estimate - Includes tabulation of all direct and indirect capital cost items for mine and all surface facilities. Normally, approximately 15 percent of the total engineering will be completed prior to making these estimates and all necessary drawings will be available for estimating the capital cost. With this much engineering information, the capital cost estimate should have an accuracy of plus 20 percent.

Operating Cost Estimate - Tabulates all operating cost items by departments and calculates the total estimated direct and indirect operating expenses for the entire operation.

Environmental Considerations - Includes federal and state environmental regulations for operations of this type, and the need for operating permits. The auxiliary equipment which is required to meet these regulations is included in the plant design and capital cost estimate.

Financial Analysis of the Project - Includes the annual gross value of all salable products, the annual sales expense, annual net revenue at the property, all direct and indirect annual operating expenses, depletion, depreciation, income before and after income taxes, annual capital outlays, cash flows by years for the life of the property and calculations for determining the rate of return, present worth, and payback time. Sensitivity analyses are made to ascertain the effect that changes in the revenue from sales will have on the rate of return. Also, sensitivity analyses are made to determine the effect that changes in the capital investment will have on the rate of return. These types of analyses will be useful to the owner in deciding whether or not to proceed with the project. Normally, unless the client requests otherwise, it is assumed that the project will be financed entirely by the owner.

Drawing List - Includes the list of all appropriate drawings which illustrate all items in the feasibility study. These drawings, which are reduced in size, are included in the Appendix.

The final feasibility report is submitted to the owner. The results may indicate that this new mining venture will be very attractive. It is assumed that the owner, after studying the report, will conclude that the engineering on the project should continue and that this work will be carried on by the same engineering company.

When the final feasibility study is finished, only about 15 percent of the total engineering work will have been completed. The next step on this project will be the detailed engineering. The engineering crews will be increased for this work and will progress uninterruptedly until approximately 50 percent of the engineering work is completed. At that time, a definitive capital cost estimate will be prepared which will have an accuracy within plus 10 percent. This will give the owner another opportunity to evaluate the project before beginning construction.

Chapter 2

MINING FINANCE

by Herbert D. Drechsler

Associate Professor
The Faculty of Commerce and Business Administration

The University of British Columbia
Vancouver, British Columbia, Canada

ABSTRACT

Mining Finance is the interrelated aspects of "determining the total amount of funds to employ in a mining firm; allocating these funds efficiently to various assets; and, obtaining the best mix of financing in relation to the overall valuation of the firm". The first section is an overview of the financing problem through examination and discussion of the Balance Sheet and Income Statement. The next section is a discussion of the nature of the criteria developed in feasibility study. The third section describes type of capital funds, and the final section describes case studies of mining finance.

INTRODUCTION

Mining finance is concerned with "(1) determining the total amount of funds to employ in the firm, (2) allocating these funds efficiently to various assets, and (3) obtaining the best mix of financing in relation to the overall valuation of the firm".* This chapter will discuss these three aspects of mining finance using various case studies and examples. A prime assumption is that the mining firm wants to maximize the market price of the firm's common shares. For government-owned firms, the objective is to maximize the value of the remaining mine reserves. These objectives are broader than the usual assumption that the firm is owned by "shareholders" and more specific than profit maximization.

Major political changes have occurred in the countries which have mines so that it is no longer a certainty that the firm has shareholders and the shares are traded on a stock exchange. Many firms are owned by government and the ownership is not subject to trading. As discussed above it is assumed that these government-controlled firms have the same objectives as traditional firms. However, financial management is necessary whether or not the firms are for profit or not-for-profit oriented. Therefore this chapter is applicable to all types of mining companies.

---

* Much of the following material has been developed from work done by Peter Lusztig and Bernhard Schwab, Managerial Financial in a Canadian Setting (Toronto: Butterworths, 1977); Ezra Solomon, The Theory of Financial Management. (New York: Columbia University Press, 1963); and V. Fred Weston and Eugene F. Brigham, Essentials of Managerial Finance, Third Edition (Ninsdale, Illinois: Dryden Press, 1974)

THE ANNUAL REPORT

This chapter will first identify elements of the firms annual report. This is done so that in the discussion on funds acquisition the reader will understand how funds are both an asset and a liability to the firm. Second is a discussion of the decision criteria used in typical feasibility studies. These criteria help evaluate the financing decision. Following that section, there will be described the types of capital acquisition funds which are used to purchase assets with which to earn income. And the fourth section will have a series of case studies with which to illustrate the varieties of long run funds for new mine or expansion investments, and some constraints. If the firms have constraints such as political forces which develop objectives different from those assumed here, then several sections in this chapter, especially related to valuation, will not directly apply.

Mining is a business and accounting is the language of business. This language has a special vocabulary which is best shown in corporate annual reports. Tables 1 and 2 following, are the principle financial statements shown in the ASARCO, Inc., 1976 annual report. Table 1 is the consolidated Balance Sheet and Table 2 is the Statement of Earnings. Accountants and financial managers often recommend reading an annual report in the following manner:

1) Read the Auditor's Report. If any unusual accounting practice has been done, or if the auditor does not completely agree with the accounting methods of the firm, this will be shown in the report.

2) Read the Notes to Financial Statements. The notes to the accounting statements explain in great detail the underlying principles used in the preparation of the key financial statements.

3) Read the Balance Sheet, Statement of Earnings, and Statement of Changes in Financial Position, and form an opinion of the firm.

4) Read the Statement by the President and Chairman of the firm. See if this statement coincides with the previously made judgement of the firm.

TABLE IA
ASARCO INCORPORATED
Consolidated Balance Sheet
As of December 31
($000,500)

Assets	1976	1975
Current Assets		
Cash	$ 30,431	$ 34,599
Accounts and Notes Receivable	104,359	102,691
Inventories	187,318	246,518
Materials and Supplies	56,103	56,929
Prepaid Expenses	1,588	1,177
Total Current Assets	379,799	441,914
Investments	422,481	378,888
Property		
Buildings and Equipment	794,753	742,702
Mineral land	146,164	142,278
Other	31,422	29,683
Total Property	972,339	914,663
Less: Accumulated Depreciation and Depletion	326,409	288,047
Net Property	645,930	626,616
Funds Committed to Construction	69,525	-
Other Assets	26,003	27,096
Total Assets	$1,543,738	$1,474,514

TABLE IB
ASARCO INCORPORATED
Consolidated Balance Sheet
As of December 31
($000,500)

Liabilities	1976	1975
Current Liabilities		
Accounts and Notes Payable		
Bank Loans	$ 6,270	$ 6,595
Trade	63,715	87,623
Other	24,103	23,109
Accrued Liabilities		
Salaries and Wages	5,163	5,822
Taxes on Income	15,897	10,442
Other Taxes	10,359	9,564
Current Portion of:		
Long Term Debt	5,040	8,567
Other	23,000	22,108
Total Current Liabilities	153,547	173,830
Long Term Debt	400,444	341,934
Deferred Credits	86,967	73,525
Other	45,393	51,264
Equity		
Preferred Stock		
Authorized - 10,000,000 shares @ no par value		
Common Stock		
Authorized - 40,000,000 shares @ no par value		
Issued - 27,678,223 share	339,239	339,239
Retained Earnings	533,852	510,260
Other	6,743	6,700
Less: Treasury Stock	22,447	22,238
Total Equity	857,387	833,961
Total Liabilities and Equity	$1,543,738	$1,474,514

TABLE 2
ASARCO INCORPORATED
Consolidated Statement of Earnings
for the Year Ended 1976 and 1975
(000's)

	1976	1975
Sales of Products and Services	$1,103,737	$1,004,638
Cost of Products and Services	960,981	898,293
Income From Products and Services	142,756	106,345
Equity in Earnings of Nonconsolidated Associated Companies	30,668	27,829
Other Income	8,676	5,682
Closing Perth AMBOY REFINERY		(20,500)
Income Before Deductions	182,100	119,356
Deductions		
Research Expense	4,416	5,282
Exploration Expense	7,185	7,104
Selling and Administrative Expenses	30,817	30,406
Corporate Taxes and Interest	1,331	2,773
Depreciation and Depletion	50,667	36,484
Interest on Debt	33,106	21,943
Total Deductions	127,922	103,992
Earnings Before Income Tax	54,178	15,364
Taxes on Income	11,865	(10,074)
Net Earnings	$42,313	$25,364
Net Earnings per Share	$1.58	$0.95
Cash Dividends Per Share	$0.70	$1.05

BALANCE SHEET

The balance sheet is a statement of a firm's financial position on one specified date. It records the firm's assets which are the resources which are expected to earn income, as well as its liabilities which are the economic obligations of the firm, and shareholders' equity which is the source of some of the assets. The name "balance sheet" is derived from the fact that the total of liabilities and equity equals assets.

In the ASARCO Notes to Financial Statements are shown the significant accounting policies. They include statements on the principles of consolidation, inventories, property, depletion and depreciation, taxes or income, copper surcharges, explanation, mine development and retirement plans. As an example, the policy on inventories is:

> Metals at primary smelters and refineries and at secondary metal plants are at last-in, first-out cost (LIFO). Metals at mines are at first-in, first-out cost (FIFO). Inventories are written down to market, if lower than cost. Primary metals sold at firm prices for future delivery are included at the sales price.

ASSETS

Assets are the income-generating portion of the balance sheet. Assets are divided into smaller categories to indicate their sources. In the case of ASARCO, the classifications are:

1) Current assets $30,431,000: These are cash plus assets that can be quickly converted to cash. In the example current assets include cash and short-term deposits, or marketable securities which are a short-term investment of funds not immediately needed.

2) Accounts Receivable $104,359,000: These are funds due from customers. This figure often includes an allowance for doubtful accounts.

3) Inventories $187,318,000: This figure includes four basic items. First are the metal inventories of smelters, refineries, and secondary metal plants. Second is the provisional cost of metals purchased for which a price has not been fixed. Third are sales prices for metals sold at firm contracts for future delivery, and fourth are mine ore inventories. Therefore this figure includes both metal, concentrates, and ore, only a part of which can be quickly converted to cash.

4) Materials and Supplies $56,103,000: This is the average cost value of normal operating supplies for all plants and offices.

5) Prepaid Expenses $1,588,000: These are short term prepayments to suppliers of operating materials.

6) Total Current Assets $379,799,000: This is the sum of the above listed assets.

7) Investments $422,481,000: This figure shows the value of ASARCO's ownership in other companies. The value is determined by the total of original cost plus the companies share in accumulated retained earnings, less dividends to ASARCO. Some of ASARCO's investments include M.I.M. Holdings Limited, Southern Peru Copper Corporation, and Revere Copper and Brass Incorporated. Investments which are intended to be converted to cash during an operating cycle are included in current assets.

8) Property $972,339,000: This figure includes tangible assets such as buildings and equipment, mineral lands, land other than mineral land, and automobiles. The value shown in the Balance Sheet is original cost or current market if it is less than cost. Maintenance and repairs are charged to earnings as they occur. These assets are depreciated at various rates, but, in general, assets at plants are depreciated on a straight line basis except those assets placed into service at primary smelters and refineries during the period 1961 and 1970 which are depreciated by accelerated methods. Depreciation and depletion at mines are generally computed on the ore reserve method.

9) Other categories: This includes funds committed to construction $69,525,000 and other assets $26,003,000. Sometimes in some Annual Reports under Other Assets is an item for deferred mine development. This account is for major development which is capitalized rather than amortized currently as an expense.

LIABILITIES

Liabilities are the firm's financial obligations to organizations or individuals outside the firm. They are listed in the balance sheet and are divided into two categories:

1) Current liabilities: these are liabilities which fall due during the coming year (or the operating cycle if longer than a year). They are accounts payable $94,088,000. Accrued liabilities (such as wages, salaries, interest) $31,419,000, and the current payments of the long term debt and reserves for plant designs. Current liabilities may also include notes payable. The total current liabilities of ASARCO are $153,547,000. The reserves listed represent the anticipated costs to be incurred after 1977 associated with the closing of the Baltimore and Perth Amboy Copper refineries and the Amarillo zinc plant.

A major consideration is that reserves are not set-aside cash. They are accounting charges. The only location of cash in the firms accounting statements is under the cash account on the assets side of the balance sheet.

2) <u>Long-term liabilities</u> $532,804,000: these are liabilities which fall due beyond one year (or the operating cycle). This includes the long term debt of the firm, and other long term accounts payable or reserves. Long Term borrowing of the firm for the purpose of developing a mine would fall into this account.

SHAREHOLDERS EQUITY

The final category is stockholders' equity $857,387,000. Equity is the excess of total assets over liabilities. In this example it is divided into preferred stock (unused); common stock of 40,000,000 shares of which 27,678,223 are issued and other capital including retained earnings. Retained earnings are not cash. They are accounting earnings which are developed from the consolidated income statement and kept within the firm.

INCOME STATEMENTS

An income statement is a record of the income and cost flows of a firm within a period of time. The net income is calculated as the excess revenue over expenses and taxes, and reflects changes in shareholders' equity. Net income also is a guide to the firm's debt-paying capacity and its ability to produce long-run earnings and dividends.

<u>Revenue</u>: The first figure on an income statement, $1,103,737, is revenue. It is the gross increase of assets that represents the total sales value of products delivered and services rendered. Revenue can be recognized at many points in the business process: when orders are secured, when the goods are manufactured, when they are delivered or when payment is received. Most accountants use the point of sale as the time to record revenue. For instance, at Craigmont Mines Limited in 1976:

Concentrate revenue is recognized at the time of sale. However, copper concentrate sales not finalized at year end have been recorded at prices estimated to be in effect on finalization dates.

Not included in the revenue are proceeds from borrowings, investments by owners, or adjustment of revenue from previous accounting periods.

EXPENSES

Expenses are deducted from revenue and are essentially costs incurred for receiving revenue. They represent a negative effect on stockholders' equity.

Cost of Production Services: $960,981. This expense in the total of direct supplies and labor necessary to produce the goods or services for sale. Indirect costs are generally shown below in Deductions.

In the Income before Deduction ASARCO labels a special Cost, Unusual Items, at a negative $20,500. In fact this is an increase in income. This account comes from a reserve account which anticipated the closing of the Perth Amboy Copper Smelter.

Deductions: $127,922. Total deductions includes the following items: Research Expense $4,416; Exploration Expense $7,185; Selling and Administrative Expenses $30,817; Corporate Taxes and Interest $1,331 (This item is not federal income tax);

Depreciation and Depletions (To be discussed further in this chapter) $50,667; and Interest on Debt $33,106.

Income Tax $11,865. Income tax is calculated from the tax statement which is not a part of the annual report. The annual report shows accounts on a "book" basis as compared with a tax basis for the government.

Depreciation Methods: Depreciation and depletion are accounting processes that convert assets into expenses. The major methods of calculating depreciation are:

a. Straight-Line

$$\text{Annual Depreciation Expense} = \frac{\text{Cost of Asset - Estimated Scrap Value}}{\text{Estimated Useful Life}}$$

b. Sum of Years Digits Q

$$\text{Annual Depreciation Expense} = \frac{1}{\left(\sum_{Q=1}^{X} Q\right)} (X - A)E$$

where: X = Estimated life of asset
A = Year of depreciation
E = Asset cost minus estimated scrap value (net cost)

c. Declining Balance

A uniform rate of depreciation (e.g. 25 percent) is applied to the declining net cost of the asset.

d. Unit of production

The net cost of the asset is divided by the number of hours, or tons, or whatever production unit is relevant. This unit of production depreciation rate is then multiplied by the number of units the machine was run or tons produced from the mine during the year. This determines the depreciation of the machine or depletion of the mine in that year.

NATURE OF DECISION CRITERIA FROM THE FEASIBILITY STUDY

The methods outlined in this section are used to rank investment proposals. Since the amount of money available for projects is not infinite, it is necessary to assess the economic worth of one investment over another. Each proposal can then be accepted or rejected on the basis of its "hurdle rate" which is the cut-off rate for decision. In all the following methods, the decision is "yes, invest" or "no, don't invest". There are no "maybe" decisions. Obviously there is no decision maker who would base an investment of hundreds of millions of dollars on a single economic criteria. There are numerous other factors to consider such as the risk propensity of the decision maker, the historic strategy and decision process of the firm, the political interrelationship between members of the decision groups, and the concern of the firm with the external world.

INTERNAL RATE OF RETURN METHOD

The internal rate of return is the rate of discount which, applied to the cash flows for investment, will yield a net present value of zero. This method calculates the discount rate which equates the present value of expected returns to the initial investment outlay. The equation for determining this discount rate r is:

$$\sum_{t=0}^{r} \left[\frac{C_t}{(1 + r)^t} \right] = 0$$

where C_t is the cash inflow or outflow and n is the number of years the flow is expected.

For example:

If the initial outlay of \$20,000 is expected to yield a cash inflow of \$6,000 per year for the next 3 years the equation would be:

$$20{,}000 = \frac{6{,}000}{(1 + r)} + \frac{6{,}000}{(1 + r)^2} + \frac{6{,}000}{(1 + r)^3}$$

Computer programmes are available to solve for r. Manual solutions require trial and error methods for solution. In the example above if

10 percent is first substituted for r, the result is \$19,894.80. If 9 percent is then substituted for r, the result is \$20,250.35 thus stradling the answer. By interpolation it is possible to find the discount rate that will equal \$20,000; i.e. 9.7 percent. The internal rate of return is then compared with a required rate of return which is the firm's weighted average cost of capital in the case of unlimited capital availability. In the usual case where capital constraints are binding, the hurdle-rate must be raised with opportunity cost the relevant criteria against which new investments are measured. In numerous conversations with managers of mining firms, the author has found that 15 percent in real terms is the common discount rate used for decision purposes.

NET PRESENT VALUE METHOD

The net present value is the sum of all future cash flows generated by a project with each cash flow discounted to the present. The discount rate used is the firm's weighted cost of capital or the more usual opportunity cost of capital as illustrated above. The equation used for calculation is:

$$\text{Net Present Value} = \sum_{t=0}^{n} \left[\frac{C_t}{(1 + r)^t} \right]$$

where C_t is the cash flow, n is the number of years the flow is expected and r is the rate of discount.

For example:

If the cash flow is \$3,000 per year for the next 5 years after an initial outlay of \$10,000 at a required rate of return of 9 percent, the Net Present Value is

$$\text{Net Present Value} = -10{,}000 + \frac{3{,}000}{(1 + .09)} + \frac{3{,}000}{(1 + .09)^2} + \frac{3{,}000}{(1 + .09)^3} + \frac{3{,}000}{(1 + .09)^4} + \frac{3{,}000}{(1 + .09)^5}$$

$$= 1668.96$$

The net present value number indicates the excess or short fall in cash flows in present value monetary terms, once financing charges at a rate of 9 percent were met. A positive cash flow indicates that a project is acceptable while a negative cash flow indicates rejection of the investment.

PAYBACK METHOD

The payback method is the period of time required for the expected after tax cash flows to recover the initial investment.

For example:

If an investment of \$20,000 has an annual cash inflow of \$6,000 per year, the payback period would be:

$$\text{Payback period} = \frac{\$20,000}{\$6,000} = 3.3 \text{ years}$$

Projects are ranked with quicker payback being preferred.

This method has several drawbacks. It does not take into account the time if cash outflows directly into the investment and it ignores returns beyond the payback period. By ignoring proposals with longer payback periods it dismisses projects such as those involved in introducing and developing new products or tapping new markets. Payback may be used as a tool for evaluating risk. The longer the payback period is, the greater the risk and the less liquid the investment.

AVERAGE ANNUAL RATE OF RETURN METHOD

This rate of return is calculated as a ratio of the average annual profits after taxes to the average book value of the investment over the life of the project.

For example:

Assume the average annual profit over a five-year period is \$3,000. The average net investment of a \$10,000 project is the difference between the beginning book value and the ending book value. The mine plant usually has no value at the end of the mine life, then the average book value of the investment is:

$$\left(\frac{\$10,000 - 0}{2}\right), \text{or } \$5,000, \text{ the}$$

$$\text{Average Annual Rate of Return} = \frac{\$3,000}{\$5,000} \times 100\% = 60 \text{ percent.}$$

Projects are ranked according to the average annual rate of return with higher percentages preferred. Managers can set a hurdle rate at any acceptable level. If the average rate of return is higher than the required percentage, the project is accepted, otherwise it is rejected. The method ignores the time flow of the profits; it is based on accounting profits rather than cash flows; and the hurdle rate is not consistent with managers.

This section has considered the investment decision of the firm. Several common decision methods and criteria have been discussed. These include the Internal rate of return, the net present value, the payback period, and the average annual rate of return. In normative terms, the internal rate of return and the net present value are conceptually better decision rules. They both include the time value of money and the pattern of cash inflows and outflows. The net present value is a theoretically better method for decision-making because it is consistent for an investment proposal.

The payback period and the average annual rate of return methods both have usefulness to the manager. They give some indication of the risk potential of the investment project and give non time-oriented decision rules.

The investment decision criteria give management an assessment of the impact of new projects or the financial statements. As such, they are dealing with forecasting the effect of present decisions. Managers utilize as many tools as possible to aid in increasing the accuracy of the forecast. This section has not dealt with uncertainty as an input in decision criteria. Managers do not make decisions in isolation from the internal and external environment of the firm,therefore the decision methods of this section must be considered interrelated with non financial project selection criteria.

MAJOR SOURCES OF FUNDS

In this section, I shall discuss the sources of funds for a mining firm. Figure I shows some of the sources for these funds. In figure 1, the classification is first by period of time between raising the monies and final repayment in the case of debt. Two periods of time are considered. Short-term financing is defined as debt maturing in one or two years or less. Long-term financing is defined as debt beyond two or more years. The time period is of an arbitrary nature, and there is much interlap between the two categories.

The firm has two prime alternative sources of financing. There are internal and external sources. The internal source of funds is the net cash inflow through sales of goods and services. This inflow is composed of retained earnings and the sum of the non-cash reductions of income before calculation of income tax. Recall from an earlier section that depreciation and depletion are not cash accounts but reflect costs of operating plant and equipment and recognition of the mine as a wasting asset. For tax purposes, these accounting costs are applied to operating income to calculate net taxable income. In the sense that depreciation and depletion are accounting costs, in the calculation, monies equivalent to these costs may be considered as a source of internal financing.

The internally generated sources of funds are adequate for providing smaller amounts of capital for day to day operations. This includes capital for financing the accounts receivable, inventories, and using cash to pay the day-to-day bills and wages. For long-term investments, these sources are inadequate. Most large mines are underinvested.

FIGURE I

Sources of Funds For a Mining Firm

- Sources of Funds
  - Short Term
    - Internal
      - Retained Earning
      - Depreciation
      - Depletion
    - External
      - Trade Credit
      - Short Term Loans
      - Commercial paper
      - Lease Financing
  - Long Term
    - Internal
      - Retained Earning
      - Depreciation
      - Depletion
    - External
      - Debt
        - World Bank
        - Export-Import Bank
        - Bonds
        - Debentures
        - Eurocurrency
        - Mortgage Bond
        - Pollution Revenue Bonds
      - Lease
      - Equity
        - Preferred
        - Common

Source: Adopted from Peter Lusztig; Bernhard Schwab, Managerial Finance in a Canadian Setting (Toronto, Butterworths, 1977) p. 209.

The full extent of the mine is unknown except on the day the mine is depleted. The size of the mining and processing plant is a function of the ore reserve size among other elements. The final ore reserve size is unknown on the day the plant is first constructed, and it is initially built smaller than optimum. As more ore reserve is located more plant is added. Investments are also needed to replace equipment destroyed in accidents or when just worn out. This investment maintains mines and processing plants at constant output with no change in cost.

Investment is needed to reduce the cost of operations through technological advance or efficiencies developed by substituting new equipment for personnel. Finally, investment is needed to produce more output as demand for mineral products grows.

Mining firms are not providing enough internally generated funds to support mining and processing operations. In the period between 1961 and 1970 and between 1961 and 1975 many firms made substantial increases in their debt positions. This increase is shown in Table 3. For the nine mining firms the debt/equity ratio was 0.03 in 1961, was 0.20 in 1970, and was 0.37 in 1975. The book value of the mining firms in the fifteen-year period of time increased 240 percent while the debt increased 2500 percent. For long term purpose, these firms borrowed heavily from banks and other locations rather than sell more shares in the financial markets. This is a strong indication that debt financing is important in the financial structure of the firm.

TRADE CREDIT

Trade credit is a form of short-term financing. When goods are delivered and not paid for at the point of delivery, trade credit exists between the supplier and the purchaser. For example, in the ASARCO Balance Sheet is shown Accounts Payable of $94,000,000. These are supplier credits to the firm. Trade credits are one of the major sources of short term financing. Trade credits are convenient and easy for mining firms to acquire. As long as the buyer maintains his credit standing, suppliers are willing to provide the financing.These credits are useful because they remain available during periods of monetary tightness or other periods when funds from other short term sources are scarce. The prime form of trade credit is open account. Under this system the supplier issues an invoice listing the item, its price, the total amount due and the credit terms. It is called open because the buyer does not sign a formal recognition of debt. The supplier's evidence of the debt is simply an accounting book entry and the purchase and shipping orders. The great use of trade credit is evidence of the trust between buyers and sellers.

SHORT-TERM LOANS

Short-term loans from banks are like trade credit in that they

TABLE 3

Debt-Equity of Some Mining Companies

	1961		1975	
	Book Value	Debt	Book Value	Debt
Copper Range	$ 51,605	$ 36,496	$105,284	$ 31,270
Johns-Manville	283,649	-	580,512	186,322
Phelps Dodge	394,528	-	893,263	522,509
Int. Metal	569,764	-	1,484,353	611,236
Newmont	112,000	-	648,331	251,437
ASARCO	381,306	-	833,961	341,934
Kennecott	748,478	6,540	1,410,423	406,387
Anaconda	814,200	65,000	1,210,918	238,279
AMAX	262,600	15,300	1,365,170	543,200
TOTALS	$3,618,129	$123,336	$8,532,215	$3,132,574

Debt-Equity ratio 1961 - 0.03 in 1970 - 0.20 in 1975 - 0.37

Source: Annual Reports,Moody's Industrial.

usually mature within a year. They are not regarded as permanent financing. Banks often stipulate that firms be out of debt for a short period each year as a guarantee that short-term loans are not static.

Short-term loans can be either secured or unsecured. Unsecured loans are those which do not require collateral. They can take the form of a line of credit in which a maximum amount that can be borrowed is specified. Loans beneath this amount are readily accepted, although lines of credit represent no legal commitment on the bank's behalf. In order to ensure a legal commitment from the bank a revolving-credit agreement must be arranged. In this agreement the bank must lend to a firm up to a maximum specified amount. Another form of unsecured short-term loan is the transaction loan which is borrowing for a single purpose. When the purpose is fulfilled or completed, the loan is repaid. Often banks require that firms maintain a minimum average balance of 15 to 20 percent of the outstanding loan in a chequing account. For the bank this is a form of security since it increases the liquidity of the firm and makes funds available if the loan payments default.

If the bank requires even more security, it asks for collateral. This security is used to guarantee that the loan is paid. If the obligations of the loan are not met, the collateral is sold and the lender receives his funds. Accounts receivable and inventory are assets which a firm may use as collateral. When using accounts receivable as security, the bank or finance company accepts those with good credit ratings. Because of the chance of fraud, the percentage amount advanced upon the face value would probably be low. Inventories are rated according to their marketability. Highly marketable inventories would receive a high percentage of their market value, whereas a highly specialized piece of equipment with low marketability would not be readily accepted as collateral. Collateral becomes necessary when the firm's credit is new or their credit rating is low. Another form of collateral is outside assets such as stocks or bonds. The percentage amount lenders are willing to pay on the market value of bonds is usually high. In addition, bonds often require compensating balances which are a requirement that the borrower maintain a fixed percentage of the loan as an interest-free deposit in the bank. Also bonds sometimes charge standby fees on lines of credit. These fees are charged on the portion of the loan not borrowed and act as a restraint on the extension of lines of credit.

An example of this type of loan is the short term debt of Hudson Bay Mining and Smelting Company, Ltd., as shown in the prospectus for \$50,000,000, 10½% Sinking Fund Debentures Due 1995 of December 9, 1975.

> The Company has a line of credit with Canadian chartered banks and also has borrowing privileges with an associated company (\$7,959,000). The loans from these sources are at interest rates equal to the prime rate in the case of the parent company and at normal commercial rates in the case of subsidiaries. The maximum outstanding

> amount under these arrangements for the eighteen months ended June, 1975, was $11,009,000. The approximate average outstanding balance for the period was $2,500,000. The average rate of interest on outstanding loans at June 30, 1975, was 9.8% and the approximate average rate throughout the eighteen months, 12.4%. There are no compensating balance requirements under these loan arrangements.
>
> The outstanding notes payable ($17,180,000) are for terms of between 30 days and six months and bear an average interest rate of approximately 7.4%. To support its commercial paper operations, the Company has a standby line of credit in addition to its open line of credit. The cost of this standby line of credit is 1/2% on the amount of the line.

COMMERCIAL PAPER

Commercial paper is a short-term, unsecured promissory note issued by major companies such as Hudson Bay Mining and Smelting Co. Limited, as above and sold to banks, insurance companies, pension funds and other business firms. Phelps Dodge Corporation lists $72,178,000 of commercial paper in the 1976 Annual Report. It is used as a supplement to bank credit and as an alternate source of funds during periods of tight money. It may mature in anywhere from 24 hours to 365 days, although most notes are 30,60 and 90 days. Advantages of commercial paper include such items as: the interest rate is often one-quarter of one percent to 2 percent below the prime rate; there is prestige associated with the acceptance of a firm's commercial paper; speed of fund acquisition; and advice from the commercial paper dealers.

Commercial paper is most appropriately used for the financing of seasonal peak requirements for funds. A problem associated with commercial paper is that the amount of funds available is limited to the excess liquidity of other firms and that in hard times, a bank is more likely to be helpful than a commercial paper dealer.

Another example of short term or medium term borrowing is illustrated in the 1976 Phelps Dodge Annual Report:

> The corporation has entered into a revolving credit and term loan agreement with various banks which permits the corporation to borrow at any time through September 1979 up to $150,000,000 at 116% of the agent bank's prime rate of interest. During the revolving period a commitment fee is payable on the unused portion of the bank's commitment at an annual rate of 1/2 of 1% and an availability fee in lieu of compensating balance is payable at an annual rate equal to 9% of the agent bank's prime rate of interest. Borrowings under the agreement which are outstanding at the end of the revolving period bear additional interest increasing from 3/8 to 5/8 of 1% per annum and are repayable from April 1981 through October 1984. There have been no borrowings under this agreement to date.

LONG TERM DEBT

Long term debt is a debt that matures after several years or more. It usually has a final repayment date as well as periodic interest payments. Collateral in the form of stocks, bonds, machinery or equipment, or the mining property including ore reserves is usually requested on these loans. Repayment is usually on a regular basis, for example, every quarter or semi-annually. Prepayment is accompanied by a penalty. The interest rate may be higher than for a short-term loan because of inflation and risk considerations and can be either a fixed rate or a rate that is variable with the bank's prime rate. As with short-term loans, the borrower may be requested to sustain a minimum average balance. Other requests from the bankers may be that net working capital be over some minimum, that changes in management be approved by the lender, and that periodic financial statements be submitted.

A typical example of a long-term note is illustrated in the 1976 ASARCO annual report. These are the 7 7/8% notes due April 1, 1994. Payable in annual installments of $4.1 million commencing in 1978. The notes to this debt indicate:

> The note agreement contains covenants relating to maintenance of certain financial ratios and limiting long term debt and declaration of cash dividends. Under the dividend limitations at December 31, 1976, $114.9 million of retained earnings were available for payment of cash dividends. However, in making dividend payments, consolidated current assets must be maintained at not less than 175% of adjusted current liabilities which amounted to $148.5 million at December 31, 1976.

In this case, ASARCO will pay an interest charge of $782.50 per annum for every $1000 of face value of debt authorized. The balance sheet limitations control the amount of debt ASARCO may borrow.

There are many ways of providing lenders security for loans. For instance, in the 1975 TECK Corporation Limited annual report is shown:

> Current bank loans aggregating $8,737,000 and the $4,100,000 term bank loan are secured by certain accounts receivable, marketable securities and investments.
>
> The $10,803,000 term bank loan is secured by a collateral fixed and floating charge debenture on the company's portion of the assets of the Newfoundland Zinc Project of approximately $11,500,000 plus specific assignment of concentrate accounts receivable and concentrate inventories from the project of approximately $1,350,000.
>
> The mortgage loan payable is secured by the company's 45% interest in the Highmont mineral properties.

SINKING FUND DEBENTURES

A debenture is an unsecured bond. Debentures are unsecured in the sense that they have no claim to any specific property but they are backed by the earning power of the firm. One restriction of a debenture often seen is that the firm shall not pledge its assets to other creditors. There are subordinated debentures which are also unsecured. Upon liquidation, these creditors have even less claim to the assets of the firm, although they are ahead of preferred and common stockholders.

Retirement of bonds or debentures can occur in several ways. First, debt can be retired by payment of face value at maturity; second, by periodic repayment of principal over the life of the debt; third, by conversion if the security is convertible; and fourth, by calling the debt before maturity if there is a call feature in the debt.

Borrowers usually require firms to provide for a sinking fund provision. This provision typically requires the firm to buy and retire a portion of the bonds issue each year. The bonds to be retired may be drawn in a lottery or the funds from the sinking fund may be used to buy bonds on the open market. An example of this are the Falconbridge Nickel Mines Limited 7.75% Sinking Fund debentures maturing February 1991:

> No portion of the principal is due on the 7.75% debentures until 1977 when the company is required to commence sinking fund repayment sufficient to retire $1,250,000 principal amount of debentures in each of the years 1977 to 1990. (From the 1976 Annual Report).

In the area of secured bonds there are mortgage bonds which place a claim upon a specific asset. Creditors usually make sure that the market value of the collateral exceeds the value of bond issue because if corporation defaults and the creditor must foreclose, he sells the collateral to recover the amount of the bond issue. If it is not fully recovered, the bondholder becomes a general creditor for the remaining amount. Falconbridge DOMINICANA, C. por A. also has 9½% mortgage bonds for company housing at the Dominican property, repayable monthly to 1993, in Dominican Republic currency.

Mortgage bonds may place severe limitation upon the borrowing power of the firm as well as requiring collateral. In the case of Falconbridge Dominicana:

> All loans are secured by a first mortgage on the assets of Falcondo (The Dominican property) and by a contract under which Falconbridge Nickel Mines Limited has agreed to buy all ferronickel of commercial value produced by Falcondo. Falconbridge is also obligated to provide 60% of all the funds required by Falcondo to enable it to meet its operating costs and debt service obligations in the

event receipts from the sale of ferronickel produced by Falcondo and other receipts are insufficient for that purpose. (From the 1976 Annual Report).

EURODOLLAR BONDS

Eurocurrency is a deposit liability, denominated in a currency banked in a country outside of the country of origin. Eurodollars are USA dollars deposited in a bank outside of the United States. Other examples are French francs in London or d-marks in France.

The sources of Eurocurrency are foreign governments or business organizations who want to hold currency outside the country of origin and banks with foreign currency outside the country of origin. The demand for Eurocurrency comes from individuals, government and organizations that need funds. Phelps Dodge in 1975 had $25,000,000 of 7½% Eurodollar Notes due in 1977.

MISCELLANEOUS BONDS FINANCING

Below are listed several alternative methods of financing which are found in annual reports. One method becoming more popular is the Air Quality Control Obligation. These are bonds issued by the town in which the mining firm has operations. The proceeds from the bonds are used for pollution control purposes. Repayment is made by the company by agreement with the town. A typical bond of this type is illustrated in the 1975 Kennecott Annual Report.

> On May 16, 1975, the town of Hurley, New Mexico issued a 7 3/8% Pollution Control Revenue Bond in the amount of $21,000,000 payable in 20 equal quarterly instalments commencing August 1, 1980 and ending May 1, 1985. Proceeds of the bond issue were made to the Company for the purpose of financing certain pollution control facilities at Kennecott's Chino Mines Division.

Purchasers of mineral products often supply funds for the equipment and development of mining properties. For instance, the Atlas Consolidated Mining and Development Corporation 1976 Annual Report shows:

> Pesos 55,406,218; Loan from Mitsubishi Metal Corporation (MMC) with interest at the rate of 8 1/6% per annum, payable through deductions from the proceeds of concentrate shipments, collateralized by a standby letter of credit of US $2,000,000 which is automatically reinstated every year for a period of 10 years to 1980, in favour of MMC.

PUBLIC SOURCES OF LONG TERM CREDIT

In the United States, the Export-Import Bank (Eximbank) is a federal agency specializing in foreign lending. This bank aids export trade through long term direct capital loans for U.S. goods or services, rebinding credits, an insurance and guarantee program for U.S. exporters and guarantees for foreign financial institutions. An example is at the Feni-Rudnici nickel property in Yugoslavia. On May 30, 1974, the Bank authorized a $38.7 million limit loan at 6% per annum and two financial guarantees for $51.6 million. The loans were to buy U.S. goods and technical services to develop an ore body, a smelter, and processing facilities.

In another case Kennecott in 1964 financed the enlargement of the El Teniente copper mine in Chile through sale of 51 percent of the corporate equity to the government and by a loan from Eximbank. The loan was to be repaid over a ten to fifteen year period and was to be unconditionally guaranteed by the government of Chile. This situation aided Kennecott in gaining compensation when the mine was nationalized.

In Canada, the Export Development Corporation (EDC) provides similar financing functions as does the Eximbank, i.e. guarantee loans, finance exporting, and insure overseas investments. For example, EDC provided $20.75 million and $57.25 million in financing the sales of Canadian goods and services to nickel mines in Guatemala and Indonesia. The loans were made on a commercial basis. The Export-Import Bank of Japan provides direct loans to exporters upon recommendation of the exporters commercial bank. Other governments such as the United Kingdom, France, Germany and Italy also provide direct loan for exported goods and services.

Several international agencies provide some financing, primarily for infrastructure in mining areas. This includes the World Bank Group composed of the International Bank for Reconstruction and Development (IBRD), the International Development Association (IDA), and the International Finance Corporation (IFC). The loans are generally made to governments for the purpose of developing infrastructure such as roads, power-generating facilities, etc.

LEASE FINANCING

One means of obtaining the use of buildings, land and equipment is leasing. In leasing, the lessor, retains title to the facilities but makes them available to a lessee in return for periodic lease payments. The lease usually is limited to a certain number of years by the economic life of the physical asset with the lessor carrying the risk of obsolescence. The lessor gains the depreciation allowance, rental income, and any residual value remaining upon termination of the lease. The lessee gains the equipment and converts the cost from capital cost to expense, for the advantages.

The two main kinds of leasing are sale and leaseback, and direct leasing. Sale and leaseback occurs when a firm sells an asset it owns to an outside party, such as a financial institution, and then leases the asset from the outside party. The kind of financial institution involved in this kind of arrangement varies with the asset to be sold and leased back. Insurance companies usually deal with real estate; commercial banks and specialized leasing companies deal with equipment.

Direct leasing involves simply renting equipment and maybe service leases or financial leases. Service leases provide the equipment as well as its maintenance. This type of lease is cancellable, that is, the lessee can return the equipment and cancel the lease before the full term of the lease has expired. Financial leases often do not include maintenance and cannot be cancelled in midstream. The lessor gains the full cost of the asset plus a financial return. Usually the firm arranges the purchase and delivery of the equipment and also arranges for the financial institution to buy the equipment. The firm and the financial institution form a lease agreement on the equipment.

COMMON STOCK

The common stockholders collectively own a firm. They are residual owners in that they are entitled to the earnings and assets of the firm only after debt and preferred stock dividends have been paid. Common stock can be considered a perpetual loan which has no maturity date. In addition, the company has no legal obligations to pay dividends to common stockholders. Common stockholders have various rights in relation to the proceedings of the firm. They have the right to vote, inspect financial records (although this privilege is limited), elect the directors, and can authorize the issue of common stock, preferred stock, bonds, debentures and other securities. Voting privileges can be given to another person by proxy.

Common stockholders may sell their shares to another individual and receive the funds. This is usually done through a stockholder through a stock market. The only change this entails is a new name on the stock certificate. There is no change in the capital or financial status of the company. However, if a company offers new common stock, the common stockholders must be offered enough shares to maintain their proportionate ownership in the firm.

Common stock may be considered authorized or issued. In the ASARCO 1976 Annual Report 40,000,000 shares without par value are authorized with 27,678,223 shares issued. Authorization indicates the maximum number of shares the company can issue. If the company amends its charter, it can authorize more shares. Increasing the authorization requires the assent of the shareholders. Unissued shares permit the management flexibility to grant stock options, pursue mergers, or raise additional capital on short notice.

Issued or outstanding shares are that portion of the authorized shares which have been sold or issued by the corporation and held outside of the firm.

Company shares can be with or without par value. ASARCO's shares are without par value while Newmont shares have a par value of $1.60. Par value is a stated monetary face value. Par value simply provides information showing the lowest possible investment behind the company's historical offering of shares. It has no value today and can confuse investors because a particular value for the stock can be incorrectly assumed. The value of a common stock is its market price.

There can be more than one class of common stock in a company. For instance, the International Nickel Company of Canada has Class A and Class B common shares.

> The two classes are interconvertible at any time and are similar in all respects, including dividend rights, except that dividends on Class B shares may be declared payable out of "1971 capital surplus on hand" as defined by the Income Tax Act of Canada.

Some classes of shares may have different numbers of votes. For example in the 1975 Teck Corporation Limited Annual report indicates:

> The Class "A" shares carry the right to 100 votes per share and the Class "B" shares carry the right to 1 vote per share; in all other reports the Class "A" and "B" shares rank equally.

Common shares can be used to raise funds. New mining firms often must resort to selling common stocks to acquire funds because the company has insufficient credit ability to to raise capital from other sources. In general, some factors to consider in selling common shares to acquire capital are:

1. Funds received are made available permanently.
2. The firm has no obligation to make dividend payments.
3. During buoyant stock market periods, common shares may be easy to sell at a relatively high price.
4. The credit standby of the firm is enhanced by the increase in equity.
5. Exiting shareholders may have their control diminished when new shares are sold.

PREFERRED STOCK

This is a debt that lies between common stock and bond or debenture holding. Preferred stock holders rank ahead of common share holders but behind the lenders in claims or the firms assets. No dividends are paid to common share holders until the preferred share holders dividends are paid. Preferred stock has voting privileges, limited to special circumstances such as when a period of time has passed without their dividend being paid.

There are advantages to the firm in using preferred stock to attain funds. Unlike debt, where lack of payments has serious consequences, there are no legal repercussions if dividends are not paid. To preferred shareholders, however, preferred stock often has provisions for carrying unpaid dividends forward from one year to the next. Before dividends are paid to common stockholders, arrears from preferred stock must be paid. A rare feature sometimes included in preferred stock is called participation. In this case the preferred stock is paid its stated dividend and common stock is paid dividends up to this amount. Any excess income remaining is split between common and preferred stockholders. Preferred stock can be retired by using the call feature, a sinking fund deposit, or converting the preferred stock into common stock.

The main advantage of preferred stock over debt is that there is no legal obligation to pay dividends. When cash becomes inadequate this can be an important difference. Preferred stock also increases the equity base from the view point of creditors and therefore helps the financial condition.

Newmont Mining Corporation illustrates some complexities in preferred shares. In the 1976 Annual Report is shown:

> Preferred stock, par value \$5.00; authorized 5,000,000 shares; issuable in series; Series A, \$4.50 Cumulative Convertible (Liquidating value \$100 per share).

Preferred shares have a par value which is meaningful. In the above case normally the liquidating value would be \$5.00 but it is stated that this value is \$100. The annual dividend is \$4.50. The cumulative feature is a protective device - all past preferred dividends must be paid before any dividends on common stock are paid. Further is the Newmont report:

> Each share of Series A, \$4.50 Cumulative Convertible Preferred Stock is convertible into 3 1/8 shares of common stock, and is callable for redemption in whole or in part at any time on at least 30 days notice at a price commencing with \$104.50 per share on January 1, 1974 and declining annually thereafter by \$0.50 to \$100 a share on and after January 1, 1983.

CASE STUDIES CAPITAL ACQUISITION

This section will illustrate with examples the varying means used to acquire investment funds for mining firms. In some inventories the actual names of the mining firm and the capital lending agent have been changed.

Case Study One

Introduction

The Borrower, X Corporation is the largest mining company in the Republic of ______________. It is engaged in the mining of copper, ore and the processing of the ore into concentrate for export.

As a result of significant new copper discoveries, X has planned a substantial expansion of its existing operations. This overall expansion program is referred to herein as the "main-stage program", consisting of:

1. Development of an open pit mine to include a 35,000 short-tons-per-day (STPD) copper concentrator (total cost, approximately $100 million).

2. Construction of a copper smelter and refinery (total cost, approximately $200 million).

The "first-stage program" to be undertaken in advance of the expansion project above, consists of the purchase of machinery and equipment at an approximate cost of $30 million. A significant amount of this equipment has already been placed on order due to long delivery lead times. The "first-stage program" is to be financed through $30 million in Eurocurrency commercial bank loans.

This financing will eventually become an integral part of the main stage financing, which will be undertaken when economic and financial conditions permit. It is anticipated that many of the banks participating in the "first-stage" financing will also join in the main-stage financing. If the $300 million main-stage program does not go forward as anticipated, the equipment purchase with "first-stage" financing will serve as normal replacements for equipment in the Borrowers existing operations.

Summary of Loan

Borrower

X Corporation

Amount

US $30,000,000

Purpose

To finance the purchase of equipment needed for the development of the open pit mine.

Latest Availability

June 30, 1975

Final Maturity

January 1, 1980

Repayment Schedule

Principal will be repayable in eight equal semi-annual installments with payments commencing July 1, 1976.

Interest Rate

2 1/4% per annum over the arithmetic mean (rounded upward to the nearest whole multiple of 1/8 of 1%) of the 6-month London interbank offered rate for relevant deposits as quoted in "Y" banks London office. Interest will be payable quarterly in arrears from the date of first drawdown.

Commitment Fee

3/4% per annum of the undrawn portion payable quarterly in arrears, commencing March 3, 1975.

Taxes

All payments of principal and interest are to be free of host country withholding or other taxes.

Reserve Requirement

The Borrower will reimburse any additional costs incurred by the lenders as a result of the imposition of any reserve requirements in respect of the loan or resulting from imposition of any interest equalization tax or other tax.

Maximum Participation

US $1 million

Governing Law

The Loan will be governed and construed in accordance with the Laws of the United Kingdom.

Other Conditions

a) The making or continuation of the loan will be conditional upon the availability of funds in the offshore currency market.

b) The Borrower is required to maintain a minimum working capital ratio of 150%. In addition there will be certain restrictions on borrowing, on the mortgaging of property whether now owned or hereafter required, the paying of cash dividends, and the acquisition of the Borrowers Capital stock.

c) Appropriate covenants relating to the activities of the Borrower to the end that should the "main-stage program" not materialize or be significantly delayed, the first-stage program will be controlled so as not to impair the financial capability of the Borrower to service the first-stage financing.

d) Managing Banks must succeed in syndicating 100% of the total financing required for the first-stage development program.

e) There shall be no material adverse changes affecting the Borrower prior to signing the loan agreement.

f) The Borrower shall have obtained all the necessary approvals from the host country government or its agencies as required.

g) There shall exist no event of default under any loans or obligations.

Case Study Two

The following case study is from a Canadian bank in a recent (1975-76) loan which provided funds for capital investment and working capital.

Summary of Loan

Borrower

LMN Company

Amount

Loan #1 US $38,000,000 project term loan

Loan #2 US/CDN $4,000,000 working capital

Purpose

To finance construction and development of mine and mill complex

To finance accounts receivable and inventory.

Availability

Drawdowns under the project term loan can be made until December 30th, 1978.

Rates of Interest

Bank's U.S. Base Rate plus 1¼% payable monthly on the basis of a 360-day year or selected maturities of London Interbank rate plus 1½%, interest payable quarterly.

For CDN dollars, the Bank's Prime Rate plus 3/4% interest payable monthly. For U.S. dollars the Bank's US Base Rate plus 3/4% interest payable monthly.

Standby Fee

To October 31, 1976, ¼% per annum; thereafter ½% per annum on undrawn portion of loan #1.

Mandatory Repayments

Eighteen equal quarterly repayments of $2,111,111 commencing the earlier of Sept. 30th, 1979 or six months following completion date.

Following completion date, mandatory repayments may be deferred during periods of force majeure of up to six months, provided such deferred repayments are made in four equal quarterly installments following the end of the force majeure period.

Final Maturity

December 31st, 1983

Mandatory Prepayments

In any fiscal year, excess cash flow (to be defined as cash generated above mandatory repayments and capital expenditures) is to be dedicated to prepayments on the Bank loan, payment of amounts outstanding under the deficiency guarantee and interest on any income debentures on a basis to be determined.

Security

Loan Agreement containing the normal positive and negative covenants (including but not limited to) ownership, operation of the project, maintenance of specific working capital level, prohibition against additional borrowings and pledging of security to others.

Other terms include:

- A first fixed and floating charge debenture on the mine-mill and mineral property.
- Assignment of proceeds to be received under the sales contract with the smelter.
- Assignment of Insurance Policy normal for operations similar in scope.
- Unconditional deficiency quarantee for up to US $10,000,000 from the smelter, deficiency guarantee to reduce pro-rata as repayments on the term loan are made.

Controls on concentrate inventories and General Assignment of Book Debts.

Conditions Precedent

Infusion by PQR Company (co-venture) or other party acceptable to the Bank of US $12,000,000 into the project (excluding present deferred development costs of approximately $4 million).

The borrower shall have entered into a long-term sales contract with the smelter with terms and conditions as outlined in the Technical & Economic Summary dated February 1976. (Not Included).

LMN Company and PQR Company have entered into a fixed price contract for construction-development of the project on the approximate terms and conditions outlined in the PQR letter. The contract is to call for the completion of the mine-mill as per design specifications and satisfactory operations on a continuous basis at acceptable product quantity and quality levels on a basis to be agreed with the Bank.

The loan Agreement and all security and supporting documents shall have been finalized and registered where necessary.

The project shall have been reviewed by the Bank's consultant and the Technical & Economic Summary shall have been confirmed to the Bank's satisfaction.

Other Terms and Conditions

Legal costs and all out-of-pocket expenses of the lender and its consultant related to the preparation of documentation to be for the account of the Borrower.

Case Study Three

The following is taken from the Financial Post, May 22, 1976:

Investment Group Tries New Wrinkle for Mine Financing

Mining exploration men for the past few years have been trying to evolve new methods of raising exploration money, faced with tight restrictions put in the way of such efforts by stock exchange and securities commissions which are proponents of the old ways of financing the penny-mine business.

The novel concept, the brainchild of the company director, attempts to reduce risks in the highly speculative mineral exploration field by mixing debt and equity to increase protection for investors.

Funds Reinvested

"We're breaking with traditional forms of mining investment in Canada to provide an investment vehicle combining safety of principal with the speculative appeal of former small exploration companies." The Director says, "Here's how the five-year plan works":

A new company, First Mineral Explorations, Inc., will offer \$10.5 million of noninterest bearing collateral trust bonds, and two million common shares, in 10,000 units of one \$1,000 bond and 200 shares. The price per unit: \$1,050.

The \$10-million capital (after underwriting and other costs) will be invested in short-term government and government-guaranteed securities as well as bank and trust company deposits.

The interest on the investment, estimated at \$1 million per year, will be used for mineral exploration.

At the end of the five-year period, the bonds are redeemable.

The investors' return will be reflected in the value of the common shares, depending on the company's exploration success.

The bonds and common shares will be transferable only in units, not separately, until a date set by the directors, not before June 30, 1978, not later than June 30, 1980.

The draft prospectus notes that "in the opinion of the directors", the \$1-million-per-year income "is adequate to enable the company to carry out a diversified, well-balanced exploration program".

It continues:

"It is anticipated that specific projects may be undertaken alone or jointly with others on a joint venture basis, and that exploration work may be done on the property of other mining companies in consideration for shares of such other companies and under agreements whereunder the write-offs for tax purposes will be available to the company rather than the company which owns the property.

Preliminary Prospectus

"As the bonds offered by this prospectus will have the benefit of a pledge of securities which will ensure their retirement in full at maturity, there will be no covenants in the trust deed securing the bonds, which would inhibit the company from borrowing moneys or otherwise financing itself for the purposes of acquiring directly or indirectly interests in properties thought to be capable of economic production and of bringing any such properties into production".

Case Study Four

The Tara lead and zinc project. (Ireland)

Financing of the Tara lead-zinc project at Navan, 30 miles north of Dublin, was arranged in two parts: US$ 100 million in bank loans and US$ 30 from the Canadian government's Export Development Corporation (EDC).

The $100 million portion was arranged by Toronto Dominion Bank and three Irish banks, Allied Irish Investment Bank, Bank of Ireland, and the Ulster Investment Bank. (The latter three provided 30% of the bank loan).

In addition, Toronto Dominion Bank Investments (UIC) Ltd. (owned by the TD bank), Citibank N.A., First National Bank of Chicago, Bank of Nova Scotia, Commerce International Trust Ltd. (owned by Canadian Imperial Bank of Commerce), and Banque Canadienne Nationale (Europe) - a subsidiary of Bank Canadiene Nationale - participated in the consortium.

These banks and buyer credit loans have been guaranteed unconditionally by Tara Exploration and Development Company which owns 75% interest in the Tara Mines Limited. The Irish government owns 25% of Tara mines. There is a standby facility of up to $15 million in the case of overruns to be provided by Bank of Nova Scotia and the TD and Commerce subsidiaries.

Case Study Five

From the New York Times, June 16, 1976

U.S. Steel

The nation's largest steel-maker faced a dilemma in its financing decision. On the one hand, its ratio of debt to equity had risen to 32 percent - considered high for an AA-rated company - making debt inadvisable.

On the other hand its stock is selling below "book" or net asset value - $52.38 compared to $59 - making equity equally unattractive. Book value is the theoretical worth of the company, based on balance sheet figures.

Even though the market value of the company can vary widely analysts still worry about stock's being sold below book value because technically that means new stockholders get a piece of the company at cut-rate prices.

Pressed for capital to finance $800 million in planned expenditures this year, the steelmaker settled on a hybrid: convertible debentures - which come close to avoiding the pitfalls of both equity and debt.

Strictly speaking, the issue is debt - the company promises to repay $400 million to lenders on July 1, 2001 while making semiannual interest payments until then. The wrinkle is that debenture holders can turn their certificates in for stock at a set price, expected to be 10 to 12 percent above the stock price on the date when the debentures are sold.

In other words, if the stock market booms and the price of U.S. Steel soars, debenture holders will have a chance to reap substantial capital gains by converting.

For this option, buyers of convertible issues will settle for a return expected to be 6 percent on the steel issue - markedly reducing the company's interest costs. But the major plus for U.S. Steel in offering convertibles at a conversion price close to book value is that debt-to-equity ratio will drop without the dilutive effects of selling stock below book.

Financial experts expect the issue to sell easily - particularly to public pension funds, which are limited by law as to how much common stock they can hold.

Case Study Six

As the final case study, a more complete project financing scheme will be presented. Much of the data has been developed by Mr. Grover R. Castle, Vice President, Petroleum and Mineral Division, Chemical Bank, New York. This document includes an addendum to provide for political risk.

Proposed 3 1/4 Year Standby Eurodollar Commitment
Followed By A 3 1/2 Year Eurodollar Term Loan
To Canadian Coal Company

Amount

$75,000,000 U.S. Dollars

Purpose

Canadian Coal Company (Coal Company) has been formed to construct and develop a new deep coal mine and preparation plant (Mine) in Canada. The Partners (Partners) in the Coal Company are listed below:

Partners	Percentage Ownership
Canadian Steel Company	35%
Canadian Mining Company	25%
Japanese Trading Company	20%
Provincial Government	20%
	100%

The Mine is estimated to have proved but undeveloped reserves exceeding 90 million tons of low sulphur medium volatile recoverable washed coal and the total projected cost of developing the mine is $100,000,000. The Proposed Loan shall finance 75% of the cost of the Mine and the balance of the cost shall be supplied by pro rata (in proportion to Percentage Ownership) equity investments by the Partners. Initial production of the Mine is expected to start in 1980 and the eventual design capacity of the Mine is planned at 3,000,000 tons per year. The coal is expected to be sold to Japanese utilities under long term contracts.

Standby Period

From September 30, 1977 through December 31, 1980

Standby Fee

1/2 of 1% per annum on the average daily unused portion payable at the end of each interest period.

Management Fee

To be determined

Borrowings

Prior to each Borrowing the Banks shall receive a certificate indicating that the proceeds of the Borrowing shall be used for the development of the mine. Also at the time of each Borrowing, the Partners shall invest cash into the Coal Company which shall be used for the development of the Mine and which shall be in an amount equal to one third of the amount of the Borrowing.

Interest Rate

(x) percent over the three, six or nine month cost of funds for Eurodollars payable at the end of each quarter.

Amortization of Term Loan

Commencing March 31, 1981, the Proposed Loan shall mature in 14 quarterly installments through June 30, 1984. The first seven installments shall each equal 3.57% of the original amount of the Proposed Loan and the next six installments shall each equal 10.71% of the original amount of the Proposed Loan. Assuming the full $75,000,000 is borrowed then the first seven installments shall each equal $2,678,571, the next six installments shall each equal $8,035,714, and the final installment shall equal $8,035,719.

Alternate Interest Rate

If any Bank is unable to quote rates to prime banks in the London Interbank Market for offering of U.S. Dollars, then such Bank shall no longer be obligated to make advances unless such Bank (in its sole discretion) and the Coal Company agree as to what interest rate shall apply to such advance.

Taxes and Governmental Action

All payments by the Coal Company (principal, interest, fees and reimbursement of expenses) will be made in U.S. Dollars free and clear of any taxes, levies, imposts and duties (except for taxes on the overall profits of the Banks). The Coal Company shall pay additional interest to compensate for any such taxes.

If any change in applicable law or regulation subjects any Bank's Loan to any U.S. interest equalization tax or any other tax, levy, impost or duty on loans to foreign borrowers, or changes the basis of taxation of payments to any Bank of principal, interest or fees payable (except for tax on overall profits of such Bank) or imposes or modifies its reserve requirements or imposes other conditions, to increase the cost of making or maintaining the Loan, the Coal Company shall pay additional interest to compensate for the resultant increase in the cost of making or maintaining the Loan.

In the event of any of the foregoing, the Coal Company shall be entitled to repay forthwith all or any part of the Proposed Loan then outstanding with accrued interest thereon and the funding loss, if any, whereupon that Bank shall be discharged from liability to make any further loans.

Prepayments

In whole or in part at the end of any interest period at the Coal Company's option without penalty. All Prepayments shall be applied against the Proposed Loan in the inverse order of maturity.

Collateral

The Proposed Loan shall be secured by a Real Estate Mortgage (Mortgage) on the Mine and Floating Charge (Floating Charge) on production from the Mine. The Debenture providing for the Mortgage and the Floating Charge shall also provide for an Assignment of the Sales Contracts (as hereinafter defined).

Other Conditions of Lending

Usual, plus prior to any borrowing hereunder:

1. All Governmental approvals and clearances shall be obtained.
2. The Banks shall be satisfied with an analysis of the Feasibility Report prepared by Canadian Engineering Company. This would include an inspection of the basic data and the Mine site and the Banks shall be satisfied that the Coal Company owns at the Mine in excess of 90 million tons of low sulphur medium volatile recoverable clean coal. The Banks shall also be satisfied with the feasibility of developing the Mine for $100,000,000.
3. The Banks shall be satisfied with the terms of the License (License), authorizing the Coal Company to exploit the coal reserves in the Mine and all consents required under the License for the Mortgage shall have been obtained from the Province of Y.

Undertakings By The Partners

The Partners shall:

A. Complete (Completion) the Mine prior to March 31, 1981. Completion shall include:
   1. The operation of the Mine for a continuous period of 120 days and during that period:
      a) the Mine shall have produced 950,000 tons of coal in accordance with a predetermined set of specifications;
      b) the cost of operating the Mine shall not exceed a level which shall be $23 per ton less than the escalated Sales Price provided in the Sales Contracts (hereinafter defined) and the cost of transporting the coal shall not exceed a level which shall be $28 per ton less than the escalated Sales Price in the Sales Contract.
   2. A requirement that all costs of Completing the Mine shall have been paid for by either the Proposed Loan or Partners Contri-

butions.

3. A requirement that on the date of Completion, no environmental suits shall be threatened or in process.
4. A requirement that on the date of Completion, Sales Contracts (Sales Contracts) covering the sale of at least 85% of design capacity shall be in full force and effect. The purchasers under such Sales Contracts shall be acceptable to the Banks and the terms of the Sales Contracts shall conform to the specifications set out below:

   Term: At least 15 years

   Price: The Base Price shall be $46 per ton and there shall be provision in the sales contract for escalation above the Base Price for all cost (of operating the Mine) increases subsequent to the date of the Loan Agreement. All payments under the Sales Contract shall be in U.S. or Canadian Dollars.

   Quality: The Sales Contracts may include a penalty on the Sales Price if the coal does not meet the following minimum standards (average of each shipload):

Moisture	7.0 maximum
Inherent moisture, air dry basis	2.0% maximum
Volatile matter, air dry basis	25% maximum
Ash, air dry basis	10% maximum
Sulphur, air dry basis	0.5% maximum
BTU, air dry basis	13,500 minimum
FSI	7.5 maximum

   Force Majeure and Events of Default: Usual provisions in normal commercial contracts, but a purchaser shall not be permitted to cancel a Sales Contract in the event that coal can be purchased at a lower price from another seller.

5. A requirement that on the date of Completion the Coal Company shall have Insurance (Insurance) on the Mine in amounts, covering risks and with underwriters all of which are satisfactory to the Banks. Such Insurance shall have a loss payable clause in favor of the Banks.

B. Represent and warrant that they own the License to exploit the coal reserves in the Mine and such license is in full force and effect; warrant that on the date of Completion all governmental clearances have been obtained; and warrant that on the date of Completion no amendments to the License have been signed since the copy that has been furnished to the Banks.

All undertakings by the Partners shall be on a several basis in proportion to their Ownership Percentage.

Covenants of the Coal Company

A. The Coal Company shall:
   1. Maintain existence, pay taxes, royalties and liabilities when due, maintain the Mine in good repair, and defend the Mine against all claims.
   2. Operate the Mine in a good and workmanlike manner.
   3. Maintain Insurance on the Mine.
   4. Furnish annual audited, quarterly interim unaudited statements and quarterly production reports.
   5. Permit representatives of the Banks to inspect the Mine.
   6. Give written notice of any claim or demand effecting title to the Mine or any adverse actions against such title.
   7. Complete the Mine prior to March 31, 1981.
   8. Pay all closing costs in connection with the Proposed Loan.
   9. Maintain working capital of at least $1,800,000 at all times.
   10. Maintain the License in full force and effect.

B. The Coal Company shall not:
   1. Incur, create or permit to exist any indebtedness for borrowed money, except the Proposed Loan.
   2. Incur, create or permit to exist any mortgage, lien or any encumbrance (including CSA's, purchase money mortgages and other title retention agreements and including any production payments) on any of its assets, whether now owned or hereafter acquired, except for the usual exceptions and the Collateral provided for herein.
   3. Guarantee, endorse or otherwise in any way become contingently liable for any indebtedness of any other party (including take or pay or working capital maintenance agreements).
   4. Incur or become liable for the payment of rent.
   5. Acquire all or a substantial part of the assets or merge or consolidate with any other entity.
   6. Sell all or a substantial part of its assets.
   7. Make any loans or advances or investments in any other entity.
   8. Make any distributions to the Partners or retire any of the partnership capital except that the Coal Company may make distributions out of 50% of Excess Cash Flow earned on a cumulative basis subsequent to the Completion of the Mine. Excess Cash Flow shall be defined as the amount by which the sum of net income plus depreciation plus all non cash charges earned in any year exceeds $24,000,000.
   9. Amend the License or the Sales Contracts.
   10. Engage in any business other than the operation of the Mine.
   11. Grant any Section 88 lien to any Canadian bank.

Events of Default

In the event:
1. Of non payment of principal or interest when due.
2. Of misrepresentation.
3. Of voluntary or involuntary bankruptcy of the Coal Company or

any of the Partners (prior to Completion only as to the Partners).
4. Of cross acceleration.
5. The Coal Company is no longer owned by the Partners in proportion to their Percentage Ownership.
6. Of final judgment against the Coal Company.
7. Of default under any Sales Contract.
8. Of default under any other provision of the Loan Agreement.
9. The License is no longer in full force and effect.
10. The Mine produces less than 1,275,000 tons on an average basis in any six month period after Completion.

Subject to approval of Counsel for the Banks as to all legal details.

Canadian Coal Company Venture
(Modifications to Financial Plan in Order to provide for assumption of Political Risk)

Other Conditions of Lending

4. The Banks shall receive a letter from the Host Country addressed to the Banks confirming that:
   a) there shall be no change in the provisions of the License (including those relating to taxes) during the payout of the Proposed Loan;
   b) the Mine shall be given its fair share of goods and services at a fair cost;
   c) the Host Country is aware of the terms of the Proposed Loan and shall not take any action that will have an adverse effect on the payout of the Proposed Loan.

Undertakings by the Partners

The Partners shall:

A. Complete (Completion) the Mine prior to March 31, 1981. Completion shall include:
   1. The operation of the Mine for a continuous period of 120 days and during that period:
      a) the Mine shall have produced 950,000 tons of coal in accordance with a predetermined set of specifications.
      b) the cost of operating the Mine shall not exceed a level which shall be $23 per ton less than the escalated Sales Price provided in the Sales Contracts (hereinafter defined) and the cost of transporting the coal shall not exceed a level which shall be $28 per ton less than the escalated Sales Price in the Sales Contract.
   2. A requirement that all costs of Completing the Mine shall have been paid for by either the Proposed Loan or Partners Contributions.
   3. A requirement that on the date of Completion, no environment suits shall be threatened or in process.
   4. A requirement that on the date of Completion, Sales Contracts (Sales Contracts) covering the sale of at least 85% of de-

sign capacity shall be in full force and effect. The purchasers under such Sales Contracts shall be acceptable to the Banks and the terms of the Sales Contracts shall conform to the specification set out below:

Term: At least 15 years.

Price: The Base Price shall be $46 per ton and there shall be provision in the Sales Contract for escalation above the Base Price for all cost (of operating the Mine) increases subsequent to the date of the Loan Agreement. All payments under the Sales Contract shall be in U.S. or Canadian Dollars.

Quality: The Sales Contracts may include a penalty on the Sales Price if the coal does not meet the following minimum standards: (average of each shipload)

Moisture	7.0 maximum
Inherent moisture, air dry basis	2.0% maximum
Volatile matter, air dry basis	2.5% maximum
Ash, air dry basis	10% maximum
Sulphur, air dry basis	.5% maximum
BTU, air dry basis	13,500 minimum
FSI	7.5 maximum

Force Majeure and Events of Default: Usual provisions in normal commercial contracts, but a purchaser shall not be permitted to cancel a Sales Contract in the event that coal can be purchased at a lower price from another seller.

5. A requirement that on the date of Completion the Coal Company shall have Insurance (Insurance) on the Mine in amounts, covering risks and with underwriters all of which are satisfactory to the Banks. Such Insurance shall have a loss payable clause in favor of the Banks.

B. Represent and warrant that they own the License to exploit the coal reserves in the Mine and such License is in full force and effect; warrant that on the date of Completion all governmental clearances have been obtained; and warrant that on the date of Completion no amendments to the License have been signed since the copy that has been furnished to the Banks.

In the event of a Political Event the Partners shall no longer be obligated to perform the Undertakings described above during the continuance of such Political Event. However, if such Political Event shall subsequently be corrected then the Undertakings hereunder shall henceforth be reinstated.

Political Event shall be defined as:

A. Any action by the Host Country (including taxation or legislation) which does not constitute the due exercise of rights under the License and which has the following consequences:
   (i) expropriation or nationalization;
   (ii) preventing the Coal Company from Completing the Mine;
   (iii) preventing the Partners from exercising their fundamental rights.

B. War, Insurrection or civil strife which prevents the Coal Company from Completing the Mine.

C. Failure of the Host Country or its Central Bank to convert local currency into U.S. Dollars in order to permit the Coal Company to make the applications on the Proposed Loan that are provided for under the amortization section of this Outline.

All undertakings by the Partners shall be on a several basis in proportion to their Ownership Percentage and the Undertakings by the Host Country shall include a waiver of sovereign immunity.

<u>Events of Default</u>

In the event:

9. Of a Political Event.

Selected References

Castle, Grover R., "Project Financing - Guidelines for the Commercial Banker", Journal of Commercial Bank Lending, April 1975, pp. 14 - 30.

Chender, Michael, "Financing New Production in the Copper Industry Calls For New Approaches", Mining Engineering, December 1976, pp. 20 - 23.

Daniels, J.D., et. al., International Business: Environment and Operations (Reading, Mass: Addison-Wesley, 1976).

Drechsler, Herbert D., "A Source of Capital: Multi-Nation Investment", Proceedings of the Council of Economics, AIME, Annual Meeting, San Francisco, California, 1972.

Drechsler, H. and Schwab, B., "Economic Evaluation of Mining Ventures with Capital Constraints: Interrelationships between Net Present Value, Payback, and Regret Due to Investment Delay", Proceedings of the IX World Mining Congress, III-2, May 1976, Dusseldorf, Germany.

Drechsler, H. and Stephenson, J., "The Effect of Inflation On the Evaluation of Mines", Canadian Mining and Metallurgical Bulletin, February, 1977, pp. 76 - 82.

Gentry, Donald W., "Buy-or-Lease Decisions for Capital Equipment", Mining Engineering, September 1975, pp. 39 - 43.

Horngren, Charles T., Accounting for Management Control, third edition, (Englewood Cliffs: Prentice-Hall, Inc. 1974).

Mackenzie, Brian W., "Evaluating the Economics of Mine Development", Canadian Mining Journal, December 1970.

Lusztig, Peter and Schwab, Bernhard, Managerial Finance in a Canadian Setting, (Toronto: Butterworth and Company, 1973).

Solomon, Ezra, The Theory of Financial Management (New York: Columbia University Press, 1963).

Ulatowski, Tomec, Frohling, Edward S, and Lewis, F. Milton, "Delaying Debt During Early Development: A New Approach to Mine Financing", Engineering and Mining Journal, May, 1977, pp. 65 - 70.

Van Horne, James C., Financial Management and Policy (Englewood Cliffs, N.J.: Prentice-Hall, Inc. 1971).

Wearly, William R., "Implications of Escalating Capital and Development Costs", Mining Congress Journal, June 1976, pp. 33-40.

Weingartner, M., "Some New Views on the Payback Period and Capital Budgeting Decision", Management Science, August 1969, pp. B-594 to B-607.

Wells, Howard M., "Decision Under Uncertainty", Journal of the South African Institute of Mining and Metallurgy, April 1976, pp. 375 - 382.

Weston, J. Fred and Brigham, Eugene F., Essentials of Managerial Finance, third Edition (Hinsdale, Ill: The Dryden Press, 1974)

Chapter 3

THE ESTIMATION OF PRELIMINARY CAPITAL COSTS

By A. L. Mular

Professor of Mineral Engineering
University of British Columbia
Vancouver, B. C.
Canada

ABSTRACT

Several capital cost estimation procedures are reviewed briefly, although the factored estimate method is discussed in preference. A table of cost versus equipment parameter data is provided, where the cost may be updated from the current value of the M&S cost index for mining and milling. Cost equations are useful to prepare computer programs which perform capital cost calculations. The use of the factored estimate method is illustrated.

INTRODUCTION

For many years mineral processing engineers have recognized the advantages of valid cost data unique to their processing field. This has led to compilations such as a handbook* for estimating major equipment costs and preliminary capital costs. The handbook was published in 1972 by the Canadian Institute of Mining and Metallurgy (1) and contains data in the form of graphs, tables and equations for the rapid estimation of the cost of an item of major equipment used in mineral processing plants. Data collected from various sources follow the equation

$$\text{Cost} = a(X)^b$$

where X is a suitable parameter such as motor horsepower and a and b

---

*Currently under revision and expansion.

are constants. Such cost data are useful to estimate the total cost of a processing plant.

The cost equations in this chapter have at least one thing in common. Each was developed from prices adjusted to a common cost index value, namely, the Marshall & Swift index (Mining and Milling) of 300. Towards the end of the chapter, the equations are employed to estimate the capital cost of a plant from a rough flowsheet. But first, before capital cost estimations are considered, the nature and utility of cost indices is reviewed.

COST INDICES

Cost indices are ratios used to estimate current prices of equipment from obsolete prices. A cost index is the ratio of costs at a particular time to costs at a specified base year. Where the price of an item at some time in the past is known, the current price is estimated from

$$\textit{cost now} = (\textit{cost then})\frac{(\textit{cost index now})}{(\textit{cost index then})}$$

Cost indices are based on mean costs over a period of time. They have an accuracy of ±10 percent and they are used with equal to or better than this accuracy when the time lapse is less than ten years. As a consequence, cost indices should be restricted to order-of-magnitude and factored estimate (ratio estimate) costing methods.

Of the various cost indices available, the following are most commonly employed in the process industries: (a) Nelson refinery construction index, (b) Chemical Engineering plant construction cost index, (c) Engineering News-Record construction index (ENR) and (d) Marshall & Swift (formerly Marshall & Stevens) cost index (M&S). These are described in Cost Engineering, October, 1968 issue, and are summarized below.

Nelson Refinery Construction Cost Index

This index is used primarily for estimations in the petroleum industry. The total index is based on material and labor in the following percentages: 24% iron and steel, 8% building material, 8% miscellaneous equipment, 30% common labor, 30% skilled labor. The base year is 1946.

Chemical Engineering Plant Construction Cost Index (CE)

This index is based on equipment, erection and installation, material, labor, engineering and supervision costs in the following percentages: 37% fabricated equipment, 14% process equipment, 20% pipes, valves and fittings, 7% process instruments and controls, 7% pumps and compressors, 5% electrical equipment, 10% structural supports, insulation and paint, 22% erection and installation labor, 7% buildings, materials and labor, 10% engineering and supervision manpower. The index is based on 1957-1959 where it has a value of 100.

Engineering News-Record Construction Cost Index (ENR)

The ENR index is based on the costs of labor and building materials in the following proportions: 25 cwt. of structural steel, 6 bbl. of Portland cement, 1.088 mfbm of 2x4 s4s lumber, and 200 hours of common labor. The base year for which the index takes the value of 100 is 1913.

Marshall & Swift Cost Index (M&S)

The M&S index has several values. The value most often used and referred to is the all-industry equipment index. This index is the average of the indices calculated for 47 different industries. Other M&S index values are calculated according to the type of industry; one of these is an M&S index for the mining and milling industry.

These indices are based on equipment appraisals, modifying factors and judgments concerning the current economic conditions. The base year is 1926.

Comparisons Between Cost Indices

Table 1 lists the ENR, M&S and CE cost index values from 1950 to 1975. Such values are taken from various journals such as *Chemical Engineering*, *Engineering News Record* and *Oil and Gas Journal*. Of interest is a graph of M&S index (mining and milling) versus year. Such a plot indicates that from 1966 to 1972, the price of an item increased on the average by about three and one half percent per year on a cumulative basis. From 1972 to 1975 this increase is closer to eight percent on the average.

Table 2 is a review of index values for 1976. Note that the M&S index for mining and milling is larger than the all industry value.

Example of Use of Cost Indices

Suppose that in June of 1967 a cone crusher was purchased for $50,000. What is the approximate cost of this unit today?

Letting "today" be September 26, 1976, then any current cost index may be chosen for the problem, although the M&S index for mining and milling is preferred. This index "today" is 487 from Table 2. June is the middle of the year so the corresponding value for 1967 will probably be representative of June. This is 264 from Table 1. Using the formula on the previous page

$$\text{cost now} = \$50{,}000 \times \frac{487}{264} = \$92{,}235$$

TABLE 1: Cost Index Values from 1950 to 1975

Year	ENR Index	M&S(all-industry)	M&S(min/milling)	CE Index
-	1913=100	1926=100	1926=100	1957-9=100
1950	510	168	171	74
1951	543	180	181	80
1952	569	181	182	81
1953	600	183	184	85
1954	628	185	187	86
1955	660	191	193	88
1956	692	209	216	94
1957	724	225	228	99
1958	759	229	234	100
1959	797	235	237	102
1960	824	238	241	102
1961	847	237	239	102
1962	872	239	240	102
1963	901	239	240	102
1964	936	242	243	103
1965	971	245	245	104
1966	1019	253	253	107
1967	1070	263	264	110
1968	1154	273	273	115
1969	1270	285	286	119
1970	1380	303	303	126
1971	1571	321	321	132
1972	1726	332	332	137
1973	1897	344	343	144
1974	2019	398	394	165
1975	2209	444	451	182

TABLE 2: Cost Index Values for Year 1976

Month	ENR Index	M&S(all-industry)	M&S(Min/Milling)	CE Index
avg	2400	472	483	192
Jan	2300	---	---	
Feb	2310	458	469	187
Mar	2317	---	---	188
Apr	2327			188
May	2357	469	479	189
June	2410	---	---	190
July	2414			192
Aug	2445	477	487	193
Sept	2468	---	---	194
Oct	2478			196
Nov	2486	485	497	196
Dec	2490	---	---	197
				198

CAPITAL COST ESTIMATION

The *total capital investment* in a mineral processing plant consists of a *fixed capital* portion and a *working capital* portion. The fixed capital cost portion is the total amount of money needed to purchase the necessary plant and auxiliary plant facilities. The working capital cost portion represents cash that must be available to begin the operation. Of major importance is the fact that the total capital investment is a sum that is separate from the *total product cost* paid out per unit time from gross income per unit time. Total product costs include *operating costs* such as direct production costs, fixed charges and plant overhead, and *general expenses* such as administrative costs, distribution and marketing costs, research and development costs and gross earnings expenses.

Fixed capital cost estimates may be based on as little prior information as plant capacity or on as much as complete process and building specifications. Out of this range, five major types of estimations have been proposed by the American Association of Cost Engineers. These are:

1. Order-of-magnitude estimate based on previous cost data. It is a ratio estimate of more than ±30 percent probable accuracy.
2. Factored estimate based on major equipment costs with a probable accuracy of ±30 percent.
3. Budget authorization estimate (preliminary estimate) based on sufficient data to permit budgeting. Probable accuracy is ±20 %.

4. Definitive or project control estimate based on almost complete data (no specifications or complete drawings). Probable accuracy is ±10 percent.
5. Detailed or contractor's estimate based on complete engineering drawings, specifications, and site surveys. Probable accuracy is ±5 percent.

Types 1., 2. and 3. of the above are preliminary (predesign) fixed capital cost estimates and require much less detailed information than types 4. and 5.

Preliminary capital cost estimates are useful to process engineers who are involved in process selection and design. Since most companies have a limit to available capital funds, an immediate assessment of initial cash requirements is provided. In situations where the total product cost of alternative processes are similar, the one with the smallest capital expenditure is selected. Thus a preliminary estimate serves to assess processing alternatives.

Type 1. estimates do not have the flexibility of the factored cost estimate (type 2.); the latter allows personal judgments to be made. Type 3. estimates require more information than either type 1. or type 2. and, therefore, are more time-consuming. When the feasibility of a process is still being evaluated, the extra accuracy of a type 3. estimate is not justified. Therefore, this chapter stresses estimates in categories 1. and 2.

Working Capital Cost Estimates

Working capital cost estimates are obtained often by taking a percentage of the fixed capital investment. Approximately 10 percent is a reasonable average.

Provided the corresponding information is available, working capital cost may be estimated from the following:

a.	Raw materials inventory (1 month's supply @ cost)	$....
b.	Materials-in-process inventory (1 month's supply @ cost)	$....
c.	Product inventory (1 month @ mfg. cost)	$....
d.	Accounts receivable (1 month @ selling price)	$....
e.	Available cash (to meet expenses of wages, raw material, utilities, supplies for 1 month @ mfg. cost)	$....
	Working Capital = a + b + c + d + e =	$....

Fixed Capital Cost Estimates

At least four ways to obtain type 1. and type 2. estimates of fixed capital requirements appear to have been used in mineral processing (2). These are the six-tenths rule, the plant cost ratio method, the equipment cost ratio method and the plant component cost ratio (factored estimate) method. To use the last three methods, the following information is helpful: a rough flowsheet showing major items of equipment and their corresponding sizes along with sufficient information to estimate plant size and complexity. Material balances and energy balances will be of value.

Six-Tenths Rule. To use the six-tenths rule, the fixed capital cost of a plant of known capacity must be available. In such cases

$$\frac{(\text{cost})_1}{(\text{cost})_2} = \left(\frac{(\text{capacity})_1}{(\text{capacity})_2}\right)^{0.6}$$

For example, suppose a 30,000 TPD concentrator cost $\$60\text{x}10^6$ to build in 1970 when the M&S mining and milling index was 303. Suppose that the cost of a similar 40,000 TPD facility is desired at the beginning of 1977. Using the above rule,

$$\begin{aligned}(\text{cost})_1 &= 60\text{x}10^6(40{,}000/30{,}000)^{0.6}\\ &= \$71.3\text{x}10^6 \text{ at a cost index of } 303\end{aligned}$$

To update the cost, the current cost index must be employed. Table 2 shows that the index at the beginning of 1977 must be close to 497. Hence,

$$\begin{aligned}(\text{cost})_{\text{now}} &= 71.3\text{x}10^6(497/303)\\ &= \$117\text{x}10^6\end{aligned}$$

The exponent, 6/10, is an average and depends on the type of plant. Such factors as type of site, economic conditions prevalent, geographic location and regional productivity are responsible for a substantial variation.

Plant Cost Ratio Method. This method requires an estimate of the cost of major items of delivered process equipment. If this cost is equal to N, then (see reference (2)):

cost of solid process plant	= 3.10xN
cost of solid-fluid process plant	= 3.63xN
cost of fluid process plant	= 4.74xN

Reasonably accurate estimates of equipment costs must be available.

For example, suppose that the current cost of major equipment items in a secondary crushing plant happen to be $\$5\text{x}10^6$. The cost of the installation is estimated as

$$\begin{aligned}\text{cost} &= 3.10\text{x}5\text{x}10^6\\ &= \$15.5\text{x}10^6\end{aligned}$$

Equipment Cost Ratio Method. Compared with the above method, more accuracy results from multiplying categories of equipment of similar nature by corresponding factors and then calculating the resulting sum. Let the cost of the ith major process unit by X_i. Then a plant cost associated with item i is equal to F_iX_i where F_i is an appropriate factor.

The cost of the total plant is calculated from

$$\text{cost} = \sum_{i=1}^{i=n} F_iX_i$$

for a plant consisting of n items of major equipment. Table 3 shows typical F_i values (2). The table may be expanded after several projects have been analyzed.

TABLE 3: Equipment Cost Ratios

Equipment Category	Factor, F_i
Bucket elevator	2.0
Mixer	2.0
Furnace	2.1
Drum Type dryer	2.2
Kiln	2.2
Conveyor	2.3
Compressor	2.3
Electrostatic precipitator	2.5
Blower, fan	2.5
Refrigeration unit	2.5
Boiler	2.8
Mills	3.0
Vacuum rotary dryer	3.2
Dry dust collector	3.5
Storage tank	3.5
Crusher	3.5
Process tank	4.1
Instrumentation	4.1
Heat exchanger	4.8
Wet dust collector	6.0
Pump	6.8
Electric motor	8.5

Plant Component Cost Ratio Method. This method provides considerable flexibility and involves a breakdown of fixed capital costs into components whose costs are a ratio of major equipment costs. Table 4 is an example of the plant component cost ratio method (factored estimate method) suggested for chemical plants (3). An alternative to Table 4 is given in Table 5 (see reference (2)).

TABLE 4: Factored Capital Cost Estimate Guide[3]

1. Purchased equipment costs from references and on current index basis..............................$000,000
2. Installed equipment costs
   a. From references and on current index basis or
   b. Item 1 multiplied by 1.43..............................$000,000
3. Process piping..............................$000,000

Type plant:	Percent of Item 2:
Solid	7-10
Solid-Fluid	10-30
Fluid	30-60

4. Instrumentation..............................$000,000

Amount of automatic control:	Percent of Item 2:
None	3-5
Some	5-12
Extensive	12-20

5. Buildings and site development..............................$000,000

Type plant:	Percent of Item 2:
Outdoor	10-30
Outdoor-Indoor	20-60
Indoor	60-100

6. Auxiliaries (e.g., electric power)..............................$000,000

Extent:	Percent of Item 2:
Existing	0
Minor additions	0-5
Major additions	5-25
New facilities	25-100

7. Outside lines..............................$000,000

Average length:	Percent of Item 2:
Short	0-5
Intermediate	5-15
Long	15-25

8. Total physical plant costs = Sum of Items 2+3+4+5+6+7....$000,000
9. Engineering and construction..............................$000,000

Complexity:	Percent of Item 8:
Simple	20-35
Difficult	35-60

10. Contingencies..............................$000,000

Type process:	Percent of Item 8:
Firm	10-20
Subject to change	20-30
Speculative	30-50
Average	30

11. Size factor..............................$000,000

Size plant:	Percent of Item 8:
Large commercial	0-5
Small commercial	5-15
Pilot plant	15-35

12. Total plant of fixed capital costs = Sum of 8+9+10+11....$000,000

TABLE 5: Plant Component Cost Ratio Method(2)

1. Purchased equipment costs from references and on current index basis..............................$000,000
2. Equipment installation (0.17 to 0.25 times Item 1).......$000,000
3. Piping, material and labor, excluding service piping (0.07 to 0.25 times Item 1)..............................$000,000
4. Electrical, material and labor, excluding building lighting (0.13 to 0.25 times Item 1)..............................$000,000
5. Instrumentation (0.03 to 0.12 times Item 1).............$000,000
6. Process buildings, including mechanical services & lighting (0.33 to 0.50 times Item 1)..............................$000,000
7. Auxiliary buildings, including mechanical services & lighting (0.07 to 0.15 times Item 1)..............................$000,000
8. Plant services, such as fresh water systems, sewers, compressed air, etc., (0.07 to 0.15 times Item 1).................$000,000
9. Site improvements, such as fences, roads, railroads, etc. (0.03 to 0.18 times Item 1)..............................$000,000
10. Field expenses related to construction management (0.10 to 0.12 times Item 1)..............................$000,000
11. Project management including engineering and construction (0.30 to 0.33 times Item 1)..............................$000,000

12. Fixed capital cost (sum of 1+2+3+4+5+6+7+8+9+10+11).....$000,000

MAJOR EQUIPMENT COSTS

Manufacturers of mineral processing equipment were contacted and prices were obtained for major items of equipment (1). Invariably, such costs were related to an equipment parameter by the equation:

$$\text{cost} = a(X)^b$$

where X is the equipment parameter and a and b are constants. Data were fitted by least squares methods. When necessary to maintain predictions within 10 percent, equations of the above form were fitted to cover different ranges of X. Thus a cost versus X curve might be described by as many as 3 or 4 equations, each with different constants appropriate to a specific range of X.

Table 6 shows cost versus equipment parameters for various mineral processing units. Ranges for X are listed. Note that a substantially more complete tabulation is found in reference (1). The equations may be extrapolated outside their ranges with extreme caution only. Clearly, the resulting cost is not a direct quote! Each calculated cost corresponds to an M&S index, mining and milling, of 300.

TABLE 6: Basic Equipment Cost Data

Item	Parameter, X	Units of X	Range of X	a	b
cone crusher	diameter of the discharge annulus	feet	2-7	4154	1.80
primary gyratory	receiver opening times mantle diameter	sq. in.	30x55-60x109	3.58	1.41
jaw crusher	area of feed opening	sq. in.	15x24-60x48	6.0	1.28
hammer mill	area of feed opening	sq. in.	6x4-46x60	217	0.67
ball mill	motor HP	HP	40-1750	3410	0.54
rod mill	motor HP	HP	50-1500	2955	0.57
cyclones	diameter	inches	6-30	132	0.91
SRL pump	capacity	USGPM	250-5000	31	0.64
dust collector	capacity	cu. ft. per min.	2000-10000	10.6	0.72
double deck vibrating screen (secondary)	screen area	sq. ft.	6x16-8x20	376	0.76
spiral classifier	spiral diameter	inches	24-78	30	1.53
table	deck area	sq. ft.	24-90	618	0.35
flotation cell	volume	cu. ft.	100-300	71	0.74
jig	capacity	ton/day	45-400	1120	0.17
steel mixing tank	volume	US Gal.	100-250,000	44	0.53
disc filter	filter area	sq. ft.	22-1800	1194	0.44
drum filter	filter area	sq. ft.	12-720	4331	0.33
thickener	diameter	ft.	10-225	147	1.38
vib. feeder	feeder area	sq. in.	2880-13824	1.28	0.99
36" belt conveyor	length between centers	ft.	200-1000	34	1.13
motor 1800 rpm	horsepower	HP	250-600	15.3	1.11

Calculation of Cost of Major Equipment

The information in Table 6 may be used to estimate the cost of equipment listed there. Suppose that the cost of five 30-inch diameter cyclones is desired. The cost equation from Table 6 is

$$\text{cost} = 132(\text{diameter})^{0.91}$$

so that cost = $132(30)^{0.91}$ = 2916 dollars per cyclone at an M&S index (mining and milling) of 300. Since the current value of this index is about 515 (September, 1977), then the current cost is 2916(515/300) or $5005 per cyclone. Since five cyclones are involved, then the total cost is $25,025.

It is apparent that, given a flowsheet which lists major equipment properly specified, capital costs based upon major equipment cost may be estimated.

COMPUTER PROGRAMS FOR CAPITAL COST ESTIMATIONS

Using cost equations fitted by least squares techniques (1) in conjunction with a factored estimate table such as Table 4, a computer program to estimate capital costs has been written (4). The program is in the BASIC language and permits direct operator-computer communication via a remote terminal. The program, which uses a "game player" approach, has been tested on the CGE Mark II time-sharing terminal system.

Figure 1 is a main flow diagram of the computer program. The user calls CAPCOST to initiate the sequence.

Program CAPCOST reminds the user that costs are based on an M&S index of 300 and requests that the current value be entered. Input and calculated information, such as the current M&S value and a calculated cost value, is stored and/or retrieved via file COSTCOM which is common to all programs. After an M&S index value is entered, CAPCOST lists the types of processing equipment available and then chains with program BRANCH.

Program BRANCH directs the computational sequence to any of eight subroutines as specified by the user. Sub-Programs CRUSH, MILLS,SIZE, SSSEP, LEACH, SLSEP, and FLOW contain operating instructions and equipment parameters. As the user follows instructions typed by the computer and enters data as requested, equipment costs are calculated. Cost data are stored and/or retrieved in file COSTCOM. After all equipment has been costed, the user directs program BRANCH to chain to program REPORT.

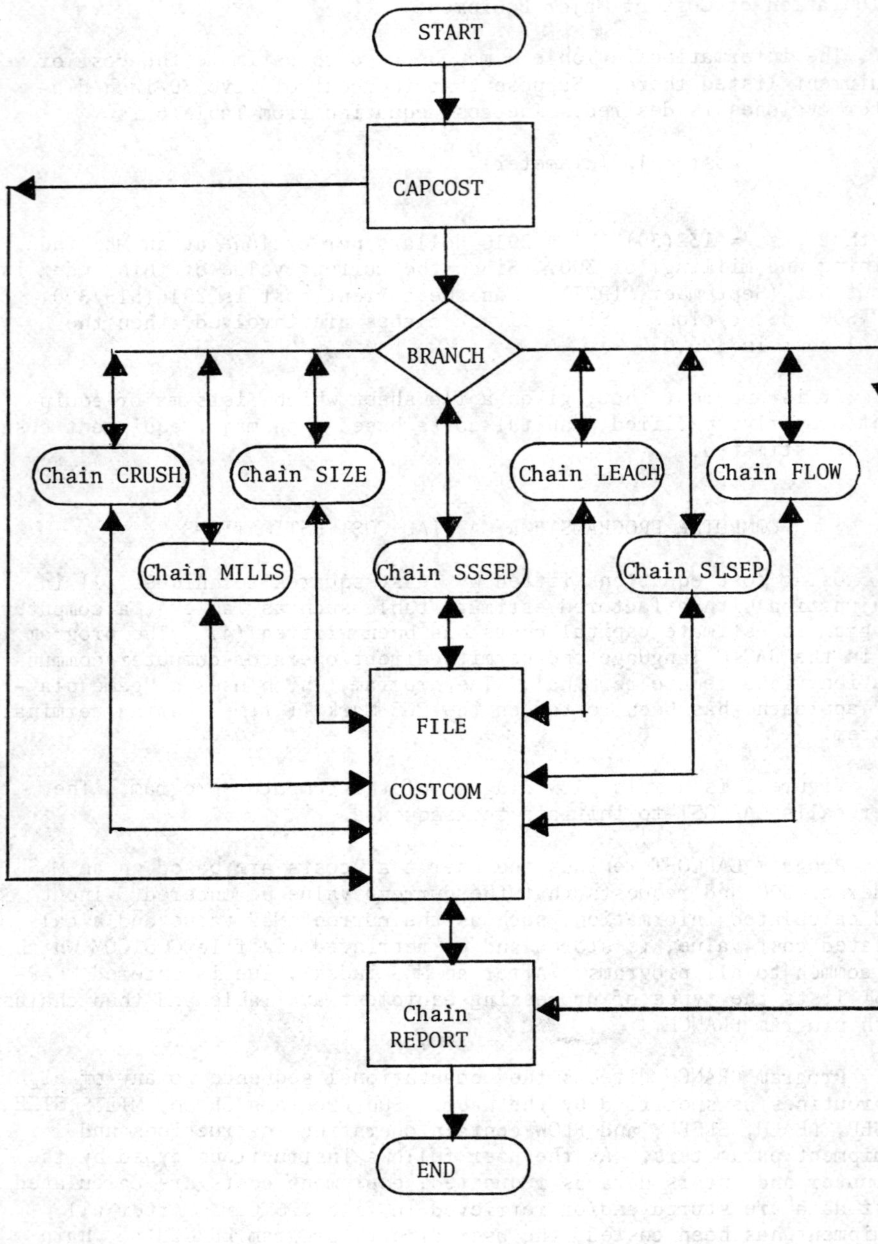

FIGURE 1: Flow Diagram of Computer Program for Capital Cost Estimations.

Program REPORT asks the user to supply any equipment costs not available in the program and then requests that cost factors be entered. A final print-out of the total capital investment is then provided.

All sub-programs require user-supplied specifications. Sub-Programs will list the specifications required, along with ranges for which equations are considered accurate. Categories of equipment are:

Sub-Program CRUSH contains crushing equipment such as cone, primary and secondary gyratory, impact, jaw and roll crushers; hammer mills and pulverizers.

Sub-Program MILLS contains grinding equipment including autogenous, ball, rod and pebble mills.

Sub-Program SIZE contains sizing and classifying devices such as cyclones, screens and classifiers.

Sub-Program SSSEP comprises solid-solid separation equipment including tables, jigs, electrostatic & magnetic separators, spirals and flotation cells.

Sub-Program LEACH includes leaching equipment and ion exchange units. Autoclaves, agitators, blenders, mixers and tanks are listed.

Sub-Program SLSEP contains solid-liquid separation devices such as filters, centrifuges, dryers, thickeners and hydroseparators.

Sub-Program FLOW includes blowers, conditioners, conveying equipment, feeders, motors, pumps and dust collection equipment.

Capital cost estimation computer programs are useful because (1) in universities the teaching of process design is facilitated, (2) in industry the process engineer rapidly obtains a capital cost estimate for preliminary feasibility investigations, and (3) in future the program will be incorporated into simulation and optimization programs to facilitate optimal flowsheet design.

EXAMPLE CAPITAL COST ESTIMATION

Figure 2 shows the flow diagram of a secondary crushing plant, where each major process unit has been numbered. This flowsheet is typical of some that are found in the literature. Several important details are missing. For example:

(1) The lengths and widths of conveyor belts are missing. This information is necessary to cost these items.

(2) The equipment necessary to effect transfer of material from ore storage onto belts, from surge bins onto screens and tertiary crushers is not listed.

(3) Motor horsepowers are missing.

(4) Information on whether the plant is to be heated and on the type of dust collection system is lacking.

(5) It is not clear whether the screens are housed separately.

Note the importance of specifying those equipment parameters, such as capacity or horsepower, which have been related to the cost

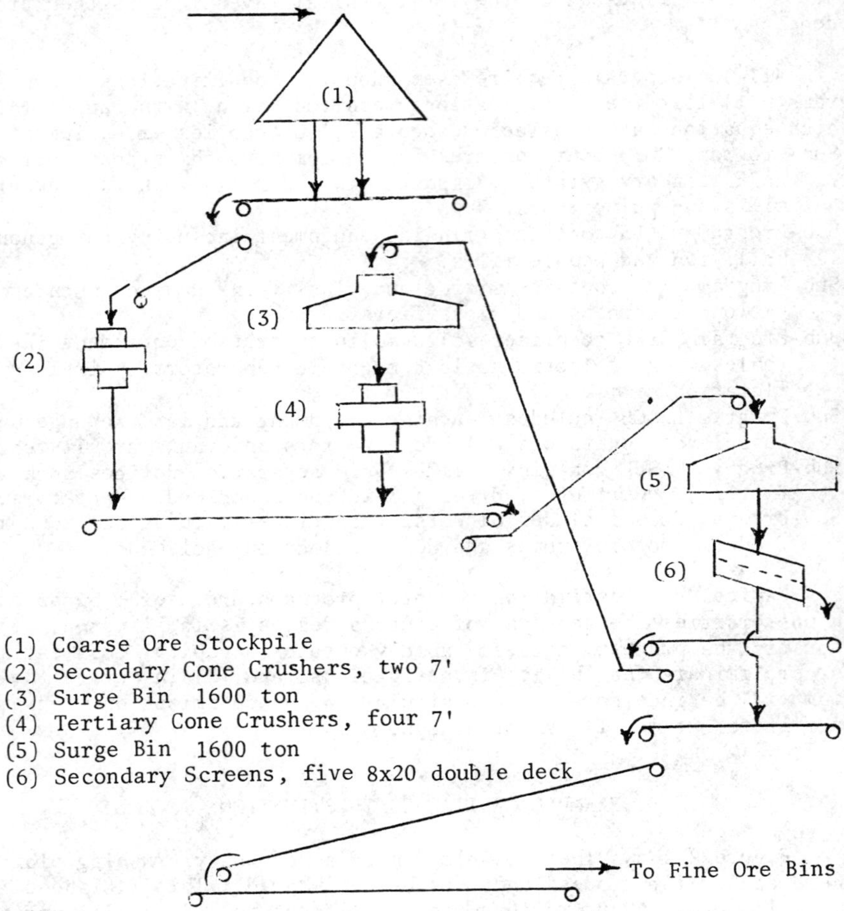

FIGURE 2: Flow Diagram of Secondary Crushing Plant

of the process unit. This requires a familiarity with cost equations or graphs that are in use.

To be somewhat more precise, the equipment list should have the following information:

Secondary/Tertiary Crusher Building
- -under coarse ore stockpile, 48-in. belt conveyor with four 40"x72" vibrating feeders, 1500 ft. of belt (two)
- -50 ft of 54" transfer belt conveyor (two)
- -7' standard cone crusher, 350 HP, (two)
- -200 ft of 72" belt conveyor
- -20,000 cu. ft. surge bin
- -vibrating feeder, 40"x72" (four)
- -7' short head cone crusher, 300 HP (four)
- -40 ft. of 54" transfer belt (four)

Screen Building/Transfer Tower
- -20,000 cu. ft. surge bin
- -vibrating feeders 80"x94" (five)
- -75 ft of 60" transfer belt
- -600 ft of 48" transfer belt
- -600 ft of 60" transfer belt

Note that belt lengths are estimates at this stage, for a preliminary estimate is all that is required. Dust collection systems may be costed on the basis of roughly 40,000 cfm.

Conveyor Belt Cost Estimate

Table 6 shows the cost equation for 36" conveyor belts. Using an M&S index of 500 as current, then the cost is found from

$$\text{cost} = (500/300)\text{x}34\text{x}(\text{length})^{1.13}$$

To estimate the cost of larger belt widths, the 0.6 rule is used. For example, the cost of a 600 ft 36" belt is about

$$\text{cost} = (500/300)\text{x}34\text{x}(600)^{1.13} = \$78{,}100$$

A 48" belt would cost about

$$\text{cost} = 78{,}100(48/36)^{.6} = \$92{,}800$$

Thus belt costs are summarized below:

Length	$ for 36" belt	Belt size	No. of belts	$ using 0.6 rule
40	3660	54 in.	4	18,700
50	4710	54 in.	2	12,000
75	7450	60 in.	1	10,100
200	22600	72 in.	1	34,300
600	78100	48 in.	1	92,800
600	78100	60 in.	1	106,000
1500	220000	48 in.	2	261,000

The approximate cost for conveyor belts is thus \$535,000 at an M&S index of 500.

Vibrating Feeder Cost Estimate

Table 6 indicates that at an M&S index of 500, vibrating feeders cost about

$$\text{cost} = (500/300)\text{x}1.28\text{x}(\text{area})^{0.99}$$

Therefore 12 feeders of 40"x72" area each will cost about \$68,100, whereas 4 feeders of 80"x94" area should be around \$73,400. The total cost is about \$142,000.

Surge Bin Cost Estimate

Assume that bins cost about the same as steel tanks, where each tank contains about 150,000 gallons (20,000 cu. ft.). From Table 6, each bin costs about

$$\text{cost} = (500/300)\text{x}44\text{x}(150{,}000)^{.53} = \$40{,}500$$

so that two bins is around \$81,000.

Vibrating Screens Cost Estimate

Table 6 shows that the cost of an 8'x20' double deck vibrating screen is around

$$\text{cost} = (500/300)\text{x}376\text{x}(160)^{.76} = \$30{,}000$$

Since there are five screens, the total cost is roughly \$150,000.

Crusher Motor Cost Estimate

From Table 6, motor costs are about $(500/300)\text{x}15.3\text{x}(\text{HP})^{1.11}$. A 350 HP motor costs \$17,000, while a 300 HP motor costs \$14,300. Total motor costs are approximately 2x17,000 + 4x14,300 = \$91,000.

Cone Crusher Cost Estimate

From Table 6, a 7' cone crusher costs $(500/300)\text{x}4154\text{x}(7)^{1.8}$ or about \$230,000. Since there are six, then total crusher costs are around $\$1.38\text{X}10^6$.

Dust Collector Cost Estimate

A 10,000 cfm collector costs $(500/300)\text{x}10.6\text{x}(10^4)^{.72} = \$13{,}400$. Employing the 0.6 rule the cost of a 40,000 cfm unit would be around \$30,800. Assume 4 of these are used so that the total cost is about \$123,000.

Total Major Equipment Costs

Major equipment costs are summarized as follows:

Item	$ Cost
Conveyor belts	535,000
Vibrating feeders	142,000
Surge bins	81,000
Vibrating screens	150,000
Crusher motors	91,000
Cone crushers	1,380,000
Dust collectors	123,000

The above table shows that the total major equipment costs are about \$2,502,000 at an M&S index of 500.

Factored Capital Cost Estimate

Once the cost of major equipment has been estimated, a preliminary capital cost may be determined. Either Table 4 or Table 5 is useful for this purpose. Note that such tables are easily modified to reflect unique cost features such as geographic location (e.g., Arizona versus Tasmania).

Using Table 5, the estimated capital cost is summarized below.

1. Purchased equipment costs.............................\$2,502,000
2. Equipment Installation (0.24 times Item 1.)........... \$600,480
3. Piping (0.16 times Item 1.)........................... \$400,320
4. Electrical (0.19 times Item 1.)....................... \$475,380
5. Instrumentation (.08 times Item 1.)................... \$200,160
6. Process Buildings (0.45 times Item 1.)................\$1,125,900
7. Auxiliary Buildings (0.12 times Item 1.).............. \$300,240
8. Plant Services (0.10 times Item 1.)................... \$250,200
9. Site Improvements (0.05 times Item 1.)................ \$125,100
10. Field Expenses (0.10 times Item 1.).................. \$250,200
11. Project Management, Engineering (0.32 times Item 1.)... \$800,640

12. Fixed Capital Cost (1+2+3+4+5+6+7+8+9+10+11)..........\$7,030,620

The estimated capital cost is calculated to be about \$7,000,000. Clearly, this figure could differ between individuals because the selection of multiplying factors depends on personal judgment.

SUMMARY AND CONCLUSIONS

By means of cost indices, major equipment costs may be updated to permit current equipment cost estimates. Such information is employed to obtain preliminary capital cost estimates for immediate assessments of processing alternatives. A computer program has been written for this purpose, where valid cost equations are employed in the program. Such programs are useful to everyone.

An example of the factored capital cost estimation method is provided. Note that equipment parameters related to cost must be used when specifying major equipment.

It must be stressed that the methods discussed in this chapter are for preliminary estimations--not for definitive ones. The methods can be in error by on the average ±30 percent. However, they are extremely versatile and their accuracy may be improved considerably with experience. Factored estimate tables may be modified to fit any desired situation.

REFERENCES

1. Mular, A. L., with E. A. Parkinson, Mineral Processing Equipment Costs and Preliminary Capital Cost Estimations, Special Volume 13, Canadian Institute of Mining and Metallurgy, Montreal, Quebec,1972

2. Balfour, R. J., and Papucciyan, T. L., "Capital Cost Estimating for Mineral Processing Plants", 4th Annual Meeting of Canadian Mineral Processors, January 25-27, 1972, Ottawa, Ontario.

3. Vilbrandt, Frank C., and Dryden, Charles E., Chemical Engineering Plant Design, McGraw Hill, N. Y., 1959.

4. Mular, A. L., with E. Zickar, "A Computer Program for Mineral Processing Capital Cost Estimation", CIM Bulletin, May, 1973.

Chapter 4

OPERATING COST

F. Milton Lewis

Roshan B. Bhappu

Metallurgical Consultant
Mtn. States Mineral Enterprises
Tucson, Arizona

Vice President & Gen. Mgr.
Mtn. States Research & Devel.
Tucson, Arizona

ABSTRACT

The methods for preparing operating cost estimates for a proposed mining venture are outlined.

The procedures for estimating the direct and indirect operating cost for proposed mining and ore processing facilities are reviewed. Profits from the by-products are credits to the total operating cost, direct plus indirect, and are subtracted to obtain the total net operating cost. The non-operating costs which are a part of the total expenses are defined but are not considered in this manuscript.

A method for developing the operating cost for a typical copper concentrator in the southwest is described in detail.

INTRODUCTION

Capital and operating costs estimates on a proposed mining project must be completed prior to calculating its profitability. Generally, the capital cost estimate is finished before the operating cost estimate. The annual estimated net revenue from sales, the annual total expenses and annual capital outlays are required for calculating the cash flows by years for the life of the project, the rate of return on investment, and the payback time. The results from these calculations will determine the viability of the project.

Since operating cost is one of the major expense items, it must be estimated prior to evaluating the proposed project. There are several methods for estimating operating cost with different degrees of accuracy. Early in a project, a less accurate estimate may be all right, since it is less expensive and the owner wants to spend minimum funds until data indicate that the project is viable. However, in this manuscript it is assumed that the preliminary study indicated that the investment on the project will be attractive and that an accurate operating cost estimate is desirable. In this type of estimate, the annual operating expense should be estimated by years for several years as the annual operating expense will probably vary. The production may not reach design capacity for some time after start-up, and the operating cost may be different by years as the mining plan may not produce uniform ore grade and tonnage year after year. These annual operating costs are dependent on the ore grade, ore to waste ratio, mine production rate and product output from the surface plants. Thus, it is necessary to estimate them for the years in which there will be planned changes in operating conditions.

In the mining industry, it is common practice to report operating cost for the mine and the concentrator in dollars per year and cost per ton of ore and for processing plants in dollars per year and cost per unit of product. Normally, operating costs are escalated from the time that the estimates are made to the time in which the start-up is scheduled and not escalated further, since it is assumed that over a period of years the rate of escalation or revenue from sales will equal the rate of escalation of operating costs.

Direct operating costs are estimated separately from the mine, concentrator and processing plants, and combined for the total direct operating cost for the project. This cost plus the indirect operating cost is the total operating cost. The charges for corporate headquarters plus the total operating cost is the total expense for operating these facilities. However, the charges from headquarters are not considered in this manuscript.

DIRECT OPERATING COST

The procedures for estimating the direct operating cost in all departments of a mining operation are similar and include the following items:

- Supervision
- Operating Labor
- Maintenance Labor
- Maintenance Overhead
- Electric Power
- Fuel
- Water
- Consumable Materials
- Maintenance Parts and Supplies
- Operating Supplies
- Charges from Central Shops

- Charges for Services from Other Departments
- Payroll Burden*
- Miscellaneous

The procedures for developing the direct operating cost for a typical copper concentrator in the southwest are explained later in this paper. Similar procedures are used for developing the direct operating cost for the mine and other processing facilities.

* In some companies, the labor burden from all departments is combined and is included in indirect operating cost. In this exercise, the labor burden for each department is included in department's direct cost.

INDIRECT OPERATING COST

Normally, the indirect operating cost is not allocated to the different departments and the following items are included in this cost:

- Salaries and Labor:
  - Administration
  - Technical Services
  - Laboratory
  - Personnel and Public Relations
  - Safety Services
  - Engineering Office
  - Environmental Expenses
  - Warehouse Expenses
  - Accounting Department
  - Security
- Payroll Burden on the Above Personnel
- Power, Heat and Maintenance for:
  - General Office
  - Laboratory
  - Warehouse
  - Field Offices
- Distribution of Electric Power
- Office Supplies
- Laboratory Supplies
- First Aid Supplies
- Communications
- Automobile Allowances
- General Manager's Expense Account
- Gifts, Donations, Etc.
- Yard Maintenance and Landscaping

- Local and State Taxes
- Hazard Insurance
- Operating Licenses
- Miscellaneous

The annual indirect cost is estimated by considering separately each of the foregoing items. The cost for each item should be estimated and, of course, the total indirect cost will be the sum of the costs for all of these items.

BY-PRODUCT CREDITS

The cost for producing the by-products is included in the operating cost. Therefore, the profits from these by-products are credits to the operating cost and these profits will reduce the total operating cost.

NON-OPERATING COST

The non-operating cost, allocated to the project from corporate headquarters may include the following expenses:

- Executive Offices
- Accounting and Auditing
- Engineering
- Research and Development
- Sales
- Financing
- Purchasing
- Legal Services
- Environmental Protection

Estimating these expenses is not a part of this paper.

DIRECT OPERATING COST FOR A TYPICAL COPPER CONCENTRATOR

Direct operating cost for a concentrator includes all operating expenses from receiving the ore at the primary crusher to delivering the concentrates to a smelter or to a rail siding for shipping to a distant smelter and disposing of tailings with a water reclaiming system. Usually, the new water is supplied at cost from a central water system and the concentrator is charged its proportional part of the total. The cost for materials and supplies includes the expense for delivering these to the concentrator plus sales or use tax.

Metallurgical data, operating schedules for crushing and concentrating plants, plot plans, a flowsheet with materials balance, and equipment list with connected horsepower, size reduction within the comminution system and hardness of the ore (Work Index) are available and are used in estimating the direct operating cost for the concentrator.

Salaried Personnel

An organization chart is prepared showing the number and title of all supervisors, staff assistants and clerks. Salaries for these personnel are estimated to be equal to the current salary rate structure in similar concentrators in the southwest.

Operating Labor

The number of operators per shift can be estimated by reviewing the plot plans and flowsheet. These will show the areas to be covered by the different operators and proximity of the operating areas. Job duties and responsibilities for each operating area are defined, and thus reasonably sized operating crews can be projected. By reviewing the operating schedules and the number of operating shifts per day, the total number of operators per day can be estimated quite accurately.

In addition to these operating crews, there is one crew of laborers on day-shift. After all vacancies are filled in the operating crews, the remaining extra trained operators are assigned to this crew along with the untrained laborers. The work requirements for laborers are reviewed and the number of manshifts to handle this work load per operating day is estimated. Thus, the size of the operating crews plus the labor crew is established. The rates for the different job classifications should be equal to average rates for similar jobs in other concentrators in the area. With this information, the daily cost for operating labor can be calculated. Normally, about ten percent is added to this cost for overtime and training operators. The expense of holidays and vacations for operators is never considered in the operating cost since these expenses are included in the payroll burden. The annual cost for operators is calculated by multiplying the cost per day by the number of operating days per year. The annual cost divided by the annual tons of ore to the concentrator gives the cost per ton of ore.

Maintenance Labor

At most mines, all maintenance crews work under the jurisdiction of a central maintenance department. However, a maintenance crew with its own supervisors is assigned to the concentrator to do the routine maintenance and repair jobs. Typical job classifications in this crew are: electrician, welder, pipefitter, repairman and helper. The rates for the different classifications should be equal to average rates in other concentrators in the area. Their wages are charged directly to the concentrator operating accounts, but the salaries for the supervisors are charged through the central maintenance office to the concentrator in the maintenance overhead expense. From the equipment list, a repair schedule is developed, based on projected life of all wear parts, the estimated man-hours required for changing out these worn parts, and the estimate of man-hours required for the preventive maintenance program on machinery, electrical components and other equipment. A certain number of man-hours must be included for repairs due to mechanical failures. These man-hours can be divided into the different job classifications and totaled for the year. From all of this information, the size of the maintenance crew and the labor cost per year are calculated.

Maintenance Overhead

The salaries of the supervisors assigned to the concentrator plus an allocation of all general expenses in the central maintenance department is charged to the concentrator. The allocation cost can be calculated by analyzing the proposed organization of the central maintenance staff and estimating the total general expense to be allocated to the departments and the percentage of the total work which will be for the concentrator. This percentage multiplied by the total general expense will be the overhead that is charged to the concentrator.

Payroll Burden

This is the expense to the company for the employees in excess of the direct payroll. The payroll burden amounts vary, but generally it is approximately 35 percent of the total concentrator payroll and includes the following items:

- Payments to employees while on vacation and for holidays
- Payment for employees to the pension plans
- Health and life insurance
- F.I.C.A.
- Workmen's compensation
- Unemployment tax
- Industrial accident expense
- Payment to union officers and stewards while attending union-company meetings

The sum of the first four items in the list averages approximately 85 percent of the total cost of the labor burden. Generally, enough information will be available for calculating quite accurately the labor burden cost.

Electric Power

Normally, the power will be delivered to a primary substation and will be billed monthly on contract rate based on demand charges plus energy charges. This power is distributed to the mine, concentrator and other facilities. Internally, the cost of the power delivered to the mill is calculated in total kilowatt-hours and total monthly cost. Prior to estimating the operating cost, the power company is contacted and furnished with data on the amount of power that will be used. A rate schedule is requested and this schedule is used in calculating power cost.

Sufficient information is available to calculate the kilowatt-hours consumed in the concentrator per ton of ore. The Bond Work Index for the ore, the size of feed, and the size of ground product in the primary grinding circuit is information that was established prior to designing the crushing and grinding circuits. Thus, the kilowatt-hours per ton of ore in the primary grinding circuit are known and this is generally over 50 percent of the total power required for the entire concentrator. The balance of the power is estimated in kilowatt-hours per ton by adding the total connected horsepower and assuming that 80 percent of this horsepower represents the amount consumed. This power is calculated in kilowatt-hours per ton of ore and dollars per year.

Fuel

In some concentrators, it is necessary to heat the pulp in certain flotation circuits. The volume of pulp was calculated in the materials balance data and the temperature of the flotation circuit was established in the test work. Thus, the amount of heat required for these circuits can be calculated and the cost can be estimated. Fuel is used for heating offices,operating stations, changehouse and bath water. This cost can be estimated.

Water

Eight to nine hundred gallons of water per ton of ore are used in the copper concentrators in the southwest. About 75 percent of this water is recycled and the remaining 25 percent is new water. The locations of the reclaimed water stations, the distance and elevations for pumping this water were determined in the design so that the quotations for correct size pumps and motors were supplied by the vendors prior to making the capital cost estimate. With these pump costs and power requirements, the cost for pumping the reclaimed water is calculated. Likewise, the sources for new water, quantity of new water and total dynamic heads on the pumps were determined. With this information, the cost for pumping the new water is calculated. The sum of the cost for recycled water and new water is the concentrator's operating water expense.

Consumable Materials

Rods and Balls. The consumption of rods and balls is proportional to power consumed in the grinding mills. The average consumptions per kilowatt-hour on porphyry ores are 0.18 pounds of rods and 0.16 pounds of balls.

Reagents. The flotation and water treatment reagents were determined by test work on representative samples. The delivered cost for these reagents is made from quotations on reagents and schedules for freight and state taxes. With information on quantities and delivered prices, the reagent cost per ton is calculated.

Maintenance Parts and Supplies

All replacement wear parts such as liners, screen cloths, conveyor belts, drive belts, slurry pump liners, impellers, etc., are classified as maintenance parts. An estimated life for these parts is determined by experiences compiled by the manufacturer and from actual records in other similar concentrators. Thus, the annual cost of these wear parts can be estimated quite accurately. All mill machinery and equipment are also classified as repair supplies. The annual cost for these supplies is much more difficult to estimate, since the preventive maintenance program is the important factor. However, a judgment estimate is made based on what is known about similar operations.

Operating Supplies

These include the light materials, such as filter cloth and frames, lubricants, tires, instrument parts, concentrator office and laboratory materials, charts, forms, etc. The annual cost of these supplies is estimated from reviewing the flowsheet and determining reasonable quantities which will be required. The delivered cost for these supplies can be determined from current prices including freight and taxes and the total annual operating supply expense is calculated.

Charges from Central Shops

The concentrator's normal and routine maintenance is taken care of by the maintenance crew assigned to the concentrator. However, there will be major repair jobs and shop work which are out of the scope of work planned for the concentrator's maintenance crew. It is difficult to estimate the cost of this work, but normally it amounts to about ten percent of the total maintenance labor cost.

Charges for Services from Other Departments

These charges are from the central laboratory, central metallurgical office, engineering office, safety office, and warehouse. In some companies, all of these charges are included in the indirect operating cost, while in other companies, these charges are treated as direct cost items and must be estimated and included in the direct operating cost for the concentrator. These may be estimated after reviewing the amount of work which will be performed by the service department.

Miscellaneous

There will be expenses which were overlooked in making the preceding calculations and other expenses which cannot be forecast. The cost for these items is grouped into a miscellaneous account. Three to five percent of the total operating cost is an ample amount to include.

Summary

The operating cost estimate is presented in a number of tables which are detailed so that they are easy to read. Normally, the following tables will be prepared:

- Table I — Summary of All Operating Costs by Items
  All expense items in dollars per year and cost per ton are tabulated. The sum of these items, which is the total operating cost, is included.

- Table II — Summary of All Operating Costs by Unit Operations
  The expense for each unit operation such as crushing, grinding, flotation, filtering, etc., are calculated and tabulated in dollars per year and cost per ton. The sum of these unit operations is the total operating cost and is the same as the total in Table I.

- Table III — Salaries
  Job descriptions, number of persons on each job, annual salaries and total annual salary expense are shown in this table.

- Table IV — Operating Day-Pay Labor
  Job classifications, number of operating shifts per week, number of manshifts per week, rate per shift including shift differentials, dollars per year and cost per ton are tabulated for each job in all of the operating crews. The labor crew is included in operating labor.

- Table V — Maintenance Day-Pay Labor
  Job classifications, number of employees on each classification, rate per shift, number of shifts per week, dollars per year, and cost per ton for the maintenance crews are shown in this table.

- Table VI — Electric Power
  The kilowatt-hours per ton of ore for each of the operating unit operations are tabulated. The total kilowatt-hours per ton, cost per kilowatt-hour, cost per year, and cost per ton are shown in this table.

- Table VII — Reagents
  The delivered cost of reagents, consumption in pounds per ton, and cost per ton are tabulated. The total cost per ton is multiplied by annual tons of ore to obtain the reagent cost per year.

- Table VIII — Maintenance Parts and Supplies
  All materials which wear normally and are replaced with new parts on a routine basis are listed with estimated life, cost of replacements parts, and cost per ton. The total is the total cost per ton for maintenance wear parts. Normally, the crusher liners and grinding mill liners are reported in pounds of wear per ton of ore and cost of liners per pound.

The other items, which are not tabulated in Tables III through VIII, are included in Table I and explained by notes following the table.

Chapter 5

OPERATING DECISIONS BASED ON CASH FLOW ANALYSIS

by J. B. Evans

Professor and Department Head
Department of Mineral Engineering
University of British Columbia
Vancouver, B. C., Canada

ABSTRACT

Maintaining overall optimum profitability at a mine site necessitates constant reassessment of operating strategies. In essence, each strategy is a particular combination of envisaged capital expenditure, income flow and elapsed time. Selecting a strategy based on profitability potential requires each contemplated strategy being subjected to discounted cash flow analysis; even with a computer such analysis can be cumbersome. The chapter discusses a 'short cut' technique which assigns profitability rankings to alternate strategies and identifies those aspects of a strategy which impact heavily on profitability. This technique can yield quick assessments of sufficient accuracy for many operating strategy decisions taken on site at on-going mines.

INTRODUCTION

A fundamental goal of any undertaking is to ensure that optimum efficiency is maintained at all times. Usually, under the free enterprise system, maintaining optimum efficiency at a mining operation is identified as ensuring that overall profitability is optimized throughout the life of the operation.

No mine operates under static conditions; at any point in time at least one of the interlocking components which control the operation is subject to change. For example, ore grades fluctuate, market conditions vary, and new technology emerges. Thus, maintaining overall profitability at a mine site necessitates continual reassessment of operating strategies.

Whenever changes in operating conditions become evident, or are anticipated, such changes must be assessed in the light of their influence on profitability, and decisions appropriate to the maintenance of optimum profitability then taken. In all instances each operating strategy option will involve its own unique combination of capital expenditure and income flow occurring within a particular elapsed time interval. Thus, the maintenance of optimum profitability demands that the specific combination of capital expenditure, income flow and elapsed time for each available operating strategy option be expressed in terms of influence on overall profitability, and the particular option which appears to hold the greatest profit potential implemented at the appropriate point in time.

OBJECTIVE

Normally an engineer at a mine site can express an operating strategy option in terms of envisaged capital outlay, anticipated operating cost and expected life expectancy. However, frequently his lack of familiarity with matters such as mining taxation, accounting procedures, concentrate marketing and the level of required profitability, prohibit him from expressing the option in terms of profitability impact.

The objective of this chapter is to describe a technique considered useful to on site engineering personnel who have limited knowledge of accounting terminology, yet are faced with making operating strategy decisions on the basis of profitability impact.

The technique is not offered as a substitute for full scale economic assessment studies of prospective mining property acquisitions or disposals. Frequently these studies involve additional variables such as divided ownership, phased transfer of title and raising capital from several sources. The technique is proposed solely as a 'short cut' method applicable to operating strategy decisions taken on a day-to-day basis at the mine site.

DESCRIPTION OF THE TECHNIQUE

As the technique involves the use of particular terminology possibly unfamiliar to engineers, certain terms require definition.

Definition of Terms

Mining Reserve	- the rock designated as mineable (as distinct from the Geologic Reserve)
Milling Reserve	- the rock which enters the mill as feed material

Mine Dilution	- that portion of the mill feed not carried in the Mining Reserve, but introduced into the mill feed in the course of extracting the Mining Reserve.
Percentage Extraction	- that percentage of the Mining Reserve which reports as mill feed.
Mill Recovery	- that portion of the valuable content of the mill feed which is recovered in the mill as concentrate.
Concentration Ratio	- the number of units of weight of mill feed material required to produce one unit of weight of concentrate. (Normally expressed on a dry weight basis).
Concentrate Moisture Content	- that portion of the concentrate produced in the mill which represents contained water.
Concentrate Loss	- that portion of the concentrate produced in the mill which does not report at the point of concentrate sale.
Net Mine Site Realization	- the sum of money received by the mine operator as a consequence of selling concentrates following settlement of all charges and deductions related to concentrate sales, such as, smelting charges, customs payments, transportation insurance premiums and freight costs.
Direct Operating Costs	- those on-site costs related to production which are permitted by the taxing authority to be classified as charges against production at the time of production.
Indirect Operating Costs	- those costs not included in Direct Operating Costs which are permitted by the taxing authority to be classified as charges against production at the time of production, e.g. Head Office costs, loading on direct labor costs such as holiday pay and other fringe labor costs, all risk insurance premiums, Director and Audit Fees. (Often called "Operating Overheads").

Capital Expenditures - those expenditures permitted by the taxing authority to be accumulated and used as offsets against taxation liabilities associated with income, e.g. Installed Cost of Plant and Equipment, and 'Capitalized Expenditures' such as interest charges paid prior to the commencement of production, exploration expenses, property acquisition payments, feasibility study costs, 'tune up' and inventory expenses.

Depreciation - the portion of the Capital Expenditures permitted by the taxing authority to be used as offsets against taxation liabilities associated with income. Usually Capital Expenditures will be assigned to specific categories which will be permitted to be 'written off' at specified rates.
Note: Frequently the depreciation rate, and the resulting Undepreciated Value of an asset bear little semblance to the physical wear and tear of the asset, or the replacement value of the asset, respectively. Never confuse 'depreciation' with physical decay. Depreciation can be viewed as an empirical allowance solely associated with taxation liability.

Depletion Allowance - an empirical allowance permitted by the taxing authority to be accumulated and used as an offset against taxation liabilities associated with income.
Note: Normally the depletion allowance for taxation purposes bears no relationship to the actual depletion of ore reserves.

Royalties - payments, usually related to production, paid to a party in accord with a contractual commitment.
Note: Royalties may or may not be an allowable taxation deduction.

Net Operating Profit - equals Net Mine Site Realization, less all Operating Costs (i.e. both Direct and Indirect).
Note: Net Operating Profit is not equal to Net Profit as reported in company financial statements.

Taxable Profit - equals Net Operating Profit less all allowable taxation allowances (i.e. Depreciation and Depletion plus other permitted deductions such as interest paid on outstanding debt).

Gross Cash Flow - equals Net Operating Profit
less (Taxes
(Royalties
(Interest on outstanding debt associated with the operation.

Net Cash Flow - equals Gross Cash Flow
less (Repayments made to reduce outstanding debt
(Capital Expenditures required to ensure continuance of the operation.

Note: Net Cash Flow is not equal to Net Profit as reported in company financial statements.

Discounted Net Cash Flow - the expression of individual net cash flows discounted in accord with a chosen time dependent discount sequence--commonly, annual net cash flows are discounted to a selected point in time in accord with a chosen equivalent compound discount criterion, e.g. $1.00 invested at 10% compound interest for seven years will amount to $1.95, principal plus accrued interest; thus $1.00 earned seven years from now is equivalent to investing $1.00 ÷ 1.95 = 53.1 cents now at 10% compound interest.

Present Value - is the algebraic sum of the Discounted Net Cash Flows.

The above definitions are general in character, sufficient in detail for the purposes of this chapter. It should be noted that variations in phraseology exist from one jurisdiction to another, consequently the reader is advised to seek confirmation of the definitions offered above before applying the terms to his specific case studies.

In essence, the profitability assessment technique proposed by this chapter utilizes a set of sequential tables (or graphs) to correlate the prime components which constitute a mining operation, converting from engineering to monetary units at the appropriate occasions. In practice the set of sequential tables would cover the ranges of values and conditions applicable to the specific operation for which the set is designed.

The Set of Sequential Tables

As a starting point it is necessary to construct tables (or graphs) to correlate engineering variables.

The Interface between the Mining and Mineral Processing Phases of a 'standard' metal mining operation are presented by Tables 1 and 2.

Table 1, "Correlation of Mining Reserve Tonnage, Amount of Mine Dilution and Milling Reserve Tonnage" is an expression of the following relationship,

Milling Reserve = (The amount of Mining Reserve actually extracted) plus (the amount of Mine Dilution introduced into the mill feed in the course of mining).

Table 2, "Correlation of Mining Reserve Grade, Mine Dilution Grade and Mill Head Grade", has been assembled in accord with the following,

Mill Head Grade = (Mining Reserve Grade multiplied by the proportion of mill feed derived from the Mining Reserve) plus (Mine Dilution Grade multiplied by the proportion of mill feed derived from the Mine Dilution material).

Mineral Processing Plant Variables are presented by Table 3, "Correlation of Mill Head Grade, Mill Recovery, Concentrate Grade and Concentration Ratio. This table exemplifies the following dry weight relationship,

$$\text{Concentration Ratio} = \frac{\text{Concentrate Grade}}{\text{Mill Head Grade multiplied by Mill Recovery}}$$

The Correlation of Concentrate Marketing and Net Mine Site Realization requires that all conditions of sale taking place beyond the mine site are known.

For the purpose of illustration, the following marketing conditions are assumed to apply,

1. Smelter Schedule for metal M, when the range of concentrate grade lies between 24 to 30% M, dry weight basis.

   Charges - \$55.00 per dry Tonne of concentrate received at smelter.

   Payment - Pay for amount of metal M received, less 1.5 Tonne units of M per Dry Tonne of concentrate received, at 90% of the Published Selling Price of M.

2. Freight Charges - From mine site to smelter - \$15 per Wet Tonne of concentrate produced.

TABLE 1. Correlation of Mining Reserve Tonnage, Amount of Dilution and Milling Reserve Tonnage

Tons taken from Mining Reserve	Mine Dilution (Weight Percent of Mill Feed) 10	15	20
1 000	1 111	1 176	1 250
2 000	2 222	2 353	2 500
3 000	3 333	3 529	3 750
4 000	4 444	4 706	5 000
5 000	5 555	5 882	6 250
6 000	6 667	7 059	7 500
7 000	7 778	8 235	8 750
8 000	8 889	9 412	10 000
9 000	10 000	10 588	11 250
	Milling Reserve Tonnage		

TABLE 2. Correlation of Mining Reserve Grade, Mine Dilution Grade and Mill Head Grade

Mining Reserve Grade, % M	Amount of Mine Dilution (Weight Percent of Mill Feed) 10			15			20		
	Mine Dilution Grade % M			Mine Dilution Grade % M			Mine Dilution Grade % M		
	0.10	0.20	0.30	0.10	0.20	0.30	0.10	0.20	0.30
1.00	0.91	0.92	0.93	0.87	0.88	0.89	0.82	0.84	0.86
1.20	1.09	1.10	1.11	1.04	1.05	1.06	0.98	1.00	1.02
1.40	1.27	1.28	1.29	1.21	1.22	1.23	1.14	1.16	1.18
1.60	1.45	1.46	1.47	1.38	1.39	1.40	1.30	1.32	1.34
1.80	1.63	1.64	1.65	1.55	1.56	1.57	1.46	1.48	1.50
2.00	1.81	1.82	1.83	1.72	1.73	1.74	1.62	1.64	1.66
	Mill Head Grade, % M								

TABLE 3. Correlation of Mill Head Grade, Mill Recovery, Concentrate Grade and Concentration Ratio

Mill Head Grade % M (Dry Basis)	Concentrate Grade % M, Dry Basis 24			26			28			30		
	Mill Recovery %			Mill Recovery %			Mill Recovery %			Mill Recovery %		
	90	92	94	90	92	94	90	92	94	90	92	94
0.80	33.3	32.6	31.9	36.1	35.3	34.6	38.9	38.0	37.2	41.7	40.8	39.9
0.90	29.6	29.0	28.4	32.1	31.4	30.7	34.6	33.8	33.1	37.0	36.2	35.5
1.00	26.7	26.1	25.5	28.9	28.3	27.7	31.1	30.4	29.8	33.3	32.6	31.9
1.10	24.2	23.7	23.2	26.3	25.7	25.1	28.3	27.7	27.1	30.3	29.6	29.0
1.20	22.2	21.7	21.3	24.1	23.6	23.0	25.9	25.4	24.8	27.8	27.2	26.6
1.30	20.5	20.1	19.6	22.2	21.7	21.3	23.9	23.4	22.9	25.6	25.1	24.6
1.40	19.0	18.6	18.2	20.6	20.2	19.8	22.2	21.7	21.3	23.8	23.3	22.8
1.50	17.8	17.4	17.0	19.3	18.8	18.4	20.7	20.3	19.9	22.2	21.7	21.3
1.60	16.7	16.3	16.0	18.1	17.7	17.3	19.4	19.0	18.6	20.8	20.4	19.9
1.70	15.7	15.3	15.0	17.0	16.6	16.3	18.3	17.9	17.5	19.6	19.2	18.8
1.80	14.8	14.5	14.2	16.0	15.7	15.4	17.3	16.9	16.5	18.5	18.1	17.7
1.90	14.0	13.7	13.4	15.2	14.9	14.6	16.4	16.0	15.7	17.5	17.2	16.8

Concentration Ratio - Tonnes Milled per tonne of concentrate produced (Dry Basis)

TABLE 4. Net Mine Site Realization

Concentrate Grade % M	Published Selling Price of M - $'s per kilogram								
	1.25 (∿56.7¢ per lb.)			1.50 (∿68.0¢ per lb.)			1.75 (∿79.4¢ per lb.)		
	Concentrate Moisture in %			Concentrate Moisture in %			Concentrate Moisture in %		
	8	10	12	8	10	12	8	10	12
24.0	179.41	170.06	178.68	229.38	229.02	228.65	279.35	278.99	278.62
26.0	201.62	201.26	200.89	256.03	255.67	255.30	310.44	310.08	309.71
28.0	223.83	223.47	223.10	282.68	282.32	281.95	341.53	341.18	340.80
30.0	246.04	245.68	245.31	309.33	308.97	308.60	372.62	372.27	371.89

Net Mine Site Realization - $'s per Dry Tonne of Concentrate produced

3. Customs and Insurance
   1.0% of (Net Smelter Return less Freight charges).

4. Concentrate Losses in Shipment

   0.3% of concentrate produced (i.e. 99.7% of the concentrate produced reports to the smelter).

5. Concentrate Moisture Content

   0.08 to 0.12 by weight (i.e. the concentrate as shipped contains 8 to 12% moisture).

6. Published Selling Price Range of Metal M

   \$1.25 to \$1.75 per kilogram (∿56 to 80¢ per lb.)

When the above assumed marketing conditions apply, the Net Mine Site Realization (expressed as \$'s received per Dry Tonne of concentrate produced at the mine site), can be determined as follows,

Net Mine Site Realization, \$'s per Dry Tonne of concentrate produced, is equal to

$$0.990\left[\begin{array}{l}0.997[(\%M-1.5)(0.9 \times \text{Selling Price of M, \$'s/kg})(10)^{1} - 55.00] \\ -\left(\dfrac{\text{Freight Charges, \$'s per Wet Tonne Concentrate shipped}}{1.00 - \text{Moisture Content, in decimal parts}}\right)\end{array}\right]$$

Table 4, "Net Mine Site Realization", presents the relationship of the variables involved when the marketing conditions as specified above are assumed to apply.

Assigning a Profitability Index to the Capital Expenditure, Net Operating Profit and Economic Life involved in an operating strategy option takes an engineer beyond his normal sphere of activity. To make this assignment the applicable taxation regime and the profitability criterion under which the mine operates must be known.

For purposes of illustration, the following taxation regime will be assumed,

1. Annual Depreciation Rate - 30% of the Capital Expenditure on a diminishing balance basis.

2. Depletion Allowance - 33-1/3% of the Taxable Profit.

3. Taxation Rate - 50% of (Taxable Profit less Depletion).

4. At all times the operation is sufficiently profitable to take advantage of any available tax deduction.

Figure 1. Pro-forma Application of Funds Statement - Economic Life Five Years

Capital Expenditure - $10 000, made at the commencement of Year 1.
Annual Net Operating Profit - $2 946

ITEM	YEAR 1	YEAR 2	YEAR 3	YEAR 4	YEAR 5	YEAR 6	TOTALS
Net Operating Profit	2 946	2 946	2 946	2 946	2 946	Nil	14 730
Less Depreciation (30%)	3 000	2 100	1 470	1 029	720	1 681	10 000
Taxable Profit	-54	846	1 476	1 917	2 226	-1 681	4 730
Less Depletion Allowance (1/3)	-18	282	492	639	742	- 560	1 577
Net Taxable Profit	-36	564	984	1 278	1 484	-1 121	3 153
Taxes at 50%	-18	282	492	639	742	- 561	1 574
Net Operating Profit	2 946	2 946	2 946	2 946	2 946	Nil	14 730
Less Taxes	-18	282	492	639	742	- 561	1 574
Net Cash Flow	2 964	2 664	2 454	2 307	2 204	561	13 154
Present Value Factors (10%)	0.909 1	0.826 4	0.751 3	0.683 0	0.621 9	0.564 5	
Present Value of Net Cash Flow	2 695	2 202	1 844	1 576	1 368	317	10 002

Thus, the Present Value of an Annual Net Operating Profit of $2 946 for five years is equivalent to $10 000; hence, a Capital Expenditure of $10 000 can be justified when the increase in Net Operating Profit is $2 946 per year for the ensuing five years.

5. Any residual undepreciated value of a capitalized asset remaining beyond the actual economic life of that asset can be claimed in full for tax purposes in the year immediately subsequent to the conclusion of the asset's physical usefulness.

Again for purposes of illustration, it will be assumed that the profitability criterion in use is as follows,

> "An investment in Capital Expenditure will be made at the mine site provided an equivalent 10% per annum compound interest return is obtainable from that investment over its anticipated life cycle".

With the applicable taxation regime and profitability criterion available, it is possible to calculate the net operating profit required to justify a capital expenditure over various time intervals. A sample calculation for a capital investment having an economic life of five years under the taxation regime and profitability criterion given above is presented by Figure 1,'Pro-forma Application of Funds Statement - Economic Life Five Years!

By completing calculations for specific economic lives of Capital Expenditures in a manner similar to that shown by Figure 1, it is possible to relate Capital Expenditure, Net Operating Profit and Economic Life in accord with the applicable taxation regime and profitability criterion.

Table 5, 'Required Total Net Operating Profit' relates Capital Expenditure, Required Net Operating Profit and Economic Life in accord with the taxation regime and profitability criterion specified above; on the proviso that the generation of Net Operating Profit is uniform throughout the life of the investment.

Should the net operating profit resulting from a Capital Expenditure be generated uniformly throughout the investment life, it is possible to assign a Present Value to each Annual Net Operating Profit.

Assuming the taxation regime and profitability criterion detailed above, then,

When the economic life is <u>one</u> year, each \$ of Net Operating Profit has a Present Value = 1 ÷ 1.181 8 = <u>84.62</u> cents.

When the economic life is two years, each \$ of Net Operating Profit has an average Present Value of 1 ÷ 1.259 3 = 79.41 cents.

But the 1st Year Net Operating Profit \$ has a Present Value = 84.62 cents, thus the Present Value of the <u>2nd Year</u> \$ is (2 x 79.41) - 84.62 = <u>74.20</u> cents.

Likewise, when the economic life is three years, each $ of Net Operating Profit has an average Present Value = 1 ÷ 1.332 4 = 75.05 cents.

But the 1st and 2nd Year Net Operating Profit $'s have an average Present Value of 79.41 cents, thus the Present Value of the 3rd Year Net Operating Profit $ must be (3x75.05) - (2x79.41) = 66.34 cents.

By extending this line of reasoning it is possible to derive a Present Value to each annual Net Operating Profit $. For the particular taxation regime and profitability criterion described above, the sequence given by Table 6, 'Present Value of Annual Net Operating Profit' is applicable when the generation of Net Operating Profit is uniform throughout the investment life.

When a mine site engineer has tabulated the operating variables in a manner similar to that given by Tables 1 through 6, covering ranges appropriate to his particular mine, he is then in a position to assess and select operating strategy options on the basis of their impact on profitability.

TABLE 5. Required Total Net Operating Profit

$$A = \frac{\Sigma \text{ Annual Net Operating Profit}}{\text{Capital Expenditure}}$$

	Total Investment Life in Years									
	1	2	3	4	5	6	7	8	9	10
A	1.181 8	1.259 3	1.332 4	1.403 2	1.472 9	1.542 6	1.612 7	1.683 8	1.756 0	1.829 0

TABLE 6. Present Value of Annual Net Operating Profit

PV of Net Op.Profit Year N = (PV Factor Year N)(Net Op. Profit Year N)

	Year in which Net Operating Profit generated									
	1	2	3	4	5	6	7	8	9	10
P.V. Factor	0.846	0.742	0.663	0.599	0.544	0.495	0.451	0.411	0.374	0.342

EXAMPLES OF THE TECHNIQUE

The following examples of typical operating strategy options encountered at a mine site are offered to demonstrate applications of the technique. In all instances the nomenclature, marketing conditions, taxation regime and profitability criterion presented by Tables 1 through 6 are assumed to apply.

Example 1. A strategy decision involving a Capital Expenditure, Increased Operating Cost and Increased Revenue.

The mine has a future life of at least eight years producing 10 000 tonnes of concentrate per year; currently the concentrate grades 28% M (dry basis), and contains 12% moisture.

A proposal to incorporate a concentrate dryer into the circuit has been made with the following parameters known

a) Installed capital cost $12 000

b) Added operating cost $0.50 per tonne concentrate (dry basis)

c) Assured reduction in concentrate moisture - from 12% to 8%.

Should this proposal be undertaken?

Solutions

Table 4 shows that reducing concentrate moisture from 12% to 8% will increase the Net Mine Site Realization by $0.73 per dry tonne of concentrate, thus the indicated increase in Annual Net Operating Profit is

$$(0.73 - 0.50)(10\ 000) = \$2\ 300$$

Then the Total Net Operating Profit increase over eight years is 8 x 2 300 = $18 400. From Table 5, the Required Total Net Operating Profit is obtained by multiplying the Capital Expenditure by the eight year life factor 1.683 8, thus Required Total Net Operating Profit increase to justify $12 000 Capital Expenditure is

$$12\ 000 \times 1.683\ 8 = \$20\ 200$$

$$20\ 200 \div 8 = \$2\ 525 \text{ per year for eight years}$$

or

P.V. of Net Operating Profit increase is 18 400 ÷ 1.683 8 = $10 930.

Therefore, the installation of the dryer is not justified!

But, by extrapolation, Table 4 indicates a reduction of moisture content from 12% to 7% would increase the Net Mine Site Realization to $0.91 per tonne of concentrate; maybe an increase in dryer fuel consumption at, say, $0.10 per tonne of concentrate would yield a concentrate containing 7% moisture? Under this circumstance, the increase in Annual Net Operating Profit would be

(0.91 - 0.60)(10 000) = \$3 100, which has a Present Value equal to

8(3 100) ÷ 1.683 8 = \$14 700.

Thus, should this dryer performance be obtainable, on a Present Value basis the installation of a concentrate dryer is equivalent to a net gain to the operation of

(14 700 - 12 000) = \$2 700 (a mill metallurgist's salary for one month?)

Example 2. A strategy decision involving Capital Expenditure, Operating Costs and Economic Life.

A small mine, milling capacity approximately 400 Tonnes per day, is reaching the end of its life. Two operating strategy options have been identified for exploiting the remaining Mining Reserve as follows,

Item	Option A	Option B
1. Tonnes available from Mining Reserve	670 000	600 000
2. Mining Reserve Grade, % M	1.80	2.00
3. Expected Mine Dilution, % Mill Feed	20	10
4. Grade of Diluting Material, % M	0.10	0.25
5. Direct Mining Cost, \$ per tonne milled	3.00	4.00
6. Milling Rate, Tonnes per year	125 000	150 000
7. Total Operating Cost excluding Mining, \$/tonne milled	4.50	4.00
8. Anticipated Mill Recovery % of M	92	90
9. Concentrate Grade % M	26	26
10. Concentrate Moisture Content, % Moisture	10	10
11. Capitalized Expenditure required, (mine development work), \$ x 10^3	1 000	1 500

If the Published Selling Price of M is expected to be in the range \$1.25 - \$1.75 per kilogram for the foreseeable future, which is the better option to take?

Solution

	Item	Option A	Option B
a)	Determine Milling Reserve (Table 1)	837 500	666 700
b)	Determine Mill Head Grade (Table 2)	1.46	1.82_5
c)	Determine Concentration Ratio (Table 3)	19.4	15.8
d)	Determine Life (Item a ÷ Item 6)	6.70	4.44
e)	Total Operating Cost, $ per T milled (Items 5 + 7)	7.50	8.00
f)	Determine Net Mine Site Realization $ per T conc. (Table 4) at Selling Price $1.25/kg	201.26	201.26
	$1.75/kg	310.08	310.08
g)	Determine Net Mine Site Realization $/T milled (Items f ÷ c) at Selling Price $1.25/kg	10.37	12.74
	$1.75/kg	15.98	19.63
h)	Determine Net Op. Profit, $/T milled (Items g - e) at Selling Price $1.25/kg	2.87	4.74
	$1.75/kg	8.48	11.63
i)	Determine Annual Net Op. Profit, $'s x 10^3 (Items h x 6) at Selling Price $1.25/kg	359	711
	$1.75/kg	1 060	1 745
j)	Present Value Factors (Table 6) (Summation Years 1 through 6 + 0.70 Year 7)	4.20	
	(Summation Years 1 through 4 + 0.44 Year 5)		3.09
k)	Present Value of Total Net Profit, $'s x 10^3 (Items i x j) at Selling Price $1.25/kg	1 508	2 197
	$1.75/kg	4 452	5 392
l)	Present Value of Option $'s x 10^3 (Item k - 11) at Selling Price $1.25/kg	508	697
	$1.75/kg	3 452	3 892
m)	Advantage Option B over Option A, $'s x 10^3 at Selling Price $1.25/kg		190
	$1.75/kg		440

This example demonstrates two operating strategies, namely

Option A - Accepting a more highly diluted mill feed from the mine, combined with reduced mill throughput rate and high mill recovery.

Option B - Higher selectivity in the mining phase, combined with higher mill throughput rate and lower mill recovery.

Example 3. A possible change in operating conditions which does not involve a capital outlay.

The nominal 12 000 tonnes per day capacity concentrator operates 360 days per year. The grinding section consists of three parallel circuits where each circuit must be shut down for relining at four monthly intervals. Each reline takes five days to complete and costs $30 000 in consumable supplies (Case A).

An alternate relining strategy (Case B) is available whereby each relining shutdown will occupy two days at six monthly intervals and cost $190 000 in consumable supplies.

Should the alternate (Case B) strategy be implemented when the following operating conditions are known?

1. Milling Reserve - at least nine years supply available at 1.2% M head grade.
2. Concentrator Performance at 1.2% M head grade,

   Recovery of M is 92% into a concentrate grading 28% M (dry basis), 10% moisture.
3. Published Selling Price of M is not expected to go below $1.50 per kg whilst the Operating Costs given below are applicable.

General Overheads, fixed at		$12 000 per day
Mining	a) Fixed	$40 000 per day
	b) Variable	$2.00 per tonne milled
Milling	a) Fixed	$24 000 per day
	b) Variable	
	Grinding Media & Power	$1.00 per tonne milled
	Mill Relining, Case A	$30 000 per reline per grinding circuit
	Case B	$190 000 per reline per grinding circuit.

Solution

At first glance the Annual Reline costs are

Case A 30 000 x 3 x 3 = $ 270 000

Case B 190 000 x 3 x 2 = $ 1 140 000

On this basis Case B is excessively expensive but the down time is less than Case A (i.e. 12 vs. 45 days per year).

Therefore, a comparison of Annual Operating Costs is in order.

Examination of Annual Operating Costs reveals the following,

Item			Case A, $\$x10^3$		Case B, $\$x10^8$
General Overhead		12 000 x 360	4 320		4 320
Mining	Fixed	40 000 x 360	14 400		14 400
	Variable	12 000x315x2	7 560	12 000x348x2	8 352
		8 000x 45x2	720	8 000x 12x2	192
Milling	Fixed	24 000 x 360	8 640		8 640
	Variable	12 000x315x1	3 780	12 000 x348x1	4 176
		8 000x 45x1	360	8 000x 12x1	96
	Relines	30 000x 3x3	270	190 000x 3x2	1 140
Total Annual Operating Cost			40 050		41 316
Total Annual Throughput in Tonnes					
		12 000 x 315) 8 000 x 45)	$4.140x10^6$	12 000 x348) 8 000 x 12)	$4.272x10^6$
Total Operating Cost \$ per Tonne Milled			9.67		9.67

Even on this examination Case B appears less attractive than Case A, by increasing the Annual Operating Cost by \$ 1 266 000.

However, as Case B yields an increase in throughput by 132 000 tonnes per year, it is necessary to examine the increase in Annual Operating Profit Case B offers over Case A, as follows

Item	Case A	Case B
1. Concentration Ratio (Table 3)	25.4	25.4
2. Net Mine Site Realization, \$ per tonne concentrate (Table 4)	282.32	282.32
3. Net Mine Site Realization, \$ per tonne milled (2 ÷ 1)	11.11	11.11
4. Total Operating Cost \$ per tonne milled (obtained above)	9.67	9.67
5. Net Operating Profit \$ per tonne milled (3-4)	1.44	1.44
6. Tonnes milled per year x 10^6 (obtained above)	4.140	4.272
7. Annual Net Operating Profit $\$x10^6$ (5 x 6)	5.962	6.152
8. Increase in Annual Net Profit (Case B-Case A) $\$ \times 10^6$		0.190

Thus case B generates a higher Annual Net Profit than does Case A, amounting to \$190 000 under the conditions assumed. This increase can be assessed as follows,

If the present operating conditions persist, then the Present Value of the Increased Operating Profit for nine years is (from Table 5)

$$190\ 000 \times 9 \div 1.756 = \text{PV \$'s } 974\ 000$$

but Case B leaves less tonnage to be processed beyong the ninth year than does Case A. If it is assumed that this tonnage $[(4.272 - 4.140)10^6 \times 9 = 1.188 \times 10^6]$ would be processed in the tenth year under Case A, then this now non-available tonnage having an Operating Profit potential of \$1.44 per tonne has a negative Present Value (from Table 6)

$$1.188 \times 10^6 \times 1.44 \times 0.342 = \text{PV } \$585\ 000$$

Then value of Case B over Case A on a Present Value basis is

$$(974\ 000 - 585\ 000) = \$\ 389\ 000$$

Following this line of reasoning it could be argued that the selection of the new relining procedure (Case B) is equivalent to receiving at least \$389 000 in Net Cash Flow at the present time.

SUMMATION

Cash flow analysis has been discussed from the viewpoint of mine site engineers charged with making operating strategy decisions on a profitability basis. In essence, the assessment technique presented is a procedure for assigning Present Value indices to estimated future costs and anticipated taxable income.

The technique cannot be applied as an overall substitute for full scale assessment of mineral property acquisition, however, many of the correlations presented by the technique can be used to advantage in such assessments.

The technique as described in this chapter utilizes a set of tables to correlate variables; alternatively the same data can be presented as a series of graphs (1).

It is noted that the data presented by Tables 1 through 4 represent valid relationships, whereas Tables 5 and 6 are mathematically rigorous only when Net Operating Profit is generated uniformly. Experience with the technique in practice has shown that fluctuations in annual Net Operating Profits, (up to ±20%), can be accommodated without invalidating the findings. (When Annual Net Operating Profits fluctuate markedly with time, regrettably it is necessary to resort to laborious Pro-forma Application of Funds compilations--access to a computer is a boon in this circumstance!)

Although specifically designed for on site operating strategy decision making, the technique can be applied to gauge the impact taxation has on a mining operation. By compiling the respective data set to yield Tables 5 and 6 for each taxation regime under consideration, it is possible to express taxation variance in engineering quantities. For example, mine 'cut off grades' can be related to taxation structures and the resulting Mining Reserves established.

REFERENCES

(1) Evans, J.B., "The Evaluation of Mining Properties - A Graphical Approach", Trans CIMM, Vol. LXIII, 1960, pp 547-553.

Chapter 6

COMPARISON OF ESTIMATED VS. ACTUAL CAPITAL COST AND OPERATING DATA FOR A COPPER CONCENTRATOR

by Thomas D. Henderson and Donald E. Crowell

Chief Metallurgist, ASARCO Incorporated
Tucson, Arizona

Director, Mineral Beneficiation Dept., ASARCO Incorporated
Tucson, Arizona

INTRODUCTION

This paper presents a "case history" of the steps taken to estimate capital and operating costs for a typical porphyry copper concentrator of ±9,070 metric tons (10,000 short tons) per day capacity. The concentrator described herein is hypothetical, but is based on experience and is, in effect, a composite of several actual projects. The numbers used are illustrative only and do not represent actual costs at any particular concentrator. These numbers are, however, representative of typical milling costs at one point in time and are sufficiently accurate to illustrate the cost estimating procedures described in this paper. It is not possible to publish actual, current milling costs and, in any event, actual cost figures quickly become meaningless in today's economic situation with various costs all rising at different rates.

The paper is divided into three parts: (1) order-of-magnitude estimate, (2) feasibility study estimate, and (3) discussion of differences between estimated and actual costs.

For the purpose of this paper, we will assume that a mining company has diamond drilled a property and broadly defined an ore body of some 45.3-million metric tons (50-million short tons) grading 0.6% sulfide copper. The company first desires an order-of-magnitude cost estimate (both capital and operating costs) to determine if the property warrants further drilling and sampling. Assuming the economic evaluation using order-of-magnitude costs is favorable, a feasibility study will be done requiring more accurate capital and operating costs.

Following is a description of the various steps taken to develop first order-of-magnitude costs and then feasibility study costs.

ORDER-OF-MAGNITUDE STUDY

Order-of-Magnitude Capital Costs

An order-of-magnitude capital cost estimate is by definition very preliminary in nature and at best only accurate to within ±25-30%. In this case the only basis for the estimate was preliminary bench-scale flotation tests on diamond drill core plus experience with similar operations. Our standard method for preparing order-of-magnitude cost estimates is to factor the major plant equipment costs.

The first step was to prepare a preliminary flowsheet. Since the preliminary flotation test work had not indicated any special problems, a conventional porphyry copper flowsheet was laid out.

A pulp and water balance was then calculated so that various equipment could be sized. There had been no test work regarding ore grindabilities, product settling rates or other design parameters at this stage, so experience factors and "rules of thumb" were used for sizing equipment and determining the pulp and water balance.

A list of major equipment was then prepared which included all equipment of the size of process pumps and larger. Vendor quotations were solicited from one or two major suppliers of each piece of equipment. These quotations were preliminary in nature and defined as being for estimating purposes only. Quotations included motors, drives, and estimated freight rates to the mine site.

The capital cost estimate was then obtained by multiplying the equipment cost by a factor of 3.5. This factor is arbitrary and ever changing with variable inflation rates and different inflation patterns. Therefore, the factor is recalculated continually as new actual cost information is gathered.

The capital cost obtained above included only the concentrator proper and did not cover power supply, assay facilities, water supply, tailing disposal facilities (aside from the tailing thickener) or other auxiliary facilities. These items were added on separately and, at this stage, were little more than "quesstimates." The repair shop, warehouse and office building served both mine and concentrator, so these estimates had to be coordinated with the mine cost estimate.

A contingency figure of 15% was added to the total capital cost estimate. This figure was not meant to be a safety factor, but actually represented many small costs which do not appear in an estimate. In this case, where a factor on equipment cost was used, the contingency could have been included in the factor. However, it is our standard practice to keep the contingency separate.

Table I gives the equipment list and calculation sheet used to obtain the capital cost.

TABLE I

Order-of-Magnitude Capital Cost Equipment List

Description	Cost
Grizzly Feeder	$ 32,600
Jaw Crusher	182,415
Vibrating Feeders (5)	17,407
Tramp Iron Magnet (2)	11,362
Weightometers (3)	16,201
Metal Detector	9,950
Screens (3)	66,785
Cone Crushers (2)	300,512
Belt Feeders (6)	33,990
Grinding Mills (3)	841,559
Horizontal Slurry Pumps	43,259
Vertical Slurry Pumps	33,919
Cyclones	30,279
Distributors	22,891
Flotation Cells (all)	213,368
Thickeners (3)	143,255
Filters (2)	60,956
RR Scale	25,930
Lime Unloading	12,809
Lime Pumps	2,128
Lime Slaker	16,700
Air Compressor	48,408
Inst. Air Dryer	2,310
Water Pumps	5,244
All Electric Motors	300,000
Total Process Equipment Cost	$2,474,237
Miscellaneous Equipment	275,000
Spare Parts	200,000
TOTAL	$2,949,237 (or $2,950,000)

Concentrator Cost (Concentrator cost equals factor of 3.5 times major equipment cost, or)	$10,325,000
Additional Items	
Power Supply	65,000
Water Supply	800,000
Auxiliary facilities (shop, assay office, warehouse, main office)	1,000,000
Tailing Dam	500,000
Sub-total	$12,690,000
Contingency at 15%	1,900,000
TOTAL MILLING PLANT COST	$14,590,000

Order-of-Magnitude Operating Costs

The direct operating cost estimate was broken down into the following categories:

Power
Operating Labor
Maintenance Labor
Operating Supplies
Maintenance Supplies
Supervision

Indirect costs were calculated as a percentage of direct costs and included taxes, vacation allowances and various labor fringe benefit costs.

The direct operating costs were estimated very crudely at this stage and were obtained mainly by comparison with similar operating plants modified as required. In this case, since we were dealing with a very conventional flowsheet, it was relatively easy to obtain comparative costs. If this had not been the case, it would have been necessary to calculate costs as discussed later under feasibility study costs.

Table 2 gives typical order-of-magnitude costs in column I.

TABLE 2

Direct Operating Costs

	Column I Order-of-Magnitude $ per metric ton	Column II Feasibility Study $ per metric ton
Power	$ 0.34	$ 0.3748
Operating Labor	0.16	0.1495
Maintenance Labor	0.09	0.0816
Operating Supplies	0.30	0.2395
Maintenance Supplies	0.15	0.1254
Supervision	0.03	0.0478
Total	$ 1.07	$ 1.0186
Indirect Costs at 30% of Direct Costs	.32	.31
TOTAL OPERATING COSTS	$ 1.39	$ 1.329

FEASIBILITY STUDY

Financial calculations using the order-of-magnitude estimates appeared attractive so money was appropriated for a detailed feasibility study. This feasibility study included four primary areas of study: (a) Various metallurgical tests to define metallurgy and develop design parameters, (b) cost estimate for concentrator and surface plant by an engineering contractor, (c) in-house studies and cost estimates of various auxiliary facilities, and (d) calculation of operating costs. These are discussed in more detail below:

Metallurgical Studies

The first step was to obtain a representative sample of the ore body. This was done by using six-inch diamond drill core which was crushed and mixed to represent the various ore types which would be mined. Samples were then used for laboratory test work to establish accurate recoveries, grades, and various design parameters. Samples were also sent to crushing and grinding equipment suppliers to determine crusher and grinding mill sizes and power requirements.

Engineering Contractor Cost Estimate

An engineering contractor was selected and company engineers worked closely with them to produce an accurate cost estimate. This estimate was based on a design criteria and 15 general arrangement drawings, plus a detailed flowsheet and pulp and water balance.

The drawings and information produced were in sufficient detail to allow the estimators to calculate labor, material, and sub-contract costs for each area of the plant. These areas included:

- Site Preparation
- Primary Crushing
- Fine Crushing
- Grinding
- Flotation
- Concentrate Handling
- Concentrator Building
- Plant Buildings
- Process Instrumentation
- Utilities
- Tailing Disposal

Indirect costs and engineering costs were factored from the total direct costs as calculated above. A 15% contingency factor was again added to the total figure. This cost figure is generally considered accurate to within ±20%.

Table 3 lists the results of the engineering contractor study.

TABLE 3

Feasibility Study Capital Cost

Engineering Contractor Cost Estimate

Item	Estimated Direct Cost
Site Preparation	$ 950,000
Primary Crushing/Stockpile	625,000
Fine Crushing	730,000
Grinding	1,300,000
Flotation/Reagent Facilities	1,550,000
Concentrate Handling	108,000
Concentrator Building	1,010,000
Plant Buildings	2,325,000
Process Instrumentation	490,000
Utilities	645,000
Tailing Disposal	735,000
Total Direct Cost	$10,468,000
Indirect Costs (including fee)	1,200,000
Engineering/Procurement	850,000
Contingency at 15%	1,878,000
TOTAL ENGINEERING CONTRACTOR ESTIMATE	$14,396,000

\- - - - -

In-house Estimates

Power Supply	73,000
Road and Railroad	220,000
Water Supply	695,000
Temporary Water and Power	60,000
Total	$15,444,000
Escalation at 6% per year for 2 years (Calculated on only one-half of total for second year.)	1,390,000
TOTAL PLANT COST	$16,834,000

In-house Studies

Various auxiliary facilities were studied in-house and cost estimates made. These consisted of:

A. Preliminary agreements were made with the power company and a cost calculated for erecting a power line to the plant site.

B. Costs for an access road and a railroad siding were calculated.

C. An engineering study was made and costs estimated for installing water wells and a pipeline to the plant site.

D. Costs were estimated for providing temporary power and water to the site for construction purposes.

The various in-house estimates were combined with the engineering contractor estimate to produce the overall project cost estimate for final economic evaluation. Mine preproduction costs were, of course, included for the evaluation.

Inflation was handled by escalating the construction cost through the construction period, in this case two years.

Feasibility Study Operating Costs

For the feasibility study phase, the proposed operating costs were studied very thoroughly and every effort made to produce the best possible estimate. The order-of-magnitude estimates were somewhat conservative in most cases, so overall operating costs were reduced slightly from the previous estimate. Direct labor costs were estimated by preparing manning tables for both operation and maintenance labor and applying current labor rates. The number of men required was determined from experience with similar operations modified as necessary to conform to our flowsheet. A typical manning chart for the flotation section is given in detail on Table 4. Similar charts were prepared for crushing (primary and secondary), grinding, concentrate handling, and tailing disposal.

Supervision costs were obtained by preparing a list of all salaried personnel required and the current salary rates. A typical example is given on Table 5.

Operating supply costs were estimated by calculating the major costs such as grinding steel and flotation reagents and estimating other costs from experience with other operations. Eighty to ninety-percent of operating supply costs will be in crushing and grinding steel consumption and flotation reagents. Table 6 gives typical operating supply costs.

TABLE 4

Estimated Total Labor - Hourly Personnel
Flotation

Title	Man Shifts Required Per Day	Hourly Rate	Total Per (1) Month
Flotation Man			
1st Shift	1	5.40	$ 1,260
2nd Shift	1	5.50	1,281
3rd Shift	1	5.58	1,302
Reagent Man	0.5	4.50	525
Laborer	2	4.05	1,890
Head Sampler	1	4.50	1,050
Sampler	1	4.28	998
Sampler Helper	1	4.05	945
Flotation Helper			
1st Shift	1	4.28	998
2nd Shift	1	4.37	1,019
3rd Shift	1	4.46	1,040
TOTAL			$12,308
			$0.0465/metric ton

(1) Assuming 29.17 days per month at 350 operating days per year.

TABLE 5

Estimated Concentrator Supervisory Costs

Title	Number of Men	Monthly Rate	Total per Month
Mill Superintendent	1	$ 1,620	$ 1,620
Asst. Mill Superintendent	1	1,440	1,440
Mill Shift Foremen	3	900	2,700
Relief Shift Foremen	1	900	900
Chief Metallurgist	1	1,080	1,080
Metallurgist	1	900	900
Mill Clerk	1	630	630
TOTAL			$ 9,270
			$0.035/metric ton
Chief Chemist	1	$ 1,080	$ 1,080
Assayer	1	720	720
Helper	1	630	630
TOTAL			$ 2,430
			$0.009/metric ton

TABLE 6

Estimated Operational Supply Costs

Area	Operation	Supplies Cost/Metric Ton
Crushing	Primary Crushing	$ .0003
	Secondary Crushing	.0011
Grinding	Primary Ball Grinding and Classification	.1258
	Regrinding and Classification	.0097
Flotation	Flotation Conditioning	.0005
	Flotation Reagents	.0806
	Sampling and Assaying	.0042
Concentrate Handling	Thickening, Filtering, and Drying-Concentrate	.0023
Tailing Disposal	Tailing Disposal	.0072
Miscellaneous	Fresh Water	-
	Return Water	-
	Lighting	.0009
	Heating	-
	Experimental	.0020
	Misc. Tools and Supplies	.0037
	Lubrication	.0011
	Instrumentation	.0001
TOTAL		$ .2395

Maintenance supplies were estimated from experience with similar operations. As a check, the commonly used figure of 5% of capital cost per year was used for comparison and came out much higher. We used our estimate since we felt this number was accurate, and operating data has proven our estimate correct. The discrepancy probably lies in our definition of "maintenance supplies." Typical maintenance supply costs appear in Table 7.

Power costs were estimated by taking installed horsepower and multiplying by a diversity factor of 0.8 to obtain operating horsepower, and electrical power consumption was calculated from this figure. Crushing, grinding and flotation power consumption can be estimated quite closely from work indices and flotation time requirements, and these three items will normally make up 85% or more of the power consumption. Typical power costs are listed in Table 8.

The individual cost items were added to give a total direct operating

TABLE 7

Estimated Maintenance Supply Costs

Area	Maintenance	Supplies Cost/Metric Ton
Crushing	Buildings & Surface	$ .0025
	Primary Crushers	.0132
	Secondary Crushers	.0269
	Conveyors	.0123
	Grizzlies & Screens	.0060
Grinding	Ore Bins & Feeders	.0015
	Primary Ball Mills	.0260
	Regrind Ball Mills	.0029
	Classifier Cyclones	.0064
Flotation	Flotation Equipment, Conditioners	.0055
	Reagent Storage & Feeders	.0011
	Product Pumps & Lines	.0050
	Water Pumps & Lines	.0013
	Samplers	.0003
Concentrate Handling	Thickeners, Filters & Dryers-Concentrate	.0014
Tailing Disposal	Tailing Disposal	.0093
Miscellaneous	Dust Collection	.0020
	Instrumentation	.0015
	Compressors & Air Lines	.0003
TOTAL		$ .1254

TABLE 8

Estimated Power Distribution and Costs

Operation	Installed(2) Horsepower	kwh/Metric Ton(3) at .8 Diversity Factor	Cost (1) $/Metric Ton
Crushing	1,000	1.63	$0.0326
Grinding	7,000	11.17	0.2234
Flot. & Regrinding	3,000	4.83	0.0966
Concentrate Handling	350	0.56	0.0112
Tailing Disposal	350	0.55	0.0110
TOTAL	11,700	18.74	$0.3748

(1) Estimated cost of power is 2.0 cents/kwh.
(2) Installed HP figure does not include duplicate equipment motors such as duplicate pumps.
(3) Overall power factor assumed as unity.

cost (see Table 2, column II). A contingency figure of zero to 10% is normally added depending on the accuracy of the cost estimate and whether or not the estimator has access to operating costs of similar operations.

Indirect costs are calculated as a percentage of direct operating costs and vary between 25% and 50%, depending on location and method of calculation.

The next section describes some typical errors which become evident when comparing estimated costs and actual operating costs.

TYPICAL DISCREPANCIES BETWEEN ESTIMATED AND ACTUAL MILLING COSTS

Supervision

Supervisory costs are generally higher than anticipated due to additions to the mill metallurgical staff. Even for a relatively straightforward porphyry copper ore, the complexity of the metallurgy can be underestimated. This necessitates more extensive bench-scale metallurgical testing and mill testing than was anticipated. During the startup and first few months of operation especially, much metallurgical work will be required to refine operating conditions such as grind, pH, reagent dosages, etc. as the types of ore being milled change with development of the mine. This type of metallurgical control testing must continue in order to maintain high metallurgical efficiency in the mill at the lowest possible cost.

Operating Labor

Manning tables for operating labor crews are prepared from a similar type of mill. There will usually be little difference between the estimated and the actual operating labor force. There might be some rearrangement of manpower between the different unit operations in the mill, but the numbers of a given job classification will remain the same.

Maintenance Labor

The requirements for the mill maintenance crew can be rather seriously underestimated; while this is partly due to an overestimate of individual productivity, the difficulty in hiring experienced mill maintenance personnel is probably more to blame for a low estimate. The turnover in this work area is also a contributing factor to the problem of acquiring and keeping experienced mill repairmen. Perhaps, more attention should be given in the future to the availability and type of experience of local mechanical personnel in the planning of a proposed mill to prevent errors of this type.

Operating Supplies

The requirements for flotation reagents are estimated from dosages used in the bench-scale flotation tests done to determine the flowsheet. The actual reagent costs might be quite close to those estimated due to higher usages of some reagents and lower ones for others. Worth mentioning here is that, for open pit mines, larger amounts of frothers than those estimated are generally required due to milling of softer ores at the startup of the mill. They tend to consume more frother than was indicated by flotation testing of diamond drill core from deeper within the ore body.

The higher frother costs are pretty much compensated for by lower usages of collectors and other reagents.

It might be found necessary after startup of the mill to use a filter aid to reduce moisture in the copper concentrate filter cake to satisfactory levels. This often is forgotten in the proposed reagent scheme. Filter aids and flocculants are two groups of reagents which might be overlooked in planning reagent requirements until deficient mill operation after startup dictates their use.

Actual grinding media costs for primary grinding and concentrate regrinding have sometimes been much higher than those estimated. This is largely due to increases in grinding ball costs between the time the estimates are made and operation of the mill commences.

The rationale for using present-day unit costs of operating supplies for estimating future operating costs is that the price of the metal produced in the mill will inflate accordingly as the operating supply costs increase. In past years, this assumption has been fairly valid; however, at today's metal prices, this is no longer true.

Maintenance Supplies

The cost of maintenance supplies is estimated by using average costs for a mill of similar type and capacity over a 5-year period. This has proven to be quite close to actual maintenance supply costs for our mills.

Power

Total power consumption is estimated by using grinding test data and installed horsepower. The power consumption estimates are only as accurate as the ore sample used in the grinding tests. Also, a few small miscellaneous power usages are often omitted from the estimates.

2

Sampling and Testing

Chairmen: D. K. Mortensen
G. V. Jergensen

Chapter 7

SAMPLING A MINERAL DEPOSIT FOR FEASIBILITY STUDIES AND METALLURGICAL TESTING

by A. J. Sinclair

Professor of Geological Sciences
The University of British Columbia
Vancouver, British Columbia, Canada

ABSTRACT

Sampling methods for feasibility studies are dependent on geological characteristics of the deposit being evaluated. A preliminary geostatistical study is required, taking into account autocorrelation aspects of variables under study in order that sample plan design can be optimized to produce grade and tonnage estimates within stipulated error limits. In general, regular sample distribution is to be aimed for, and regular sample lengths (from drill core or cuttings, for example) are preferred in order to facilitate statistical analysis. Sample plans are modified and improved (either technically or economically), as a sampling program proceeds and additional information is obtained.

Bulk samples for metallurgical testing can be evaluated in terms of the variance shown by all such samples in a deposit and the internal variability of grade. Attention to these sources of variability will ensure proper evaluation of pilot test results from bulk samples. In addition, a general mineralogical study of a deposit is an important, often ignored, requirement in a more general but semi-quantitative evaluation of the representativeness of a bulk sample; and, moreover, provides information useful in optimizing plant design.

INTRODUCTION

Sampling methodology and evaluation, so essential to a feasibility study, is a topic that cannot be considered in all its complexity in a brief chapter such as this. It is not possible here to document the variety of techniques used to evaluate every type of ore body. In particular, it is difficult to consider sampling aspects of large porphyry-type deposits on one hand and on the other hand to treat in the same detail some of the specialized sampling procedures used for pegmatitic masses or uranium deposits.[1,2] Instead, the author has chosen to approach the problem of sampling in a more general context, beginning with a short introduction to sampling methods and subsequently, emphasizing evaluations of procedures and sampling results.

The fundamental purpose of a sampling program is to delimit an orebody and to characterize its grade distribution. In more practical terms this means that the locations and assay values of samples must be convertible into grade and tonnage estimates. Such estimates may be required on several scales, e.g. global reserves or local reserves, and the error limits deemed essential to these estimates will have an overpowering effect on sampling procedure. For this reason considerable emphasis in the following account of sampling is placed on quantitative analysis that leads to meaningful error estimates.

Bulk sampling procedures, that supply material for pilot tests, will not be discussed in detail. Instead, emphasis will be placed on procedures that minimize the random and systematic errors inherent in bulk samples in practice.

SAMPLING METHODS

Samples and specimens of ores and waste must be distinguished. Specimens of ore or gangue are chosen routinely by geologists to exemplify particular features of mineralogy, texture, structure or other geological relationships, and need not be representative in any way of ore grade of a larger volume of material. In fact, depending on their nature, specimens can be strongly biased. Samples, on the other hand, are rock or aggregate taken in such a way that they are representative of a larger volume of material, especially with regard to one or more particular variables. Many sampling procedures have been developed that differ in detail, but for discussion purposes those important to our topic can be grouped generally as follows:[3]

A: Channel, chip and grab samples, and

B: Long hole drill samples.

Sampling errors are strongly affected by the general method of sampling and the detailed methodology.

Channel samples are those in which _all_ solid material in a rectangular prism (or channel) or rock extending across a mineralized face is collected as chips and fine matter. _Chip_ samples are those in which rock chips are collected along a line (at regular intervals with a predetermined spacing) or in some well-defined pattern from a face or other surface. _Grab_ samples are collected with relatively little effort compared to channel and chip samples either systematically or randomly from broken or solid ground. They should be without bias and thus represent a higher level of confidence than do specimens although the term "grab" is commonly used incorrectly as being essentially synonymous with specimen. Grab samples usually have a high error and find little application in most feasibility studies. Both channel and chip samples are particularly suited to outcrops, trenches, pits, stripped areas and, where appropriate, underground workings. Originally, channel and chip sampling involved hand use of a sampling moil. In most cases, modern methodology almost universally incorporates the use of small, hand-held drills to speed the sampling procedure. The desirability of channel or chip samples for a particular mineral deposit depends very much on the character of the mineralized zone being sampled and the required accuracy of the results. Both of these aspects require thorough consideration and in some cases procedures must be re-evaluated as information from check samples is obtained. In well stratified zones, be they alternating beds or mineralogical zones in a vein, it is obviously best to use channel samples with channels oriented perpendicular to layering. In general, chip sanples are more appropriate where such directional heterogeneities do not exist. The main problem with both chip and channel samples results from different physical properties of ore and gangue minerals. In particular, soft ore minerals may break off easier than hard gangue minerals such that grade will be overestimated. This problem can be minimized if wide channels or large chips are taken, or, if separate samples are taken from soft and hard zones. Where several samples must be averaged to produce a single value it is normal procedure to estimate a weighted average with each sample being weighted by a length (or area) to which it applies. Obviously, length of a sample zone is an essential piece of sampling information. Surface samples from trenches, pits, outcrops and stripped areas are commonly taken as chips or channels. Care must be taken to recognize the presence of weathering products in such case.

Drill-based sampling methods are now used routinely, particularly for evaluation of large mineral deposits where abundant data are required from what would otherwise be inaccessible parts of a deposit. All of rotary, percussion, auger and diamond drills are used depending on particular circumstances. Rotary, percussion and auger drills bring fragmental and/or finely ground or loose material to surface at or near the collar of the drill hole. This loose material is sampled by a tested, systematic procedure that provides an acceptable reproducibility. Normally such sampling is repeated at regular intervals to provide analytical data for the entire length of the hole. One of the main difficulties with these drilling approaches to sampling is the possibility of loose or soft material high in the hole sloughing as the hole is deepened, thus affecting assays thought to represent lower depths.

Diamond drilling is a routine method of sampling both from the surface and underground. As a rule, a more-or-less continuous cylindrical core of rock is produced for the total length of bit penetration through rock. One of the principal advantages of this core is that it permits <u>direct</u> examination of inaccessible mineralized zones. The core is split in half, lengthwise, one half representing a sample to be sent for assay, the other half being retained for geological information, particularly mineralogical study. In rare cases such as asbestos deposits the entire core is destroyed in the process of determining accurate recoverable fibre content; consequently, thorough detailed logging should precede such determinations. The main problems encountered with diamond drilling as a sampling technique are a) lost core, b) preferential grinding of soft materials from the core, c) lost sludge, and d) wandering of drill holes. Lost core refers to core that is finely ground, possibly due to its locally fractured nature or extreme, short range differences in hardness. The effects of lost core on assays generally can be evaluated by independent analysis of both sludge and core with the production of a weighted assay. A similar approach can be used to examine the possibility of preferential grinding of some ore minerals relative to gangue, a situation that is common where soft ore minerals are highly concentrated locally in hard tough gangue (molybdenite veins in granite rocks, for example). Drill holes can wander off course to some extent, particularly in layered sequences containing rocks of highly variable competency or in very incompetent rocks. A few orientation tests for dip (weak hydrofluoric acid in a test tube) or direction (tropari compass) will show whether or not significant wandering has occurred. Lost sludge may not be critical where core recovery is high.

In all drill-based sampling techniques the length of sample to be assayed is important. In North America, lengths of 5 or 10 feet (1.5 or 3 m) are common during the exploration stage, whereas during exploitation the bench height (say 45 feet or 15 m) may be the sample length. Sample length is obviously dependent on a particular situation. However, one cardinal rule to bear in mind is that a common sample length throughout will provide the most meaningful data base for statistical analysis. In the early stages of evaluating a deposit the uniform sample length should be as short as is economically and practically feasible. Hence, if for geological reasons it is desirable to have an odd sample length of 3 feet (1 m) to evaluate a dike in a porphyry deposit for example, then the next sample should be 7 feet (2 m) so the two samples can be combined to produce a standard length of 10 feet (3 m). If the dike in the preceding example were 55 feet thick (17 m) it probably could be avoided by selective mining, in which case 10-foot (3 m) samples need not be maintained throughout the entire thickness.

ACCURACY AND REPRODUCIBILITY OF SAMPLING PROCEDURES

Accuracy of a sampling procedure is a measure of how closely the procedure approaches a true value.[4] Analytical accuracy is generally measured by analysis of standards either synthetic or natural but having chemical characteristics similar to the mineral matter in question. Accuracy can be a function of the quantity of the element being analyzed. Hence, if a wide range of assay values are being obtained accuracy should be tested at two or more abundance levels. Where differences in accuracy exist a linear interpolation generally can be assumed in practice.

Where standards are not available or are difficult and costly to obtain it is common practice to accept the average result of several analytical methods as a "true" value against which to estimate accuracy. Accuracy can be expressed as a variance (S_{ac}^2), a standard deviation (S_{ac}) and in some cases as two standard deviations ($2S_{ac}$).

$$S_{ac}^2 = \frac{\Sigma(x_i - x_t)^2}{n - 1}$$

where x_i is the ith value, x_t is the true value and n is the number of determinations. In each case it is important to state clearly how accuracy is recorded. Many assaying procedures attain an accuracy of a few percent for $2S_{ac}$ expressed as a percentage of the "true" value.

Reproducibility is a measure of the inherent variability within a total sampling procedure and need bear no relation to accuracy. For example, a particular sampling and analytical procedure may consistently give more-or-less uniform results for duplicate samples but be significantly different than the true value. Such a situation indicates a bias in the procedure, a consistent error by an operator reading a dial for example.

Reproducibility is a function of 1) local sampling variability, 2) handling variability, and 3) analytical variability, and can be represented in equation form as

$$S_r^2 = S_s^2 + S_h^2 + S_a^2$$

where S_r^2 is reproducibility as a variance, S_s^2 is sampling variability, S_h^2 is variability arising from physical handling, and S_a^2 is the variability of analytical techniques. Local sampling variability is a function of the inherent variability in material being sampled as well as the size, shape and orientation of the sample itself. In many cases where grades are strongly autocorrelated over a range (distance) that is long relative to the sample dimension, a practical increase in sample size will not reduce the local sampling variability appreciably. On the other hand, a short range to autocorrelation may mean that an increase in sample dimension will reduce the local sampling variability appreciably as described by David and Dagbert.(5) Variability due to physical handling includes errors that arise in coning and quartering, riffling, repeated handling, etc., that is, those physical manipulations involved in reducing a sample to an appropriate size for analysis. Obviously much care is required to minimize the potential for both random errors and bias. A rigid systematic approach, once proven, will generally produce a very low error level routinely. Analytical variability includes the sum of all errors inherent in chemical and/or machine-oriented analytical procedures. Normally, in most tested analytical methods, the analytical error is small relative to other sources of error, particularly sampling error.

In any sampling program it is important to know the magnitude of errors that arise at various levels within the program and regularly to monitor possible variations in such errors. Evaluation of errors at different levels can be attained by duplicate sampling, careful splitting of duplicate samples at various levels in the sampling procedure, with duplicate splits carried through the remainder of the procedure and each split analyzed in duplicate. Such a procedure allows variances to be tested and compared by a hierarchical analysis of variance as described by Griffith[6], among others.

SAMPLING AGGREGATES AND BROKEN GROUND

For crushed, granular or fragmented rock and aggregates, a comprehensive sampling theory has been developed by Gy[7,8] who advances the equi probability theory, that is, the case in which fragment or grain types are distributed throughout a mass in unbiased fashion. In such a case the inherent sampling error is given in somewhat simplified form by

$$S^2 = \frac{\rho d^3}{M}$$

where ρ is a constant characteristic of a particular material, d is the screen diameter (in cm) that retains the coarsest five percent of the granular or fragmental material, and M is the weight of the sample (in gms). The constant ρ is defined by Gy[8] as

$$\rho = \frac{1-a}{a} \left[(1-a)\,\rho_a + a\rho_w\right] l\ f\ g \ = m\ l\ f\ g$$

where m is a mineralogical parameter, ρ_a is the specific gravity of the ore mineral(s) of interest, ρ_w is the spacific gravity of non-valuable minerals (i.e. waste), a is the average volumetric proportion of valuable minerals, f is a shape factor that for crushed ore is commonly 0.5, g is a grain size distribution factor generally taken to be 0.25, and l is a liberation factor determined as a function of d/d_o as described by David[9]. d_o is the liberation size of the valuable minerals.

In general ρ is fixed for a particular material, therefore, to reduce inherent sampling error it is necessary either to increase the mass (i.e. size) of the sample or to further crush and homogenize the material being sampled. It is apparent that because d is raised to the third power, a small decrease in d will have a more pronounced effect in decreasing precision. For example, if d were cut in half, the sampling error as a variance would be reduced to one-eighth its former value.

This approach to sampling applies to granular, fragmented and crushed material at various scales and the Gy basic formula can be used as a guide to sampling procedures for such complex tasks as the evaluation of grade of waste dumps, tailing piles, placer deposits and so on. A slide rule has been developed for rapid application of the formula with a description in English by Ottley.[10]

SAMPLING PLANS

Sampling plans are to some extent determined by factors that limit access to particular sample sites. Here we will ignore such problems and discuss sampling plans in a more idealistic manner. In practice, many sample plans are the result of a combination of common sense, empirical rules and practical factors all of which are difficult to quantify. Consequently, it is difficult to evaluate the errors inherent in sampling plans based on such an approach. There are three basic types of sample patterns to which reference commonly is made--random network, stratified random network and regular network. Conceptually, these patterns can be viewed respectively as (1) a block in which each of n samples is located randomly, (2) a block divided into n smaller blocks with a single, randomly located sample within each of the smaller blocks, and (3) a block divided into n smaller blocks with a single sample located centrally within each of the smaller blocks.

For a uniform number of samples (n) the random and random stratified sample networks have inherent errors of $S^2(0/V)/n$ and $S^2(0/v)/n$ respectively, both terms referring to the error (variance of a point sample (0) that is extended to apply to a volume (v or V). In this case V refers to the large block and v to the small blocks. For autocorrelated variables Matheron[11] has shown that $S^2(0/V)$ is always larger than $S^2(0/v)$, which means that some regularity to a sampling plan is desirable. In fact, the error is diminished even more if a sampling plan for an autocorrelated variable is perfectly regular, although this is difficult to prove simply without an extensive geostatistical argument. Nevertheless, intuitively it is apparent that the best position for a single sample as an estimation of a small block is at the centre of that small block. Furthermore, for variables that are strongly autocorrelated the improvement in error of a centrally located sample over a randomly located sample becomes very pronounced. Thus, it is apparent that for variables such as grade, thickness, metal accumulation, etc., that are in general autocorrelated and are used routinely and necessarily in ore reserve estimations, regular sampling plans are to be desired rather than irregular sampling plans.

Departures from a regular drilling pattern commonly are necessary for other reasons than simple grade estimation. For example, fairly specific drill sample locations that depart from a regular pattern may be needed to evaluate geological complexities that affect ore continuity. Thus, although a regular sample pattern is desirable because it provides minimum error for estimates based on a particular number of drill holes (sample sites), it should not be adhered to blindly in the face of other demands on a drilling program.

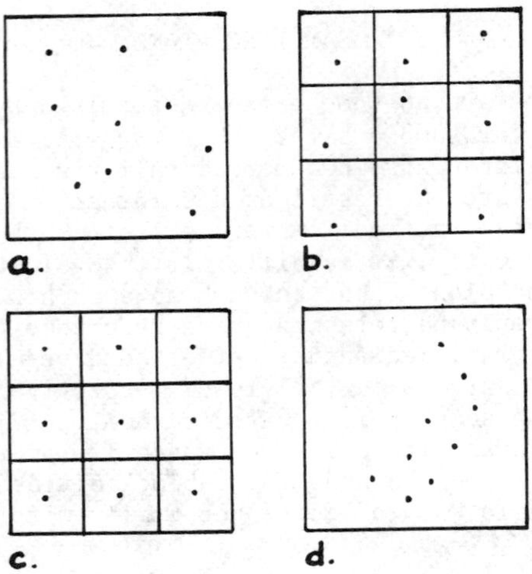

FIGURE 1: Sampling Patterns. (a) random sample locations in a block V, (b) stratified random sampling, with each of n samples located randomly in small blocks v where V = nv, (c) regular sampling pattern with all samples located at centers of small blocks v and (d) biased sampling pattern.

ORE RESERVE ESTIMATIONS

The principal purpose of a sampling program is to provide basic data with which to estimate ore reserves, including grade and tonnage. In general, these two variables are estimated independently and both estimates have associated with them errors that are dependent on the number and locations of sample information.

Traditionally ore reserves have been classified as measured, indicated and inferred (proven, probable and possible) as outlined in Table 1. This grouping is empirical and is not particularly rigid, leading to subjective applications and inconsistencies. Other descriptive classifications of some importance to feasibility studies are (a) insitu or geological reserves and (b) mining reserves. Geological reserves are defined on the basis of geological structure or extent of a mineralized zone and commonly do not take into account the fact that some parts of the zone may not lend themselves to extraction because of physical reasons. In addition, geological reserves may not be based on a cut-off grade. Mining reserves, on the other hand, represent

TABLE 1:

ORE RESERVE CLASSIFICATION

Category		General Description
U.S.A	Canada	(after Patterson, 1959)
Measured	Proven	Ore for which tonnage is computed from dimensions revealed in outcrops, trenches, workings and drill holes and for which the grade is computed from the results of detailed sampling.
Indicated	Probable	Ore for which tonnage and grade are computed partly from specific measurements, samples or production data and partly from projection for a reasonable distance on geologic evidence.
Inferred	Possible	Ore for which quantitative estimates are based largely on broad knowledge of the geological character of a deposit and for which there are few sample measurements.

a later stage in the evolution of ore reserve estimations in which mining and milling procedures are incorporated in the distinction between ore and waste. Parts of the deposit may therefore not be accessible to the extraction machinery, and some parts that are accessible may be too low grade. Thus, in calculating mining reserves it is essential to specify cut-off grade.

Cut-Off grade is the lowest grade of mineralized material that qualifies as ore in a given deposit, i.e., material of the lowest assay that is included in an ore estimate (12). Estimation of cut-off grade, although a complex economic problem (13) beyond the scope of this chapter, can be viewed simplistically as follows:

$$t_c = C_1 + \frac{C_2}{P} - D$$

where t_c is cut-off grade (\$/tonne), C_1 represents operating costs (\$/tonne), C_2 is annual fixed costs (\$), P is annual production (tonne) and D is the difference between average mining costs per tonne and the cost of mining marginal ore (\$/tonne). Note that t_c is defined in terms of variables that can fluctuate dramatically; thus, the cut-off grade is itself a variable that can fluctuate with time and its variations should be monitored. If t_c is transformed to metal content or grade rather than dollars per tonne its practical use as a selection for distinguishing ore-grade samples from waste is more apparent.

Methodology In Ore Reserve Estimates

Having determined a cut-off grade, sample information must now be used to define ore reserves. Traditional practice in North America has been to use one or more of several empirical techniques such as those described by McKinstry (3) or Patterson (14). The methods depend on arbitrary choices as to area or volume of influence of an individual sample, that is, subjective weighting factors in the calculation of average grade. Procedures applied to specific mineral deposits are modified empirically in each mineral deposit as production experience is gained.

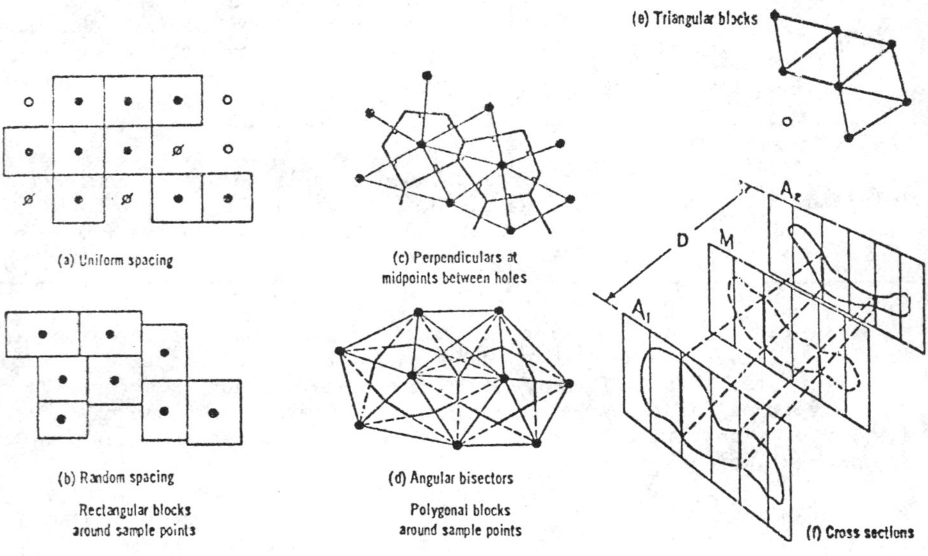

FIGURE 2: Various empirical patterns for obtaining sample weights for tonnage and grade calculations, after Patterson (14).

Recently these empirical methods have been computerized, particularly some of the more esoteric "polygon of influence" techniques for drill core samples with non-uniform orientations. O'Brian and Weiss (15) discuss general aspects of computer-related ore estimation techniques. The computer-based approach has also led to the application of various statistical techniques to ore reserve estimation problems. One of the most widely used methods is trend surface analysis as described by Phillips (16), where mathematical surfaces, generally polynomials, are fitted to spatially distributed assay values as a means of averaging and interpolating grades. The method apparently cannot be used to extrapolate sample data as shown by Link et. al. (17). The principal advantage of a statistical approach such as trend surface analysis is that in addition to being amenable to computer treatment it provides confidence limits to estimates. Unfortunately, difficulties can arise in the application of statistics of random independent variables to ore estimation problems where the variables commonly show significant autocorrelation, that is, they are not random (18).

Grade is generally a regionalized rather than a random variable. Matheron (7) has developed an extensive theory of regionalized variables with the principal advantages that it is a general approach applicable to any mineral deposit and the fact that meaningful errors can be calculated for all grade and tonnage estimates. The fundamental tool of Matheron's approach to geostatistics (the theory and application of regionalized variables especially as applied to ore estimation problems) is the variogram which is defined as half the expectation of the squared difference in values between two samples situated at x and x + h in a direction α,

$$\gamma_\alpha(h) = \frac{1}{2n} \sum_{i=1}^{n} ((x) - (x+h))^2$$

so that $\gamma_\alpha(h)$ is half the variance (error) if a value at point x is used as an estimator of (is extended to) point x+h.

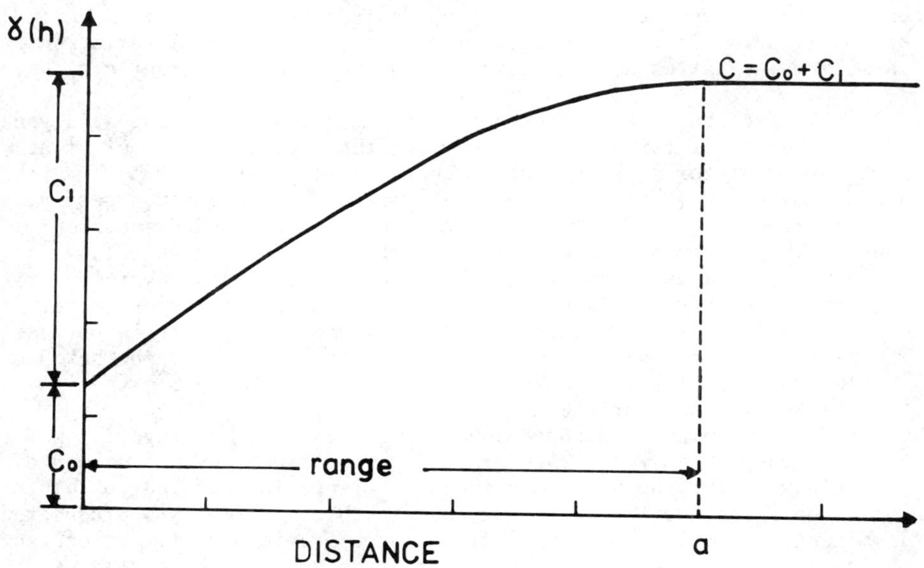

FIGURE 3: Smooth curve approximation of saw-tooth curve of a real experimental variogram.

Figure 3 shows an ideal variogram, a spherical (or Matheron) model in which the saw-tooth curve (not shown) of a real experimental variogram is approximated by a smooth curve of the form

$$\gamma_\alpha(h) = C_o + C_1\left(\frac{3h}{2a} - \frac{h^3}{2a^3}\right)$$

where C_o is an underlying random component (nugget effect), h is sample spacing (lag), a is the range beyond which samples are independent and C_1 is the sill value of the structured part of the variable in question.

As a rule $\gamma_\alpha(h)$ is an increasing function of h up to some distance a, the range, beyond which it is relatively constant (sill value C_1). By comparison, the semi-variogram for a random variable will fluctuate about a constant value for <u>all</u> sample separations. The range represents the average distance beyond which samples are uncorrelated (i.e., are independent); thus, it is an important parameter in the design of sampling plans. The range can also be thought of as the average dimension of "structures" within the data-mesh; hence, the term "structural analysis for determination and evaluation of variograms to describe autocorrelation of regionalized variables".

If a uniform range exists for all directions the data are said to be isotropic. If range, a, depends on direction, α, the data are anisotropic. Theoretically all variograms begin at the origin. In practice, many display a discontinuity, C_o (nugget effect), at the origin due to the presence of a structure with range shorter than the closest sample spacing for which $\gamma_\alpha(h)$ values can be estimated. A further complication is the proportional effect in which local variations in sill value varies as a function of the mean value of the local data.

All of these factors are taken into account in arriving at a general mathematical formula that describes the average character of a semi-variogram for a set of data. These models, the details of which are discussed in (7), are an integral part of formulae for determining two important types of variance--extension variance and dispersion variance. The difference between the two may be described as follows:

(1) Suppose we have 10 samples within a block of ore and want to estimate grade of the block and the probable error of the estimate. The extension variance is a measure of the uncertainty in our estmate, arising from the fact that we are <u>extending</u> a value based on 10 small samples to the entire block; hence, the term extension (or estimation) variance.

(2) Suppose we plan to produce 100 tons per day from a single stope. What variations in average grade might we expect over a period of 30 days? This problem concerns the "dispersion" of 100-ton blocks (1 day's production) within a 3000-ton block (1 month's production).

In mining operations both types of variance are important. Extension variances tell us <u>how well</u> we know our grade and tonnage estimates on global and local scales. Dispersion variances can provide important to the mineral processor regarding the range of fluctuations of ore quality to be expected over various time periods (because differently sized blocks are mined over different time periods).

In addition to such obvious practical applications, geostatistics provides the possibility of forecasting the effects that additional information will have in reducing extension variances; in other words, of increasing the confidence with which estimates are known. This arises because estimation variances depend only on the variogram (known) and the geometric pattern of samples. A variety of alternative sample patterns may be assumed and corresponding errors evaluated. Techniques of this sort provide the basis for optimizing costly pre-exploitation drilling programs.

Numbers of Samples Required

In any sampling program it is essential to have some general appreciation of the optimum number of samples to provide sound tonnage and grade estimates without the excessive time and expense involved in over-sampling. Without the use of a statistically based approach such a decision is either empirical or guesswork. Even with a statistical approach it is necessary to incorporate assumptions into the decision process. In general, it is essential to have some appreciation of the certainty with which an estimate is to be made--this decision is routine in any statistically oriented procedure. For example (see Figure 4), in an extended geostatistical study of the Ladner Creek gold deposit Montgomery and others (19) showed that 32 drill holes provided an in situ tonnage estimate with a relative standard error of about 11 %. In order to reduce the relative error in tonnage to about 5 %, an additional 22 drill holes were required. These were located such that the total 54 drill holes approximated a regular square sample network with spacings 100 ft (30 m) apart. As outlined previously, a regular sample pattern provides grade estimates with minimum error for a set number of samples. It is equally apparent that the error of estimates will decrease as the number of samples per unit volume increases. Hence, both aspects, pattern and number of samples, must be taken into account in optimizing a sampling program to produce estimates with a prestated error. This type of approach applies on any scale and depends on establishing an appropriate variogram model for a deposit.

In some cases, as in the early stages of exploration, most samples are fairly widely spaced, as might be the case in evaluating porphyry-type or flat-lying sedimentary deposits. In some situations the spacing of samples may exceed the range of autocorrelation; hence, statistics of random variables would prevail. Provided the mineralized zone is continuous between sample points, an average grade and errors can be determined by standard averaging techniques. Furthermore, tonnages can be correspondingly determined. However, as the sample spacing becomes less than the range (distance over which autocorrelation exists) the use of classical statistical methods applicable to random variables will provide false information. In the Ladner Creek gold example, the effect of thickness had to be considered. Using a geostatistical approach involving a variogram analysis, average thickness of the gold-

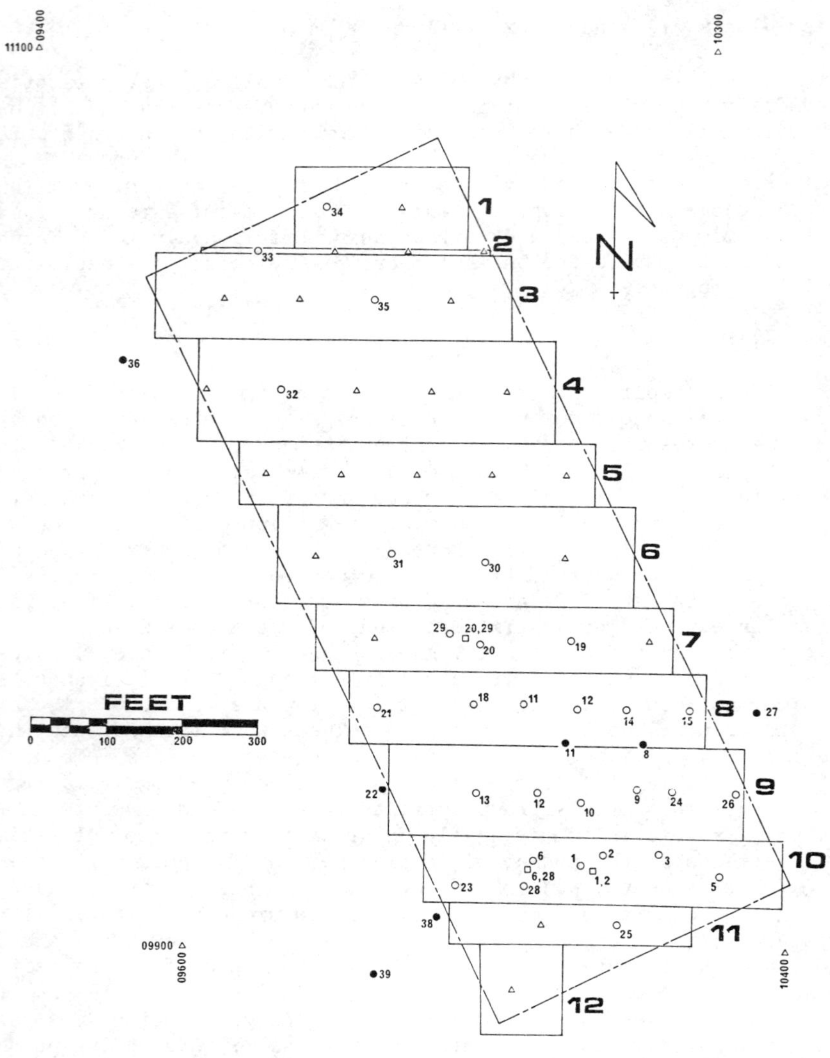

FIGURE 4: Plan of drill hole locations for Ladner Creek gold deposit after Montgomery et. al. (19). The large rectangular area (dash-dot line) represents lateral limits to zone for which ore reserve estimates were made. Numbered open circles are existing drill hole intersections used in reserve estimations. Open squares are collar positions of inclined holes. Open triangles are proposed sites of an additional 22 vertical drill holes, which provides information sufficient to permit a reduction in tonnage and grade errors by about 50 percent.

bearing zones was estimated to be 98.7 ft (30 m) with a relative standard error of 10.0 % based on 32 existing drill holes. If statistical methods for random variables were used a mean thickness of 96.1 ft (29.2 m) would be calculated with a standard deviation of 52.21. Thus, the standard error of the mean,(52.21/32) = 9.23, to give a relative standard error of (9.23/96.1)x 100 = 9.61%, compares favorably with the estimate based on the autocorrelation study. However, to reduce the error to 5% requires a standard error of 4.8 which in turn requires a total of 118 randomly located drill holes (i.e., 86 additional drill holes as opposed to the 22 which the geostatistical study showed would reduce the relative error in thickness to 4.8%). It is apparent that ignoring the autocorrelation and relying on statistics of independent variables could lead to financial disaster. However, examples exist in which the statistics of random variables can be applied to grade or tonnage estimates. The necessary methodology is given by Wolfe and Niederjohn (20) whose problem was to determine tonnage of a buried, flat-lying clay layer with a 95% confidence that the tonnage estimate is correct to within 10%. A preliminary drilling campaign of 10 holes, 7 of which intersected clay, provided statistics substituted into the Stein equation to estimate the number of additional holes required in clay to produce the required accyracy. Thus

$$n_2 = \frac{4t^2s^2}{d^2} - n_1$$

where n_1 is the number of holes previously drilled, t is obtained from t-tables to correspond with the stipulated probability level, s is the standard deviation of the variable in question (thickness), d is the permissible range of the variable (taken as a percentage of the mean value). In a firstapplication of this equation an additional 47 holes (n_2) in clay were calculated as being essential. The initial random sampling indicated that 70% of the holes were in clay; hence, an additional 68 holes are required in reality. Note that in this example no account has been taken of the error in delimiting the horizontal extent of the clay. Only the error in thickness has been considered.

Errors In Tonnage and Grade Estimates

Errors in grade estimates are the routine errors calculated by geostatistical methods in the extension of a limited amount of sample data to a volume of mineralized rock. Consequently, these errors are referred to as extension variances or errors. This type of error has been mentioned and has been detailed by Matheron (11).

Errors in tonnage estimates arise from two sources, namely, (a) error in the factor used to convert volume to tonnage and (b) error in some part of the volume estimation procedure. In many cases too little attention is paid to the tonnage conversion factor and its variability, which generally arises from the heterogeneous distribution of minerals with different specific gravities. In general, the conversion factor for short tons is equal to 32/(s.g.), where s.g. is the specific gravity of the ore. Suppose that the central part of a layered massive sulfide deposit consists of 10% sphalerite (s.g. = 4), 45% pyrrhotite (s.g. = 4.6), and 45% pyrite (s.g. = 5.1). The s.g. of the ore would be 4.77 (assuming no porosity) and the tonnage conversion factor is then 6.71. If the fringes of the same deposit consist of 60% argillite (s.g. = 2.7), 30% pyrite, 4% sphalerite and 6% galena (s.g. = 7.6), the ore would have an s.g. of 3.77 to give a tonnage conversion factor of 8.49--quite different from the previous factor. This example, though extreme, emphasizes the reason that ore density should be monitored by appropriate sampling to permit honest comparison of tonnage estimates with production. Even a change in the pyrite content of an argillite of from 2% to 10% increases the s.g. from 2.75 to 2.94, i.e., a 7% difference that would be reflected entirely in tonnage estimates.

Errors in volume estimates arise from errors in delimiting ore in place and errors in determining thickness of the ore. In many cases, both of these errors are negligible. This is particularly so if an ore zone has wide, gradational boundaries. Where relatively sharp boundaries occur the error in delimiting lateral boundaries, called the error in surface estimation, is given by (11):

$$\frac{\sigma^2}{s^2} = \frac{1}{N^2}((N_2/6) + .061(N_1^2/N_2)) \qquad N_2 \leq N_1$$

where σ^2 is surface variance, s is the surface area estimate, N_1 and N_2 are half the number of ore-waste unit sampling cell lengths in direction 1 and 2 respectively with directions defined such that N_2 is less than or equal to N_1, and N is the sum of N_1 and N_2.

Thickness of a deposit may introduce an error component in a volume estimate. This error may be calculated in a manner analogous to the error for grade, but based on an autocorrelation study of the thickness variation.

In general, the surface and thickness errors can be assumed as independent, so their squared relative errors can be added to provide a single estimate of the error variance for volume (21). For tonnage calculations, errors in conversion factors are independent of thickness to a first approximation. Thus errors in tonnage estimates can be expressed as (21):

$$\frac{\sigma_T^2}{T^2} = \frac{\sigma_s^2}{s^2} + \frac{\sigma_t^2}{t^2} + \frac{\sigma_f^2}{f^2}$$

where T refers to tonnage estimate, s is surface area estimate, t is average thickness f is the volume to tonnage conversion factor and the subscripted variances are self explanatory.

BULK SAMPLING FOR METALLURGICAL TESTING

A large bulk sample must be taken for pilot plant tests for the purpose of measuring the recoverable values in an ore and developing an optimal beneficiation procedure. In general, such a sample (or samples) should be representative of the total material to be tested so that potential problems in recovery can be foreseen and dealt with in advance. Unfortunately, bulk samples are not always representative. Plant recovery may not meet original expectations because of unexpected changes in the character of ore mined from one part of an orebody compared with another part. This is a common problem, since pilot tests are conducted during the early evaluation of an orebody when only limited parts of the deposit are accessible for bulk sampling. Availability of only parts of a deposit for bulk sampling is thus a practical matter of economics that is unlikely to dissappear. However, there are two important lines of study that help in the evaluation of bulk samples, namely, (a) a thorough understanding of grade variability of bulk samples and (b) a mineralogical study of the ore.

Variability of Grade of Bulk Samples

Grade variability of a bulk sample can be viewed in two different ways. In the first place it is important to know how representative the bulk sample is relative to the deposit as a whole. This can be evaluated by comparing the mean value of a bulk sample with the mean value of the deposit. It is probable that such a comparison can be made in advance of actually extracting the bulk sample on the basis of preliminary sampling information. However, another aspect that can be investigated is the variance of all possible samples the size of the bulk sample but within the limits of the ore deposit. Matheron (11) shows that this variance is given by

$$s^2(v/V) = F(V) - F(v)$$

where $s^2(v/V)$ is the variance of all possible bulk samples of size v within the ore deposit (V), F(V) is the average variogram value for two points each of which takes all possible positions within the ore deposit (V) and F(v) is the average variogram value for two points that describe all possible positions within the bulk sample (v). If $s^2(v/V)$ is too high the size of the bulk sample will have to be increased from v to a new value.

From another point of view it is essential to have a bulk sample in which the internal variability is representative of all such possible samples in the deposit. This internal variability is F(v). It must be evaluated as to its acceptability, perhaps by comparing the average expectation of F(v) with variability determined on the basis of a sampling campaign on the bulk sample itself.

F(v) and F(V) can be estimated for a deposit only after a variogram model has been established by means of graphs and equations given by David (9).

Mineralogical Sampling of A Mineral Deposit

A mineralogical study is broader than simply reporting on mineral species, and includes a thorough textural and structural description of the ore. Unfortunately, despite their relatively low cost and their value as described by Gaudin (22), mineralogical studies are sometimes omitted. Their main advantage is that relatively small specimens and samples from split drill core can provide direct observation of trends or abrupt changes in mineralogical characteristics that otherwise might not be taken into account in mill design or in estimates of recovery used for feasibility studies. This is not the place to dwell on details of the wide variety of mineralogical techniques that are available but their importance when directed towards the evaluation of a bulk sample relative to an entire mineral deposit cannot be overemphasized. Excellent accounts of the advantages of appropriate mineralogical studies of ores are given by Cadwell (23), Aletan (24), Petruk (25) and Kraft (26).

An often-ignored aspect of a mineralogical examination is the investigation of the mode of occurrence of important trace elements be they benefical or deleterious. For this purpose mineralogical techniques may be supplemented by appropriate chemical tests, electron microscopy and the electron microprobe. Abnormally high traces of Hg in massive sulfide deposits are commonly shown to occur mainly in ZnS. In porphyry deposits, Rh occurs in solid solution with molybdenite. The association of the precious metals must be established.

Routine mineral examination should include identification of both gangue and ore minerals and the occurrence and associations of both should be described in a thorough and systematic manner. Samples and specimens should be from locations more-or-less evenly distributed throughout the deposit. Proportions of all mineral species should be noted. Textural data must of necessity include routine information about average grain size, grain size distribution, grain shape, nature of gangue-ore boundaries, and types of intergrowths. Edwards (27) discusses aspects of both ore and gangue minerals important to mill design. Amstutz (28) has presented a useful empirical classification of locked particles that provide a basis for evaluating ore textures. Much of the data from such investigations may be illustrated graphically by means of Vandeveer diagrams (29).

SUMMARY

The ultimate aim of exploration sampling is to provide grade and tonnage estimates with acceptably low errors for the purpose at hand. Sampling procedures must be adapted to each particular mineral deposit. Quality control of sample method, handling procedures and analytical technique should be maintained by a rigorous procedure that tests accuracy and reproducibility and monitors errors.

Sample density is controlled by allowable error limits. A regular disposition of samples should be maintained to provide minimum error for a specified number of samples in a particular block of ore. Whereever possible a uniform sample length should be maintained, at least until an acceptable model for the autocorrelation function (variogram) has been obtained.

Error estimation involves a thorough geostatistical analysis that takes into account the autocorrelation of the regionalized variables (grade, thickness, metal accumulation, etc.) in question. A complete description of these techniques is beyond the scope of this chapter but the methods are fundamental to much of what has been described.

Tonnage estimates are subject to errors which can be minimized by appropriate attention to a sampling plan design. In particular, the volume-tonnage conversion factor can be the source of considerable error in tonnage estimates, a fact not always recognized and a variable rarely monitored by appropriate sampling and measurement of specific gravities. Errors in estimates of lateral limits and thicknesses of mineralized zones are in some cases very large and their values must be correctly estimated.

Bulk samples for beneficiation tests are not always as representative as might be hoped. This is a practical difficulty that can be minimized by examination of grade variability within the bulk sample, as well as the grade variability of all such theoretically possible bulk samples within the mineral deposit. Furthermore, a thorough mineralogical study of a deposit will provide a qualitative estimate of the representativeness of the bulk sample and will signal unexpected variations in the nature of the ore.

REFERENCES

1. Rowe, R. B., 1953, Evaluation of pegmatitic mineral deposits; Can. Min. Metall. Bull., p. 700-705.
2. Norton, J. J. and Page, L. R., 1956, Methods used to determine grade and reserves of pegmatites; Min Eng., April, p. 401-414.
3. McKinstry, H. E., 1948, Mining Geology; Prentice-Hall, N.Y., 680 p
4. Hill, W. E., Jr., 1975, The use of analytical standards to control assay projects; Assoc. Expl. Geoch., Spec. Publ. #2, p. 651-657.
5. David, M., and Dagbert, M., 1977, Lakeview revisited: Variograms correspondance analysis--new tools for the understanding of geochemical data; Assoc. Expl. Geochem., #2, p. 163-181.
6. Griffith, J. C., 1967, Scientific method in analysis of sediments; McGraw-Hill, N. Y., 508 p.

7. Gy, P., 1967, L'echantillonage des minerais en vrac; Tome I, revue de l'Industrie minerale, Saint-Etienne, 186 p.
8. Gy, P., 1968, Theory and practice of sampling broken ores; Can. Inst. Min. Metall., Spec. Vol. 8, p. 5-10.
9. David, M., 1977, Geostatistical ore reserve estimation, Elsevier Press, Amsterdam,
10. Ottley, D. J., 1966, Pierre Gy's sampling slide rule; Can. Min. J., Vol. 87, no. 7, p. 58-82.
11. Matheron, G., 1971, The theory of regionalized variables and its applications
12. American Geological Institute, 1957, Glossary of geology and related sciences, 325 p.
13. Henning, U., 1963, Calculation of cut-off grade; Can. Min. J., No. 2, p. 651-657
14. Patterson, J. A., 1959, Estimating ore reserves follows logical steps; Eng. Min. J., Sept. p. 111-115.
15. O'Brian, D. T. and Weiss, A., 1968, Practical aspects of computer methods in ore reserve analysis; Can. Inst. Min. Metall, Vol. 9
16. Phillips, R. J., 1968, Trend surface analysis in mining, Proc. Sym. on Decision-Making in Min. Expl., Western Miner Press, p. 28-33
17. Link, R. F., Koch, G. S. Jr and Gladfelter, G. W., 1964, Computer methods of fitting surfaces to assay and other data by regression analysis, USBM, R.I. 6508
18. Matheron, G., 1967, Polynomial interpolation procedures, Can. Inst Min. Metall., Vol. 60, p. 1041-1045
19. Montgomery, J. H., Sinclair, A. J., Symonds, D. F., and Giroux, G, 1977, Geostatistical study of the Lakner Creek gold deposit of Carolin Mines Ltd., Western Miner, March, p. 11-20
20. Wolfe, J. A. and Niederjohn, J. A., 1962, Statistical control of an exploration program; Min. Eng., Nov., p. 54-58
21. Sinclair, A. J. and Deraisme, J. R., 1974, A geostatistical study of the Eagle copper vein, northern British Columbia; CIM Bull., vol. 67, p. 131-142
22. Gaudin, A. M., 1939, Principles of mineral dressing, McGraw-Hill, N. Y.,
23. Cadwell, E. P., 1954, Color microscopy for the mill man, Eng. Min. Jour., vol. 160, p. 81-93.
24. Aletan, G., 1960, The significance of microscopic investigation in the course of beneficiation of the Brunswick ore, CIM Bull, Dec.
25. Petruk, W., 1968, The application of ore microscopy to mineral beneficiation, CIM Bull. June, p. 63
26. Kraft, M., 1967, Ore dressing microscopy as applied to concentration products of the Pb-Zn ores of the Brander Revier, Freiberg mining district, Germany; Can. Min. J., July, p. 52
27. Edwards, A. B., 1947, Textures of the ore minerals, Austr. I.M.M.
28. Amstutz, G. C., 1960, How microscopy can increase recovery in your milling circuit; Mining World, Dec., p. 19
29. Robertson, F., and Vandeveer, P. L., 1952, A new diagrammatic scheme for paragenetic relations of the ore minerals; Econ. Geol, vol. 47, p. 101-105

Chapter 8

ANAMAX MINING COMPANY

OREBODY SAMPLING AND ANALYTICAL METHODS

by

E. E. Krist, Chief Metallurgist

and

W. H. Sweetser, Sulfide Mill Metallurgist

ABSTRACT

Sampling of an orebody has a number of purposes. In the initial phase, it is necessary to define the limits of the orebody and mineral content to the extent that it can be stated whether it is, in fact, an economic orebody. Also, samples must be generated so that geological and metallurgical evaluations can be made. Finally, when sufficient information concerning the geology, mineralogy, and metallurgy are known, a mining method and plan can be developed and the metallurgical plants constructed to process the ores as they become available[1].

With today's large, complex open pit copper mines, single sampling methods would be inadequate. Diamond drilling, percussion drilling, underground shafts and drifts all play a part in the total orebody definition. At Anamax Mining Company's Twin Buttes mine, sampling and orebody definition will eventually lead to the existence of three metallurgical processing plants and, perhaps, an underground mining operation. A 30 000 tons per day sulfide concentrator was placed in operation in September 1969 and a 10 000 tons per day oxide leaching plant was placed into service in May 1975.

The aim of this presentation is to historically discuss the orebody sampling that has been done at Twin Buttes, briefly describe the orebody, and discuss the analytical methods currently used to assist the geologist, metallurgist and mine planner in their respective evaluations.

SULFIDE OPERATION

Throughout the 1950's, Banner Mining Company acquired extensive holdings at Twin Buttes where there existed a history of copper mining dating back to the early 1900's[2]. The Anaconda Company obtained an option agreement with the Banner Mining Company in March 1963 and began an extensive drilling program[3]. By April 1964 the orebody was partially outlined and The Anaconda Company exercised its option to proceed. Both surface drilling and an underground operation proceeded rapidly and in March 1965, the decision was made to develop an open pit at Twin Buttes based on a copper and molybdenum sulfide flotation operation. At this point in time, there existed no economic method for processing the extensive overlying copper oxide ores.

Before proceeding to the sampling that was done during this period, it should be pointed out that because of the 400 to 600 feet of alluvial overburden above waste rock and the orebody, unique mining methods were developed and placed into operation to integrate the production of ore and the construction of a plant to process sulfide copper ores. Regretfully, it is beyond the scope of this paper to regress to the challenging and successful mining methods and planning that have been employed at Twin Buttes.

During the initial period, an underground shaft was driven to confirm surface drilling and to generate sufficient samples for a pilot plant program. Because of the complexity of the ores, a pilot plant campaign was considered mandatory. Summarized sampling accomplished during this period is as follows:

Surface Drilling	
Period	3/63 to 6/67
Type of Drill	Truck Mounted, Diamond Drill
Number of Holes	193
Depth Range	700 to 2 500 feet
Total Footage	321 377 feet

Underground Excavation	
Period	7/64 to 7/67
Shaft Depth	956 feet (to the 2 400 foot level)
Drifts and Cross Cuts	16 835 feet
Tons Hoisted	
Alluvium & Waste Rock	8 885 tons
Drifts & Cross Cuts	142 442 tons
Stopes & Raises	27 852 tons
Total Tonnage	179 179 tons

Underground Drilling	
Type of Drill	Column Mounted, Air Driven, Diamond Drill
Number of Holes	73
Hole Inclination	Usually 15° to 60° up or down
Range	300 to 800 feet
Footage Drilled	45 339 feet

With respect to sampling the Twin Buttes orebody, inclined holes are preferred, when possible, in that the ore zones lie in vertical beds and better information is gained when crosscutting these zones.

The pilot plant was a converted operating mill[4]. Banner Mining had a 1 000 tons per day mill located at Mineral Hill which is approximately 7 miles north of Twin Buttes. This plant, which processed ore from Banner's Palo Verde underground mine, was shut down in May 1963. In July 1964 the decision was made to convert this mill to a pilot plant, which was done, and in February 1965 the first ores were processed and tested by the Anaconda Extractive Metallurgical Research Department.

The pilot plant processed 250 tons per day, and, during the conversion, the crushing plant was modified to act as a bulk sampling plant for the underground ores. During the two-year pilot plant campaign, which terminated in April 1967, a total of 93 793 tons of ore were sampled and tested, 85 978 tons of which came from the Twin Buttes underground operation.

The design parameters were thus developed which allowed the construction of a 30 000 tons per day copper and molybdenum sulfide extraction facility. The first milling division was placed on-stream in September 1969.

Since this initial thrust into the orebody, there have been a total of 618 diamond drill holes driven with an average range of 800 to 2 000 feet, and an additional 964 percussion drill holes driven with a range of 200 to 800 feet[5]. In general, holes were placed on 200 or 400 foot centers. In addition, a second exploratory underground operation was undertaken to obtain information with respect to a high grade zone. This operation will be further described later in this presentation. Core samples and pulp samples for chemical analysis have been saved and stored for future reference, if required. When blasting waste rock, sulfide or oxide ores, each rotary blast hole is assayed and, if in a doubtful ore zone, tested metallurgically. All rotary blast holes are sampled and assayed for the extremely important purposes of grade control of the sulfide, oxide and waste rock leaving the mine. In a given year, 15 000 to 20 000 individual samples will be prepared, assayed and tested, if necessary, from rotary blast holes alone.

OXIDE PLANT

Although oxide ores were the first ores to be mined from the open pit at Twin Buttes, at that point in time there was no known economical process to recover the contained copper. Conventional sulfuric acid leaching could not be used because of the excessively high requirement of 250 pounds of acid per ton of ore; consequently, as oxide ores became available, they were stockpiled in three distinct grade classes. By the time an extraction plant was put into operation in

May 1975, over 20 million tons of oxide ore had been mined and put into stockpile. This made access to representative ore samples for pilot plant testing quite easy.

Investigations of processes to treat oxide ores were started in the spring of 1963 by the Anaconda Extractive Metallurgical Department[6]. After a large number of processes to treat the refractory oxide ores were investigated, ammonia leaching became the most promising. A detailed laboratory investigation of this process was started in early 1968 and progressed through miniplant and one ton per hour pilot testing. With the prospect in 1970 of the availability of large quantities of inexpensive sulfuric acid due to newly created pollution laws in Arizona, a reevalution of the acid leach process was started in mid-1970 and was piloted on a 20 tons per day scale plant from November 1970 to October 1971. It was this pilot campaign that generated the design parameters for the construction of the 10 000 tons per day agitated leach, CCD, solvent extraction, electrowinning plant which went on-stream in May 1975.

In the fall of 1972 Amax was considering a possible merger with Banner Mining Company. To assure themselves of ore reserves, 143 percussion holes and 24 rotary holes were drilled in the oxide stockpiles. In addition, 5 diamond drill holes were drilled in the orebody which allowed laboratory flotation work to be done. In July 1973, after the merger between Banner and Amax, a partnership between Amax and Anaconda was formed to operate Twin Buttes under the name of Anamax Mining Company.

During the oxide pilot plant campaign, uranium values were detected in the leach solutions but, due to the relatively low uranium price, extraction was not pursued at the time. However, in 1975 the selling price of uranium began to increase, and when viable quantities of uranium were encountered in the existing solutions, a test program was initiated in May 1976. By July 1977, the parameters were available with which to design a plant. Because the chemistry of uranium extraction is well known, a pilot scale program was not considered necessary. An extraction facility which will utilize the Eluex process would consist of extraction of uranium onto a strong base resin, elution with sulfuric acid, concentration using ammine solvent extraction, and final precipitation to yellow cake using ammonia.

The importance of saving reject core or pulp samples can now be stressed. This is done at Twin Buttes, and, when it became apparent that uranium might become economically viable, it was possible to obtain samples for amenability testing for both the ores in stockpile and also for the oxide ores yet to be mined without additional drilling. The Anaconda Metallurgical Research Department conducted 133 amenability tests on samples from the stockpiles and an additional 112 tests on the reject core or pulp samples yet to be mined.

HIGH GRADE UNDERGROUND PROJECT

Slope stability problems occurred on the southwest side of the existing open pit, and in this area at depth there exists substantial tonnage of excellent grade sulfide ores; therefore, thought was given to developing this reserve by underground methods which could provide a potential alternative to open pit mining[5]. This could lengthen the mine life by adding reserves beyond economic pit limits, and this possibility continues to be evaluated.

In March 1975, surface drilling began on this project and is continuing. In May of the same year, a decline was initiated from an elevation of 2 250 feet in the existing open pit. This decline was at 12.5 percent until the ore zone was reached, at which point an incline of 2 percent was maintained. The initial underground development work in the ore zone was completed in July 1977. Nearly all surface and underground diamond drill holes were angle holes which, again, provided better information when confronting the vertical ore zones. Summarized statistics at the time of this writing are as follows:

Diamond Drill Angle Holes	
Number	65
Total footage	84 071 feet
Underground Excavation	
Footage to Ore Zone	3 839 feet
Inclination	- 12.5%
Drifts, Crosscuts in Ore	2 514 feet
Inclination	+ 2.0%
Tonnage in Ore	40 432 tons
No. of Rounds in Ore	485

The crushing and bulk sampling plant for the first underground operation, originally installed at Mineral Hill, was dismantled in 1975. Consequently, it was necessary to construct a new bulk sampling facility at Twin Buttes to obtain a representative sample of each round as it was mined from the underground operation. This facility consists of a 20 by 36 inch primary jaw crusher, a 10 by 16 inch jaw crusher, a 18 by 16 inch roll crusher, and a No. 2 Denver cone crusher which makes the final reduction to 10 mesh. Because the ore was clean, non-oxidized sulfide ore, only laboratory flotation tests were conducted on the samples and no pilot plant work was done.

TWIN BUTTES OREBODY

No attempt will be made to fully describe the Twin Buttes orebody in that this was recently done by J. Kelly of the Anamax staff[7]. A dikelike mass of quartz monzonite porphyry several hundred feet wide extends in a southeasterly direction under 400 to 600 feet of alluvium. The porphyry intrudes Paleozoic limestones, siltstones and Mesozoic clastic rocks. The limestones, siltstones, and quartzites are locally warped, contoured and disrupted by faulting. The quartz monzonite porphyry is very weak in mineral content. Because of this, the Twin Buttes orebody cannot be considered a true "disseminated copper porphyry." It is the silicated, garnetized and chloritized limestones and siltstones that show the strongest copper mineralization.

Oxidation of the orebody has altered the ore to a significant degree and the depth of oxidation is extremely variable. The copper minerals in the oxide zones are principally chrysocolla and cuprite with lesser amounts of azurite, malachite and brochantite. There is also a widespread occurrence of black amorphous "wad" high in copper. The "wad" is extremely refractory and is not responsive to either flotation or leaching techniques. Medmontite, a copper-bearing clay, is not uncommon and native copper occurs as fine stringers near the base of the oxide zone.

Although leaching and secondary enrichment is highly erratic, there are zones where secondary copper sulfides have substantially enhanced the grade of sulfide ore. Secondary chalcocite and covellite coatings on primary sulfide persist to considerable depth below the oxide zone.

Chalcopyrite is the most abundant primary sulfide mineral with only small quantities of bornite and primary chalcocite. Molybdenum is irregular but widespread. Tungsten values are also widespread in small quantities in the form of powellite and scheelite. Locally, sphalerite can occur in considerable quantity. Small quantities of silver exist and only trace quantities of gold.

ANALYTICAL METHODS

The analytical methods required for analysis of today's massive orebodies are as varied as the sampling techniques themselves. The methods developed for any particular property must be a balanced blend of speed, throughput, and accuracy. While the geologists and metallurgists would like instantaneous results accurate to 6 places, the experienced chemist knows that, generally, the more accuracy required, the longer the analytical procedure.

Most common mining analyses fall into one of two broad categories: First is the tried and proven chemical (wet) analysis, and second is the relatively new family of atomic methods[8]. While the atomic methods are, on the average, 5 to 20 times faster than wet methods, the excellent accuracy and reliability of the chemical methods insure that they will be a major portion of all mining analyses for some time to come.

Atomic Methods

The energy required to eject an electron from its normal orbital position is unique for each element, and once ejected, the electron emits a characteristic amount of energy as it returns to its normal orbit. All of the atomic analysis systems rely upon this phenomenon with primary differences being in the methods of electron excitation and energy level detection. Studies involving atomic spectra methods for chemical analysis were known in the 1920's; however, development of commercial equipment did not begin until the 1950's. It was immediately realized at that time that the speed and flexibility of the atomic methods far outweighed the capital equipment costs and lack of accuracy. The common atomic analysis methods fall into the following 3 categories: (1) emission spectrograph, (2) x-ray fluorescence, and (3) atomic absorption.

Emission Spectrograph. The source of energy for electron excitation is generally a direct current arc in which the sample material is fused. The generated light is passed through a defraction grating and a photosensitive negative (or detector) is exposed to the resulting spectrum. Analysis is made from characteristic emission lines on the developed negative. The emission spectrograph is only accurate to 10-25% of the material present; however, it has the ability to assay for all the heavy elements in a single burn and is extremely useful in accurately detecting trace elements. While the Twin Buttes chemical laboratory does not have a spectrograph, during the initial orebody development stages, use of spectrographic analysis was common. Then when interest in uranium and other associated valuable metals began in early 1976, use of spectrographic analysis became quite useful.

The typical emission spectrograph costs approximately $100,000.

X-ray Fluorescence. Probably the most common of the quick analysis techniques employs use of x-ray fluorescence. In this technique, an x-ray beam is used for electron excitation and the emitted energy detected by a crystal coupled to a photomultiplier tube. By adjusting the angle of the crystal deflection, the desired element can be analyzed. Many of the on-stream systems are currently generating up to 500 determinations per hour with accuracies in the 4-10% relative error range. This high speed analysis makes the x-ray unit extremely desirable; however, the accuracy is highly dependent upon mineralogy of both the desired elements and the host rock. If the orebody mineralogy is highly variable, x-ray calibration is extremely difficult to maintain.

At startup, the Twin Buttes concentrator was equipped with an on-stream x-ray system which analyzed 36 mill streams every 20 minutes. Due to the extremely erratic nature of the Twin Buttes orebody, accuracy of the system never reached acceptable levels and the system was abandoned in 1972. The chemical laboratory has a laboratory model x-ray unit which has run continuously since 1965. This unit currently analyzes approximately one third of the over 30 000 determinations done each month.

A standard laboratory model x-ray unit costs $100,000 to $125,000 while the sophisticated on-stream systems demand $250,000 to $300,000.

Atomic Absorption. During the past 15 years atomic absorption has advanced from a novel new instrument to the mainstay of many modern chemical laboratories. Atomic absorption is somewhat different from the other two techniques in that the energy absorbed by neutral atoms is measured, whereas in the other methods, the energy which is emitted by excited atoms is measured.

The sequence of atomic absorption is as follows: A solution containing an element of interest is sprayed into a flame where droplets of solution are volatilized. The solution components are broken down into clouds of neutral atoms. A light beam from a hollow cathode lamp constructed of the element of interest passes through the cloud where the neutral atoms absorb the light.

Since the specific cation to be assayed must be in solution, the atomic absorption method is much slower than x-ray. However, digestion of the sample eliminates virtually all of the matrix interference found in solid sample spectra methods. This results in extremely accurate readings, provided that the sample is diluted to the required concentration (generally less than 10 ppm). In most instances accuracy is limited to the dilution accuracy.

During the past 8 years the Twin Buttes use of atomic absorption has increased from 4 000 to 12 000 determinations per month. A top-line atomic absorption unit can be purchased for approximately $20,000.

Wet Methods

Although many of the techniques of wet methods are tedious and time consuming, the necessary accuracy cannot be reached with the faster atomic methods. Perhaps the best feature of wet analysis is the adaptability to different ore types. Every orebody has its own unique minerals and gangue materials which contain interferring ions. It is the singular ability of the analytical chemist to nullify the effect of these interferences which has made the wet analysis the trusted standard it is today. At Twin Buttes wet methods still account for approximately 1/3 of all determinations and are used for all metallurgical accounting and final product assays.

ACKNOWLEDGMENTS

The authors are grateful to Anamax Mining Company for the permission to publish this paper and to co-workers within the Anamax, Anaconda and Amax organization who contributed to the presentation. The writers are especially indebted to the Anamax Geology Department for the assistance which they gave.

REFERENCES

[1]Thomas, L. J., An Introduction to Mining, Hicks Smith & Sons, Sydney, 1973.

[2]Kalt, William D., Jr., "Awake The Copper Ghosts!", The History of Banner Mining Company and The Treasure of Twin Buttes, Banner Mining Company, 1968.

[3]Knaeble, J. B., "Development of the Twin Buttes Mine for Production," AIME reprint No. 70-AO-58, February 1970.

[4]Moore, J. E., "Final Report, Mineral Hill Pilot Plant, 1964-1967," Internal Reporting by the Anaconda Extractive Metallurgical Department.

[5]Kelly, J. L., Barter, C. F., and Wylie, E. T., Internal Anamax Communication.

[6]"Summary Report, The Twin Buttes Oxide Orebody Acid Leach Process Study," Internal Reporting by the Anaconda Extractive Metallurgical Department, December 1971.

[7]Kelly, J. L., "Geology of the Twin Buttes Copper Deposit, Pima County, Arizona," AIME reprint No. 76-I-3, February 1976.

[8]Liebhaesky, H. D., X-Ray Absorption and Emission in Analytical Chemistry, John Wiley & Sons, 1960.

Chapter 9

METALLURGICAL TESTING PROCEDURES

Richard L. Coleman

Director of Milling
Noranda Mines Limited
Toronto, Ontario

ABSTRACT

Metallurgical testwork to develop a flowsheet is described, beginning with background information required right through to a final process flowsheet. Both laboratory and pilot plant testing are reviewed with emphasis on the former.

INTRODUCTION

The amount and type of metallurgical testing to be performed on any mineralized samples will vary with each circumstance. Very little work may be necessary when developing flowsheets for new processing facilities to treat ore from an enlarged existing plant or a new adjacent orebody while conversely, years of research may be required for a complex ore far from other plants and test facilities or in a completely different environment. This testing is an on-going activity from discovery to eventual abandonment.

The initial research program in the metallurgical laboratory develops the general flowsheet which may be verified in the pilot plant depending on the complexity disclosed. Also, at this stage, the profitability of the project may be confirmed or placed in doubt before expenditures have become excessive. The flowsheet used for design and construction can be rearranged in degree following the commencement of production during the tune-up phase and may take years to optimize.

INFORMATION REQUIRED

Location

The metallurgist can proceed with his laboratory and pilot plant testing without knowing the location of the deposit and details of the topography, climate, water resources and transportation facilities. However, although these data may be generally of interest during the research program, they are essential to the flowsheet development and, therefore, he should not proceed too far without a knowledge of the geography and site characteristics.

As far as the bench scale testing is concerned, one very important local feature which will influence the metallurgical results is the quality of water available while the quantity is important in flowsheet development. Due to climate, geography, economics or local regulations, it may not be possible to obtain adequate fresh water for processing purposes and, hence, the majority of the required research may have to be conducted using the available saline or recycle water.

A thorough knowledge of the physical factors in the vicinity of the deposit is essential for flowsheet development. As examples, the snowfall affects the compactness of the plant, the terrain affects the amount of pumping, the remoteness affects the choice of equipment types, sizes and country of manufacture, the water temperatures affect the metallurgy and the manpower supply and quality affect the degree of sophistication and the approach to labour requirements.

The Deposit

In the majority of cases, the metallurgist's initial contact with the deposit of interest is through the exploration arm of his company or by letter from other companies when he is asked to do some research on this "copper" ore, that "lead-zinc" find or those "uranium" claims, etc. By implication, the research route is narrowed down to the one or more metals mentioned in the request but in fact, this is simply the start of information gathering. A knowledge of the geology is helpful and of the mineralogy may be essential. If the mineralogical data is not available, the research team will have to arrange for adequate studies by others or in their own laboratories. Studies should be conducted on specimens as coarse as practical because of the potential for heavy media or gravity concentration.

The features of interest from a geological or mineralogical viewpoint will include, but not be limited to, studies of the minerals present, degress of interlocking, extent of dissemination, size distribution of economic minerals and amount of alteration.

Samples

What samples are available for testing purposes and which samples are the most suitable for the research contemplated? There is no single answer to these questions as each case is different but one can be more confident of the results if a wide variety of samples rather than a large quantity of a limited number of types is provided. As the various stages of the research are contemplated, the metallurgist may be provided with surface hand samples, diamond drill core, trench samples or samples from stockpiles. As mine development progresses, bulk lots for the laboratory, the pilot plant and/or autogenous grinding tests should become available. Initial quantities preferred for bench tests range from 1 to 3 kilograms depending on complexity anticipated while the pilot plant may require hundreds of tons.

With respect to samples, two very important prerequisites to a good research program are that they should be representative and fresh. Split diamond drill core is most frequently available today but the quantities are limited because the geologists will keep a portion, various government agencies may require samples and portions may be kept in reserve. Hence, the research team may expect to see only 1/4 to 1/3 of the whole. Individual diamond drill hole samples should be kept separate at least until the geologist and the metallurgist confirm their representativeness and calculate the correct combining ratios. Initial tests may be done on a composite portion and later confirmed on the individual samples or the reverse order may be recommended when known differences are recognized early.

Samples of ores which are known to degrade, oxidize or otherwise change on contact with air or water should be sealed in plastic bags, possibly stored in strong metal containers and occasionally refrigerated in the case of unstable sulphides. These ores should not be crushed finer than necessary to facilitate handling and storage until required for the specific tests planned. When asking for "coarse" samples, the metallurgist should stress the point that he wants a complete sample including coarse and fines but that the amount of breakage should be limited to yield a convenient top size - typically 15 to 20 cms.

The obtaining of representative samples does not stop with the initial research programs because the metallurgist will require samples throughout the operating life of the concentrator to monitor, improve or revise the efficiency of the flowsheets, equipment and reagents. Thus, the importance of good sampling techniques never diminishes.

Mining

It is essential that the flowsheet development engineers know the proposed mining rates, methods, sequences and mine equipment sizes. All of these features affect the choice of mill equipment, operating hours in crushing, storage capacity, manpower requirements and the location of the concentrator. It makes an appreciable difference to a development and design engineer whether the ore is loaded from a 15 yard shovel, an orepass or a scooptram and is delivered to the crusher by a 100 ton truck, an underground train or a front-end loader and whether it will be wet, dry or full of snow.

The research metallurgist will be particularly interested in the mining sequence because the treatment methods in the early production phases may be significantly different than in later periods due to differences in response to beneficiation techniques for the various types of ore in the same deposit. Alternatively, the economics of the overall operation may favour the treatment of higher grade ores than the mine average in the early years of production.

Markets and Specifications

One cannot assume today that any beneficiation product can be sold. All consumers of concentrates and all secondary processing plants are encountering increasing regulatory and quality control restrictions and it is axiomatic that the buyers are more quality conscious when they purchase concentrates. The research engineer must be aware of the current and probable concentrate specifications and particularly the undesirable impurities so that minimum quality objectives can be established and premium quality products may be targetted if warranted.

The securing of a market for the mine products must be another prerequisite during flowsheet development, design and construction because the future for certain products and for marginal specification concentrates may be strong today but soft tomorrow unless a satisfactory contract can be obtained. Projected sales may be such that the size of the initial plant proposal in the flowsheet development stage may be adequate for a few years but may fail to produce products which will meet market requirements later so that provision for alterations, additions or expansion may be wise. Saleable products may have to be stored for long periods pending shipment (e.g. potash because of seasonal demand or base metal concentrates in the arctic due to open navigation seasons) during which time, compaction or degradation may occur and it may be necessary to investigate possible preventative or remedial action.

Environmental Guidelines

A thorough awareness of present, proposed and potential restrictions which might be imposed on the operations due to water and gas emissions, space limitations, noise, aesthetics and waste disposal is recommended in the early stages of the research. It is probable that the metallurgists will have to provide the backup data required for permit applications before any disturbance of the area will be allowed.

In addition, one may expect to encounter legislative requirements which would dictate alternate approaches to minerals beneficiation such as the replacement of one reagent for another more acceptable to the authorities. Simultaneously, a statement of fact may be required to show that the flowsheet proposed is the most practical route, the most economic and the technically correct approach.

Literature Review and Practice

Although the majority of companies contemplating new ore processing facilities would have people with experience in the specific field of interest, very few engineers today can keep abreast of all technical developments even in their own fields. Therefore, time spent in literature reviews on the subject or process of interest can be time well spent. In-house knowledge may not be disseminated as widely as it should be particularly with large companies and, therefore, good communications and exchanges of information are to be promoted always. At the same time, the most successful developers will study industry practice to take advantage of its experience and modernization trends and arrange visits to processing facilities similar to those proposed. Fortunately, the mining-metallurgical industry has been relatively free with its exchange of ideas and, for this reason, progress in minerals beneficiation has been more rapid than in some other fields.

LABORATORY TESTING

The types of tests to be recommended can be selected only after much of the suggested prior information mentioned above is available. There are a great number of tests available to the metallurgist but no ore would need to be subjected to all of them. Experience will dictate those which should be employed and problems encountered will force attention to other less-common techniques.

Bench testing of ore samples is one of the most important steps in the long series of metallurgical and economic studies involved in bringing new mines into production. If these investigations are not thorough, many questions and problems that develop during the research programs, in mill design engineering and later during operations can be traced to inadequate or improperly interpreted laboratory test data. Experimental programs should not be terminated until a satisfactory general flowsheet with all requirements from crushing to tailing disposal and concentrate load-out are met.

The laboratory testing does not terminate when the operations commence. On a regular routine basis, bench tests, using a standardized laboratory procedure, can be used to determine the differences in ore types, response, reagents and flowsheet changes required during the operating life of the orebody. They also can be used to develop or test new techniques or processes prior to the full scale pilot trials. If favourable results can be obtained in the laboratory under controlled conditions, the flotation engineer gains assurance that "it can be done" and thus new, improved, achievable objectives can be established for the operations.

The pilot plant concentrate production can be used by the potential buyers for their own metallurgical research. Finally, laboratory tests and pilot plant programs can be used to advantage to train mill operators - particularly those responsible for flotation.

Crushing

The first step in minerals beneficiation of an ore is normally crushing which has been defined "as the operations required to reduce run-of-mine ore to grinding feed size".[1] It would therefore follow that the first metallurgical testing should be conducted on run-of-mine ore but, due to its size, this would be rather impractical. However, there is little need to test large rocks because smaller competent rocks upon examination and under test conditions can show similar characteristics. Good relationships between rock sizes, tonnage rate and equipment sizes have been established in the industry.

Crushing tables supplied by the manufacturer normally can be used directly to size crushing equipment for a given flowsheet. However, if the ore contains a large percentage of primary fines, wet sticky clay or is extremely competent, it may be prudent to conduct crushing tests in the pilot plant on representative samples of the material. These tests will identify crusher power requirements or unusual screen surface area for the ore in question.

It is seldom necessary for the owners of a new mine to invest large sums of money in sophisticated laboratory crushing facilities for the limited number of tests which may be required to prove the crushing characteristics of the ore. All of the major equipment manufacturers have research facilities for determining crushing requirements, their personnel have experience and they welcome the opportunity to work on new ores. They can do crushability tests to determine the work index of the ore and the amount of work required as well as the compressive strength to determine the necessary strength in the crusher. However, the ore dressing research staff must be able to evaluate the results from the crusher manufacturer because, in many instances, they have the final responsibility for flowsheet design and development.

Nevertheless, at least one small jaw crusher and a small cone crusher should be considered standard laboratory equipment for ongoing operations. The jaw crusher procedure should be standardized and may even be calibrated against a "known" ore to compare breaking characteristics.

Following crushing in the laboratory, test sample lots should be riffled, preweighed and the weight adjusted to obtain standard sample lots. One of these lots should be set aside for a comprehensive physical analysis followed by a chemical analysis possibly after the initial grinding test. One should crush no more of an ore than can be satisfactorily tested on the bench scale within one month - particularly when treating high sulphide ores.

Grinding

Although metallurgical testing may not start with grinding,

intensive research must be conducted in this area since the power and media consumptions in grinding in an operating plant are frequently the most significant operating cost items in minerals beneficiation. The purpose of comminution is to affect adequate liberation of the desirable components from each other and from the waste so that subsequently, they may be separated efficiently by flotation or other processes. Grinding is the initial key variable to define and refine. Exceptions to this statement may be noted in the processing of potash and other non-metallics, in handling extremely complex sulphide ores and in some of the less common ore treatment plants where expensive or large quantities of reagents are more costly than grinding. Therefore, as power, energy and material costs escalate, it behooves the metallurgist to develop the most economical comminution techniques.

In the early years of this modern milling era, the operators had but one choice of grinding media and method, i.e. balls in ball mills. Today, there is a wide range to select from. Broadly speaking, this range includes fully autogenous, semi-autogenous, pebble, rod and ball milling and it is the responsibility of the metallurgical research team to choose the most efficient type or combination of types. Grinding may be performed wet or dry, although, in the base metal, uranium and iron ore industries, wet methods are usually preferred for a number of reasons, chief of which may be the ease of handling, the possible effects on metallurgy and the environmental advantages.

The geologists, mineralogists and miners can assist the metallurgists to do the preliminary analyses on the ores to determine the potentials for some form of autogenous grinding. They will note whether the ore is hard, soft, highly fractured, massive, weathered or blocky and many other significant features. If at least part of the grinding may be accomplished in this manner, the number of steps, the number of pieces of processing equipment including conveyors and the consumption of purchased grinding media could be lowered. While appreciably lower capital costs in the crushing plants will be partially offset by the higher costs of the grinding equipment, the reduction in operating costs may make the use of autogenous grinding units very attractive. Recently, methods[2] have been proposed which may permit early assessment for feasibility of autogenous grinding from diamond drill core.

It is common practice to conduct the initial grinding tests in laboratory rod mills or ball mills. For these tests, the ore must be crushed to 6.5 mm (1/4 inch) or less and a uniform quantity of ore used for each test. To evaluate the results, most research establishments use the procedure developed by F.C. Bond in the Allis Chalmers laboratories[3]. This method permits the calculation of a "Work Index" which may be defined as the kilowatt-hours required to reduce one ton of ore from infinite size to 80% passing 100 microns. The formula to calculate the amount of power required in grinding uses this work index as a basis and may be written as follows:

$$W = 10Wi\left(\frac{1}{\sqrt{p}} - \frac{1}{\sqrt{f}}\right)$$

Where W = Work input - Kw Hr per ton of ore
Wi= Work index
p = size in microns through which 80% of the product passes
f = size in microns through which 80% of the feed passes.

Equipment suppliers and commerical laboratories are available to perform these tests but most company research facilities should be outfitted similarly because grinding tests will be required throughout the life of the orebody to confirm grind requirements and prepare test lots for flotation, leaching or other research.

It is suggested that a typical laboratory rod mill might be fabricated from 20 cm (8 inch) diameter schedule 40 type 316 stainless steel pipe and would be about 38 cm (15 inches) long. See Figure No. I. The stainless steel grinding rods for this size of mill may be a graduated charge from 25 to 10 mm diameter (1 inch to 1/2 inch) but variations in size are not essential. The Allis Chalmers Company uses a 30 cm diameter x 61 cm long (12 inch x 24 inch) rod mill with 30 and 45 mm rods (1 1/4 inch and 1 3/4 inch). A similar but smaller 30 cm x 30 cm (12 inch x 12 inch) ball mill with 40, 30, 25 and 20 mm balls (1 1/2, 1 1/4, 1 and 3/4 inch) is used for their ball mill work index calculations. Stainless steel is suggested because the abrasion and corrosion will be negligible, it is easy to clean out and there will be no rust to influence the subsequent metallurgical results. However, mild steel or carbon steel test mills and rods or balls may be desirable for certain ores to duplicate plant flotation or leaching practice when soluble iron has a bearing on the metallurgy or the solution used might react with the commercial mill or media. These influences are claimed to have been demonstrated recently in a research program by Canmet.[4]

Other sizes of rod mills or ball mills can be used but it is suggested that, regardless of the choice, the same mill should be used on all ores and for all tests. This will permit continuous comparisons with past research and with known ores.

The 20 x 38 cm laboratory mill mentioned above is suitable for testing 2000 gram lots of most normal specific gravity ores. If the same quantity of sample and pulp density are used for all tests in the same mill, the results obtained when changing the grinding time will be analysed readily since the length of time required to grind the sample is proportional to the work input.

It is impractical or at least extremely difficult to measure the grinding power consumed by laboratory mills directly because of the high proportion of the power which is consumed in drive inefficiencies and fixed load characteristics. Without very sophisticated instruments, it is difficult to tell the difference between a lab mill which

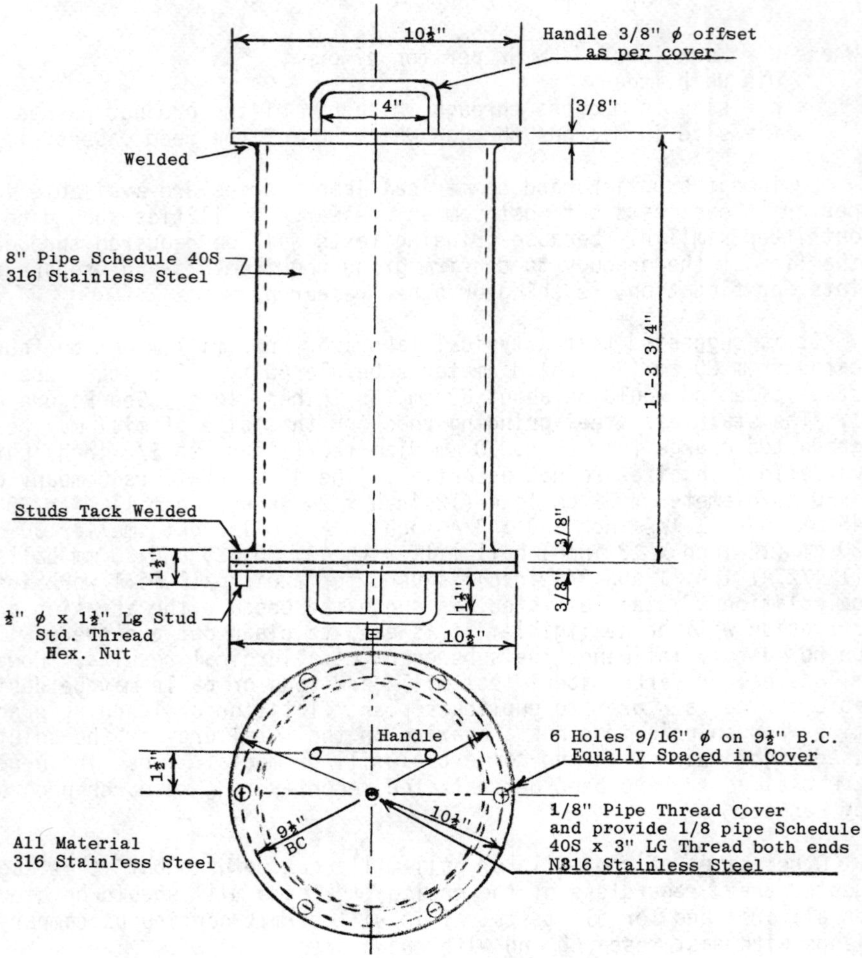

FIGURE 1 - Laboratory Rod Mill

is loaded with steel and pulp or one with steel alone. Fortunately, it is not necessary to measure this power because alternate means to determine the work index are available.

Where there are no previous records of grinding mill power calculations, it may be necessary to consult with a commercial or an associate's laboratory personnel to have the initial work index determinations done on your ore in their laboratories and compared with reference ores. A second method is available when an operating grinding mill is processing an ore which can be tested in the laboratory. By producing the same product screen analysis in the laboratory, a direct comparison between the products in the plant and the laboratory permits calculations of relative work performed. Alternately, when your own research staffs have established work indices during previous projects, they will have historical data upon which to base their present project. The research team will find that the straight lines shown on log-log plots of the results from the various ores tested will be useful for comparisons.

The amount of grinding required for any project is established by the efficiency or adequacy of the initial and subsequent separating stages. Hence, once the correlation between grind and recovery has been established, the economic power requirements can be estimated. The most suitable method of grinding which might be with steel or autogenously can then be determined.

If the indications during the early physical examinations of the ore samples did not reject the possibility of autogenous grinding, various tests to determine the feasibility of this process should be planned. This means moving out of the laboratory and into the pilot plant where grinding units are larger. Relatively speaking, there are very few mining companies who have in-house pilot plant autogenous grinding equipment and, therefore, this type of testing will be done in commercial laboratories. In most cases, the metallurgists and the owners of the new plant should insist on piloting before proceeding with full scale plant design because, despite the advances in the technology, there are greater risks in choosing the autogenous grinding route than in steel grinding. To date, no successful autogenous grinding plants have been built without some pilot plant research and the most successful applications have been the result of the most extensive research.

In the pilot plant, it will be possible to determine if some steel is required to augment the ore as grinding media and to determine if pebble milling is viable. Since the media in primary autogenous mills and some pebble mills must come from crushing plants, the flowsheet development of the latter cannot proceed until answers to grinding questions are obtained.

If some form of autogenous grinding is not the chosen route, grinding with steel is the principle alternate available but even then, there are at least two options open - rods and balls or slugs - particularly for the primary stage. There are few laboratory tests which can be conducted to assist in choosing between rod milling or ball milling and, hence, many practical factors influence the ultimate choice.

Rod mills may have been termed "fine crushing machines" but two directly opposing statements about their operating costs have been made.[5] In the earlier days of this century, one frequently heard that "it is cheaper to crush than to grind" but today, with the tremendous increase in grinding mill sizes and power utilization, it is doubtful that the statement is still true. Since a rod mill will accept a coarser feed than a ball mill, it fits in naturally as the intermediate comminution step. However, rod mills, like crushers, must be operated under controlled conditions to obtain maximum efficiency and are more difficult to operate than ball mills.[1] Thus, the ultimate choice between a rod mill or ball mill for the primary grind may depend on the operator's or designer's experience as much as the metallurgist's research.

Regrinding

The amount of regrinding of intermediate concentrate or middling products and its location in the flowsheet can be determined only during the testing of the separating techniques developed by the research staff. The need for regrinding is more widespread today because smelter contracts are more restrictive than they used to be and thus, better concentrate grades are required. Calculations for sizes of mills and power inputs will be influenced by the work index, the degree of liberation required, circulating loads, surface freshening or polishing desired, viscosity, chemical interactions and industry or company experience. Therefore, although the amount of regrinding outwardly might appear to be routine and the Bond Work Index can be utilized, the final choice may involve as much reference to industry practice and judgement by the designers as by the theoretical laboratory results. Fortunately, the quantity to be reground is normally appreciably less than the tonnage of primary ore and, hence, the incorporation of a few capacity safety factors is not liable to represent appreciable increased capital or operating costs.

Classification

Although the hydrocyclone has earned a prominent place in the minerals beneficiation field, it should not monopolize the classification step. There is still room for the mechanical spiral and rake classifiers in specific areas despite the fact that they may require more space during installation and more maintenance during operation.

Testing of cyclone classification units on a bench scale is difficult because the equipment openings suitable for laboratory samples and quantities are too small for passage of the normal sized products. The research may define the desirable cyclone splitting points but the metallurgist will have to rely on experience to determine sizes, types, openings, pressures, etc. to use in selecting cyclones. Fortunately, the commercial laboratories and equipment suppliers have gained enough experience in the past 20 years that, with their computerized programs, they can make reliable recommendations based on the screen analyses, specific gravities and densities desired.

Mechanical classifiers are utilized today in specialized circumstances or by operator preference. They may be used for coarse particle separation where power requirements for cyclone operation would be very high and where their efficiency is preferred.

Flotation

If comminution is accepted as the key cost factor in milling, flotation can be termed the key metallurgical or efficiency factor for all base metal and many precious metal, ferrous, non-metallic and solid fuel beneficiation plants. The range of tests available is endless and should be used throughout the life of the orebody because unlike crushing, grinding, gravity concentration, thickening, etc., most changes in flotation flowsheets are relatively easy to make during operation without significant cost as new reagents become available or new flow arrangements are deemed necessary.

As flotation is still largely an art, flotation tests must be planned by experienced flotation engineers on the basis of actual experimental results with less consideration to theory than most other metallurgical processes require. No testing program can be successful unless the test data are correctly interpreted and succeeding tests carefully planned so that all interesting avenues are investigated.

Typical flotation tests are conducted in laboratory flotation cells with nominal capacities from 50 to 2000 grams and pilot plant flotation cells from 6 to 75 litres. Some research groups have had to make their own ultra-small cells when very high ratios of concentration were involved. However, scale-up from these models to plant equipment sizes is not practical.

It is convenient to use the laboratory flotation cell as a conditioner should conditioning be a variable to be tested. Most of these cells also have provision to add a gas (air or other) as required.

Keen observation of all details during actual flotation testing is essential. Although most of the prior and subsequent steps can be carried out by competent technicians, the actual froth color and removal should be observed by experienced flotation engineers. It is not acceptable practice to rely solely on inanimate product weights and assays recorded after the test runs.

Many articles and books have been written about flotation testing techniques. Macdonald and Brison have written about their own methods and include a good summary of reference articles.[6] The variables to be tested in flotation will include but not be limited to the following:

Reagent Types
Reagent Quantities
Pulp Density
Aeration and Conditioning
pH Regulation
Retention Time
Temperature
Water Quality
Sample Aging.

It has been found that, as these variables are changed, the results obtained will be analysed most readily if plotted on graphs and/or listed in tables. Tables should show results against one variable only and graphs with two or three only to be most easily interpreted. However, in certain instances, a series of statistical design tests to evaluate the rank and importance of a number of variables under controlled conditions may be recommended.

Reagent Types

It is beyond the scope of this article to list the reagents available other than by classification because there are so many, new ones are continually appearing and common ones are becoming obsolete, non-available or non-economic. Suffice to say that the experienced research metallurgist will know which reagents to start with and the inexperienced will find numerous references to suitable ones in the literature. At this stage in the research, it is probably more important to find a reagent combination that ensures the success of start-up rather than one which is the "ultimate" - a condition which is truly never obtained. In certain instances - e.g. potash - suppliers are actually "customizing" reagents to fit the specific ore. The classifications of reagents to consider will include collectors or promoters, frothers, modifiers or alkalinity regulators, dispersants, emulsifiers, depressants, activators, protective colloids and flocculants. The major commercial laboratories and reagent suppliers have extensive literature on their products and frequently provide samples for research purposes.

Reagent Quantities

An integral part of reagent testing is to vary the quantities used. Previous research, plant practice or the literature will provide a good starting point from which the metallurgist should vary up and down the scale to determine the optimum.

Minor or major affects may be observed or show up in the assaying of products whether dealing with starvation quantities or large doses. Since relatively small quantities of ore are to be tested, the amount of reagent to be used will be very small and hence, most reagents should be diluted or emulsified prior to their addition to laboratory flotation cells.

Pulp Density

Throughout the testing of ores and their processing by mineral dressing techniques, the density of the slurry will be an important factor which the research metallurgist and operator will want to control. Before the slurry density can be determined, the specific gravity of all products of interest will have to be ascertained. The diamond drill core samples are convenient for making these determination.

When the gravity has been determined, the pulp density tables can be used for the specific gravity of interest or the density can be calculated using the formula:[7]

$$D = \frac{P}{W + C} \times 1000$$

Where: D = density of pulp in grams per liter
P = sum of the solids to solution ratio
W = weight of water in pulp at stated percentage solids
C = a constant determined by dividing 1.00 by the specific gravity of the dry solids in the feed.

Viscosity of the slurry, intensity of agitation, mineral crystal shape and size, specific gravity of solids and solution, etc. all have a bearing on the choice of the most suitable slurry density.

Although the pulp densities that are experienced in the laboratory are lower than in plant practice, they must be controlled, varied and recorded in all research programs. In primary grinding with laboratory rod mills, the densities of low specific gravity ores are generally about 50 percent solids by weight and with higher specific gravities may be as high as 65 percent solids. With viscous ores, densities will be less than 50 percent solids and even may be lower than 40 percent as recently demonstrated by Claridge and Tenbergen.[8] Laboratory regrinding in ball mills will range frequently between 20 and 35% solids principally because it is difficult to thicken intermediate flotation products.

Due to the necessities to wash grinding media, transfer pulps, raise cell volumes to provide adequate froth depths and to wash down flotation cell walls during every individual test, laboratory flotation pulp densities are low. Common figures are 15 to 25 percent solids and frequently lower figures will be recorded, particularly in cleaning.

Aeration and Conditioning

The separation by differential flotation of base metal minerals from each other and from a sulphide gangue can be improved frequently by 15 to 20 minutes of aeration prior to flotation. This fact may not be recognized or appreciated ordinarily in bench tests because the action of transferring of pulps from laboratory mills to buckets and to flotation cells and the agitation in the cell prior to froth removal introduce appreciable air to the pulp. A frequent cause of the discrepancy in metallurgical response of high sulphide ores between the laboratory and the operating concentrator can be traced to the relative degrees of oxidation or oxygen demand in the two. Whether or not aeration is critical to good mineral separation can be demonstrated by conditioning in the laboratory flotation cell and confirmed in a 10 cm diameter x 2 meter long (4 inch x 78 inch) pipe mounted in the vertical position and with provision to add plenty of air at the bottom.

Conditioning without specifically adding air may be essential or desirable for various lengths of time for all ores and therefore should be tested.

pH Regulation

The effects of pH variations in flotation can be very marked in some instances while in others, no detectable trends are apparent. However, there are few ores that can be completely beneficiated by flotation in the ambient pH of the local waters. Therefore, a planned program with modifiers such as lime, soda ash, ammonia and/or sulphuric acid always must be included in the research agenda.

Unfortunately, simple pH measurements alone are not enough with some high sulphide ores because the presence of Ca^{++} or Na^{+} ions or both may be more significant than pH. Certain ores - e.g. copper, lead, zinc - may require both soda ash and lime at specific points in the flowsheet for optimum grade and recovery control. Similarly, the dispersing, flocculating or buffering effects of these alkalis may be highly significant despite a low or high pH. As with many other flotation reagents, the rates and points of additions of these chemicals can be critical and important. In addition, changes from acidic to basic conditions and the reverse - particularly with some high sulphide ores and in differential separation circuits - may be necessary to obtain the desired metallurgical results.

Retention Time

The total amount of time required in flotation or in the various steps contemplated can vary from the one minute range to one hour or more depending on the types of minerals, circulating loads, the response to reagents and the stage involved such as roughing vs. cleaning. The time factor helps to establish recovery, grade and cut-off between gradient assays for subsequent processing, conditioning within cells, etc. and is an essential parameter to control in all tests. It is very easy to extract products at specific time intervals which, on analyses and with graphical plots, will indicate cut-off points to commence recovery of the second mineral or the need for alterations to affect additional recovery prior to advancing to the next step.

Temperature

A few ores will show improved efficiencies with elevated or depressed temperatures and some particularly complex ores and non-metallics show marked changes with temperature variations. Therefore, occasionally temperature related tests should be performed and, even for simple ores, the operating temperatures should be recorded. As examples, warmer solutions may be required in chalcopyrite-galena or sphalerite-pyrrhotite separating circuits while cooling can be beneficial in potash flotation.

Water Quality

The difference in response of certain ores to separating techniques by flotation particularly in cleaning steps can be pronounced when comparing results of tests using fresh water vs. other types of water. Therefore, a metallurgist's research should be done with fresh water _and_ the type of water he may expect to be provided with in practice. Recycle water from within the concentrator or from the tailing dam, treated effluent, mine water and/or sea water may be the available fluid which will have to be considered for most of the mills of the future and even for the older mills to meet environmental guidelines. It may be necessary on occasion to make up synthetic solutions similar to the water expected at the mine site for testing purposes although the results might be less dependable and thus should be confirmed with some quantity of the available water obtained from the proposed concentrator site.

Sample Aging

Samples from the majority of ores under examination could be treated today, six months from now or two years from now with little or no difference in metallurgical response. However, there are types such as the complex sulphides with a high pyrrhotite content or the degradeable ores such as the phosphates which should be tested fresh from the mine or kept in plastic bags and/or under refrigeration until required. In addition, samples of these unstable ores should not be reduced in size by crushing or grinding until the research staff is prepared to conduct the flotation tests. In some cases, the only available samples may be old and therefore, the research would be conducted, of necessity, on this material. However, prior to design, confirmatory work should be done on fresh samples.

Cycle Tests

The number of batch tests required will vary from a few on the simple ores to perhaps hundreds for some complex ores. When middling products contain significant quantities of the metals to be recovered, cycle testing permits more accurate recovery calculations by reducing the percentage of middlings that arbitrarily must be assigned to final concentrates or tailings. This type of research is normally best performed by two or more technicians because a number of flotation cells and grinding mills may be required during the test and simultaneously will help to confirm that the technique proposed can be duplicated by others. Variability within a team strongly suggests flowsheet weaknesses.

The flowsheet, reagents, times, etc. which have been developed during batch tests are used to design a flowsheet for cycle tests. In these tests, intermediate products from one batch test are saved and reintroduced at the appropriate point in the subsequent batch test. The procedure is repeated until equilibrium as determined by weights and analysis is reached. Figure II shows some suggested typical standard laboratory grind and flotation flowsheets for batch and cycle tests. It is common to see six batch tests used in one cycle test. Dilution becomes a problem in these tests and must be controlled. Dewatering between steps might be necessary but the solutions should be reused as appropriate dilutants and drying should not be practiced unless it is proposed for the full scale plant.

When equilibrium has been established, the results of the last two tests may be used to calculate the projected results for a full scale operation. If equilibrium has not been reached, it cannot be assumed that the intermediate products will end up in the desirable concentrates in the ratio indicated by the overall test because some of these products may not be upgradeable. The more complex the separation, the more caution the research metallurgist should be in his predictions. See Table 1 for an example.

Sand-Slime Separation

Sand-Slime separating techniques may be advantageous at any point in a flowsheet between crushing and tailing disposal so that the experienced metallurgist will look for the possible applications. Although strictly not "sand-slime separation", a classification step may be incorporated in the crushing plant to separate coarse and fine fractions so that a clean sand may be leached by acid, base or cyanide.[9] If the slimes were permitted to stay with the sand, solution channeling might occur, it might be impossible to dewater the mixture, the liquid-solid separating technique might be so inefficient that the product from subsequent precipitation would be unacceptably low in grade or reagent consumptions would be excessively high.

Alternately, during or immediately after crushing, there might be

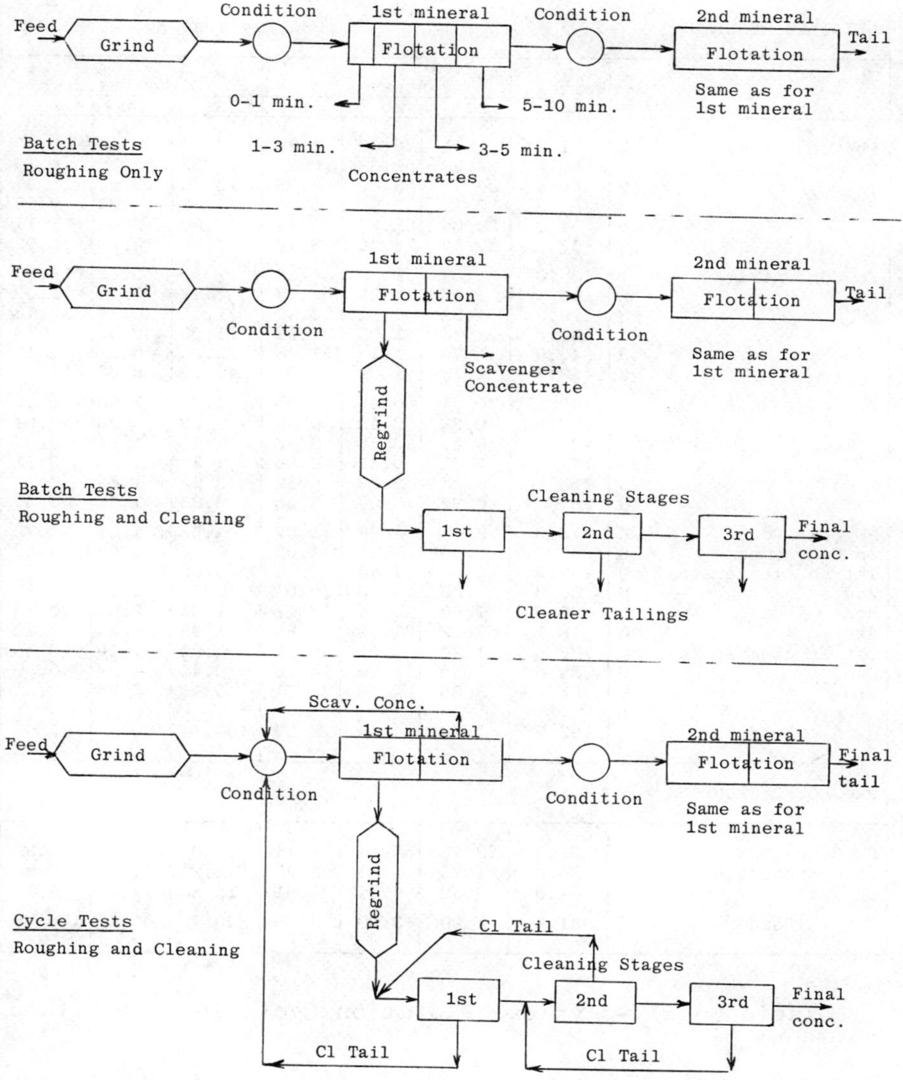

FIGURE II - Standard Laboratory Grind and Flotation Flowsheets

DETAILED RESULTS								Cycle Test No: Van 325
Sample		Weight Grams	Assay %			Distribution %		
			Ag	Pb	Zn	Ag	Pb	Zn
Pb Concentrate	1	27.8	31.04	72.06	5.46	7.19	8.15	0.30
	2	56.2	24.89	58.33	8.21	11.53	13.16	0.19
	3	59.8	24.79	56.63	8.95	12.22	13.66	1.04
	4	54.6	25.11	55.90	8.25	11.38	12.37	0.88
	5	54.9	23.06	51.90	8.74	12.39	13.61	1.11
	6	57.8	23.82	54.30	8.14	11.38	12.69	0.92
Zn Concentrate	1	104.6	1.62	0.99	54.18	1.39	0.42	11.03
	2	113.8	1.57	0.92	55.37	1.42	0.42	12.30
	3	116.3	1.51	0.91	53.47	1.45	0.43	12.13
	4	131.8	1.40	1.03	52.56	1.53	0.55	13.58
	5	140.9	1.73	1.18	51.46	2.01	0.67	14.13
	6	130.7	1.73	1.24	55.04	1.87	0.65	14.02
Zn Rough Tail	1	590	0.21	0.42	0.45	1.03	1.00	0.52
	2	726	0.32	0.43	0.45	1.92	1.26	0.64
	3	721	0.21	0.52	0.48	1.25	1.51	0.67
	4	741	0.21	0.52	0.56	1.29	1.55	0.81
	5	763	0.32	0.52	0.62	2.02	1.60	0.92
	6	765	0.32	0.57	0.60	2.03	1.76	0.90
Pb Scav Conc	6	73.6	2.48	3.60	12.02	1.50	1.07	1.72
Zn Scav Conc	6	53.4	1.08	1.36	4.02	0.48	0.29	0.42
1st Pb cleaner tail	6	108.1	3.66	7.18	13.18	5.55	5.44	4.83
2nd	6	28.3	5.92	19.10	13.04	1.37	2.16	0.71
3rd	6	28.0	7.98	23.04	13.11	1.85	2.60	0.72
4th	6	16.1	9.59	21.93	12.83	1.28	1.42	0.40
1st Zn cleaner tail	6	168.5	1.29	1.51	8.49	1.80	1.10	2.80
2nd	6	21.5	2.38	3.05	14.10	0.42	0.26	0.59
3rd	6	21.0	2.80	2.90	24.71	0.48	0.24	1.01
Head		5963.7	2.03	4.16	8.59	100	100	100
PROJECTED RESULTS								
Pb Concentrate		61.3	23.44	53.14	8.44	75.0	85.0	6.3
Zn Concentrate		135.8	1.73	1.21	53.25	12.2	4.3	88.0
Zn Tail		764.0	0.32	0.54	0.61	12.8	10.7	5.7
Head		961.1	1.99	3.99	8.55	100	100	100

Table No. 1 - Typical Flotation Cycle Test Results

advantages to separate the coarse from the fines by screening or sand-slime techniques to permit heavy media or gravity beneficiation of the coarse fraction.

Many researchers and operators have found it advantageous to treat sand and slime portions of an ore separately by flotation. There may be instances where the processing should be specific for each fraction throughout the concentrator and others where separation would take place after some initial beneficiation and then probably only the coarser portion would be treated to provide extra recovery.[10] However, the majority of ores do not require sand-slime separation for optimum results.

Even in dewatering steps, there may be merit in sand-slime separation to provide more capacity in existing or proposed equipment when the indicated settling rates of the fractions are so different that unusual action is required.

Gravity and Heavy Media Concentration

Beneficiation of ores by gravity concentration methods has been practiced longer than by flotation and doesn't appear to be losing ground. In fact, the two techniques are frequently complementary when the advantages of both are utilized.[11] In general, gravity beneficiation is most efficient on coarse materials while flotation is more suitable for fine sizes from 35 mesh (420 microns) but occasionally even from 8 mesh (2380 microns) in potash flotation down to 10 microns.

Nature frequently forms orebodies wherein the specific gravity of the desirable components is sufficiently different from the gangue or each other that gravity can be utilized to affect separation. Some indications of this difference may be obtained by physical examinations, by microscopy or even plain visual means but, more frequently, it may be necessary to test the ground ore in various heavy liquid media. The differences may be so pronounced (e.g. galena vs. silica) that recovery of coarse fractions may be simple or they may be so slight (e.g. salt vs. potash) that the separation may appear to be impossible. However, both types of ore mentioned above and many others are being beneficiated in gravity and heavy media circuits today.

In the laboratory, differences in specific gravity which might be taken advantage of in mineral processing can be tested by superpanners, infrasizers, tables, spirals, elutriators, heavy media liquids, cyclosizers, classifiers, centrifuges, jigs, etc. Some of this equipment may not be too useful except in the laboratory or pilot plant but it is used to determine the gravity characteristics so that the proper process equipment may be selected.

The heavy liquids used in the laboratory may be aqueous solutions of inorganic salts such as zinc chloride or heavy organic liquids such as carbon tetrachloride and other halogenated hydrocarbons. These can be adjusted to give a full range of densities from S.G. 1.0 to 4.0. Unfortunately, most of these heavy liquids are expensive and toxic and thus have not been used in quantity yet on a production scale.

In industrial practice, finely ground heavy minerals such as galena, magnetite, barytes, ferrosilicons, etc. may be uniformly suspended in a fluid by gentle agitation and the resultant pulp has many of the characteristics of a true heavy liquid. The ferrous media - magnetite and ferrosilicon - are more common in commerce because they can be recovered efficiently and cleaned by magnetic separators.

Leaching

Laboratory leaching procedures include studies of dissolving metal ions with an acid, a base or cyanide. Although not so common today, gold amalgamation may also fall in this category. The testing may involve field tests on ore in place, broken ore in stopes, ore and waste in dumps or crushed ore in vats and laboratory tests of ground ore, concentrates or calcines in agitators. The technique may be used to effect the selective solubilization of a desirable component such as copper, uranium or gold or the purification of a product to remove impurities such as MgO from zinc concentrates, copper from molybdenum concentrates or iron from silica sands.

Very little laboratory testing can be done on ore in place so that the research team will have to depend on descriptions of the deposit from the geologist, mineralogist and miner, tests on coarse rock samples and knowledge of the characteristics of the ores of the region involved. Leaching tests on rock samples of any size from 20 cms to 1 cm can be simulated in columns of adequate diameter and height. The metallurgist should choose a large enough diameter that there is no adverse "wall effect". Non-corrosive materials of construction such as asbestos or plastic pipe should be utilized for the columns so that maintenance in the lab will not be high and undesirable constituents of the pipe will not enter the leachant.

The bottom of the column should be fitted with a suitable filter media. A positive circulating, corrosion-proof pump and reservoir system should be employed - usually using plastic tubing for solution transfer. The most important variables to monitor when leaching ores are time, temperature, feed size, additive requirements, density and leachant concentration and consumption vs. extraction of desirable and undesirable components.

Agitation leaching of ore, concentrates or calcines may be easier to test than static leaching but liquid-solid separation of the residues may be a problem and should be included in the testing. Consumptions of reagents may be higher than in static column tests so that the extra operating costs would have to be justified by higher recoveries. One is not liable to discover that elevated temperatures and/or pressures would be economic today with high energy costs in treating low grade ores by leaching but this statement may not be true with some high grade ores or when treating high value metals such as uranium or higher grade materials such as concentrates or calcines.

Certain additives such as lime during precious metal cyanidation and oxidants during uranium leaching will be important reagents to monitor.

Calcining or Roasting

It is not the intent of this article to discuss pyro-metallurgical extraction or refining. However, calcining prior to concentrating by ore dressing methods is practiced occasionally in order to alter the mineral surface to make it amenable to minerals beneficiation techniques.

The term roasting is used rather loosely here because the temperatures involved may not accomplish complete sulphur or carbon dioxide elimination and may not be designed to do so. Simple surface alteration such as the heating of copper concentrates to destroy the surface layer of reagents to permit molybdenum-copper mineral separation or adequate roasting to permit efficient cyanide leaching of an arsenical gold ore may be the objective of these tests.

Laboratory calcining equipment is not common and indeed it may be difficult to justify such an expenditure since the number of times it would be needed may be very limited. However, most researchers can gain access to an assay office muffle furnace and at least the preliminary tests can be conducted with these furnaces. Most extensive research may have to be done by commercial laboratories or in a pilot plant.

Ion Exchange[12]

Ion exchange may be defined as the reversible interchange of ions between solid and liquid phases in which there is no permanent change in the structure of the solid. With few exceptions, the great utility of ion exchange rests with the ability to use and re-use the ion exchange materials. The inorganic exchangers which were common prior to 1940 have been almost completely displaced by the resinous types because of the latter's physical and chemical stability. Uranium is commercially extracted from leach liquors by an ion exchange and then concentrated as well as purified by elution with a solution of nitric acid or ammonium nitrate. However, in a number of areas, it is expected that the use of nitrates will be discouraged in the future for environmental reasons and, thus the operators may change to sulphates or chlorides.

The two principle techniques of contacting solutions with ion exchange resins are known as the batch method and the column method. The column method may be further subdivided as to whether the resin bed is fixed or moving and whether the solution and regenerant solution flow past the resin in the same relative direction or in opposite directions. While it is occasionally convenient on a small scale to use a resin batchwise to check some features, it is generally impractical to regenerate the resin for reuse with this method.

Commercially, the packed bed-downflow type of column has had wide acceptance but the introduction in recent years of the fluidized bed, upflow, ion exchange systems is expected to have a pronounced affect in the industry of the future. In addition to the advantage of the continuous or semi-continuous nature of these columns, the designers and manufacturers hope to handle solutions with an 8 to 10% solids content thus simplifying liquid-solid separations.

Before testing may proceed, the metallurgist will want to know what type of resin to use, i.e. whether cationic or anionic. In uranium recovery systems (anionic), there are strong base resins and weak base resins. The former are less selective but todate, the latter have been of low specific gravity and not suitable for upflow columns.

The variables to be determined in the laboratory are the type of resin, resin mesh size, degree of cross-linkage of the resin, temperature, flow rates, resin selectivity, regenerant concentration, absorption and elution curves.

The 100 ml graduated burette is a readily obtainable piece of laboratory equipment which can be converted to an ion exchange column. However, to eliminate the disadvantages of this size of burette, the use of a 25 mm diam. x 2 meter high (approx. one inch by 78 inch) column should be considered. See figure No. III for an example. The solutions to be treated are pumped continuously through the bed until the ion that is being removed is detected in the column effluent. Loading is then discontinued and the bed is regenerated to the proper form. The choice of regenerant chemical will depend on the ionic species to which it is desired to convert the resin and its cost. In the laboratory, the research personnel should run enough cycles to be sure there is no poisoning of the resins by thionates, silica, heavy metals, etc.

Solvent Extraction

The solvent extraction process, as applied in uranium or copper extraction plants, is in many ways analagous to the resin ion exchange process and indeed is often referred to as liquid ion exchange. The dissolved metal from the leaching steps is extracted from the clarified pregnant solution by an organic agent, the two phases - organic and aqueous - allowed to separate and the purified or concentrated metal stripped from the organic into a second aqueous phase permitting the barren organic to be recycled to the extraction step. The metal of interest is then recovered from the aqueous phase by conventional means such as precipitation, electrolysis or other.

The development of a liquid ion exchange process begins at a bench scale level using separatory funnel techniques. It is common to make a mixture of 5% by volume organic extractant, 2.5% isodecanal and 92.5% kerosene and then to contact the aqueous metal-pregnant solution with the extraction phase to 4 volumes of feed solution. A good indication of phase disengagement time and emulsion tendencies can be gained through such experiments as can data showing the distribution of the desired metal and impurities between the phases under various extraction conditions. The existance of a tendency to emulsify or form a third phase may be proven only by long continuous runs.

The effects of varying organic phase composition, extraction reagent, additive, carrier, temperature, contact time and any tendency for the organic phase to become poisoned should be investigated. Poisoning can be determined by batch stripping studies in the separatory funnels with separation and reuse of the organic phase over and over again so as to determine any change in its metal loading capacity.

Ultimately, the procedure should be tested with a continuous flow solvent extraction circuit having a number of mixing and phase disengagement stages in both the extraction and stripping steps. To determine the relative extraction in a single stage, the percentage extraction is related to the distribution coefficient as shown in the equation:

$$\%\ \text{Extracted} = \frac{PK}{1 + PK} \times 100$$

Where: P = phase ratio
and K = distribution coefficient.

A second important use of the distribution isotherm is in predicting the number of counter-current stages which would be required to achieve any given extraction of the described metal. This is done by use of a McCabe Thiele diagram.[13] A popular arrangement for the laboratory circuit consisting of four extraction and three stripping stages is shown in Figure No. IV. This type of laboratory equipment is available commercially.[14]

Magnetic Separation

Magnetic separation utilizes the force of a magnetic field coacting with some other force to produce differential movements of mineral grains through the field.[15] Most imporant magnetic separations in practice are those involving iron ores, pyrrhotite from pentlandite and recovery of magnetite in a heavy media plant but there are numerous small scale applications. Laboratory equipment includes

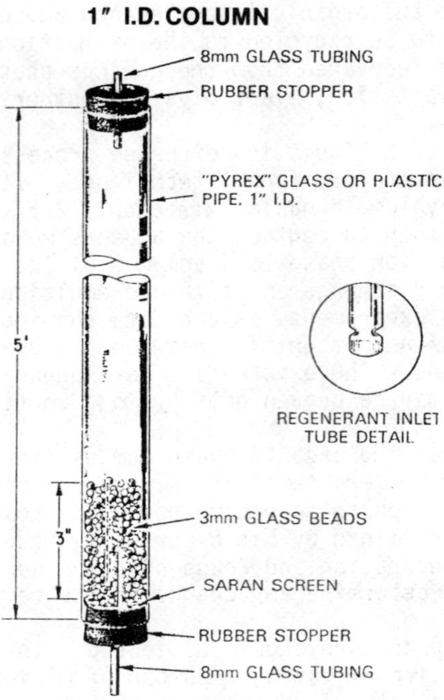

FIGURE III - Laboratory Ion Exchange Column
Source: Dow Chemical Company[12]

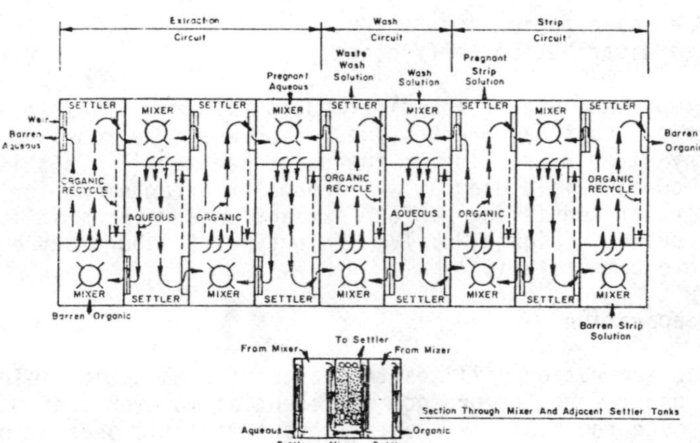

FIGURE IV —Typical mixer-settler solvent extraction system.
Source: W. C. Hazen.[13]

Davis tubes, drum separators, hand magnets, etc. and tests may be conducted wet or dry. Roughing and cleaning steps may be necessary. In some instances, roasting may be desirable to convert iron bearing materials into a magnetic state and in a few cases, demagnetizing may be necessary to prevent agglomeration.

Electrostatic Testing

Electrostatic separation utilizes the force of an electric field coacting with some other force to produce differential movement of mineral gains.[15] The basis is found in the differences in interfacial resistance offered by different minerals to passage of electrons there through. Modifying factors are specific gravity, size, shape, surface condition and purity of the minerals and electrical attributes of the separator.

Various makes of separators are available on the commercial market. Electrostatic separations which are applied in practice in the minerals beneficiation field, are rutile from zircon, vermiculite from gangue and upgrading of some phosphate flotation concentrates. Relatively few laboratories would expect to conduct electrostatic separations and hence, few have the necessary equipment. Since its use is limited, narrow and specialized, no effort to describe laboratory procedures will be included in this chapter. However, more information on the subject may be noted in Chapter 21.

Fluorescence and Radioactivity

The degree that a mineral is fluorescent or radioactive may be used in the laboratory, principally for mineral identifications, but seldom used today for mineral separation in practice because of labour costs and environmental hazards. However, machines have been developed in Africa which, it is claimed, rely on photometric or radioactive principles to beneficiate ores.[16]

Microscopic Studies

In the technical role, microscopy involves the investigation of ore lumps, crushed ore and mill products in order to identify the mineral constituents of the ore and to determine their physically significant features, which include shapes, chemical properties, mode of occurrence, mutual intergrowth, grain sizes, possible alteration, oxidation and weathering products.

Mineral grains down to 150 microns occasionally can be recognized by the naked eye. This is frequently the grain size at which the art of flotation is most efficient while comminution is not too expensive. With the aid of a good microscope, however, grain sizes down to one micron in diameter can be distinguished and classified. In addition, it is claimed that a new image analyser[17] provides accurate quantitative mineralogical analysis of finely ground ores.

A trained mineralogist - microscopist can provide effective guidance to the metallurgist during the latter's research. However, regrinding requirements, poor concentrate grades or high tailings assays confirmed by the microscopist may be as much the fault of poor laboratory techniques as they are the lack of liberation or degree of association. Therefore, the microscope must not be used as a means to find an excuse for poor separation technique and must be used with caution.

Screen Analyses

No laboratory, whether used solely for research or for operation monitoring, will be complete without minimum sets of testing sieves and a sieve shaker. Throughout the research programs, the metallurgist will want to analyse the products with screens to determine how much work has been done or how much more work there is to do to permit the separation objective. In early research, a correlation between the chemical and screen analyses will be essential to determine where the values, contaminants and waste portions are concentrated but as the work progresses, the chemical analyses may not be required as frequently.

Knowing in which size fractions the metals of interest are concentrated, the metallurgist can formulate ideas of subsequent treatment such as regrinding or sand-slime separation, etc. Frequently, graphical plots of size analyses vs. other variables of interest will show sharp or adequate breaks in the curves to indicate the point at which a certain technique should be discontinued and a new attack planned.

However, screen analyses also can be misinterpreted or applied incorrectly. High chemical assays of the coarse fractions of a tailing may be the result of improper flotation conditions as frequently as they may be due to true middlings while mineralogical analyses may not be suitable since many middlings will float. Similarly, high metal values in the slime portion of a tailing may be a result of poor flotation practice. Thus, re-examinations of the grinding conditions periodically during flotation research are recommended.

Pumping

All too frequently, pumping tests have been neglected by the research staffs and omission of these studies has caused problems later. Although the input by operators and designers has prevented many major errors in layouts of new facilities, the failure to appreciate the potential problems may have caused excessive maintenance or exceptional amounts of line blockages. Within the plants, the lack of prior pumping tests may not be critical, but if any long distance pipe lines are to be contemplated, a series of pipe line and pump loop system tests should be considered mandatory.

A simple line loop system as shown in Figure No. V, is easy and inexpensive to construct, although more extensive systems may be recommended for very specific problems. The quantities of sample or products of interest for pumping generated in a laboratory may not be adequate for loop tests and, thus, the tests may have to be delayed until the pilot plant can produce adequate quantities.

Although, todate, a single formula has not been developed which permits proper consideration of all factors known to influence slurry flow characteristics, a commonly used formula to scale-up from pump loop systems to process piping follows:

$$V_c D = \sqrt{\frac{D}{d}} \times V_c d$$

Where: D = Diameter of large pipe
d = Diameter of small pipe
$V_c D$ = Critical velocity in large pipe
$V_c d$ = Critical velocity in small pipe

This should yield conservative results and therefore should be modified for long lines.

As with a number of other aspects of metallurgical ore dressing research, there are a number of commercial laboratories and pump manufacturers in North American who are fully equipped to do pumping tests. Therefore, it is not necessary to set up one's own loop systems unless the problem is expected to be very complex or the tests prolonged.

This research should be designed to monitor and compare line velocities, slurry densities, viscosities, settling characteristics, pressures, abrasion, degradation of product friction losses, critical carrying velocities, screen analyses and specific gravity.

Thickening and Settling

The capacities of thickeners and tailings dams are inherently related to the settling rates of the material to be densified. Since both units are inclined to be very expensive items, it well behooves the research staff to test the settling characteristics extensively.

Initial tests should be done on the material as it comes from the process. They may be done in 1000 c.c. graduated cylinders where the interface may be visible and measured so that a rate per hour and an area can be calculated. A safety factor of 33% should be used for these results for scale-up. Confirmatory tests should be done in a 7 to 10 cm (3 to 4 inch) diameter pipe which is 2 to 3 meters (6 to 9 feet) long, capped at one end and tapped at regular intervals for sample extraction. The results of these tests should be multipled by 120% for safety. The area required may be calculated from the

A - Pump & DC Motor	H - Magnetic Flow Meter	Q - Suction Tank
B - Drain	J - Heat Exchanger	R - Pressure Gage
C - Rubber Pinch Valve	K - Valve Pinch	S - Vent
D - Plug Valve	L - Swing Valve	T - Diff. Pressure Gage
E - Pressure Tap	M - Weigh Tank	U - Slurry Trap
F - Pyrex Glass Pipe	N - Scale	V - Suction Pressure
G - Rubber Hose	P - Control Tank	γ - Angle of Item F

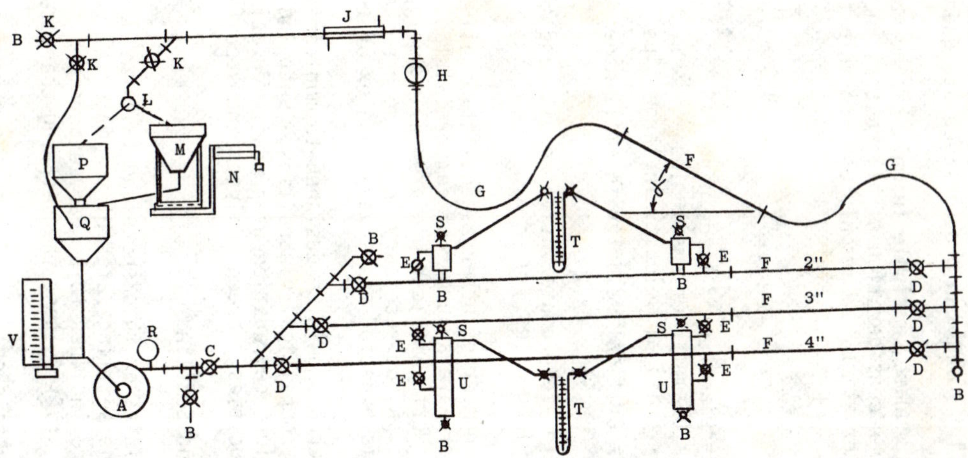

FIGURE V - Slurry Test Loop[18]

formula:

$$A = \frac{1.333\,(F - D)}{R \times g}$$

Where: A = area in square meters/Tonne/24 Hours
F = initial density (parts water to solids by weight)
D = final density - from thickener
R = rate of settlement in meters/hour
g = specific gravity of solution.

Confirmatory tests should be done at various feed dilutions because the settling characteristics can be appreciably different. Also, since temperatures have a pronounced affect on settling rates, routine tests should be conducted at the temperature expected in practice. Previously heated pulps after cooling frequently settle faster than unheated pulps and, therefore, if heat is involved in the process, the researcher may want to take particular note of that fact.

Normally, the settling tests will be done without settling aids. Hence, if the results are average or reasonable, a measure of safety is available when operations begin, upsets are encountered and the operator finds that settling aids have to be used. However, some materials such as clays or fine precipitates are very difficult, if not impossible, to settle without flocculants and, under these circumstances, the next step must be conducted using the aids available. Although reasonable rates may be demonstrated in the laboratory, using these aids, it would be wise to use an additional safety factor of 20%.

There are recorded instances where it would appear that settling or flocculating in tailings dams is extremely slow so that large basins and/or long period of time are required to get satisfactory effluents. Due to the high volumes which must be handled, it is doubtful that it would be economic to add flocculants to speed up the tailing settling rates.

Filtering

Throughout the laboratory testing, the research staff will be monitoring the results by collecting, filtering, drying and assaying intermediate and final products. These preparatory steps on samples will give the metallurgists a good idea about the product filtration characteristics so that if anything unusual is noted, special attention should be paid to dewatering. Even if no particular problems show up in these intermediate tests, the filtration tests are simple and should be conducted using a simple vacuum disc fitted with various types of appropriate filter cloths.

Significant factors affecting filtration are screen analysis, mineral shape, viscosity and density of pulp, temperature, amount of vacuum available, pH and presence of materials such as clay which tend to blind filter cloth. Although good filter aids are available commercially today, it would not be deemed wise normally, even if the laboratory results were encouraging, to depend on them solely when scaling-up to plant scale equipment. The aids could be termed the safety factors to be used later if necessary.

There are a number of types of filters being manufactured and since their ability to handle various qualities of materials is drastically different, significant economics have been realized after appropriate testing. The four principle types are drum (with inside or outside cloth), agitated disc, horizontal table and pressure filters.

Laboratory vacuum filters will yield filter cakes more closely approaching those of production vacuum filters than laboratory pressure filters - particularly with reference to moisture content. Filtrates should occasionally be saved so that flotation tests may be made with them because most significantly, some of the modern filter aids in use prevent their immediate reuse without aging.

Tests should also be done to test the type of filter cloth to use because the physical characteristics of the solids and the chemistry of the solutions encountered today are extremely varied so that many of them may require special cloths. Even for ordinary concentrates, significant reductions in moisture content can be demonstrated with some of the synthetic or specially woven cloths.

Chemical Analyses

Throughout ore dressing laboratory and plant research projects, the most significant and common control feature is the chemical analysis of products. In certain instances, or with a number of specific ores, physical and/or visual characteristics may be very important but the chemistry is the one most frequently used.

Detailed chemical analyses will be required for most of the types of ore to be tested, all the final concentrates produced and the majority of the waste products discharged. It will not be necessary to assay for every metal in all tests since most routine work can be monitored by watching one or two elements. However, early in the program, the detailed head analyses will establish the basic data and raise the warning flags to guide the metallurgists should any potential objectionable elements show up. The concentrate analyses in detail will be required for the sales staff because all smelting and refining contracts take note of the contaminants as well as the desirable metals. Similarly, more and more detailed waste product analyses are being required by regulatory authorities as more ions and anions are stated to be potential environmental hazards which must be recognized and perhaps controlled.

Rapid control chemical analyses are required during the testing because subsequent test programs are dependant on the results of the current tests. Assay laboratories can become overloaded by samples from research personnel so that it is essential to properly plan the essential amounts and types of chemical analyses.

Effluent Treatment

Almost all new concentrators will plan on using significant quantities of recycle waters and many of the older plants which have not been using return solutions are, or will be, investigating the possibilities. In those areas where the annual evaporation exceeds the precipitation, there is possibly no net effluent and, therefore, treatment for environmental reasons is not required. However, when the net precipitation is positive or when the make-up exceeds the sum of the losses, various water treatment techniques may be required to meet imposed guidelines and the cost may be high.

Even though the water balance may be theoretically satisfactory in a number of plants, some discharge to the local streams may be necessary because fresh water may have been essential within the process for efficient separation in certain steps, for equipment cooling or for domestic purposes. In other instances, internal recycle cannot be used immediately without treatment and thus, these solutions may have to be dispatched to the tailing dam for aging.

Therefore, during laboratory research, solutions which are liable to be recirculated must be collected and their use as make-up throughout the circuits tested. The quality must be assessed so that, if discharge to the surroundings is contemplated, treatment techniques can be developed if the need is indicated. The acid-producing and thiosalt generating potentials when treating sulphide ores must be investigated so that, if detected in the laboratory, remedial action can be planned for the operating system. The chemical reactions within a tailing impoundment system may change drastically seasonally. In winter in northern climates, the ice cover and cold temperatures will stop or slow down oxidation. In the summer, entire ponds can turn acid very quickly and the effluents will have to be treated prior to reuse or discard.

Environmental Research

Control of the physical environment is more a problem for the designer, the mechanical engineer or the mill superintendent than it is for the research metallurgist. Nevertheless, the metallurgist, during his research, should be aware of potential environmental or health hazards which may occur. His analyses of the ores, concentrates and tailings and his knowledge of how his proposed processes might cause problems are essential inputs to the overall plans. Dust quality, wet or dry processing, fume potentials, degree of solution corrosion and reagent toxicity are a few of the important features to study.

A field that will require more and more research is that of noise and thus, hearing protection. So much of present day mill equipment presents a noise hazard above the currently acceptable 80 to 85 decibel level, that studies of the machine characteristics, location in the flowsheet, isolation and insulation become integral parts of the planning program.

Equipment

Most experienced designers, operators and researchers are familiar with the types or makes of equipment which are common in the mineral dressing field. When this is the case, their experience plus the manufacturer's literature will permit suitable choices. However, as the metallurgist investigates new fields and as technology introduces new equipment, testing of these machines is recommended. If neighbours or associates are using the suggested item, their performance data may be adequate for your purposes, but if not, it may be necessary to arrange for trials in some other parts of your organization or elsewhere in the industry.

Seldom can production size equipment be tested in the laboratory and thus, models may have to be tried if available. Scale-up from these small units can be simple or a serious problem so that thorough studies are recommended for unproven or unfamiliar machines.

Materials of construction must also be tested by research staffs. This is particularly true when severe corrosion or abrasion problems are indicated.

REPORTS

Format

The best metallurgical research in the world will lose much of its value if not presented in a form that is readily understood and useful to a wide variety of professionals and executives. It is well to remember that the report that is generated will be used by the manager, geologist, miner, pyrometallurgist, designer, financier, sales agent, regulatory agencies and a host of other interested parties. Therefore, the summaries, conclusions, tables, graphs, flowsheets, etc. should not be prepared in a manner which only the ore dressing metallurgist would understand. "Executive" summaries serve a very useful purpose. However, the reports must be technical in nature and hence, to be fully useful, should contain considerable scientific terms, technical language, metallurgical details and data.

The entire research program should be planned in depth. This does not imply that results and conclusions will be pre-judged, but rather that a minimum of a certain number and types of tests will be undertaken regardless of early success or failure. At the same time, it would be rare that some of the planned tests did not prove or indicate that additional research is required.

Prior to embarking on the research program, the staff should know what format of report will be used. Standard forms for screen analyses, flotation tests, tables, etc. are common in most laboratories because they permit similar repetitive tests in identical order and reduce the possibilities of forgetting important procedures. Frequently, the final report will have an appendix with all the detailed test results which will be useful to the detailed analyst even in the handwritten state.

It is not expected that everyone who is interested in the work will have the knowledge, desire, time or need to study all the detailed data and hence, summaries, conclusions, purposes, proposals, etc. should be prepared to make up the principle portion of the report. Also, engineers and metallurgists should remember that the value of the report diminishes rapidly if the English composition, grammar and spelling is poor.

Since there is such a variety of metallurgical ores and tests which can be done on them, the form of the report will change with each program and from company to company. Nevertheless, there are common features in all of the better reports. Although the order may change, good reports will contain, but not be limited to:

1. A title page including distribution
2. An index
3. An introduction
   origin of samples
   purpose of research
4. Brief summary of best practical results
5. Conclusions
6. Recommendations for further work and/or flowsheet development
7. Detailed summaries of research
   Step by Step and material balances
8. Cycle test results
9. Tables
10. Graphs
11. Flowsheets
12. Appendices - detailed test results
    Related information
    Inclusion of unsuccessful tests may be as meaningful as some of the good ones.

FLOWSHEET DEVELOPMENT

Pilot Plant

Not all ores must be tested in a pilot plant after the bench scale research is completed. There will be instances where the laboratory results are very good and reproducible, where the ore treatment is simple or where the ore under test is nearly identical to the ore being processed in an operating concentrator. Under these circumstances, the costs of a pilot project might not be justifiable although the research staff may want to confirm their work by processing a bulk sample in the existing circuit.

However, more and more ores are being tested in the pilot plant and, for complex ores, for ores with indicated problems or uncertainties and for ores with minute quantities of recoverable values, the cost of confirmatory tests in a pilot plant is warranted and recommended. If autogenous grinding or prototype equipment is proposed for the new plant, it should be tested in the pilot plant also.

The flowsheet for the pilot project should be based on the results of the best series of cycle tests from the laboratory. However, one must be prepared to modify the flows and reagents in the pilot stage. The quantities to be processed will depend on the capacity of the pilot plant comminution units more frequently than on any other equipment sizes or numbers. The collection of bulk samples from the mine development which will be the usual feed for the pilot project should be observed by the metallurgical research team if possible. They should be delivered in advance of scheduled start-up but not so early that sample deterioration might be a factor. As with all research, the bulk sample must be representative or the results will be of less than full value.

Direct scale-up from pilot plant to production is relatively straight forward.

Process Flowsheet

The process flowsheet has been indicated by the bench scale testing and confirmed by the pilot plant tests. Therefore, the preparation of a proposed production flowsheet is rather simple. However, the numbers, types, sizes and layouts of equipment will be judgement factors dependant on experienced personnel, power requirements, manufacturer's specifications, laboratory retention times and other physical characteristics. Safety factors must be considered but not over-emphasized to the detriment of excess capacity or cost. Provision for expansion may be important if management deems it pertinent. Some of the features which haven't been mentioned in the previous parts of this article may be noted below.

The first size-reduction operation after mining normally will be done in a primary crusher regardless of whether the subsequent step is crushing or grinding. The size of the crusher frequently is determined by physical factors beyond the control of the metallurgical research staff. The factors, although numerous, may be summarized as blasting techniques, size of loading equipment, capacity of hauling or hoisting machines, location of orebody (underground or near surface), mining methods and rate. This information must come from the mine planning group and discussions on their recommendations are beyond the scope of this chapter.

However, the size of primary crusher does affect the product size and this is significant to the flowsheet development team. If the ore is sufficiently competent to consider any degree of autogenous grinding, the product discharged from the first crusher may become grinding media and, therefore, extremely important. This feature has been discussed in more depth under grinding research.

There are three principle types of primary crusher on the market today. They are gyratory, jaw or impact. Normally, the gyratory crusher would be chosen for large tonnages or large rock, the jaw for intermediate tonnages and sizes or for underground service and the impact crusher for specialized duty such as clayey material, less abrasive rock, product size control, etc.

The choice of units for secondary-tertiary crushing may be a little more complex than for primary crushing but the manufacturers' tables are of great value. Wet or sticky ores may require special treatment or types of crushers and these varieties should have been tested before the final selections were made. Poor flow characteristics, tendencies to pack, variable abrasiveness, etc. can be serious factors if overlooked. Primary and secondary crushers are operated usually in open circuit while tertiary crushers are open circuited about 50 percent of the time and closed for the balance.

Screens are used to reduce the load on the crusher in secondary and tertiary plants and to control the final product size particularly in closed circuits. Screen size selection is largely based on empirical data from the manufacturers, but some testing is recommended for those ores which have unusual physical factors, such as stickiness, preponderence of fines, abrasiveness, considerable flake portion or excessive water or snow content.

The selection of grinding equipment has been reasonably well dealt with under the testing procedures. However, because mills are very expensive items, it is worthwhile emphasizing again that thorough research, experience and judgement are required when proposing grinding mills.

In practice, pulp densities during continuous operation are more readily controlled than in the laboratory and, in general, they will be somewhat higher. In grinding, they usually vary between 65% solids in ball and pebble milling to 85% solids in rod mill grinding of high specific gravity sulphides. The more common ores are ground between 70% and 80% solids. Rougher flotation will most efficiently be conducted between 25% and 50% solids and the usual range is from 35% to 45% solids. Cleaner flotation may range from 35% down to a few percent but more frequently, if controllable, will be between 10% and 25% solids.

Adjustments may have to be made to the above density ranges when floating in solutions other than water such as brine or when floating very coarse particles. Certain solids such as clay, etc., when mixed with fluids, will tend to form viscous slurries which must be diluted to permit separation by flotation.

The effect of various densities does not stop after grinding and flotation because the influence is great in thickening, settling and filtering.

In choosing flotation cells, most research, design and operating personnel have found that a factor of at least two, and in the case of gold ores may be as high as five, should be used when scaling up from the figures for retention and flotation times developed in the laboratory.

Considering pumping tests alone is not enough to determine pump or line sizes for the full scale plants because the majority of concentrator products contain various quantities of air or froth. Therefore, after the basic engineering capacities have been decided upon, most operators have found it necessary to multiply the resultant by a "froth factor". This will vary from 1.0 for fine tailings to 3.5 for an extremely frothy middling or concentrate.

Conversely, the friction loss of pipe lines handling slurry or pulp tends to be less than one would expect. It is good practice to use a C factor of 140 for in-plant pumping systems with short pipe runs. The C 140 factor will prevent over calculating system heads and consequent over pumping by running out on constant speed pump curves. On long distance schemes where reduced flow has a more pronounced effect on stable operations - e.g. tailing disposal - where the TDH consists primarily of friction head, a good engineering approach is to use C - 130[19].

Most other scale-up and safety factors have been included in the discussions of testing procedures.

SUMMARY

Metallurgical testing and corresponding flowsheet development involve acquisition of certain basic information, a laboratory testing phase, the preparation of reports and, finally, flowsheet development.

Basic information includes information relevant to location, the nature of the ore body, sampling, marketing and specifications, environmental guidelines, and a review of literature and current practice.

The laboratory testing phase can be substantial and may cover the following: crushing, grinding, regrinding, classification, flotation with a study of reagent types and quantities, pulp density, aeration and conditioning, pH regulation, retention time, temperature, water quality, sample ageing and cycle tests. Further testwork could include: sand-slime separation, gravity and heavy media concentration, leaching, calcining, solvent extraction, ion exchange, magnetic and electrostatic separation, microscopic studies, pumping, thickening, filtering, effluent treatment and environmental research. Equipment is finally selected.

Flowsheet development may rely on pilot work; the final result is a process flowsheet.

REFERENCES

1. Lewis, Coburn and Bhappu, - Comminution: A Guide to Size-Reduction System Design, Mining Engineering, September and November, 1976.

2. MacPherson, Arthur L., - A Simple Method to Predict the Autogenous Grinding Mill Requirements, AIME, 76B327, 1976.

3. Bond, F.C., - Crushing and Grinding Calculation, Canadian Mining and Metallurgical Bulletin, July 1954, 46, No. 507, pp 466-72, Trans., 1954, LVII, 286-92.

4. Bruce, R.W., - The Effect of Grinding Media on the Selective Flotation of a Copper-Zinc Ore, Canadian Mineral Processors, January 1977.

5. Steane, H.A., - Developing and Implementing Mineral Processing Flowsheets for New Mining Ventures, McGill University, November, 1974.

6. Macdonald, L.D. and Brison, R.J., - Applied Research in Flotation, Froth Flotation, 50th Anniversary Volume, AIME, 1962, pp. 298-327.

7. Denver Equipment Company - Bulletin No. S1C-B4.

8. Claridge, P.G. and Tenbergen, P.A., - Pipe Ore Processing Developments in the Thompson Mill, Canadian Mineral Processors, Ottawa, January, 1975.

9. Dayton, Stan - Mining Program aims at Growth Opportunities, E & M.J., August 1974.

10. Anderson, Royal P. and Tanner, John A. - Design Considerations for Mineral Plant, Minerals Processing, May 1973.

11. The Sullivan Concentrator - Denver Trefoil, March-April 1955.

12. Wheaton, R.M. and Seamster, A.H., - A Basic Reference on Ion Exchange, The Dow Chemical Company, Reprint from Encyclopedia of Chemical Technology, Vol. II, 1966.

13. Hazen, Wayne C., - Solvent Extraction Techniques, Denver Trefoil, Aug. - Sept. - Oct. 1963.

14. Lewis, Clifford J., - Liquid Ion - Exchange Processes, Chemical Engineering, July 1965.

15. Taggart, Arthur F., - Handbook of Mineral Dressing, 1954, pp. 13-01 and 40.

16. Goode, J.R., - Recent Coarse Ore Sorting Advances, Canadian Mining Journal, June, 1975.

17. Petruck, William, - The Application of Quantitative Mineralogical Analysis of Ores to Predict the Optimum Grind, Canadian Mineral Processors, Ottawa, January, 1976.

18. Morris Machine Works, - Baldwinsville, New York.

19. Horne, Charles A., - A Guide to Good Slurry Pumping Practice, Canadian Mining Journal, November - December 1976.

Chapter 10

METALLURGICAL FLOWSHEET DEVELOPMENT

by Robert A. Wilson and Harmel A. Dawson

Consulting Engineer, The Galigher Company
Salt Lake City, Utah

President, Dawson Metallurgical Laboratories Inc.
Murray, Utah

ABSTRACT

First the basic steps to determine the values present and the minerals involved which are the basis for the metallurgical testing are presented in this paper. Next, choosing the process or processes indicated for treating an ore are discussed. This is followed by the test procedures used, the results of which will indicate amenability to treatment. These test procedures have been developed, used and proven effective in laboratory evaluations relative to plant processing. Special, uncommon processing is beyond the scope of the paper. Evalutation of test results and incorporation into the preliminary and final flow sheet for plant design are discussed. The value of a flow sheet, with material and water balances, to plant design is shown.

INTRODUCTION

This paper seeks to clarify the development of a satisfactory metallurgical plant flow sheet. Theoretical aspects of the various processes used have been left to numerous other technical publications.

The philosophy behind this presentation can be summarized in the following statements:

1. A few simple, logical steps recognizing the fundamentals involved will permit the development of an optimized flow sheet from both a financial and metallurgical point of view.

2. There is more than one way to do almost everything. Therefore, the objective should be to bring the best possible accommodation to the total system or project.

3. Objectives need to be clearly defined and accommodated in order of their priority.

4. In the real world there are very few 'absolutes'. Therefore, the possibility of bringing to any new project or flow sheet all of the knowledge future experience will bring, is so remote it can be dismissed as impossible. As Taggert's Handbook has so aptly put it, "Adequate margins of safety, suitable means of adjustment over a wide range and due pessimism as to the accuracy of the results implied" are all in order.

Laboratory investigations for the development of a process flow sheet on any mining property is an important step in determining the economic viability of a mine. If the values present cannot be economically extracted to give marketable products, then it is not an ore, regardless of the analyses obtained. Of course, the results of analyses and testing can be only as good as the sample used in the evaluation. Studies made on a sample that does not represent a mineable area could be worse than no investigation, because it could lead to false conclusions. Even though the sample tested is considered representative, additional sample should be taken from each area of the mine and testing conducted on each to determine their amenability to the process developed.

In the opening of any new property a preliminary metallurgical evaluation would be recommended in addition to analytical and microscopic studies. The results of this evaluation could be of real value in further development. When sufficient reserves have been developed, then additional testing is required for the final flow sheet and plant design.

Initially the laboratory investigation should be limited to known commercial processes. The use of new, untried procedures could be costly as well as potentially disastrous. If the ore body will not respond to standard processes and has the potential, then the cost of developing and implementing a new process could be considered.

The steps used in the laboratory testing should be made knowing that each is reproducible in plant operation using available equipment. A process requiring the development of a new type of equipment should be considered only if necessary.

If the process developed is complex, then consideration should be given toward having the flow sheet checked in a second laboratory. The ability of the second laboratory to reproduce the results, as well as other imput, would give assurance to ultimate operation.

Without the development of an economic process for treatment of an ore, all the geologic and mine development is of questionable value.

DETERMINATION OF VALUES PRESENT

Knowledge of the type, character and quantity of the metal or mineral values present in an ore body is necessary before considering laboratory procedure used for process development. The analyses required could be quite simple to very complex depending on the ore in question.

Visual and Microscopic Examination

Initially, visual examination of hand samples directly or under a microscope (or hand magnifier) will show the major type of, as well as character of, the mineralization. Much can be learned by grinding or pulverizing the sample and producing an array of the minerals by specific gravity using a vanning plaque. By examining the array under a microscope, indications as to the quantities of each type of mineral and size of liberation can be obtained. In many cases mineral types not readily identified on the hand samples can be identified on the plaque.

As testing progresses further microscopic studies may be desired. These could include sizing of ground ore and obtaining mineral counts on each size fraction under the microscope. This would show the portions of each type of mineral present and give indication as to grind required for mineral liberation. Microscopic examinations of selected thin sections and/or mounted polished sections could aid in identifying particular beneficiation problems.

The extent of the microscope studies depends on the ore. In many cases a series of laboratory separation tests - gravity or flotation - will rapidly define grind for optimum product grade and mineral recovery.

Analyses

On most ores (other than those containing gold and silver values) initial microscopic observations will indicate the major values for analysis. However, minor constituents may be missed. Therefore, X-ray or arc spectrographic analyses will show the metals present and if they are major or minor constituents. In general, quantitative spectrographic analyses are only approximate. Chemical or fire (or special assay techniques, if required) are necessary to obtain accurate analysis. The spectrographic procedures for analysis can be effective if standardized against similar ores of known metal content. However, this would not be possible on the testing of an ore from a new property.

Atomic adsorption methods for analyses are effective when operated by a qualified technician, but too often it is operated without proper safe-guards and controls. Consequently, the results are often questionable. If atomic adsorption is used, periodic checks using standard chemical or fire procedures are required.

Metal content (assay) is not a measure of metal recovery. Mineral type may determine extraction. Therefore, analyses to show state of oxidation may be required - such as the acetic acid soluble analyses to indicate oxide copper, lead and zinc values.

On ores where the values are present as metals (not as compounds) sample preparation and subsequent analyses require special treatment. In many cases it is not possible to obtain dependable accurate analyses. One piece of metal into or out of the sample to assay can effect the results appreciably. The effects of the metal particles can be reduced (but not eliminated) by carefully pulverizing a weighed portion of ore and screening out the metallics, which are assayed separately and then pro-rated assays combined. Often it is more effective to make the required metallurgical separation, assay the products and calculate the feed analysis. The whole sample containing metallics is consumed in the assay and weight of metal only reported. This can be converted to gm/1000 kg, (oz. 1T.) or per cent weight as required.

CHOICE OF METALLURGICAL PROCESS OR PROCESSES

General Processes

In minerals beneficiation studies the processing generally consists of a combination of unit operations which include in a normal Sequence:

1. Crushing
2. Grinding
3. Classification (sizing)
4. Concentration procedures
   Sizing, such as slime removal
   Gravity - heavy media, jigs, tables
   Flotation
   Magnetic
   Leaching
   Hand Sorting
5. Liquid-Solid Separations
   Thickening
   Filtering
   Screening
   Drying

Considerations in Making Choice

The types of unit operations to be considered for treating any ore is influenced by combination of:

1. Character of ore and value
2. Desired products and marketing
3. Location and size of ore body

It has been considered best metallurgical practice to remove the values as soon as they are liberated consistent with overall metallurgical results. Values may be lost by fine grinding as well as increased grinding costs are incurred. However, with present day flotation procedures, overgrinding is not as great a problem. In many cases it has been found that the primary slimes, inherent in the ore are the main cause of slime losses.

The flow sheet considered initially for each ore must be based on the type of separation that appears to be most effective, tempered by products required for market and the overall economics determined by recoverable values, size of ore body, and location relative to market.

Some industrial minerals may just require crushing and sizing prior to marketing, or may require screening to reject particle sizes that are sub-grade. It may be that the differences in hardness or particle shape of minerals aid in making the separation by particle size. There are some ores in which washing and separation by some form of classification (cyclone, rake classifier, settling tank) give a marketable coarse or slime product.

If liberation of the mineral particles occurs at relatively coarse size and there is a marked specific gravity difference, then gravity concentration processing (heavy media, jigging, and tabling) can be used. In gravity separation processes (other than heavy media which should be limited to 8 mesh in standard units and 40 mesh with cyclones) the separation is a function of particle mass, and as the particle size is reduced the mass difference is reduced. Therefore, the separation efficiency reduces as the particle size is reduced. However, gravity separation is generally not too effective where the minerals are close to the same specific gravity and the separation may result in an up-graded bulk product that requires further treatment. This procedure has been effectively used as an initial up-grading and is followed by other treatment such as flotation.

Flotation has proven to be the most universal tool, although not a cure-all, for up-grading and selective separation of minerals. After crushing and grinding to effect liberation as well as reduce to a size for effective flotation (generally minus 35 mesh), a wide range of mineral separations have to be effected by the use of the proper regulating and collector reagents.

Hydrometallurgical processes (leaching) are the most economic for extraction of some values. Many gold and silver ores are readily leached by cyanide and some ores handled by heap leaching procedures. Oxidized copper, as well as mixed oxide-and-sulfide ores, are being treated with dilute sulfuric acid leaching (including heap leaching). Most uranium is extracted by leaching - sulfuric acid or sodium bicarbonate - on high lime ore.

The ultimate flow sheet depends upon the overall economics. A flow sheet that gives the best overall metallurgy may not be best because of high capital costs relative to the size of the ore body. Possibly the location of the plant relative to the market may dictate type of operation. Being located close to, or allied with, a smelter may allow for the production of a lower grade product with increased recovery. On the other hand, as isolated property may necessitate a high-grade product at the expense of recovery. A small ore body may require a simple flow sheet to be economic.

TESTING PROCEDURES

The analyses along with the microscopic observations generally will indicate the type of processing that may be effective and dictate the initial testing procedures. Where mineral particle separations are involved, often effective preliminary evaluations can be made by the use of the vanning plaque on the test products. Careful ovservations of the minerals on the plaque will indicate if the separation is effective as well as give a good idea as to the amount and type of mineral locking. An experienced person using a vanning plaque can in many cases rapidly evaluate test results which then allows the technician to continue testing without waiting for analyses. Where high ratios of concentration are obtained, particularly on gold and silver ores, the use of the plaque with a microscope on the concentrate can show the character of mineralization which is not otherwise readily discernible. Of course, the ultimate measure of the separations is dependant on analysis of each product and calculation of a material balance.

Observations made during the testing are as important as the analyses and material balance in the development of a process. Therefore, by delegating the physical testing to an unqualified technician without proper supervision important physical characteristics of an ore could be overlooked.

Sampling and Sample Preparation

Assuming that a representative sample has been obtained in the laboratory, then preparation in the laboratory for analysis and testing should be conducted based on the ore type and projected test program. If coarse screening, washing, gravity separation or leaching tests are anticipated, then the initial crushing would be dependant on the coarsest particle size desired. Only a portion would be crushed and sampled for analysis. Careful mixing prior to each stage of splitting is necessary to insure representative samples. Experience has shown that crushing to minus 20 mesh is sufficient prior to cutting out 100-150 gram portions to be pulverized for analysis. Sample mixing and splitting techniques are thoroughly covered in most mineral processing text books.

In preparation of samples for flotation testing, crushing through a fixed screen size using standard procedure prior to grinding makes possible a comparison of grind characteristics. Where batch laboratory grinds are employed, the crush should be fine enough to insure that the coarse particles will readily grind. Experience has shown that a 20 mesh crush will allow an effective grind without excessive grind time.

Sizing and Classification Separations

The initial examination of an ore may indicate that the values could be concentrated in certain size fractions or that removal of certain size fractions would aid in the ultimate processing. Assay screen analyses would aid in this evaluation. If appreciable slimes are present, repeated washing and decanting over a sieve (100 to 400 mesh) followed by drying may be necessary prior to screening. The fines (slimes) should be filtered, dried and analysed with the other screen fractions.

An assay screen analysis after crushing or grinding may show that up-grading is possible in certain size fractions because of differences in the breaking characteristics of the minerals.

Possibly the valuable portion of the ore is the coarse or fine fraction, such as in glass sand or clay recovery, then blunging followed by screening or wet classification steps should be considered. A first step would be to slurry the ore at high solids in a high-speed mixing device, such as a laboratory flotation machine then dilute, settle and decant several times over a fine sieve. Based on these results, decanting tests simulating counter-current plant operation could be employed or tests conducted using laboratory cyclone units.

The sizing and classification separation steps may produce a final product or may be the first stage of up-grading in a flow scheme.

Gravity Concentration Separation

In considering any type of gravity concentration, initially testing is required to determine the particle size for liberation. Visual inspection gives an indication as to size, but testing is required to demonstrate the product grade and values lost in tailings. If the gangue constituents are relatively low in specific gravity (such as silica, silicates or limestone), then heavy liquid separation using tetrabromoethane (2.96 specific gravity), methyene iodide (3.31), bromoform (2.89), chlorinated hydrocarbons (1.4-1.6), or mixtures could be employed. After first wet desliming and drying, the unsized ore is separated at the desired specific gravity. The products are washed with solvent to remove the heavy liquid and dried. An assay screen analysis of the sink and float screen products would indicate the size of liberation. Next, in order to verify these indications as well as evaluate the product grade and recovery to be anticipated, a sample of ore should be crushed to size of apparent economic liberation and retested.

On a majority of the ore samples the results of these heavy liquid separation tests would determine the amenability to gravity separation and the type of unit that should be considered. If up-grading can be accomplished on particles coarser than 8 mesh, then standard heavy media separation may be considered. Heavy media may be used down to about 40 mesh using cyclones; however, the use of jigs may be more economic in this size range. Depending on the separation involved, jigs can effectively make separation on particles as coarse as 0.10 meter (4 inches) on coal separation down to 100 mesh on some heavy mineral separations. Although the jig can effect separations over a wide range of particle sizes, selection of the type required for the desired separation is important. Tables will make separations on particles as coarse as 0.01 meters (0.4 inch) for coal down to 200 mesh on heavy minerals. However, effective separations require treatment of sized fractions because separation is based on mass of a particle and not specific gravity alone. Spirals can be used on minus 10 to 20 mesh material to the 100 mesh range but are also limited to sized fractions.

On most ores the heavy liquid separation testing is sufficient to evaluate jig operation. But further laboratory evaluation may be required where mineral separations of very high gravity minerals are considered. Laboratory test jigs are available; however, in the use of these units it is difficult to obtain a balanced operation which will give meaningful results. If necessary, hand jigging can be employed. In hand jigging the crushed ore is placed in a sieve (down to about 48 mesh). The screen is moved up and down in water to simulate jig action. Then the light fractions are removed using a spoon or spatula. A number of passes are required on the tailing and concentrate in order to optimize grade and recovery.

Initially tabling results can be simulated by a combination of treatment with a gold pan and a vanning plaque. It is important to realize that the fines should be handled separate from the coarse fraction. Therefore, an initial sizing or classification step is required. As with the hand jigging, a number of passes on each product is required to optimize results. Ultimately a laboratory test table should be used to verify results. If unsized feed is used, the tailing should be deslimed and fines retreated. By first making a broad cut on the table a final tailing is obtained with a rougher concentrate. The rougher concentrate is then re-tabled to give the final grade product and a middling. The results of the testing would show the balance of grade and recovery as well as show the possible effects of middling particles.

In treating of sand deposits, such as beach sand and placer deposits, a number of gravity units have been developed such as variations of the pinched sluice and various types of riffles. Each has its application in special areas of treatment. Laboratory testing of placer deposits is not possible because the values are so low that they may or not be present in a selected sample. Pilot plant testing at the property is required. Then, if desired, laboratory testing could be conducted on the up-graded product to determine the most economic procedure for final recovery of the values.

Gravity concentration may be only one stage of up-grading in a process. It may be used to recover a fraction as a high-grade product such as the use of a jig in closed circuit with a grinding section where the final recoveries are made by flotation. Gravity treatment may be used as the initial stage of up-grading such as treatment of low-grade sulfide ore by heavy media separation prior to flotation. If the mineral selectivity by flotation is not possible, then, tabling of the flotation concentrates has been used to obtain required concentrate grades. Tabling is used for final up-grading of scheelite (tungsten) concentrates where the limestone gangue cannot be depressed during flotation.

Flotation Separation

Historically flotation procedures have been developed for the separation of a wide range of minerals as well as other substances and is the basis of our present mineral industry. Flotation is effected by the use of a collector reagent which will selectively coat a type of mineral particle so that it will attach to air bubbles generated in a flotation machine and be carried to the surface where it flows from the machine with the froth. Modifying agents are added to act as selective activators, selective depressants, pH regulators, or to reduce the harmful effects of colloidal material or soluble salts, and often one reagent may perform several functions. Frothing agents are used which will give a stable, but brittle froth, to carry the collected particles from the flotation machine.

An explanation as to the use and effects of the various types of collectors, frothers and modifying agents is available in a variety of publications and are not considered in this article. Laboratory test procedure will be discussed which will aid in the development of a process flow scheme which is a simple and straightforward as possible.

The initial choices of collector reagents are dependant on the type of mineral to be floated and are of three basic types:

Type	Class	Example
Anionic for Sulfide	Sulphydric	Xanthates, Thiophosphates
Anionic for Oxides and Carbonates	Cargoxylic	Fatty Acids, Fatty Acid Soaps
Cationic for Silicates and Acidic Minerals	Cationic	Alkyl Amines

The laboratory procedure used in the initial flotation testing will be dependant on the mineralization present and is governed by personal experience and/or processes used for treating similar ores as discussed in the literature. Before starting testing, a sample of ground ore should be slurried in water, a natural ph obtained and then, if necessary, adjusted to the desired ph with soda ash, lime or other reagent as required. This will give a measure of the reagent to be added to the ball mill.

In the initial laboratory test on any ore the grind time is always a question. Therefore, it is estimated based on past experience and is generally close enough on most ores to take the slurry directly to the preliminary flotation test. If necessary, grind tests can be conducted prior to flotation testing. Determination of the grind can be obtained by a screen analysis of a sample taken directly from the flotation cell prior to testing or by re-combining pro-rated portion of test products. The Tyler Screen Series is usually employed.

Water soluble reagents used in the testing are generally added as 0.5 or 5% solutions, depending on the quantity required. Liquid reagents, used without dilution, are added drop-wise using a calibrated dropper. Dry reagents are weighed and directly added. In flotation processing reagents may be added in the ball mill, to pulp conditioners prior to flotation, just prior to flotation, during flotation, or to conditioners prior to a second stage of flotation. The point of reagent addition is important. In the initial testing, observations made of froth character, selectivity and tailings values will dictate stage additions of reagents. The use of vanning plaque and microscope on the concentrate and tailings is of real value.

Batch testing is used for the development of the basic process. Open circuit operation is used in batch testing; that is, the rougher concentrate and rougher tailing (final tailing) are first obtained by flotation of the ore slurry from the ball mill and then the rougher concentrate is retreated in the flotation machine several times in order to give the final cleaner concentrate and cleaner tailings. All products, including the cleaner tailings, are filtered, dried, weighed and assayed in order to abtain a material balance. The number of cleaner stages may vary from one to eight or nine depending on the mineral separations being made. Reagents are added to each cleaner to aid in selectivity and recovery. The value of dilution water as a reagent during cleaner stages should not be overlooked. It aids removing entrained gangue as well as diluting reagents to give improved selectivity.

In plant operation the flotation section is generally operated in closed circuit with the cleaner tailings flowing counter to the concentrates. The first cleaner tailings are combined with the feed to rougher flotation, the second cleaner tailing combined with the rougher concentrate to the first cleaner, etc. Closed circuit operation can be simulated in the laboratory by conducting a series of batch tests (at least four) in which the cleaner tailings are recycled to subsequent tests in the manner that is anticipated in plant operation. Often the effects of cleaner tailing recycle can be evaluated with fewer cycles if the combined cleaner tailing (1st, 2nd, 3rd, etc.) are added to the rougher flotation. If only minor values are reporting to the cleaner tailings, closed circuit testing may not be necessary because whether these values report to the concentrate or tailings will not have a marked effect on the economics.

As a flow scheme is developed a variety of circuits may be considered in order to optimize product grade and recovery. However, if possible, a simple, straight forward, procedure should be used. Many flow schemes used in operating plants have developed over a long period and attempting duplication in the laboratory could be difficult and the results meaningless. The laboratory procedures should be kept as simple as possible and such that the results can be interpreted into plant operation.

After an effective basic reagent combination has been obtained, then the testing can be conducted to optimize grind to rougher flotation. In this evaluation, rougher flotation only is conducted using grind times selected to give a coarser and finer grind than used in the initial testing. In these tests a sand slime separation is made on the rougher tailing and each assayed to determine the portion of the values lost in the sand fraction. Depending on the results of these tests one or two additional tests would be required to optimize grind. In the primary grind study the possibility of improved recovery by sand scavenger flotation should be considered. Improved recovery at a coarse grind may be possible if the rougher tailing is deslimed and the sand portion refloated (sand scavenger flotation) to recover a middling which can be reground and cleaned to a marketable product.

If after optimizing the rougher flotation recovery, cleaning stages do not give the desired grade product with recovery and mineral locking is indicated as the problem; then regrind testing should be considered. Ultimately a variety of regrind circuits may be considered. However, initially thickening or desliming the rougher concentrates by decanting over a 325 mesh sieve, and regrinding prior to cleaning would be indicated. The regrind can be best accomplished by using small diameter (0.0125 meter) steel balls in the grinding mill. As with the primary grind evaluation, a series of tests varying the regrind should be conducted.

In flotation systems the interdependence of the variables can make optimizing of a flow scheme difficult and often testing to obtain exact circuit conditions is not realistic because minor changes in ore grade or character will effect the balance. The process developed should be such that it will handle the ore variations and can be operated in a plant with a minimum of operator control. Optimum conditions, not maximum, should be the processing goal. However, until an acceptable concentrate grade with recovery has been obtained in the laboratory, a flow scheme has not been developed. Knowledge of the effects of each reagend added is important in adjusting quantities to improve results. Published literature covers to a large extent the effects of each; however, much is dependant on the experience of the operator.

Although there is a scientific basis for flotation separations, it is still an art, but experience has shown that on most ores, laboratory testing is sufficient for the development and design of good operating flow schemes. Pilot plant testing cannot be justified in most cases.

Hydrometallurgical Separation

The recovery of values from some types of ore is best and most economically obtained by leaching with aqueous agents. Leaching of gold and silver ores with sodium cyanide solutions is a well-known procedure that has been employed for many years and is effective for extracting values from gravity or flotation concentrates as well as ores. Oxide as well as mixed oxide plus sulfide copper ores have been economically leached with dilute sulfuric acid in dumps, heaps, vats and agitated tanks, and has proven to be the best method for handling low-grade ores. Leaching is the basis for recovering virtually all the uranium and associated vanadium from their ores. Dilute sulfuric solutions are used unless it is a limestone ore, and then sodium carbonate-bicarbonate leaching is used. Some uranium ores require oxidizing treatment as well as high temperature autoclave treatment to effect a leach.

With a recent emphasis on environmental problems from smelter operations much emphasis has been directed toward hydrometallurgical extractions of a variety of metals from concentrates as well as ores. Although much progress has been made, these processes have not been used to any extent to replace present smelter operations which are faced with high costs for controlling their stack gases and fumes.

Laboratory test procedures for cyanide leach evaluations on gold and silver ores are straight forward. The variables to be considered are grind, leach time, cyanide concentration and per cent solids. On a new ore the natural ph of a water slurry should be determined as well as the quantity of lime to bring the ph to the desired level for cyanidation (10.0 - 11.0). Initial amenability testing can be conducted by adding the cyanide after grinding, but ultimately testing is necessary to determine the value of cyanide in the grinding circuit. Cyanide consumptions on ore can vary over a wide range depending on the mineralization and will have to be evaluated during testing. For gold ores cyanide concentrations of 0.1% NaCN or less can be effective, whereas, silver ores may require 0.2 to 0.4% or more, and on concentrates 1.0% or more should be considered. The cyanide leaching should be conducted with adequate air contact and can be readily accomplished with an agitator in a beaker or in a bottle on rotating rolls.

The basic amenability to cyanide leaching can be obtained where the ore is given a 24-hour leach after a standard minus 48 to 65 mesh grind. Periodic sampling of the solution is necessary to assure the presence of free cyanide and lime throughout the test. Then a series of tests are necessary to evaluate (in turn) cyanide concentration, grind, leach time and possibly per cent solids during leach. If adequate extractions can be obtained on coarse ore (0.01 meter or coarser) and it appears that solution flow through a column is possible, then heap leaching may be considered. Indications as to an ores' amenability is dependant on results obtained from a large test heap. Where heap leaching is being considered, a very conservative approach is recommended.

The traditional procedure for recovering the gold and silver values from solution is by cementation using metallic zinc or aluminum (Merrill-Crowe Process). Recently considerable success has been obtained where the values are stripped from solution on to activated charcoal. They are removed from the char with a hot caustic cyanide solution and recovered by electrolysis. Heat treatment is required to reactivate the charcoal before recycle. These processes are described in literature and the charcoal process has been developed by recent work at the U. S. Bureau of Mines.

An oxide copper analyses on an ore is indicative of the ease that the ore can be leached. Some types of mineralization, such as the carbonate and oxides, readily leach; silicates are more difficult, and sulfides require oxidation before leaching can be accomplished. Chalcopyrite ($CuFeS_2$) is difficult to leach. Dilute sulfuric acid is generally used and ferric iron used for oxidation. Ammonium salt leaches can be employed with a high lime ore, but the high cost of the leaching agent generally is prohibitive. In heaps and dumps reoxidation of the ferrous iron can be accomplished by bacteria. However, in the laboratory oxidants are required, such as permanganate, manganese dioxide or hydrogen peroxide. Sodium hypochlorite would be effective but the effects of the chloride ion on the leach should be considered.

The variables to be considered in copper leaching are acid concentration (generally measured by pH), degree of oxidation (measured by E.M.F. in millivolts), leach time, temperature and grind. The acid concentration should be maintained as low as possible throughout the leach to keep leaching of other minerals to a minimum, and to end up with low free acid for recovery of copper from solution. Initial laboratory testing is generally conducted at room temperature for 4 to 8 hours with stage additions of sulfuric acid to maintain about a 1.5 pH on the agitated slurry. Titration of the liquor at the end of the leach with a base will give a measure of free acid. Because of metal salts in the solution, the base used for titration should be added to the first permanent precipitate formed. The results of the initial test will dictate the variables to be considered in subsequent testing. Where a relatively high acid concentration is required at the end of a leach, two-stage counter current leach may be considered. The relatively high acid solution from the tailing (2nd stage leach) could be added to the fresh ore and leach without acid addition (1st stage) to give a low acid pregnant liquor.

The copper can be recovered from the filtered and clarified liquor by cementation or by solvent extraction procedures which are well known and covered in many publications.

In the acid leach of uranium ores the same variables as for copper ores are considered and laboratory testing follows the same general procedure. However, some uranium ores require leaching in autoclaves in order to obtain the desired temperature and oxidizing conditions. Although oxygen-enriched air or pure oxygen will speed up leaching in an autoclave, the corrosive effects on equipment made its use impractical.

Where high acid consumptions are obtained because of limestone in the ore, then sodium carbonate-bicarbonate leaches can be considered.

The leached uranium can be recovered from a ground slurry using ion exchange resin (R.I.P. or resin-in-pulp procedure), or can be recovered from the filtered and clarified liquor by ion exchange resin and/or solvent extraction. The special techniques for these procedures are beyond the scope of this article.

Grindability

The optimum grind for recovery of the values from an ore is most readily determined during the process development testing and the grinding time in the laboratory ball mill relative to the fineness of grind is indicative as to the hardness of the ore. Manufacturers of ball mills have equipment to develop the grind work index (energy per ton of ore to grind) in order to size the equipment. The results should be considered in light of the sample submitted and the testing should be conducted on more than one sample from a mining property.

An indicative measure of work index is possible where a standard crush and grind are used in the laboratory. Plots are made of grind work indices as determined by the manufacturer vs. the plus 150 and 200 mesh per cent weight screen fractions on the ground ore. The lines generated by connecting the plots can be used for estimating the work indices of other ores. Although not infallible, the procedure has proven to be a quick method for estimating grind index.

Liquid-Solid Separations

A majority of the metallurgical beneficiation processes are conducted in aqueous slurries. Therefore, the liquid-solid separations on the final products are a very important consideration and should not be overlooked in the flow sheet design. There have been cases where an effective metallurgical process was developed but could not be used because the liquid-solid separation could not be economically effected. Procedures used are dependant on the character of the product and the moisture content ultimately desired. One-unit operation or a combination of unit operations may be required.

The basic unit operations used are:

1. Screening

2. Cycloning
3. Thickening
4. Centrifuging
5. Filtering
6. Drying

Screening is employed to dewater relatively coarse products such as coarse coal and sand products. Dewatering screens alone are effective to give the desired product on coarse coal, but thermal driers may be required on sand products after screening.

Hydro-cyclone separations are used on intermediate size coal particles, sandy materials, or used in conjunction with thickeners for gravity and flotation products. This special type of application of the cyclone should be carefully considered based on the character of the material to be separated.

Thickening devices, based on the settling of solids from a slurry by gravity, may be as simple as a settling pond or a cone bottom tank to a unit that has rakes to carry the thickened product to a draw point and with weirs to remove solution. Thickeners are used to increase solids content prior to filtering or centrifuging and are used in washing solutions from solids in hydro-metallurgical operations such as counter-current decantation (C.C.D.) employed in cyanide leaching.

Laboratory procedures have been developed for determining the thickener area required to effect a separation. The two methods generally used are the Kynch and the Coe-Clevenger. There are advocates for each method or variations thereof. However, with the most diligent testing and evaluation, there is still some question as to the accuracy of the results. Adequate safety factors should be applied.

The thickener studies are generally conducted by obtaining settling rates for each unit of time in a graduated cylinder and intercepting the curve generated by plotting time vs. settled distance. There are a variety of metal salts and organic compounds that are used to cause flocculation of the solids and increase settling rate. Although there are some guides as to the type or types to be used; much is dependant on the experience of the technician and what actually does the job.

Centrifuges are effectively employed for dewatering of solids that have only a minor amount of very fine particles (slimes). The unit gives a rapid separation, resulting in a product with relatively low moisture content. However, the operating and maintainence costs should be considered before placing the unit in a flow scheme. Although equipment can be obtained for centrifuging tests, the ultimate evaluation and sizing should be determined by the manufacturer.

Filtration is widely used for removal of a majority of the liquid from a slurry. Depending on the character of the material moisture contents as low as 8 to 10% are possible but may be as high as 30 to 35% on fine coal or a material high in clays. A selection of vaccuum drum, disc, or belt; or pressure units are available. Each has its area of application and should be considered based on the character of the material and separation desired. Although one type of unit appears to be theoretically effective, testing is required for final evaluation.

Filter tests can be conducted in most laboratories to obtain indications as to the type and size of filter unit required and can be used in the initial design studies. However, before the final selection and sizing is made, manufacturers should be consulted and, if necessary, conduct testing.

Further reduction of moisture in the filter or centrifuge product may be required for marketing or shipping of a material and then thermal drying is considered. Estimates as to the energy type and size of unit required for a given tonnage and moisture content can be calculated from information shown in publications. Much is dependant on the experience of the engineer. Recommendations from qualified people may be desired.

EVALUATION OF TEST RESULTS

A majority of the evaluations of test results are made as an investigation progresses and can be readily made based on the product grades and recoveries, as well as impurities obtained. Through the years investigators have developed calculations which show relative degrees of selectivity and give a measure for evaluating results. In the development of the initial flow scheme visual comparisons are sufficient. Ultimate evaluation may require further calculations and graphs to optimize flow scheme procedures. Care should be taken in making comparisons to be sure that the variables considered can be logically compared. If more than one change is made in a series of tests, then care should be taken to be sure that the other variable or variables do not invalidate the comparison.

A common method of comparison is based on per cent metal or mineral recoveries. However, where test results on ore samples of varying grades are compared, the per cent recovery measure is not adequate. With similar product grades, a comparison of the tailing assays may be more valid. On most ores the tailing assays do not change to a marked degree with changes in feed grade where an effective procedure is employed. The ultimate economic evaluation should be based on quality and quantity of the product produced irregardless of the feed grade and per cent recovery.

Where indicated optimum metallurgical results have been obtined, then relatively simple calculations can be made to estimate the net value for similar ore. By subtracting the average units (tailing assay x per cent weight) in the tailings from the feed assay, the units recovered are obtained. The per cent weight product can be obtained from the product grade and units recovered (product assay ÷ units recovered). The product value per ton times the per cent weight less the mining and milling costs will give the indicated net value per ton of ore. Graphs can be generated to show the net value for a range of ore grades and could be a real asset in evaluating a mining property or as a guide in mining.

FLOW SHEET LAYOUT

Assuming metallurgical testing has been competently done on a sufficient number of samples to provide a basis for a workable flow sheet, the next logical step is to translate this information into a suitable flow sheet.

The same data can result in many variations depending on the preference of those doing the work, but some general comments may be helpful in guiding those making the interpretation and those charged with approving or modifying the finished flow sheet.

1. Keep the flow scheme as direct and simple as possible.

2. Provide reasonable safety factors so the 'abnormal' ore can be treated without serious economic consequences due to low capacity or poor recovery.

3. Crushing is cheaper than grinding. Crush to the finest practical size depending on ore characteristics. Finished products in 10 - 16 mm. (3/8 - 5/8 inch) size range are preferred, providing clay content and/or moisture content do not interfere.

4. Provide adequate <u>live</u> ore storage at as fine a crushed product as practical.

5. Provide means of blending feed to the plant.

6. Avoid single point feed openings from bins which tend to allow segregation, particularly in coarser ore sizes.

7. Closed circuit crushing is generally preferred over open circuits.

8. Grinding is generally cheaper in a ball mill than in a rod mill. Rod mills do their best grinding from 16 - 22 mm. (5/8-7/8 inch feed to 10 - 14 mesh size range. Ball mills, particularly large diameter mills, can accept feed up to 19 - 13 mm. (3/4 - ½ inch) and grind to a finished product more economically than a rod mill - ball mill combination.

Rod mills can accept relatively coarse feed of 100 - 25 mm. (4 - 1 inch) and are best applied when the ore is of such a nature that crushing to minus 25 mm. (1 inch) is not practical.

9. Provide at least three conditioning units to prevent short circuiting. Pulp which arrives at the concentration process, flotation for example, without being properly conditioned generally yields poor metallurgy and therefore poor economics.

10. Flotation circuits should be matched to the grinding units used. Generally flotation circuits with 12 to 18 or more cells in length will give superior metallurgy and economics to short strings of large cells.

In the authors' view the use of large cells should be confined to plants of sufficient size to require a minimum of 12 cells per row for roughing circuits in base metal and most non-metallic circuits. Coal and some non-metallic circuits generally have shorter roughing circuits only because of the relatively lower values of the products being handled.

Cleaner circuits should generally be of sufficient number of cells to produce cleaner tailings which can be discarded without serious metallurgical losses. It is often advantageous to open-circuit the cleaners. This technique is difficult or impossible if insufficient cleaner cells are provided.

11. Liquid-solid separation can generally be accomplished in a number of ways. In the authors' view the use of thickeners ahead of filtration is preferred to direct filtration except in coal and some non-metallics. Thickeners should be of sufficient size to provide satisfactory overflow and surge storage of concentrates during minor filter repairs.

Filters should be of a type and size to provide sufficient capacity for anticipated fluctuations in the amount of concentrate produced at an acceptable moisture. Marginal or undersize equipment in the liquid-solid separation process step can be a major error.

12. The authors' preference for a finished flow sheet is a pictorial schematic diagram which illustrates clearly the process steps, the recommended equipment and the tonnages of solids, liquids and pulps which must be dealt with.

The authors recognize that a final flow scheme requires the expertise of many specialists, especially in large plants and complex processes. In such instances the translating of metallurgical test data to final flow sheet is a complex problem and is seldom the work of only the metallurgical engineer.

3

Crushing and Grinding

Chairmen: L. E. Lipphardt
C. B. Dugger, Jr.

Chapter 11

CRUSHING

by Jerome C. Motz

Product Manager, Crusher Department
Process Machinery Division, Rexnord Inc.
Milwaukee, Wisconsin

Primary Crushers

The types, sizes and number of crushers employed in a complete reduction system will vary with such factors as the volume of ore to be processed, the size of the mine—run lump material, the hardness of the ore bearing rock and the size and dimension required for the final product.

Basically, for primary crushing there are only three distinct types of crushing equipment. These are the Jaw crushers, Gyratory crushers and Impact crushers, and they each have their own distinctive operating characteristics.

Jaw crushers operate by squeezing the rock between the fixed and the movable sides of a tapered cavity. Variations of pitch and swing have been tried, but most Blake-type machines have a crushing angle of about 27° between swing and stationary jaws.

The principle of which Gyratory crushers work is very simple. If a Cone is mounted on the upper end of a vertical shaft and the top of the shaft is held stationary while the lower end is rotated eccentrically, the Cone will also swing eccentrically. If the Cone is enclosed in a suitable housing, it will swing toward and away from the housing walls as it rotates. If the Cone and the housing walls are sufficiently strong and heavy, anything caught between them will be crushed.

Besides the increased efficiency developed by the continuous crushing action and curved crushing faces, Gyratory-type crushers have other advantages:

1. The largest unrestricted feed opening available in comparison to other crusher types.
2. A high range of sizes and capacities where rates of between 600 and 6,000 tons per hour are required
3. Generally, no feed control is necessary. Direct dump trucks of 150 to 200 tons capacity are now being used in the larger installations.

Impact crushers accomplish material breaking and reduction primarily through impact action of the material with fixed or free-swinging hammers revolving about a central rotor.

Product gradation is a function of rotor speed and friability of the material. An increase in rotor speed or material friability leads to an increase in the proportion of fines in the product; conversely, a decrease in rotor speed or material friability will decrease the proportion of fines in the product.

The principle of impact crushing is the total use of rotor energy in the reduction of the material. Kinetic energy is imparted to a particle of the material, and this energy is used chiefly to break down the bonding of the particle structure. The particle breaks along its planes of weakness producing a cubical natural grain-shaped product as the result. The end product is free from weak zones. The impact factor is equal to the particle kinetic energy and varies as:

$E = \frac{MV^2}{2}$ when E = Energy, V^2 = Speed Squared, and M = Mass Weight

The velocity is fixed by the peripheral speed of the rotor; therefore, particles of a material with a common impact value will break according to their initial mass. A large piece will receive a severe impact and will break more readily than a small one. The latter receives only a light impact and will tend to pass through with minimal further breakage. If two materials of different impact value are fed to a machine, even though they may be particles of the same mass, one will be reduced more than the other.

Benefits in using primary Impact-type crushers, as compared to compression crushers, are lower installed capital costs per ton of capacity, much greater capacity weight for weight of comparable Jaw and Gyratory crushers thus reducing installation cost and raising the feasibility for mobile units, production of more cubical product and generally, a finer product gradation. This may reduce the need for secondary crushing units.

Due to the relatively high maintenance costs generated through welding of hammers and the internals of the Impact unit when crushing moderately abrasive rock types, thorough testing and evaluation of the feed material is recommended by the manufacturer. Refer to Tables 1 through 5 for capacities and product analyses for primary-type crushers.

Table 1

JAW CRUSHERS **CAPACITY**

JAW CRUSHER CAPACITY

TPH

DISCHARGE SETTING (INCHES)	1236	1836	2136	2436	2540	3040	3240	3648	4248
1½	40								
2	55	55	55						
3	85	90	90	90	145				
4		125	125	125	185	185	185	275	
5			155	155	240	240	240	350	
6			195	195	290	290	290	425	425
7					340	340	340	475	475
8					390	390	390	525	525
9								575	575
10								650	650
12									800

Capacities listed are average based on dry, free crushing rock at 100 lbs. per cubic foot. Tonnage will vary depending on size of feed, rate of feed, operating conditions, breaking characteristics of the rock, type and condition of crushing members (jaws), speed of crusher, and horsepower available. Listed capacities may vary ±25%.

Crusher cannot be operated economically where ratings are not specified.

Table 2

JAW CRUSHERS **GRADATION**

PRODUCT ANALYSIS

This chart gives an average analysis of stone crushed by an overhead eccentric jaw crusher. It does not imply any guarantee of product sizing, since all stone varies from area to area and local conditions are never constant.

OPEN CIRCUIT
Percent of Each Size in the Product of a Crusher

Size Range	Crusher Setting to Give 15% Recirculation																	
	¾	1	1¼	1½	1¾	2	2¼	2½	2¾	3	3½	4	5	6	7	8	9	10
+10″																		15
-10″ +9″																	15	32
-9″ +8″																15	29	9
-8″ +7″															15	20	6	4
-7″ +6″														15	16	9	5	4
-6″ +5″													15	13	9	6	4	3
-5″ +4″												15	10	8	7	6	5	4
-4″ +3½″											15	9	8	8	6	5	4	3
-3½ +3″										15	8	8	8	6	6	5	5	4
-3″ +2¾″									15	7	6	6	5	4	3	3	3	2
-2¾″ +2½″								15	7	7	7	6	5	4	3	3	2	2
-2½″ +2¼″							15	6	6	6	6	4	3	3	3	2	2	2
-2¼″ +2″						15	7	7	7	7	6	5	5	4	4	3	2	2
-2″ +1¾″					15	8	8	8	7	6	6	6	6	5	3	3	2	1
-1¾″ +1½″				15	8	7	7	7	7	6	6	6	5	4	3	2	1	1
-1½″ +1¼″			15	9	9	8	8	7	6	6	6	5	4	3	3	2	2	1
-1¼″ +1″		15	12	11	11	10	9	9	8	7	5	4	3	3	2	2	1	1
-1″ +7/8″		7	7	6	6	6	5	5	5	5	4	3	3	3	2	2	1	1
-7/8″ +3/4″	15	8	8	7	7	7	6	5	4	3	3	3	3	2	2	1	1	1
-3/4″ +5/8″	10	10	7	7	6	6	6	5	4	4	4	3	3	2	2	2	2	1
-5/8″ +1/2″	11	10	9	8	8	6	5	4	4	4	3	3	2	2	2	1	1	1
-1/2″ +3/8″	13	10	8	8	7	6	6	6	5	5	4	4	4	3	2	2	2	1
-3/8″ +5/16″	7	6	5	4	4	4	3	3	3	2	2	2	1	1	1	1	1	1
-5/16″ +1/4″	7	5	5	5	3	3	3	3	3	2	2	2	1	1	1	1	1	1
-1/4″ +4M	8	7	5	4	4	4	3	3	3	3	3	2	2	2	2	1	1	1
-4M +8M	13	9	7	6	5	3	3	2	2	2	1	1	1	1	1	1	1	1
-8M	16	13	12	10	7	7	6	5	4	3	3	3	3	3	2	2	1	1

EXAMPLE: Jaw crusher for 1½″ with 15% recirculation – find 1½″ at top and follow down verticle column.

+1 1/2″	-15%	7/8″ to 3/4″	-7%	3/8″ to 5/16″	-4%
1 1/2 to 1 1/4″	-9%	3/4″ to 5/8″	-7%	5/16″ to 1/4″	-5%
1 1/4″ to 1″	-11%	5/8″ to 1/2″	-8%	1/4″ to 4M	-4%
1″ to 7/8″	-6%	1/2″ to 3/8″	-8%	4M to 8M	-6%
				-8M	-10%
				TOTAL	100%

Table 3A

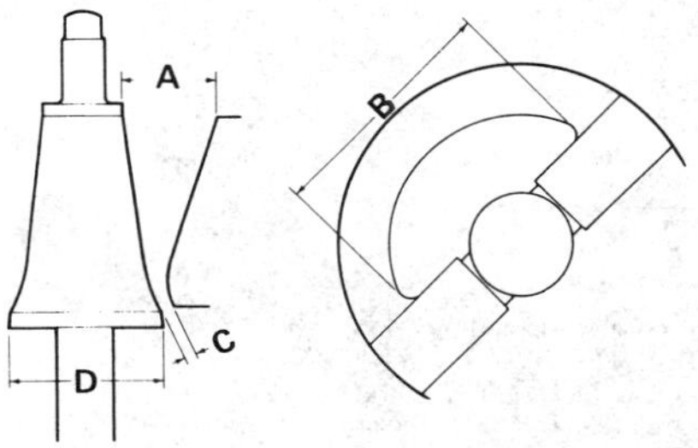

SELECTION DATA: The condensed information on these pages is only a guide in determining the size and operating data of a Nordberg Primary Gyratory Crusher for a particular application. Crusher capacities in these tables are given in short tons (2000 lbs.) per hour and are based on average operating conditions when handling a dry friable material weighing 100 lbs./cu. ft. (crushed). Factors which may affect capacity include hardness, toughness, fracture pattern, moisture content, size gradation of the feed material and method of feeding.

The horsepower column represents the maximum rated horsepower required for the crusher. These ratings do not refer to the horsepower required for throws and capacities listed. In general, the recommended duty and capacity of the crusher can be determined by knowing the feed, throw, speed and product desired. Horsepower for specific applications can be determined on following pages.

CAPACITIES IN TONS PER HOUR (2000 Lbs.)

Size	A x B Feed Opening (Inch)	D Head Diameter (Inch)	Gyra per Min.	Cntr. Shaft RPM	Eccentric Throw Inch	Max. HP	Open Side Discharge Setting – C												
							3 1/2	4	4 1/2	5	5 1/2	6	6 1/2	7	7 1/2	8	8 1/2	9	10
42-70 HD	42 x 114	70	137	514	3/4	200				525	625	730	835	925	1020				
					1	250				700	830	975	1110	1230	1360				
					1 1/8	270				785	936	1100	1250	1385	1530				
					1 1/4	300					1040	1220	1390	1540	1700				
42-70 XHD	42 x 114	70	137	514	3/4	240			430	525	625	730	835	925	1020				
					1	320				700	830	975	1110	1230	1360				
					1 1/8	360				785	936	1100	1250	1385	1530				
					1 1/4	400					1040	1220	1390	1540	1700				
48-75 HD	48 x 133	75	140	514	1	300				1175	1250	1330	1440	1565	1700	1840			
					1 1/4	375					1567	1660	1800	1955	2125	2300			
					1 3/8	410					1720	1880	1980	2150	2340	2530			
					1 1/2	450						1995	2160	2350	2550	2760			
54-75 HD	54 x 144	75	140	514	1	300					1300	1400	1510	1630	1775	1730			
					1 1/4	375						1750	1885	2040	2220	2410			
					1 3/8	410						1925	2075	2240	2440	2655			
					1 1/2	450							2265	2445	2660	2895			
54-80 HD	54 x 145	80	140	514	1 1/4	400						1825	1950	2170	2420	2650	2900		
					1 3/8	425						2010	2145	2385	2660	2915	3190		
					1 1/2	460							2340	2605	2905	3180	3480		
					1 5/8	500							2535	2820	3145	3445	3770		
60-90 XHD	60 x 165	90	115	514	1 3/8	550							1955	2130	2420	2660	2900	3145	
					1 1/2	600							2130	2390	2640	2900	3160	3430	
					1 5/8	650								2580	2860	3140	3425	3715	
					1 3/4	700								2775	3080	3385	3635	4000	
60-102 HD	60 x 174	102	115	514	1 1/2	600								3140	3500	3900	4270	4640	5400
					1 5/8	650								3400	3790	4225	4625	5025	5850
					1 3/4	700									4085	4550	4980	5415	6300
					2	750									4665	5200	5695	6185	7200
60-102 XHD	60 x 174	102	115	514	1 1/2	750								3140	3500	3900	4270	4640	5400
					1 5/8	800								3400	3790	4225	4625	5025	5850
					1 3/4	875									4085	4550	4980	5415	6300
					2	1000									4665	5200	5695	6185	7200

Table 3B - pg. 1

HORSEPOWER REQUIREMENTS *

This method employs a common factor known as the work index which is the total work input in kwhr per short ton required to reduce a given material from theoretically infinite particle size to 80% passing 100 microns or 67% passing 200 mesh.

Material: (Notes 1 & 4) ________ Wt./Cu. Ft.: (Note 1) ________

Sp. Grav: (Note 2) ________ Av. Impact: (Note 2) ________

Work Index = $2.59 \frac{\text{Av. Impact}}{\text{Sp. Grav.}}$ = ________ ; ROM (Run of Mine) or Scalped

Feed Size = 80% Passes (Note 3) ________ In. F = ________ microns

Prod. Size = 80% Passes (Note 4) ________ In. P = ________ microns

(In calculating F & P, 1" = 25,400 microns)

HP/Short Ton = $\frac{WI \times 13.4 (\sqrt{F} - \sqrt{P})}{\sqrt{F} \times \sqrt{P}}$ = ________ HP/Short Ton

Total HP = Capacity in Short Tons/Hr. (Note 1) x HP/Short Ton = ________ HP

NOTE 1 – Refer to Nordberg Bulletin 318C for crusher capacities. Interpolate for specific bulk density since the values are based on 100 lbs./cu. ft.

NOTE 2 – The average impact, specific gravity and the resulting work index can be determined at the Nordberg Process Machinery Test Center if the samples are provided.

NOTE 3 – If the feed analysis is not available, experience indicates that 80% of the feed size will pass a square opening equal to from half to two-thirds of the crusher receiving opening.
Feed Size = 80% Passes ________ In, F = ________ microns

NOTE 4 – Product size can be determined with reasonable accuracy from Product Gradation Curve Tables A & B.
Prod. Size = 80% Passes ________ In, F = ________ microns

MATERIAL FRACTURE CHARACTERISTICS

Example:

Assuming the feed to be run of quarry soft limestone (WI-8.5) including fines to a 54-80 Nordberg Gyratory Crusher set at 6" O.S.S.

Table "A" shows 90% of product should pass a 6" square opening in flat testing screen. Follow the 90% horizontal line within the 6" screen size opening on table "B". This line will yield the approximate screen analysis shown on the right.

Square Screen Opening In Inches	Approx. % Passing
8	98
4	66
3	50
2	34
1-1/2	29
3/4	16
1/2	12

TABLE A
Percent of Product Passing a Square – Opening Equal to Open Side Setting of Crusher

Work Index Guide	Material Characteristics	Run of Mine or Quarry	Scalped Feed: Crusher Product Only	Scalped Feed: Crusher Product And Grizzly Thrus Recombined
5.0 to 10.0	Soft	90	85	88
	Soft (Spongy)	85	82	85
10.0 - 13.0	Average	90	85	85
	Average (Slabby)	85	82	85
13.0 & Up	Hard (Brittle)	90	85	85
	Hard (Tough)	82	75	80
	Hard (Slabby)	75	70	75

*From Fred C. Bond, "Third Theory of Comminution"

Table 3B - pg. 2

PRIMARY CRUSHER PRODUCT

TABLE "B" – PERCENT PASSING

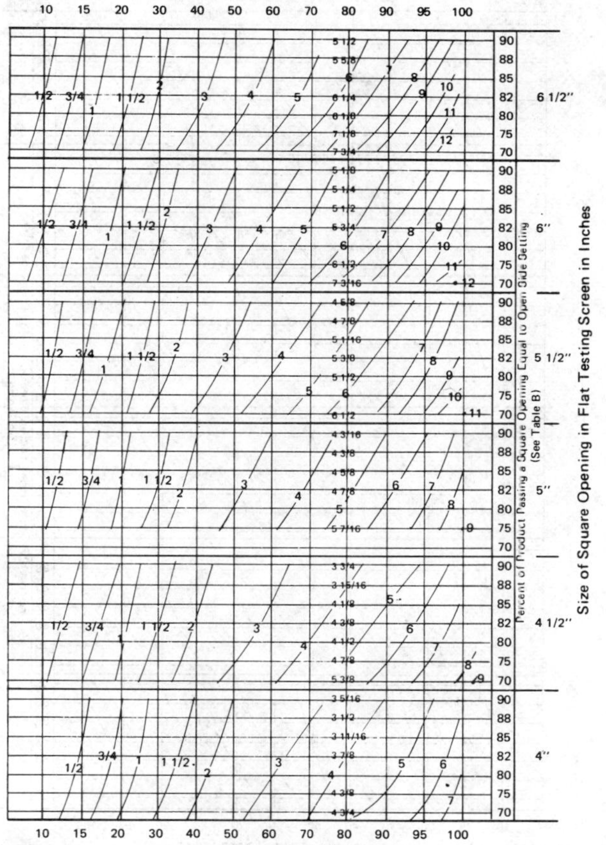

The above product gradations which represent square screen openings in inches yield the approximate screen analyses and should be used in conjunction with table "A".

Table 3B - pg. 3

PRIMARY CRUSHER PRODUCT

Table "B" – Percent Passing

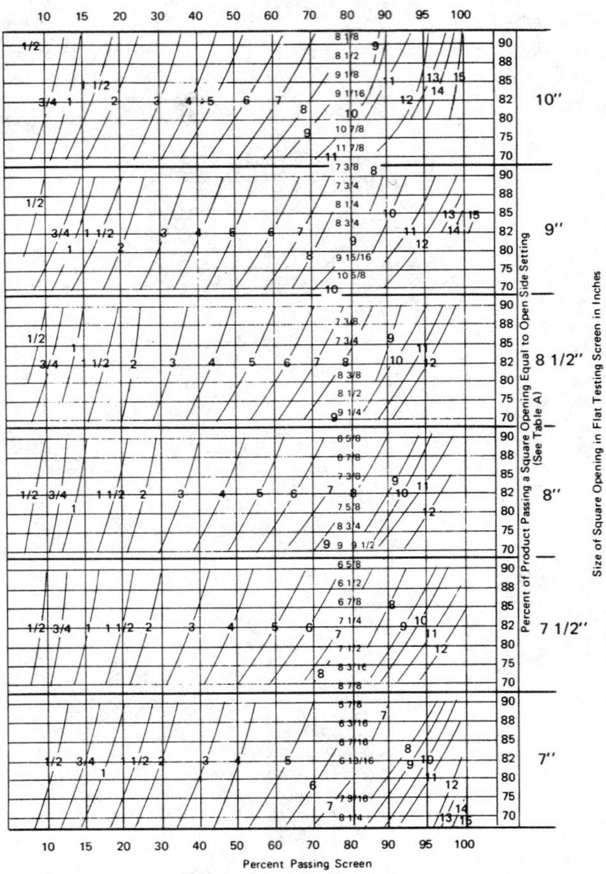

The above product gradations which represent square screen openings in inches yield the approximate screen analyses and should be used in conjunction with table "A".

Table 4

Impact Crusher

DYNAPACTORS CAPACITY

DYNAPACTOR CAPACITY

Capacities listed are average based on dry, free crushing limestone at 100 lbs./cu. ft. Actual tonnage will vary with size of feed, rate of feed and general operating conditions.

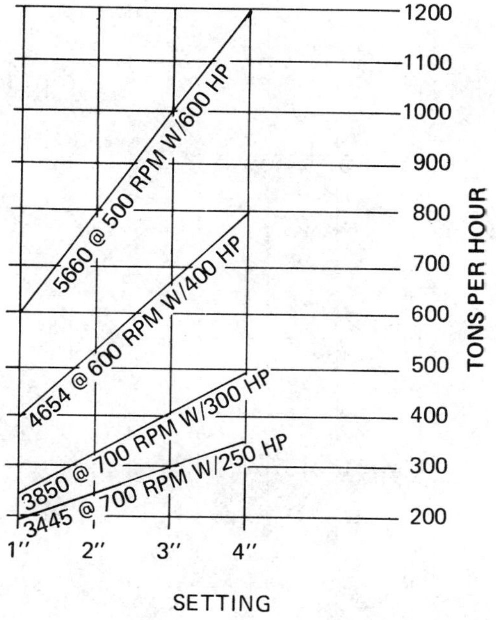

DYNAPACTORS Table 5 **GRADATION**

Impact Crusher

DYNAPACTOR CRUSHER

PRODUCT ANALYSIS

OPEN CIRCUIT

Percent of Each Size in the Product of a Crusher

SIZE RANGE		CRUSHER SETTING IN INCHES																							
		1"	2"	3"	4"	5"	6"	1"	2"	3"	4"	5"	6"	1"	2"	3"	4"	5"	6"	1"	2"	3"	4"	5"	6"
	+10"						3					5	10				2	10	19				5	15	25
-10"	+9"					2	4				2	2	3				3	5	5				4	6	6
-9"	+8"					3	4				3	3	5				5	6	5			1	6	7	7
-8"	+7"				2	4	5			2	4	6	5			3	6	6	6			6	7	8	6
-7"	+6"				5	6	5			3	5	7	6			7	8	7	5			8	8	8	6
-6"	+5"			4	6	6	6		2	5	7	7	7		2	10	8	6	6		5	10	11	6	5
-5"	+4"			6	7	7	7		3	8	9	7	7		8	[illegible]	8	6	6		10	13	9	6	5
-4"	+3 1/2"		2	5	5	4	4		3	6	5	4	4		7	[illegible]	5	4	4		8	7	4	3	3
-3 1/2"	+ 3"		5	5	5	5	4		5	6	4	4	3		7	5	3	3	3	1	8	5	4	3	3
-3"	+2 3/4"		3	4	3	3	3		4	3	3	3	3		4	3	3	3	2	2	5	3	2	3	2
-2 3/4"	+2 1/2"		3	3	3	3	2	1	4	3	3	2	2	2	4	3	2	3	2	3	5	2	2	2	2
-2 1/2"	+2 1/4"		4	4	3	3	3	2	4	3	2	3	3	3	4	2	3	2	2	4	4	3	2	2	2
-2 1/4"	+2"		4	4	3	3	3	3	4	3	3	3	3	5	4	3	3	2	2	5	5	3	3	2	2
-2"	+1 3/4"	2	5	4	4	3	3	4	5	4	4	3	3	6	5	3	3	3	3	8	4	3	3	3	3
-1 3/4"	+1 1/2"	4	4	5	4	4	4	5	5	4	3	4	3	7	4	4	3	3	3	8	4	3	3	3	3
-1 1/2"	+1 1/4"	7	6	5	5	4	4	7	6	5	5	4	4	9	5	4	3	4	3	11	4	3	3	3	3
-1 1/4"	+1"	8	6	6	5	5	4	8	5	5	4	5	4	8	5	5	4	3	3	8	5	5	4	3	3
-1"	+7/8"	5	4	3	3	2	3	5	3	3	3	2	2	5	3	3	3	2	3	4	3	2	2	2	2
-7/8"	+3/4"	5	4	4	3	3	3	5	4	4	3	3	3	4	3	3	2	2	2	4	3	3	2	2	2
-3/4"	+5/8"	6	5	3	3	3	3	5	5	4	4	2	2	5	3	3	3	3	2	4	4	3	3	2	2
-5/8"	+1/2"	6	5	4	4	3	3	6	4	4	4	3	3	5	5	3	3	3	2	5	3	3	3	2	2
-1/2"	+3/8"	7	6	5	5	5	4	7	6	5	4	4	3	6	5	4	4	3	3	6	5	3	2	3	2
-3/8"	+1/4"	10	8	6	6	4	4	9	8	5	4	4	4	7	6	5	4	3	2	7	5	4	3	2	1
-1/4"	+4M	5	5	4	3	4	3	6	4	3	3	2	2	7	4	4	2	2	2	5	3	2	1	1	1
-4M	+8M	15	10	7	5	4	3	10	7	5	3	3	2	9	5	3	3	2	2	6	3	2	1	1	1
-8M		20	11	9	8	7	6	17	9	7	6	5	4	12	7	5	4	4	3	9	4	3	3	2	1
Crusher Size and R.P.M.		3445 @ 600-850 R.P.M. 3850 @ 600-725 R.P.M.						3445 @ 450-600 R.P.M. 3850 @ 500-600 R.P.M. 4654 @ 550-650 R.P.M.						3850 @ 400-500 R.P.M. 4654 @ 450-550 R.P.M. 5660 @ 450-600 R.P.M.						4654 @ 350-450 R.P.M. 5660 @ 350-450 R.P.M.					

This chart gives an average analysis of stone crushed by a single rotor, two hammer impact crusher. It is based on dry, free crushing, medium hard lime stone. It does not imply any guarantee of product sizing, since all stone varies from area to area and local conditions are never the same.

SECONDARY, TERTIARY AND FOURTH STAGE CRUSHING

Cone Crushers

The Cone crusher is used as a secondary, tertiary and fourth stage crusher in hard rock applications. The Cone crusher was invented by Mr. Edgar B. Symons in the mid-1920's after spending considerable time in the development of secondary crushing machinery. The Cone crusher proved to be remarkable due to its radical departures from the normal principles of a Gyratory crusher operation.

In previous Gyratory machines the crushing process comprises reduction by pressure between two confining faces and a subsequent freeing movement during which the material settles by gravity until it is caught and subjected to further pressure and again released. In the normal Gyratory the crushing stroke or travel of the head has usually an important bearing on the size of the finished product, although this factor is subject to modification when a parallel sizing zone is adopted.

The movement of the head in the Cone crusher is similar to that in the ordinary Gyratory with the exception that, towards the bottom of the Cone, it travels through a much greater distance and gyrates much faster. This large stroke, combined with the flat angle of the Cone, causes the stone to follow a zigzag path through the crusher. The rock entering at the top drops upon the head and is then elevated and crushed against the bowl liner. As the head recedes in its gyration, faster than the effect of gravity on the crushed particles, the crushed rock drops vertically down towards the head to be met on the next gyration by the upcoming face, and again is crushed against the bowl liner.

The long movement changes the crushing stroke from pressure to impact, and the increased clearance on the freeing stroke allows the stone to fall away vertically after each impact. The wide gap at the end of the opening movement also permits the stone to move freely for the greater part of the time occupied in passing through the crushing zone. The periods of confinement are therefore limited.

The angle between the faces of the bowl liner and the mantle decreases gradually, and these faces are parallel near the outlet for a sufficient distance to ensure that the head must make one or two gyrations before the material can pass through the sizing zone, which again means that it is the opening on the closed side of the crushing head which regulates one dimension of the product size. As in the slow-speed Gyratory crusher, the volume transported through each vertical position in the Cone crusher cavity is directly related to the length of the crushing stroke, sometimes called eccentric throw, but more accurately defined as head displacement.

The machine departs from the usual design in that there is no upper bearing for the Gyratory shaft and no spider to bridge over the feed opening. The head is supported in a part-spherical seating and the tapering main shaft is actuated by an eccentric. The absence of the spider ensures a more even distribution of the feed to all parts of the crushing chamber.

The crushing bowl, inverted over the Cone and screw-threaded to permit vertical adjustment inside an adjustment ring, is held down to the main frame by a series of powerful springs which allows the bowl to rise when tramp iron or other uncrushable material enters the crushing cavity.

It is very desirable with this machine that the feed be controlled and the material be evenly distributed around the circumference if optimum efficiency is to be obtained. A feed-distributing and regulating plate is therefore mounted on the top of the main shaft with a vertically adjustable spout above it.

CONE CRUSHER APPLICATION

Crushing Stages

To reduce run-of-mine to mill feed requires reduction in a series of stages. The physical size and power requirements of a crusher capable of reducing hard rock vary depending on the application.

Even when a crusher has the capability of achieving a high reduction ratio it is normally more efficient to run the crusher at a mid-setting, as shown in the manufacturers' tables, rather than choosing the closest setting which offers the greatest reduction ratio. The main consideration at each stage for maximum production is efficient power draw.

There is an optimum setting for each crusher and an optimum number of stages required for maximum plant production based on the individual characteristics of the material being crushed. Overloading the crusher does not increase production, but instead is counterproductive and will decrease the lives of the crusher components. Ideally, the top size feed should receive four to five impact blows during its progress through the crushing chamber. This is a combination of reduction at the upper zone of the liners as well as the parallel zone. The crusher should be fed so as to be operating at or near continuous full load horsepower capability. Operating the crusher at too narrow a setting decreases capacity and creates high wear. Too wide an opening, in proportion to top size feed, will prevent crushing in the upper zone and will develop excessive power draw. Power drawn per ton of crusher feed is not in itself a measure of productivity. Efficient use of power through proper application of the cavity, with respect to feed and product requirements, will determine the optimum production per horsepower drawn. From the following crusher application descriptions, it can readily be seen that each crusher has its own designed level of reduction. The range of feed sizes and product sizes, for each crushing stage, somewhat overlaps into the succeeding and preceding stages. This allows for flexibility in adjustment at each stage for optimum production of the entire system. This also presents the option of eliminating a stage when it is advantageous.

Primary Crusher

The Primary crusher's main purpose is to reduce large fragments of blasted or natural rock down to a size suitable for handling by transfer equipment and the secondary stage crusher. Feed opening and product size ranges for the various models of Gyratory crushers are as follows:

Maximum Feed Opening -- 1500 MM to 750 MM (5 ft. to 2-1/2 ft.)

Recommended Minimum Product -- 162 MM to 100 MM (6-1/2 In. to 4 In)

Preferred Reduction Ratio -- 8:1

Standard (Secondary) Cone Crusher (Refer to Fig. 1)

The Standard Cone crusher is normally applied as a secondary crusher in a multi-stage crushing circuit. The small diameter feed distributor and the wide throat opening at the top of the liners enable the Standard Cone crusher to accommodate the larger feed produced by the Primary crusher.

Feed opening and product size ranges for the various models of the Standard Cone crusher are as follows:

Maximum Feed Opening -- 625 MM (25 In.)

Recommended Product Range -- 100 MM to 19 MM (4 In. tc 3/4 In.)

Preferred Reduction Ratio -- 6 or 8:1

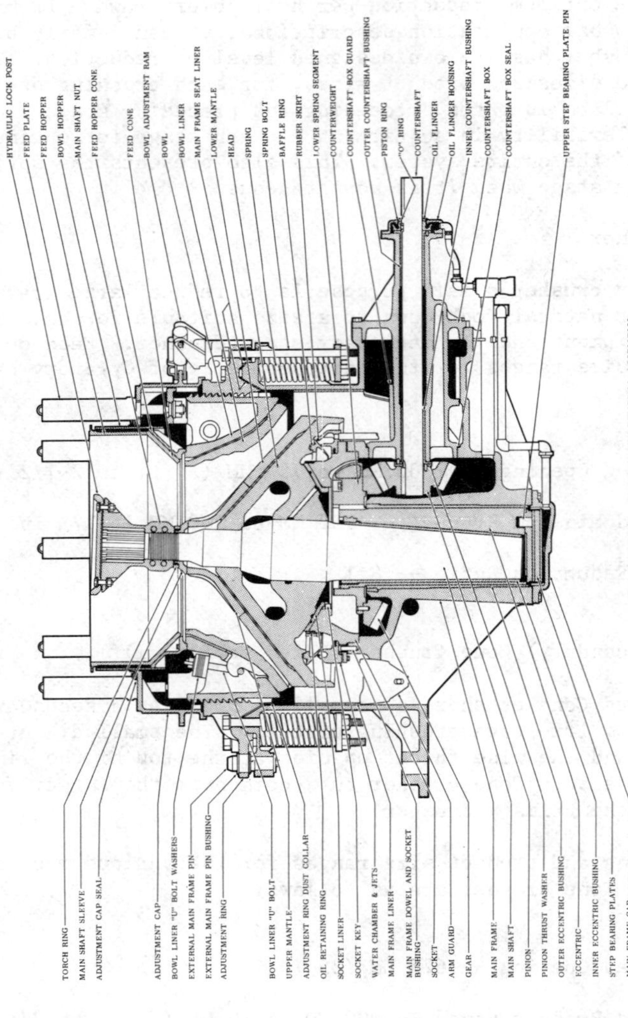

7 FT. STANDARD HEAVY DUTY GENERAL ASSEMBLY
(water seal-typical arrangement 5½ ft. & 7 ft. extra heavy duty standards)

Figure 1

The feed to the Standard Cone is normally scalped if over 10 to 15 percent of the desired crusher product is present in the new feed. Since the range of feed size will often vary from approximately 300 MM to 5 MM (12 In. to 2 In.) in a given installation, the crushing cavity must be selected to reduce these sizes efficiently. Where the majority of the top size of the feed is only 200 MM (8 In.) the crushing cavity would be modified to reflect this smaller feed size. To meet these variations in size of feed and desired product size, Standard Cones can be equipped with various designs of fine, medium, coarse, or extra coarse crushing cavities.

In selecting the crushing cavity, care is taken to obtain a design that will permit the top size of the feed to enter the crusher in sufficient volume throughout the entire wear life of the liners. Since the most efficient crushing action takes place when the feed is reduced in a series of four or five blows as it passes through the cavity, it becomes important to select a design that will effect some reduction in the upper portion of the cavity as well as the lower or parallel sizing zone. In other words, too large an opening for the feed will prevent crushing in the upper zones and can develop excessive power draw. Since the crusher feed rate is best governed by the power draw, an incorrect cavity can reduce the crusher capacity and, on occasion, create unnecessary maintenance. On the other hand, if the cavity will only accept the feed when the liners are new, but as the liners wear the feed opening closes off reducing the feed rate, a more efficient cavity is required. Since no two ores or crushing applications are exactly the same, a large number of cavity designs have become available through the years. Therefore, the optimum in cavity design is one which permits close-to-the-rated power draw throughout the entire wear life and results in a scrap loss of between 25 and 40 percent, depending on crusher type and size.

With the proper cavity design, maximum efficiency will be obtained if the feed is evenly distributed around the entire crushing cavity. Normally, the wide feed size range required in the Standard or secondary crushing stage is such that the crushing cavity is required to have a receiving opening considerably larger than the average portion of the feed. For best results, the feed should be regulated based on the crusher motor kilowatt draw. When handling the harder materials, it is likely that the crushing cavity in the Standard Cone crusher will not be entirely filled up or "choked". Depending on the size and type of secondary crusher, normal kilowatt fluctuations during crushing may approach 50 to 70 kilowatt, depending on the feed analysis and method of feed control. However, maximum capacity will be obtained under these conditions when cavity design and proper feed distribution methods are observed.

Short Head (Tertiary) Cone Crusher (Refer to Fig. 2)

The Short Head Cone crusher is normally applied as a third stage crusher in plants designed for three or four stages of crushers. However, the wide range of crushing cavities available in the Short Head crusher permits its installation as a secondary crusher in some cases.

Feed opening and product size ranges for the various models of Short Head Cone crushers are as follows:

Maximum Feed Opening -- 250 MM (10 In.)

Finished Product Range -- 25 MM to 3 MM (1 In. to 1/8 In.)

Reduction Ratio -- 4 or 6:1

The feed to the Short Head Cone crusher is normally screened prior to the crusher to remove the finished product sizes and to provide void space for the crushed particles produced in the cavity. Maximum production will be obtained when the crusher operates at or near full horsepower load continuously. To achieve this condition, plant design, feed distribution and the type of crushing cavity are factors which warrant considerations. The schematic arrangement shown in Figure 3 permits the crusher to operate with a controllable feed rate. The circuit shown in Figure 3 offers the optimum in crusher feed control and automation. The new feed is withdrawn from the bin directly to the crusher assuring a steady, controllable feed rate. In a properly fed crushing cavity the material is distributed in a manner which utilizes the entire circumferential area of the receiving opening. It is equally important to have the finer gradations in the feed interspersed with the coarse feed.

To meet variations in feed size and product requirements, the Short Head Cone can be equipped with various designs of fine, medium, coarse or extra coarse crushing cavities. Essentially, the same considerations for selecting the crushing cavity should be used in this stage as was described for the Standard Cone crusher.

When the feed material is relatively nonabrasive, the crushing cavity selection is usually made to permit a condition where the entire cavity is filled up or "choked". A certain degree of self-regulation takes place under these conditions.

When crushing a material which is both hard and abrasive, it is possible that the full motor power can be drawn without having the

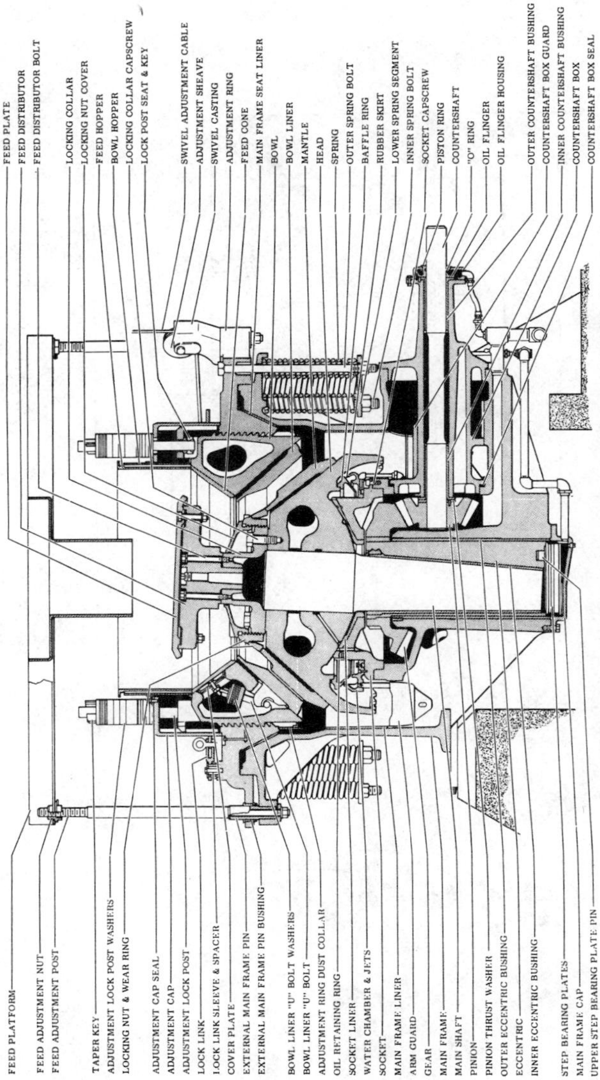

7 FT. SHORT HEAD EXTRA HEAVY DUTY GENERAL ASSEMBLY
(water seal - typical arrangement 5½ ft. & 7 ft. heavy duty short heads)

Figure 2

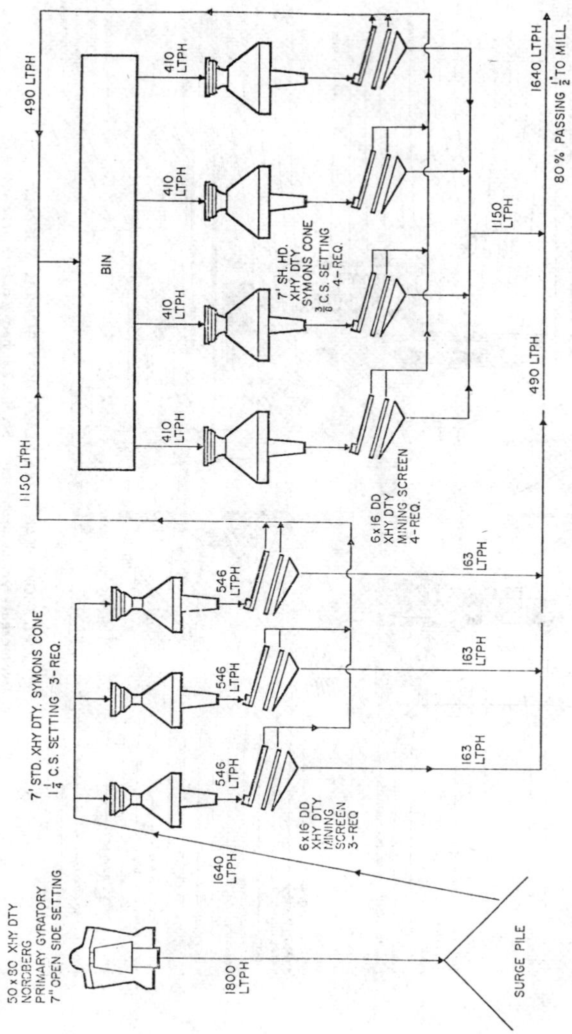

3-STAGE CRUSHING FLOWSHEET
FOR ROD MILL FEED

Figure 3

crushing cavity entirely filled up or "choked". As the crusher liners wear, the condition will develop where a "choked" cavity can be observed.

The normal 10 to 30 kilowatt fluctuations in the tertiary position are much less than those observed in the secondary due to better control of feed consist and control.

Refer to Tables 6 through 8 for capacities and product analyses for secondary and tertiary Cone crushers.

GRADATION **CONE CRUSHERS**

Table 6

SCREEN ANALYSES
SYMONS CONE CRUSHER
"OPEN CIRCUIT"

PRODUCT	CLOSED SIDE SETTING OF CRUSHER										
SIZE	2½"	2"	1½"	1¼"	1"	7/8"	3/4"	5/8"	1/2"	3/8"	1/4"
+ 4½"	2%		These average table values will vary with the method of feeding, selection of crusher cavity, the weight, cleanliness & moisture contents of the material and its fracture pattern. Accurate values should be established by actual testing.								
-4½" + 4"	4%										
-4" + 3½"	6%	3%									
-3½" + 3"	12%	5%									
-3" + 2½"	16%	13%	4%								
-2½" + 2¼"	9%	10%	5%	2%							
-2¼" + 2"	7%	10%	7%	4%							
-2" +1¾"	8%	10%	11%	6%	3%						
-1¾" + 1½"	7%	10%	13%	12%	6%	4%					
-1½" + 1¼"	7%	9%	14%	16%	12%	7%	4%				
-1¼" + 1"	5%	8%	13%	17%	19%	17%	12%	6%			
-1" + 7/8"	2%	3%	6%	7%	11%	12%	11%	6%	3%		
-7/8" + 3/4"	2%	3%	5%	7%	9%	12%	13%	12%	6%		
-3/4" + 5/8"	2%	3%	4%	7%	10%	11%	14%	16%	12%	4%	
-5/8" + 1/2"	11%	2%	4%	5%	8%	10%	12%	17%	19%	12%	
-1/2" + 3/8"		11%	3%	4%	6%	8%	12%	14%	20%	24%	8%
-3/8" + 5/16"			11%	2%	3%	4%	4%	7%	10%	14%	13%
-5/16" + 1/4"				11%	2%	3%	4%	5%	8%	12%	19%
-1/4" + 3/16"					11%	12%	2%	4%	6%	12%	20%
-3/16" + 6M							12%	3%	4%	7%	12%
-6M + 10M								10%	12%	6%	11%
-10M + 14M										9%	4%
-14M + 28M											13%
-28M											
	100%	100%	100%	100%	100%	100%	100%	100%	100%	100%	100%

CONE CRUSHERS Table 7 **GRADATION**

SCREEN ANALYSES
"CLOSED CIRCUIT"

These average table values will vary with the method of feeding, selection of crusher cavity, the weight, cleanliness & moisture contents of the material and its fracture pattern. Accurate values should be established by actual testing.

PRODUCT SIZE	TOP SIZE OF PRODUCT			
	3/4"	5/8"	1/2"	3/8"
+ 5/8"	4%			
- 5/8" + 1/2"	12%	6%		
- 1/2" + 3/8"	24%	19%	8%	
- 3/8" + 5/16"	14%	15%	13%	4%
- 5/16" + 1/4"	12%	16%	19%	12%
- 1/4" + 3/16"	12%	15%	20%	24%
- 3/16" + 6M	7%	11%	16%	24%
- 6M + 10M	6%	7%	12%	21%
- 10M + 14M	2%	3%	3%	4%
- 14M + 28M	7M	8%	3%	4%
- 28M			6%	7%
	100%	100%	100%	100%

CONE CRUSHERS

Table 8A

CONE CRUSHER CAPACITY CHARTS

Cone crusher capacity charts are developed for use as an application tool to properly utilize the machine's capabilities when operated in closed circuit. Symons cone crushers have 3 separate and distinct capacities (see diagrams on following page). The difference between these is the point in the circuit at which the measure is taken in either open circuit or closed circuit operation The crusher is one component of the circuit. As such, its performance is in part dependent on the proper selection and operation of feeders, conveyors, screens, supporting structure, electric motors, drive components and surge bins. Where used, attention to the following factors will enhance crusher capacity and performance.

1. Proper selection of crushing chamber for material to be crushed.
2. A feed grading containing proper distribution of the particle sizes.
3. Controlled feed rate.
4. Proper feed distribution 360° around the crushing chamber.
5. Discharge conveyor sized to carry maximum crusher capacity.
6. Properly sized scalping and closed circuit screens.
7. Automation controls.
8. Adequate crusher discharge area.

The following factors will detract from crusher capacity and performance.

1. Sticky material in crusher feed.
2. Fines in crusher feed (smaller than crusher setting) exceeding 10% of crusher capacity.
3. Excessive feed moisture.
4. Feed segregation in crusher cavity.
5. Improper feed distribution around circumference of crushing cavity.
6. Lack of feed control.
7. Inefficient use of recommended connected horsepower.
8. Insufficient conveyor capacity.
9. Insufficient scalper and closed circuit screen capacities.
10. Insufficient crusher discharge area.
11. Extremely hard or tough material.
12. Operation of crusher at less than recommended full load countershaft speed.

The capacities shown in the tables are based on results obtained from thousands of installations worldwide, crushing the broadest range of ores, rocks and minerals to determine effect of individual conditions. Consult Rexnord Process Machinery Division.

USING THE CHARTS ON THE FOLLOWING PAGES

To determine crusher capacity for proposed installation, determine whether crusher is to be operated in open circuit or closed circuit, as shown in schematic on following page.

For Open Circuit Operation:

A. Capacities shown in column listed as "Capacity 3" lists the capacities of all size short head crushers in tons per hour passing through crusher. Recommended closed side setting for each crusher size and type of cavity is listed from left to right.

B. To determine exact tonnage, using known data of proposed feed size and tonnage, select correct size machine and proper type of crushing schematic on following page.

For Closed Circuit Operation

A. Column listed as "Capacity I" shows capacity at various settings of each size. Crusher to produce a net finished product passing the effective square opening of the screen.

B. Using known data on proposed feed size and tonnage required, select correct size machine and proper type of crushing cavity.

Crushing Cavity Selection:

For **preliminary** selection of crushing cavity, proposed feed grading top size should pass a square opening not to exceed 90% of "D" dimension at open side.

Note: Capacities shown in column listed as **"Capacity 2"** are provided to assist in the proper sizing of conveyors, feeders and screens. These capacities are based on an average screen efficiency of 90%.

CONE CRUSHERS

Table 8B

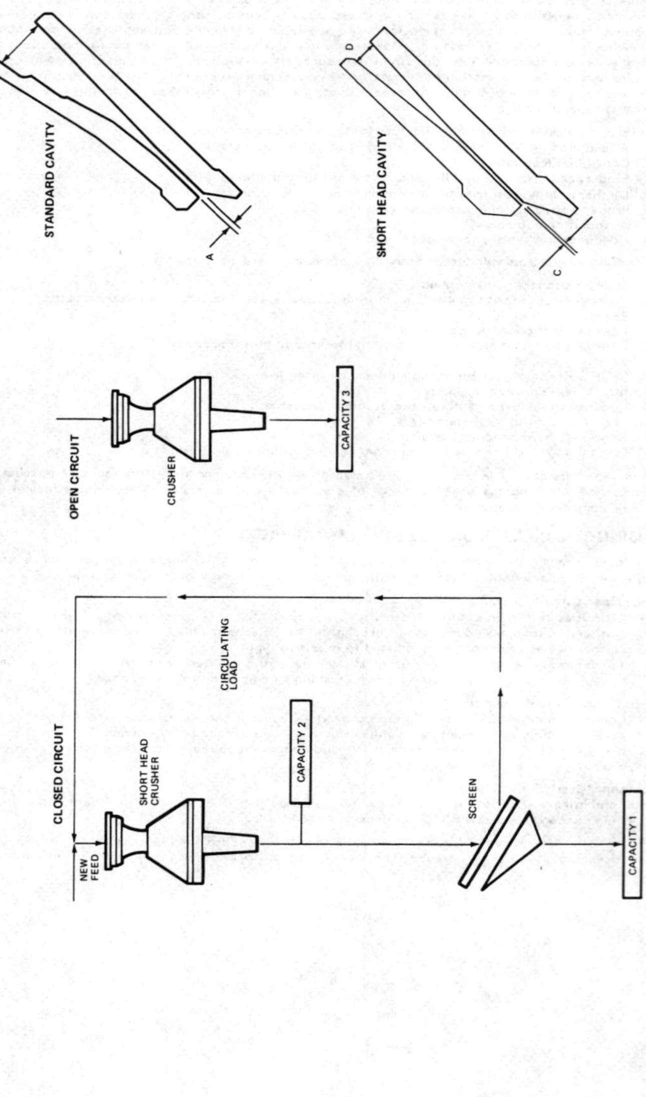

Table 8C

STANDARD SYMONS CONE CRUSHERS—OPEN CIRCUIT / Cavities Feed Openings Product Sizes Capacities*

Size	Type of Cavity	Recommended Minimum Discharge Setting A	Feed Opening With Minimum Recommended Discharge Setting A		Capacities in Tons (2000 lb.) Per Hour Passing Through the Crusher at Indicated Discharge Setting A										
			B Closed Side	B Open Side	1/4" (6 mm)	3/8" (10 mm)	1/2" (13 mm)	5/8" (16 mm)	3/4" (19 mm)	7/8" (22 mm)	1" (25 mm)	1 1/4" (32 mm)	1 1/2" (38 mm)	2" (51 mm)	2 1/2" (64 mm)
2 ft.	Fine	1/4"	2 1/4"	2 3/4"	18	20	25	30	35	40	45	50	60		
	Coarse	3/8"	3 1/4"	3 3/4"		20	25	30	35	45	50	60	75		
	Extra Coarse	1/2"	3 1/2"	4"			25	30	40	50	55	70	80		
3 ft.	Fine	3/8"	3 3/8"	4 1/8"		40	50	60	70	75	80				
	Coarse	1/2"	4 7/8"	5 11/16"			50	60	75	90	100	120	140		
	Extra Coarse	3/4"	6 7/8"	7 1/2"					75	90	100	120	140		
4800	Fine	3/8"	5"	5 5/8"		70	90	110	130	140	150	170			
	Medium	1/2"	6 1/8"	6 3/4"			100	110	140	150	175	200	220		
	Coarse	3/4"	7 3/8"	8 1/4"					145	165	190	220	270	320	
	Extra Coarse	3/4"	8 15/16"	9 3/4"					145	175	200	230	280	330	
5100	Fine	1/2"	4 1/2"	5 3/8"			120	140	150	160	175	200			
	Medium	5/8"	7 3/8"	8 1/4"				140	175	195	190	240	265		
	Coarse	3/4"	9"	10"					175	210	225	265	325	385	
	Extra Coarse	1"	10 3/8"	11 1/4"							240	275	335	395	
5 1/2 ft.	Fine	5/8"	7 1/8"	7 3/4"				180	200	235	275	300	350		
	Medium	7/8"	8 5/8"	9 1/2"						275	300	375	400	450	
	Coarse	1"	9 7/8"	10 7/8"							300	375	450	500	700
	Extra Coarse	1 1/2"	13 1/2"	14 1/2"									450	500	800
7 ft.	Fine	3/4"	10"	11"					370	400	500	620	750		
	Medium	1"	11 1/2"	12 3/4"						500	600	750	800	1100	
	Coarse	1 1/4"	13 1/2"	14 7/8"								750	850	1200	1400
	Extra Coarse	1 1/2"	16 3/4"	18 1/8"									850	1200	1400
10 ft.	Fine	7/8"	12 1/2"	14"						900	1300	1650			
	Medium	1"	15 1/2"	17"							1400	1650	1800		
	Coarse	1 1/2"	18 1/2"	20"									1900	2500	3000
	Extra Coarse	1 1/2"	24 1/2"	26"									1900	2500	3000

CONE CRUSHERS

Table 8D

SHORT HEAD CAPACITIES

EFFECTIVE SQUARE OPENING OF SCREEN CLOTH ON CLOSED CIRCUIT SCREEN.

CLOSED SIDE SETTING in shaded areas, recommended to control the circulating load in closed circuit operation, normally less than closed circuit screen opening.

CAPACITY 1: Net finished product (screen undersize) CAPACITY 2: Total TPH passing through crusher, net finished product plus circulating load. CAPACITY 3: Approximate TPH passing through crusher in open circuit operation. See circuit schematic.

Size	Cavity	Minimum Discharge Setting C	Feed Opening D At Min. Discharge Setting C: D Closed Side	D Open Side	3/16": Closed side setting	1	2	3	1/4": Closed side setting	1	2	3	3/8": Closed side setting	1	2	3	1/2": Closed side setting	1	2	3	5/8": Closed side setting	1	2	3	3/4": Closed side setting	1	2	3	1": Closed side setting	1	2	3
2 ft.	Fine	1/8"	3/4"	1 3/8"	1/8"	10	20	19	3/16"	13	26	23	1/4"	15	28	25	5/16"	25	30	26	3/8"	30	40	35								
	Coarse	3/16"	1 1/2"	2"	3/16"	10	26	23	3/16"	13	26	23	1/4"	15	28	25	3/8"	25	40	34	7/16"	35	45	40								
3 ft.	Fine	1/8"	1/2"	1 5/8"	1/8"	20	50	48	3/16"	30	58	54	1/4"	40	60	58	5/16"	55	75	64	7/16"	60	80	70	5/8"	65	85	80				
	Medium	1/8"	1"	2"	1/8"	20	50	48	3/16"	30	60	58	1/4"	40	62	60	3/8"	55	80	70	1/2"	60	90	77	5/8"	70	95	80	3/4"	80	100	85
	Coarse	1/4"	2"	3"					1/4"	35	75	68	1/4"	45	75	68	3/8"	60	90	77	1/2"	90	105	95	5/8"	75	110	100	3/4"	100	120	105
4800	Fine	3/16"	1 7/8"	3"	3/16"	30	80	72	3/16"	40	80	72	1/4"	65	90	81	3/8"	85	120	102	1/2"	100	130	111								
	Medium	5/16"	2 1/2"	3 1/2"									1/4"	65	105	95	3/8"	85	130	111	1/2"	100	130	115	5/8"	125	145	135	3/4"	150	180	165
	Coarse	1/2"	2 3/4"	4"													3/8"	90	135	115	1/2"	115	175	149	5/8"	145	200	170	3/4"	170	210	180
5100	Fine	3/16"	1 1/8"	2 1/2"	3/16"	45	100	90	3/16"	55	100	90	1/4"	80	110	99	3/8"	105	150	128	1/2"	125	160	136								
	Medium	3/16"	1 5/8"	3"	3/16"	45	105	92	1/4"	55	120	108	1/4"	80	120	108	3/8"	105	160	136	1/2"	125	170	145	5/8"	150	180	155				
	Coarse	5/16"	2 3/4"	4"									5/16"	80	130	115	3/8"	110	165	140	1/2"	140	205	175	5/8"	175	240	192	3/4"	200	240	210
	Extra Coarse	5/8"	4 3/4"	5 1/2"																	5/8"	140	210	180	5/8"	175	240	200	3/4"	200	250	220
5 1/2 ft.	Fine	3/16"	1 3/8"	2 3/4"	3/16"	65	130	124	3/16"	90	130	124	1/4"	135	200	180	3/8"	175	230	196	1/2"	210	240	204								
	Medium	1/4"	2 1/8"	3 1/2"					3/16"	90	165	155	1/4"	135	200	180	3/8"	175	230	196	1/2"	210	250	213	5/8"	245	270	216	3/4"	280	300	240
	Coarse	3/8"	3 3/4"	5 1/4"									3/8"	140	210	220	3/8"	175	250	220	1/2"	220	310	280	5/8"	250	350	300	3/4"	320	360	320
	Extra Coarse	1/2"	6"	7 1/4"													1/2"	175	260	230	5/8"	230	350	290	5/8"	250	350	305	3/4"	320	360	325
7 ft.	Fine	3/16"	2"	3 3/4"	3/16"	120	240	230	3/16"	160	240	230	1/4"	280	390	350	3/8"	315	436	390	7/16"	360	450	383	5/8"	420	500	400				
	Medium	3/8"	3 7/8"	5 3/4"									3/8"	280	430	385	3/8"	315	440	395	1/2"	360	500	425	5/8"	420	550	470	3/4"	500	570	520
	Coarse	1/2"	5"	7"													1/2"	315	500	468	1/2"	360	500	468	5/8"	450	560	500	3/4"	550	600	560
	Extra Coarse	5/8"	6 1/4"	8 1/4"																	5/8"	360	620	510	5/8"	450	620	510	3/4"	550	650	600
10 ft.	Fine	1/4"	3"	5"									1/4"	500	800	720	3/8"	670	1000	850	7/16"	850	1100	935	5/8"	900	1200	960	3/4"	1150	1450	1160
	Medium	1/2"	4"	6"													3/8"	670	1000	850	7/16"	850	1100	935	5/8"	900	1200	960	3/4"	1150	1450	1160
	Coarse	5/8"	7"	9"																	7/16"	850	1100	935	5/8"	950	1300	1040	3/4"	1250	1550	1350
	Extra Coarse	3/4"	8"	10"																					5/8"	950	1300	1040	3/4"	1250	1550	1400

CONE CRUSHERS

Gyradisc® (Quarternary) Stage (Refer to Fig. 4)

The Gyradisc is applied as a fourth or final stage crusher. The Gyradisc normally handles feed of less than 50 MM (1 to 1/2 In.).

Feed and product size ranges for the Gyradisc are as follows:

Feed Size -- 75 MM (3 In.) Maximum

Finished Product Range -- 12 MM (1/2 In. to #20)

Gyradisc crushers are specialized reduction machines. They are designed to economically produce large quantities of cubical product from stone, gravel, ores, and nonmetallic minerals. The feed for Gyradisc crushers usually has a top size not larger than 75 MM (3 In.) for the larger units and 30 MM (1-1/4 In.) for the smaller size crushers.

Gyradisc crushers differ from conventional cone-type crushers because the comminution of material is achieved by a reduction process called Inter-Particle Comminution. This principle of reduction utilizes a combination of impact and attrition of a multi-layered mass of particles.

For proper operation, the crusher always operates with a choke level in the cavity. A controlled feed rate permitting a constant power draw is required for an efficient circuit.

Through proper application of the Gyradisc crusher, mill feeds of 100 percent passing 3, 4, and 6 mesh are attainable. The crushing action of the Gyradisc crusher results in circulating loads of less than 50 percent of the new feed rate when closed circulating on 3 mesh, to 150 percent when closing the circuit on 6 mesh.

One 84" Gyradisc crusher is rated to produce 180 tons per hour of 3 mesh screen undersize. The flow sheet, Figure 5, shows the equipment required to produce the finer mill feed.

Using the arrangement shown in the flow sheet, mill feed rates are increased by a minimum of 20 percent over the rates possible when feeding the nominal 12 MM (1/2 In.) size. The work index and the final grind, however, directly influence the degree of feed rate improvement.

Refer to Tables 9 through 11 for capacities and product analyses for Gyradisc crushers.

® Gyradisc is a registered trademark of Rexnord, Inc.

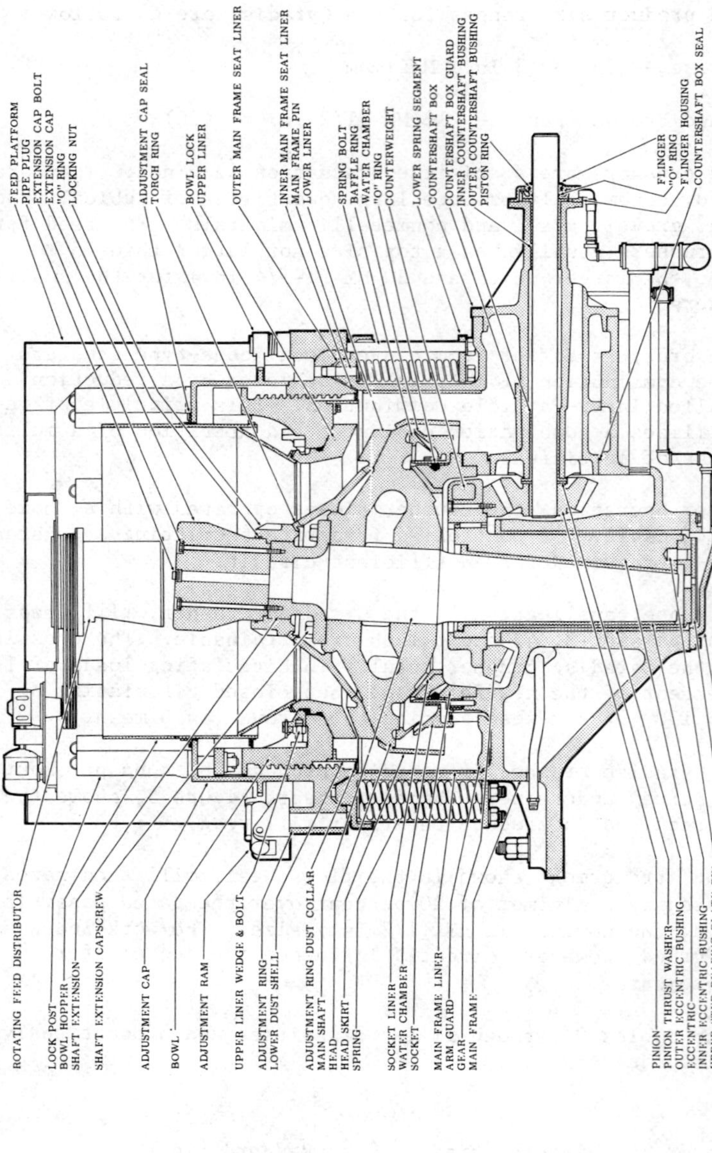

84" GYRADISC
(water seal - typical arrangement)

Figure 4

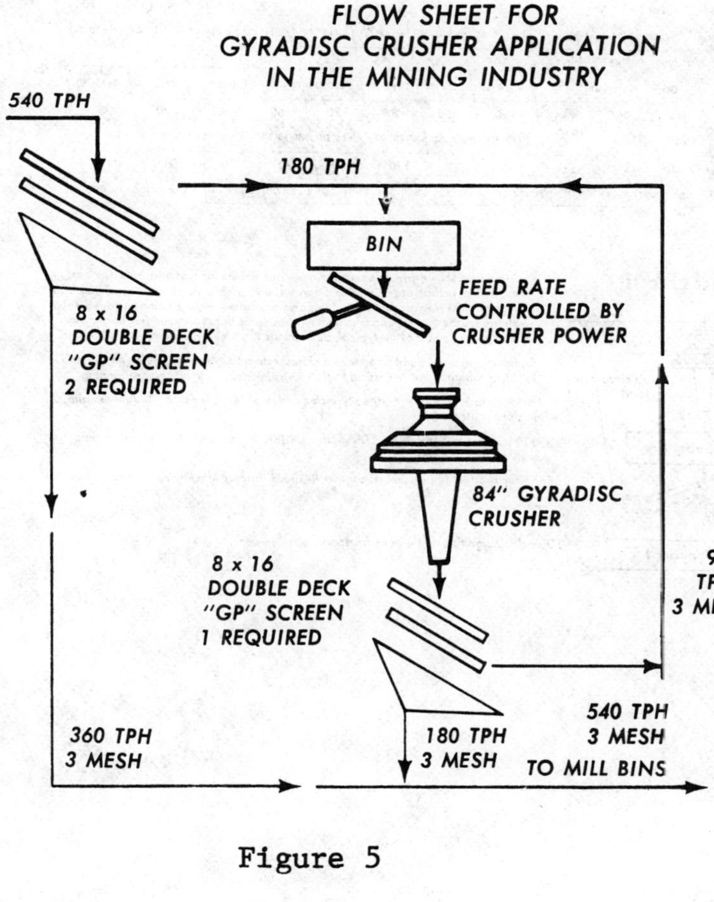
FLOW SHEET FOR
GYRADISC CRUSHER APPLICATION
IN THE MINING INDUSTRY
540 TPH
180 TPH
BIN
FEED RATE
CONTROLLED BY
CRUSHER POWER
8 x 16
DOUBLE DECK
"GP" SCREEN
2 REQUIRED
84" GYRADISC
CRUSHER
8 x 16
DOUBLE DECK
"GP" SCREEN
1 REQUIRED
90
TPH
3 MESH
540 TPH
3 MESH
360 TPH
3 MESH
180 TPH
3 MESH
TO MILL BINS

Figure 5

Table 9

Capacity table based on closed circuit operation

GYRADISC CRUSHER DESIGNATION		HEAD DIAM. INCHES (MILLI-METERS)	CAPACITY 1 TOTAL TPH THRU CRUSHER	CAPACITY 2 IN TONS (2000 LBS.) PER HOUR (SCREEN UNDERSIZE) — RECOMMENDED SCREEN SIZE									
ENGLISH	METRIC			.375" (9mm)	.250" (6mm)	.185" (5mm) U.S. NO. 4 4 MESH	.131" (3mm) U.S. NO. 6 6 MESH	.093" (2mm) U.S. NO. 8 8 MESH	.065" (1.6mm) U.S. NO. 12 10 MESH	.046" (1mm) U.S. NO. 16 14 MESH	.0328" (.833mm) U.S. NO. 20 20 MESH	ASTM C-33 SPEC SAND
36" GD	900 GD	36" (914mm)	65-75	60	50	35	30	23	18	15	10	20-25
				4'x12' SD	4'x12' SD	4'x14' SD	5'x14" DD	5'x14' DD	6'x14' DD	6'x20' DD	6'x20' DD	
48" GD	1200 GD	48" (1219mm)	115-130	105	80	55	45	40	30	25	17	34-40
				4'x12' SD	5'x14' SD	6'x16' SD	6'x20' DD	6'x20' DD	8'x20' DD	8'x20' DD	2 Req. 6'x16' DD	
66" GD	1650 GD	66" (1676mm)	190-210	160	120	85	70	60	45	40	25	55-60
				5'x12' SD	6'x20' SD	8'x20' SD	2 Req. 6'x16' DD	2 Req. 6'x16' DD	2 Req. 6'x20' DD	3 Req. 6'x16' DD	2 Req. 8'x20' DD	
84" GD	2100 GD	84" (2134mm)	265-300	250	180	130	100	90	70	60	40	80-70
				6'x14' SD	8'x20' SD	2 Req. 6'x20' SD	2 Req. 6'x20' DD	2 Req. 6'x20' DD	2 Req. 8'x20' DD	2 Req. 8'x20' DD	4 Req. 6'x16' DD	

Closed circuit

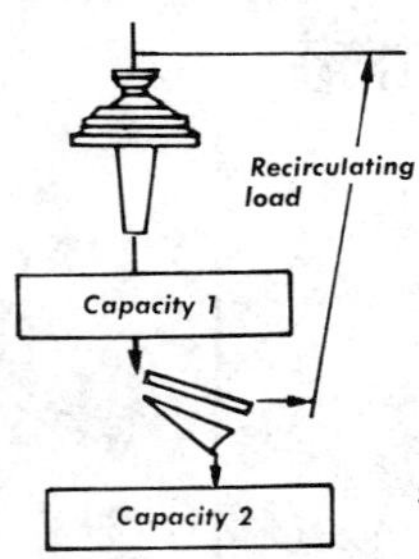

Capacities shown are neither maximum nor minimum, and are based on results secured in actual practice. The figures shown apply to short tons of material weighing 100 pounds per cubic foot, and are based on a properly graded feed with the crusher consuming a minimum of 80% of the recommended horsepower. A feed size of minus 3/4" is typical, but a minus 1 1/2" feed can be readily accepted.

Optimum capacity is obtained with a closed side setting ranging between 1/4" and 1/2". The setting is not necessarily related to the closed circuit screen opening.

Factors which may affect capacity include, size of feed, friability, toughness and moisture content.

Figures shown under column "TPH THRU CRUSHER" are included for calculating screen areas.

For finer product requirements, contact Rexnord Process Machinery Division.

SD and DD signifies single deck or double deck screen.

PRODUCTION CURVES **GYRADISC CRUSHERS**

Table 10

SIEVE SIZE (INCHES)	MESH (TYLER)	U.S. NO.	CUMULATIVE PER CENT RETAINED		
			NEW FEED	CRUSHER PRODUCT BEFORE SCREENING	FINISHED PRODUCT THRU SCREEN
1.48 (38mm)					
1.06 (27mm)					
.742 (19mm)					
.525 (13mm)			47	1	
.371 (9mm)			81	7	
.263 (7mm)	3		91	21	
.185 (5mm)	4	4	97	36	
.131 (3mm)	6	6	99	51	19
.093 (2mm)	8	8		62	42
.065 (1.6mm)	10	12		71	55
.046 (1mm)	14	16		77	64
.0328 (.833mm)	20	20		83	73
.0232 (.589mm)	28	30		86	78
.0164 (.416mm)	35	40		89	83
.0116 (.294mm)	48	50		92	88
.0082 (.208mm)	65	70		93	89
.0058 (.147mm)	100	100		95	92
.0041 (.104mm)	150	140		96	94
.0029 (.073mm)	200	200		97	95

Feed: Copper Ore
Product: Ball Mill Feed

STANDARD SCREEN ANALYSIS

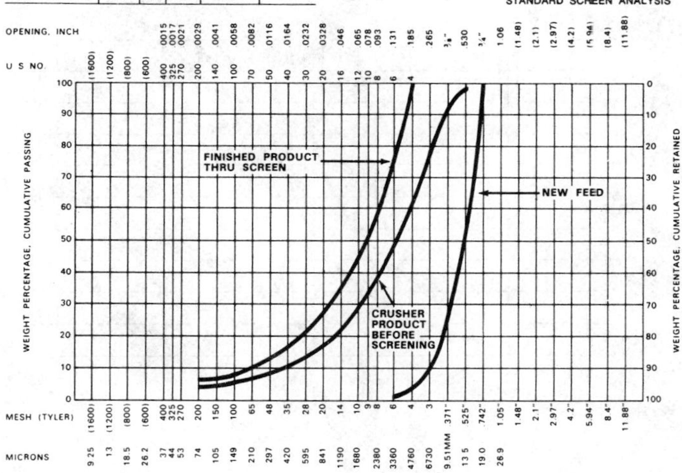

GYRADISC CRUSHERS **PRODUCTION CURVES**

Table 11

SIEVE SIZE (INCHES)	MESH (TYLER)	U.S. NO.	CUMULATIVE PER CENT RETAINED		
			NEW FEED	CRUSHER PRODUCT BEFORE SCREENING	FINISHED PRODUCT THRU SCREEN
1.48 (38mm)					
1.06 (27mm)					
.742 (19mm)			1		
.525 (13mm)			8		
.371 (9mm)			21	8	
.263 (7mm)	3		49	16	
.185 (5mm)	4	4	72	27	
.131 (3mm)	6	6	84	42	
.093 (2mm)	8	8	92	55	19
.065 (1.6mm)	10	12	97	65	39
.046 (1mm)	14	16		72	52
.0328 (.833mm)	20	20		77	59
.0232 (.589mm)	28	30		81	65
.0164 (.416mm)	35	40		85	70
.0116 (.294mm)	48	50		87	75
.0082 (.208mm)	65	70		90	80
.0058 (.147mm)	100	100		92	85
.0041 (.104mm)	150	140		94	89
.0029 (.073mm)	200	200		96	92

Feed: Iron Ore
Product: Ball Mill Feed

STANDARD SCREEN ANALYSIS

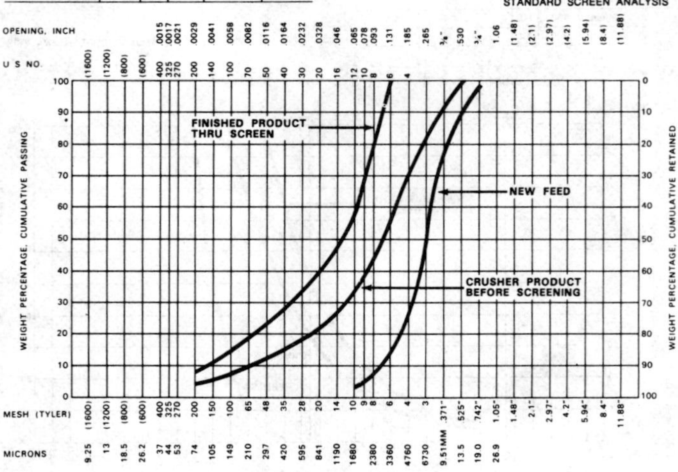

FINE CRUSHING PLANT FLOW SHEETS

The trend in design of fine crushing plants is in the direction of the horizontal arrangement with all crushers on a single floor. Conveyors from the coarse ore bin or stockpile feed scalping screens are usually double decked. When the feed to the fine crushing plant contains less than 15 percent passing 19 MM (3/4 In.) and less than three percent moisture, the scalping screens ahead of the secondary crushers can be eliminated without adverse effects to the crusher.

This practice has been followed successfully in several Minnesota taconite installations. The secondary crusher product is screened and conveyored to bins ahead of the tertiary crushers. The tertiary crushers operate in either open or closed circuit with a variety of arrangements available in positioning the tertiary screens. The screens can be placed above or below the tertiary crushers or in a separate screenhouse. With the advent of automatic crusher feed control systems it becomes important to reduce the reaction time between the feed control mechanism, such as a vibrating or belt feeder, and the crusher to a minimum.

Where secondary and tertiary crushers are stacked with a scalping screen between the two, surge and feed control for the tertiary crusher becomes more difficult. This results in lower efficiency from a power utilization standpoint. The initial cost of construction for this type of plant is usually lower than the aforementioned horizontal layout.

The positioning of the secondary and tertiary crushers depends on a variety of factors including terrain, moisture level in the ore, storage capacity both before and after fine crushing, and beneficiation steps.

Usually the fine crushing plant is located some distance from the primary with an intermediate ore storage facility between. The operation of the fine crushing plant is independent, to a large degree, of the performance of the primary crusher.

A great deal of thought has been given to the problem of attaining maximum efficiency in this type of operation. Efficiency is measured by the work of reduction on the ore per unit of time and by percent running time. The work of reduction is represented by tonnage of throughput and degree of size reduction.

To attain the maximum work of reduction, each crusher must be kept at full capacity. The capacity may be limited volumetrically in the case of softer ores, or by the horsepower that can safely be applied to the crusher in the case of harder ores.

The use of hydraulic mechanisms for monitoring and controlling fine crusher settings from a central control room gives the operator a means for maintaining overall crushing plant efficiency. All that is needed for complete automatic control is the proper coupling of the power draft and volumetric controls of one stage of crushing to the positioning mechanisms of the preceding stage to insure the best balance between successive stages of crushing.

Figures 6 through 9 show typical crushing plant flow sheet arrangements.

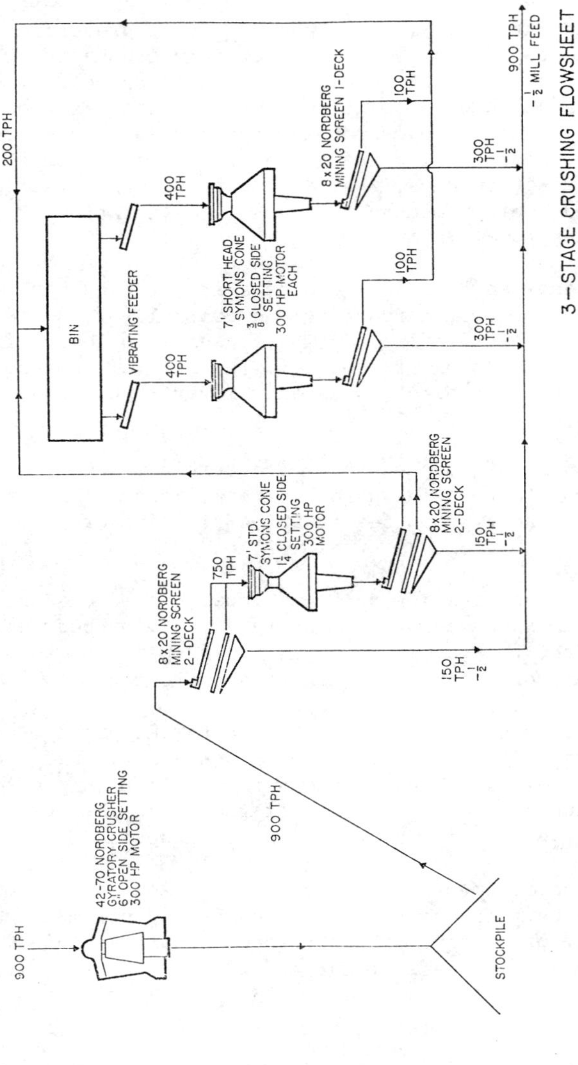

Figure 6

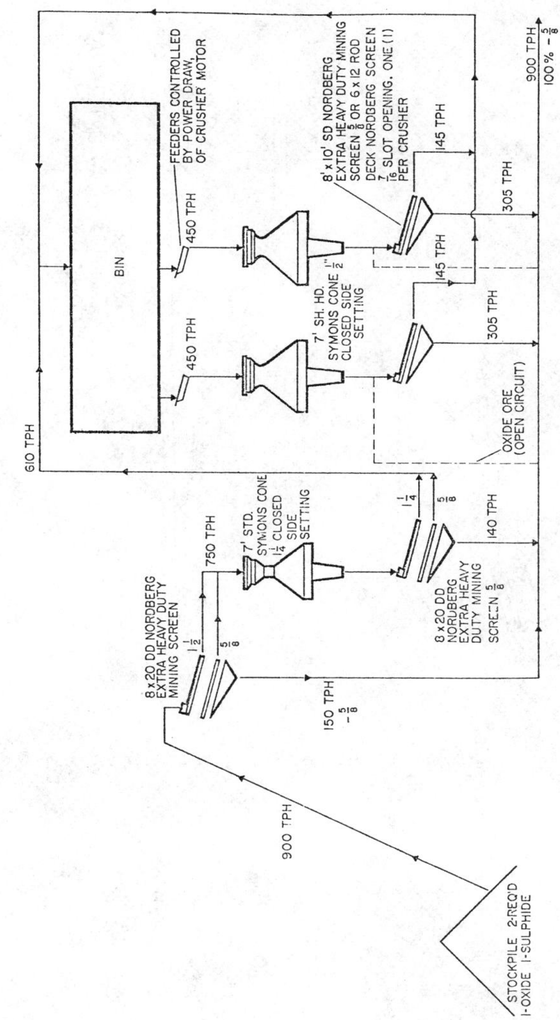

Figure 7

SECONDARY AND TERTIARY
FLOWSHEET FOR ROD MILL FEED

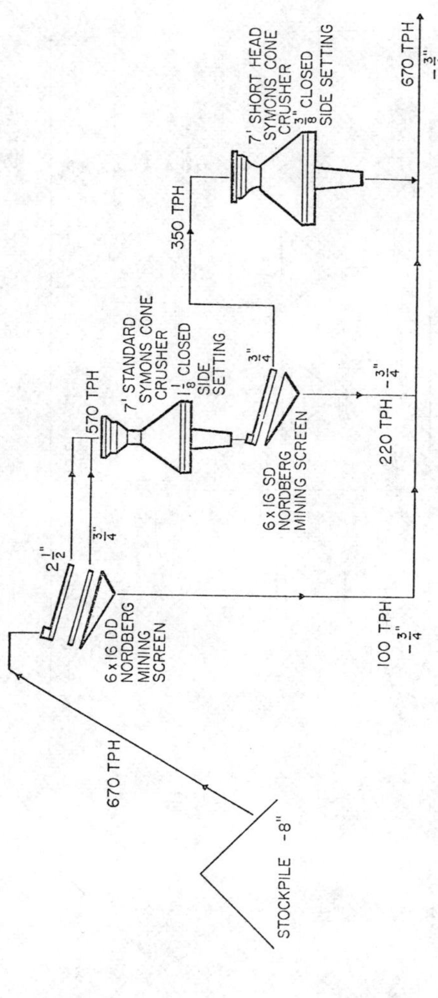

FLOWSHEET FOR SECONDARY AND TERTIARY CRUSHING

Figure 8

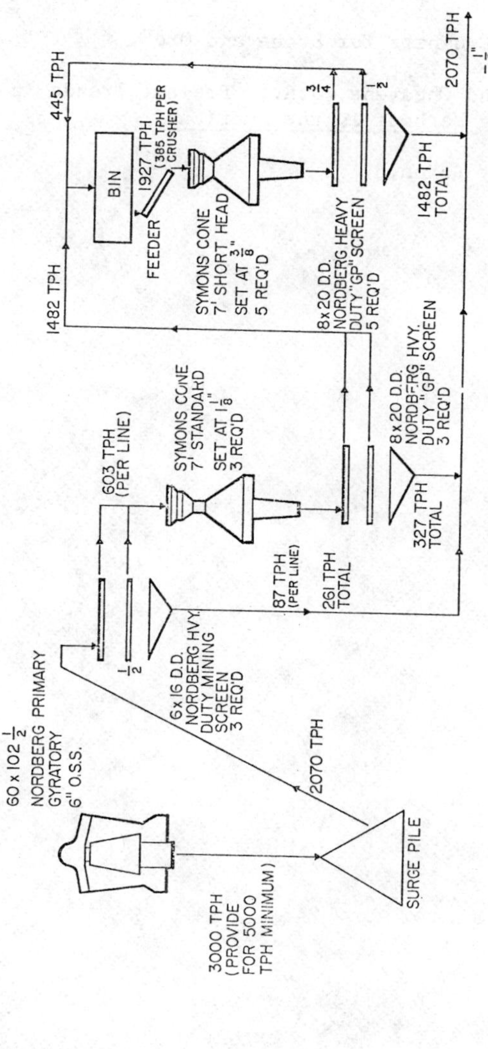

Figure 9

3 STAGE CRUSHING FLOWSHEET
FOR SINGLE STAGE BALLMILLING

REFERENCES

1. Miller, T.W., "Crushers for Stone and Ore".

2. Weiss, Norman and Cheavens, J.H., "Present Trends in Mill Design", Milling Methods in the Americas.

3. Data supplied by Rexnord.

Chapter 12

ROD AND BALL MILLS

By: Chester A. Rowland, Jr. and David M. Kjos, Mgr.

Chester A. Rowland, Jr.
Sr. Process - Project Engineer
Process Development Engineering
Mining Systems Division
Allis-Chalmers Corporation

David M. Kjos, Mgr.
Grinding Mill
& Special Eqpt. Application
Mining Systems Division
Allis-Chalmers Corporation

INTRODUCTION

Mineral ore comminution is generally a feed preparation step for subsequent processing stages. Grinding, the fine product phase of comminution, requires a large capital investment and frequently is the area of maximum usage of power and wear resistant materials.

Grinding is most frequently done in rotating drums utilizing loose grinding media, lifted by the rotation of the drum, to break the ores in various combinations of impact, attrition and abrasion to produce the specified product. Grinding media can be the ore itself (autogenous grinding - primary and secondary), natural or manufactured non-metallic media (pebble milling) or manufactured metallic media - steel rods, steel or iron balls. This chapter covers rod and ball mills which utilize manufactured metallic grinding media. Fig. 1 shows the cross section of an overflow rod mill, Fig. 2 an overflow ball mill, and Fig. 3 a diaphragm (grate) discharge ball mill.

MILL DESIGN

The interior surface of rod and ball mills exposed to pulp and/or grinding media are protected from wear and corrosion by rubber, metallic or a combination of rubber and metallic wear resistant materials.

Rod and ball mills essentially draw constant power, thus are well suited for use of synchronous motors with power factor correction capabilities as drive motors. A net of approximately 120 to 130 percent of running torque is required to cascade the charge in these mills. The pull-in torque is about 130 to 140 percent with the pull-out torque to keep the motor in-step (in-phase) generally in excess of 150 percent. When rod and ball mill are started across-the-line the starting and pull-in torques result in inrush currents exceeding 600

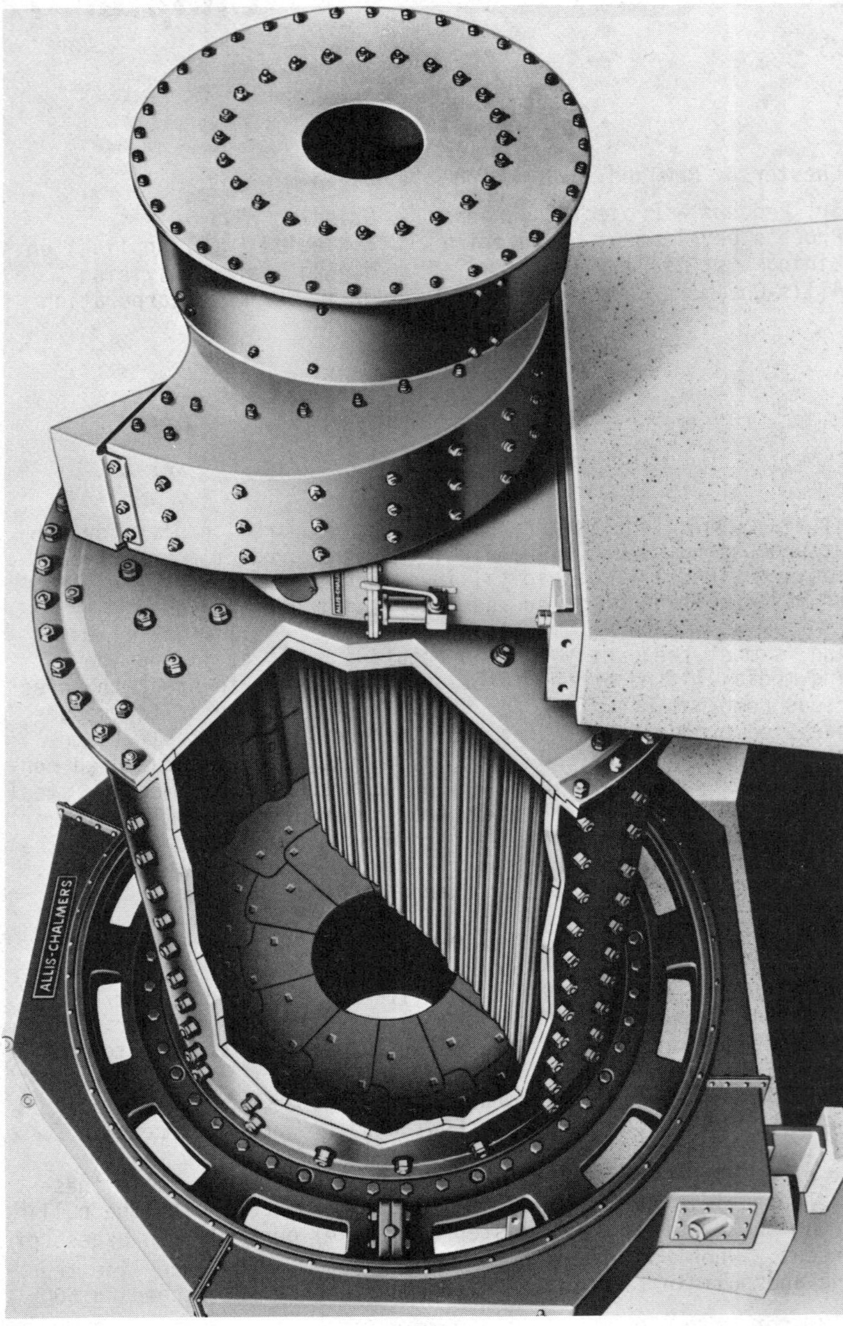

Fig. 1 - Overflow Rod Mill

Fig. 2 - Overflow Ball Mill

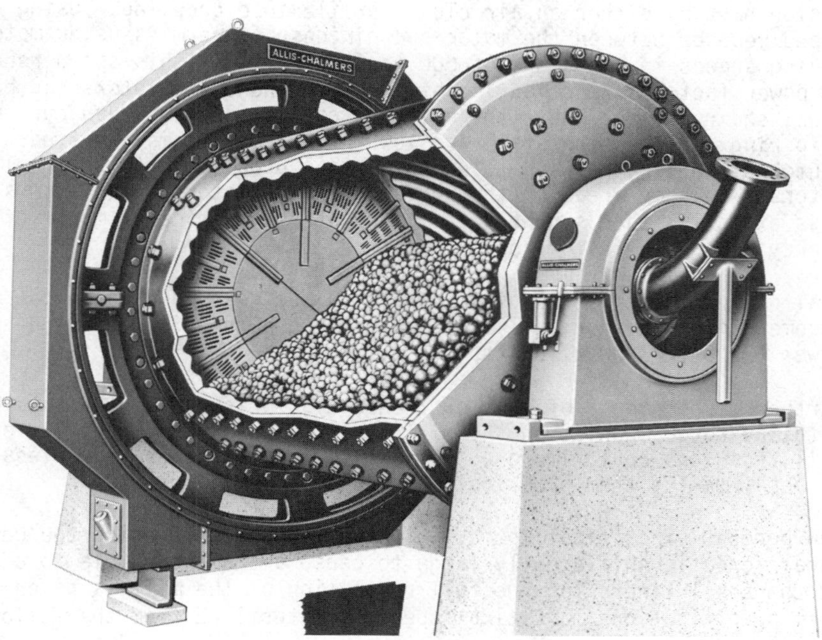

Fig. 3 - Diaphragm Ball Mill

percent which result in possibly high voltage drops. To deliver 130 percent starting torque to the mill the motor design must take into account the maximum anticipated voltage drop. Motor torque decreases as the decimal fraction of the voltage available squared. E.g., a motor rated 160% starting torque with a 10% system voltage drop will deliver 160% x $\left(\frac{100\%-10\%}{100}\right)^2$ or 129.6% torque to its output shaft.

When it is not possible or practical to start a fully loaded synchronous motor across-the-line it is possible to utilize the motor's pull-out torque to start the mill. By using a clutch, normally an air clutch, between the motor and the mill, the motor is brought up to synchronous speed before the clutch is energized. If the motor has an adequate amount (175 or greater) of pull-out torque the pull-out torque starts the mill without major disruptions on the electrical system.

Since the energy release at initial cascade of the mill charge is an inverse function of acceleration time, a minimum acceleration time of 6 to 10 seconds or more is recommended to prevent damage to the mill or the mill foundation.

Economics at the time of plant design and mill purchase determine the drive to be used. The simpliest drive is the low speed synchronous motor with speeds in the range of 150 to 250 RPM connected to the mill pinionshaft by either an air clutch or flexible coupling. Using a speed reducer between the motor and pinionshaft permits using motors having speeds in the range of 600 to 1000 RPM. In this speed range, if power factor correction is not required induction motors can be used; squirrel cage where there is no restriction on inrush current; slip ring where a slow start and low inrush current is required. Air clutches can also be used to ease starting problems with squirrel cage motors. In some areas of the world induction motors and starters are less expensive than synchronous motors at a sacrifice of motor efficiency and power factor correction.

Dual drives, that is two pinions driving one gear mounted on the mill, become economical for ball mills drawing more than 3500 to 4000 horsepower (2600 to 3000 kilowatts).

Further developments of the low frequency, low speed synchronous motors with the rotor mounted on the mill shell or an extension of one of the mill trunnions could improve the cost picture for these "gearless drives", making them practical for large ball mills.

The percent of critical speed, which is the speed at which the centrifugal force is sufficiently large to cause a small particle to adhere to the shell liners for the full revolution of the mill is given in mill specifications. Critical speed is determined from the following:

$$Cs = \frac{42.305}{\sqrt{D}} \qquad (1)$$

Where D is mill diameter inside liners specified in meters.
Cs is critical speed in RPM.

When D is specified in feet:

$$Cs = \frac{76.63}{\sqrt{D}} \qquad (1a)$$

Peripheral speed, which doesn't influence mill power but is a factor in liner wear and to an extent media wear, has to be considered in mill design. It can be determined by the following either as meters per minute or as feet per minute.

$$Mp = \pi DN \qquad (2)$$

where

Mp = Peripheral speed.
D = Diameter inside liners.
N = Mill speed in rpm.

To obtain some balance of critical speed and peripheral speed as mill diameters increase, the average recommended speed as percent of critical speed reduces as shown in Table I. These are guide lines for initial plant design. Actual speeds may differ from these to suite specific ore and economic conditions that apply to the specific plant.

ROD MILLS

To prevent most conditions leading to rod charge tangling, the generally recommended relationship of rod length to mill diameter inside liners is 1.4 to 1.6. When this ratio becomes less than 1.25 the risk of tangling increases rapidly. For rod mills larger than 3.8 meters (12.5 feet) in diameter rod availability and quality have to be considered. Table II gives rod length to mill diameter ratios for the larger diameter rod mills.

6.8 meters (20 feet) is a practical limit on the length of good quality rods (that is rods that will stay straight in the mill and will break into pieces that will discharge from the mill when worn). This length is a function of rod quality and production limits imposed by the suppliers. The mill length inside end liners measured along the surface of the shell liners should be 0.1 to 0.15 meters (4" to 6") longer than the rods, so that the rods will fit in the length of the grinding chamber without tipping or laying across the charge. A slope as steep as possible should be used for rod mill head (end) liners to prevent unsupported ends of rods from protruding from the charge and being broken under impact from other rods.

TABLE I

AVERAGE % OF CRITICAL SPEED

MILL DIAMETER INSIDE LINERS		% OF CRITICAL SPEED	
METERS	FEET	ROD MILLS	BALL MILLS
0.91-1.83	3-6	76-73	80-78
1.83-2.74	6-9	73-70	78-75
2.74-3.66	9-12	70-67	75-72
3.66-4.57	12-15	67-64	72-69
4.57-5.49	15-18		69-66

TABLE II

ROD MILL DIAMETER - ROD LENGTH

MILL DIAMETER INSIDE LINERS		ROD LENGTH L = 1.25 D		ROD LENGTH L = 1.4 D	
METERS	FEET	METERS	FEET	METERS	FEET
3.81	12.5	4.76	15.6	5.33	17.5
3.96	13	4.95	16.2	5.54	18.2
4.11	13.5	5.14	16.9	5.75	18.9
4.27	14	5.34	17.5	5.98	19.6
4.42	14.5	5.53	18.1	6.19	20.3
4.57	15	5.71	18.8	6.40	21.0
4.72	15.5	5.90	19.4	6.61	21.7
4.88	16	6.10	20.0	6.83	22.4
5.03	16.5	6.29	20.6	7.04	23.1

The rod specifications given in Table III can be considered as a minimum specification. Better rod quality, which reduces breakage, allows wearing the rods to a smaller size and which can reduce rod operating costs are available. The better quality rods are generally recommended when using 100 mm (4") diameter rods and/or the larger diameter rod mills.

The feed end of rods wear into a long tapered "spear-shaped" profile, while the discharge ends wear into more of a conical shape. Approximately the middle two thirds of the rod length eventually wears into an eliptical shaped section. Small pieces of broken rods can accumulate in the mill before being discharged. The tapered wear and accumulation of broken rods reduces the bulk density of the mill charge, and thus mill power. The rod charge bulk density given in Table IV can be used to determine the power a rod mill with a worn-in charge should draw. Bulk density is a variable subject to care given a rod charge, and experience indicates mill diameter also has an effect on bulk density of the worn-in charge. The larger the diameter of the rod mill the less practical "culling" of the charge becomes.

Rod mills normally carry from a 35 to 40% by mill volume rod charge, they can carry up to a 45% charge. The limits on charge level are: keeping the feed end trunnion open so feed will go into mill, and keeping the rod charge low enough so rods will not work into discharge end trunnion opening, where they can tip and cause rod tangling.

Rod mills are normally fed by spout feeders as shown in Fig. 4. A minimum head of 1.5 meters (5 feet) above the mill center line to the bottom of the feed hopper to which the feeder is attached is required to get the proper flow of feed into the mill.

Heavy duty single wave shell liners cast of either alloy steel (manganese steel is not recommended) or wear resistant alloyed cast iron are most frequently used in rod mills. The number of lifters to the circle is usually equal to approximately 6.6 D in meters (for D in feet divide 6.6 D by 3.3). These liners have 65 mm (2.5") to 90 mm (3.5") high waves above 65 mm to 75 mm (3") liners. Rubber backing can be used between the liners and shell to protect the shell from washing and corrosion. However, with rubber backing care must be taken with the liner bolt specifications and sealer assembly to assure the liners will stay tight and not move on the shell. This creates leaky liner bolts and causes the bolt holes in the shell to wear into an elongated shape. There are modifications such as the two piece liner - lifter design that can be used instead of the single wave liner. Rubber shell liners have been successfully applied in the smaller diameter rod mills running at slow speeds. When using rubber liners care must be given to using good quality rods and culling broken and thin rods from the charge. Rubber liners can help reduce the noise level emanating from a rod mill.

TABLE III

MINIMUM ROD SPECIFICATIONS

Grinding mill rods should be hard enough to remain straight throughout their entire life, yet they cannot be so brittle as to break up at coarse sizes.

When rods are too soft, they are subject to bending in the mill. Bending causes premature breakage and tanglement of rods. Tangled rods make mill cleaning difficult and hazardous, and cause costly downtime.

Material of the following chemical analysis is recommended:

CHEMICAL ANALYSIS

Carbon	0.85 to 1.03%
Manganese	0.60 to 0.90%
Silicon	0.15 to 0.30%
Sulphur	0.05% Max.
Phosphorous	0.04% Max.

PHYSICAL REQUIREMENTS

Rods should also have the following physical requirements:

. Rods are to be special commercial straightened.

. Rods are to be hot sawed to length where mill (steel) facilities permit. If hot sawing is not possible, use an abrasive cutting wheel or machine cut both ends to proper length.

. All grinding mill rods should be 152 mm (6 inches) shorter in length than the working length of the rod mill.

TABLE IV

BULK DENSITY WORN-IN ROD CHARGES

		BULK DENSITY KG PER CUBIC METER	BULK DENSITY LBS PER CUBIC FOOT
NEW RODS		6247	390
WORN-IN CHARGE			
MILL DIAMETER			
METER	FEET		
0.91-1.83	3-6	5847	365
1.83-2.74	6-9	5766	360
2.74-3.66	9-12	5606	350
3.66-4.57	12-15	5446	340

Fig. 4 - Spout Feeder

End liners are generally a thick, smooth liner cast of alloy steel. Impacting from the rod charge, which has a lateral movement in the mill, requires great caution in using wear resistant cast iron end liners. Rubber liners can be used with caution as they can be subject to damage from the sharp ends on worn rods. Except when using rubber liners there should be a rubber backing between the head liners and the heads. End liners should be smooth with no waves or lifters as these can disrupt rod action and cause rod tangling.

Rod mills can be equipped with trommels to remove broken pieces of rods and tramp oversize from the rod mill discharge. The discharge end of a rod mill can be enclosed in a housing which will help contain the noise and splash coming from the mill. A door should be provided at the end of the housing which can be opened for charging rods. Sufficient clear space at the discharge end of the mill should be allowed for charging rods. See Fig. 5.

The following equation is used to determine the power that a rod mill should draw.

$$Kw_r = 1.752\ D^{1/3}\ (6.3-5.4\ V_p)\ fC_s \quad (3)$$

where

Kw_r = Kilowatts per metric tonne of rods (1000 kg).
D = Mill diameter inside liners in meters.
V_p = Fraction of mill volume loaded with rods.
fC_s = Fraction of critical speed.

In terms of mill diameter in feet and rod charge in short tons (2000 pounds) the equation becomes:

$$Kw_r = 1.07\ D^{1/3}\ (6.3-5.4\ V_p)\ fC_s \quad (3a)$$

Table V lists many of the common size rod mills giving speed, loading and power data. The power is in horsepower at the mill pinionshaft. For different length rod mills power varies directly as rod length. For difference between new and worn liners increase power draw by 6%, and adjust for bulk density per Table IV.

Wet grinding rod mills are normally used in minerals processing plants. Experience with dry grinding generally indicates many difficult problems and should be avoided except where absolutely necessary; in which case the problem should be referred to the mill manufacturers for recommendations.

The various rod mill manufacturers have different equations for determining the power rod mills draw, but all come close to the same calculated power draw.

Fig. 5 - S N & R Mission Plant

TABLE V

ROD MILL POWER AT MILL PINIONSHAFT (HORSEPOWER)

ROD MILL DIAMETER		ROD MILL LENGTH		ROD LENGTH (L)		L/D	MILL SPEED			BULK DENSITY ROD CHARGE		ROD CHARGE WEIGHT						MILL POWER			DIA (D) INSIDE NEW LINERS	
												METRIC TONNES % VOLUMETRIC LOADING			SHORT TONS % VOLUMETRIC LOADING			% VOLUMETRIC LOADING				
M	FT	M	FT	M	FT		RPM	% C_s	FPM	kg/m^3	lb/ft^3	35	40	45	35	40	45	35	40	45	M	FT
0.91	3.0	1.22	4	1.07	3.5	1.4	36.1	74.5	284	5847	365	1.0	1.13	1.27	1.1	1.25	1.4	7	8	8	0.76	2.5
1.22	4.0	1.83	6	1.68	5.5	1.57	30.6	74.7	336	5847	365	2.25	2.58	2.9	2.48	2.84	3.2	23	25	26	1.07	3.5
1.52	5.0	2.44	8	2.29	7.5	1.67	25.7	71.2	363	5847	365	6.91	7.95	8.89	7.62	8.76	9.8	57	61	64	1.37	4.5
1.83	6.0	3.05	10	2.90	9.5	1.73	23.1	70.7	399	5847	365	13.1	15.0	16.8	14.4	16.5	18.5	114	122	128	1.68	5.5
2.13	7.0	3.35	11	3.20	10.5	1.62	21.0	69.9	428	5766	360	20.0	22.8	25.6	22.0	25.1	28.2	181	194	204	1.98	6.5
2.44	8.0	3.66	12	3.51	11.5	1.53	19.4	69.3	457	5766	360	29.0	33.2	37.4	32.0	36.6	41.2	275	295	310	2.29	7.5
2.59	8.5	3.66	12	3.51	11.5	1.44	18.7	69.0	470	5766	360	33.0	37.7	42.5	36.4	41.6	46.8	318	341	359	2.44	8.0
2.74	9.0	3.66	12	3.51	11.5	1.38	17.9	67.5	470	5766	360	36.0	41.1	45.5	39.7	45.3	50.1	344	369	388	2.55	8.35
2.89	9.5	3.96	13	3.81	12.5	1.41	17.4	67.6	483	5606	350	42.7	48.8	54.9	47.1	53.8	60.5	416	446	470	2.70	8.85
3.05	10.0	4.27	14	4.11	13.5	1.44	16.8	67.0	493	5606	350	51.5	59.0	63.8	56.8	65.0	70.3	507	544	572	2.85	9.35
3.20	10.5	4.57	15	4.42	14.5	1.47	16.2	66.4	501	5606	350	61.4	70.1	78.9	67.7	77.3	87.0	609	653	687	3.00	9.85
3.35	11.0	4.88	16	4.72	15.5	1.50	15.9	66.8	517	5606	350	72.5	82.8	93.5	79.9	91.3	103	735	788	829	3.15	10.35
3.51	11.5	4.88	16	4.72	15.5	1.43	15.5	66.6	528	5606	350	79.7	90.7	103	87.8	100	113	819	878	924	3.31	10.85
3.66	12.0	4.88	16	4.72	15.5	1.37	15.1	66.4	538	5606	350	82.7	99.8	112	91.1	110	123	906	972	1023	3.46	11.35
3.81	12.5	5.49	18	5.34	17.5	1.48	14.7	66.0	547	5446	340	104	119	134	115	131	148	1093	1173	1234	3.61	11.85
3.96	13.0	5.79	19	5.64	18.5	1.50	14.3	65.6	555	5446	340	120	137	154	132	151	170	1264	1356	1426	3.76	12.35
4.12	13.5	5.79	19	5.64	18.5	1.44	14.0	65.5	565	5446	340	130	148	166	143	163	183	1385	1486	1562	3.92	12.85
4.27	14.0	6.10	20	5.94	19.5	1.46	13.6	64.9	570	5446	340	147	169	190	162	186	209	1580	1695	1783	4.07	13.35
4.42	14.5	6.10	20	5.94	19.5	1.41	13.3	64.6	579	5446	340	159	181	204	175	200	225	1715	1840	1935	4.22	13.85
4.57	15.0	6.10	20	5.94	19.5	1.36	13.0	64.3	586	5446	340	171	194	219	188	214	241	1853	1988	2091	4.37	14.35

BALL MILLS

Being free of the limits imposed on rod mills by the rods, ball mills have more variations in length to diameter ratios, ranging from L/D ratios of slightly less than 1:1 to some greater than 2:1. There are no fixed rules on the proper L/D ratios to use as these vary with the circuit used, ore type, feed size and overall grinding requirements. Table VI gives some rough guide lines showing, based upon past experience, the general L/D ratios used in the application of ball mills.

Grinding balls can be made of forged or cast steel or cast iron. The quality depends upon the source of supply. While not always true, frequently the better quality balls are forged steel. Generally balls are spherical shaped, but they can be in various cylindrical, conical or other irregular shapes. Balls vary considerably in hardness with soft balls having Brinnell hardnesses in the range of 350 to 450, and the hard balls having hardnesses in excess of 700. A rule of thumb subject to arguments is: "the harder the ball the better its life" (provided it is not too brittle and breaks or becomes too highly polished and too smooth to nip the material being ground). Local economics and the specific grinding application generally are the deciding factors in selecting the balls to use. The balls giving the lowest operating cost and best performance are generally selected. This need not be the lowest priced balls available nor the ones giving the lowest wear rate, but can be a compromise between the two extremes.

Balls should be solid with a reasonably uniform hardness thru the entire ball. They should wear in a relatively uniform pattern. An indicator of good ball wear is when the worn balls discharging from the mill are around 16 millimeter (5/8") or smaller in size and are polygon shaped having as many as 8 to 12 surfaces, which can be slightly concave. Evidence of broken balls is found when pieces of balls are being discharged, some as circular discs, some as half rounds, some crescent shaped. Pieces of worn or broken balls with holes in them indicate poor quality balls with sand inclusions and/or blow holes and/or hollow centers.

For calculating the power that a ball mill will draw, forged steel and cast steel balls are assumed to weigh 4646 kilograms per cubic meter (290 pounds per cubic foot) with cast iron balls weighing 4165 kg per m^3 (260 pcf).

Ball mills normally carry a ball charge occupying from 40 to 45% of the mill volume, but can carry up to a 50% or slightly higher charge. Fig. 6 shows the relationship of mill power and volumetric loading. For plant capacity and design purposes, ball mills are frequently selected based upon carrying 40% ball charges with the mills and drives designed to carry higher charges if required.

TABLE VI

BALL MILL L/D RATIO - APPLICATION GENERAL GUIDELINES

FEED 80% PASSING SIZE	TOP BALL SIZE		L/D RATIO
MICROMETERS	MILLIMETER	INCHES	
5,000 to 10,000	60 to 90	2.5 to 3.5	1:1 to 1.25:1
900 to 4,000	40 to 50	2.5 to 2	1.25:1 to 1.75:1
Fine Feeds - Regrind	20 to 30	3/4 to 1-1/4	1.5:1 to 2.5:1

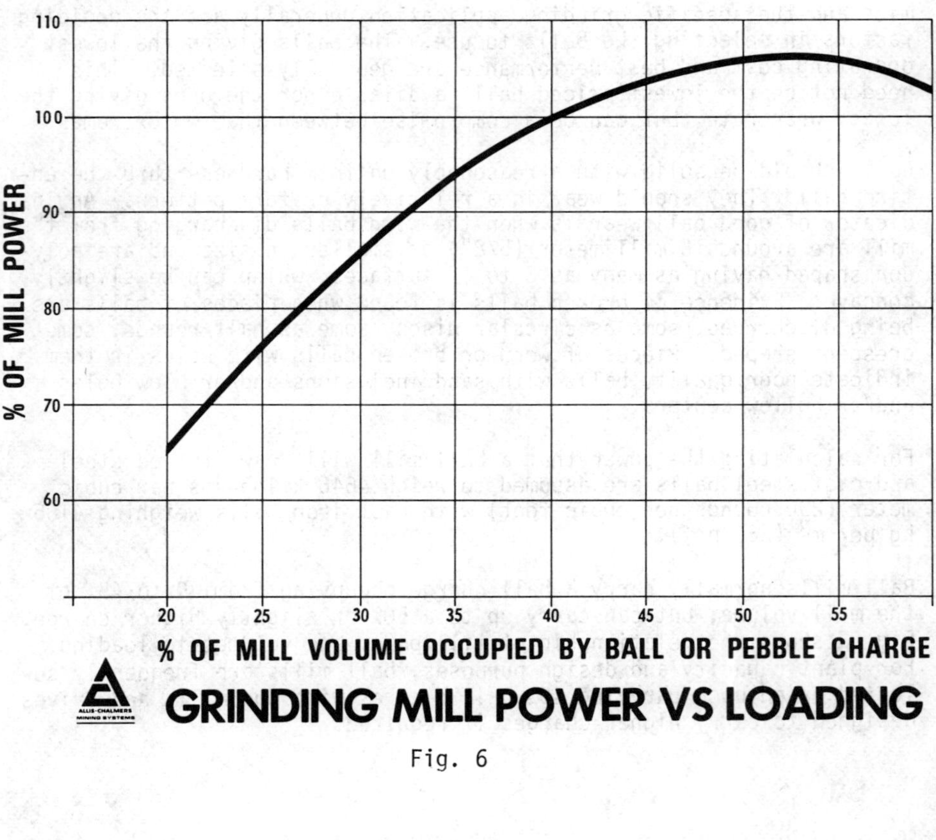

Fig. 6

When a 45 to 50% ball charge is to be carried, generally a double scoop feeder as shown in Fig. 7 is used. This is a more expensive feeder than a spout feeder as shown in Fig. 4, which can also be used to feed ball mills. Plant design, capital cost and operating costs all influence ball mill feeder selection. When a ball mill is closed circuited with a rake or spiral classifier, a scoop feeder is required to feed the circulating load into the mill. With cyclone classifiers spout feeders can be used. Spout feeders allow arrangement where the classifier underflow flows by gravity into the spout feeder hopper. Therefore, the cyclone classifiers must be installed high enough to obtain the head required for this flow into the hopper. This can result in high pumping heads and pump power to pump the mill discharge to the cyclone classifiers. Fig. 8 shows a single stage ball mill installation using double scoop feeders with cyclone classifiers installed at about the horizontal centerline of the mill. Depending upon the radius and width of the scoops and the capacity (including circulating load) double scoop feeders consume from 25 to 30 kilowatts (20 to 40 horsepower). This arrangement reduces considerably pumping head and power and must be balanced against higher maintenance cost for the scoop feeder.

Make-up grinding balls are fed to the mill as required thru the mill feeder with the mill in operation. Balls may feed directly through a spout, but should <u>not</u> be fed into a scoop box because of possible jamming and serious mill damage. Scoops usually have a central ball feed pipe or a small charging drum to accomplish this.

There are many different designs and styles of ball mill liners. As with grinding balls local economics and ultimately operating costs determine the best design and material to use. The initial set of liners is rarely the final design selected. Based upon individual experience, mill superintendents develop perferences for liner designs. The following is given as a guide line for the initial set of liners.

A. For 60 mm (2.5") and smaller top size balls for cast metal liners use double wave liners with the number of lifters to the circle approximately 13.1 D in meters (for D is in feet divide 13.1 D by 3.3). Wave height above the liners from 1.5 to 2 times the liner thickness. Rubber liners of the integral molded design follow the cast metal design. If using the replaceable lifter bar design in either metal or rubber the number of lifters should be about $3.3\pi D$ in meters (for D is in feet divide $3.3\pi D$ by 3.3) with the lifter height above the liners about twice the liner thickness.

   The use of double wave liners, particularly when using 50 mm (2") or larger balls, may show a loss of 5% or so in the mill power draw until the waves wear in and the balls can

Fig. 7 - Single or Double Scoop Feeder

Fig. 8 - Duval Sierrita Plant

nest between the lifters. When liners, and double wave liners in particular, wear with circumferential grooves, slipping of the charge is indicated, and this warns of accelerated wear. When the top size ball is smaller than 60 mm (2.5") and mill speed is less than 72% of critical wear resistant cast irons can be used. For other conditions alloyed cast steel is recommended.

Rubber liners are well suited to this same area and not only reduce operating costs but can reduce noise levels.

B. Single wave liners are recommended for larger size balls (60 mm/2.5" and larger). The number of the lifters to the circle equals approximately 6.6 D in meters (for D is in feet divide 6.6 D by 3.3). The liners are from 50 to 65 mm thick (2" to 2.5") with the waves from 60 to 75 mm (2.5" to 3") above the liners. The replaceable lifter bar design made of either metal or rubber in about the same design proportions can be used. There could be a loss in power with rubber particularly if the mill speed is faster than about 72% of critical speed and the ball size is larger than 75 mm.

Because of the impacting from the large balls, single wave liners for ball mills are usually made from alloyed steels or special wear-resistant alloyed cast irons. Because of the difficulty of balancing growth and wear with work hardening manganese steel is used infrequently and then with extreme care to allow for growth.

C. End liners for ball mills conform to the slope of the mill head and can be made of rubber, alloyed cast steel or wear resistant cast iron. To prevent racing and excessive wear end liners for ball mills are furnished with integral radial ribs or with replaceable lifters or with both.

D. When a grate discharge is used the grates and wear plates are normally perpendicular to the mill axis while the discharge pans conform to the slope of the mill head. The grates and wear plates are normally made from alloyed wear resistant cast steel or rubber. They are ribbed to prevent racing and excessive wear. The dischargers and pans are generally made from either wear resistant cast iron or rubber, or wear resistant fabricated steel.

Slot plugging can be a problem in grate discharge mills. Whether the grates are made of metal or rubber the slots should have ample relief tapered toward the discharge side. Total angles 7 to 10 degrees (3.5 to 5 degrees per side) are commonly used. Metal grates often have a small lead-in pocket or recess which can fill in with peened metal rather than have the slot peen shut. With the proper combination of metal internals and rubber surfaces, rubber grates have flexibility that tend to make them self cleaning and yet not fail due to flexing.

E. Except when using rubber liners, the mill surfaces are covered with a protective rubber or plastic material to protect the surfaces from pulp racing and corrosion. This is done in wet grinding mills. Since dry grinding mills get hot due to heat from grinding generally rubber liners and rubber materials cannot be used.

The following equation is used to determine the power that wet grinding overflow ball mills should draw.

$$Kw_b = 4.879\ D^{0.3}\ (3.2\text{-}3\ V_p)\ fC_s \left(1 - \frac{0.1}{2^{9-10\ fC_s}}\right) + S_s \qquad (4)$$

where

Kw_b = Kilowatts per metric tonne of balls (1000 kg).
D = Mill diameter inside liners in meters.
V_p = Fraction of mill volume loaded with balls.
fC_s = Fraction of critical speed.
S_s = Ball size factor.

In terms of mill diameter in feet and power per short ton at (2000 pounds) of ball charge Equation 4 becomes:

$$Kw_b = 3.1\ D^{0.3}\ (3.2\text{-}3\ V_p)\ fC_s \left(1 - \frac{0.1}{2^{9-10\ fC_s}}\right) + S_s \qquad (4a)$$

For mills larger than 3.3 meters (10 feet) diameter inside liners the top size of the balls used affects the power drawn by the mill. This is called the ball size factor S_s.

$$S_s = 1.102\ \left(\frac{B-12.5\ D}{50.8}\right) \qquad (5)$$

where

B = Ball size in millimeters.
D = Mill diameter inside liners in meters.
S_s = Kilowatts per metric tonne of balls.

In terms of ball size in inches and mill diameter in feet and power per short ton of ball charge equation 5 becomes:

$$S_s = \frac{B-\frac{3D}{20}}{2} \qquad (5a)$$

To determine the power that a wet grinding, low level grate discharge mill should draw multiply Kw_b by 1.16 and for a dry grinding, full grate discharge mill multiply by 1.08.

For special applications such as cement raw materials, bauxite in caustic solutions and other clay-like materials consult the mill manufacturers, since these materials affect the power drawn by ball mills.

Table VII lists essentially "square" ball mills giving mill speed as percent of critical, weight of a 40% ball charge, top ball size and calculated power draw. Ball mill power changes in direct proportion to mill length. The power is horsepower at the mill pinionshaft inside new shell liners. Increase power for worn shell liners by 6%. There are indications that rubber liners may cause from a 5 to 10% loss in mill power.

The various ball mill manufacturers have different equations for determining the power ball mills draw, but all come close to the same calculated power draw.

ORE TESTING FOR MILL SELECTION

After the grind requirements are established, testing for the selection of comminution circuits and mill size can be initiated and can include the following:

- Primary Autogenous Media Competency
- Primary Autogenous and Semi-Autogenous Pilot Plant
- Secondary Autogenous (Pebble) Testing
- Impact Crushing - Bond Work Index
- Rod Mill Grindability - Bond Work Index
- Ball Mill Grindability - Bond Work Index
- Abrasion Index
- Crushing, Grinding and Concentration Pilot Plant

This discussion will be limited to testing for selection of rod and ball mills. Rod and ball milling, using grinding media of known quality, are well established and require much simplier grinding test programs than required for the selection of autogenous grinding mills and circuits.

Samples for grindability tests should be crushed to about minus 1". The samples can be taken from adits, pits, etc., in the ore body or they can be crushed drill cores. While it is helpful in anticipating variations in feed rates to know the grindability of the various ore types in a deposit, for mill selection grindability tests should be run on composite samples representative of the blended mill feed. It is infrequent that the grindability of the composite or blend is the same as that obtained by mathematically making a proportional blend based on grindability tests performed on each component. Differential

TABLE VII

BALL MILL POWER AT MILL PINIONSHAFT (HORSEPOWER)

BALL MILL DIAMETER		BALL MILL LENGTH		BALL SIZE		MILL SPEED			BALL CHARGE WEIGHT						MILL POWER						DIA (D) INSIDE NEW LINERS	
									METRIC TONNES % VOLUMETRIC LOADING			SHORT TONS % VOLUMETRIC LOADING			OVERFLOW DISCH. % VOLUMETRIC LOADING			GRATE DISCH. % VOLUMETRIC LOADING				
M	FT	M	FT	MM	IN	RPM	% C_s	FPM	35	40	45	35	40	45	35	40	45	35	40	45	M	FT
0.91	3.0	0.91	3.0	50	2.0	38.7	79.9	304	0.68	0.77	0.87	0.75	0.85	0.96	7	7	7	8	8	9	0.76	2.5
1.22	4.0	1.22	4.0	50	2.0	32.4	79.1	356	1.77	2.02	2.28	1.95	2.23	2.51	19	20	21	22	24	25	1.07	3.5
1.52	5.0	1.52	5.0	50	2.0	28.2	78.1	399	3.66	4.19	4.71	4.03	4.61	5.19	42	45	47	49	52	54	1.37	4.5
1.83	6.0	1.83	6.0	50	2.0	25.5	78.0	441	6.56	7.50	8.44	7.23	8.27	9.30	80	85	89	93	99	103	1.68	5.5
2.13	7.0	2.13	7.0	50	2.0	23.2	77.2	474	10.7	12.3	13.8	11.8	13.5	15.2	137	145	151	158	168	175	1.98	6.5
2.44	8.0	2.44	8.0	50	2.0	21.3	76.1	502	16.2	18.6	21.0	17.9	20.5	23.1	215	228	237	249	265	275	2.29	7.5
2.59	8.5	2.44	8.0	50	2.0	20.4	75.3	513	18.5	21.1	23.8	20.4	23.3	26.2	250	266	277	290	308	321	2.44	8.0
2.74	9.0	2.74	9.0	50	2.0	19.7	75.0	526	23.5	26.9	30.2	25.9	29.6	33.3	322	342	356	373	397	413	2.55	8.5
2.89	9.5	2.74	9.0	50	2.0	19.15	75.0	541	26.4	30.1	33.9	29.1	33.2	37.4	367	390	406	425	453	471	2.74	9.0
3.05	10.0	3.05	10.0	50	2.0	18.65	75.0	557	32.7	37.3	42.0	36.0	41.1	46.3	462	491	512	535	570	593	2.89	9.5
3.20	10.5	3.05	10.0	50	2.0	18.15	75.0	570	36.1	41.4	46.5	39.9	45.6	51.2	519	552	575	602	640	667	3.05	10.0
3.35	11.0	3.35	11.0	50	2.0	17.3	72.8	565	43.0	49.2	55.4	47.4	54.2	61.0	610	649	676	708	753	784	3.17	10.4
3.51	11.5	3.35	11.0	50	2.0	16.75	72.2	574	49.1	54.0	60.8	54.1	59.5	67.0	674	718	747	782	832	867	3.32	10.9
3.66	12.0	3.66	12.0	50	2.0	16.3	71.8	584	56.4	64.4	72.5	62.2	71.0	79.9	812	864	900	942	1003	1044	3.47	11.4
3.81	12.5	3.66	12.0	50	2.0	15.95	71.8	596	61.4	70.2	79.0	67.7	77.4	87.1	896	954	993	1040	1106	1152	3.63	11.9
3.96	13.0	3.96	13.0	50	2.0	15.60	71.7	607	72.3	82.7	92.6	79.7	91.1	102	1063	1130	1177	1233	1311	1365	3.78	12.4
4.12	13.5	3.96	13.0	64	2.5	15.30	71.7	620	78.2	89.4	99.8	86.2	98.5	111	1189	1266	1321	1379	1469	1532	3.93	12.9
4.27	14.0	4.27	14.0	64	2.5	14.8	70.7	623	90.7	104	117	100	115	129	1375	1464	1527	1595	1699	1771	4.08	13.4
4.42	14.5	4.27	14.0	64	2.5	14.55	70.8	635	98.0	112	126	108	123	139	1492	1588	1656	1730	1842	1921	4.24	13.9
4.57	15.0	4.57	15.0	64	2.5	14.1	69.8	638	113	129	144	124	142	159	1707	1817	1893	1980	2107	2196	4.39	14.4
4.72	15.5	4.57	15.0	64	2.5	13.85	69.8	648	121	138	155	133	152	171	1838	1956	2037	2132	2264	2363	4.54	14.9
4.88	16.0	4.88	16.0	64	2.5	13.45	68.9	651	137	157	179	151	173	194	2084	2217	2309	2417	2571	2678	4.69	15.4
5.03	16.5	4.88	16.0	64	2.5	13.2	68.7	659	146	167	188	161	184	207	2229	2370	2468	2585	2750	2863	4.85	15.9
5.18	17.0	5.18	17.0	75	3.0	13.0	68.7	670	165	189	212	182	208	234	2595	2764	2883	3010	3206	3344	5.00	16.4
5.33	17.5	5.18	17.0	75	3.0	12.7	68.1	674	176	201	226	194	221	249	2750	2929	3053	3190	3397	3542	5.15	16.9
5.49	18.0	5.49	18.0	75	3.0	12.4	67.5	678	197	225	253	217	248	279	3077	3276	3414	3569	3800	3961	5.30	17.4

grinding occurs so that each component can be ground to a different size consist than the composite which has been ground to the required size.

Depending upon the size of the deposit, capacity of the concentrator, life of the operation, complexity of the ore deposit, complexity of the extraction or subsequent processes, etc., testing can range from one series of grindability tests to a large number of grindability tests and even possibly pilot plant testing. Often pilot plant testing is done primarily for metallurigical testing with grinding used for feed preparation.

Frequently pilot plants are built using available or used equipment with the grinding circuit oversize for the concentration section and as such the two are not balanced operations. Caution, therefore, should be used in selecting the test data to be used to select mills and circuits for the plant. The data should be selected from tests run <u>specifically</u> to obtain grinding data. This data should include:

a. Mill Size
b. % volumetric loading in mill occupied by media.
c. Type of circuit (open or closed) and flowsheet diagram.
d. Any concentration steps in the grinding circuit.
e. Size and type of classifiers.
f. If cyclone classifiers: size, rated capacity, power draw and speed of cyclone feed pump.
g. Size analysis circuit feed.
h. % moisture in mill feed.
i. Feed rate to the circuit, dry basis.
j. Size analysis and pulp density of:
    1.) Circuit product.
    2.) Discharge from each mill.
    3.) Feed to each classification stage.
    4.) Oversize from each classification stage.
    5.) Fines from each classification stage.
k. Power drawn by each mill (motor input).
l. Motor and drive efficiency of each mill.
m. Size of and type of grinding media used in each mill.
n. Speed in rpm for each mill.
o. Liner design and condition in each mill.
p. Media wear rate.
q. Liner wear rate if tested long enough to obtain.
r. % circulating load for each stage that is close circuited.

During grinding tests obtain samples of mill feed for grindability tests so that work indices calculated from the pilot plant data can be compared to grindability test results. Operating work indices can be obtained using pilot plant data in the following equation:

$$W_{io} = \frac{W}{\frac{10}{\sqrt{P}} - \frac{10}{\sqrt{F}}} \qquad (6)$$

where

W_{io} = Operating Work Index.
W = Kwh per ton (can be metric, standard or long).
P = Product size which 80% passes in micrometers.
F = Feed size which 80% passes in micrometers.

The application of grinding circuit and equipment-related factors discussed later in this chapter are applied to W_{io} to put it on the same basis as grindability test results. This allows a direct comparison of pilot plant test results and grindability test results. In addition to these, also, be sure to apply motor and drive efficiency factors so that the pilot plant mill power data is refered to the mill pinionshaft or to the mill shell (measured power data is generally electrical energy into the motor). For comparison with grindability test results W_{io} has to be placed on the basis of standard tons (907.4 kg).

With the difficulty in obtaining accurate pilot plant data, particularly power data, Bond closed circuit grindability test results often give the more accurate data for selecting rod and ball mills. Grindability test work should span the feed and product sizes of the proposed grinding circuit. The grindability test work generally recommended for primary milling circuits includes:

A. Bond rod mill grindability tests at 10 or 14 mesh for Work Index.

B. For each ball mill grinding step, a Bond ball mill grindability test for Work Index at one mesh size coarser than the desired 80% passing size and at the mesh size of or just finer than the 80% passing size.

C. If 50 mm x 75 mm (2" x 3") ore lumps are available an impact crushing Work Index test.

D. If 30 mm x 20 mm (1-1/4" x 3/4") ore is available an abrasion index test.

For regrind ball milling or ball milling of rougher concentrates produced in the grinding circuit run ball mill grindability tests as outlined in B above. Since the Bond grindability test requires a ratio of reduction of about 6:1 to obtain accurate results it may be necessary to run the test at a finer size than required by the specified grind or even a special test will have to be run.

This will give a good cross section of the grindability of the ore and will allow for accurate calculations of the grinding power required.

The balance of the discussion in the chapter will be an example demonstrating the selection of a primary grinding circuit and a regrind circuit where all of the ore is ground to the required product size in the primary circuit. For grinding circuits where concentration is included in the circuit the basic approach is the same as given in the example, considering each stage as a separate entity and adjusting for new feed rates and feed size (which could be different than the rate and product size from the preceding stage).

PROBLEM

Select rod mills, ball mills and pebble mills as required for the following circuits.

- Rod Mill - Ball Mill
- Single Stage Ball Mill
- Rod Mill - Pebble (Secondary Autogenous) Mill
- Primary Autogenous or Semi-Autogenous Mill - Ball Mill
- Regrind Ball Mill

Feed rate to the primary mill circuit is 500 metric tonnes per hour including factor for availability. Feed rate to regrind mill is 40 metric tonnes per hour.

Rod mill feed and feed for single stage ball mill will be prepared with closed circuit crushing. The feed sizes for the various mills will be:

- Rod Milling: minus 25 mm 80% passing 18 mm.
- Single stage Ball Mill: minus 12 mm 80% passing 9.4 mm.
- Ball Mill and Pebble Mill following Rod Mill and Ball Mill following Primary Autogenous or Semi-Autogenous Mill: minus 2 mm 80% passing 1.2 mm.
- Regrind Ball Mill: 80% passing 210 micrometers.

The circuits are all wet grinding type. All ball or pebble mills are closed circuit with the exception of the regrind mill which will be open circuit for this example.

Pebble size for pebble milling plus 30 mm minus 70 mm with a pebble consumption of 30 metric tonnes per hour which is 6% of the circuit production rate.

The specified grinds are: primary grinding circuit 80% passing 175 micrometers, regrind circuit 80% passing 45 micrometers.

Bench scale grindability test results to be used for grinding power calculations. The test results to be used in the example are:

Impact Crushing Work Index	11.5	
Rod Mill Grindability Test at 10 mesh	13.2	(W_i)
Ball Mill Grindability Test at 65 mesh	11.7	(W_i)
Ball Mill Grindability Test at 100 mesh	12.1	(W_i)
Ball Mill Grindability Test at 325 mesh on regrind mill feed	14.0	(W_i)
Abrasion Index	0.215	(A_i)

EQUATIONS USED TO DETERMINE GRINDING POWER

This first step in selecting grinding mills is to determine the power needed to produce the desired grind. The basic equation used for this is the Bond Equation.

$$W = \frac{10\,W_i}{\sqrt{P}} - \frac{10\,W_i}{\sqrt{F}} \qquad (7)$$

where

W = kwh per short ton.
W_i = Work Index.
P = Product size in microns which 80% passes.
F = Feed size in microns which 80% passes.

The power determined from equation 7 is for the following specific conditions.

A. Rod Milling - wet, open circuit grinding in a 2.44 meter (8') diameter inside liners rod mill.

B. Ball Milling - wet, closed circuit grinding in a 2.44 meter (8') diameter inside liners ball mill.

C. Power calculated is the power required at the pinionshaft of the mill, which includes mill bearings and gear pinion losses but does not include motor losses or losses in any other drive component, such as reducers and clutches.

The feed for Bond Grindability Tests is:

Rod milling ore crushed to minus 13,200 micrometers (0.530") or finer.

Ball milling ore crushed to minus 3,350 micrometers (6 mesh) or finer.

which have been used to establish optimum rod and ball mill feed sizes.

There are eight efficiency factors to be applied to the calculated grinding power to allow for variations from the specified conditions and optimum feed sizes.

EF_1 Dry Grinding.

EF_2 Open Circuit Ball Milling.

EF_3 Diameter Efficiency Factor.

EF_4 Oversized Feed.

EF_5 Fine Grinding in ball mills to product sizes finer than 80% passing 200 mesh (75 micrometers).

EF_6 High or low ratio of reduction rod milling.

EF_7 Low Ratio of reduction ball milling.

EF_8 Rod Milling.

The multipliers for the efficiency factors are determined by the following:

EF_1 - Dry Grinding - for the same range of work, dry grinding requires 1.3 times as much power as wet grinding.

EF_2 - Open Circuit Grinding - when grinding in open circuit ball mills, the amount of extra power required, compared to closed circuit ball milling, is a function of the degree of control required on the product produced. The inefficiency factors for open circuit grinding are given in Table VIII.

EF_3 - Diameter Efficiency Factor - using the base mill diameter of 2.44 meters (8') inside liners, the diameter efficiency factor can be calculated from the following:

When D is in meters:

$$EF_3 = \left(\frac{2.44}{D}\right)^{0.2} \tag{8}$$

When D is in feet:

$$EF_3 = \left(\frac{8}{D}\right)^{0.2} \tag{8a}$$

Table IX gives a tabulation of the EF_3 factors for some of the more common mill diameters in both the imperial and metric measuring systems.

TABLE VIII

OPEN CIRCUIT INEFFICIENCY MULTIPLIER

PRODUCT SIZE CONTROL REFERENCE % PASSING	INEFFICIENCY MULTIPLIER
50	1.035
60	1.05
70	1.10
80	1.20
90	1.40
92	1.46
95	1.57
98	1.70

TABLE IX

DIAMETER EFFICIENCY MULTIPLIERS

MILL DIAMETER INSIDE SHELL		MILL DIAMETER INSIDE LINERS		DIAMETER EFFICIENCY MULTIPLIER
FEET	METERS	FEET	METERS	
3.0	0.914	2.6	0.79	1.25
3.281	1.0	2.88	0.88	1.23
4.0	1.22	3.6	1.10	1.17
5.0	1.52	4.6	1.40	1.12
6.0	1.83	5.6	1.71	1.075
6.562	2.0	5.96	1.82	1.06
7.0	2.13	6.5	1.98	1.042
8.0	2.44	7.5	2.29	1.014
8.5	2.59	8.0	2.44	1.000 Base
9.0	2.74	8.5	2.59	0.992
9.5	2.90	9.0	2.74	0.977
9.843	3.0	9.34	2.85	0.970
10.0	3.05	9.5	2.90	0.966
10.5	3.20	10.0	3.05	0.956
11.0	3.35	10.5	3.20	0.948
11.5	3.51	11.0	3.35	0.939
12.0	3.66	11.5	3.51	0.931
12.5	3.81	12.0	3.66	0.923
13.0	3.96	12.5	3.81	0.914
13.124	4.00	12.62	3.85	0.914

It should be noted that for mills where the diameter inside liners is larger than 3.81 meters (12.5') that the diameter efficiency factor does not change and remains 0.914.

NOTE: In selecting mills for new operations, where this factor is less than 1.0 it is sometimes neglected and is used as a safety factor. In the example it will be applied.

EF_4 - Oversized Feed - when being fed a coarser than optimum feed, this factor applies to rod milling and ball milling. However, the most frequent use is found in conjunction with single stage ball milling. This is the one efficiency factor that is related to Work Index as is seen in the following equation:

$$EF_4 = \frac{R_r + \left[W_i - 7\right]\left[\frac{F - F_o}{F_o}\right]}{R_r} \qquad (9)$$

$$\text{where } R_r = \text{Ratio of reduction} = \frac{F}{P} \qquad (10)$$

F_o = Optimum feed size

$$\text{Rod milling:} \quad 16{,}000 \sqrt{\frac{13}{W_i}} \qquad (11)$$

$$\text{Ball milling:} \quad 4{,}000 \sqrt{\frac{13}{W_i}} \qquad (12)$$

When available, use the Work Index from a grindability test at the desired grind for W_i in equation 9. For equation 11, use either the Work Index from an impact test or a rod mill grindability test, whichever is higher. For equation 12, use the Work Index from a rod mill grindability test, since this more represents the coarse fraction of the feed; if not available then use the ball mill grindability test results.

EF_5 - Fineness of Grind Factor - this applies to fine grinding when the 80% passing size of the product is finer than 75 micrometers (200 mesh). The equation to determine this is:

$$EF_5 = \frac{P + 10.3}{1.145\ P} \qquad (13)$$

EF_6 - High or Low Ratio of Reduction Rod Milling - the equation to be used, unless R_r is between $R_{ro} = -2$ and $R_{ro} = +2$ is:

$$EF_6 = 1 + \frac{\left(R_r - R_{ro}\right)^2}{150} \qquad (14)$$

$$\text{where } R_{ro} = 8 + \frac{5L}{D} \qquad (15)$$

L = Rod Length

This factor always applies to low ratios of reduction but its application to high ratios of reduction is not always needed, but should be used for mill size selection whenever W_i from the rod mill and ball mill grindability tests exceed 7.0.

EF_7 - Low Ratio of Reduction Ball Mill - the need to use this factor does not occur very often as it only applies to ball milling when the Ratio of Reduction is less than 6. This shows up particularly in regrinding concentrates and tailings. The equation for this is:

$$EF_7 = \frac{2\ (R_r - 1.35) + 0.26}{2\ (R_r - 1.35)} \qquad (16)$$

EF_8 - Rod Milling - a study of rod mill operations shows that rod mill performance is affected by the attention given to preparation and feeding a uniform top size feed size to the mill and the care given to maintaining the rod charge. This efficiency factor has not been definitely determined. In selecting rod mills based upon power calculated from grindability tests, the following procedure has been recommended:

1) When calculating rod mill power for a rod-milling-only application, use an inefficiency factor of 1.4 when the feed is to be prepared with open circuit crushing, and use 1.2 when the feed is to be prepared with closed circuit crushing. The mill diameter, low or high ratio of reduction, and oversize feed factors also must be applied to the calculated grinding power.

2) When calculating rod mill power for a rod mill-ball mill circuit, do not allow for improvement in the ball mill performance due to receiving rod mill feed. If the rod mill feed is produced with open circuit crushing, apply a 1.2 inefficiency factor to the power calculated for the rod milling stage only. If the rod mill feed will consistently be the same size, such as produced with closed circuit crushing, do not apply a rod mill inefficiency factor. The mill diameter, low or high ratio of reduction, and oversize feed factors should be applied to the calculated grinding power.

GRINDING POWER CALCULATIONS AND GRINDING MILL SELECTIONS

The following demonstrates the use of the Bond Work Index Method to determine the power required to produce the desired grind. After the grinding power has been determined by this or other methods, the mill(s) that will draw the required power can be selected. For final mill size recommendations all the process design data and controlling economic and geologic factors should be given to grinding mill manufacturers and their recommendations obtained.

With the growth of mill sizes and changing economic situations new factors influencing grinding power calculations and mill size selection are becoming known and more will become known. The grinding mill manufacturer's are a good source for the application of this continually growing technology including the practical application of the newer approaches and math modeling being developed thru academic research.

A. Rod Mills

$F = 18{,}000$ micrometers.
$P = 1{,}200$ micrometers.
$W_i = 13.2$

$$W = \frac{10 \times 13.2}{\sqrt{1{,}200}} - \frac{10 \times 13.2}{\sqrt{18{,}000}} = 2.83 \text{ kwh/s. ton}$$

Efficiency Factors:

EF_1 does not apply.

EF_2 does not apply.

EF_3 determine after power calculations is completed.

EF_4 Feed size is coarser than 16,000 micrometers.

$$R_r = \frac{18{,}000}{1{,}200} = 15.0$$

$$F_o = 16{,}000\sqrt{13/13.2} = 15{,}878$$

$$\frac{15 + (13.2-7)\left(\frac{18{,}000-15{,}878}{15{,}878}\right)}{15} = 1.06$$

EF_5 does not apply.

EF_6 will not apply as ratio of reduction will be within $R_{ro} \pm 2$.

R_{ro} determined after mill size selection.

EF_7 does not apply.

EF_8 The rod mill feed will be prepared by closed circuit crushing and the rod mill will be in a rod mill-ball mill (or pebble mill) circuit with no intermediate concentration stage so no EF_8 factor need be applied. If it were just a rod milling circuit or if there were an intermediate concentration stage between the rod and the mill a 1.2 factor would apply.

Conversion short ton to metric tonne 1.102
Kilowatts to horsepower 1.341
2.83 x 1.06 x 1.102 x 1.341 = 4.43 Hph/metric tonne
4.43 x 500 = 2215 HP

Referring to Table V two mills will be required. The preliminary rod mill selection would be a 3.66 meter (12 foot) inside shell 3.46 meter (11.35 foot) diameter inside new shell liners. Referring to Table IX the EF_3 (Diameter Efficiency) factor is 0.931.

$$2215 \times 0.931 = 2062 \text{ HP}$$

Referring to Table V the 3.66 m x 4.88 m rod mill with 4.72 m (15.5 ft.) long rods calculates to draw 972 HP when carrying a 40 percent rod charge with a worn-in bulk density of 5606 kg per cubic meter (350 pounds per cubic foot). 1031 HP is required. Therefore, increase mill length by 0.3 meters (1 foot).

$$\frac{972}{4.72} \times 5.02 = 1034 \text{ HP}$$

Therefore, use two 3.66 meter (12 foot) diameter inside shell 3.46 meter (11.35 foot) diameter inside new shell liners by 5.18 meter (17.0 foot) long overflow rod mills with a 40 percent by mill volume rod charge with 5.02 meter (16.5 foot) long rods.

$$R_{ro} = 8 + \frac{5 \times 5.20}{3.46} = 15.25$$

$$R_r = 15.0$$

Therefore, EF_6 assumption is confirmed.

These mills are required to prepare ball mill feed.

With pebble milling the pebble portion of the product does not go thru the rod mill thus the rod mill feed rate is reduced by 30 metric tonnes per hour (6% of 500 metric tonnes per hour).

$$\frac{2062}{500} \times 470 = 1938 \text{ HP}$$

$$1938 \div 2 = 962 \text{ HP}$$

Therefore, use two 3.66 meter (12 foot) diameter inside shell 3.46 meter (11.35 foot) inside new shell liner by 4.88 meter (16 foot) long overflow rod mills with a 40 percent by mill volume rod charge with 4.72 meter (15.5 foot) long rods.

$$R_{ro} = 8 + \frac{5 \times 4.72}{3.46} = 14.8$$

$$R_r = 15.0$$

Therefore, EF_6 assumption is also confirmed here.

B. Ball Mills: Rod Mill-Ball Mill Circuit

F = 1,200 micrometers.
P = 175 micrometers.
W_i = 11.7 (the grindability work index at 100 mesh did not show any unusual characteristics that would require special consideration).

$$W = \frac{10 \times 11.7}{\sqrt{175}} - \frac{10 \times 11.7}{\sqrt{1,200}} = 5.47 \text{ kwh/s. ton.}$$

The only efficiency factor that applies is the diameter efficiency factor EF_3.

$$5.47 \times 1.102 \times 1.341 \times 500 = 4039 \text{ HP}$$

By referring to Table VII it is obvious the mill is larger than 3.81 meters (12.5 feet) diameter inside liners so the EF_3 per Table IX is 0.914.

$$4039 \times 0.914 = 3692 \text{ HP}$$

Since 2 rod mills are required use 2 ball mills (1 ball mill per rod mill). This is the simpliest circuit to operate and control or automate and represents a lower capital cost than the one rod mill two ball mills of the same diameter circuit.

$$3692 \div 2 = 1846 \text{ HP}$$

Referring to Table VI the general guidelines indicate the ball mills should have an L/D in the area of 1.5. Referring to Table VII a 4.12 meter (13.5 foot) by 3.96 meter (13.0

foot) overflow ball mill with a 40 percent by mill volume ball charge, new shell liners and 64 mm (2.5") diameter balls draws 1266 HP.

$$\frac{1846}{1266} = 1.46$$

$$3.96 \times 1.46 = 5.78 \text{ meters (18.96 feet)}$$

Therefore, use two 4.12 meter (13.5 foot) diameter inside shell 3.93 meter (12.9 foot) diameter inside new liners by 5.79 meter (19.0 foot) long overflow ball mills with a 40 percent by volume ball charge.

For lower operating costs, slightly better efficiency, and better mill availability the current practice favors overflow ball mills however, there are some operators that prefer grate discharge mills. Referring to Table VII a 3.96 meter (13.0 foot) diameter by 3.96 meter (13.0 foot) diaphragm ball mill with a 40 percent by mill volume ball charge, new shell liners and 50 mm (2") diameter balls draws 1311 HP.

$$\frac{1846}{1311} = 1.41$$

$$3.96 \times 1.41 = 5.58 \text{ meters (18.3 feet)}$$

Therefore, use two 3.96 meter (13.0 foot) diameter inside shell 3.78 meter (12.4 foot) diameter inside new liners by 5.79 meter (19.0 foot) long diaphragm (grate) discharge ball mills with a 40 percent by volume ball charge.

C. Ball Mills: Single Stage

The feed to the standard Bond ball mill grindability test is minus 6 mesh. Thus, the coarser fraction of a minus 1/2" single-stage ball mill feed is not included in the feed to the grindability ball mill. The minus 1/2" feed to a standard Bond rod mill grindability test, however, does include the coarse fraction of a single-stage ball mill feed. To obtain the complete grindability profile (W_i vs size) of an ore when selecting a single-stage ball mill, it is advisable to run both rod and ball mill grindability tests.

If there is a difference in the work indices obtained from the rod mill and the ball mill grindability tests, which frequently occurs, then, particularly if the rod mill test work index is higher, a two-step calculation should be made to determine the required grinding power. The rod mill work index should be used to calculate from the plant ball mill feed size to 80% passing 2100 microns. The calculation from 2100 microns to the desired product size is made using the work index from the ball mill grindability test. The sum of these two gives the total uncorrected power per ton required for grinding.

F = 9,400 micrometers.
P = 175 micrometers.
W_i = Rod mill test 13.2.
W_i = Ball mill test 11.7.

Step one:

$$W = \frac{132}{\sqrt{2,100}} - \frac{132}{\sqrt{9,400}} = 1.52 \text{ kwh/s. ton}$$

Step two:

$$W = \frac{177}{\sqrt{175}} - \frac{117}{\sqrt{2,100}} = 6.29 \text{ kwh/s. ton}$$

Total $\overline{7.81}$ kwh/s. ton

7.81 x 1.102 x 1.341 x 500 = 5766 HP, uncorrected.

Efficiency Factors:

EF_1 does not apply.

EF_2 does not apply.

EF_3 Mills will be larger than 3.81 meter in diameter so use 0.914.

EF_4 Feed is coarser than 4000 micrometers.

R_r = 9,400 ÷ 175 = 53.7

$F_o = 4,000 \sqrt{13/13.2} = 3970$

$$\frac{53.7 + (11.7-7)\left(\frac{9,400-3,970}{3970}\right)}{53.7} = 1.12$$

EF_5, EF_6, EF_7 and EF_8 do not apply.

5766 x 1.12 x 0.914 = 5903 HP

Use 2 mills

$$\frac{5903}{2} = 2952 \text{ HP}$$

Per Table VI the L/D should be around 1.25. Referring to Table VII a 5.03 meter (16.5 foot) diameter by 4.88 meter (16.0 foot) overflow ball mill with a 40 percent by mill volume ball charge, new shell liners and 64 mm (2.5") diameter balls draws 2370 HP.

$$\frac{2952}{2370} = 1.25$$

$$4.88 \times 1.25 = 6.1 \text{ meters (20.0 feet)}$$

Therefore, use two 5.03 meter (16.5 foot) diameter inside shell 4.85 meter (15.9 foot) diameter inside new liners by 6.1 meter (20.0 foot) long overflow ball mills with a 40 percent by volume ball charge.

Referring to Table VII for sizing a grate discharge mill a 4.72 meter (15.5 foot) diameter by 4.57 meter (15.0 foot) grate discharge ball mill with a 40 percent by mill volume ball charge, new shell liners and 64 mm (2.5") diameter balls draws 2269 HP.

$$\frac{2952}{2269} = 1.30$$

$$4.57 \times 1.3 = 5.94 \text{ meters (19.5 feet)}$$

Therefore, use two 4.72 meter (15.5 foot) diameter inside shell 4.54 meter (14.9 foot) diameter inside new liners by 6.1 meter (20.0 foot) long diaphragm (grate) discharge ball mill with a 40 percent by volume ball charge.

D. Ball Mill: Following Autogenous or Semi-Autogenous Primary Mill

If the product size from the primary autogenous or semi-autogenous mill is the same as from a rod mill, the ball mill calculation and size selection is the same as covered under section B above. If the ball mill feed size is different from this, the same procedure as covered by either section B or C (whichever applies) should be used to determine grinding power, and ball mill size selection.

E. Pebble Mill: Rod Mill Pebble Mill Circuit

The calculation for determining grinding power for Pebble milling (secondary autogenous) can be the same as for ball milling from rod mill product size to the desired specified size, neglecting the diameter efficiency factor if less than 1.0.

$$5.47 \times 1.102 \times 1.341 \times 500 = 4039 \text{ HP}$$

To this add the power required to wear the pebbles down to rod mill product size (pebble mill feed size).

F = 70,000 micrometers.
P = 1,200 micrometers.
W_i = 13.2

$$W = \frac{10 \times 13.2}{\sqrt{1,200}} - \frac{10 \times 13.2}{\sqrt{70,000}} = 3.31 \text{ kwh/s. ton}$$

The inefficiency factor to allow for the inefficient use of power in wearing down from pebble size to rod mill product size is 2.0.

$$3.31 \times 2.0 \times 1.102 \times 1.341 \times \frac{30}{500} \times 500 = 294 \text{ HP}$$

$$\begin{array}{r} 4039 \\ \underline{294} \\ 4333 \text{ HP} \end{array}$$

Select two 2200 HP pebble mills. For specific sizing refer to mill manufacturers for recommendations as they have proprietary equations for calculating mill power draw taking into account the various ore media and pulp factors involved.

F. Regrind Ball Mill

F = 210 micrometers.
P = 45 micrometers.
W_i = 14.0

$$W = \frac{10 \times 14.0}{\sqrt{45}} - \frac{10 \times 14.0}{\sqrt{210}} = 11.21 \text{ kwh/s. ton}$$

Efficiency Factors:

EF_1 does not apply.

EF_2 Many regrind operations are closed circuit, but assume this one is open circuit and 80 percent passing grind will be the controlling point. Refer to Table VIII. The EF_2 factor is 1.2.

EF_3 Because balls will be smaller than 40 mm (1.5") and other minor factors neglect EF_3 unless mill diameter is less than 2.44 meter (8.0') diameter inside liners.

EF_4 does not apply.

EF_5 Grind is 80 percent passing 45 micrometers.

$$\frac{45 + 10.3}{1.145 \times 45} = 1.07$$

EF_6 does not apply.

EF_7 $R_r = 210 \div 45 = 4.67$ which is less than 6.

$$\frac{2.0\ (4.67-1.35) + 0.26}{2.0\ (4.67-1.35)} = 1.04$$

EF_8 does not apply.

Referring to Table VI the L/D can be between 1.75 and 2.0 or even greater. Referring to Table VII a 3.05 meter (10 foot) by 3.05 meter (10 foot) overflow ball mill with a 40 percent by mill volume ball charge, new liners and 50 mm (2") balls draws 491 HP.

Using equation 5 there will be a loss of 0.55 Kw (0.74 Hp) per metric tonnes of balls.

$$37.3 \times 0.74 = 28 \text{ HP}$$
$$491-28 = 463 \text{ HP}$$

$$\frac{884}{463} = 1.91$$

$$3.05 \times 1.91 = 5.83 \text{ meters (19.1 feet)}$$

Therefore, use one 3.05 meter (10.0 foot) diameter inside shell 2.89 meter (9.5 foot) diameter inside new liners by 5.79 meter (19.0 foot) long overflow ball mill with a 40 percent by volume ball charge. For rubber liners add 10% or 0.58 meters (approximately 2 feet) to the length.

G. Motor Selection

In all cases the mills should be driven by a large enough motor to allow the mill to operate with a 45 percent by mill volume charge with new liners and at least a 35 percent charge with worn liners. It may be desired to utilize more of the available mill volume as the liners wear so a higher charge can be specified for worn liners. Specifications can also call for the drive and motor to be rated to allow using pinions with one less and two more teeth, thus allowing for changing mill speed if it is found necessary to balance circuit, increase capacity, suit changing ore characteristics, etc.

SELECTION OF GRINDING MEDIA SIZES AND ESTIMATING STEEL CONSUMPTION

The equation for selection of the largest diameter rod for the initial charge and for the make-up charge is:

$$R = \left(\frac{F^{0.75}}{160} \sqrt{\frac{W_i \times S_g}{(\% \; C_s) \sqrt{3.281 \; D}}} \right) \times 25.4 \qquad (17)$$

R = Diameter of rod in millimeters.
F = Feed size 80% passes in microns.
W_i = Work Index.
S_g = Specific Gravity.
C_s = Critical Speed.
D = Diameter inside shell liners in meters.

With R in inches and diameter (D) in feet equation 17 becomes:

$$R = \left(\frac{F^{0.75}}{160} \sqrt{\frac{W_i \times S_g}{(\% \; C_s) \sqrt{D}}} \right) \qquad (17a)$$

Table X gives the equilibrium start-up rod charge for top rod sizes from 125 mm (5") to 65 mm (2.5").

The equation for selection of the largest diameter ball for the initial charge and for the make-up charge is:

$$B = \left[\left(\sqrt{\frac{F}{K}} \sqrt[3]{\frac{S_g W_i}{(\% \; C_s) \sqrt{3.281 \; D}}} \right) \right] \times 25.4 \qquad (18)$$

B = Diameter of ball in millimeters.

NOTE: Except for K which is given below all other terms the same as for equation 17.

TABLE X

START-UP EQUILIBRIA GRINDING ROD CHARGES, PERCENT WEIGHT

Make-Up Rods Fed Sizes, MM=R	125 5.0"	115 4.5"	100 4.0"	90 3.5"	75 3.0"	65 2.5"
125 (5.0")	18					
115 (4.5")	22	20				
100 (4.0")	10	23	20			
90 (3.5")	14	20	27	20		
75 (3.0")	11	15	21	33	31	
65 (2.5")	7	10	15	21	39	34
50 (2.0")	9	12	17	26	30	66
TOTAL Pct	100	100	100	100	100	100

TABLE XI

START-UP EQUILIBRIA GRINDING BALL CHARGES, PERCENT WEIGHT

Make-Up Balls Fed Sizes, MM=B	115 4.5"	100 4.0"	90 3.5"	75 3.0"	65 2.5"	50 2.0"	40 1.5"
115 (4.5")	23.0						
100 (4.0")	31.0	23.0					
90 (3.5")	18.0	34.0	24.0				
75 (3.0")	15.0	21.0	38.0	31			
65 (2.5")	7.0	12.0	20.5	39	34		
50 (2.0")	3.8	6.5	11.5	19	43	40	
40 (1.5")	1.7	2.5	4.5	8	17	45	51
25 (1.0")	0.5	1.0	1.5	3	6	15	49
TOTAL Pc.	100.0	100.0	100.0	100	100	100	100

Ball Mill K Factor

Mill Type and Grinding Circuit	Steel or C.I. Balls K
Wet-Overflow-Open Circuit	350
Wet-Overflow-Closed Circuit	350
Wet-Diaphragm-Open Circuit	330
Wet-Diaphragm-Closed Circuit	330
Dry-Diaphragm-Open Circuit	335
Dry-Diaphragm-Closed Circuit	335

With B in inches and diameter (D) in feet equation 18 becomes:

$$B = \sqrt{\frac{F}{K}} \sqrt[3]{\frac{S_g W_i}{(\% C_s) \sqrt{D}}} \qquad (18a)$$

Table XI gives the equilibrium start-up ball charge for top ball sizes from 115 mm (4.5") to 40 mm (1.5").

These two equations give the largest diameter of the grinding media required. Since the calculated size is not always an available standard size select the nearest larger size available. Actual operating experience may dictate a change from the calculated size. Theoretically it is always advisable to use a graded charge as a replacement charge. Using a graded charge often is not practical. The loss in efficiency by not using a graded charge generally can not be measured. In some cases, it is only necessary to add the largest size media calculated as make-up. Operating results will indicate the necessity of using more than one size of media in the make-up charge.

The best figures for media and liner consumption come from actual operating experience and as operating data is generated this should be used to establish wear rates.

Actual testing in laboratory scale equipment to determine the abrasion characteristics of an ore is difficult, and available tests are guides for estimating purposes but are not completely accurate. One abrasion test measures the weight loss of a steel paddle continuously impacting falling ore particles for a prescribed time period under standard conditions. From this is developed a measurement called an abrasion Index, Ai. From plant data, emperical equations correlating with Ai: were developed to be used to estimate rod, ball, mill liner and crusher liner wear rates. These equations are:

Wet Rod Mills:

$$\text{Rods, kg/Kwh} = 0.175\,(Ai - 0.020)^{.2} \qquad (19)$$

Liners, kg/Kwh = $0.175\ (Ai - 0.015)^{.3}$ (20)

Wet Ball Mills:

Balls, kg/Kwh = $0.175\ (Ai - 0.015)^{1/3}$ (21)

Liners, kg/Kwh = $0.013\ (Ai - 0.015)^{.3}$ (22)

Multiply equations 19, 20, 21 and 22 by 2.2 to get pounds per kilowatt.

These formulas give estimates of wear rates which can be used as a guide. They are subject to such things as mill speeds, percent volumetric loading, alloy of grinding media and liners, operating practices, etc.

The reference information used to prepare this was:

Bond F. C. — "Crushing & Grinding Calculations", Reprint from British Chemical Engineering. Allis-Chalmers Publication 07R9235B.

Rowland C. A. — "Grinding Calculations Related to the Application of Large Rod and Ball Mills", Allis-Chalmers Publication 22P4704.

Bond F. C. — "Grinding Ball Size Selection", Mining Engineering May, 1958.

Rowland C. A. & Nealey R. C. — "Application of Dry Grinding Rod Mills", Transactions of Society of Mining Engineers of A.I.M.E.

Chapter 13

AUTOGENOUS GRINDING FROM TEST WORK TO PURCHASE OF A COMMERCIAL UNIT

by Arthur R. MacPherson and Robert R. Turner

Vice President and Consulting Metallurgist,
Canadian Bechtel Limited
Vancouver, British Columbia

Vice President, Research, Aerofall Mills Limited
Mississauga, Ontario

ABSTRACT

This chapter reviews the steps which should be taken from the time it is decided to consider autogenous grinding in the development of a new ore deposit or the expansion of an existing plant up until the purchase of the new autogenous mills. In this case the term autogenous includes semi-autogenous, secondary pebble mills, and lump milling as well as fully autogenous grinding systems. Also, both wet and dry grinding alternatives have been included.

Before the test program it must be decided how many bulk samples are required to properly assess the autogenous grinding possibilities of the deposit. Also, a decision must be made on where and how the samples are to be ground i.e. pilot mill at site or in established location, wet and or dry grinding, etc.

In the test program itself, certain fundamental data must be obtained. The testing should include enough work to determine whether fully autogenous or semi-autogenous grinding should be used and what type produces best power efficiency. Also, all the other pertinent data should be collected such as power; tonnages; feed, charge, and product sizing; circulating loads, etc. If a secondary mill is required to produce a final grind the decision on whether a pebble or ball mill should be used may require additional test work. Finally, the concentration of the mineral may play an important part in selection of the circuit and should become a part of the final testing once the grinding has been established.

To establish what type of mills might satisfy the requirements produced by the test plant, certain scale up's must be made such as mill power, steel wear, single or two-stage grinding, type of secondary mills.

To further ensure understanding of ore deposit and its grinding characteristics, some further small scale work should be carried out using small samples and mills to produce comparable grinding data. Also, such tests as media competency and work indices should be completed.

The grinding configuration should then be established from all the test work that has been produced. One of the last decisions is to arrive at the type of autogenous mill itself and the various possibilities of mill diameters and lengths to produce the required power.

After the flow sheet has been established and preliminary layouts have been made, the specifications for the mills are prepared and issued to vendors of such equipment. After the tenders for supply of equipment has been received and comparitive bid analyses are made the mills are ordered from the successful supplier.

INTRODUCTION

The term "Autogenous Grinding" has considerably different meaning to various writers and manufactuers. The terms in this presentation are therefore clearly defined.

Autogenous Grinding applies where run-of-mine or coarsely crushed ore is ground in a mill without the addition of other grinding media.

Semi-Autogenous Grinding applies where run-of-mine or coarsely crushed ore is ground in a mill with the addition of steel balls as additional grinding media besides the ore itself.

Pebble Milling applies where previously ground or finely crushed ore is further ground in a mill using pebbles as the grinding media. The pebbles may be of selected size from the ore itself, from screens in the crushing circuit, or from a primary autogenous mill. Pebbles may also be selected material from outside the orebody and may be either natural or manufactured pebbles. In general, the pebble mill replaces the ball mill in such a circuit.

Lump Milling - This term applies where coarse ore is used to grind the same ore in a mill where most of the ore has passed through the fine crushing plant and only the coarse media is screened out before or at an early stage in the crushing. In general, a lump mill replaces an autogenous mill but does require, in addition, the fine crushing circuit. To differentiate it from normal pebble milling it might be termed primary pebble milling.

Usually autogenous or semi-autogenous grinding is used in Primary Grinding or the first stage of grinding in any concentrator. Pebble milling is usually used for Secondary Grinding, that is the bulk of the feed has previously been ground in a primary milling stage.

The purpose of this chapter is to define the steps which must be taken to determine the design of an autogenous grinding circuit beginning with the ore deposit. This can be either a new deposit or a deposit from which ore is already being processed by another system of crushing and grinding and for which an expansion is now contemplated using autogenous grinding.

Both wet and dry systems of autogenous and semi-autogenous grinding are in normal use throughout the metallurgical industry today.

The general procedure in obtaining samples from an orebody is very similar for both systems, but details of testing, scale-up and mill design are somewhat different.

SAMPLES FOR TEST WORK AND SELECTION OF TEST PLANT

The collection of bulk samples of ore from a property in order to test the method is one of the first considerations when investigating an autogenous grinding system for that particular ore. If the property is in production, operating on a conventional system, and expansion using autogenous grinding is considered, the problem of obtaining proper samples is usually not a major one. In this case the usual procedure would be to install a small test unit at the operating mine and run the pilot mill on normal mill feed. This provides almost immediate comparative data with the actual operating plant and decisions on the mill expansion can usually be readily made. (Examples: Palabora, Republic of South Africa, and Pima, Arizona, U.S.A.)

However, if there is no existing plant and it is a new orebody, often in an isolated area, the obtaining of bulk samples for autogenous mill tests can be a problem. The location of the pilot plant must first be considered. Bringing a pilot plant into an isolated area, and supplying personnel for its operation can be a time consuming and expensive procedure. On the other hand obtaining bulk samples amounting to several hundred tons and transporting them to an existing test facility may also be expensive and time consuming.

It is generally considered that it is better to obtain the bulk samples and ship to an existing facility. The work at an existing facility is usually better supervised and is carried out by a trained crew who are experienced in such programs. After carrying out the program at an established test facility and it is decided to proceed with autogenous grinding, there may then be a case for installing a test plant at the mine. This test plant can be operated by the new mine's technical staff and used as a training facility to become familiar with the type of plant and to further test the ore from the new mine. Consequently as the commercial plant may be in the design stage at that time, modifications can be made to the design based on new data from the mine site pilot plant.

In deciding on how many samples and from what areas these should be collected for the test work, there should be close co-operation between the geologists and the metallurgists. In many cases the collection of bulk samples is left to the field crew and they obtain samples where they can be most readily obtained. The varying characteristics of the orebody such as weathering, grade variations, faulting, folding, rock types, etc., should be studied by competent geologists and explained to the metallurgists. Small batch samples from drill cores should be processed by the metallurgists and variations in metal recovery and liberation examined and related to the expected average results of the orebody.

The decisions can then be reached as to how many bulk samples and from what areas they should be obtained. In general, the more complex the orebody appears to be the more samples may be required to cover the variations.

In an orebody which appears to be very uniform, one bulk sample may be sufficient to provide data for plant design, whereas in others many samples may be necessary. A close study of the geology of the orebody and of preliminary metallurgical results is the key to this decision.

It costs considerable money to obtain bulk samples, often entailing shaft sinking, declines, adits, trenches, or samll open pits. If the decision is made by the technical staff that a sample from a certain area of the orebody is necessary then every effort must be made to procure that sample. Mine development and process plant construction are very expensive in today's markets and every effort must be made to limit the risk by careful test work. On the other hand needless expensive test work and the collection of many samples may become a disincentive to further development of the project. There is always risk involved and the important thing is to get experienced advice when the orebody is complex.

A minimum sized bulk sample for treatment in a 5 ft (1.5 m) dia. test mill is about 25 tons. If continuous processing is an essential part of the program then this sample should probably be 50 tons. The samples should be run-of-mine, that is they should be prepared in a normal manner by drilling and blasting with top size limited to 8" or 10" (200 to 250 mm).

Satisfactory results for scale-up have been obtained using mills as small as 1.5 metres in diameter. Most test mills used in laboratories or permanent test facilities are in the order of 1.5 to 2 meters in diameter with lengths from 0.6 to 2 metres. Pilot plants established in field locations have tended to be somewhat larger, varying in diameter from 2 up to 4 metres. Indeed in one case a mill of approximately 7 metres was used as a pilot plant and this became the first unit of a multi-unit commercial plant. An essential part of a good test plant is to obtain good grinding power data. The grinding power per ton has been found as reliable a method as is known to arrive at sizing for the commercial mill.

A study of 15 properties where the ore has been tested in wet autogenous mills indicated the following:

Test plant at mine site - 9
Sample taken to established test facility - 6

However, many of the properties studied were close to existing mines or research facilities. Of the really remote mining properties only 3 out of 9 such properties had pilot plants at the site. Of the 15 properties looked at in this study 14 of them now have operating plants. Of 8 dry mill installations in North America, all began with laboratory testing in 1.7 m dia. mills, 2 following with a larger dia. pilot mill on site, one with a production sized unit.

TEST PROGRAM

Autogenous grinding has attained sufficient prominence, and has proven advantages with regard to both capital and operating cost, to warrant an investigation being made for any new project under consideration.

Existing and proven alternatives may involve wet or dry grinding, autogenous or semi-autogenous operation, open or closed circuit and single or two stage systems. Preliminary testing will eliminate some of these alternatives, thorough testing being conducted on those remaining to fully evaluate the benefits available from the process.

Wet or Dry Grinding

Although it is generally considered that a wet process plant normally uses a wet grinding system, dry grinding as typified by the Aerofall mill system may provide features advantageous to the process. Many developing properties will warrant testing, at least of a preliminary nature, by both methods before a major program is initiated.

In areas where water is unavailable or in short supply, or when a dry ground product is required, dry autogenous grinding should be a first consideration. In wet process plants the current high cost of the fuel required for ore drying becomes a limiting factor, 3 percent moisture being the probable economic limit, unless a source of dry waste heat is available or an offsetting metallurgical benefit is obtained. If required, the mill is capable of accepting ore containing in excess of 20 percent moisture and producing a dry ground product.

*Other similar systems include Koppers-Allis Chalmers.

Mill Feed

Further to the comments regarding sample selection, the ore treated in an autogenous or semi-autogenous mill should be naturally graded in size, as produced by a primary crusher. The optimum ore size, with which maximum power draught is obtained is given by:

$$d = 0.95\ D^{2/3}$$

where, d = size that 80 percent of the rock passes, inches
D = mill dia. inside shell, ft.

The ore sample should be thoroughly blended so that a consistently sized material is fed to the mill, or the sample may be screened into two or more fractions that will be re-combined to give a consistent blend. The formula can be used as a guide for primary crushing.

Test Procedure

The primary purpose of testing is to determine the capability of the mill grinding to a specific product size, or to various product sizes, under optimized conditions.

Normal procedure involves feeding the ore to the mill continuously until steady state conditions are established at the prescribed mill load level. This steady state refers not only to a balanced input-output rate but also to a stable chemical and physical condition. It is the nature of the mill to accumulate the harder fractions of the ore in the mill load and it may require 6 to 12 hours of operation before such stability is attained. In wet mills it is necessary to attain a constant discharge pulp density which can further extend the pre-run time.

It should be recognized that this steady-state mill load is the grinding media of the mill. Once it is established care should be taken against loss of the charge so that subsequent changes can be made with that ore without repetition of the full pre-run procedure.

In addition to capacity and grind, additional data should be collected to permit calculation and scale-up of mill power, steel wear, and to specify and layout of the peripheral equipment. These data include:

1. Flowsheet;
2. Ore type, sizing, spec. gravity, moisture;
3. Feed tonnage;
4. Mill charge: weight, volume and sizing;
5. Mill discharge: volume and density;
6. For dry mills: airflow volume, density and mill pressure drop;
7. Mill power: gross and net;
8. Mill speed: r.p.m.;
9. Product: sizing and distribution;

10. Circulating load: tonnage and density;
11. Steel wear: balls, liners, grates. (Commercial Mill data probably more accurate. See Table 3).

When sizing the plant equipment from the collected data, it should be recognized that the test mill capacity is a maximum, cbtained under controlled and optimized conditions. The average performance in a commercial plant, subject to ore variation, classifier fluctuation and operating techniques will be less than optimum for which a minimum allowance of 10 percent should be made in selecting the equipment.

Open or Closed Circuit

The grind should be varied from a coarse open circuit grind, normally conducted with a trommel or scalping screen, to the finished grind required for concentration, usually performed in closed circuit with screens or classifiers. A study of the test results will indicate the type of grind best suited to the treatment of the ore under investigation, by power efficiency, magnitude of circulating loads, feed rates, or from metallurgical results.

TABLE 1

TYPICAL VARIATION IN GRIND DATA

Closed Circuit Screen	Feed Rate Lbs/Hr	Net kWh/ Ton	80% Passing Size In Microns	Calc. Bond W.I.
4 Mesh	1300	11.7	300	20.1
48 Mesh	1050	15.5	53	11.3
100 Mesh	900	16.2	35	9.6

Assumed feed size 80% passing 150,000 microns

Note: It is obvious from this case that the mill must be operated to produce a grind of at least minus 48 Mesh in order to achieve high power efficiency.

Autogenous or Semi-Autogenous

Very few ores will grind with high power efficiency or a full autogenous basis. Therefore, at an early stage tests should be conducted using ball charges of varying weight and ball size to determine the maximum power efficiency. This, along with estimated ball consumption, can be used for economic studies to determine whether the commercial mills should be designed with or without a ball charge. The selection of the proper ball charge usually comes with experience from operation of many test programs. The examination of the mill charge and the circulating load of sand or pebbles is an important part of the assessment. Data from a typical test program is shown in Table 2.

TABLE 2

FULLY AUTOGENOUS OR SEMI-AUTOGENOUS

Ball Charge Lbs	% Mill Vol.	Feed Rate Tons/Hr	Net KWh/Ton	KWh/Ton -100 Mesh
0	Nil	2.46	8.18	6.788
860	2.3	3.49	7.46	4.750
1650	4.3	5.21	5.67	2.654
2552	6.7	5.64	5.50	2.868

Note: From this work the most favorable ball charge would be in the order of 5% mill volume.

Power on Commercial Mills

It has been found that as a general rule the power requirements to grind a ton of ore in a test mill to a given size can usually be duplicated in a large commercial mill. This assumes that the test work is carried out under carefully controlled conditions with good power measurements. It is generally found that in the small test mill the load should be maintained between 25% and 30% of mill volume. The total power drawn under these conditions can be used for scale up to larger mills and to obtain unit power requirements per ton of ore treated.

Various researchers and manufacturers have used many formulas for arriving at the total power requirements of a commercial mill. Many of the early wet autogenous mills were underpowered and had to operate at lower load levels particularly if a ball charge had to be added to obtain the throughput. Such a situation increases both the liner and ball wear. It is therefore essential to have enough installed power on the commercial mill.

In general, most scale-up formulas use an exponent of the mill diameters and a direct comparison of mill lengths to arrive at a multiplying factor for the scale-up of the test mill power. The exponent used for mill diameters has varied from 2.4 to 3.0 as a study of the literature will reveal.

For wet mills an exponent of 2.65 should be used to arrive at the power to drive the commercial mill provided that good power measurements have been taken in the test mill at a known charge level between 25 and 30%. The power on the test mill is always given in net kW and the power on the commercial mill is generally measured as gross power. Therefore, using the formula:

$$P_C = P_T \frac{L_C}{L_T}\left(\frac{D_C}{D_T}\right)^{2.65} \Big/ 0.95$$

should give the required power of the mill motor on the commercial mill where,

P_C = Gross Power in kW Commercial Mill
P_T = Net Power in kW - Test Mill
D_C = Nominal Diameter Commercial Mill
D_T = Nominal Diameter Test Mill
L_C = Nominal Length Commercial Mill
L_T = Nominal Length Test Mill

Gross Power: Power input to mill motor.

Net Power: Power actually consumed in grinding after deducting mechanical and electrical losses.

This formula has been used to check the actual power obtained on commercial mills against the calculated figures. These cases were taken from actual test work where it is known that the work was carefully carried out and correctly supervised. In most cases the calculated figure from test work is generally conservative or safe for the design of commercial power requirements.

To find the nominal capacity of the mill in tons per hour 90% of the gross power should be divided by the net grinding power per ton obtained in the test mill.

Analyses of the data from operating dry Aerofall mills shows that mill power varies on the diameter raised to the power of 2.8 and on the effective length. Specifically, the power drawn by mills of standard design treating normal ores under optimized conditions may be calculated from:

$$P = 0.000174 \times W \times D^{2.8} \times L$$

where P = motor input killowatts for direct driven mills; W = bulk density of the mill load, lbs per cubic ft; D = mill diameter inside shell liner, ft; L = mill length inside liners, ft.

The calculated power requirements of a commercial mill should also be compared with operating mills with corrections for differences in size and in bulk density of the mill loads.

All manufacturers have their own methods for arriving at mill power. It is therefore important to submit the test data to the manufacturer to verify that the mill will have the correct installed motor size.

In summary;

1. Take careful power measurements on the test mill with the load in mill between 25 and 30% of total mill volume.

2. Scale-up by formula to obtain mill motor size for commercial mill.

3. Compare this power requirement with mills of comparable size using correction for difference in bulk density of the mill load or the load plus ball charge.

4. Obtain nominal capacity of the mill by dividing 90% of the mill input power by the net kWh per ton obtained in the test work.

5. Submit test data to mill manufacturer for verification that the commercial mill is correctly sized and powered to produce the required tonnage.

Mill Speed

The speed of wet autogenous and semi-autogenous mills has been varied over the range of 60 to 90 percent of critical speed, the normal range being in the order of 70 to 75 percent, calculated inside the mill shell.

The dry Aerofall mill is run at 84.5 percent of critical speed, calculated inside shell, in order to optimize the condtions required for efficient grinding and for air-sweeping of the mill.

Steel Wear

Ball wear rate, a measure of the abrasiveness of the ore, can be measured in pilot plant testing provided that sufficient tonnage is processed to give a measureable loss of weight. Data may be obtained from the loss of weight of the total ball charge or by the use of tared test balls included in the mill ball charge, from which ball consumption may be calculated.

Ball consumption in dry mills can be predicted with accuracy, but in wet mills the effect of corrosion makes prediction more difficult; comparison with existing mills treating similar ores may provide conformatory data.

Liner consumption may be estimated from the measured ball wear rate or by actual measurement of the weight loss of the lining during testing. Again, confirmatory data may be obtained from existing operations.

TABLE 3

BALL AND LINER CONSUMPTION

Wet:	Ball Charge % Mill Vol.	Balls Lbs/Ton Feed	Liners Lbs/Ton Feed	kWh/Ton	lb/kWh Balls	lb/kWh Liners
Lornex	3-8	0.62	0.0967	5.8	0.107	0.017
Pima (Published)	8	0.8	0.157	7.9	0.101	0.020
Island Copper	-	(2.25	(0.34	21.45) single	0.105	0.016
(Published)	-	(1.6	(0.19	17.2) two stage	0.093	0.011
Similkameen	8.5	0.85	0.26	19.5	0.040	0.013
Dofasco (Sherman)	-	-	0.294	12.2	-	0.024
Savage River (Published)	-	-	0.09	12.6	-	0.007
Dry:						
Hanna Mining, National	11.9	1.07	0.36	10.2 (Primary)	0.105	0.035
Somina, Mauritania	5.0	0.45	0.50	4.8	0.094	0.104
Ass. Port, Cement, Weardale	11.6	0.09	0.02	4.6 (Primary)	0.020	0.004
I.O.C. Carol Lake	-	-	0.09	1.6 (Fully Autog.)	-	0.056
Brit. Ind. Sand Ltd.	1.2	0.03	0.05	0.6	0.050	0.083

TABLE 4

TYPICAL EXAMPLES OF AUTOGENOUS GRINDING PLANTS

Single Stage

Wet	Dry
Hibbing Taconite, U.S.A.	I.O.C., Labrador City, Canada
Palabora, S.A.	B.I.S., U.K.$^{(3)}$
Q.C.M. Mt. Wright, Canada	J & L (Benson) U.S.A.
Q.C.M. Lac Jeannine, Canada	LKAB, Sweden
Wabush, Canada	S.L.D.C., Sierra Leone
Rana Gruber, Norway	Somima, Mauritania
Bong, Liberia	Union Carbide, (Uravan), U.S.A.
	St. Marys Cement, Canada
	Intalco, U.S.A.

Two-Stage

Wet	Dry
Griffith Mine, Canada	Hanna Butler, U.S.A.$^{(1)(4)}$
Empire Mine, U.S.A.	Coldstream Copper, Canada$^{(1)}$
Tilden Mine, U.S.A.	M.T.D. (Mangula) S. Rhodesia$^{(1)(4)}$
Aitik, Sweden	St. Lawrence Cement, Canada$^{(1)}$
Sherman, Canada	A.P.C.M. (4 plants) U.K.$^{(1)}$
Savage River, Australia	Dyckerhoff Cement, Germany$^{(1)}$
Lornex, Canada$^{(1)}$	Cements Lafarge, France$^{(1)}$
Similkameen, Canada$^{(1)(2)}$	Cementerie d l'est Alergien$^{(1)}$
Pima, U.S.A.$^{(1)}$	Norcem, Norway$^{(1)}$
I.O.C. (Sept Isles) Canada$^{(1)(2)}$	
Island Copper, Canada$^{(1)(2)}$	
Pilot Knob, U.S.A.$^{(1)(4)}$	
National Steel, U.S.A.$^{(1)}$	

Notes: (1) Semi-autogenous
(2) Originally designed as single stage
(3) Converted to wet mill
(4) Concentration between grinding stages

Single or Two-Stage Grinding

Although there are single stage grinding plants in operation, in which the primary mill produces the finished ground product, these are normally confined to coarser grained ores. Several plants, initially commissioned with single stage mills have been converted to two-stage systems, to obtain more tonnage or greater flexibility in treating ore of variable characteristics, for the minimum capital expenditure.

In general the primary autogenous mill should be considered as a coarse grinding unit reducing the ore to its natural grain size. If finer grinding is required this will be done more efficiently in a regrind mill in which the media is sized as required by the feed. As the autogenous mill tends to grind to grain size it may provide an opportunity for a concentration stage between the grinding circuits thereby reducing the power requirements of secondary grinding.

Pebble Mills as Secondary Grinding Mills

A great many autogenous grinding plants have been built using pebble mills as the secondary grinding mills. In the test work on the primary mills it may often be noted that the ore breaks down readily to a minus 75 mm size and at that point well rounded and tough pebbles form in the charge and the grinding rate of the primary mill is greatly reduced. These small well rounded tough pebbles are normally broken by using a ball charge in the primary mill. However, in many cases these pebbles can be removed from the primary mill through pebble ports in the mill grates and used as grinding media in secondary pebble mills. If used in semi-autogenous mills this procedure requires a means of separating out the larger balls that escape with the pebbles and returning them to the primary mill. If there is an excess of pebbles generated beyond the requirements of the pebble mill, these may be crushed before returning to the primary mill. This process usually improves the performance of the primary mill and saves money in the operation of secondary grinding. Pebble mills, being larger, cost more money than ball mills to install and require slightly more power, but the grinding media cost saving often more than offsets the extra capital and power costs. The effect on the primary mill of removal of pebbles for grinding media in the secondary mill should be studied as part of most preliminary testing programs.

The pebbles produced from primary mills are generally the hardest part of the ore and make much better grinding media than pebbles screened from a crusher product.

Media Competency and Work Indices

One difficulty found in scaling-up from test to commercially sized mills is commonly relegated to the area of media competency. The impact forces in a large diameter mill are greater than those in the smaller test mill and, consequently, there is an increased rate of breakage in the larger mill. This is normally counter-acted by feeding the larger mill with larger sized ore, as indicated previously. In spite of this, situations arise as in the case of certain Labrador iron ores, where the larger mills operate autogenously with larger ore and good power efficiency although the test mill required a substantial ball charge to achieve a similar result. It might well be argued that the discrepancy is not one of compentency but is due to failure to remove the ground material from the mill as fast as it is produced, thus limiting the feed rate and resulting in over-breakage of the media sized ore. The result is a mill "choked" with fine material. If the means for removal were improved, the larger mill with a ball charge might show greater efficiency than the test mill.

A truly non-competent ore might best be described as one that breaks down readily to small shattered pieces, resulting in a mill load composed of material too small to act as media for its own reduction.

If a ball charge is required in the test mill it is reasonably certain that the ore is competent as feed for a larger mill. If no ball charge is required, competency tests may be run to evaluate the ore for autogenous grinding. A test procedure has been developed by Allis-Chalmers in Milwaukee where large sized lumps of ore are run in a mill for a definite number of revolutions, then the charge screened to determine the break-down rate from coarse to smaller sizes. In addition, comparisons may be made with existing plants treating similar ores autogenously or arrangements may be made to test run the material in that larger mill.

Bond Work Index tests should be run on the ore, rod mill indices for coarse, rod and ball mill indices for fine grinding. This gives a base point for comparison with the autogenous grinding work. If fully autogenous grinding indicates a poor power efficiency as compared to the work index, then probably semi-autogenous grinding is required. In the semi-autogenous test work the ball charge should be adjusted until the mill efficiency is close to power requirements as indicated by the Work Index.

Small Mill Tests on Cores or Small Samples to Determine Variations in Grindability in Various Parts of the Deposit

It has been found that most large deposits can have a wide variety of ore types and with a wide variation in autogenous grinding characteristics. It has been found impracticable to take enough bulk samples from the orebody and run them through a pilot mill to determine the full range of grindability. Consequently many autogenous plants have started and found that wide swings in grinding rates have affected both the grinding mill operation and the follow-up concentration processes.

For this reason considerable work has been carried out to develop a method of treating small samples from the orebody to indicate the probable changes in grindability. The best samples from an orebody are usually the drill cores on which grade and tonnage calculations are based. If these cores could be used to determine variations in grinding characteristics, then the mining plan development could be based both on grade and grinding characteristics.

The mill used for this correlation work was an 18" (450 mm) diameter Aerofall mill. This is a dry air swept mill similar to large commercial mills. The ore or core samples are fed to the mill by an automatic feed control system and the feed size is normally minus 1¼ inch (32 mm) in size. A constant weight and size of balls is used as grinding media.

The products from the mill are collected dry and weighed. The feed size and product size are established and with known power and feed rate a Bond Work Index calculation is made. This results in a Work Index figure that is called the Autogenous Work Index. This Autogenous Work Index has been established for many ores that are also being processed in commercial autogenous grinding plants. Consequently by comparing the Autogenous Work Index of new ore with an ore being treated in a commercial plant, the requirements of a full scale autogenous plant to treat the new ore can readily be established. Also, by grinding a variety of samples from the orebody in the 450 mm mill the complete range of grindability that might be expected can be established. Correlation with the bulk samples treated in a larger test plant can also be established and thus determine what allowance should be made in the commercial plant for the differences established by the small mill from a greater number of samples.

TABLE 5

TYPICAL RESULTS OF 18" MILL TESTS ON DRILL CORES AND BULK SAMPLES

	18" MILL AUT. W.I.	PILOT MILL kWh/TON
Bulk Sample (Hard)	33.57	8.57
Bulk Sample (Soft)	18.90	5.14
Core Sample Softest	14.64	
↑	15.44	
	25.17	
↓	38.58	
Hardest	34.51	
* Selection of Commercial Plant	22.0	6.5

* Based on plus or minus 20% to cover all but the hardest or softest samples.

The mining plan is usually developed with the object of keeping the grade to the plant more or less uniform for reasonable lengths of time. To the design equation we can now add variations in grindability. The grindability factor may not be important enough to worry about or it may present special problems of blending and/or stockpiling.

The samples selected for testing in the 450 mm mill should be composited from cores or development samples, taking into account features such as:

1. Geological changes;
2. Grade changes;
3. Location in deposit, i.e., near surface, at margins of the ore zone, in depth, etc.

The 18" (450 mm) mill will ensure there will be no grindability surprises in the orebody and at the same time reduce the number of bulk samples which may have to be run through the test plant.

ESTABLISHMENT OF GRINDING CIRCUIT

When the data from the test work is finally collected, a decision must be made on the grinding circuit that best suits that particular ore. In addition, with test work on cores and the development of the Autogenous Work Index, direct comparisons can be made between successful commercial plants grinding ore of the same general characteristcs.

The circuits which should be examined are:

A. Wet or Dry Grinding

B. Single Stage - 1. Fully Autogenous
   - 2. Semi-Autogenous

C. Two-Stage - 1. Fully Autogenous Primary - Pebble or Ball Mill Secondary
   - 2. Semi-Autogenous Primary - Pebble or Ball Mill Secondary

The test work normally will supply the answers that will eliminate some of the possible combinations. These include:

1. Inefficient power consumption with fully autogenous, will limit consideration to semi-autogenous or pebble removal.

2. Single stage grinding tests that require extremely high circulating loads and high power requirements to produce the required product fineness would indicate grinding should be explored.

3. Build up of small hard pebbles in the primary mill: Fully autogenous probably not suitable. Try pebble removal or semi-autogenous.

4. No good development of tough pebbles in primary mill: Probably requires hard ore addition or a ball charge.

It should now be possible to compare ores that have approximately the same grindabilities. Consequently, an operating plant can be determined that has had good success in grinding an ore of similar characteristics. This operating experience can often indicate the proper grinding circuit for the new plant.

Other factors which should be examined include:

- Accessibility of Property - Road, rail, etc. - limitations.
- Power availability and cost.
- Water availability and cost.
- Steel media and liner costs.
- Grinding media competency.
- Type of mining and handling system (underground mining and conveyor hoisting).
- Blending requirements - expansion possibilies.

Accessibility of Property

Many new mines are located in isolated areas and transporation between the port or rail head may only be over a poor road or a rail link with limited carrying capacity. Some mines now pump their product to the shipping point so road connections may be fairly primitive. There can often thus be limits to the size or weight of equipment which can get to the property.

Power Availability and Cost

This is becoming one of the real critical items in any grinding evaluation. With the increased cost of all forms of fuels, the cost of power has escalated more rapidly than most other basic items such as steel. Grinding is probably the major user of power in any ore concentration plant, consequently careful examinaton of the means of reducing power consumption is essential.

Water Availability and Cost

In many areas water is scarce or expensive to procure. Evaporation losses from reclamation systems may be high. Dry autogenous grinding may prove to have an economic advantage.

Steel Costs

This item ties in with accessibility of the property and expensive or difficult transporation may govern the decision to eliminate the use of steel as grinding media. At the same time, if grinding is eliminated many of the steel mill liners can be replaced with rubber with much less weight of material to be handled.

Grinding Media Competency

This is a special test but if there are indications that the media might break down in a mill it may preclude the use of fully autogenous grinding. However, there are notable exceptions, for instance, Labrador iron ores break down very rapidly in a media competency test but in practice are ideal ores to grind fully autogenously. If the mill is designed as a fully autogenous mill, sufficient mill power should be installed so that the use of a ball charge would be possible (if necessary) at some future date.

Type of Mining and Handling System

If it is known that the ore to be ground will come from an underground system and will be hoisted and handled by conveyor systems then it is obvious the mill feed will be relatively fine. Therefore there is less chance that a fully autogenous grinding system will be completely successful and perhaps a semi-autogenous system should be investigated.

Blending Requirements

If it is found from the test work that wide fluctuations in grinding rates can be expected from an ore, then probably a semi-autogenous system can provide more flexibility and uniformity of throughput than a fully autogenous plant. However, if the ore has particularly wide ranges of grindability then some form of blending should be considered. Forms of blending which can be utilized are shovel control in the pit or partial stockpiling of one type of ore.

Expansion Possibilities

It is normally much easier, quicker and less expensive to expand a single stage grinding plant by addition of secondary units than to expand the same plant by addition of more lines. If the plant starts out with two-stage grinding there is no alternative, if expansion is required, but to add further complete lines. If it is known when a plant is built that modest expansion is contemplated within a few years then it may be advantageous to build a single stage grinding plant to start with and allow room to add secondary grinding at a later date. The test work should show the possibilities for such a plant design.

Examples

Similkameen	- 3 single stage primary mills	- tonnage 15,000 tpd
	Add 2 second stage ball mills	- tonnage 24,000 tpd
Island Copper	- 6 single stage primary mills	- tonnage 30,000 tpd
	Add 3 second stage ball mills	- tonnage 44,000 tpd
Dunbar Cement	- 1 primary, one regrind mill	- tonnage 3,220 tpd
	Add second regrind mill	- tonnage 3,920 tpd

If there is no clear cut evidence from the test work then the decision must be made on the basis of capital and operating costs.

Some factors which are self evident in capital cost studies include:

1. It is more expensive to install the same grinding horsepower in a multi-unit plant using all primary mills when compared to two-stage grinding. This is because primary mills require expensive feeding, classification and ore storage systems.

2. Pebble mills are more expensive than ball mills for the same power requirements as they require a larger plant area.

3. Semi-autogenous mills are cheaper to install because the mill will draw more power than the same size mill used as a fully autogenous mill. Consequently more tons per hour from an equal size mill.

4. Install the largest mills as are proven in commercial use without mechanical problems. If necessary, install slightly larger units than those used successfully if it means elimination of one or more lines in a large plant.

Factors which affect operating cost studies:

1. Pebble mills are usually less costly to operate than ball mills, particularly if good quality pebbles are obtained from the primary mill and help to improve the efficiency of the primary mill operations by their removal.

2. Semi-autogenous primary mills are often cheaper to operate than fully autogenous mills because of higher power efficiency, This is not always true as the primary fully autogenous may have high power efficiency which is not improved by the addition of a ball charge.

3. It is usually cheaper to operate a single stage grinding system as maintenance and operating labour are generally lower than for a two-stage system.

So, in summary, the decision on the configuration of the grinding system in a commercial installation involves:

1. A careful analysis of all test data.
2. A comparison of the new ore with known ores of the same grindability and the types of plants used successfully to treat these ores.
3. Other factors such as mine location, etc.
4. A capital cost comparison.
5. An operating cost comparison.

LONG OR SHORT PRIMARY MILLS

Wet autogenous mills have been built in a variety of shapes with diameter : length ratios varying from 3 : 1 to less than 0.5 : 1. In dry Aerofall mills the length is a constant function of the diameter, giving ratios of 0.3:1 for smaller mills to 0.23:1 for the largest mills.

The expensive part of building large diameter mills is the cost of the large diameter ring gear and the machining, casting, etc., of large diameter mill ends. Consequently if the same mill horsepower can be obtained by lengthening a mill it is usually less expensive to obtain the power requirements by lengthening the mill than it is to add to its diameter.

However, there are other considerations which must enter into the equation. These are:

A. The grinding media must be competent enough to be distributed throughout the mill length so that efficient grinding can be carried out throughout the mill length.

B. Classification is an important function of any grinding mill. It is obvious in a long grinding mill that fine material produced in the feed end of the mill will be in the mill much longer than fine material produced in the discharge end of the mill. Material not removed from the mill immediately after it reaches liberated size uses power and also will become finer as it passes through the mill. Consequently long mills will:

   Other things being equal, use more power per unit of finished product produced and will tend to produce more slimes.

C. Although dry grinding mills add to their cost through the use of smaller L:D ratios, this is compensated for by:

(i) Scale-up at a higher exponent of the diameter.

(ii) Over-grinding is minimized by removal of the product by air sweeping.

(iii) Reduced steel consumtpion as smaller ball charges are normally required and corrosion is eliminated.

It is generally considered that a coarse grained ore such as Labrador iron ores containing specular hematite and magnetite should be ground in as short a mill as possible. These mills use very little grinding power and produce liberated coarse grained iron minerals with very little sliming.

Ores that are to be concentrated by flotation should not be ground in a circuit that produces excessive slimes. Copper ores with chalcopyrite and bornite mineralization are representative of this type and slightly longer mills than the iron ore mill should be used to reduce circulating loads and take advantage of the competent grinding media.

Some particular ores such as gold, where extraction is by leaching, can be ground in longer mills but with a penalty of excessive power. However, circulating loads are reduced to a minimum and the capital cost of the plant could be decreased by using the longer mill.

Therefore, the decision regarding the length of a mill can come from the following considerations:

- Type of ore and mineralization.
- Type of concentration.
- Cost of power.
- Capital cost comparisons.
- Risk involved - the longer the mill the greater the possibility of media deficiency or segregation.
- Grinding media competency - the larger the diameter of the mill the greater the impact forces and the possibility of rapid breakdown of coarse ore.

DEVELOP FLOWSHEET

Once the decision on the grinding configuration for the plant has been reached then the flowsheet to treat the required tonnage can be finalized. The working flowsheet will show the expected mill flows, densities, sizings, and anticipated metal grades. This information, along with mill power and probable mill dimensions, will then be available to prepare a mill specification which will be sent to various mill manufacturers.

PREPARATION OF SPECIFICATIONS FOR SUBMISSION TO VENDORS OF AUTOGENOUS MILLS

Once the grinding configuration is established and the likely size of mills has been selected the next step is to prepare specifications for submission to suppliers of such equipment. In many cases the power requirements only are given to a supplier and the suppliers sizes the mill.

It costs manufacturers much time and money to prepare proposals. Consequently these may be preliminary to broadly arrive at prices for use in feasibility studies. For final tendering and purchasing it should be clearly stated that this is the final price and that the equipment will be purchased from the successful bidder.

It is not ethical for engineering firms to allow bidders to alter their prices after this proposal is made unless there is a change in scope or requirement for some part of the program. In this case all bidders must be given the same privilege unless it is for some part which one bidder has or has not included and clarification is required from that one bidder.

Feasibility studies include an estimate of the total cost of a project usually based on quotations on major equipment and latest information on bulk materials. If, after careful analysis of these figures by the owner, a decision is made to proceed with the venture there may be a period of from 3 to 12 months or more between the first proposal and the final tendering. Quite often the feasibility study points to changes in the overall project which must be made before it will be acceptable as a financial risk by the owner and his financial backers. This revision may:

- Increase or decrease the size of the project.
- Decrease expected plant performance to reduce initial capital expenditures by elimination of equipment.
- Increase size of equipment to reduce the number of lines and thus overall capital cost.

In general, the prices received on the preliminary bidding will be somewhat higher than those obtained when the project is actually proceeding. However, this is not always true as it depends on the shop availability and the amount of inflation between the first bid and the final tenders. Large grinding mills and the gears can only be produced in a limited number of shops in the Western World. If these shops are not busy then better prices are usually obtained than when there is a good backlog of orders for all the shops.

The specification for final feasibility or plant construction will give certain specific information to the vendor, including:

1. Pertinent test work with a full description of the type of ore to be processed.

2. Number and type of mills required for the commercial plant.

3. Tons per hour and sizing of the feed, circulating load and product.

4. Probable horsepower of the drive motor and type.

5. Requirements for ball charge, if any.

6. Expected rotational speed - % critical.

7. Special requirements such as type of bearing, location of ring gear, etc.

8. Location of plant site and access.

9. Required delivery date.

The vendor will be required to supply the following information:

1. Price - normally liners, gear and drive, taxes and freight would be listed as separate items. Probably at this time a firm price will not be given but will be based on escalation of labour and materials based on a certain index. If prices are firm they will only be held for a limited time.

2. Expected delivery dates.

3. Warranties.

4. Patent infringement.

5. Payment schedule.

6. Shipping information. Size and weight of largest pieces.

7. Technical Information

   (a) Liner design - drawing - weight - material of construction - size of bolts.

(b) Drive arrangement - gear - pinion - reducer - inching device.

(c) Mill Ends and Trunnions - Material of construction - weight - number of pieces.

(d) Mill Bearings - type - drawing - lubrication system.

(e) Mill Shell - thickness - number of pieces - material.

(f) Discharge Detail - trommel, etc.

(g) Feed Chute - design - materials, etc.

8. Other Information

- Assembly drawing.
- What type of shop assembly or erection if anticipated.
- Lubricants required.
- What erection supervision can be supplied in the field.
- Recommended spare parts.
- Rotation reversal recommended or possible.
- Jacking cradle requirement.

SELECTION OF EQUIPMENT FROM TENDERS

Once the engineers receive the tender documents from all bidders, they are expected to prepare a bid analysis which can be used to make a recommendation to the client, and the client can use this for study purposes to verify the engineers' recommendations. Quite often at this stage it is necessary to contact the supplier to obtain clarification on certain items in his bid and/or additional prices for items not included in the original bid.

The final decision on the placing of the order is usually determined by:

(a) Price

(b) Expected delivery

(c) Warranties

(d) Past history of supplier - does he adhere to delivery dates, prices, quality, etc.

(e) Materials for construction

(f) Adherence to original specifications

(g) Bearing and lubrication details

(h) Owner preference

A great many of the decisions are made on price and this is probably the most important factor. However, many owners have a strong preference for a certain supplier and may decide that even if it costs somewhat more they wish to purchase this equipment.

Once a tender or bid is accepted a purchase order would be issued by the owner to the successful supplier.

In general, warranties are usually given by the manufacturer to replace or repair any defective part which fails within one year of first operation. The replacement part is supplied at the manufacturing plant and must be transported and erected by the owner. Manufacturers of such equipment normally do not accept any damages for loss of production or consequential damage. If these are required by an owner they can usually be procured by insurance. The supplier's representative is usually present during startup operations to ensure the equipment is properly used and any repairs made are a legitimate back charge to the supplier. Once the mill is accepted the supplier should receive any holdback payments which are defined in the original purchase agreement.

REFERENCES

1. Bachman, W.D., Last, A.E., and Nabbs, S.W., "Autogenous Grinding of Disseminated Copper Ores:, AIME, Preprint No. 70-B-71.

2. Bergstedt, L., Boliden Aktiebolag, "Some Basic Factors Influencing the Use and Optimization of Autogenous Grinding", AIME, March 1977.

3. Bond, "An Expert Reviews the Design and Evolution of Early Autogenous Grinding Systems", Engineering and Mining Journal, 1964

4. Bruce, "The Effect of Grinding Media on the Selective Flotation of Copper-Lead-Zinc Ores", AIME, 1976.

5. Coyle, D., Cleveland Cliffs Iron Co., "Progress Report on Autogenous Grinding", 1963.

6. Dor, A., "Primary Autogenous Grinding Concentrators - Recent Trends and Development", AIME, 1965.

7. Dor, A., "Design and Operation of Primary Grinding Mills", 1968.

8. Fagerberg, B., and Ornstein, H., "Grinding Tests on Magnetic Ores with Aerofall Mill and with Wet and Dry Rod Mills:, IMPC, London, April 1960.

9. Fahlstrom, G., Boliden, "Autogenous Grinding of Base Metal Ores at Boliden, Aktiebolag", 1973.

10. Fleck, R.G., and Durocher, R.E., "A Progress Report on the Aerofall Mill:, Mining Congress Journal, December 1965.

11. Hartzog, L.D., Ray Mines, "Semi-Autogenous Grinding of Ray Mines Diversion Sulphide Ores", 1973.

12. Hellyer, W.C., Newmont, and Campbell, R.A. Michigan Technological University, "Control of an Autogenous Grinding Circuit by Means of a Crusher", 1970.

13. Jackson, O.A.E., "Autogenous and Composite Load Milling Practice at the South African Gold Mines of Union Corp.", Symposium of Autogenous Grinding, 1963.

14. MacPherson, A.R., Canadian Bechtel Limited, "A Simple Method to Predict the Autogenous Grinding Mill Requirements for Processing Ore From a New Deposit", SME/AIME, 1976.

15. Mamen, C., Iron Ore Company of Canada, Carol Project, Canadian Mining Journal, Nov. 1962.

16. Marechal, B., "The Effect of Rotational Speed in Autogenous Grinding", Mining and Minerals Engineering, Oct. 1967.

17. McDermott, W.F., Lipovetz, G.J., and Peterson, H.R., Pickands-Mather & Co., "The Dollars and Sense of Autogenous Grinding", AIME, 1972.

18. McManus, J.R., Lornex, "Semi-Autogenous Grinding of Cu/Mo Ore at Lornex", 1974.

19. Meaders, R.C., and MacPherson, A.R., "Technical Design of Autogenous Mills", AIME, 1964.

20. M.T.D. Mongula Ltd., "Deco Trefoil", Bulletin No. M4-B109, Nov-Dec, 1960.

21. Paterson, J.G., and Frame, G.A., Inco, "Grinding Practices at Frood-Stobie Mill", CIMM Meeting, 1971.

22. Pena, Fred, Cyprus-Pima Mining Company, "Cyprus-Pima Pilot Plant Predictions vs Actual Plant Results", S.M.E. Fall Meeting, 1974.

23. Rowland, C.A., "Why Intermediate Autogenous Grinding", AIME, 1964.

24. Rule, W.T.,"Dry Autogenous Grinding at Butler Taconite", 29th Annual Mining Symposium, Duluth, Minn., January, 1968.

25. Smith, A.R., and King, G.A., "Four Years Operating Experience with the Aerofall Mill at British Industrial Sand Ltd.," Eighth Commonwealth Mining and Metallurgical Congress, Australia, 1965 Vol. 6, Proceedings, General.

26. Vincent, J.D., and Weyler, P.A., Climax Molybdenum Co., "Development of Henderson Mill Design", AIME, 1972.

Chapter 14

DIGITAL SIMULATION: AN AID FOR MINERAL PROCESSING PLANT DESIGN

A.L. Mular* and J.A. Herbst**

*Professor of Mineral Engineering
University of British Columbia
Vancouver, B. C.
CANADA

**Associate Professor of Metallurgical Engineering
University of Utah
Salt Lake City, Utah
USA

ABSTRACT

Digital simulation is defined and simulation techniques are reviewed. Typical methodology involves the selection of models, the adjustment of raw data, the selection of a fitting strategy, the assessment of model accuracy in a statistical sense, and the development of an executive simulation structure. Simulations serve as aids for process engineers in evaluating the feasibility of a proposed flowsheet and the effect of a modification to an existing flowsheet. In addition, simulation plays an important role in off-line optimization and in the evaluation of alternative circuit control strategies. Several examples are presented to illustrate these uses for simulation including applications in crushing circuits, grinding circuits, and flotation circuits. It is the opinion of the authors that, within the next decade, mineral processing plants will be studied routinely via digital simulation during the preliminary design stage. Simulation will constitute an important component of the design decision making process.

INTRODUCTION

Digital computers became available for process control and data logging in the nineteen fifties (1). As a consequence, mineral engineers were encouraged to initiate programs which, over the years, have led to valid digital simulations of several plant circuits for various purposes (2,3,4,5,6,7,8,9,10). Initial efforts were directed toward the development of process models either from laboratory data or from data acquired around full scale production units. This work provided accurate quantitative descriptions of several important unit operations including crushing, grinding, classification and flotation and in many instances provided valuable insight into underlying process mechanisms. In recent years, executive structures have been sought to allow the simulation of entire circuits. Although generalized structures are available (11,12), these have usually been adapted to meet simulation needs for specific types of circuits. Fortunately these simulators have been flexible enough to permit a variety of applications in mineral processing.

The purpose of this chapter is to define digital simulation, to review simulation methodology, and to describe current applications in plant design.

DIGITAL SIMULATION DEFINED

The methodology utilized to obtain a mathematical representation of an operating plant or portion thereof (such as a grinding circuit or a filtration circuit) for computer study is referred to as digital simulation. The word digital is used to make it clear that computations are made with a digital rather than an analog computer. The mathematical representation of the plant or circuit of interest is converted to a computer program. The program is termed a simulator. It will have at least two essential parts: a main program, termed an executive, and a set of subprograms or subroutines*** containing mathematical representations or models of each process unit in the circuit to be simulated.

Executives

An executive structure is, in effect, a flexible mathematical flowsheet. It receives information about input streams to a process unit (subroutine)--thus the flow of solids and water to a rod mill are inputs--and then commands the subroutine containing the model of the process unit to calculate output information. Output information is used either as input to other process units or printed as final output. The executive directs the flow of masses, compositions and other properties of process streams to their appropriate locations. Such structures must have the flexibility to describe mathematically a variety of complex plant flow diagrams as selected by the engineer.

Models

Most of the subroutines controlled by the executive are collections of mathematical equations which describe pertinent behavior characteristics of process units of interest in the simulation. The set of equations used to represent a specific process unit is called a model.

***Special subroutines termed utility routines, e.g., convergence routines, optimization routines, and statistical routines, are also frequently employed.

Thus it may be necessary to have models of various crushers, bins, screens, rod mills, ball mills, flotation cells, pumps, pump boxes, thickeners, filters, and so on. In each case the model must describe selected responses of a particular process unit to changes in operating variables, e.g., feed rate, feed size distribution mineralogical composition, percent solids, and/or to changes in design variables, e.g., equipment dimensions, power draft, etc. Depending upon the purpose of the simulation these models may be dynamic (time dependent) or steady state (time independent) in nature.

Models for process units may be broadly classified as either theoretical or empirical. Theoretical models are based on process mechanism and are in principle valid over a full range of values of operating variables. They require substantial time and talent to build. On the other hand, empirical models involve simple polynomial equations fitted to data obtained by direct experimentation around process units. Such models apply over the range of values of operating variables examined experimentally. Usually, far less expense is involved. Frequently, models used in practice are the result of a combination of theoretical and empirical approaches. Even the so-called "theoretical" models of mineral processing are phenomenological in nature, i.e. they derive their form from theory but model parameters (or constants) must be determined experimentally.

All models must be tested and proven valid before they can be used. Considerable time in sampling, data analysis and computational effort may be expended before this stage is reached. Valid models are then programmed for use as subroutines to the executive portion of a simulator.

SIMULATION METHODOLOGY

A general procedure for developing a digital simulation may be given as follows: (1) define the purpose of the simulation, (2) select those process units which must be modeled, (3) develop statistically valid models, (4) select an executive structure and incorporate models as subroutines, and, finally, (5) test the simulator and modify as necessary.

Purpose of Simulation

The purpose of a simulation must be clearly defined to ensure that models are of the correct form. For example, suppose that a grinding circuit simulation is desired such that alternative control strategies may be compared with each other. In this situation, models must be time dependent. Relationships must be found between circuit variables to be controlled and those to be manipulated. In addition, the controllers of interest to a particular strategy must be simulated and time delays, sensors, and final control elements must be included in suitable fashion. Often, the nature of the models required for control strategy evaluation will not be the same as that of models required for other purposes such as evaluating new circuit arrangements or optimizing the steady-state performance of an existing circuit. In some instances simulators may have to serve several purposes, which may require the development of complex models with a high degree of flexibility.

Selection of Process Units to Model

The purpose of the simulation will govern the selection of process units to be modeled. Of course, a general, multipurpose simulation would obviate the need for a choice. Unfortunately, general simulators of this type are expensive to develop. Their flexibility may not justify their cost. On this basis, the following rule of thumb is suggested: "The simulator developed should not contain superflous models". For example, suppose that a steady state simulation of a primary grinding circuit must be developed to compare alternative circuit configurations. For this purpose, time delays, pump models, pump box models need not be included because no change in the primary grinding response, particle size distribution occurs in these units. Process units selected for modeling might be a rod mill, a ball mill and cyclone classifiers. Operating variables which might be incorporated include water and solids flow to the rod mill, water flow to the pump box, water and solids flow to the ball mill, the number of cyclones, vortex and spigot diameters, and the flow of water and solids through the cyclones. These variables may be manipulated during the model building stage. If equipment sizes are not to be varied in the evaluation of circuit alternatives, then the influence of such factors as mill diameters, mill lengths, mill speeds and cyclone diameters which are not easily altered for testing in the plant may not be considered during model development.

Development of Statistically Valid Models

Model building plans have been reviewed (13,14) and several texts dealing with this topic are available (15,16). The following systematic procedure has been useful: (1) specify the process responses to be predicted by the model, (2) list independent variables and specify their ranges for study, (3) select an appropriate form of mathematical model, (4) based on the form of model selected, devise a data acquisition plan in which the independent variables are manipulated in a systematic fashion to allow the determination of adjustable model parameters from experimental responses, (5) adjust the resulting data statistically if necessary, (6) fit the pre-selected model to the adjusted data, and (7) test the adequacy of the fitted function and modify as needed until a valid model is obtained.

Selection of Responses(s). Responses (i.e., dependent variables of interest to the engineer) are selected such that they 1) satisfy the overall objectives of the simulation and 2) can be measured or calculated from experimental data for the purpose of model building and verification. Thus, if a simulator is required to evaluate the influence of fineness of grind on steady state rougher flotation performance, appropriate model responses to consider would include valuable mineral recoveries on a size fraction by size fraction basis. Recoveries are not directly measurable but do satisfy criterion 2) since they can be computed from various combinations of measured assays and mass flows for the feed, concentrate, and tail.

Selection of Independent Variables and Ranges. Independent variables are those which influence process responses. Ideally, these variables can be manipulated at will. They must at least be measurable or calculatable from suitable measurements of other variables. Ranges for study must be chosen to correspond to ranges of practical interest in the plant. Constraints on levels of independent variables must be determined (see reference 14).

Selection of Mathematical Model(s). A literature search should be conducted to help determine an appropriate form of models(s). Several alternatives may be available. These alternatives should be categorized according to expected accuracy, and ease of estimation and simulation. If model forms are not available from any source (including from the evaluation of old data, a dangerous practice), then the form of model that might fit the data may be assumed. Select the simplest form of model which will satisfy the objectives of the simulation.

Data Acquisition Plans. Independent variables judged to influence responses should be deliberately altered in a systematic way. Such deliberate interference in process operation is necessary, because normal operation may not include a range of variable settings that influence significantly the measured responses. A good data acquisition plan must be devised. It should involve the simultaneous manipulation of all independent variables both individually and jointly. Moreover, the influence of process error on measured responses must be determined to permit an statistical assessment of the quality of fit of a model. Plans must minimize the effort to obtain a valid model. The minimum number of runs (a run is the procedure of adjusting variables to a set of predetermined values, holding them there and finally measuring the response) must at least equal the number of coefficients in the assumed model. The nature of the experimental plan also will depend upon the form of model selected in a previous step.

It is important to recognize that if a plan cannot be followed exactly due to process restrictions, it is not a serious problem. The importance of a plan is associated with deliberate, simultaneous and joint changes. A variety of efficient plans are available (14,15,16).

Statistical Adjustment of Raw Data. Raw data, particularly if acquired around full scale production equipment, are subject to several sources of error attributable to process fluctuations, sampling, off calibration instruments, laboratory analysis technique and so on. Frequently data adjustment procedures (17,18,19) are required to smooth such data and obtain mass balance closure. The best data adjustment procedures assume that all measured data are in error and adjust all values within the experimental error associated with each measured data point. The adjusted data, provided the adjustments are in accord with the least squares criterion (19), can then be employed for model building. Thus a set of widely varying circulating loads calculated from individual screen fractions measured from cyclone products can be reduced to one reliable values by proper data adjustment. The resulting "average" circulating load can be used for grinding model development.

Fittings Models to Acquired Data. Fitting a model to data is essentially a "french curve" technique, except that a mathematical equation is used in lieu of a pencil line. The "best" values of coefficients in the model equation must be found. Often, the best values are taken to be those which minimize the objective, ϕ, given by the following expression:

$$\phi = \sum_{i=1}^{k} \frac{(Y_i - \hat{Y}_i)^2}{S_i^2} \tag{1}$$

where Y_i is the response measured at the ith level of independent variable settings (considering a total of k), S_i^2 is the variance associated with the determination of the ith response, and $\hat{Y}_i$ is the response predicted from the assumed model for any specified set of the coefficients.

The selection of a mathematical strategy to minimize equation 1 depends upon whether the models linear or nonlinear in the parameters and upon the assumptions regarding error variance. When the variance is taken as constant and the model is linear in the coefficients, the method of least squares,linear regression may be used (20). For nonlinear models, a variety of alternative fitting strategies are available (14,16).

Testing the Adequacy of a Fitted Model. Although "best values" of model coefficients can always be determined, there is no guarantee that the fitted model accurately describes the process. After fitting,the model may be simplified statistically (why use a complex one if a simpler version can be found), the residuals, $Y_i - Y_i$, associated with the simplified model examined for abnormalities, and the simplified model checked for lack of fit. Confidence limits can then be placed on predicted responses and estimated coefficients. Finally, the model is tested for accuracy against the process unit. If rejected, an alternative model is selected and fitted to the data. Eventually, a valid model will be found.

It is important to remember that all models, whether empirical or theoretical, should be tested in the above fashion. There are few, if any, theoretical models of process units that are complete. Depending upon the purpose of the simulation, various degrees of empiricism will always be necessary. After all, the immediate requirement is a model that always really works over practical operating ranges in the plant.

Choosing an Executive Structure

There are two basic types of simulator structures available for engineers (12). The first relies on the near linear nature of circuit balances. Flows are calculated by a simultaneous solution of balance equations. The second is based on the sequential solution of material balance equations at each process unit. In this procedure selected streams are cut and an iterative calculation is carried out until component mass flows in cut streams match trial values. A recent review of the latter structures is available (21). PACER (22) and GEMCS (23) typify the generalized structures capable of dealing with complex recycle streams. Such programs have not been widely used by mineral engineers to date. Most often, the executive structure has been written specifically for the circuit of interest. This is probably the best approach at this time. Eventually, structures specific to mineral processing will be available for general usage (12,44).

The computer language in common use is FORTRAN IV and enriched versions thereof. A language useful for conversational programming is BASIC.

Flexible executive structures are written such that models of process units are introduced easily as subroutines. A given model may be called upon again and again by the executive. Sometimes, a model may differ from another only in the values of constants, so that the same subroutine may represent two or more different process units.

Testing the Simulator

After the simulator has been written it must be tested for typical inputs. In some instances it may also be desirable to extrapolate outside of the ranges used to develop models. Sometimes, greater flexibility for the executive may be desired. Furthermore, the models may not extrapolate well, so that modifications are required. In some cases, convergence problems arise during iterative calculations. Such problems must be resolved before the simulator is judged acceptable.

SIMULATORS AS PLANT DESIGN AIDS

Digital simulators may be employed to (1) evaluate the feasibility of a proposed flowsheet, (2) investigate the effect of modifications to an existing flowsheet, (3) assist in obtaining new knowledge about the behavior of a complete circuit, and (4) to form the basis for off-line optimization of circuit performance. Other applications include the comparing of alternative control strategies and the estimation of process equipment capital costs. The remainder of this paper is devoted to the description of example applications of circuit simulators in each of these areas of design.

Flowsheet Feasibility

The feasibility of a proposed flowsheet may be investigated with a simulator so long as the process unit models contain certain basic scale-up design information. The models involved can be developed from laboratory data, pilot scale data and/or full scale production data. In almost all instances, however, these models must include the dependence of process unit responses on design variables to allow an extrapolation to be made from the scale of the data used for model development to the scale of the proposed flowsheet.

Models Developed from Laboratory Data. Relatively few example applications are available of this type simply because accurate scale-up information for large extrapolations of process model parameters is often non-existent. A notable exception involves the use of a general grinding circuit simulator called ESTIMILL to predict pilot scale grinding circuit performance from laboratory scale batch data.

ESTIMILL is a general purpose grinding program which permits simulation and parameter estimation for batch, locked-cycle, open circuit continuous (at steady state) and closed circuit continuous (at steady state) behavior (24). Subroutines currently in use include a ball mill, a classifier, a parameter estimator and a statistical assessor. Figure 1 shows a program flow chart for ESTIMILL. The subroutines SIMUL, MILL, JAY, SZETA and BALPHA contain a population balance grinding model and an arbitrary classifer (see APPENDIX). In this case selection functions (S_i) and breakage functions (B_{ij}) of the form

$$S_i = S_1 \exp\left(\sum_{j=1}^{r} \zeta_j (\ln(\sqrt{X_i X_{i+1}}/\sqrt{X_1 X_2})^j \right) \quad (2)$$

$$B_{ij} = \alpha_1 (X_i/X_{j+1})^{\alpha_2} + (1-\alpha_1)(X_i/X_{j+1})^{\alpha_3} \quad (3)$$

are used to describe the grinding kinetics. In equations 2 and 3 α_i and ζ_i are constants, X refers to particle size and r is an integer valued constant. The effect of material transport through a continuous mill is described with a flexible mixers-in-series residence time distribution.

The balance of the subroutines of ESTIMILL are written for parameter estimation and statistical assessment of goodness of fit. The fitting strategy (i.e., parameter estimator) utilized by ESTIMILL is one which minimizes the weighted sum of squares objective function (13,27,28):

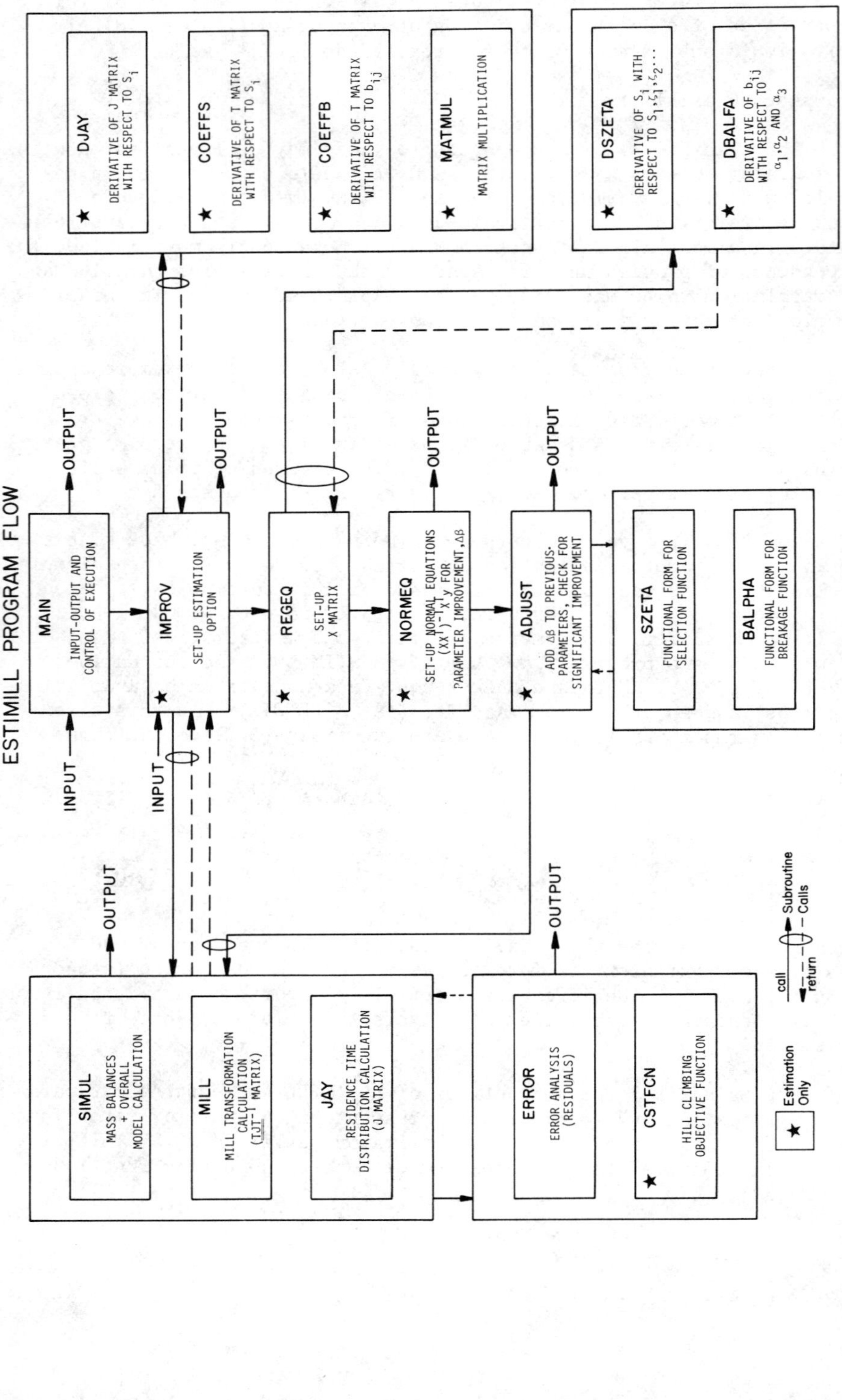
ESTIMILL PROGRAM FLOW
MAIN
INPUT-OUTPUT AND CONTROL OF EXECUTION
IMPROV
SET-UP ESTIMATION OPTION
REGEQ
SET-UP X MATRIX
NORMEQ
SET-UP NORMAL EQUATIONS $(X X')^{-1} X' y$ FOR PARAMETER IMPROVEMENT, $\Delta\beta$
ADJUST
ADD $\Delta\beta$ TO PREVIOUS-PARAMETERS, CHECK FOR SIGNIFICANT IMPROVEMENT
SZETA
FUNCTIONAL FORM FOR SELECTION FUNCTION
BALPHA
FUNCTIONAL FORM FOR BREAKAGE FUNCTION
DJAY
DERIVATIVE OF J MATRIX WITH RESPECT TO S_i
COEFFS
DERIVATIVE OF T MATRIX WITH RESPECT TO S_i
COEFFB
DERIVATIVE OF T MATRIX WITH RESPECT TO b_{ij}
MATMUL
MATRIX MULTIPLICATION
DSZETA
DERIVATIVE OF S_i WITH RESPECT TO $S_1, \zeta_1, \zeta_2 \ldots$
DBALFA
DERIVATIVE OF b_{ij} WITH RESPECT TO α_1, α_2 AND α_3
SIMUL
MASS BALANCES + OVERALL MODEL CALCULATION
MILL
MILL TRANSFORMATION CALCULATION (TJT^{-1} MATRIX)
JAY
RESIDENCE TIME DISTRIBUTION CALCULATION (J MATRIX)
ERROR
ERROR ANALYSIS (RESIDUALS)
CSTFCN
HILL CLIMBING OBJECTIVE FUNCTION
INPUT
OUTPUT
Estimation Only
call
return
Subroutine Calls

FIGURE 1

$$\phi = \sum_{i=1}^{n} w_i (m_i - \hat{m}_i)^2 \tag{4}$$

where w_i is a weighting factor, m_i is measured weight percent retained in the size interval X_i to X_{i+1}, and $\hat{m}_i$ is the weight percent retained in the size X_i to X_{i+1} predicted from the model. Best values of constants (parameters) in the model are determined by a modified Gauss-Newton procedure (29). The stopping criterion involves either an F-ratio test or a comparison of residual sums of squares with the error variance.

In the particular application of ESTIMILL described here, batch data obtained for the wet grinding of minus 10 mesh limestone in a 10 inch x 11.5 inch laboratory mill was used to estimate the adjustable parameters in equations 2 and 3. These parameter values were, in turn, scaled up to predict pilot scale performance in a 30 inch x 16 inch continuous open circuit mill using specific power information and the following approximations (25,26):

a) Selection functions are proportional to the specific power draft of mill, i.e.,

$$S_i = S_i^E \left(\frac{P}{H}\right) \tag{5}$$

where P is the net power draft of the mill, H is the mass hold-up of material in the mill and the S_i^E values are constants termed specific selection functions.

b) Breakage functions are independent of mill design and operating variables, i.e.,

$$B_{ij} = \text{invariant} \tag{6}$$

Thus from a knowledge of S_i values and the specific power draft of the 10 inch mill the specific selection functions, S_i^E, were calculated according to equation 5. Subsequently S_i values for the 30 inch mill were predicted from a reapplication of equation 5 using the appropriate P/H value for this mill. The breakage functions for the 30 inch mill were taken to be identical to those estimated in the 10 inch mill as suggested by equation 6. The residence time distribution for the 30 inch mill was assumed to conform to that of a perfect mixer.

In order to predict the open circuit continuous performance of 30 inch mill, the scaled parameter values were used in conjunction with ESTIMILL operated in the simulation mode. The resulting predictions were checked by experimental measurements for identical conditions to those simulated. Figure 2 shows a comparison of the predicted and experimental response. Similar accuracy of prediction was obtained for simulations of closed circuit pilot scale grinding.

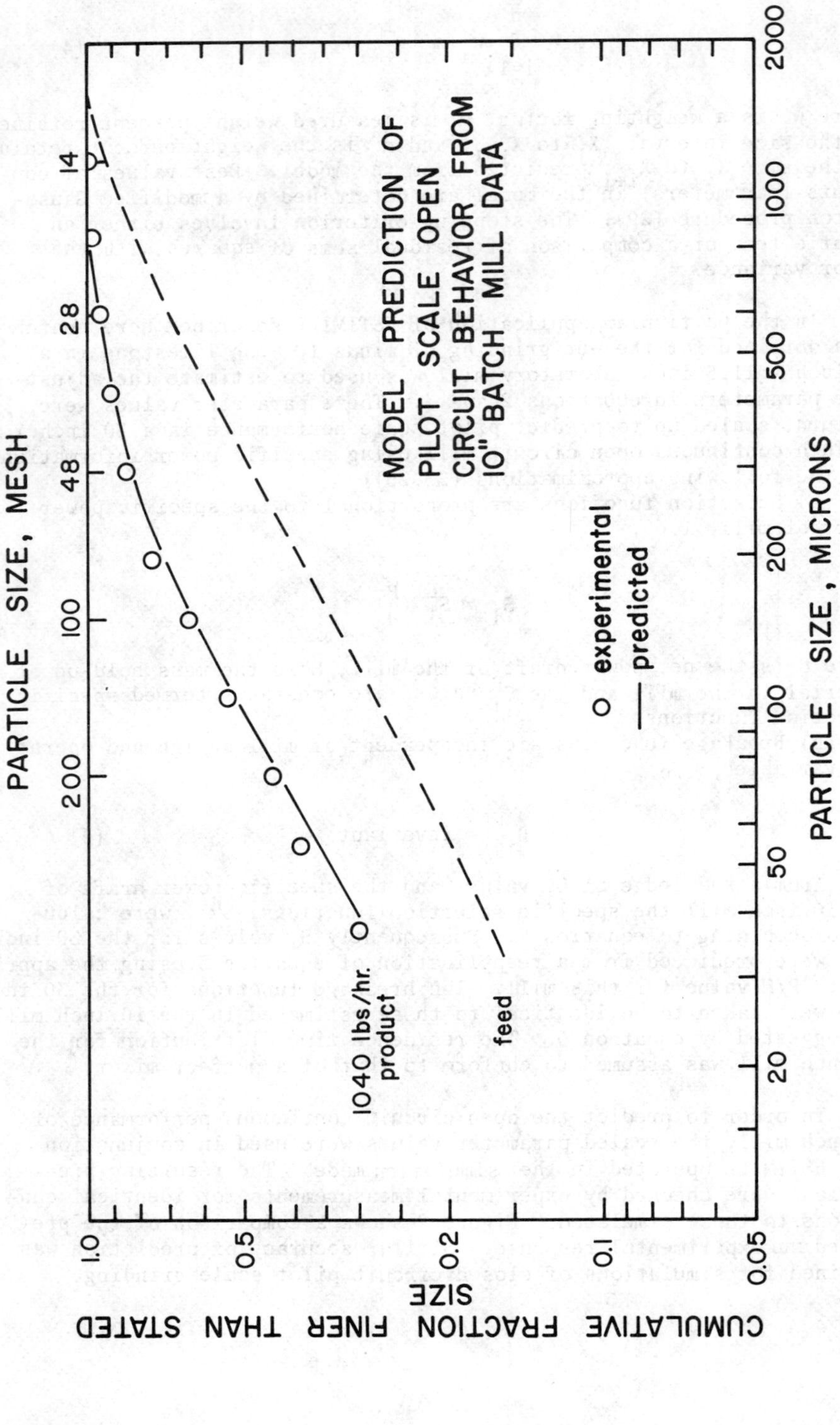

FIGURE 2: Pilot scale grinding simulation with ESTIMILL (26)

Models Developed from Pilot Plant Data. An example of a simulator that depends on pilot/laboratory data for inputs is the NIM simulator for flotation circuits (12,30). The flotation cell model developed from pilot data involves a froth transmission coefficient, flotation rate constants associated with mineralogical components as a function of particle size, bubble surface area per unit volume of pulp, the fraction of bubble area that is free from solids expressed as an average and the pulp retention time. A separate computer program has been developed to allow the estimation of model parameters using a modified Gauss-Newton procedure similar to that used in ESTIMILL. This model has been incorporated into a simulator that is based on the material balance equations for particles of each type (size/composition) in the circuit. Any number of roughers, cleaners and scavenger cells with variable volumes and inter-stage water additions can be described with this simulator. The simulator has been tested in both pilot scale and production scale circuits and has been judged useful for feasibility studies and for studies to compare alternative circuit configurations.

Models Developed from Production Plant Data. A steady state simulation has been developed (7) for a dry cement grinding circuit located in the Canada Cement Lafarge Ltd. plant at St. Constant, Quebec. The production circuit was sampled and residence time distributions (RTD) were determined experimentally by fluorescein tagging. The RTD was found to correspond to ten fully mixed tanks in series, with a mean residence time of 5.58 minutes. The ball mill model employed was effectively a population balance model. The air classifier model selected was of log normal form.

The simulator was used to predict how design schemes for different operating conditions (same size mill) might influence product size distribution and circuit capacity, when the discharge diaphragm was assumed to be redesigned to preserve mill filling at different circulating loads. The conclusions reached were: the increase of circuit output from the breakage viewpoint due to separation and rejects recycle was slight, and improved separator efficiency to produce the same product would not result in significant increase in output.

Flowsheet Modification

Figure 3 shows a grinding circuit that existed at Mt. Isa Mines Ltd., located in Australia. In 1966, this circuit was simulated at steady state (2). To describe the rod and ball mills mathematically, matrix models were derived and fitted to raw data obtained around the operating circuit. A novel cyclone model, consisting of 4 equations, was developed. The models were used as subroutines to a simulator structure written in FORTRAN. Process units were switched around "on paper" until the best circuit arrangement was discovered (the best one was the one with the highest capacity per KWH to obtain a desired fineness). Figure 3 shows the re-arranged circuit, where up to 15 percent increase in efficiency was observed. The most recent versions of the

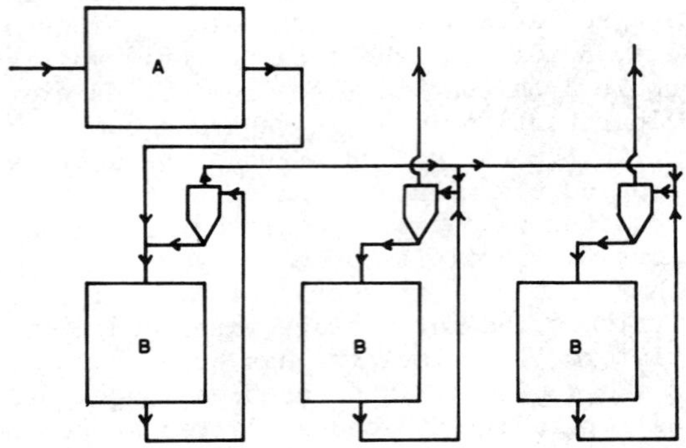

A - 9 FT. X 12 FT. ROD MILL

B - 10.5 FT. X 10 FT. BALL MILLS

— Flowsheet of the M.I.M. Grinding Circuit Prior to Rearrangement

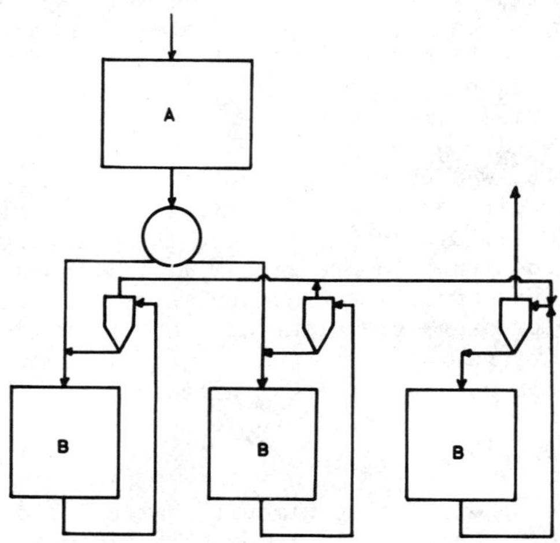

A - 9 FT. X 12 FT. ROD MILL

B - 10.5 FT. X 10 FT. BALL MILLS

— Mt. Isa Circuit Rearranged

FIGURE 3: Grinding Simulation at Mt. Isa (2)

grinding models initially used at Mt. Isa are given in the APPENDIX of this chapter.

A more recent application of these grinding and classification models involved data obtained as part of a routine data acquisition program at Gibraltar Mines Ltd., located near Williams Lake, B.C. The Gibraltar circuit consists of three identical lines of the type shown as circuit A, Figure 4. Discharge from a 13.5 x 20 ft rod mill (2500 HP) joins discharge from a 13.5 x 20 ft ball mill (2500 HP) at the pump box where water is added. A 16 x 14 x 34 SRL-C pump transmits slurry to a cyclopac comprised of six 30-inch cyclone classifiers in parallel. Cyclone underflow returns as ball mill feed, while overflow goes to flotation. The matrix models of the rod and ball mill and the cyclone model employed for the Mt. Isa Mines grinding circuit were fitted to the Gibraltar data. The fitted models (Note: the original cyclone equations were modified for Gibraltar data) were programmed and a simulator was developed to permit the comparison of several alternative circuit configurations (6). Three of the circuits compared are shown schematically in Figure 4. Conclusions reached were: Circuit A, the existing circuit, should use 4 cyclones in parallel, circuit B shows little advantage over circuit A at equivalent capacities, and circuit C produces about three percent more material of finished size at equivalent capacities. Thus circuit C should be studied by means of more accurate simulations.

Increasing Process Knowledge

The Broken Hill South Ltd. concentrator in Australia was sampled over a 4 hour period at points in the grinding and flotation circuits. Size fractions of each composite sample were analyzed for Pb, Zn and gangue minerals to permit a comparison throughout both circuits. The grinding circuit was simulated using population balance grinding models and an experimental classification matrix. The simulator suggested that the grinding circuit has reserve capacity. The flotation circuit simulation, where the simple probability model was employed (32), suggested that half the Pb losses in the final Zn concentrate and one third the Pb losses in final tails were due to slow floating Pb minerals. These were mainly ultra fines. Reference 4 contains all details.

To improve upon the available knowledge which deals with crushing and screening operations, the Brenda Mines Ltd. secondary crushing plant, located in British Columbia, was simulated (31). Figure 5 is a simplified flowsheet of the circuit. Feed from the coarse ore stockpile enters two 7-ft standard cone crushers (secondaries) in parallel. The discharge joins discharge from four 7-ft short head crushers in parallel (tertiaries) and the combined flow constitutes the feed to five 8 x 20 ft. double deck vibrating screens in parallel. The coarse screen rejects are returned as feed to the tertiary crushers, while the fines are sent to fine ore bins. A primary fines stream from the primary screening plant is likewise sent to the fine ore bins. Acquisition of plant data was in accord with two level factorial designs

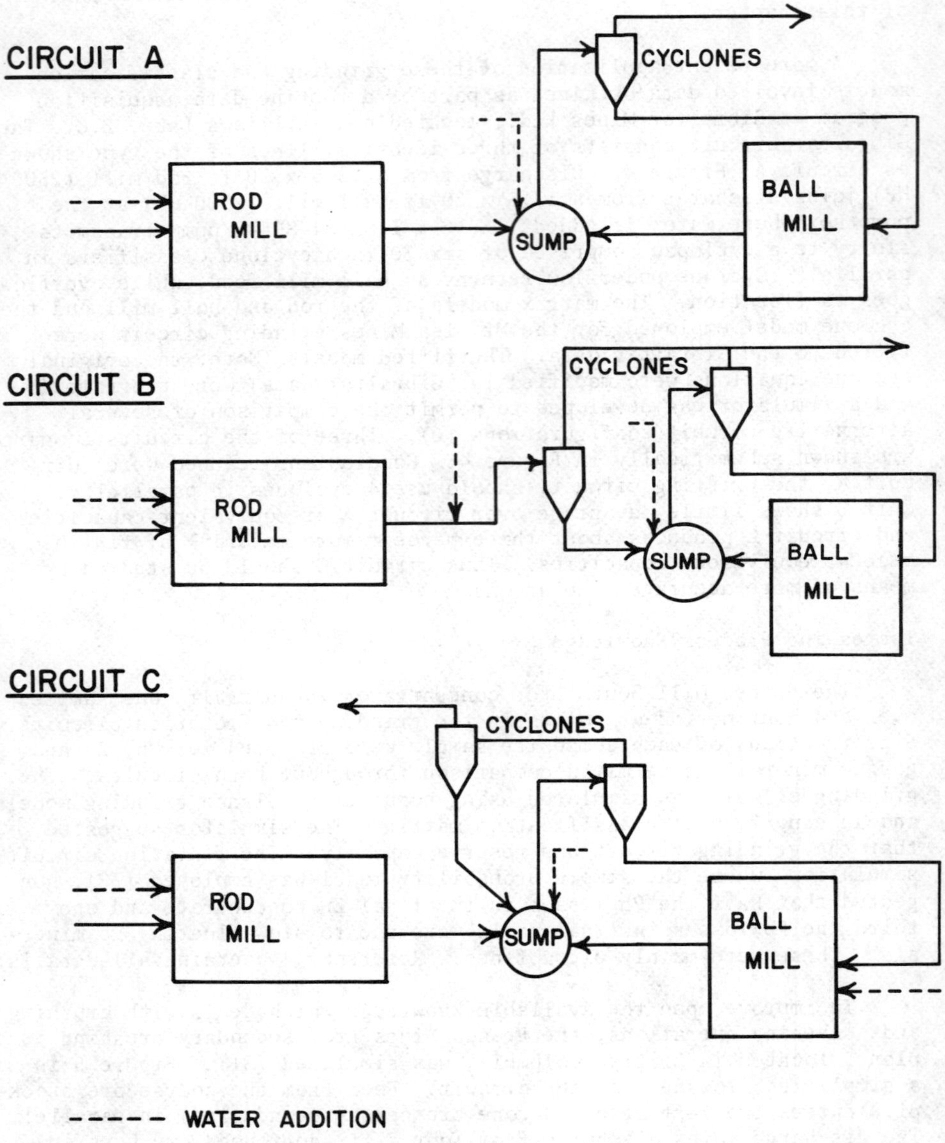

FIGURE 4: Alternative Grinding Circuit Configurations (6)

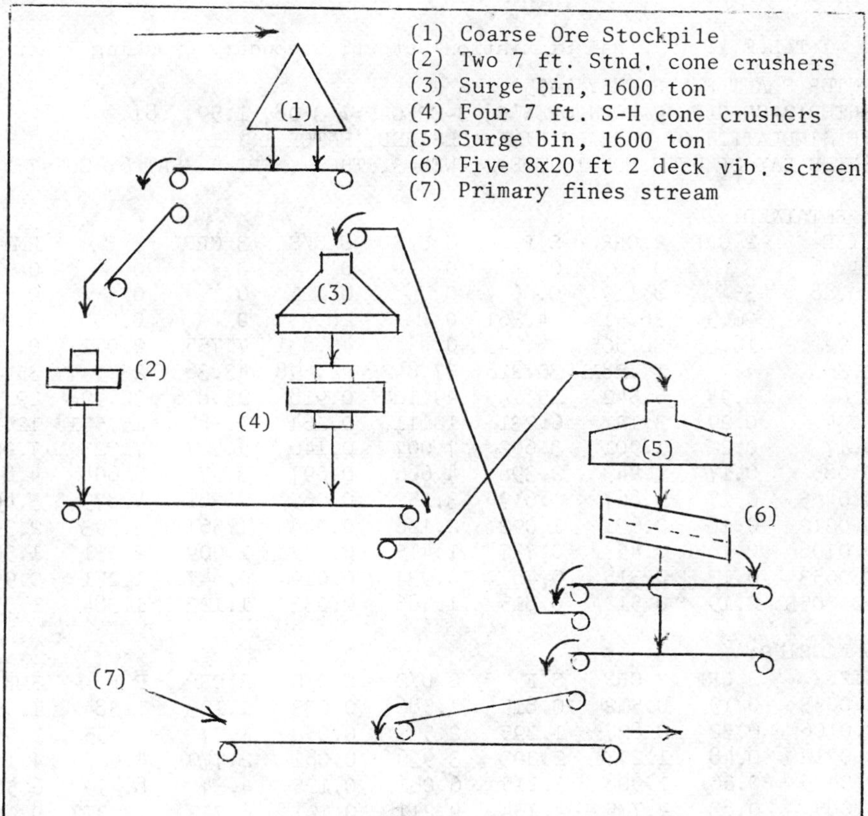

FIGURE 5: Rough Flowsheet, Secondary Crushing Plant

(13). Sampling was conducted as close to steady state as possible. Samples were screened by means of a standardized procedure and the resulting data, where necessary, was adjusted by means of a procedure that assumes all measured values to be in error (19). The crusher models employed are modifications of matrix models used to describe Mt. Isa Mines Ltd. crushers (8). On the other hand, the screening model was derived from small particle statistics. The models as developed are listed in the APPENDIX of this chapter. Table 1 lists typical output from the simulation.

To acquire output from the simulation for study, a 2-level factorial design was employed, along with a set of arbitrary intermediate values for factor levels. Five variables were manipulated, namely, plant tonnage, percent +1 inch in the plant feed, secondary close side setting, tertiary close side setting and the screen opening on the

TABLE 1: Typical Simulation Output; Secondary Crushing Plant

```
ENTER PLANT FEEDRATE:? 1455
ENTER:2-CR GAP, 2-SCRN OPEN'G, 3-CR GAP:? 3.08, 1.59, .81
IF SIMULATION OF PRIMARY FINES DESIRED, ENTER 1:? 1
ENTER DAY OF WEEK:MON=1, TUE=2, WED=3, THU=4, FRI=5, SAT=6, SUN=7:? 3
```

% RETAINED:

SIZE	2 CRF	2 CRP	S F	S U/S	S O/S	3 CRP	P F	RMF
21.76	3	0	0	0	0	0	0	0
10.88	35	0.119	0.06	0	0.121	0	0	0
5.44	36.5	28.512	14.361	0	28.935	0	0.008	0.001
2.72	18.35	38.505	23.243	0	46.831	7.755	2.071	0.375
1.36	4.7	17.281	30.226	37.842	22.498	43.363	23.547	35.254
0.68	0.94	6.542	15.145	29.164	0.918	23.876	32.498	29.768
0.34	0.29	3.121	6.231	12.11	0.264	9.387	15.571	12.737
0.17	0.22	1.901	3.602	7.007	0.146	5.327	7.323	7.064
0.085	0.17	1.243	2.398	4.666	0.097	3.57	5.604	4.836
0.0425	0.17	0.867	1.619	3.15	0.065	2.382	4.879	3.463
0.0212	0.18	0.631	1.092	2.126	0.044	1.561	3.593	2.391
0.0106	0.16	0.45	0.727	1.415	0.029	1.009	2.251	1.567
0.0053	0.13	0.315	0.48	0.934	0.019	0.647	1.271	0.995
-0.0053	0.19	0.512	0.815	1.586	0.033	1.123	1.384	1.55

% PASSING:

SIZE	2 CRF	2 CRP	S F	S U/S	S O/S	3 CRP	P F	RMF
0.0053	0.19	0.512	0.815	1.586	0.033	1.123	1.384	1.55
0.0106	0.32	0.827	1.295	2.52	0.052	1.77	2.655	2.545
0.0212	0.48	1.277	2.023	3.936	0.081	2.779	4.905	4.111
0.0425	0.66	1.908	3.115	6.061	0.125	4.34	8.499	6.503
0.085	0.83	2.775	4.734	9.211	0.191	6.722	13.377	9.965
0.17	1	4.018	7.132	13.876	0.287	10.292	18.981	14.801
0.34	1.22	5.92	10.734	20.884	0.433	15.619	26.304	21.865
0.68	1.51	9.04	16.964	32.993	0.697	25.006	41.876	34.602
1.36	2.45	15.582	32.109	62.158	1.616	48.881	74.373	64.369
2.72	7.15	32.864	62.336	100	24.113	92.245	97.921	99.623
5.44	25.5	71.368	85.579	100	70.944	100	99.992	99.999
10.88	62	99.881	99.94	100	99.879	100	100	100
21.76	97	100	100	100	100	100	100	100
43.52	100	100	100	100	100	100	100	100

OPERATING CONDITIONS

TONS:	1455	1455	2888.75	1455	1433.75	1533.75	321.71	1776.71
+1":	92.85		37.66415				75.88658	

```
CRUSHER CURRENTS: Secondary = 23.17   Tertiary = 39.64
SETS:   2 CR SET = 3.08 SCRN OP = 1.59  3 CR SET = 0.81
%-1/2 INCH IN FEED = 2.45        %-1/2 INCH IN SCREEN U/S = 62.15761
%-1/2 INCH RATIO, SCREEN U/S TO PLANT FEED = 25.37045
CIRCULATING LOAD RATIO = 0.9853952
```

lower deck of the screens. Plots of the resulting data, where dependent variables of interest were percent -1/2 inch in fines, plant reduction ratio, circulating load ratio, and screen efficiency, show that the simulation behaves as might be expected, qualitatively, except for the effect of tertiary close side setting on circulating load ratio. A minimum is observed. This is attributed to more near-screen-size material being produced at a certain close side set. It therefore seems that there is a "best" close side setting at a given screen opening. Coincidently, the relationships developed between crusher current and operating variables suggest that more horsepower for the tertiary crushers may be beneficial.

Off-Line Optimization of a Grinding Circuit

Off-Line optimization techniques (33,34) reduce considerably the effort required to find optimum sets of operating conditions from mineral processing circuits. A general procedure will involve the following steps (34): (1) development of a process simulator for optimization purposes, (2) an evaluation of the nature of constraints on the process, (3) the choice of an objective function, (4) the selection of an off-line strategy, and (5) a comparison of results with the process. This procedure is demonstrated below using the Gibraltar simulator to optimize grinding circuit performance.

Construction of Grinding Circuit Simulator. The grinding circuit simulator was developed from Gibraltar Mines data. The flowsheet of the circuit is shown in Figure 4, circuit A. Models employed for the simulator are described in the APPENDIX. Models were connected in such a way that the following variables could be manipulated by the program user: F = TPH of rod mill feed solids, P_r = percent solids in the rod mill discharge, P_c = percent solids in the cyclone feed, N = number of cyclones in the cyclopac, V = diameters of vortex finders, S = diameters of spigots, P_b = percent solids in the ball mill discharge provided that the cyclone underflow percent solids at steady state is greater. The program works as follows: solids and water enter the rod mill and undergo comminution to produce a rod mill discharge. The rod mill discharge joins ball mill discharge (equal to zero in the first cycle) plus pump box water to bring cyclone feed up to the desired percent solids. The cyclone feed splits according to the cyclone model, with the underflow solids fractions and water serving as ball mill feed added to ball mill feed water (note--if the percent solids in cyclone underflow is less than that desired in ball mill discharge, ball mill feed water is not added). The ball mill feed solids undergo breakage by means of the ball mill model to produce weight fractions of ball mill discharge solids. These are added to corresponding fractions of rod mill discharge solids. Water flows are likewise summed. The procedure is then repeated until the cyclone overflow solids discharge rate is equal to the rod mill feed rate of solids. Typical output consists of a total water and solids balance around the circuit, size distributions at every key location, and information about the cyclone classifier such as inlet pressure.

Grinding Circuit Contraints. The circuit simulator permits the user to adjust 7 variables, excluding the rod mill feed size distribution, either directly or indirectly (i.e., the number of cyclones is altered in a direct manner whereas the percent solids in rod mill discharge is altered by water addition to the feed to obtain the percent solids desired). Explicit constraints such as that on P_b are taken into account by the simulator. Thus, the user may "hope" for a ball mill discharge percent solids of 80, but he must be satisfied with whatever is necessary to balance circuit water. In consultation with manufacturers and operators, the following implicit constraints were selected: $9 < V < 11$, $5 < S < 7$, $76 < P_r < 82$, $56 < P_c < 66$, $76 < P_b < 82$, $4 < N < 7$, and $520 < F < F_u$. The upper limit, F_u, is the rod mill overload tonnage, while the lower limit was judged to be the safe operating tonnage. To constrain these variables between the limits shown the following transformation was employed (14):

$$V_i = L_i + ((U_i - L_i)\exp(X_i))/(\exp(X_i) + \exp(-X_i)) \quad (7)$$

where V_i, the transformed variable, serves as input to the simulator, X_i is the value of the variable that an optimization strategy must use, U_i is the upper constraint and L_i is the lower constraint. The value of F_u was calculated from an equation derived by regression analysis (35) as:

$$F_u = 931 - 53.34B_i - .8609R + .7981P \quad (8)$$

where B_i is the plant work index (kwhr/ton), R is the weight percent coarser than .75 inch in the rod mill feed, and P is the 80 percent passing size in cyclone overflow. At 82% solids, the ball mill overload tonnage was 3400 TPH, at which point steel balls were judged to be floating out the discharge trunnion. A roping constraint on the cyclone was calculated from graphical data (36), so that

$$U_c = 1.1\, N \exp(0.928S + 0.0688P_u - 3.5) \quad (9)$$

where P_u is the cyclone underflow percent solids. If the underflow tonnage, U, exceeds U_c a roping condition is imminent.

Choice of Objective Function. The objective could be stated as follows: maintain the percent +48 mesh in cyclone overflow at 20 for a given number of cyclones. Find the best set of manipulatable variables that permit this without exceeding circuit constraints. Clearly, several other objectives are possible, such as maximize rod mill feed rate by manipulating adjustable variables but do not exceed a critical size in cyclone overflow.

The objective function was constructed in such a way that an optimization strategy could minimize it to obtain the solution. Thus

$$Y = (C - 20)^2 + P_1 + P_2 \tag{10}$$

where Y is the objective function, C is the percent +48 mesh in cyclone overflow calculated by the simulator, P_1 is a ball mill tonnage penalty and P_2 is a roping penalty. Functional forms for P_1 and P_2 were:

$$P_1 = (B/B_{max})^3 - 1$$
$$P_2 = (B/U_c)^3 - 1 \tag{11}$$

where B is the value of underflow tonnage at any current stage in the simulator. As long as $B < B_{max}$ and/or $B < U_c$, P_1 and P_2 are set to zero. Otherwise, Y is penalized as it enters the constrained region in cubic fashion.

Optimization Strategy. An off-line optimization strategy is a mathematical procedure which finds a unique combination of adjustable process variables that maximize or minimize an objective function. Aspects to consider prior to final selection of a strategy have been discussed (37). Because the objective function is formulated to approach a parabolic minimum, a modified Simplex direct search strategy that minimizes was chosen for the task (33). Figure 6 shows a simplified diagram of the final computer program employed. In this case the optimizer acts as the main program and the grinding circuit simulator as a subroutine.

Results of Off-Line Optimization Study. Table 2 shows how the number of cyclones will influence the objective. Note that to maintain 20% +48 mesh in the cyclone overflow as more and more cyclones are placed in parallel, the circuit capacity must decrease and the ball mill feed tonnage must increase. More water must be added to the ball mill as well, since the % solids decreases correspondingly. The X_{50c} size (see the APPENDIX) decreases somewhat, along with the overflow percent solids. Other adjustable variables remain essentially constant.

Other Uses for Simulation in Design

Simulators have been utilized for several purposes not emphasized in previous sections. Thus process control strategies may be compared by means of simulators and cost analyses for design may be conducted to maximize profitability.

Comparing Process Control Strategies. Specific digital computer control schemes may be developed off-line and alternatives may be compared. A study of this kind was performed on a grinding circuit at the Anamax concentrator, Sahuarita, Arizona (38). A control objective was to obtain the finest grind at a rod mill feed rate specified by the operator. Dynamic models were developed for the rod and ball mills

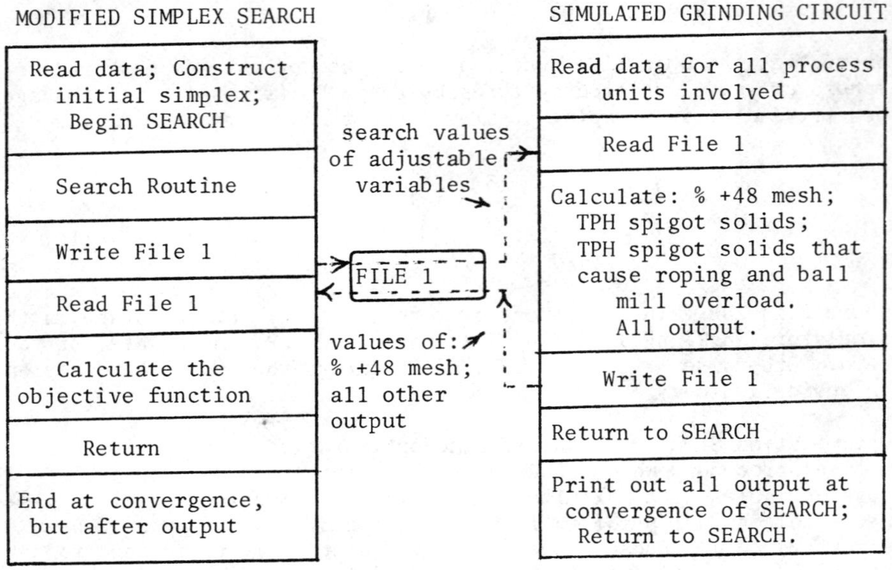

FIGURE 6: Simplified Flow in Off-Line Optimization Program

(using simplified population balance kinetics with perfect mixing), the cyclone classifiers, the pump and pump box, and the P+I controllers including measuring elements, transducers and final control elements. A simulator was constructed such that the control program to be studied could call the appropriate subroutines in accord with the "block diagram" of the control strategy. One conclusion reached from this study was that, for a fixed rod mill feed rate, one should add as much water as possible to the pump box without roping the cyclones or disrupting high/low level alarms. Thus, for a given feed rate of solids to the rod mill, one should observe the finest possible cyclone overflow. However, the optimum fineness will require a different flow rate of feed slurry to the cyclones. Hence, for a given percent solids in the cyclone feed, the simulator will permit the user to select the rate at which to increment the pump discharge set point. Further details may be found in reference 38.

Recently a similar simulator involving more detailed grinding

TABLE 2: Influence of Number of Cyclones on Off-Line Optimization Study.

Adjustable Variable:	Number of Cyclones Used 4	5	6	7
Rod Mill Feed Rate, TPH	585	580	577	574
Cyclone Feed Percent Solids	63.3	63.5	63.5	63.5
Overflow Percent Solids	40.0	39.8	39.1	38.5
X_{50c} size, micron	183	182	180	178
Vortex diameter, inch	10.5	10.5	10.5	10.5
Spigot diameter, inch	6.5	6.4	6.4	6.4
Inlet pressure, PSI	8.6	6.6	5.5	4.7
Rod Mill Percent Solids	80.5	80.7	80.7	80.7
Ball Mill Percent Solids	78.0	76.4	75.3	74.4
% +48 mesh in overflow	20	20	20	20
Ball Mill Feed Rate, TPH	1800	2029	2284	2525

models and variable hardness ore has been developed for the purpose of evaluating various types of optimizing control strategies (45). In this case the objective was taken to be maximum mill throughput at a fixed product size. Simulations performed for a single stage grinding circuit have resulted in the following tentative conclusions:

1) The use of a measured rather than an inferred particle size improves the quality of control,

2) Of possible feedback control strategies the manipulation of cyclone feed sumpwater to control particle size and the manipulation of fresh feed rate to control circulating load results in the best performance, and

3) Model based feedforward control in which an on-line grinding model is used to compute optimal controller settings seems to out perform all feedback strategies.

These conclusions are currently being tested in a 0.5 tph computer controlled grinding circuit.

Cost Analysis and Optimum Profitability. Simulators have been constucted to perform capital cost estimations (39). Cost equations were developed (40) and these serve as subroutines to an executive written in conversational mode for terminal usage. A factored estimate table is employed to obtain a preliminary capital cost estimate of a processing plant. The M&S index (for mining and milling) is utilized to update an estimated cost of major equipment.

By combining capital with operating cost estimations, profitability may be maximized by means of techniques described in a chapter of this volume entitled "Operating Decisions Based on Cash Flow Analyses". Simulators may be written for this purpose.

SUMMARY AND CONCLUSIONS

Digital simulation has been defined and corresponding methodology employed for this purpose has been reviewed. To develop a digital simulator the purpose must be defined, models must be developed, an executive structure must be written and, finally, the simulator must be tested. A good model building plan is an essential step. Responses and factors associated with process units are selected for study. Models are statistically fitted to data acquired on-line and adjusted. Fitted models must be statistically simplified and tested for lack of fit. Finally, models are compared with the operating process unit.

Simulators have been employed for flowsheet feasibility studies, for comparing modifications of flowsheets, for gaining process knowledge, for off-line optimization, for assessing alternative control strategies and for capital cost estimation.

The modern mineral processing engineer should be aware that models of various types now exist for a large number of unit operations including crushing, rod milling, ball milling, cycloning, flotation, heavy media and magnetic separation, thickening and filtration. In addition, generalized executive structures of the type required to integrate these unit operations into an overall circuit simulator are available. The existence of both models and executive structures suggest that process engineers should be in a position to develop specialty simulators for a variety of circuit types. In many instances simulators which are already available such as UBC's crushing circuit simulator, Utah's grinding circuit simulator and the NIM's flotation circuit simulator may be directly applicable to solution of plant design problems.

Judging by the amount of research effort which has recently been devoted to modeling and simulation in mineral processing, it seems that accurate digital simulation of all major types of circuits is a foreseeable goal. Within the next decade, a mineral processing plant will be studied routinely via digital simulation during the preliminary design stage. Simulations will provide important input for major design decisions.

REFERENCES

1. Savas, E.S., Computer Control of Industrial Processes, McGraw-Hill, New York, 1965.

2. Lynch, A.J., et. al., "Developing the Optimum Performance of a Multistage Grinding Circuit", I.M.M. London, 76, 1967, p. c169.

3. Brookes, G.F., et. al., "Control and Modelling of a Grinding/Classification Circuit", Proc. 9th Commonwealth Mining and Metallurgical Congress, vol. 3, 1969, p.701.

4. Cameron, A.W., Kelsall, D.F., et. al., "A Detailed Assessment of Concentrator Performance at Broken Hill South Limited", Proc. Aus. I.M.M., 240, 1971, p.53.

5. Freeh, E.J. and Horst, W., "Analysis and Control of Grinding Circuits: I and II", SME-AIME Preprint 70-B-27 and 70-B-28, Denver Annual Meeting, 1970.

6. Mular, A.L., et. al., "Digital Simulation of the Gibraltar Grinding Circuit", Ann. General Meeting of C.I.M., paper 71, Vancouver, 1973.

7. Austin, L., et. al., "Steady State Simulation of a Cement Milling Circuit", Int. J. Min. Proc., 2, 1975, p.127-150.

8. Whiten, W.J., "The Simulation of Crushing Plants with Models Developed Using Multiple Spline Regression", J.S.A.I.M.M., May, 1972, p.257.

9. King, R.P., "The Simulation of Flotation Plants", SME-AIME preprint 74-B-25, Dallas Meeting, 1974.

10. Herbst, J., et. al., "Development and Use of Lumped Parameter Models For Continuous Open/Closed Circuit Grinding Systems", Trans. I.M.M., C77, 1968, p.193.

11. Crowe, C.M., et. al., Chemical Plant Simulation, McMaster University Bookstore, Hamilton, Ontario, 1969.

12. King, R.P., "The Use of Simulation in the Design and Modification of Flotation Plants", Chapter 32 in Flotation, Ed. M.C. Fuerstenau, AIME, 1976, vol.2.

13. Mular, A.L. and Bull, W.R., Mineral Processes: Their Analysis, Optimization and Control, Dept. of Mineral Engineering, U.B.C., Vancouver, 1969.

14. Mular, A.L., "Empirical Modelling and Optimization of Mineral Processes", Mineral Science and Engineering, 4, no.3, 1972, p.30.

15. Mular, A.L., and Bacon, D.W., Statistical Design of Experiments and Process Optimization in Metallurgical Engineering, Dept. of Metallurgical Engineering, Queens University, 1967.

16. Himmelblau, R.M., Process Analysis by Statistical Methods, Wiley, New York, 1970.

17. Smith, H.W., and Ichyen, "Computer Adjustment of Metallurgical Balances", CIM Bulletin, Sept. 1973, p.97.

18. White, J.W., "Mill Modelling and Data Analysis", College of Mines, University of Arizona, 1975, p.3.

19. Mular, A.L., et. al., "Mass Balance of a Grinding Circuit", CIM Bulletin, December, 1976.

20. Draper, N. and Smith, H., Applied Regression Analysis , Wiley, New York, 1966.

21. Motard, R.L., et. al., "Steady State Chemical Process Simulation", AIChE, 21, #3, 1975, p.417.

22. Mosler, H., "A Digital Computer Executive Program For Process Simulation and Design (PACER), M.Sc. Thesis, Purdue University, 1964.

23. Toong, T. and Johnson, A., "The Modular Approach to System Design and Analysis", CGE Information Systems, Toronto, 1968.

24. Herbst, J.A., Rajamani, K. and Kinneberg, D., "ESTIMIL: A Program for Grinding Simulation and Parameter Estimation with Linear Models", Dept. of Metallurgical Engineering, University of Utah, Salt Lake City, 1977.

25. Herbst, J.A., et. al., "Population Balance Models For The Design of Continous Grinding Mills", 10th IMPC, London, 1973.

26. Siddique, M., "A Kinetic Approach to Ball Mill Scale-Up for Dry and Wet Systems," M.S. thesis, University of Utah 1977.

27. Mular, A.L., "The Determination of Selection Elements and Lumped Parameters For Grinding Mills", CIM Bulletin, 63, 1970, p.820.

28. Klimpel, R.R., and Austin, L.G., "The Back Calculations of Specific Rates of Breakage", Int. J. of Min. Proc., 4, 1977, p.7.

29. Herbst, J., et. al., "An Approach to the Estimation of Parameters of Grinding Models From On-Line Measurements", 3rd European Symposium on Comminution, Canvas, 1971, Dechema Monographien, 69, 1972.

30. King, R.P., et. al., "A Pilot Plant Investigation of A Flotation Model", Report 1573, National Institute For Metallurgy, Johannesburg, South Africa, Feb., 1974.

31. Hatch, C.C. and Mular, A.L., "Digital Simulation of a Secondary Crushing Plant", to be presented at Annual General Meeting of CIM, Vancouver, April, 1978.

32. Kelsall, D.F., Private Communication.

33. Mular, A.L., "Optimization in Flotation Plants", Chapter 31 in Flotation, Ed. M.C. Fuerstenau, AIME, 1976, vol.2.

34. Mular, A.L., "Off-Line Optimization of A Grinding Circuit", Presented at joint meeting, Montreal Branch of CIM and Computer Appl. and Proc. Control of CIM, McGill University, March 21, 1975.

35. INAMDAR, A., "Some Process Improvements on Gibraltar Grinding Circuit", Presented at Vancouver Branch Meeting, Canadian Mineral Processors, Vancouver, 1974.

36. Tarr, D.T., "Practical Application of Liquid Cyclones in Mineral Dressing Problems", Krebs Engineering, Menlo Park, Calif., 1965.

37. Mular, A.L., "The Selection of Optimization Methods For Mineral Processes", Special volume 12, CIM, Montreal, 1971, p.442.

38. Peterson, J.W., White, J.W., and E. Krist, "Off-Line Development of A Minicomputer Process Control Program For An Industrial Wet Grinding Circuit", Trans. AIME, 260, 1976, p.355.

39. Mular, A.L. and Zickar, E., "A Computer Program For Mineral Processing Capital Cost Estimation", CIM Bulletin, May, 1973.

40. Mular, A.L., Mineral Processing Equipment Costs and Preliminary Capital Cost Estimations, Special vol. 13-R, CIM, Montreal, 1978.

41. Mular, A. L., Progress Report on "Digital Simulation of the Brenda Concentrator", Dept. of Energy, Mines and Resources, Ottawa, Ontario, 1977.

42. Plitt, L. R., "A Mathematical Model of the Hydrocyclone Classifier", CIM Bulletin, December, 1976, p. 114.

43. Lynch, A. J., and Rao, T. C., "Modelling and Scale Up of Hydrocyclone Classifiers", Proceedings of the 11th IMPC, Cagliari, 1975

44. White, J.H., "A Systems Study of Grinding as Applied to Mill Scale Operations", NSF/RANN Final Report GI 38067, 1976.

45. Herbst, J.A., and Rajamani, K., "Models for the Design and Control of Tumbling Mills", University of Utah Research Report, 1977.

APPENDIX

Simulators mentioned in the text have employed models of ball mills, rod mills, cyclone classifiers, cone crushers and vibrating screens. These models are summarized below.

Ball Mill Models

Linear lumped parameter ball mill models have been widely accepted and are in common usage. An alternative ball mill model, namely the stage of grinding (or matrix) model, has been utilized in some cases.

Linear Lumped Parameter Ball Mill Models. If the kinetics of breakage in a ball mill are linear, the batch response for a set of n size fractions $m^T(t) = (m_1(t), m_2(t), m_n(t))$ is given by (10):

$$\underline{m}(t) = \underline{\underline{T}}\,\underline{\underline{J}}(t)\underline{\underline{T}}^{-1}\underline{m}(0)$$

where

$$T_{ij} = \begin{cases} 0 & i > j \\ 1 & i = j \\ \sum_{k=j}^{i-1} \dfrac{b_{ik}\,S_k}{S_i - S_j}\,T_{kj} & i > j \end{cases}$$

$$J_{ij} = \begin{cases} \exp(-S_i\,t) & i = j \\ 0 & i \neq j \end{cases}$$

and S_i gives the fractional rate of breakage (Selection Function) for material in the ith size interval (X_i to X_{i+1}) and b_{ki} gives the fraction of primary breakage fragments from the ith size interval which reports to the jth interval (Breakage Function).

For continuous grinding if material transport through the mill is size independent, the steady state mill discharge size distribution $\underline{m}_{MD}$ can be obtained in terms of a weighted average of batch responses for a feed $\underline{m}_{MF}$ where the weighting function is the material residence time distribution for the mill, i.e.,

$$\underline{m}_{MD} = \int_0^{\infty} \underline{\underline{T}}\,\underline{\underline{J}}(t)\underline{\underline{T}}^{-1}\underline{m}_{MF}\cdot E(t)\,dt$$

A flexible form of the residence time distribution is the mixers-in-series model with mean residence time τ:

$$E(t) = \frac{j^j (t/\tau)^{j-1} \exp(-\,jt/\tau)}{\tau\,\Gamma(j)}$$

This RTD model yields a plug flow response for $j \to \infty$ and a single perfect mixer response for $j = 1$. In this case the expression for $\underline{m}_{MD}$ given above becomes:

$$\underline{m}_{MD} = \underline{\underline{T}}\,\underline{\underline{J}}_c(\tau)\underline{\underline{T}}^{-1}\underline{m}_{MF}$$

where

$$J_{c_{ij}}(\tau) = \begin{cases} (1 + S_i\tau/j)^{-j} & i = j \\ 0 & i \neq j \end{cases}$$

For open circuit grinding the mill discharge is equal to the circuit product, $\underline{m}_P = \underline{m}_{MD}$, and the mill feed is equal to the fresh feed, $\underline{m}_F = \underline{m}_{MF}$, therefore

$$\underline{m}_P = \underline{\underline{T}}\,\underline{\underline{J}}_c(\tau)\underline{\underline{T}}^{-1}\underline{m}_F$$

Further for closed circuit grinding a material balance around the classifier yields expressions for the circuit product.

For standard closed circuit

$$\underline{m}_P = [\underline{\underline{I}}-\underline{\underline{C}}]\,\underline{\underline{L}}\,[\underline{\underline{I}}-\underline{\underline{C}}\underline{\underline{L}}]^{-1}\underline{m}_F$$

For reverse closed circuit

$$\underline{m}_P = [\underline{\underline{I}}-\underline{\underline{C}}][(\underline{\underline{I}}-\underline{\underline{L}}\underline{\underline{C}})^{-1}\underline{\underline{L}}\underline{\underline{C}}+\underline{\underline{I}}]\underline{m}_F$$

where $\underline{\underline{L}} = \underline{\underline{T}}\,\underline{\underline{J}}_c(\tau)\underline{\underline{T}}^{-1}$ and $\underline{\underline{C}}$ is an arbitrary classifier matrix.

These equations form the basis for simulation with the ESTIMILL program.

Stage of Grinding Model. The latest version of the stage of grinding (matrix) model may be written as follows for a ball mill (41):

$$z = a_o + a_1 W + a_2/Q + a_3 W_b \qquad v-1 \leq z \leq v$$

$$\underline{P}_z = (v - z)\underline{P}_{v-1} + (z+1-v)\underline{P}_v$$

$$\underline{Y}_j = \underline{\underline{B}}\,\underline{\underline{S}} + \underline{\underline{I}} - \underline{\underline{S}}$$

$$\underline{P}_{v-1} = \underline{Y}_1\underline{Y}_2 \ldots \underline{Y}_{v-1}\,\underline{F}$$

$$\underline{P}_v = \underline{Y}_v\,\underline{P}_{v-1}$$

where z is the number of stages, a_o, a_1, a_2, and a_3 are constants, W is the work input to the mill in KWHr/ton, Q is the volume flow rate of feed slurry, W_b is the water flow rate to the ball mill, $\underline{P}_z$ is the predicted product size distribution vector, $\underline{F}$ is the feed size distribution vector, $\underline{\underline{B}}$ is the breakage matrix and $\underline{\underline{S}}$ is the selection matrix. The term v is an integral stage.

Rod Mill Models

To describe grinding in rod mills, several have used the linear lumped parameter grinding model. However, linear breakage processes do not appear reasonable for such machines. One alternative has been a stage of grinding version, which is written as (2):

$$zQ = K \qquad v-1 \leq z \leq v$$

$$\underline{P}_z = (v - z)\underline{P}_{v-1} + (z+1-v)\underline{P}_v$$

$$\underline{Y}_j = (\underline{\underline{I}} - \underline{\underline{C}})\,(\underline{\underline{B}}\,\underline{\underline{S}} + \underline{\underline{I}} - \underline{\underline{S}})(\underline{\underline{I}} - \underline{\underline{C}}(\underline{\underline{B}}\,\underline{\underline{S}} + \underline{\underline{I}} - \underline{\underline{S}}))^{-1}$$

$$\underline{P}_{v-1} = \underline{Y}_1\,\underline{Y}_2 \ldots \underline{Y}_{v-1}\,\underline{F}$$

$$\underline{P}_v = \underline{Y}_v\,\underline{P}_{v-1}$$

where symbols have meanings similar to the matrix ball mill model, except that in the above, K is a constant and $\underline{\underline{C}}$ is the classification matrix.

Cyclone Classifier Models

Cyclone models developed for the Mt. Isa Mines grinding circuit (2) are reported below with minor modification (6). Recent developments should be reviewed by the interested reader (42,43).

Capacity equation:

$$Q = k_1 V P^{.5} (W_f)^{.125}$$

Water split equation:

$$H_o = k_2 + 1.1 H_f - 4.3 S$$

X_{50c} equation:

$$\ln(X_{50c}) = k_3 + .047(P_f) - .47S + .14V + .015P$$

Efficiency equation:

$$\ln(1 - Y_{cx}) = (-\ln 2)((x/x_{50c})^{1.38} - .096)/.904$$

where Q is volume flow rate of slurry, k_1, k_2, k_3 are constants, V is vortex diameter (in), P is inlet pressure (psi), W_f is feed percent water by weight, H_o is mass flow of water to overflow (tph), H_f is mass flow of water in feed (tph), S is spigot diameter (in), X_{50c} is the size at which Y_{cx} is .5 (microns), P_f is the feed percent solids, and Y_{cx} is the weight fraction of a narrow size range of feed that reports to underflow after correction for water bypassing of solids.

Crusher Models

Models developed (31) for secondary and tertiary crushers are modifications of those utilized for Mt. Isa Mines crushers (8).

Secondary Crusher Model. This model is of the form

$$\underline{P} = (\underline{\underline{I}} - \underline{\underline{C}})(\underline{\underline{I}} - \underline{\underline{B}}\,\underline{\underline{C}})^{-1}\,\underline{F}$$

$$\underline{\underline{B}} = \alpha \underline{\underline{B}}_1 + (1 - \alpha)\underline{\underline{B}}_2$$

$$\alpha = a_o + a_1 G + a_2 T + a_3 S$$

$$y_i = 1 - (1 - (x_i/x_j)^{\gamma})^{\beta}$$

$$Y_i = \exp(-(x_i/2)^{\delta})$$

$$c_{ii} = (F_i - F_{i+1})/(x_i - x_{i+1})$$

$$F_i = k_1 - (1/3)(k_1 - k_2) \qquad x_i < k_1$$

$$F_i = x_i - (1/3)((x_i - k_2)^3/(k_1 - k_2)^2) \qquad k_1 \leq x_i \leq k_2$$

$$F_i = x_i \qquad x_i > k_2$$

$$k_1 = c_o + c_1 G + c_2 G^2 + c_3 S$$

$$k_2 = b_o+b_1G+b_2G^2+b_3S$$

where $\underline{P}$ is the product vector, $\underline{\underline{I}}$ is an identity matrix, $\underline{\underline{C}}$ is a classification matrix with diagonal elements c_{ii}, $\underline{\underline{B}}$ is the breakage matrix, $\underline{\underline{B}}_1$ is a normalizable component of $\underline{\underline{B}}$, and $\underline{\underline{B}}_2$ is a non-normalizable component of $\underline{\underline{B}}$. Elements of B_1 are found from y_i which is a cumulative fraction finer than x_i from the breakage of x_j. Elements of B_2 are found from Y_i which is a cumulative fraction coarser than x_i. $\alpha,\gamma,\beta,\delta$, a_o, a_1, a_2, a_3, b_o, b_1, b_2, b_3, c_o, c_1, c_2, c_3 are constants, G is crusher close side setting, T is crusher feed rate, and S is the percent +1 inch in the crusher feed. Spline functions were judged unnecessary for the above model.

Tertiary Crusher Model. This model is similar to that for the secondary crusher, except that:

$$\alpha = d_o + d_1T + d_2/(1 + d_3(T-d_4)^2)$$

$$\beta = g_o + g_1\exp(-g_2\alpha^n)$$

$$k_1 = b'_o + b'_1S$$

$$k_2 = c'_o + c'_1G + (c'_2/(1+c'_3(G-c'_4)^2))^m$$

where symbols not previously defined are constants.

Vibrating Screen Model

The vibrating screen model has been derived (31). The derivation is summarized below.

Let a vibrating screen be fed at steady rate with F/T particles per unit time, where the particles are described by a distribution of sizes. In one retention time period, T, F particles (not involved in bypassing) are fed. Consider a narrow size fraction of these particles. Let their number, of geometric mean size X_i, be F_i, where $F = \Sigma F_i$. After time T, U_i particles of size X_i have passed through the screen, while $F_i - U_i$ particles have been retained as part of the oversize stream. The number of particles, U_i, that pass through depends on particle dimensions, particle orientation, size of screen opening, crowding, mechanical factors such as slope, throw and speed, and the value of T. One thing for certain is that a given particle either passed or did not pass through the screen, although, because size is measured with standard sieves, that particle cannot be distinguished individually within the geometric mean size range measured by X_i. Let each feed particle in the size range be tagged with an identifying number, Z_j. If the particle passes through the screen, assign a value 1 to Z_j. Otherwise the value of Z_j is zero, i.e., if there are $F_i = 8$ particles the following may occur:

Z_1 Z_2 Z_3 Z_4 Z_5 Z_6 Z_7 Z_8 (identifying numbers)
1 0 1 1 0 0 0 0

This means that particles Z_1, Z_3 and Z_4 passed through the screen, while the others did not. It is obvious that there are a large number of distinct sequences that may occur. Alternatively, group F_i particles according to whether they pass the screen or do not as follows:

Pass (1)	Do Not Pass (0)
Z_1 Z_3 Z_4	Z_2 Z_5 Z_6 Z_7 Z_8

Again note that a large number of sequences is possible, including all particles pass, no particles pass, etc. There are $F_i!$ ways to write the Z_i identifying numbers on a line, as the first position may be chosen in F_i ways, the second in F_i-1 ways and so on. But many of these sequences are not distinguishable because of the way particle size is measured. For example, interchanging the order Z_1 Z_3 Z_4 to Z_3 Z_4 Z_1 is not a distinguishable sequence. Of interest are the number of distinct sequences that are possible; not the permutations. Each distinct sequence can be permuted in $U_i!(F_i - U_i)!$ ways. If W_i is the total number of distinct sequences, then

$$W_iU_i!(F_i - U_i)! = F_i!$$

so that

$$W_i = F_i!/(U_i!\ (F_i - U_i)!)$$

There are W_i distinct ways that the F_i particles might distribute to the undersize stream. If we now cover the complete range of sizes involved in the feed, then the total number of distinct ways of distributing feed particles to the undersize stream is W, where

$$W = \prod_{\text{all } x} W_i = \prod_{\text{all } x} (F_i!/(U_i!(F_i - U_i)!)) \qquad (1)$$

Take natural logs of both sides of equation 1 so that

$$\ln(W) = \sum_{\text{all } x} (\ln(F_i!) - \ln(U_i!) - \ln((F_i - U_i)!)) \qquad (2)$$

To find the most probable way to distribute particles to the undersize, make $\frac{\partial \ln(W)}{\partial U_i}$ a maximum, subject to the following conditions:

(a) $U = \Sigma U_i$ = constant at steady state, where U is the total number of particles which enters with feed but could report to underflow, and (b) $V = \Sigma U_i V_i$ is the total volume of U particles = constant and V_i is the average volume per particle of the ith size fraction. Using the approximation that $\ln(j!) = j\ln(j) - j$, equation 2 may be written as

$$\ln(W) = \sum_{\text{all } x} (F_i \ln(F_i) - (F_i - U_i)\ln(F_i - U_i) - U_i \ln(U_i)) \qquad (3)$$

The LaGrange multiplier method may be used to find a maximum subject to conditions (a) and (b). Thus:

$$\frac{\partial(\ln(W) + \alpha(U - \Sigma U_i) + \beta(V - \Sigma(U_i V_i)))}{\partial U_i} = 0 \qquad (4)$$

where α and β are the multipliers. On substituting equation (3) in (4) and taking partial derivatives,

$$\ln(F_i - U_i) - \ln(U_i) - \alpha - \beta V_i = 0$$

which may be rearranged such that

$$Y_i = \frac{F_i - U_i}{F_i} = \frac{1}{1 + \exp(-\alpha \ -\beta V_i)} \qquad (5)$$

where Y_i is the weight fraction of the ith narrow size range of feed that is reporting to the oversize stream on the assumption that the feed does not bypass. When $Y_i = .5$, $V_i = V_{50}$ so that $-\alpha = \beta V_{50}$. Hence equation 5 becomes:

$$Y_i = \frac{1}{1 + \exp(\beta(V_{50} - V_i))} \qquad (6)$$

With shape factors constant and with V_{50} and V_i proportional to the cube of the geometric mean size, equation 6 becomes:

$$Y_i = \frac{1}{1 + \exp((X_{50}^3 - X_i^3)/k)} \qquad (7)$$

where k is a new constant.

Bypassing may occur due to fine particles sticking to coarse wet particles. In addition, dusting may be a problem so that dust particles are not involved in the screening process described. In this case, a weight fraction, C_i, of a narrow size range of feed must be accounted for. This is accomplished as follows:

$$y_i = C_i + \frac{1 - C_i}{1 + \exp((X_{50}^3 - X_i^3)/k)} \qquad (8)$$

where y_i is the actual weight fraction of a narrow size range of feed that reports to the oversize stream. For the model developed, the value of C_i was found to be essentially zero (the ore was dry and the dusting problem was not severe). In addition, the parameters X_{50} and k depended on operating variables in the following way:

$$X_{50} = a_o + a_1 d + a_2 d^2 + a_3 S^2 + a_4 T^7$$

$$k = b_o + b_1 d^2 + b_2 S + b_3 T^2$$

where a_o, a_1, a_2, a_3, a_4, b_o, b_1, b_2, b_3 are constants, d is screen opening, S is the percent +1 inch in the feed, and T is the screen feed rate of solids.

4

Screening-Classification-Gravity Separation

Chairmen:, G. M. Meisel
C. F. Lagergren

Chapter 15

SELECTION GUIDELINES FOR SIZE AND TYPE OF VIBRATING SCREENS IN ORE CRUSHING PLANTS

by K. G. Colman
Sr. Application Engineer

W. S. Tyler, Inc. Screening Division
Combustion Engineering, Inc.
Mentor, Ohio

INTRODUCTION

This text is intended to provide the operator and turn-key contractor with the basic data for sizing vibrating screens to be used in ore crushing plants.

The basic capacities, modifying factors, etc. used in the area requirement calculations were developed empirically and modified to their present values based upon years of practical applications.

It must be pointed out that screening is much more of an art than an exact science, however, the information presented here should provide for reasonably accurate screen sizing when applied to metallic ores.

TYPES OF SCREENS

There are two types of screens which can be used for strictly sizing applications. These are two bearing and four bearing. Both of these are inclined, circle throw machines.

Two Bearing

The motion of a two bearing screen is generated by unbalanced weights attached to the vibrator shaft. The body of the machine is usually supported by steel coil springs (see Fig. 1.)

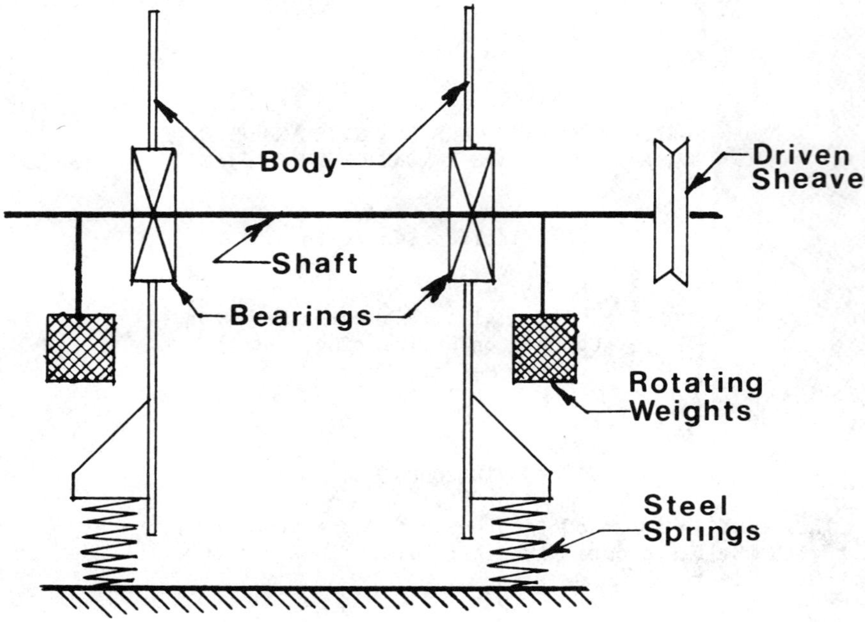

FIG. 1 TWO BEARING SCREEN DIAGRAM

The main advantage of the two bearing screen is the ability to change the amplitude of the motion by adding and subtracting weights if the application should be changed.

The biggest disadvantage in ore applications is the fact that should there be any buildup of wet, sticky ore on the vibrating body, the amplitude of motion will be reduced because of the increased weight. This reduction in amplitude will result in slower rate of travel of material and reduced vibrating "G" force leading to lower screening capacity and increased deck blinding tendencies.

Also the two bearing screen will transmit dynamic loading to the supporting structure when running. This will affect the structural design criteria for the proposed building.

Four Bearing

The motion of a four bearing screen is created by eccentrics machined into the vibrator shaft, in principle, like the eccentrics of an automobile crankshaft. This provides a positive stroke.

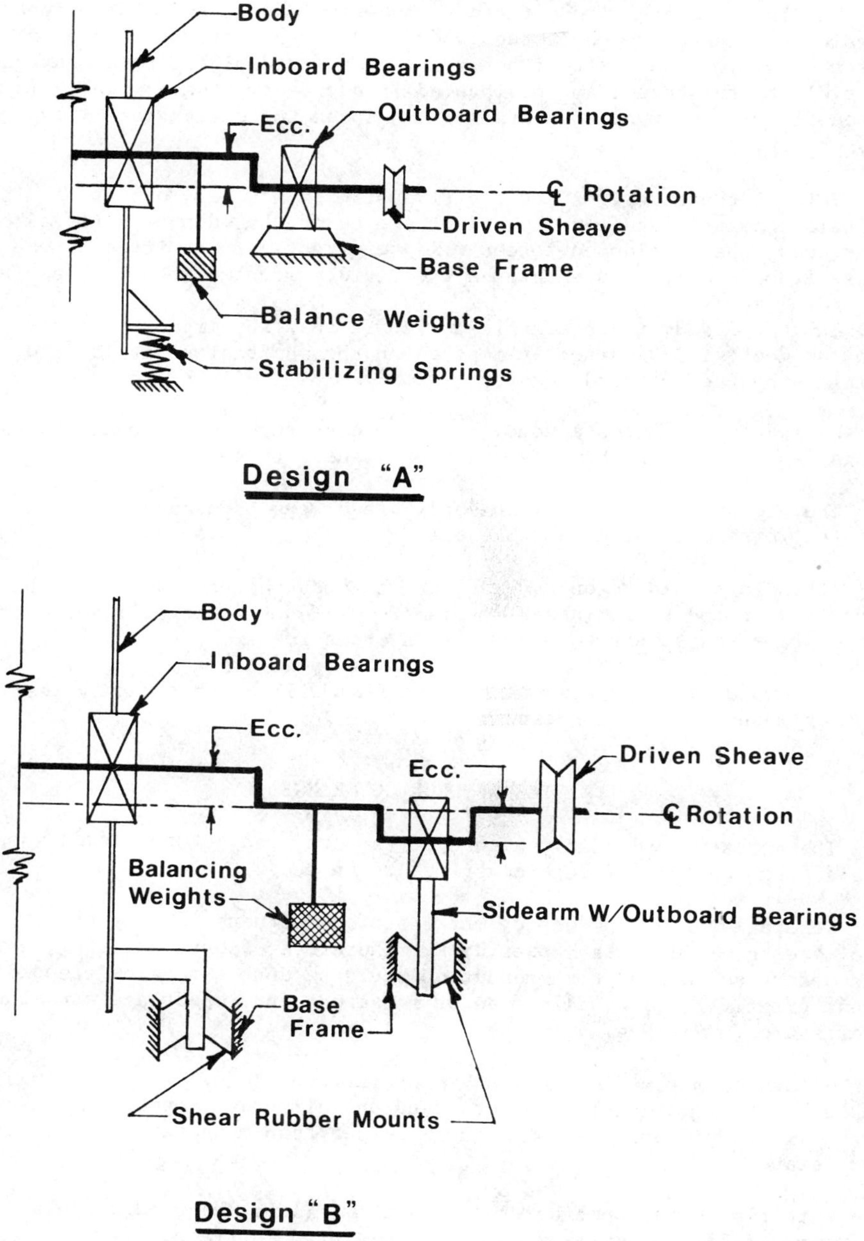

FIG. 2 FOUR BEARING SCREEN DIAGRAMS

The four bearings are mounted on the same shaft, two at the inboard position fastened to the screen body, and two at the outboard position at the ends of the shaft either mounted in fixed pillow blocks attached rigidly to the base frame or mounted in a sidearm assembly which is free-floating on rubber shear mounts. These two designs are illustrated in Fig. 2.

Both of these designs are positive stroke, however, design "A" is a single eccentric shaft and should there be any buildup of material on the body, the unbalanced forces will be directly transmitted to the base frame and structure through the rigidly mounted outboard bearings.

Another disadvantage of this design is that any impact of material on the deck will be directly imposed on the shaft since it is held rigidly by the outboard bearings.

Design "B" utilizes a double eccentric shaft with the outboard bearings, as well as the body, supported by shear rubber mounts.

The sidearm assembly and the body are always opposed to one another so the forces cancel each other out.

If there should be any buildup on the body the unbalance will be absorbed in the rubber shear mounts and will not be transmitted to the base frame and structure unless it is excessive.

Any impact load imposed by the material will be absorbed by the rubber mounts, not the shaft.

TYPES OF DECK COVERINGS

The abrasive and sticky nature of the ore must be considered when selecting the type of deck coverings to be used.

Generally, the top deck of multi-surface screens is only there to relieve the load on the lower deck. Therefore, accurate sizing is not necessary so, a good recommendation would be double tapered stepped cast grizzly bars, usually made of manganese for impact and abrasion resistance. See Fig. 3.

Alternate materials such as hi-chrome iron (26%c_r) are sometimes used but this material does not stand up well under impact conditions since it is too brittle but is much better than manganese for abrasion resistance.

A fairly normal opening would be 39 mm. (1.50 in) used above a 11.1 mm. (7/16 in) or 12.7 mm. (1/2 in) bottom surface.

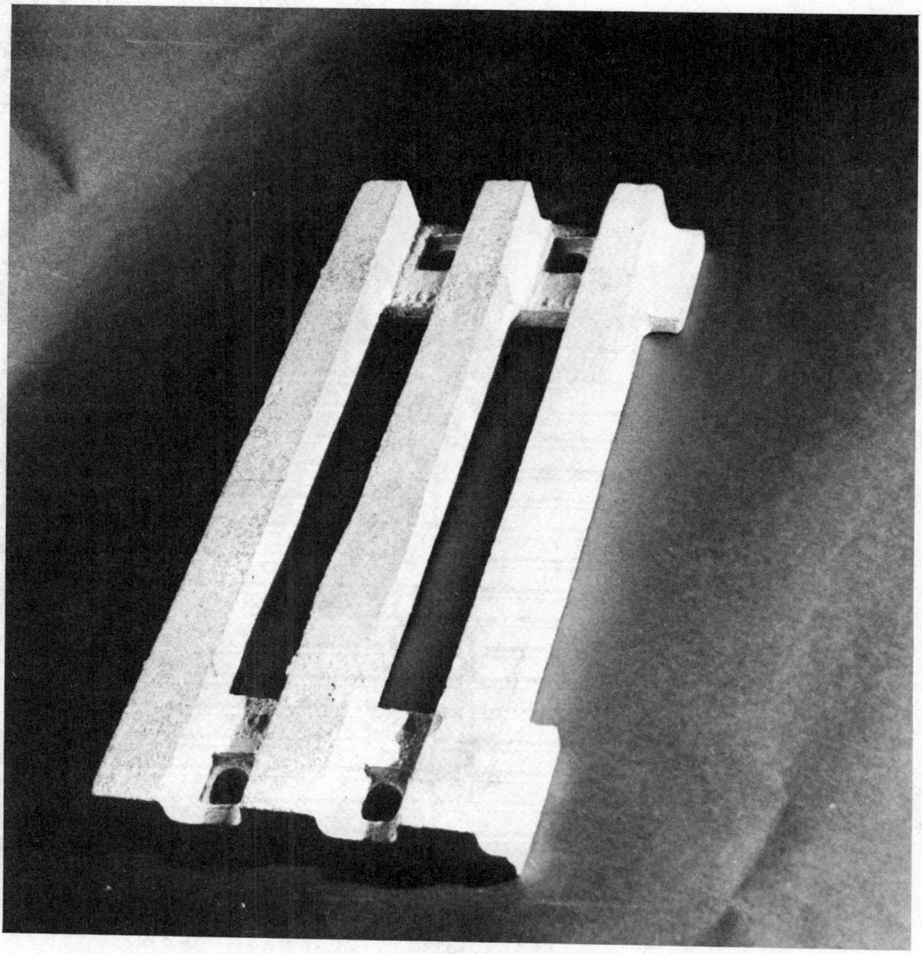

FIG. 3 DOUBLE TAPERED CAST MANGANESE GRIZZLY BAR SECTION

There are a couple of deck covering configurations used on the bottom or sizing surface. The long slot, or Ty-Rod type of woven wire cloth, such as shown in Fig. 4, provides the advantages of both increased throughput capacity as well as anti-blinding action as a result of secondary vibrations in the long wires.

FIG. 4 LONG SLOT (TY-ROD) WOVEN WIRE CLOTH

Another type of sizing deck used extensively is a rod deck as pictured in Fig. 5. The deck is constructed from individual rods of high carbon, oil tempered steel with a hardness of 46-48 Rockwell "C" which are held in rubber end holders and intermediate spacers.

FIG. 5 ROD DECK

The rubber allows the rods to move with a secondary vibration to help alleviate buildup of sticky ore and relieve wedge blinding.

This type of deck gives reasonably good life as well as high capacity. Fig 6 illustrates a two surface crushing plant screen equipped with cast grizzly bars on the top deck and a rod deck on the bottom, a fairly typical arrangement used in taconite plants.

FIG. 6 TWO DECK SCREEN WITH CAST GRIZZLY BAR TOP DECK AND ROD DECK ON THE BOTTOM

SPECIAL FEATURES

When writing bid specifications for vibrating screens, there are several features which you should strongly consider including.

Very often, if these features are not spelled out by the buyer, it will be difficult to evaluate bids from suppliers since some may include these options and some may not.

Liners

Screens equipped with liners, wherever possible, on wearing surfaces will help to avoid costly replacement of major screen components and loss of production.

Liners should be applied to feed plate, side plates above the deck, deck support members, vibrator housing, and discharge lips.

Materials of liner construction can be various grades of abrasive resistant steel, rubber, or polyurethane.

Dust Enclosures

The present day dust emission control regulations essentially dictate the use of some sort of dust enclosure on the screen.

The most common is the stationary air seal type such as pictured in Fig. 6. This type provides access to the screen side plates, quick removable top covers for ease of inspection and deck replacement, and an air exhaust outlet.

The live cover, or vibrating type, although less costly has several disadvantages.

(1) Flexible connections must be provided at the feed and discharge points and the air exhaust.

(2) Increased vibrated weight will reduce the bearing life.

(3) Covers cannot be removed for inspection while screen is running.

Roll-Out Carriages

Several of the modern plants now in operation have crushing plant screens that are equipped with roll-out carriages.

This feature provides for the absolute minimum of downtime when a screen needs maintenance by rolling it out and rolling in a spare standby screen. This spare screen generally serves a bank of screens since normally only one screen would be taken out of service for maintenance at a time.

The screen that was removed can then be taken to the maintenance shop or other area out of the production area and whatever work that needs to be done can be done without the immediate pressure of downtime causing an emergency type fix rather than a well organized, complete maintenance program.

SIZE SELECTION

The generally accepted method of screen area selection is the throughput method which is based upon the amount of material (STPH) that will pass through 0.0929 m^2 (1.0 ft^2) of a screen deck having a specific opening.

The total area "A" can be defined by the formula:

$$A = \frac{\text{STPH of underside in feed}}{C \cdot \text{Bulk Density} \cdot \text{(Modifying factors F,E,S,D,O,W)}}$$

Where C is the Basic Capacity from Fig. 7, bulk density is expressed as a decimal found by dividing the density of the feed material (kg/m^3) by 1602 kg/m^3 (100 lb/ft^3) and the modifying factors are found in appropriate figures following in this text.

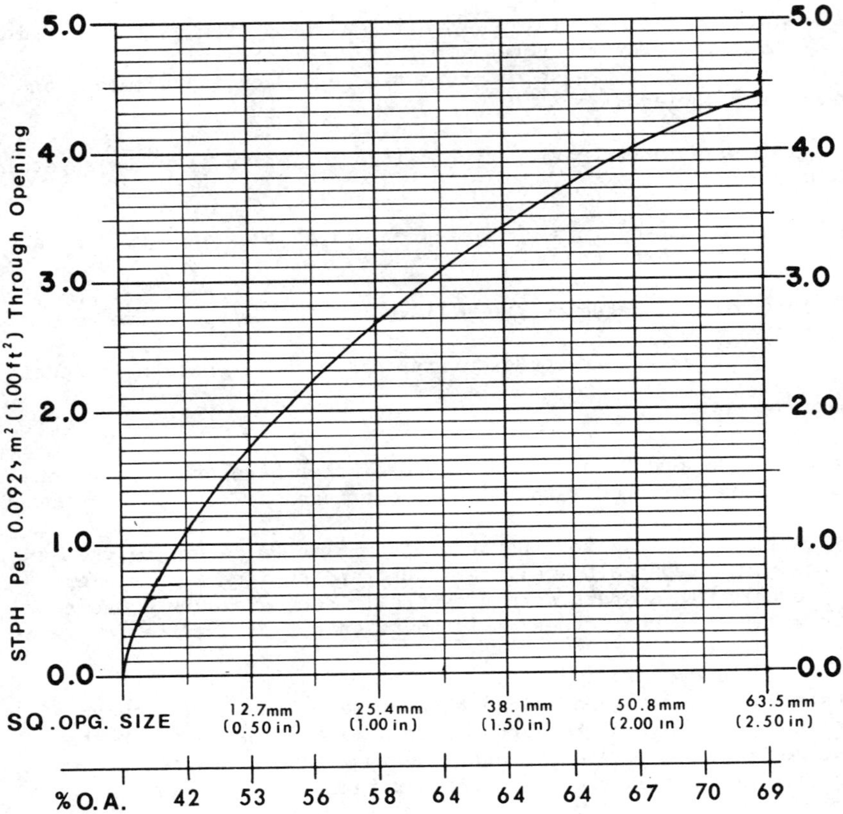

FIG. 7 BASIC SCREEN CAPACITY
For Ore having 1602 Kg/m^3 (100#/ft^3) Bulk Density

Basic Capacity

The curve representing the value of C for the various openings is based on a material with a bulk density of 1602 kg/m^3 (100 lb/ft^3).

Since most metallic ores have similar screening characteristics, the C value for any ore can be determined by a simple ratio of densities.

However, all materials do not have the same properties or screening characteristics so do not attempt to use the capacities given in Fig. 7 for materials such as coke, sand & gravel, etc., because the results will be in error. These other materials have their own specific capacity curves so use Fig. 7 only for metallic ores.

Modifying Factors

There are many variables and interrelationships of these variables which affect the screenability of a particular material but we will apply only those which significantly affect the sizing calculations for ore screens.

Fines Factor - F. The fines factor is a measure of the amount of material, in the feed to a screen deck, which is less than 1/2 the size of the opening in the deck.

It is a measure of the difficulty of screening when compared to 40% fines (F = 1.00) which is the value used for the Basic Capacity (C) in Fig. 7.

The values for various percentages of fines are given in Fig. 8.

The important thing to remember is to use the percent of half-size coming to the particular deck being analyzed, expressed as a percentage of the feed to that deck, i. e., with top deck 38 mm. (1.50 in.) opening, bottom deck 12.7 mm. (0.50 in.) opening, feed analysis 70% passing 39 mm., 50% passing 12.7 mm., 35% passing 6.3 mm., the fines factor for the bottom deck would be 35/70 = 50%.

Efficiency Factor - E. The efficiency of separation is expressed as the ratio of the amount of material that actually passes the opening divided by the amount in the feed that should pass.

"Commercially perfect" screening is considered to be 95% efficiency, therefore the factor for 95% is 1.00 as indicated in Fig. 8.

For scalping and crusher relief applications, an efficiency of 80-85% is generally acceptable.

%*	FACTOR	
	Fines F	Efficiency E
0%	.44	
10%	.55	
20%	.70	
30%	.80	
40%	1.00	
50%	1.20	
60%	1.40	
70%	1.80	2.25
80%	2.20	1.75
85%	2.50	1.50
90%	3.00	1.25
95%	3.75	1.00

*Interpolate for percentages not shown

FIG. 8 FINES & EFFICIENCY FACTORS

Slot Factor - S. The slotted opening factor compensates for tendency of more particles to pass through the screening medium due to a fewer number of cross bars or wires obstructing the particles path through the deck.

The values for various slot configurations are shown in Fig. 9.

Typical Deck Preparations	Length/Width Ratio	Slotted Opening Factor S
Square and Slight Rectangular Openings	<2	1.0
Rectangular Openings Ton-Cap	>2 but <4	1.15
Slotted Openings Ty-Rod	>4 but <25	1.2
Parallel Rod Decks	>25	SP 1.4 * RA 1.3

*SP = Slots Parallel to Flow, RA = Slots Right Angles to Flow

FIG. 9 SLOTTED DECK OPENING FACTORS

Deck Factor - D. The deck factor (Fig. 10) allows for the fact that stratification does not take place at the extreme end of the deck so the bulk of the fines do not fall through until the material has traveled a short distance down the deck.

Therefore, except for the top deck, the entire feed does not land on the feed end which results in some of the screening area being ineffective.

Deck	Deck Factor D
Top	1.00
2nd	.90
3rd	.80

FIG. 10 DECK FACTOR

Open Area Factor - O. The basic capacity curve shown in Fig. 7 is based upon a square opening woven wire deck with an open area as indicated under the opening size.

When using deck coverings having an open area significantly different from that shown, the factor is obtained by the ratio of the open area of the deck used to the standard open area given in Fig. 7 for the opening being used.

For example, if for a 24 mm separation a deck with 36% open area were used, the factor would be 36 ÷ 58 (Fig. 7) or 0.62 and alternately, if a deck with 72% open area were used, the factor would be 72 ÷ 58 or 1.24.

Wet Screening Factor - W. The wet screening factor is applied when the screening is being done with the aid of water being sprayed on the material being screened.

The benefit derived from spraying water on the feed varies with the opening size so the factor is selected from Fig. 11 by referring to the opening column and reading the factor in the "W" column.

It is also important to note that the advantage of using sprays can only be realized if the correct amount of water is used.

The recommended volume for efficient wet screening is 11.35 liters/minute (3 GPM) minimum to 18.92 liters/minute (5 GPM) per 0.765 cubic meters (1.0 cu yd) of feed material.

SIZE OPENING (Square)	W
1/32" or less	1.25
1/16"	3.00
1/8" & 3/16"	3.50
5/16"	3.00
3/8"	2.50
1/2"	1.75
3/4"	1.35
1"	1.25
+2"	1.00

FIG. 11 WET SCREENING FACTOR

To illustrate the selection procedure, the following sample problem will be used.

Feed: 300 STPH Iron Ore, 2082 kg/m^3 (130 lb/ft^3), 8% H_2O, with following sieve analysis:

	Cumulative % Passing
38 mm (1-1/2")	100.00
25 mm (1")	98.00
19 mm (3/4")	92.00
12.5 mm (1/2")	65.00
6.3 mm (1/4)	33.00

Separation required: 12.7 mm (1/2"), dry screening.

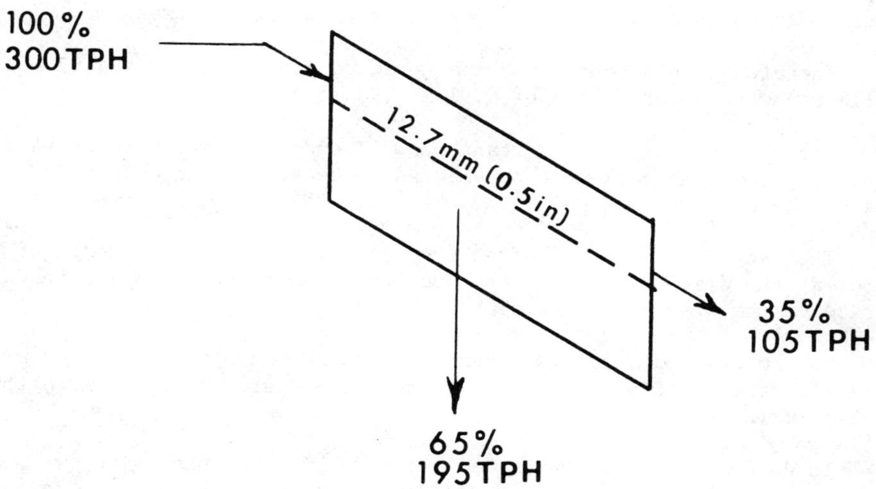

Solution

$$A = \frac{195\ \text{TPH}}{C \cdot \text{Density} \cdot F \cdot E \cdot D \cdot S}$$

Where C = 1.7 T/.09 m^2; Density = $\frac{2082}{1602}$ = 1.30, F = 0.86 (33%), E = 1.00 (95%), D = 1.00 (Top), S = 1.2 (Ty-Rod Cloth).

$$A = \frac{195\ \text{TPH}}{1.7 \times 1.3 \times 0.86 \times 1.00 \times 1.00 \times 1.2} =$$

$$\frac{195\ \text{TPH}}{2.28\ \text{TPH}/.0929\ m^2} = 7.95\ m^2\ (85.5\ ft^2).$$

The 7.95 m^2 (85.5 ft^2) represents the net effective screen area required so to allow for the space taken up by tension rails, center clamps and openings that are blocked by cloth support bars, add 10% to the net area and use the total figure for the selection of screen size.

In this case, 7.95 m^2 + 0.79 m^2 = 8.74 m^2 (94.0 ft^2).

Now, a standard commercial size screen with at least an area of 8.74 m^2 (94.0 ft^2) must be selected, keeping in mind that the length to width ratio should be at least 2:1 for effective screening.

Therefore, this tentative size would be 1.83 m (6 ft) x 4.87 m (16 ft) with a total area of 8.91 m^2 (96 ft^2).

The next step is an important one. The depth of bed of material passing over the deck must be checked to be sure it is within acceptable limits.

The recommendation for effective screening is that the depth of bed at the discharge end of the deck not be more than 4 times the size of the opening in the deck.

This means that, for the example which has an opening of 12.7 mm (1/2"), the depth of bed at the discharge end should not be more than 50.8 mm (2 in).

To determine the depth of bed refer to Fig. 12 which gives values, for various width screens, of STPH/cm depth of bed based on a rate of material travel of 18.29 m/min (60 ft/min).

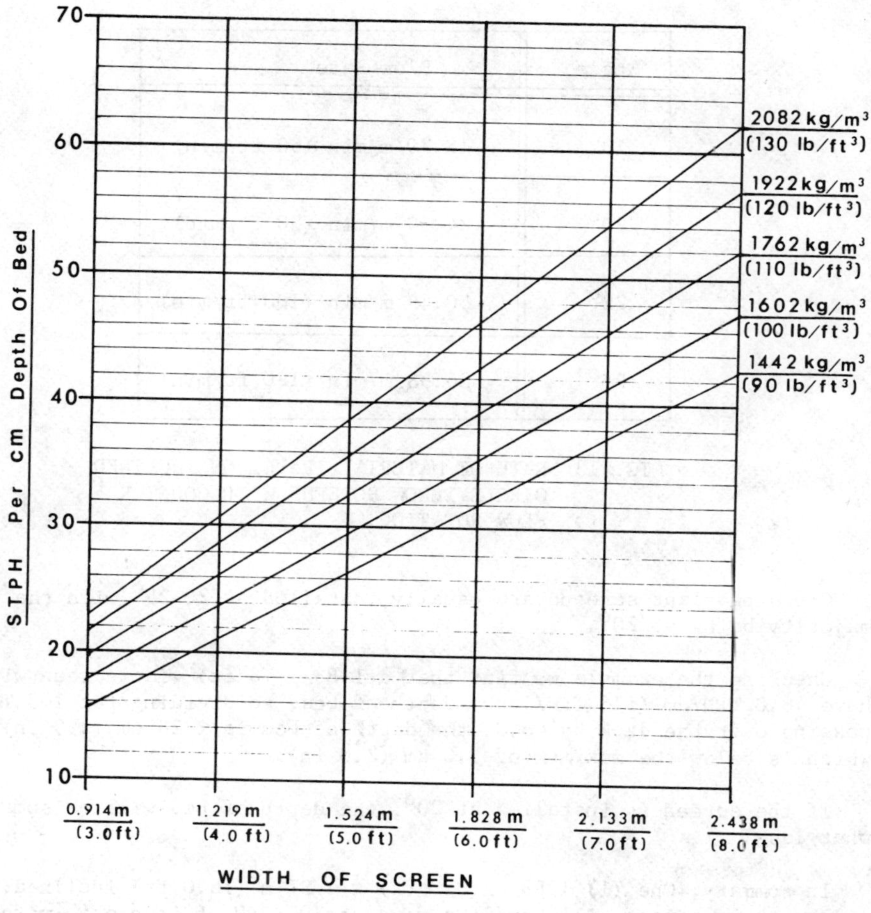

FIG. 12 DEPTH OF BED CHART BASED ON 18.29 m/min (60 ft/min) Flow Rate

Rates of travel for various angles of installation are given in Fig. 13 for inclined circle throw screens. Adjust the depth of bed obtained from Fig. 12 by the ratio of 18.29 m/min (60 ft/min) divided by the rate of travel of your tentative installation angle.

Angle	Flow Rate
18°	18.29 m/min (60 ft/min)
20°	24.39 m/min (80 ft/min)
22°	30.48 m/min (100 ft/min)
25°	36.58 m/min (120 ft/min)

FIG. 13 RATE OF MATERIAL TRAVEL ON INCLINED CIRCLE THROW SCREENS WITH COUNTER FLOW ROTATION.

Crushing plant screens are usually installed at 20-25° with the majority being at 20°.

Checking the example we find that a 1.83 m (6 ft) wide screen will have 46.0 TPH/cm (116 TPH/inch) depth of bed, so dividing the 105 TPH passing over the deck by 46.0, the depth of bed is 2.28 cm (.90 in) which is below the maximum of 5.0 cm (2.0 in).

If the screen is installed at 20°, the depth of bed will be somewhat less.

In summary; One (1) 1.83 m (6.0 ft) x 4.87 m (16.0 ft) inclined, circle throw screen will handle a feed of 300 TPH of iron ore, with a density of 2082 kg/m^3 (130 lb/ft^3) and having a sieve analysis of 100% -38 mm (1-1/2"), 65% -12.7 mm (1/2"), 33% -6.3 mm (1/4") and make a 95% efficient removal of -12.7 mm (1/2") material.

When calculating double deck screen areas, the important thing to remember is to treat each deck as a separate problem.

Of particular importance is that the fines factor be based on the percent of material that is half-size in the feed to the specific deck being calculated, not the original feed, unless it is the top deck.

RECIRCULATED LOADS

When the crushing circuit includes a closed circuit in one of the crushing stages, the approach to screen selection is slightly different.

The critical factors needed for selection are: (1) the total feed rate to the screen including recirculated load; (2) sieve analysis of this total feed.

It is vital to determine what the recirculated load is because the tentative crusher selection may not be adequate thereby restricting the system from achieving its desired capacity.

There are several complex formulas in existence for determining the recirculated load but the following method is straightforward and logical.

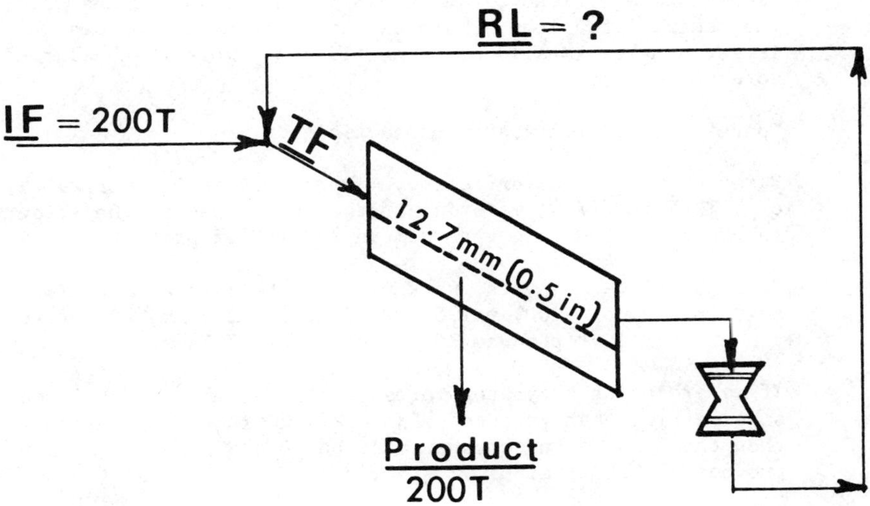

FIG. 14 CLOSED CIRCUIT SCREENING

IF Sieve Analysis

Size	% Passing
38 mm (1-1/2")	100.00
25 mm (1")	98.00
19 mm (3/4")	92.00
12.7 mm (1/2")	65.00
6.3 mm (1/4")	33.00

Problem: Given the flow diagram and Initial Feed (IF) sieve analysis shown in Fig. 14, determine the recirculated load (RL).

1. The first step is to assume a screening efficiency. The higher the efficiency, the larger the screen but the smaller the recirculated load. This can be an advantage because a larger crusher usually costs considerably more than a larger screen.

   Therefore, we will assume 95% efficiency.

2. Based on 95% screen efficiency, there must be $\frac{200}{0.95}$ or 210 TPH of -12.7 mm (1/2") material in the total feed to the screen in order for the screen to remove 200 TPH of product.

3. From the sieve analysis, we know that the Initial Feed (IF) contains (0.65 · 200) = 130 TPH of minus 12.7 mm, therefore the crusher must produce 70 TPH of minus 12.7 mm.

4. If we are using a gyratory crusher at 12.7 mm (1/2") closed side setting that produces 75% -12.7 mm in the discharge, then the feed to the crusher must be $\frac{70}{.75}$ = 93 TPH in order to produce 70 TPH of minus 12.7 mm.

5. Therefore, the Recirculated Load <u>RL</u>, 93 TPH, plus the Initial Feed <u>IF</u>, 200 TPH, equals the Total Feed <u>TF</u> or, in this case, 293 TPH.

6. The next step is to select the screen area using the method previously described and based on 293 TPH total feed having a sieve analysis obtained by proportionately combining the Initial Feed analysis and the crusher discharge analysis.

Some circuit designs will have the recirculated material going back through the same crusher as the initial feed. In this case, the procedure for determinging the recirculated load is the same except that the sieve analysis of the initial feed and the recirculated load are the same, so a combined analysis is not necessary.

SUMMARY

Screens are a very vital and critical piece of equipment in any ore processing plant; therefore, careful attention should be paid to selecting screens of adequate size and type.

Any tendency to skimp or economize in this area can result in serious bottlenecks in material flow that could prevent the plant from reaching its anticipated production rate.

By utilizing the information presented here, you should be able to assure that the recommendations given by screen suppliers, in response to your bid specifications, are adequate but not excessive. Should you find any discrepancies, you will be able to discuss them with the screen supplier intelligently and using his language.

Chapter 16

APPLICATION AND SELECTION OF SPIRAL CLASSIFIERS

by Raymond E. Riethmann and Beuford M. Bunnell

Senior Applications Engineer: Classifiers and Furnaces, Mine and Smelter Corporation, Denver, Colorado

Market Planning Manager, Mine and Smelter Corporation Denver, Colorado

ABSTRACT

The spiral classifier was originally developed for closed circuit grinding It has since been applied very successfully to other classification duties where a two-product size split is required. Inherent operational qualities of this classifier make it ideally suited for sizing, washing, or dewatering materials such as ores, phosphate rock, sands, coal, oyster shells, solar and potash salts. Operating characteristics, flexible design and guidelines for proper application are presented.

WHAT IS A SPIRAL CLASSIFIER

A spiral classifier consists of an inclined tank in which one or two spirals revolve slowly and freely without touching the sides or bottom of the tank. The machine makes a classification of solids according to size and/or specific gravity through the different settling rates of the particles in suspension in the pool.

It is a simple, rugged and easily maintained mechanism for continuously decanting water and fine particles in suspension, termed "overflow", and simultaneously draining and conveying the coarse particles, termed 'rake' or 'sand', out of the classifier.

DESCRIPTION OF EQUIPMENT

A spiral classifier consists of an inclined tank enclosing one or more revolving helixes, commonly called "spirals". The tank is equipped with an overflow weir and an overflow box for collecting the overflow product which generally consists of fine solids and water, although under certain conditions very nearly clear water may constitute the overflow. The settled product, commonly termed "rake", is discharged at the upper end of the tank by the revolving spiral.

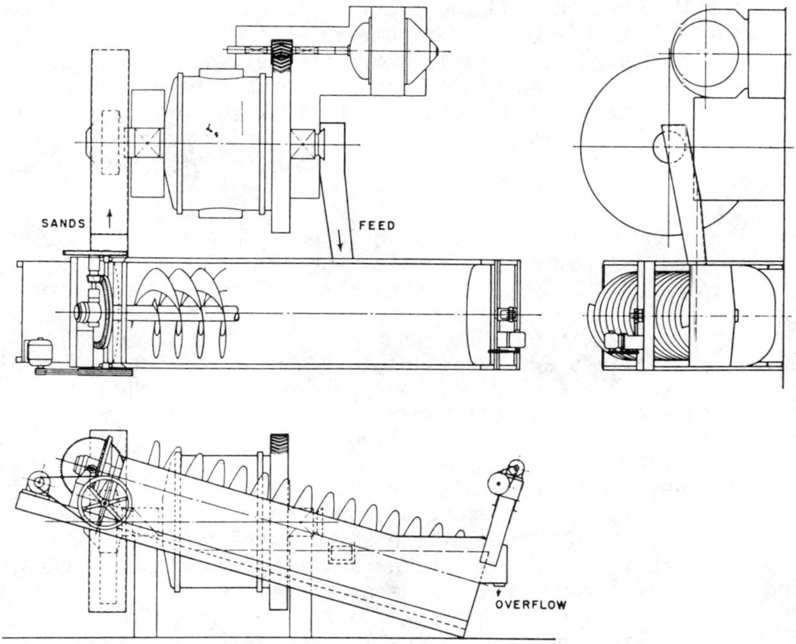

This illustrates a typical closed circuit arrangement of a Simplex Akins Classifier and a Marcy Ball Mill. Note the steep slope permitted by the Akins in making the closed circuit possible with no auxiliary equipment.

The raking mechanism is made up of double or single spirals mounted on a shaft, spirals being continuous from the overflow weir to a point above the feed entrance. The double and single spiral construction are termed double pitch and single pitch, respectively.

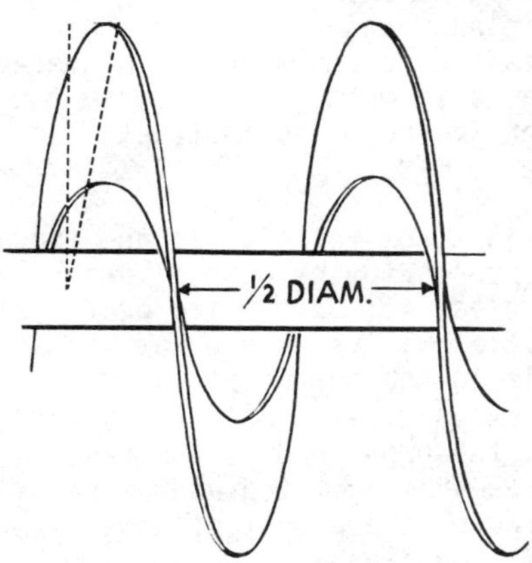

Single Pitch Akins Spiral

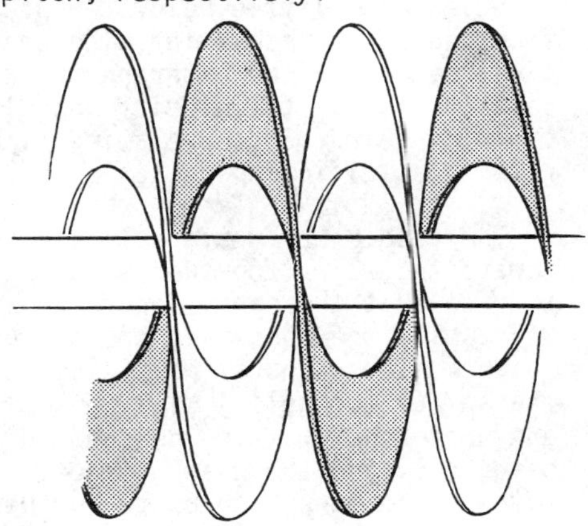

Double Pitch Akins Spiral

The feed enters through a feed opening at one, or, in some cases, both sides of the tank, and the overflow is discharged at the lower end over an adjustable weir. In some cases the feed and overflow points are reversed. The heavier, or coarser, solids settle to the bottom, and in simplex machines are advanced by the action of the spiral along the bottom and one side of the tank out of the settling pool to the point of discharge. In the case of duplex machines, the spirals are operated toward each other so that the sands are conveyed up the center of the tank between the spirals.

The lower end of the spiral is equipped with a lifting mechanism to remove the spiral from settled solids for ease of starting and inspection. A spiral drive is so arranged that the spiral may rotate and be lifted or lowered simultaneously.

Spiral classifiers are furnished in two general types: The high-weir type in which the spiral is not completely submerged at the lower end; and the submerged-spiral type in which the spiral is completely submerged to provide a larger settling pool.

Spiral classifiers are generally set at a tank inclination of from 3 to 4 in/ft., although for special conditions these limits may be exceeded. The operating speed of the spiral varies generally between 10 RPM for small units to 2 RPM for large units, but these limits are exceeded for special conditions. The power required will depend upon the size of the classifier and the rake load. It varies from 1 HP on the small simplex sizes to 25 HP on the larger, duplex sizes.

HOW IS THE CLASSIFICATION PROCESS ACCOMPLISHED

Classification at a particular particle size, called the "separation size", is accomplished by gently introducing a slurry of feed solids and liquid (usually water) into the classifier 'pool'. Minimal turbulence at the entry point promotes predictable settling rates; therefore, the use of a feeder box is best. For the same reason the feed should enter at or a little below the pool level and on the side opposite from the coarse conveying side on simplex machines. The depth a particle settles after entering the classifier depends on its hindered settling rate, its velocity and distance it must travel toward the overflow-weir. By correlating time and velocity against settling rate, a specific separation can be performed.

The feed entry must be located so that the velocity of the liquid toward the weir, together with the distance to the weir allows proper time so that the specified fine particles will be carried over the weir while the specified coarse particles will settle below the weir height. These coarse particles settle to the bottom of the pool where the spiral (or helix) will convey these particles up the inclined tank and where they will be discharged out the upper end of the tank. While being conveyed the coarse fraction (rake) is agitated and washed in a counterflow manner so as to further reduce the amount of undersize fines carried out with the coarse rake fraction.

DISCUSSION OF OPERATING PARAMETERS

Spiral classifiers are mechanical gravity separators and hence depend on the difference in settling rate between coarse and fine and/or heavy and light gravity particles in the pool. The variables easiest for an operator to control in an existing classifier installation are feed rate and turbulence. Concentration of solids in the feed and spiral rotation speed affects both capacity and turbulence.

Some fundamental laws of classification that apply are:

1. Coarse particles have a relatively faster settling velocity than fine particles of the same specific gravity.
2. Heavy gravity particles have a relatively faster settling rate than light gravity particles of the same size.
3. Porous and/or irregularly shaped particles will generally settle slower than would be predicted for spherical impermeable particles of the same size classification and specific gravity.
4. Settling rates of solid particles become progressively slower as the viscosity or density of the fluid medium increases.
5. There is a point (called critical dilution) where lowering the density or viscosity by adding more liquid creates a velocity effect toward the weir which overcomes the increase in settling rates; thereby coarsening the separation.
6. Conversely at the critical dilution point less liquid will cause a viscosity and buoyancy effect (hindered settling) which will also coarsen the separation.

The control of liquid into classifiers is important to the spiral classifier operator, although not nearly as critical as with hydrocyclones, since separation into overflow and rake products is made largely by the buoyancy, viscosity and degree of agitation in the pool.

Lower water concentration creates lower velocity of the overflow toward the weir. This results in high overflow solids capacity per unit width of the weir. However, the resultant high percent solids results in a narrow spread of hindered settling rate of the particles. This makes for a "dirty" separation with a fairly high percentage of misplaced particles.

High concentration of water past the critical dilution has the same effect. However, if the weir width is increased and the feed entry geometry is proper the critical dilution liquid concentration may be increased. Under these circumstances a much sharper separation (lower percentage of misplaced particles) can be obtained.

Water in the overflow varies considerably. Often it is dictated by the next processing step. This could be flotation requiring say 30% solids (70% H_2O) in the feed to the flotation cells. Often the total water to the classifier is specified and the overflow water is the amount left after deducting the water in the rake product. Sometimes there are no restrictions and the application engineer is free to specify the desirable amount of water necessary for the separation.

Total water to the classifier is usually introduced with the feed solids. However, if there is reason to spray wash the rake product on the drainage deck, this water is added to the feed water. Also, for various reasons, auxiliary water is sometimes added to the classifier, which reports to the overflow.

Water in the rake product varies to some degree. Depending on the coarseness of the solids and the specific gravity of the particles, it can vary from say 15% for fairly coarse solids of high specific gravity (such as iron ore concentrate) to say 25% for fairly fine solids of low specific gravity (such as coal). In most cases this provides a rake product dry enough for transporting on a conveyor belt.

Conditions affecting the amount of water in the rake product:

1. Specific gravity of the particles
2. Size of the particles.
3. Amount and position of spray washing on the drainage deck above the emergence of the solids from the pool, if any.
4. Slope of the drainage deck.
5. Length of the drainage deck.
6. Spiral RPM (transporting speed).
7. Special design features, such as serrated outer wearing shoes, a "dam angle" at the bottom of the tank at the discharge lip or shortening the spiral at the discharge end.

SPIRAL CLASSIFIER APPLICATION CONSIDERATIONS

Overflow

As the feed slurry enters the classifier pool, it immediately moves toward the overflow weir at a velocity dependent on the overflow volume and the width of the weir. These conditions determine the overflow weir crest.

During this travel, the solids particles all settle downward at a settling rate dependent on their shape, their size, specific gravity and percent solids of the fluid.

Experience has proven that if a particle settles to a depth of two times the weir crest by the time it reaches the weir, it will probably not overflow and will report in the rake product. If it has not settled this amount, it will probably report in the overflow.

As the overflow percent solids increase, settling rates for the particles decrease. Also, for the various size particles the settling rates variation is less. Therefore, sharpest separations can be obtained in dilute overflows (if less than critical dilution) and as the overflow becomes thicker the separations become less sharp, which is called a dirty separation.

All the above is based on comparatively free travel of the material toward the weir in a quiescent pool. Pool turbulence excessive enough to disrupt free settling will result in a dirty separation. Therefore, the feed material must flow into the pool somewhat below the surface as gently as possible, and the spiral RPM must be slow enough to create no disruptive turbulence.

Quality of separation under all the above variations of conditions is difficult to predict, and can only be determined by past experience.

Rake: If the classifier is being used to close circuit a ball mill the length of the slope of the tank as well as the distance between the feed opening and the sand discharge must be compatible with the closed circuit diagram. The dryness of the rake product is not important as it is usually introduced into the mill scoop feeder box by gravity from a chute and utilizing flush water. However, the tank slope is also a function of the spiral raking capacity and a pool length must be acceptable from an overflow capacity consideration.

The above photograph shows a typical Marcy Ball Mill-Akins Classifier closed circuit installation.

If a classifier is not close circuiting a mill, the rake material is usually transported on a conveyor belt, and must be sufficiently dry for this purpose. The proper dryness of the rake material is a function of the size and nature (drainability) of the material, the speed of the spiral, the slope of the tank and the length of the drainage deck above the pool. Of course, if spray washing on the drainage deck is required, the length of the drainage deck must be increased. Normal logic and experience are extremely valuable in these determinations.

Over the years it has been determined that the minimum practical classifier slope is 3" per foot, and is used for the finest size separations (such as 200 mesh). This balances the necessary long pool length for maximum overflow capacity and necessary flat slope for the efficient raking of the fine rake material with a tank and spiral length that are not excessive.

Also, it has been determined that the maximum practical slope is 4" per foot, and is used for the coarsest size separations (such as 20 mesh). This results in a comparatively short tank and spiral while still allowing sufficient pool length for obtaining adequate overflow capacity and gives good raking capacity for the coarse material.

HISTORICAL DEVELOPMENT OF CURRENT DESIGN

The first Akins Classifier (and to my knowledge the first spiral classifier) was a deep rectangular trough containing a steel conveying spiral extending throughout its length. It was set along side a grinding mill at a slope. The lower end was at a height to accept the mill discharge and additional water to form a pool. A weir was provided to overflow the small size finished particles. The upper end was at a height to allow the larger size particles that needed more grinding to be raked out of the pool, up the tank bottom where they were drained and then flushed into the mill feed box. Although no longer used, this machine is termed a <u>"Low Weir"</u> type classifier.

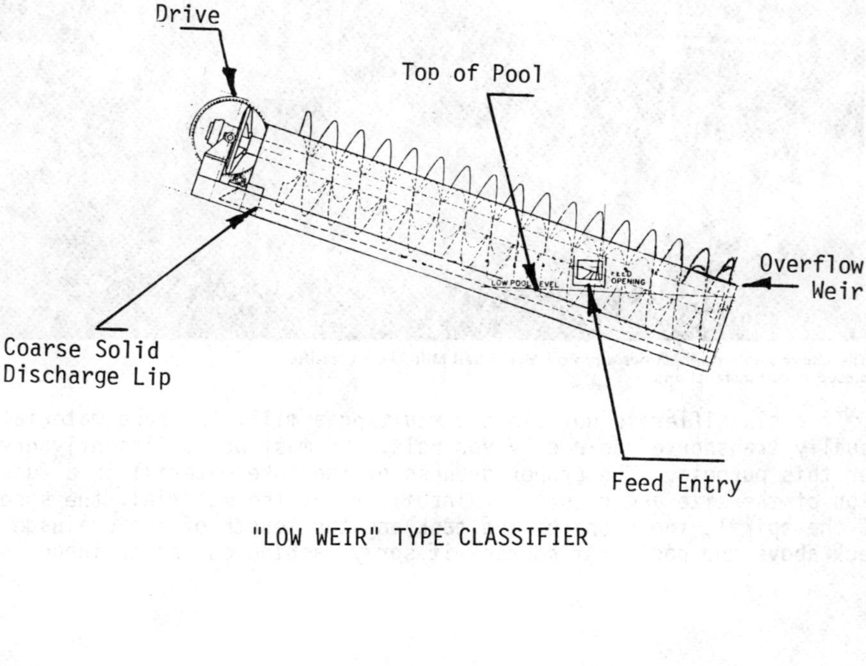

"LOW WEIR" TYPE CLASSIFIER

This principle was acceptable, but many problems developed. The spiral flights cut off the pool flow from feed entry to the overflow weir. Therefore, the tank sides were built up in the pool region which allowed a larger pool, and the overflow weir was now higher. This was called a "High Weir" or type "H" classifier. As the increase in height of weir was not significant, the problem was only partially solved. It was obvious that by allowing a generous clearance between the inside of the flights and the spiral shaft, the pool could flow through this clearance toward the weir.

DOUBLE PITCH SPIRAL WITH CLEARANCE FOR POOL FLOW

As the spiral load was not great near the overflow weir, the pool end flights could be cut for larger clearance in this region. To hold the flights in place, arms projecting from the shaft were required.

Basically, this is the principle of the "High Weir" classifier to-day. Due to the size of the pool, separations are fairly coarse (20 mesh through 65 mesh).

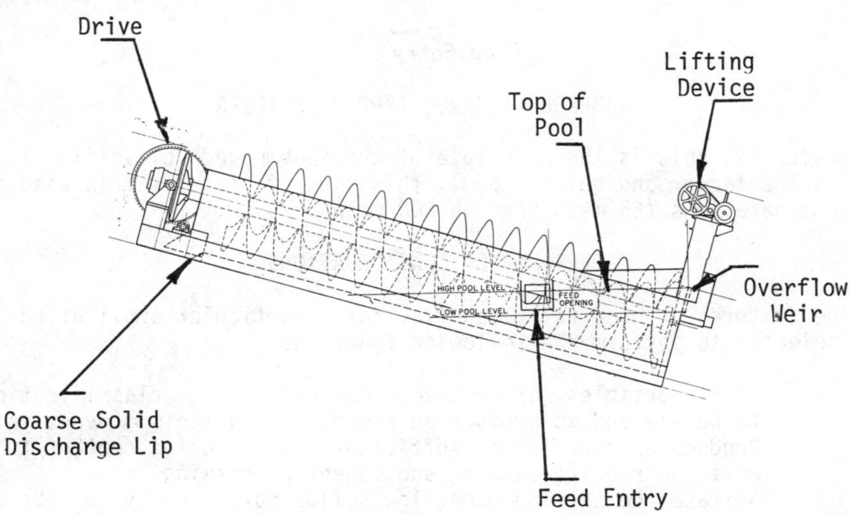

"HIGH WEIR" TYPE CLASSIFIER

As ore beneficiation progressed, it became evident that finer separations were required. For this a considerably larger and more quiescent pool was necessary. Therefore, the tank sides in the pool region were built up even higher than before to result in a considerable increase in pool size. The overflow weir was high enough to result in complete submergence of the spiral for a considerable distance in front of the weir. This was called the "Submerged" or type "S" classifier. This resulted in a large quiet pool for finer size separations and a greater pool volume which dampened the effect of feed solids volume fluctuation. Naturally, the tanks and spirals were longer to allow good drainage of the coarse particles conveyed from the pool to the upper end of the classifier.

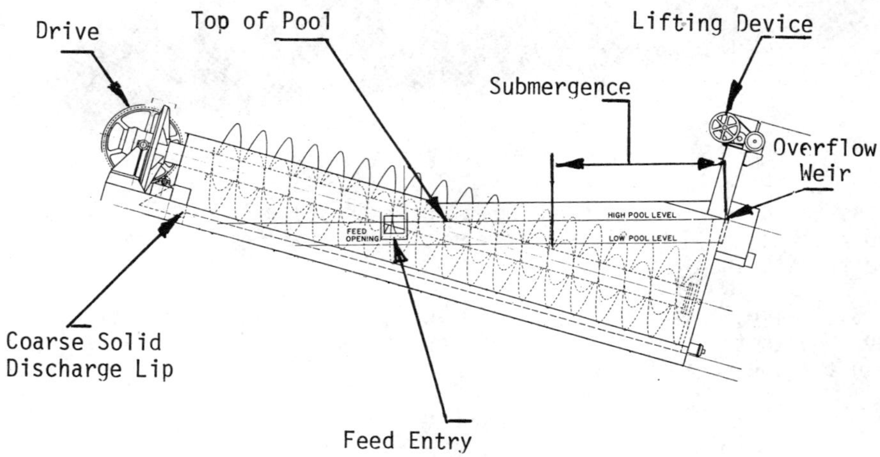

"SUBMERGED SPIRAL" TYPE CLASSIFIER

Basically, this is the principle of the "Submerged" classifier today. Due to the larger and quieter pool, this type of classifier is used for finer separations (65 mesh through 200 mesh).

PERFORMANCE CONSIDERATIONS

The features of a spiral classifier for a particular application are selected to perform the following functions:

1. Allow particles larger than a desired size of classification to settle and to produce an overflow with minimum oversize.
2. Produce an overflow of sufficiently high solids content to meet the requirements of subsequent processing.
3. Agitate the settled underflow solids sufficiently to liberate entrapped undersize so it may report in the overflow.

4. Drain and remove the underflow solids (rake) from the settling pool.

A great many factors determine efficiency of classification, and experience with actual operating data is extremely important.

The following parameters govern the size and type of classifier required for a specific application:

Feed solids tonnage, specific gravity and dry bulk density.
Feed solids particle size distribution.
Feed solids concentration in the feed slurry.
Feed slurry liquid and specific gravity.
Required classification and settling rate of feed particle of required separation size in feed slurry liquid.
Solids concentration required in the overflow.
Solids concentration required in the rake.
Feed temperature.
Feed pH and corrosion with respect to mild carbon steel.

The dryness of the rake product is a function of the particle sizes and drainability, spiral speed, drainage deck length, classifier tank slope, and settled product rate. The rake capacity for a classifier is related to the specific gravity of the particles, spiral diameter, spiral speed, pitch of the helix, slope of the machine and drive horsepower.

Overflow capacity of a classifier is dependent on overflow velocity, particle settling rates, turbulence and distance the particle must travel from the feed point to the overflow weir.

By proper design of the feed entry, pool area, volume and geometry, the spiral classifier has the ability to handle feed volume fluctuations for short periods of time without upsetting the system. This material process system dampening effect means that sophisticated and costly control equipment are not usually required. Since the overflow and rake product are easily observed the need for control instrumentation is often eliminated.

Table I gives the average raking and overflow capacities for spherical sand particles with furnished horsepower for simplex Akins Classifiers with straight sides operating at conservative spiral speeds at normal slopes.

The listed rating capacities and drive motor horsepower are based on double pitch spirals, except for the twelve and sixteen inch spirals which are single pitch. Capacities and horsepower requirements will vary as follows:

Raking capacity: Approximately double for duplex classifiers, inversely proportional to change in slope and proportional to spiral speed.

Drive motor horsepower: Porportional to change in rake capacity.

Where applicable, classifier tanks can be flared at the deep pool end to increase overflow capacities by the following rates:

	Simplex	Duplex
Type S	1.60	1.30
Type H	1.30	1.15

TABLE I: CAPACITY TABLE

In the table below are average raking and overflow capacities of Akins Classifiers to be used only as a general guide.

Spiral Dia. In.	Spiral Speed RPM	Drive Motor H.P.	Raking Capacity Tons Per Hour	OVERFLOW CAPACITY-TONS PER HOUR OF SOLIDS											
				TYPE "H"					TYPE "S"						
				Percent Solids					Percent Solids						
				40	35	30	25	20	25	22	19	16	13	10	7
				Limiting Mesh					Limiting Mesh						
				20	28	35	48	65	48	65	80	100	150	200	325
*12	20	1/2	1.76	10.5	6.38	3.97	2.37	1.26	8.59	4.97	2.64	1.55	.566	.213	.055
*16	14.8	3/4	3.12	30.4	18.5	11.4	6.88	3.60	12.5	7.26	3.88	2.27	.849	.320	.073
24	10.1	2	13.4	52.0	31.5	19.6	11.8	6.17	24.1	14.0	7.49	4.35	1.63	.613	.146
30	8.1	2	21.1	85.1	51.6	32.0	19.2	10.1	33.6	19.4	10.5	6.04	2.26	.880	.220
36	6.6	3	30.1	99.2	60.1	37.4	22.4	11.8	55.0	31.9	17.1	9.91	3.71	1.41	.348
42	5.8	3	41.2	132	80.2	49.8	29.9	15.7	71.7	41.5	22.3	12.9	4.84	1.87	.457
48	5.0	5	53.6	175	106	66.1	39.7	20.9	85.5	49.5	26.6	15.4	5.76	2.21	.531
54	4.4	5	67.8	203	123	76.3	45.8	24.1	116	67.2	36.1	20.9	7.81	3.01	.732
60	4.0	7-1/2	83.7	263	160	99.1	59.5	31.3	133	76.9	41.3	23.9	8.95	3.44	.842
66	3.7	7-1/2	101	303	180	112	67.2	35.4	159	91.6	49.4	28.6	10.7	4.11	1.01
72	3.3	10	120	324	196	122	73.3	38.5	182	105	56.5	32.7	12.2	4.69	1.13
78	3.0	10	141	398	241	150	90.1	47.4	209	121	65.0	37.7	14.1	5.41	1.32
84	2.8	15	164	493	299	186	111	58.7	238	138	74.1	42.9	16.0	6.16	1.50
Tank Slope in in./ft.			3	4		3-3/4		3-1/2		3-1/2		3-1/4			3

*Raking and overflow capacities are listed for products of 2.65 specific gravity, and vary proportionally with same.

APPLICATIONS

Although the spiral classifier was developed for close circuiting with wet process grinding mills, it has since been applied very successfully to other classification duties where a two-product size split is required. Because of the inherent operational qualities of this classifier, it is ideally suited for sizing, washing, or dewatering ores, phosphate rock, sands, coal, oyster shells, solar and potash salts, as well as many other specialized applications.

Installation Washing Oyster Shell

Spiral classifiers are utilized to good advantage in the Phosphate field. Generally the duty is dewatering and desilting at 65 mesh to 150 mesh.

Desliming Phosphate Rock

In iron ore plants, spiral classifiers are used to dewater tails, dewater plant wash-down, dewater slurries from spray water at troublesome dusting points such as pellet screens, conveyor transfer points and loading areas. One 66" duplex is receiving all the plant spills, overflows, etc., which was being sent to tails. The classifier overflows the minus 20 mesh, which is now being sent to tails and the spirals rake over 150 TPH of plus 20 mesh which is added to plant production.

OTHER APPLICATIONS ARE LISTED BELOW

Iron Ore:	Heavy media separators and densifiers for upgrading coarse ore. Dewatering and desliming concentrates.
Copper Ore:	Dewatering cyclone underflows.
Molybdenum Ore:	Closed circuiting finish grinding mills.
Potash Ore:	Closed circuiting grinding mills and classifying cyclone underflow.
Silica Sand:	Washing, desliming, dewatering fine feed.
Uranium Ore:	Closed circuiting grinding mills. Acid or alkaline counter washing.
Oyster Shells:	Washing and dewatering dredged shell.
Sand & Gravel:	Washing, desliming, dewatering.
Chemical Crystal:	Counter washing.
Salt:	Heavy media separators and densifiers for upgrading. Classifiers for washing and dewatering.
Manganese, Fluorspar, Garnet:	Heavy media separators and densifiers.

DEFINITIONS

Classification: Separation of solid particles into size or weight fractions by differential settling through a fluid.

Overflow: A fine fraction along with the liquid medium which is taken off by flowing over the discharge weir.

Rake: Partially drained fraction containing the settled coarse material. Sometimes also termed the "sand" or "underflow".

Separation Size: In closed circuit grinding a size specified by some point in the overflow screen analysis: Example 1-1/2% on 65 mesh.
In other cases this may refer to the smallest size fraction which cannot be found in the rake product or, conversely, the largest size fraction which cannot be found in the overflow.
In the case of hydrocyclones 'separation size' often refers to the size fraction which splits half in the overflow and half in the underflow.

Wet Classification: The art of separating the solid particles in a mixture of solids and liquid into fractions according to particle size or density by methods other than screening.

NOTE: Although this chapter discusses spiral classifyers specifically, the principles apply to rake and drag classifyers also.

Chapter 17

THE SELECTION OF CYCLONE CLASSIFIERS, PUMPS AND PUMP BOXES FOR GRINDING CIRCUITS

A. L. Mular* and Norman A. Jull**

*Professor of Mineral Engineering
University of British Columbia
Vancouver, B. C.

** President
Technequip Limited
Toronto, Canada

ABSTRACT

Fundamentals of cyclones are reviewed briefly, where a typical cyclone is considered. Graphical cyclone data available in the literature have been converted to mathematical equations for convenience. Preliminary selection and design procedures for both cyclones and pumps are provided, and a sample calculation for a hypothetical grinding circuit is covered in some detail. In general, cyclone and pump suppliers should be consulted for definitive design purposes. Some cyclone suppliers currently employ digital computers for definitive design to ensure greater accuracy.

INTRODUCTION

Hydrocyclones have been in use for many years to classify solids suspended in liquids, where applications have ranged from removal of sand in hot catsup to closed circuit grinding of ores (1). Process engineers and/or plant operators faced with a cyclone application invariably ponder the question "What size is required, how many should there be and is the pump size correct"?

The purpose of this chapter is to review, briefly, cyclone fundamentals and selection, pump and pump box selection along with piping and distributors, and sizing methodology that may be employed for preliminary design in closed circuit grinding operations.

HYDROCYCLONE CLASSIFIERS

Figure 1 depicts the cutaway view of a "typical" cyclone*. Feed slurry, either pumped or flowing by gravity, enters the inlet to a cylindrical chamber tangentially and under pressure. Particles larger than those desired in the overflow (vortex finder), ideally, remain close to the wall and travel downwardly in a spiral path and out the underflow (apex). Normally, both overflow and underflow streams discharge at atmospheric pressure. At the velocities used, turbulent mixing inside the cyclone is a hazard which is minimized by smooth surfaces and gradual changes in cyclone shape.

Most of the volume of slurry entering the cyclone departs through the vortex finder; this "average" medium is viscous and drags particles out of the feed in an inward-upward direction. Each particle resists the drag, because of a centrifugal force acting on that particle in an outward radial direction. The force is proportional to particle mass; higher forces drive smaller and/or light particles from the overflow stream. Since small cyclones have a small radius and since accelerative forces vary inversely with cyclone diameter,then a smaller cyclone makes a finer separation (2).

The underflow stream consists of coarse particles whose voids are filled with water and fines having characteristics analogous to the overflow stream. Thus a cyclone with a "thick" overflow will have thick overflow slurry filling the void spaces. When more water is added to the feed so that the overflow becomes more dilute, then the by-passed overflow slurry in the voids is correspondingly more dilute and the underflow is "cleaner".

The Typical Cyclone

A typical cyclone, where cyclone diameter is defined as the approximate inside diameter of the cylindrical feed chamber, has an inlet area at the point of entrance into the feed chamber of 6% to 8% of the cross sectional area of the feed chamber. Often, the inlet nozzle at the entrance is rectangular (1).

The vortex finder extends below the feed entrance to minimize short circuiting of coarse particles to overflow. The vortex finder diameter, defined as the inside diameter at the point farthest from the top of the cyclone, is approximately 35% to 40% of the cyclone diameter. On the other hand, the cylindrical section located between the feed chamber and the conical section is the same diameter as the feed chamber.

The conical section has an included angle of about 12° for cyclones smaller than 10 inches; for larger cyclones the cone angle is approximately 20°.

* Courtesy of Krebs Engineers, Menlo Park, California, 94025

FIGURE 1

CYCLONE CUTAWAY

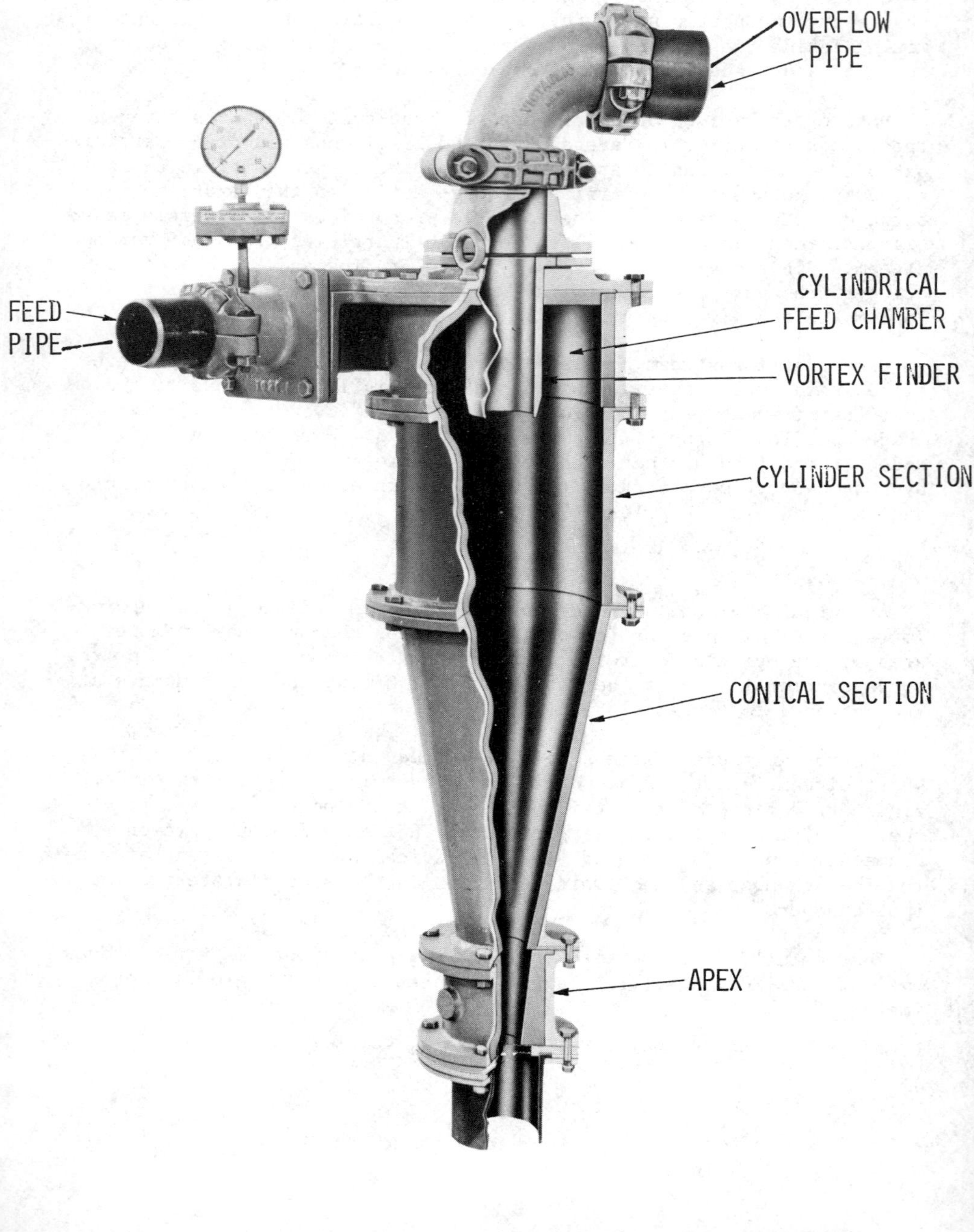

The apex, the point of highest wear, has a diameter usually not less than one-fourth that of the vortex finder, but there is no absolute lower limit.

Materials of Construction For Cyclones

While many cyclones used in chemical and paper industries have been molded of plastics or cast in hard ceramics and metals, the mineral industry has utilized cyclones mainly built of steel shells lined with various elastomers. By far the greatest number has had molded replaceable liners of pure gum rubber for rapid maintenance and assurance of good internal geometry. The smooth surfaces which molding can provide resist abrasive wear particularly well.

Where the ore is unusually abrasive, replaceable liners of ceramics are used. Their higher cost is, under those conditions, returned by longer life. An advantage of replaceable liners is the ability to try liners of different materials in different parts of the cyclone. Sections near the apex often suffer greatest wear from coarse particlea and can be protected with the more exotic materials. Upper sections use less costly materials.

Cyclone Performance and Separation Size

In recent years, the trend has been to describe cyclone performance graphically for each narrow size fraction of feed solids. The weight fraction of a narrow size range of feed recovered to underflow is plotted against the geometric mean size of the fraction. For example, if the feed contains 16 tonnes per hour (TPH) of -48 +65 mesh material, and if 12 TPH report to underflow then the weight fraction is $12/16 = 0.75 = Y_x$ recovered to underflow and of geometric mean size $\sqrt{300 x 212} = 252$ micron = X. Y_x may be determined for each size X. Figure 2 shows Y_x plotted against X.

By convention, a portion of each feed size fraction of interest is assumed to enter the underflow by short circuiting in direct proportion to the fraction of feed water reporting to underflow. For the above example, if 25% of the feed water reports to underflow, then $0.25 \times 16 = 4$ TPH have short circuited so that only $16 - 4 = 12$ TPH of the feed fraction is capable of being classified. Of the 12 TPH actually reporting to underflow, the amount actually there due to classification is $12 - 4 = 8$ TPH. The ratio $(12 - 4)/(16 - 4) = 0.67$ is Y_{cx} which is referred to as the corrected recovery of the 252 micron fraction. A Y_{cx} may be calculated for each size fraction of interest. The corresponding curve is shown in Figure 2. Note that the point at which $Y_{cx} = 0.5$ is associated with a special size, namely, X_{50c}. Computing X/X_{50c} gives a new scale for the abscissa and permits comparing the shapes of corrected recovery curves calculated from cyclone operating data, since they all pass through the common point $Y_{cx} = 0.5$, $X/X_{50c} = 1$.

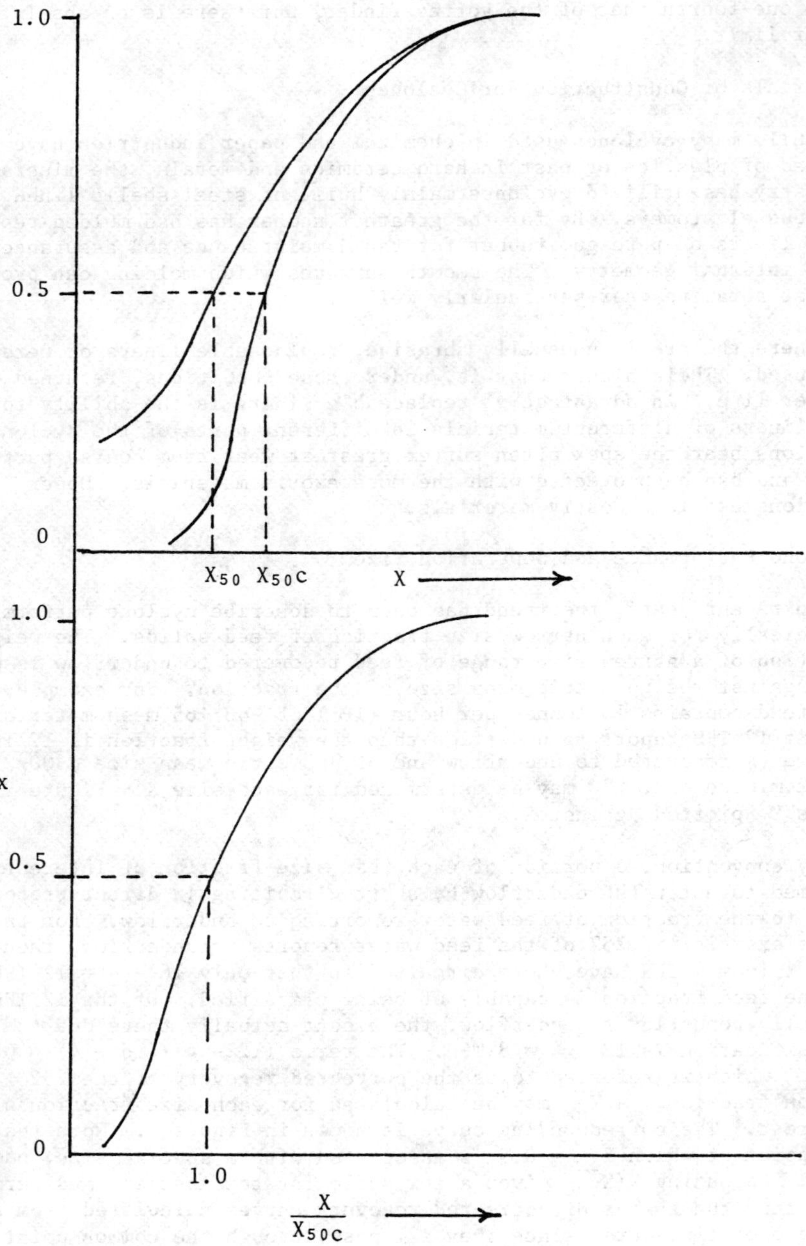

Figure 2: RECOVERY CURVES

The steeper the curve, the closer the classification approaches a theoretical screen separation.

X_{50c} has been observed to depend upon operating variables (3), and may be used as a measure of separation size. Classically, separation size is defined as the particle size, $X_{1,3}$, which constitutes 1% to 3% of dry solids weight in the overflow, while essentially all particles larger than this size report to underflow along with some fines. The ratio $X_{1,3}/X_{50c}$ varies between 1.5 and 2.5 in open-circuit cycloning such as desliming and degritting; it can be greater in closed-circuit grinding cyclones.

Influence of Cyclone Geometry and Operating Variables on Separation

In a cyclone, variables associated with cyclone geometry (design variables) and operating variables interact with each other. With this in mind, their influence on separation, starting with a set of base conditions, can be discussed. Base conditions are as follows: feed liquid is water at $20^{o}C$, feed solids are of specific gravity 2.65, feed concentration is less than 1% solids by weight and cyclone inlet pressure is 10 PSI.

Variables Associated with Geometry. Variables included in this category are: cyclone diameter, vortex finder diameter, inlet area, apex diameter, cylindrical section length and cone angle.

CYCLONE DIAMETER. Larger cyclones tend to separate at a coarser size than do smaller ones, because larger ones generate much smaller accelerative forces (10 times gravity versus 4000 times for small cyclones). Of course, each size produces a range of such forces, but the force is approximately in proportion to the inverse of cyclone diameter. A recent series of investigations (1) verify that X_{50c} is proportional to D^n for the typical cyclone under base conditions, where D is cyclone diameter and n is a constant.

VORTEX FINDER DIAMETER. The diameter of the vortex finder is one of the more important variables. For cyclones of fixed diameter and at constant pressure, the vortex may be altered to influence X_{50c}. The larger the vortex finder, the coarser the overflow. Vortex finder length should terminate below the feed inlet nozzle and above the top of the cylindrical section (4). Outside this range, X_{50c} tends to become larger. It has been noted (2,4) that the separation size, $X_{1,3}$ is approximately proportional to the diameter of the vortex raised to a power.

INLET AREA. The inlet area determines the entrance velocity and is one of the factors that govern the tangential velocity at various radii. In consequence, the transition radius between free and forced vortices is affected. While maintaining base conditions in a typical cyclone, an increase in inlet area involves an increase in flowrate of feed. Decreasing the inlet area will slightly increase pressure drop at similar capacities. A rectangular shape at the feed entrance is judged superior to other shapes (1).

APEX DIAMETER. Determination of the optimum apex diameter seems difficult at best. This diameter determines both solids capacity and and underflow percent solids. However, at least with apex diameters over 3 inches as used for cyclones in many grinding circuits, solids capacity changes more rapidly with apex diameter than does percent solids. This is especially true for higher specific gravity solids.

The central air core of the cyclone, inside the annulus through which solids are discharging, is possibly a small part of the cross-sectional area in apexes larger than about 2 inches. The air core becomes unstable and pinches shut when the cyclone "ropes". Roping is a condition where the apex becomes overloaded with coarse solids or when the underflow is inadvertently throttled, so that coarse particles are forced into the overflow stream--clearly an undesirable situation. The rope condition is hard to detect in large cyclones. At low pressures the appearance of the underflow does not change dramatically between roping and spraying. At roping, the underflow stream is uniformly thick and tramp oversize can be detected in the overflow. For normal operation, the air core can be felt with the fingers.

Operators who object to recirculating any more slimes than can be avoided will use apexes as small as will pass the underflow solids without roping. Figure 3 shows recommended apex diameters versus underflow solids capacity at various specific gravities of ore. Diameters much smaller than suggested may be detrimental, although the curves are approximate. Field data shows that underflows higher than 60% solids by volume, without roping, are not uncommon in conjunction with high overflow percent solids. The phenomenon is attributed to filling up of the void space between coarse underflow particles with fine slurry of high percent solids.

Figure 4 is intended only to show approximate limits on underflow percent solids to avoid roping. Note that higher values of underflow percent solids can be attained, without roping, when the cyclone is operated so as to have a high overflow percent solids.

An alternative to Figure 3 has been calculated from graphical data supplied in reference 4, such that:

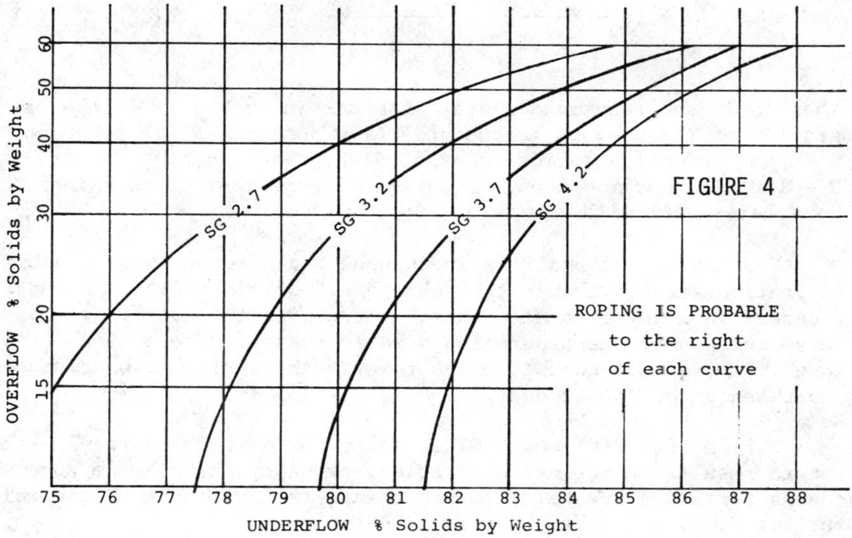
FIGURE 4
OVERFLOW % Solids by Weight
UNDERFLOW % Solids by Weight
SG 2.7
SG 3.2
SG 3.7
SG 4.2
ROPING IS PROBABLE
to the right
of each curve
15
20
30
40
50
60
75
76
77
78
79
80
81
82
83
84
85
86
87
88

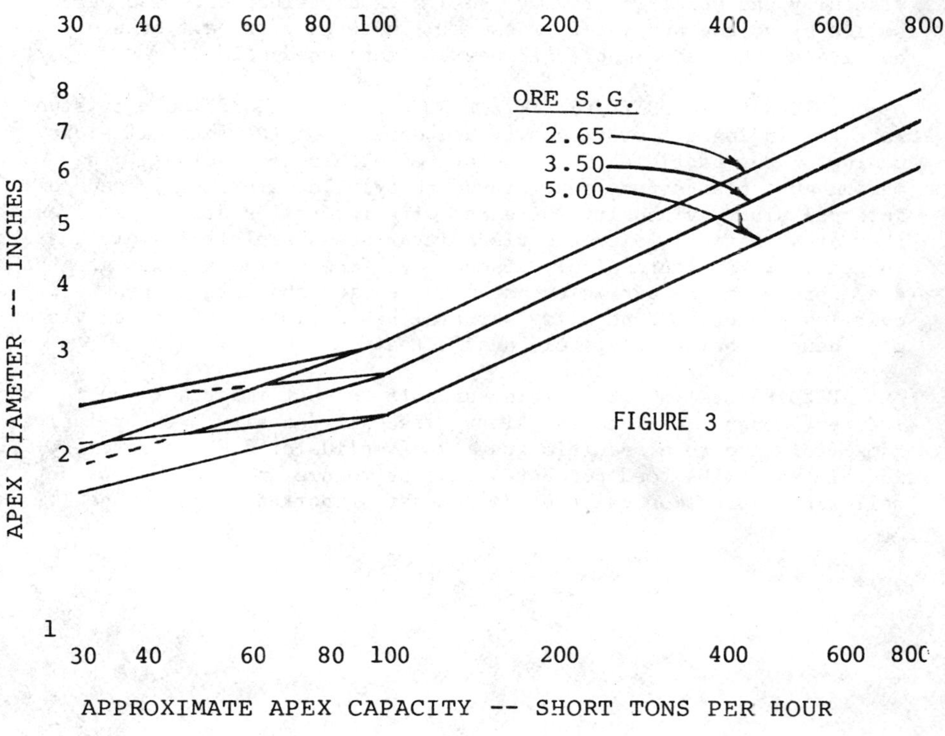
ORE S.G.
2.65
3.50
5.00
FIGURE 3
APEX DIAMETER -- INCHES
APPROXIMATE APEX CAPACITY -- SHORT TONS PER HOUR
30
40
60
80
100
200
400
600
800
1
2
3
4
5
6
7
8

$$S = 4.16 - \frac{16.43}{(2.65 - \rho + \frac{100\,\rho}{p_u})} + 1.10 \ln(U/\rho) \qquad (1)$$

where S is the recommended apex diameter in inches, ρ is the specific gravity of the ore, p_u is the underflow percent solids by weight and U is the underflow solids tonnage, STPH. For example, suppose p_u = 79%. U = 150 STPH and ρ = 2.65. Equation 1 shows that S = 3.7 inch. This correlates well with Figure 3.

CONE ANGLE. A small included cone angle tends to decrease the separation size, although the "sharpness" of classification may be influenced in a detrimental manner. Increasing the cone angle has a reverse influence; the separation size increases. The action of the cone is to squeeze coarse solids towards the center to obtain a concentrated underflow product.

CYLINDRICAL SECTION LENGTH. An increase in cylindrical section length results in a finer separation, probably because the zone where coarse particles are being forced toward the axis by the cone wall is further removed from the vortex finder.

Operating Variables. The separation size is influenced by many variables, including particle shape and specific gravity, internal slurry viscosity and specific gravity, feed size distribution, feed percent solids by volume and inlet pressure. These factors have been discussed (2,4). The more important ones are mentioned below.

INTERNAL VISCOSITY/DENSITY OF SLURRY. It is difficult to separate the influences of viscosity and density of the internal slurry medium on the particles being separated within the cyclone. The medium must bear strong resemblance to overflow slurry. In general, internal slurry viscosity increases with internal slurry density such that at a critical point the viscosity rises sharply for small changes in density. An increasingly viscous overflow stream applies greater drag forces on the particles, drawing heavier and larger ones into the overflow product. Fine slimy material has a similar effect on viscosity, density being relatively unaffected.

FEED PERCENT SOLIDS. This variable is most important and is an indirect measure of internal slurry viscosity/density. The relative X_{50c} (relative to a specific gravity of solids of 2.65) denoted by C_1, is related to the feed percent solids by volume in accord with the following expression calculated from data reported in reference 1:

$$C_1 = \exp(-.301 + .0945V - .00356V^2 + .0000684V^3) \qquad (2)$$

where exp means the base e, and V is the volume percent solids.

INLET PRESSURE. Increasing the volume throughput to a cyclone consumes more energy, as indicated by the increasing pressure drop from feed inlet to overflow. This energy gives the slurry an angular velocity, which increases dramatically as the flow spirals inward from the wall to the central vortex. The velocity creates centrifugal forces which may be represented by a single vector directed radially outward. On the other hand, particles are carried to the overflow by drag forces, for the majority of the feed volume spirals inward. These forces may be represented by a single vector directed radially inward. Figure 5 is a sketch that shows how the two vectors oppose each other. Changes in volume throughput, which alter the magnitudes of both vectors along with pressure drop, have small effect. The influence is greatest on the centrifugal force vector, so that the net effect of increasing the volume flow rate is to cause coarser particles to stay nearer the wall and report to underflow. The weakness of the effect is such that the pressure drop must increase by a factor of 4 to make the cyclone separate a $\sqrt{2}$ mesh size finer.

The influence of pressure drop on the relative X_{50c} size, C_2, is calculated from graphical data in reference 1 to be:

$$C_2 = 2.0(\Delta P)^{-.3} \qquad (3)$$

where ΔP is in PSI.

In some cases, products are not discharged at atmospheric pressure. If discharge pressure in the overflow is positive, a greater portion of feed volume is forced to the underflow, reducing drag toward the overflow and thus making a finer separation.

SPECIFIC GRAVITY OF SOLID. The centrifugal force, which acts on particles and opposes the drag of the overflow stream, depends on particle mass that is related to particle size and specific gravity. For this reason, cyclone overflow contains fine heavy particles along with coarse light particles. The effect on the relative X_{50c} size, C_3, is related to the specific gravity, ρ, of solids in water as follows (1):

$$C_3 = \sqrt{\frac{1.65}{\rho - 1}} \qquad (4)$$

FEED SIZE DISTRIBUTION. The influence of changes in feed size distribution may be understood, qualitatively, by recognizing that a coarse feed without fines will coarsen the separation; a fine feed without coarse particles is conducive to a finer separation. Attempts have been made to account for a feed size effect on separation size(5).

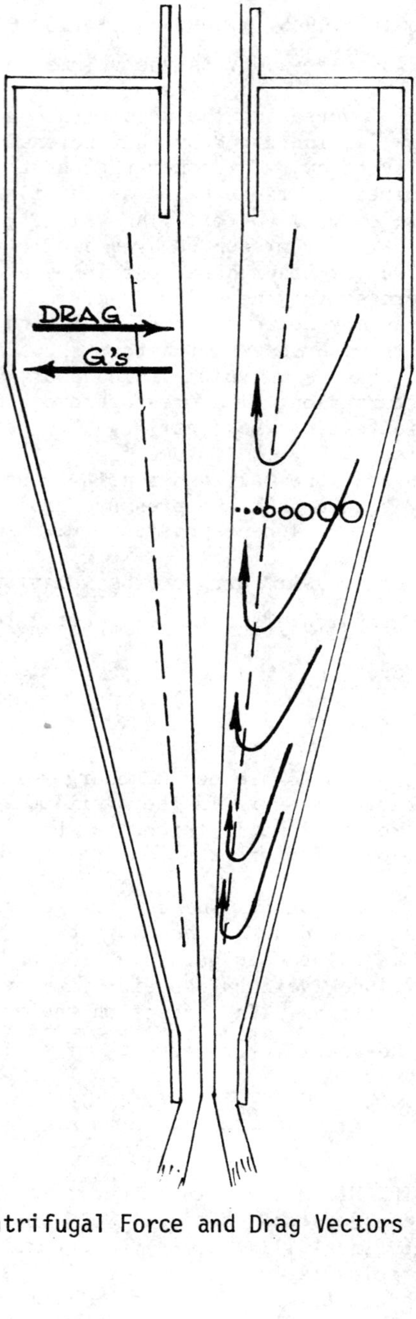

FIGURE 5: Centrifugal Force and Drag Vectors in Cyclones

PUMPS, PUMP BOXES, LINES, DISTRIBUTORS

Centrifugal pumps are commonly employed to handle slurry being fed to cyclones. The task of selecting the pump, pump box, piping, and distributor system should not be delegated by the designers of a grinding circuit. The designers have determined the range of tonnage and circulating load that is desired and are aware of the physical limitations on handling, piping and location. They must be involved in the corresponding selection. Otherwise, it may cost the owner money and the operator grey hair.

Pump Construction Materials

A variety of construction materials are available. Both rubber-lined and cast hard-alloy construction is seen in the wet ends of slurry pumps feeding cyclones. In primary grinding, where large, sharp-edged particles will be present, rubber might be expected to have a short life at the velocities which exist inside even a large pump. However, only experience can provide specific guidelines for a choice.

Information Required For Pump Selection

For pump selection, the following information is required:

(a)
- Determination of velocity head and friction head losses at a given flow rate.
- Determination of pressure drop through the cyclone (s) at the same flow rate.
- Determination of static head from the pump centerline to the cyclone overflow.

(b)
- Determination of characteristic pump curves at various pump speeds.

Information in (a) will provide a system curve which is a plot of total dynamic head required at different flow rates, while information in (b)--generally available from pump manufacturers--will provide a pump characteristic curve which is a plot of total dynamic head delivered by a pump at various flow rates and impeller speeds. These curves permit a selection of the size of the pump and its speed. With this information, the power requirement may be determined from the power curves supplied by the manufacturer.

System Curve. To plot this curve, the Bernoulli Theorem is applied from the pump centerline to the cyclone overflow. The total dynamic head required may be determined from:

$$H_s = \frac{.78\ v_s^2}{2\ g} + \frac{v_i^2}{2\ g} + \frac{2.31\ \Delta P}{\rho}\left[(\rho - \frac{.01p_f(\rho-1)}{}\right] + f_sL_s + f_dL_d + h_d - h_b \quad (5)$$

where g = 32.2, v_s is the slurry velocity at the pump suction line entrance, v_i is the slurry velocity at the cyclone feed inlet, ΔP is the pressure drop across the cyclone, ρ is the specific gravity of the ore, p_f is the cyclone feed percent solids by weight, f_s is the friction loss rate in the suction line, L_s is the length of the suction line, f_d is the friction loss rate in the discharge line from the pump centerline to the cyclone inlet, L_d is the equivalent length of the discharge line, h_d is the vertical height from the pump centerline to the cyclone overflow and h_b is the vertical height from the pump centerline to the top of the slurry in the pump box. Remember that slurry velocity = flow rate of slurry divided by cross-sectional area at the point of interest.

Slurry velocities must be high enough to prevent settling of coarse solids in the line which leads to the cyclone inlet. Cyclone inlet pressure must be 3-5 PSI for proper cyclone operation. When lines are vertical and radial distributors are used (see section on distributors in this chapter), flow velocities need not be high. Use of velocities typical of horizontal pipes is inappropriate; such values can be divided by 2 or 3 with complete safety, as reference to settling velocity data will show. Practical experience shows that a flow of 1 m/s (3.05 fps) of slurry containing 19 mm (3/4 inch) particles is entirely trouble-free. Benefits of large diameter piping are long life and reduced pump speed, which the reduced friction factors permit, saving maintenance dollars and downtime.

Friction loss rate factors have been determined in several ways; one method uses the Williams and Hazen formula (6,7):

$$f = .2083\ \left(\frac{100}{C}\right)^{1.85}\ \left(\frac{Q^{1.85}}{d^{4.8655}}\right) \quad (6)$$

where f is the friction loss per 100 ft. of pipe, d is the internal pipe diameter in inches, Q is flow in USGPM and C is 140 for new steel pipe and 100 for very old pipe. Most often, C is taken as 140. Other equivalent factors are required for elbows and bends. Note that using equation 6, developed for horizontal lines, gives factors that may be higher than actually required.

The cyclone inlet pressure is estimated from graphical information shown in reference 1 for a typical cyclone (remember, the vortex finder diameter is about 0.4 times the cyclone diameter) as:

$$\Delta P = \frac{2Q^2}{D^4} \tag{7}$$

where D is cyclone diameter in inches, Q is flow in USGPM and ΔP is in PSI. Equations 5,6, and 7 are useful for preliminary design primarily. Cyclone/Pump manufacturers should be consulted for more definitive design calculations.

Pump Characteristic Curve. Pump manufacturers supply pump curves, which are plots of total dynamic head or a related head that the pump will deliver at various flow rates and impeller speeds. Normally, these curves are for pumping water, so that a correction factor is necessary. This factor has been recommended (8) in the form:

$$R = 1 - .000385(\rho - 1)(1 + \frac{4}{\rho})p_f \ln(X_5/22.7) \tag{8}$$

where ρ is the specific gravity of the solids, p_f is the feed percent solids by weight and X_5 is the particle size in microns which 50% by weight would pass and 50% by weight would be retained. Equation 8 is less reliable above about 60% solids.

If H_p is the total dynamic head for water as supplied by the manufacturer, then

$$H_d = R\ H_p \tag{9}$$

where H_d is the delivered head.

In general, it is best to consult with pump suppliers for definitive pump design and selection.

Lines and Pump Boxes

Often, pump experts recommend a bottom horizontal discharge followed by a short, high-velocity, horizontal run of pipe to minimize recirculation of coarse particles inside the pump. A long-radius bend then connects to the larger vertical pipe. Use of rubber-lined pipe is probably good economy where the pipe would be difficult to replace.

Small, high-velocity piping connected to few cyclones (which then require a high pressure drop to pass the flow) results in a steep slope on the system head curve. An example is shown in a subsequent section, where the range in flow at fixed pump speed is forced to become narrow as a consequence. Hence there is important flexibility inherent in using vertical piping (velocity and friction loss may be low as a result) and more cyclones to keep pressure drop down. Also, a lower pump speed and a lower inlet pressure at the cyclone will cut maintenance costs.

In primary grinding circuits, it seems clear that the modest increase in capital cost for large feed piping and more cyclones will be repaid in a short time through reduced operation cost. In fine grinding circuits, such as for regrinding, higher velocities and cyclone pressures do not result in serious wear problems of equivalent magnitude.

If overflow piping is lower than the feed entrance to the cyclone, a siphon may be established such that the overflow is coarsened. A large vent pipe at the high point of the overflow pipe may prevent it.

As shown in a later section, deep pump boxes are of value. The design stage is the time to appreciate this; provide space for it. The cross-sectional area of the box should not be too large, for then the sides will slope steeply to prevent solids from hanging up and sloughing off intermittently into the pump inlet. Feed slurry should be introduced so it slides as smoothly as possible down the sloping side to minimize aeration; never let the feed plunge into the box from a long fall through the air. Air seriously disturbs classification and can exacerbate any tendency for pump surging which is likely to occur when the slurry level gets too close to the pump inlet.

Feed Slurry Distributors

An important consideration given a cyclone system is how slurry is to be distributed to each cyclone in parallel. If the separation is critical, care must be exercised. In such cases (1), a manifold in which all cyclones are fed radially from a central mixing chamber is recommended. A properly designed radial feed distributor provides each cyclone with slurry of identical properties in equal amount. An in-line manifold system (one in which feed is removed by cyclones as the slurry moves along a pipe) frequently fails to provide a representative sample of feed slurry to each cyclone. In some installations, the last cyclone receives a higher concentration of coarse and heavy particles. In others, this expected result does not occur but rather the first cyclones may get coarser feed. Use of reducing pipe diameters along the manifold and 45° takeoffs instead of 90° still fails to cure the problem in many cases.

Feed distributors and product collection launders can be designed to avoid high velocities and heavy impact loads from coarse material. When this is the case, life of elastomer (usually gum rubber) linings is measured in years. If the design is poor, with only initial price in mind, wear is more rapid and cost of maintenance can be high due to awkward size and inaccessibility.

SELECTION OF CYCLONES AND PUMPS FOR GRINDING CIRCUITS

Figure 6 is a sketch of a classical grinding circuit and is shown below. Before aspects of cyclone/pump selection are discussed, it is useful to review certain characteristics of cyclone operation in a grinding circuit.

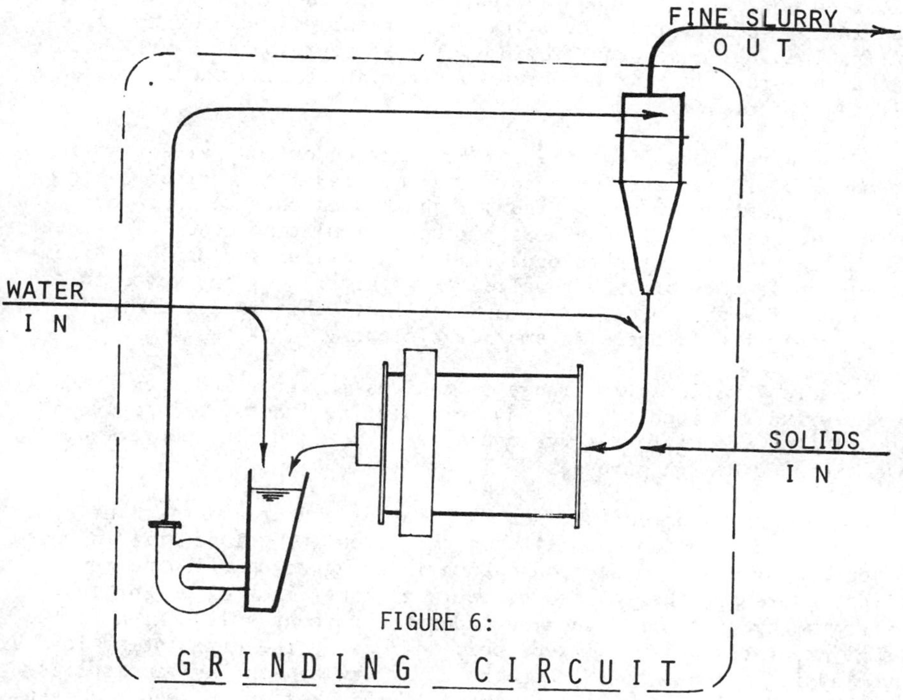

FIGURE 6:
GRINDING CIRCUIT

Cyclone Operation in Closed Circuit

While analyzing what happens within the grinding circuit, keep in mind that a closed circuit is a unit operation by itself. Like any other process unit "what goes in must come out" and, unlike a thickener, there is little inventory. Changes in input show up as output from the circuit, i.e. as cyclone overflow. Input rate of dry solids is usually changed only gradually; water addition is easier to change. The slurry, comprised of solids plus all the water added (including pump gland water along with any water from other sources of this kind), appears as cyclone overflow. Thus overflow percent solids is directly controllable via the water valve (s), and can never be higher than the percent solids entering the circuit from a previous process step.

Adding water (the most potent control of cyclone separation for the short term) has the immediate effect of producing a finer overflow. However, the weight recovery of solids to underflow will increase--in other words, the circulating load increases. This thick underflow of larger volume , added to fresh feed, raises the initially lowered cyclone feed percent solids and thereby restores the original separation. The final result is lower overflow percent solids and a higher circulating load. The new equilibrium value of overflow size will be finer than before only if the increased circulating load has increased the grinding efficiency.

If constant overflow size and constant underflow percent solids are assumed, a characteristic curve may be drawn for a given cyclone treating a particular ore. Figure 7 shows such curves, where overflow percent solids is plotted versus percent circulating load. The curves show an inverse relation to each other. A smaller cyclone has a comparable curve more or less parallel to a larger one, but moved in the direction of higher overflow percent solids. This is attributable to higher centrifugal forces in smaller cyclones.

Figure 7 reflects the range of choices in water addition rates to the grinding circuit at various circulating loads. This information, plotted for different cyclone sizes, is useful for operation, selection and design purposes.

If underflow percent solids is allowed to decrease (such as by opening up an adjustable apex) then the circulating load will increase without water addition and consequent lowering of overflow percent solids. However, the tonnage of recirculating fines rises and most operators object to it. Keeping underflow percent solids high is the normal practice. With a fixed apex, increasing the circulating load by water addition tends to thicken the underflow by sending more solids down through the same apex diameter. If this has an influence on separation size it is in a coarser direction, opposing the effect of water addition which is in a finer direction.

The cost-benefit relationship of higher circulating loads is not studied regularly in full-scale plants. Hopefully, operators who possess parallel, identical grinding circuits will publish such studies. Costs for pumping power and for maintenance of pumps and cyclones will rise with larger circulating loads.

Once an economically optimum circulating load is chosen, along with the corresponding overflow percent solids which a particular cyclone is capable of at that grind and underflow density, control of grind may be accomplished by tonnage variation to balance changes in ore hardness and/or in fresh ore feed size distribution. Attempts to

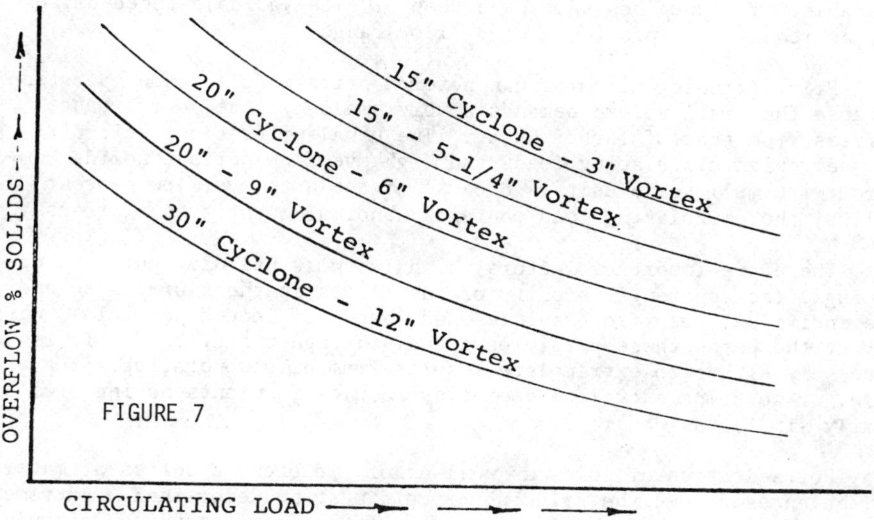

FIGURE 7

maintain the grind on a harder ore at the same tonnage by adding water (and thus increasing the circulating load) may not pay if prior decisions are well made.

In the above paragraphs, it has been assumed that percent solids in the mill, size and quantity of grinding media, mill speed and available power have all been maintained at optimum values.

Raw Data Acquisition For Selection

The size at which economic liberation of ore minerals occurs must be determined in the laboratory. This size is most often reported as an 80% passing size in microns. An estimate of the finished overflow size, $X_{1,3}$, or the X_{50c} size must be obtained. A knowledge of the minerals present, their relative amounts and individual specific gravities, is important in predicting behavior of the mixture in a cyclone.

Data on energy requirements for size reduction influence cyclone selection indirectly. If the installed grinding power is more than needed to grind the tonnage, low circulating loads can be expected unless the operator accepts overgrinding. A lower circulating load implies fewer cyclones and feed density will be lower for the same overflow percent solids. Later, it is almost certain that the available power will be utilized to grind increasing tonnage, so circulating load will tend to rise. This widens the range of operating conditions that the cyclone and pump must cover. Provision for more cyclones, with inlet shutoff valves, may be needed to keep pump head reasonable. The pump box should be deep or else variable-speed drives may be required to provide a wider flow range.

Pilot grinding circuits can never duplicate full-scale operation because the small volume demands a very small cyclone; performance differs from that of larger units. The usual pilot closed circuit has very high circulating loads and high overflow percent solids compared with full scale units. Because of the high overflow percent solids, the circulating load contains abnormal quantities of fines.

There are important factors which lab work has difficulty in defining: the apparent viscosity of the slimes in the slurry feed and the individual specific gravities and shapes of locked particles, which affect the paths these particles travel through the cyclone. It is necessary either to extrapolate results from pilot operation, if available, or to compare cyclone operating results in plants having ores as nearly similar as possible.

For selection of pumps as well as of cyclones, a series of material balances around the grinding circuit must be determined. Extreme ranges to be expected should be covered for: circuit tonnage, circulating loads, overflow percent solids; from this will emerge the range of: pumping volumes, slurry specific gravities, percent solids by volume. The task is simplified by use of a form such as the one illustrated in Tables 1 and 2, where Table 1 is in international units. The form utilizes any available data. The rest is derived by working up the columns or downward.

The typical cyclone should be considered for preliminary design purposes. Remember that the pressure drop at the inlet is 10 PSI.

At this stage, suppose that the following basic information is available:

252 kg/s (1000 STPH) dry feed ore, S.G. = 3.2
Flotation feed to be 80% -150 micron at 35% solids by weight
Rod mill discharge fed directly to ball mill, 80% -1000 micron
Bond work index estimated at 21.5
Ball mills will draw 11950 kw (16000 HP) with optimum liners, speed and ball load

Predicted power consumption at 1000 STPH from Bond (9):

$21.5((10/150^{.5}) - (10/1000^{.5}))\ 1000 = 10756$ kw which is

90% of the 11,950 kw available.

TABLE 1. Sample Material Balance Form

$$\frac{\text{kg/s}}{\text{S.G. x 1000}} = \text{m}^3/\text{s} \qquad \text{S.G.} = \frac{\text{kg/s}}{\text{m}^3/\text{s x 1000}}$$

	kg/s	S.G.	m^3/s	
solids	252	4.0	0.063	
liquid	378	1	0.378	
OVERFLOW SLURRY	630	1.43	0.441	
% solids/wt.	40		14.3	% solids/vol.
(200%CL) solids	504	4.0	0.126	
liquid	126	1	0.126	
UNDERFLOW SLURRY	630	2.5	0.252	
% solids/wt.	80		50	% solids/vol.
solids	756	4.0	0.189	
liquid	504	1	0.504	
FEED SLURRY	1260	1.82	0.693	
% solids/wt.	60		27.3	% solids/vol.

$$\text{OVERFLOW SOLIDS x } \frac{\%\text{CL}}{100} = \text{UNDERFLOW SOLIDS}$$

$$\text{FEED ALWAYS} = \text{OVERFLOW + UNDERFLOW}$$

$$\text{CIRCULATING LOAD \%CL} = \frac{100.\%\text{R}}{100-\%\text{R}}$$

$$\text{RECOVERY TO UNDERFLOW \%R} = \frac{100.\%\text{CL}}{100-\%\text{CL}}$$

TABLE 2. Sample Material Balance Form

$$\frac{\text{STPH} \times 4}{\text{S.G.}} = \text{USGPM} \qquad \text{S.G.} = \frac{\text{STPH} \times 4}{\text{USGPM}}$$

	STPH	S.G.	USGPM	
solids	1000	4.0	1000	
liquid	1500	1	6000	
OVERFLOW SLURRY	2500	1.43	7000	
% solids/wt.	40		14.3	% solids/vol
(200%CL) solids	2000	4.0	2000	
liquid	500	1	2000	
UNDERFLOW SLURRY	2500	2.5	4000	
% solids/wt.	80		50	% solids/vol
solids	3000	4.0	3000	
liquid	2000	1	8000	
FEED SLURRY	5000	1.82	11,000	
% solids/wt,	60		27.3	% solids/vol

$$\text{OVERFLOW SOLIDS} \times \frac{\%\text{CL}}{100} = \text{UNDERFLOW SOLIDS}$$

$$\text{FEED ALWAYS} = \text{OVERFLOW} + \text{UNDERFLOW}$$

$$\text{LONG TPH} \times 1.12 = \text{SHORT TPH}$$

$$\text{METRIC TPH} \times 1.1023 = \text{SHORT TPH}$$

$$\text{CIRCULATING LOAD } \%\text{CL} = \frac{100.\%\text{R}}{100-\%\text{R}}$$

$$\text{RECOVERY TO UNDERFLOW } \%\text{R} = \frac{100.\%\text{CL}}{100+\%\text{CL}}$$

From Figure 4, note that at an overflow percent solids of 35% by weight, the recommended underflow percent solids is approximately 80%. This information will permit the estimation of the cyclone feed volume percent.

Calculation For Cyclone Selection

Unless operation begins with reduced ball loads to reduce power consumption, circulating loads will be low. Assume 100% initially, rising to 400% later when the full power is utilized for added grinding work. Using the material balance form (see Tables 1 and 2) with an 80% solids by weight for underflow, calculate:

Circulating load	100%	400%
Underflow dry solids, kg/s(STPH)	252 (1000)	1008 (4000)
Cyclone feed volume, m^3/s(USGPM)	.0788 (10928)	.315 (17678)
Feed percent solids by volume	22.9%	35.4%

It is necessary to design for the highest feed percent solids expected; otherwise the selected cyclones would be incapable of producing the desired grind at full capacity. Initially, overflow percent solids will be higher than the desired 35% but it may be diluted for flotation.

Assume that the grinding power is distributed over 4 ball mills, so that the feed rate per mill will be 63 kg/s (250 STPH). Each mill will draw 2988 kw (4000 HP). The above table can be rewritten as:

Circulating load	100%	400%
Underflow dry solids, kg/s(STPH)	63 (250)	252 (1000)
Cyclone feed volume, m^3/s(USGPM)	.0197(2732)	.0788(4420)
Feed percent solids by volume	22.9%	35.4%

Cyclone Diameter Selection. Graphical data reported in reference 1, when converted to equation form, shows that:

$$X_{50c} = \frac{13.2\ D^{.675}\ \exp(-.301+.0945V-.00356V^2+.0000684V^3)}{(\Delta P)^{.3}\ \sqrt{\rho - 1}} \tag{10}$$

For preliminary design purposes, the X_{50c} should be near the 80% passing size, so that $X_{50c} \simeq 150$ micron. If the pressure is taken as 10 PSI, the "typical" cyclone pressure, then the only unknown in equation 10 is D, the cyclone diameter. From equation 10, D ≃ 17 in. A 20 inch cyclone is close enough, although this choice places limits on vortex and apex diameters to be compatible with the typical cyclone.

Estimating The Number of Cyclones. There are 4 grinding lines and each line requires one or more cyclones. Each group of cyclones must handle 2732 to 4420 USGPM of slurry at 35.4% solids by volume (63.7% by weight).

The volume of slurry that the typical cyclone will handle is estimated from equation 7, so that

$$Q = 0.7\ \sqrt{\Delta P}\ D^2 = 0.7\ \sqrt{10}\ (20)^2 = 885 \text{ USGPM}$$

per cyclone. This suggests that each grinding line should have a minimum of (2732/885) ≃ 3 cyclones and a maximum of (4420/885) ≃ 5 cyclones. Obviously, this choice will depend upon whether vortex and apex diameters and inlet nozzle area have the proper slurry handling capacity such that the desired X_{50c} size, the overflow percent solids and the underflow tonnage is attainable in operation. For definitve design, the cyclone supplier should be consulted.

Vortex Diameter/Apex Diameter/Inlet Area. Vortex diameter, apex diameter and inlet area must be selected, bearing in mind that these variables influence pressure drop and hence Q in equation 7. The typical cyclone has a vortex diameter that is about 0.4 times the cyclone diameter. Therefore, the vortex size is approximately 0.4 (20) = 8 in. Suppliers offer various types of 20 inch cyclones with recommended ranges for vortex diameter, which results in reasonable flexibility.

The apex diameter must be such that roping does not occur, in spite of "typical cyclone" relationships between apex diameter and vortex finder diameter. The fraction of feed solids reporting to underflow in a closed grinding circuit is often greater than that found in open-circuit classification. Equation 1 may be used to estimate the apex diameter that should provide the required underflow capacity per cyclone:

$$S = 4.162 - \frac{16.43}{\left(2.65 - 3.2 + \frac{320}{80}\right)} + 1.10\ \ln(200/3.2)$$

which gives S ≃ 4.0 inches at 400% circulating load and five cyclones.

Inlet area for the typical cyclone is about 7% of the cross-sectional area of the feed chamber. Consequently $0.07 \times \pi \times (20)^2/4$ is approximately 22 square inches, the approximate required nozzle area.

Final Decision. A decision should be made in consultation with a cyclone supplier. Graphical data of the type shown in Figure 8 will be supplied for final decision making.

Figure 8 shows overflow percent solids versus percent circulating load for various sizes of cyclones analogous to Figure 7. From the curves it is apparent that the middle curve meets the design specifications; the lower curve might be selected if a compromise is decided upon. The upper curve provides leeway for unusually high circulating loads but the underflow will contain more fines due to the higher overflow percent solids. In this example, it would be desirable to keep pressure drop in the 6 to 7 PSI range for long life of pumps and cyclone liners/apexes.

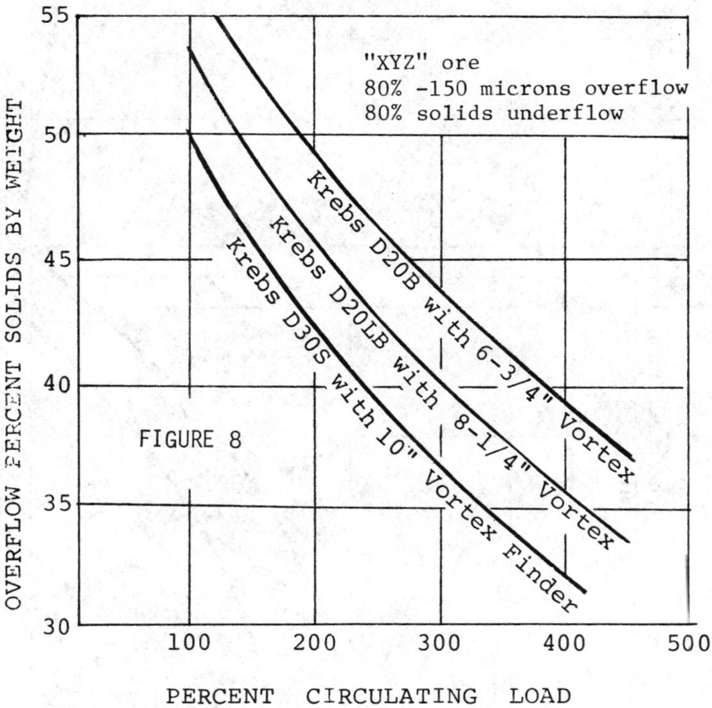

FIGURE 8

Pump Selection

Figures 9 and 10 show pump/system curves plotted with their origins in the upper left corner. Thus, head increases downwardly. Two system curves are plotted on each figure, one curve represents low pump box level and the other represents high pump box level. The flow rate range lies between the points where the system curves and pump curves intersect. If the RPM of the pump increases, the pump curve displaces downwardly in parallel. For a given pump and flow rate, the head is proportional to the square of the pump speed.

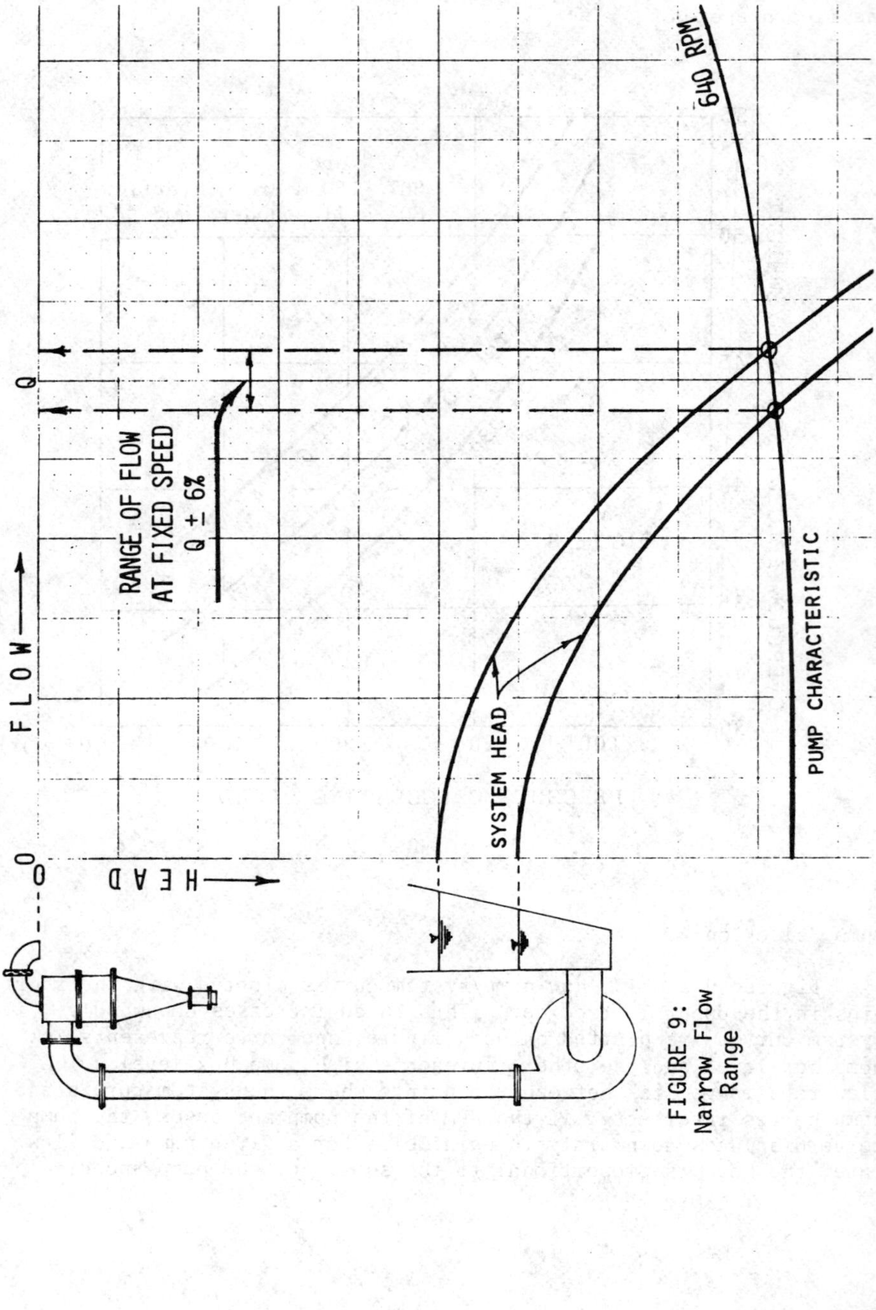

FIGURE 9: Narrow Flow Range

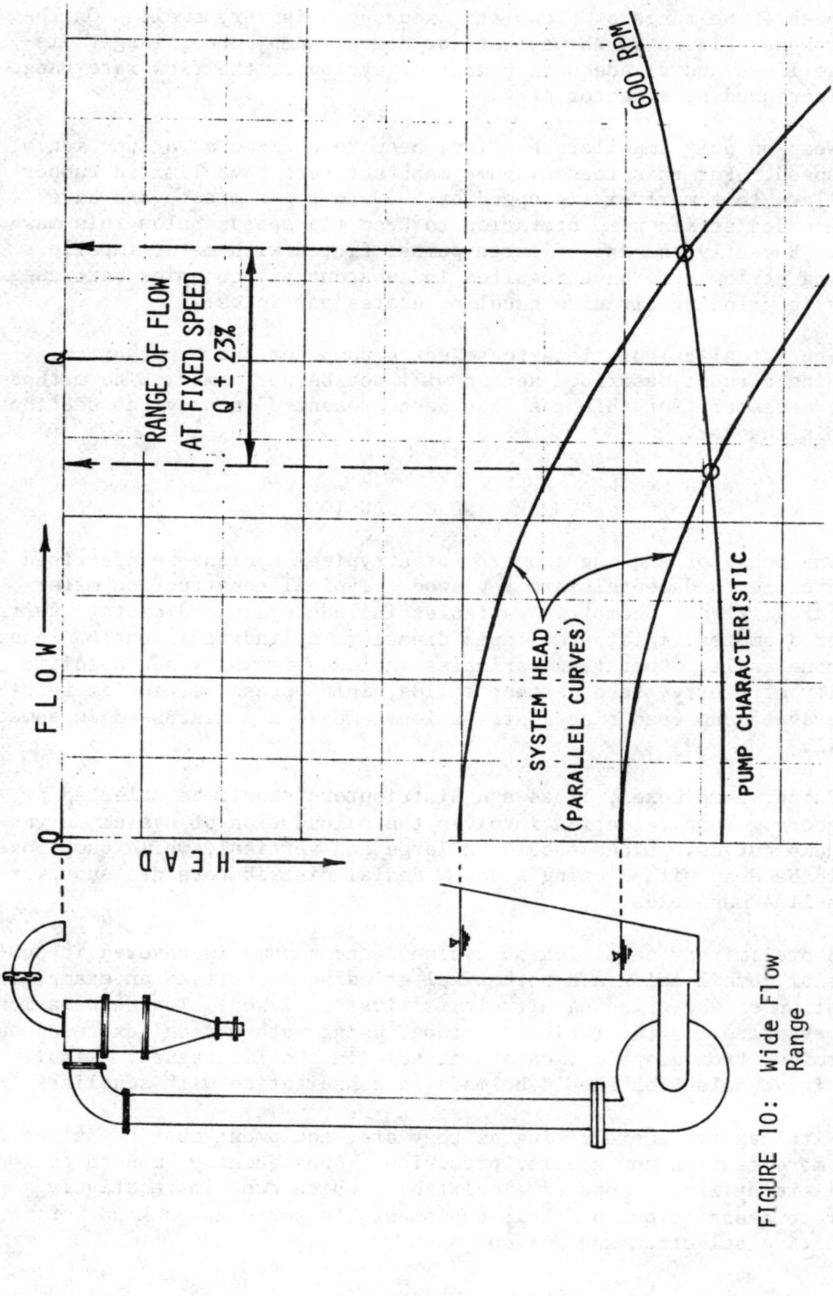

FIGURE 10: Wide Flow Range

Figure 9 illustrates a situation where the pump box is not very deep; discharge lines are of small diameter and there are insufficient cyclones. The range of flows at fixed speed is very small. On the other hand, Figure 10 shows that for deeper pump boxes, larger discharge lines and an adequate number of cyclones, the flow rate range has increased by a factor of 4.

Wear on pump impellers has long been recognized as a function of tip speed. For this reason, pump manufacturers have limited rubber impellers to a maximum tip speed of 5000 feet per minute and have recommended series pump operation to keep tip speeds below this maximum. Recently the use of large pumps, fitted with metal impellers running at low RPM, have resulted in substantial economies particularly in grinding circuits handling coarse particles.

The actual calculations to select a pump for the hypothetical grinding circuit described herein will not be performed. The methodology necessary for this task has been presented in previous sections of this chapter.

SUMMARY AND CONCLUSIONS

The behavior and the geometry of a typical cyclone is described and cyclone fundamentals are reviewed. Typical construction materials are listed. Geometric variables include cyclone diameter, vortex finder diameter, inlet area, apex diameter, cylindrical section length and cone angle. Operating variables include viscosity and specific gravity of slurry, feed percent solids, inlet pressure, solids specific gravity and feed size distribution. These are discussed in some detail.

Pumps, pump boxes, lines and distributors should be selected by designers. Pump selection involves the calculation of system curves and pump curves. Lines should be large and vertical, while pump boxes should be deep with sloping sides. Radial distributors are superior to in-line manifolds.

A preliminary selection of cyclones and a pump is covered in substantial detail using a hypothetical grinding circuit as an example. Operation of the grinding circuit is first reviewed. The calculations are performed for the typical cyclone, using mathematical expressions determined from graphical data available in the literature. Final, definitive selection should be made in consultation with suppliers.

With capital costs rising as they are, equipment must be selected with more caution and greater precision. Consequently, modern cyclone suppliers utilize computer simulations, which contain relatively accurate descriptions of their equipment, to serve as back up for definitive selection and design.

REFERENCES

1.-------, "The Sizing of Hydrocyclones", Krebs Engineers,1205 Chrysler drive, Menlo Park, California, 94025, 1977.
2.Jull, Norman A., "Parameters For Cyclone Selection", Annual Meeting, Canadian Mineral Processors, Ottawa, January, 1972.
3.Lynch, A. J., and Rao, T. C., "Modelling and Scale-Up of Hydrocyclone Classifiers", 11th Int. Min. Proc. Congress, Cagliari, 1975.
4.Tarr, D. T., "Practical Application of Liquid Cyclones in Mineral Dressing Problems", Fall Meeting AIME, Phoenix, October, 1965.
Tarr, D. T., "The Influence of Variables on the Separation of Solid Particles in Hydrocyclones", Minnesota Section, AIME, Duluth, 1972. (contains comprehensive bibliography on hydrocyclones)
5.Lynch, A. J., Technical Report, July 1 - December 31, 1974, Julius Kruttschnitt Mineral Research Centre, University of Queensland.
6.---------, The Transportation of Solids in Steel Pipe Lines, Colorado Research Foundation, Inc., 1963.
7.Horne, C. A., "Slurry Pumping Practice", Annual Meeting, Canadian Mineral Processors, Ottawa, 1976.
8.Cave, I., "Effects of Suspended Solids on the Performance of Centrifugal Pumps", Proc. Hydrotransport, 4, organized by BHRA Fluid Engineering, 1976.
9.Bond, F. C., Allis Chalmers Bulletin 07R9235B, 1961.

Chapter 18

PROCESS DESIGN, SCALE-UP AND PLANT DESIGN FOR GRAVITY CONCENTRATION

by Chris Mills

Industrial Minerals Consultant
Princeton, New Jersey

ABSTRACT

The role of gravity separation in the mineral industry is briefly considered in perspective. The mineralogical and laboratory approaches to gravity process design are examined with emphasis on process technology selection. The equipment available for small scale and pilot testwork is discussed, particularly in regard to scale-up and the acquisition of reliable, accurate data.Transposition of performance data from pilot scale to plant scale is explained. The available types of full scale gravity separation equipment are discussed with emphasis on large high-capacity units. Plant layout and selection of ancillary equipment are considered.

INTRODUCTION

As an introduction to this chapter, it is pertinent to put gravity separation, as a collection of process stages, in perspective. This has been done at least once before by Aplan (1), and figures for slightly more recent times are shown below. These figures are in part abstracted from Minerals Yearbook 1974 (2), and in part taken from the author's knowledge of gravity separation in the United States. These figures are at best very approximate, but they do serve to give some idea of the magnitudes involved. One can see from Table 1 that for the year 1973 in the USA, about 313 million tonne of ore were treated by gravity concentration to produce concentrates with a value of about 3900 million dollars. This compares with an estimated 400 million tonne of ore treated by froth flotation in the same period with concentrates valued at about 1500 million dollars.

Table 1. Ore tonnages treated by gravity separation in the USA in 1973 (excluding DMS)

Ore type	Ore processed (million tonne)	Approximate value of products (million $)
Coal	228	2900
Iron ore	50	900
Mineral sands	23	72
Gold	9	2
Tungsten	3	37
TOTAL	313	3911

As can be seen from the above figures, coal represents the bulk of the ore treated by gravity separation in the USA, with iron ore representing a major portion of the balance. The major units of gravity separation equipment used in the USA are jigs, wet shaking tables, spirals, sluices and (more recently) cones. Dense media separation (DMS) is also used extensively, but this process will not be covered in this chapter and figures for gravity concentration by DMS have accordingly been omitted from Table 1.

It is generally thought that gravity separation is applicable only to coal separation and a few obscure separations where flotation has failed. This view is rather erroneous at the present time, and will become more so in the future. The reasons for the latter statement are threefold. In general, gravity separation has a lower installed cost per tonne of throughput than flotation for any given job, and usually has a lower installed power requirement per tonne of throughput. Gravity separation does not use expensive reagents, the cost of which (for flotation) is continually spiralling upward. With the exception of slime disposal (common to flotation), the environmental impact of gravity plant effluent is considerably less than that for flotation, due to the absence of organic chemicals and their reaction products.

Although not the subject of this chapter, brief mention must be made of the two processes that are often complimentary to gravity separation, electrical and magnetic separation. Both of these processes share most of the cost benefits of gravity separation, and also the non-pollutant advantages.

Examination of the literature of the last ten years shows the remarkable neglect of gravity separation. This neglect is reflected by the sparse research into the subject, the almost complete lack of teaching of the subject in US mining schools and the suspicious attitude of major US mining houses toward any suggestion of the use of gravity separation when flotation has any conceivable chance of success. This area could be expanded ad infinitum, but it is more appropriate to examine the subject carefully and to try to dispel some of the myths.

The genuine key to effective gravity separation is a complete understanding of the mineralogy of the ore under consideration. Figure 1 shows the size range applicability of unit processes in gravity separation. In addition, it shows the size range applicability of some other relevant processes. In addition to the particle size constraints, of course, there are constraints on specific gravity differentials and liberation, both of which will be dealt with later. However, consideration of Figure 1 in combination with a knowledge of the ore mineralogy will immediately indicate whether gravity separation is a viable processing route. In addition, the chart in Figure 1 will indicate what other process technology is available as an alternative to gravity separation. Naturally, the experienced mineral engineer will be able to determine a great deal from the mineralogical analysis. Assuming that gravity separation technology is appropriate, the course of action to be followed is usually straightforward.

LABORATORY TESTWORK

The first step at the laboratory level should always be heavy liquid analysis of the ore to be fed to the gravity separation plant. Now, the standard sink-float separation at s.g. 2.95 is not usually very illuminating (the coal industry has for many years been far ahead of the metalliferous industry in this regard, using sequential sink-float testing (3,4)). What must be done is a complete sequential sink-float analysis of the proposed plant feed.

One of the major problems in this area is ignorance of the available techniques. 1,1,2,2-tetrabromoethane and bromoform (and occasionally methylene iodide) are the heavy liquids with which most mineral processing engineers are familiar. These are of little use for heavy liquid analysis in gravity separation, as most minerals concentrated by gravity methods have specific gravities much higher than the three liquids mentioned. Clerici solution (an aqueous solution of thallous formate and thallous malonate), however, is of great use since its maximum s.g. (at 95°C) is 5.2. Clerici solution is much less hazardous to handle than is generally thought, and when properly handled is much safer than the halogenated hydrocarbons. Clerici solution should never be heated in

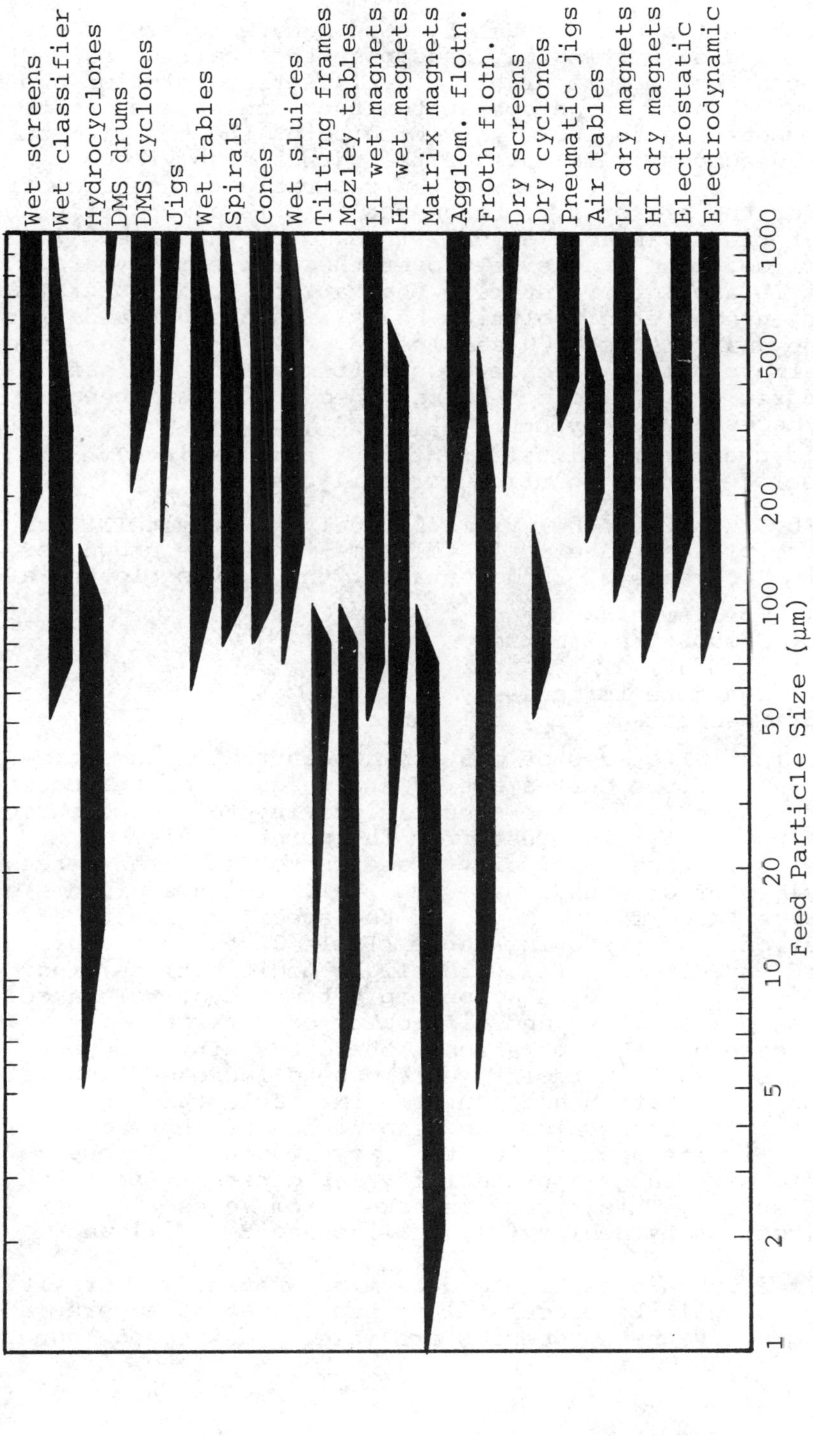

Figure 1. Size Range Applicability of Commercial Unit Processes

a sand bath (as recommended by some manufacturers) since this method of heating may decompose the thallium salts, whereupon they become highly toxic. Clerici solution should always be heated in an open water bath since this limits the temperature to 100°C (at which point the thallium salts do not decompose). Because of the rather high viscosity of Clerici solution, it is often desirable to use a centrifuge for the higher s.g. range but this is a low price to pay for the advantages gained. This author has used Clerici solution extensively over the last eight years, and it is a point of interest that at least one Australian company uses Clerici solution heavy liquid separation for routine analysis (c. 200 samples/day). In the latter case no health problems have developed (because of the safe procedures used), and no expensive equipment has been found necessary. A hypothetical example will be given later in this chapter to illustrate the principles involved in the use of Clerici solution for analysis.

The results of heavy liquid analysis of metalliferous ores are plotted in exactly the same manner as shown for coal in Reference 3. Fundamentally, the curves plotted are

1. Specific gravity
2. Cumulative % sinks
3. Cumulative % floats
4. Characteristic assay
5. Specific gravity tolerance

In the initial stages of gravity concentration investigation, all of the curves are of about equal importance as will be shown later. The specific gravity tolerance curves are of particular interest from the point of view of process selection. Basically these curves show the percentage weight of ore head that would fall within a given s.g. tolerance (e.g. ±0.05, ±0.1) of the specific gravity of separation. The following table (Table 2, taken from Leonard and Mitchell (3) with slight modifications) indicates the approximate relationship between s.g. tolerance, process applicability and difficulty of separation.
In the case of DMS separations, the difficulty has been reduced somewhat in recent years due to increased sophistication of operation and control. The table takes no account of particle size, and experience is therefore required in its application to heavy liquid test results. However, size range applicability may generally be taken from Figure 1. This figure is taken from Jones (5), with modifications by Henley (6), Collins and Read (7) and Mills (8).

It is thus possible to determine whether wet gravity separation will be successful in beneficiating an ore, and which gravity unit processes are likely to succeed, purely on the basis of heavy liquid analysis over the appropriate specific gravity range.

Table 2. Approximate relationship between s.g. tolerance, process applicability and difficulty of separation

Weight % within ±0.10 gravity of separation	Degree of difficulty expected	Gravity process recommended	Type
0 - 7	simple	almost any process	JIGS, TABLES, SPIRALS, SLUICES, CONES, DMS, VANNERS, ETC.
7 - 10	moderately difficult	efficient process	
10 - 15	difficult	efficient process - good operation	
15 - 20	very difficult	very efficient process - expert operation	DMS
20 - 25	exceedingly difficult	very efficient process - expert operation	
Above 25	Formidable	Limited to a few exceptionally efficient processes - expert operation	DMS with close control

In fact, the range of possible gravity unit processes can be further narrowed based upon the estimated projected plant throughput. Moncrieff and Lewis (9) have recently compared the feed size range, unit capacity, capital cost and capital cost/unit throughput for various gravity unit processes. Examination of their data, for example, indicates that although the installed cost/tonne of feed is very low for a Reichert cone (when compared with other unit processes), the capacity of a Reichert cone is about 70 tonne/h. If the proposed plant will have a capacity of 15 tonne/h (300 tonne/24h - not unusual for a small cassiterite, wolframite or tantalite plant), then a cone would obviously be too large and spirals would be a more appropriate choice in terms of capacity.

A brief recapitulation is in order at this stage. From a knowledge of the ore mineralogy and heavy liquid analysis of the ore, the following have been determined:-

1. The feasibility of gravity separation

2. The probable quantity and quality of gravity

separation products

3. The most appropriate unit process(es) for gravity separation

Now, it is worth reflecting at this point on the small-scale equipment available for miniature tests and the mode of gravity separation achieved with each. Figure 2 shows the action of various devices upon a family of particles. For completeness, perfect screening and density separation are included.

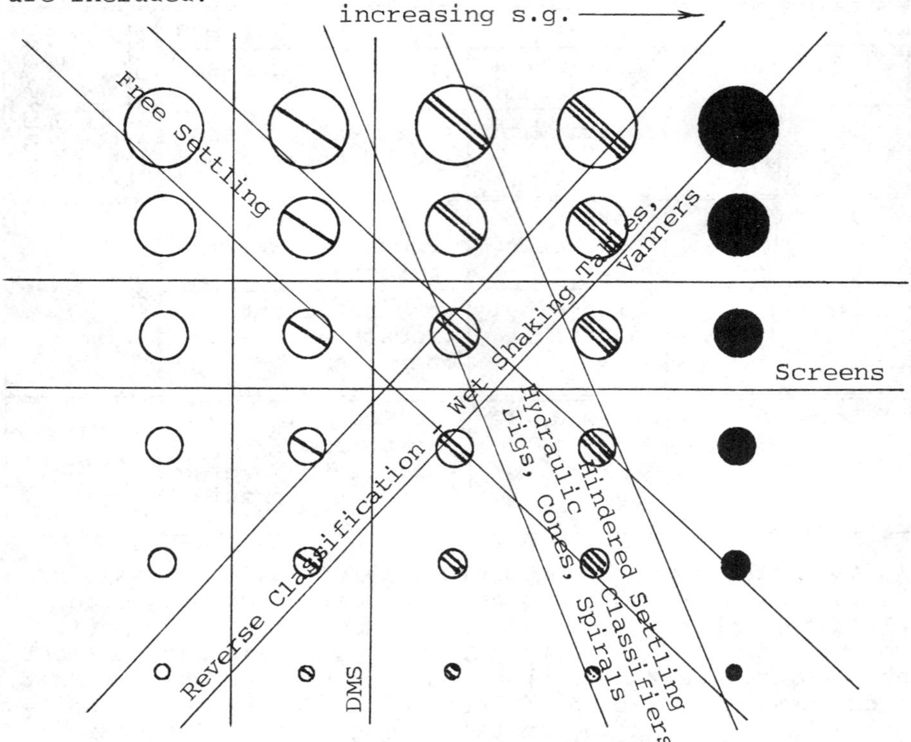

Fig. 2. Separation effects of various devices with regard to particle size, specific gravity and mass.

At the smallest scale, say a few grammes, laboratory test screens, heavy liquids and the micropanner are the available processes. The micropanner has the flexibility to operate as both a film sizing device and a mass classification device. It is thus appropriate to test any ore (presumably amenable to gravity separation) on a micropanner to test the hypothesis that gravity separation is appropriate. While the micropanner does not perform in exactly the same manner as most gravity separation devices, it will rapidly confirm the applicability of such methods.

The next stage is to scale-up the data to laboratory scale, which means the production of larger feed samples. Feed preparation must also be considered. This may include grinding to liberation size, desliming, screening to remove unwanted oversize, hydraulic classification or a combination of the above. Desliming is usually the one feed preparation step necessary for efficient gravity separation in all cases.

The equipment available for laboratory testwork will now be considered. Spirals and cones, however, will be discussed later as their throughput is so large that they should really be considered as pilot plant equipment.

Where the feed is to be processed on film sizing devices such as shaking tables and vanners, hydraulic classification is indicated. For laboratory scale batch hydraulic classification (about 10 to 15 kg or so), a classifier such as the Deister laboratory classifier is recommended. These units produce data that can be directly scaled to pilot plant or full scale operation with very reproducible results. Three products, each of about 5 kg, from 15 kg of classifier feed are sufficient for laboratory wet shaking table tests. For laboratory wet table testwork, the Deister 15-S, Wilfley 13A or Wilfley 13B is the most useful (the 13B has a slightly larger deck than the 13A - the capacity of the 13B being comparable to that of the 15-S). A partially riffled deck (the most commonly used) gives a product that is film sized (middlings are coarse heavies with fine lights), while a fully riffled deck approximates mass classification (middlings are coarse lights with fine heavies). Figure 2 refers to partially riffled decks which are the standard production types in the industry. In general terms, a Deister 15-S or Wilfley 13B has a nominal capacity of about 100 kg/h. Thus a 15 kg sample fed to a laboratory classifier to produce three products of about 5 kg each will produce enough table feed for a minute or so of table running time. The products of such a run may always be combined and rerun.

For laboratory scale vanner testwork Bartles (Carn Brea) manufacture an ideal unit that is suitable for the treatment of samples in the 200 g to 5 kg range.

For film sizing devices such as partially riffled wet tables and vanners scale-up to larger units is very straightforward. Capacity is roughly proportional to surface area. A full-size Deister wet table deck (such as one deck of a triple-deck '999') has sixteen times the area of a 15-S laboratory table, so that if laboratory tests indicate a capacity of 100 kg/h then the full size unit will generally give a comparable separation performance at a throughput of 1600 kg/h.

The most common laboratory scale jig is the Denver

No.1M which is suitable for batch samples of about 2 kg, and by virtue of its transparent hutch gives rapid indication of feed jiggability. It is, however, rather difficult to scale up from the 1M, and if scale up data is needed then the larger Denver 4"x6" "Selective" Mineral jig should be used. This jig is designed to operate on unclassified feeds and scale up to larger units is made on the basis that capacity is proportional to the area of the jig bed.

With all of the abovementioned equipment, testwork should be carried out on a batch basis with measurement of all pertinent operating parameters. There are three common methods of running laboratory tests of this type. The one most favoured by the author is the steady state multiple product test. This is particularly applicable to tables. Feed is supplied to the unit until an apparent steady state is reached, whereupon simultaneous samples of six to ten fractions of the products are taken for weighing and analysis. This method gives an excellent picture of the distribution of values in the separator, and more importantly the distribution of locked particles and true (as opposed to mechanical) middlings. A method preferred by some workers is the middlings extinction type test. Here the concentrate and tailings are continuously collected, while the middlings are continuously returned to head feed. This method may be appropriate for fully liberated ores such as mineral sands, where middlings are predominantly mechanical, but it gives an incorrect view of the disposition of the true locked middlings as it forces them to report with either the concentrate or the tailings. This may be particularly misleading when the middlings fraction really needs to be reground to achieve further liberation.Finally, there is the simple two or three product test. This is almost unavoidable with laboratory jig tests, but at the laboratory stage of gravity separation testwork it should really not be used if avoidable. Three products (concentrate, middlings and tailings) can always be composited from the samples from a multiple product test. A classic case of this is the tabling of material containing monazite, zircon, staurolite, garnet and tourmaline (known in the industry as "zircon magnetics") for monazite recovery. What often appears to be a good concentrate cut visually, often is not upon analysis. The multiple cut system allows the construction of a composite concentrate following product analysis. This reduces the overall number of tests required and alleviates considerably the frustration of the mineral engineer.

PILOT PLANT TESTWORK

As mentioned earlier, Reichert cone and spiral testwork should be considered as pilot plant testwork because of the high capacity of such units. For all practical purposes, this is also true of sluices and jigs. The quantity of feed required for a cone test is about 5 tonne (10),and that for a spiral test is about 0.75 to 1 tonne. Since a 5 tonne sample is required for a cone test and the cost of preparation of a 5 tonne sample may be high (as is the cost of cone testwork), it is wise to determine whether the proposed operation is large enough to justify a cone. As a general guideline, the smallest operation that can really justify a cone is about 50 tonne/h (1000 tonne/24h) cone feed. Below this throughput it is probably more sensible (if the feed particle size is in the range suitable for a cone) to consider 2-start spirals as more appropriate.Both cones and spirals have the significant advantage that scale up is purely numerical as full-sized units are always used for testwork.

The question of wet gravity separation pilot plant testwork is always controversial, particularly if a continuous pilot plant is envisaged. Because of the high unit capacities of jigs, cones, sluices and spirals, it is difficult to justify a continuous pilot plant unless:-

1. It is absolutely essential because of the marginal results so far.
2. The envisaged plant is large enough that a continuous pilot plant could be built at the mine site (or adjacent to it), and serve eventually as part of the full scale plant. Pilot plant concentrate therefore having net positive value, and may be needed for further testwork.
3. The pilot plant equipment needed is mostly already available to a company so that capital costs are low,and the required feed is readily available.
4. Other considerations of a non-technical nature.

Examples of the above might be:-

1. A low-grade scheelite ore where tests indicate possible difficulty in producing a marketable concentrate.
2. A large low-grade ilmenite/zircon deposit where plant feed is estimated at 1500 tonne/h, and electrical and magnetic separation testwork must be done on gravity concentrate.
3. Projected plant will treat tailings from existing ilmenite dry mill to produce a zircon product. Feed is currently produced tailings and pilot plant can be readily fitted into existing ilmenite wet mill.
4. A company is diversifying into new minerals (e.g.

a lead-zinc company going into wolframite/tantalite) and is unfamiliar with the process technology. Board feels that pilot plant demonstration is necessary before approving finance.

Continuous gravity pilot plants can be very expensive if they incorporate large jigs (such as the IHC), sluices or cones, since they are basically small production units in their own right. With the possible exception of large mineral sands and iron ore deposits, such pilot plants are very difficult to justify.

Assuming for the time being that a pilot plant is necessary, let us consider some of the problems that may be encountered. With the exception of Case 2 above, a compromise will always be necessary. Batch pilot plants are not nearly as useful as continuous pilot plants, but often they may be the sole option. Small continuous wet gravity are quite rare. However, the author has designed two such plants in the US, both of which operated extremely satisfactorily. One of these pilot plants contained eleven unit processes, excluding feeding and pumping, with four separate recirculating flows. However, the surveillance and manpower required were completely out of proportion to the size of the plant.

Spirals can be substituted for cones, if necessary, since their metallurgical performance is not radically dissimilar to that of cones. This allows a considerably lower head feed rate, and a rougher-scavenger-cleaner cone plant can be approximated by a similar spiral plant (e.g. a 2-start rougher, a 2-start scavenger and a 1-start cleaner). Flows can be judiciously split (with rotary splitters) to reduce flows to succeeding stages. For example, cleaner spiral concentrate might be divided into three parts. Two parts would be discarded, while the remaining part would flow to a hydraulic classifier (4-spigot), which would in turn feed four laboratory wet shaking tables. However, the problems associated with the operation of such items as laboratory shaking tables on a continuous (24h/day) basis have to be experienced to be believed.

Probably the most difficult problems with wet gravity pilot plants are those of (small scale) pumping and water balance. These are also the two most critical areas of full scale wet gravity plant design, and will therefore be considered later. Small scale wet gravity plants are much more difficult to keep in balance than full size plants, and it is therefore advisable to use variable speed pumps and oversized peripheral discharge sumps. Where full size equipment can be used throughout a continuous wet gravity pilot plant, many of the difficulties disappear and the operation approximates that of a full size plant.

In summary, it is not nearly as simple to build a wet

gravity pilot plant as it is to build a flotation pilot plant. It is always possible to design and build a wet gravity pilot plant, but many compromises are usually necessary. Every aspect of such a plant requires careful and accurate (and often ingenious) design. The most important question is always "Will the results justify the cost, or will batch tests give almost the same information at considerably less expense?". There is, of course, no general answer to this question - each case must be examined on its own merits.

WET GRAVITY PLANT DESIGN

In the case of all gravity concentration devices, feed preparation is not only desirable but almost essential. For unit processes utilising film sizing (reverse classification), such as the standard partially riffled wet table, Bartles-Mozley table or vanner, hydraulic classification (hindered settling) is the appropriate feed preparation. For the less sensitive separators that approach mass classification such as the jig, Reichert cone, sluice and spiral, the screening removal of oversize coupled with desliming is usually adequate feed preparation. It should be noted that at least two very large gravity plants built within the last ten years have partially or totally disregarded this latter point. The oversize category is not limited to mineral particulate matter, but includes organic material such as tree roots. The Reichert cone is particularly sensitive to fine tree roots, which readily block the annular splitter slots. Spirals are very sensitive to "pea-sized" mineral matter, as such particles tend to settle in the turns of the spiral and divert slurry flow. The metalliferous mineral engineer may baulk at the prospect of wet screening 1200 tonne/h of solids at 20 mesh, but such a step has proven necessary in at least two large mineral sands plants. Such screening can currently be undertaken economically with commercially available screening equipment.

Almost without exception, wet gravity separation equipment is extremely sensitive to the presence of excessive slimes. While a small amount of minus 400 mesh material is generally acceptable, quantities in excess of 5% should be avoided. More than 10% minus 400 mesh causes serious separation problems for two reasons. The first, and probably the most serious, is that slimes increase the viscosity of the slurry and hence reduce the sharpness of the separation. The second problem is one of operator vision where split points depend upon good visibility (such as wet tables and spirals). Slimes can usually be dealt with by cyclones, hydrosizers or dewatering screws. The choice will depend upon the friability of the valuable mineral(s) and upon the economics of the operation. Cyclones cost less per unit throughput than hydrosizers and dewatering screws, but generate more interparticulate abrasion. Slime problems are usually directly interrelated to water supply problems, as water balance is always an important criterion in wet gravity plants. Usually a considerable percentage of plant water usage is recycled plant water, and appropriate steps must be taken to minimise slime build-up in the recirculated water supply.

If one examines the few gravity plants still in existance from the last century, one aspect of plant design is immediately apparent. Practically all plants were built on the side of a hill, cliff or mountain to allow maximum utilisation of gravity. With modern day ores this may seem equally appropriate, but such is true only in exceptional cases. The old plants mentioned had, in general, coarse mineral liberation and only moderate recovery of fines. Most plants had few, if any, middlings cuts because of the difficulty of recirculating middlings uphill. The low recoveries were of little consequence when one considers the economics of the times. Such older plants, in addition, did not have access to high efficiency rubber lined centrifugal pumps and to hydrocyclones. When considering the plant design of modern day gravity units, the lessons of the old-time milling engineers should not be forgotten - i.e. when feasible, use natural forces to conserve energy. There has been a tendency to build gravity plants on completely level bases over the last few decades, and with increasing power costs a compromise is obviously required.

Let us consider a common example. The use of large spiral plants has been fairly commonplace. The design of these plants (which are usually three-stage units) has generally been similar. The plants tend to be of the three-level type. On the base or datum level are all sumps and pumps; on the level above, the spirals; on the uppermost level, the pulp distributors for the spirals. This may be justified in the case of floating plants, since stability is important. It cannot usually be justified, however, for land based plants where there should be two or more levels of spirals. An arrangement might be roughers on an upper level with scavengers on another, lower level. This two-level spiral arrangement would thus have a considerable energy saving over a single-level one, as can be shown by simple calculations. The two-level spiral plant will be somewhat higher than a single-level one, but its building cost will not differ very much as it covers considerably less floor space. This argument is less valid for Reichert cone plants, since a single cone unit is nearly five times the height of a single spiral unit, but should nevertheless be considered.

Where the gravity separation device has moving parts (particularly reciprocating parts), the cost of supporting such machines above ground or datum level may be quite substantial. The problem here is one of structural vibration and potential structural resonance. Shaking tables and vanners are the obvious unit processes for ground level installation, since their motion is horizontal. Jigs are less critical in this regard as their motion is vertical. Suspended tables such as the Deister '999' (triple deck) and the Bartles-Mozley (multiple deck) have considerable advantages from the plant construction viewpoint as their contribution to building vibration is very small to negligible. From a pure structural standpoint it is hence indicated that units with no moving parts should occupy the highest levels of a plant, while those with the greatest horizontal vibration should occupy the base or ground level. In all cases the economics must be optimised - consideration being given to both plant construction cost and energy conservation.

It must be borne in mind that the reduction of slurry pumping to a minimum not only reduces energy consumption, but also reduces slimes generation. This is relevant to our subject as many minerals concentrated by gravity are notoriously friable. This brings up the matter of pump selection which, while not the subject of this chapter, deserves certain comments. If the minerals to be concentrated are very friable, then the following generalisations are appropriate. Slurry pump discharge velocities should be as low as possible while maintaining sufficient velocity for solids suspension. This has the additional advantage of reducing wear on pump parts. In as much as many ores processed by gravity concentration are low grade, with the bulk of the feed as silica or silicious gangue, the selection of abrasion resistant pumps is almost mandatory. This author has a preference for rubber lined pumps for gravity plants. The vertical type of pump, such as that made by Sala, has the dual advantage of being glandless and of having a sump integral with the pump. However, the complex water balances of gravity plants often require the use of peripheral discharge sumps or tanks in order to increase pulp density, and horizontal pumps perform better under such circumstances.

The last paragraph alluded to one of the most important considerations in the design of a gravity plant. This is the problem of water balance. Almost all gravity separation devices have an optimum feed pulp density (e.g.cones 60% solids; spirals, 30% solids; tables, 25% solids), and these devices are quite sensitive to variations from these optimum figures. Proper feeding of gravity equipment thus requires accurate pulp density control. In addition, unless a plant has unlimited fresh water supply, it is necessary to recycle large quantities of water (much of which is contaminated with slimes). This requires adequate settling ponds or thickeners, fine wet screens, hydrocyclones or dewatering screws (screw or spiral classifiers).

Let us first consider the subject of pulp density control. Effective pulp density control is essential throughout gravity plants, but nowhere is it more important than on the plant raw feed. This is true regardless of the primary concentration step, but it is probably most critical on Reichert cone concentrators because of the high pulp density required for feed. As noted earlier, the optimum feed pulp density for cones is about 60% solids. While there are many claims regarding the tolerance of cones to pulp density variations, most cone separations fall off rapidly in efficiency if the pulp density variation is greater than ±5% solids. While ±5% may be rather empirical, it happens to apply well as a tolerance guide for almost all gravity concentration devices and is useful as a practical and engineering guide. Cones, spirals, tables and most other gravity devices require water additions at some stage of the process, and the combined pulp density of the products is thus always less than the pulp density of the feed.

The manner in which pulp density control is attained in a gravity plant depends upon several considerations. The first is the size of the plant. In small plants it may be possible to control pulp density by some manual method, but any plant treating more than about 25 tonne/h should have automatic pulp density control on the new feed. Much has been written on this subject, and the choice of type will depend upon the nature of the plant. The most foolproof method of pulp density control at this time appears to be the use of an electromagnetic flowmeter and nuclear (γ-ray) density gauge to determine pulp density, with feedback to increase or decrease water addition to the feed sump. While such a control system may appear to be expensive, its long term financial effectiveness is almost unquestionable.

In the plant itself several options are available for pulp density adjustment. Obviously pulp dilution (density reduction) is simply a matter of water addition. Where the desired increase in density is small (say 20 to 30% solids) the cheapest and most effective method is the use of peripheral discharge sumps. Where more than one of these units is used, all of the overflows should be directed to a central large sump of the same design or to a thickener. This prevents material loss since there is always some solids carry-over from overflow sumps. Where the increase in pulp density is substantial (say from 25 to 60% solids) there are two basic choices available. The less expensive is the use of hydrocyclones, while the more expensive alternative is to use thickeners. Where very high pulp densities must be attained (e.g. 80% solids for scrubbing or attritioning), the alternatives are dewatering screws (spiral classifiers), dewatering screens (such as the Derrick-Linatex) or in extreme cases, centrifuges. Centrifuges are rarely justified because of their high operating cost on even moderately abrasive material. Of the density adjustment methods mentioned above, two are particularly useful (although expensive) since they add surge capacity to a plant. These two are thickeners and dewatering screws.

There remain a small number of ancillary items that deserve mention. Pulp distributors for gravity plants should be as simple and effective as possible. Rotary distributors tend to cause intermittent flow to separating units if the number of units is large. The stationary type of distributor (Humphreys or Mineral Deposits) is the most effective where a large number of units must be fed. For small divisions, such as a three-way split to the three decks of a Deister '999' wet table, the rotary type of splitter made by Deister is more appropriate.

Multi-spigot hydraulic classifiers for feed preparation for wet tables and the like come in two varieties. The completely automatic type, such as the Stokes, needs less attention than the manually controlled Deister or Denver, but costs considerably more. Whichever type is chosen, it is essential that the water additions (called "hydraulic water") are from a steady head source and that the feed rate is uniform and of a sufficiently high pulp density (50 to 60% solids).

Choice of filters for gravity plants depends upon the nature of the final products. Often gravity plant concentrates are sufficiently porous that horizontal disc filters can be used successfully. These are less expensive to operate than conventional drum or disc filters.

Adequate automatic sampling is essential in wet gravity plants. The number of sampling points will depend upon the type of plant.

Finally, it should be noted that all launders should be rubber lined, as should all wear points, and that right angle bends in pipework should be avoided completely.

WET GRAVITY SEPARATION EQUIPMENT

As in all areas of mineral dressing, the tendency in gravity separation in recent years has been towards larger and larger units. In an excellent recent paper Terrill and Villar (11) summarised the trends in gravity separation plant design and it would be pointless to duplicate such a fine article. In the following, the large scale gravity units will be enumerated, and additional reading material suggested.

The largest commercially available jig is the IHC - Cleaveland 25, a circular jig of diameter 7.5 m with a nominal capacity of 150 to 300 m^3/h. This unit has been described by Zaalberg (12) and Ziegler (13). There are at least eighteen of these units in operation in the far east. For coal cleaning, Zimmerman (14) has described the Batac jig which is an improved Baum type jig.

For high capacity treatment of finer sized low grade ores, the Reichert cone is becoming an industry standard. It is well described in papers by Ferree (15,16), Giffard (17), Graves (18) and Nel (19). A new type of sluice called the Wright impact plate gravity concentrator (20) has recently become available and may offer some of the advantages of the Reichert cone.

For smaller scale plants, and for concentrate cleaning in cone plants, the spiral still appears to have no competitor for efficiency, ease of operation and economy. The 2-start units economise on space and are offered by Mineral Deposits, Humphreys and Reading U.S.. Again see Giffard (17), Graves (18) and also Li (21).

The highest capacity wet shaking table on the market is the Deister '999' triple-deck which is seeing ever-increasing use. Tiernan (22) has detailed the Deister table performance on US coals.

In the very fine particle size range, the Bartles-Mozley table has overcome the problem of the large areas normally needed for film sizing by using a large number of stacked decks. The unit is well described by Mozley (23) and Burt and Ottley (24).

REFERENCES

1. Aplan,F.F.,"Gravity Concentration",<u>Research Needs in Mineral Processing</u>, Somasundaran,P. and Fuerstenau,D.W., Eds., National Science Foundation, New York, 1976,193p.

2. Staff, United States Bureau of Mines, <u>Minerals Yearbook 1974 Vol.1</u>, United States Government Printing Office, Washington,D.C.,1976.

3. Leonard,J.W. & Mitchell,D.R.,(Eds.),<u>Coal Preparation</u>, 3rd Edn.,AIME, New York, 1968,pp.4-23 to 4-30 and pp. 18-3 to 18-29.

4. Symington,R., Higginbotham,G.H. & Armstrong,F.,"The Performance of Coal preparation plant",<u>Proc. 2nd. Symposium on Coal Preparation</u>, University of Leeds, 1957,pp.353-377.

5. Jones,M.P.,"Mineral Dressing Tests on the Extraction of

Columbite and other Heavy Minerals from the Odegi Younger Granite",Records of the Geological Survey of Nigeria,1967,pp.36-59.

6. Henley,K.J.,"The Quantitative Mineralogical Evaluation of Alluvial Ores",Australian Mineral Development Laboratories Bulletin, Vol.14, 1972,pp.21-31.

7. Collins,D.N. & Read,A.D.,"The Treatment of Mineral Slimes",Minerals Science and Engineering,Vol.3, No.2, 1971,pp.19-31.

8. Mills,C.,"The Recovery of Cassiterite Slimes by Oil Phase Extraction",M.Sc. Thesis, Colorado School of Mines, 1974, 122pp.

9. Moncrieff,B.A. & Lewis,P.J.,"Treatment of Tin Ores", Transactions Institution of Mining and Metallurgy, Vol.86, 1977, pp.A56-A60.

10. Mineral Deposits Limited,"Reichert Cone Concentrators", Company bulletin No.RCC.1,1975,12pp.

11. Terrill,I.J. & Villar,J.B.,"Elements of High-Capacity Gravity Separation", The Canadian Mining and Metallurgical Bulletin, Vol.68, No.757, 1975,

12. Zaalberg,P.H.A.,"Offshore Tin Dredging in Indonesia", Transactions Institution of Mining and Metallurgy, Vol.79, 1970, pp.A86-A95.

13. Ziegler,R.B.,"Dredge Mining Methods and Equipment for the Exploitation of Alluvial Mineral Deposits", Paper presented at Placer Exploration and Mining Short Course, University of Nevada, Reno, 1976, 38pp.

14. Zimmerman,R.E.,"Batac Jig, a new improved Baum type Jig", Mining Congress Journal, Vol.60, No.5, 1974, pp.43-49.

15. Ferree,T.J.,"An Expanded Role in Minerals Processing seen for the Reichert Cone", Mining Engineering, Vol.25, No.3, 1973, pp.29-31.

16. Ferree,T.J.,"Introduction to the Reichert Cone", Paper presented at AIME-MBD annual meeting, Colorado Springs, Colo., 1972, 15pp.

17. Giffard,P.J.,"Technological Development for Sands Mining", Australian Mining, 1972, pp.61-62.

18. Graves,R.,"Gravity Separation Technology by Mineral

Deposits Ltd., Australia", Mines Magazine, Vol.62,No.2, 1972.

19.Nel,V.,"Palabora's new Heavy Minerals Plant adds Uranium Concentrate to the Recovery list", Engineering and Mining Journal, Vol.173, No.11, 1972.

20.Reading U.S.,Inc.,"The Wright Impact Plate Gravity Concentrator", Company bulletin IPC 475, 1975.

21.Li,T.M.,"Startup of Manchester Mine and Mill boosts U.S. Production of Primary Ilmenite", Engineering and Mining Journal, Vol.174, No.12, 1973, pp.71-75.

22.Tiernan,C.H.,"Production of Low-Ash Coking Coal by Tabling with Special Reference to U.S. Practice", The Canadian Mining and Metallurgical Bulletin, Vol.66, No.732, 1973, pp.84-91.

23.Mozley,R.H.,"A Gravity Concentrator for Fine Minerals", A Technical Conference on Tin, The International Tin Council, London, 1967, pp.79-98.

24.Burt,R.O. and Ottley,D.J.,"Fine Gravity Concentration using the Bartles-Mozley Concentrator",International Journal of Mineral Processing, Vol.1, 1974, pp.347-366.

APPENDIX: EXAMPLE OF HEAVY LIQUID ANALYSIS

Let us consider a hypothetical gravity project and examine the various phases of development. The example chosen is a low grade scheelite bearing contact metamorphic skarn, assaying 0.3% WO_3.

Mineralogical investigation has shown that the scheelite is nominally liberated at 48 mesh (300 μm), but that scheelite contained in the minus 400 mesh (38 μm) fraction is inter-grown in a complex manner with several other minerals, notably oxides of iron.

Market considerations require that a salable product must contain at least 20% WO_3, and reserves suggest that a 100 tonne/h plant is the most appropriate size.

At this point the single most important question is - can a concentrate assaying 20% WO_3 be produced from this ore by gravity separation? The question really must be answered as inexpensively and accurately as possible before the project progresses any further.

The best approach, and certainly the least expensive, is heavy liquid analysis. A representative head sample is wet ground to liberation size (300 μm), and deslimed at 38 μm. Sequential heavy liquid separation is then performed on the ground deslimed head sample. In this hypothetical example, the specific gravities used are 2.7, 2.8, 2.9, 3.0, 3.4, 3.8, 4.2, 4.6 and 5.0. Starting at s.g. 2.7, the floats are sequentially removed, weighed and assayed for WO_3. Table 3 shows the possible results of such a procedure. There are several possibilities for the combination of liquids needed for the sequential separation. Probably the most cost-effective is to use diluted TBE for the separations at 2.7, 2.8 and 2.9, and Clerici solution for the remaining separations. Since 88.1% of the sample would float at s.g. 2.9, a fairly large sample could be used for the heavy liquid analysis - 400g would be appropriate. In this case only 47.6g would remain for the Clerici solution separations and the smallest sample for assay would be 0.8g (sinks at s.g. 5.0). A brief inspection of the data in Table 3 indicates that plotting the entire cumulative % sinks, cumulative % floats, characteristic assay, specific gravity and s.g. tolerance curves would not be particularly illucidating, because of the range of values involved. This would be true of almost any low grade ore. However, the information required can be extracted by examining a small portion of the curves, and these portions can then be drawn to a much larger scale than is usual. Figure 3 shows the cumulative % sinks curve and the specific gravity curve in the range of interest, and Figure 4 the cumulative % floats curve.

Table 3 shows that a concentrate assaying 20% WO_3 can be produced from the ore. However, the table also

TABLE 3. Hypothetical Results of Heavy Liquid Analysis

Specific Gravity	SINKS						FLOATS			Characteristic Assay	
	% wt	Cum %wt	% WO_3	WO_3 units	Cum WO_3 units	Cum % WO_3	Cum %wt	Cum WO_3 units	Cum % WO_3	% WO_3	Av. Cum % wt. of sinks
+5.0	0.2	0.2	42.5	8.500	8.500	42.5	99.8	21.348	0.214	42.5	0.10
-5.0+4.6	0.3	0.5	26.4	7.920	16.420	32.8	99.5	13.428	0.135	26.4	0.35
-4.6+4.2	0.3	0.8	12.0	3.600	20.020	25.0	99.2	9.828	0.099	12.0	0.65
-4.2+3.8	0.4	1.2	0.92	0.369	20.389	17.0	98.8	9.459	0.096	0.92	1.00
-3.8+3.4	2.3	3.5	0.588	1.352	21.741	6.2	96.5	8.107	0.084	0.588	2.35
-3.4+3.0	4.0	7.5	0.321	1.284	23.025	3.1	92.5	6.823	0.074	0.321	5.50
-3.0+2.9	4.4	11.9	0.300	1.320	24.345	2.05	88.1	5.503	0.062	0.300	9.70
-2.9+2.8	19.6	31.5	0.085	1.666	26.011	0.83	68.5	3.837	0.056	0.085	21.70
-2.8+2.7	58.5	90.0	0.062	3.627	29.638	0.33	10.0	0.210	0.021	0.062	60.75
-2.7	10.0	100.0	0.021	0.210	29.848	0.298	-	-	-	0.021	95.00
	100.0	-	0.298	-	-	-	-	-	-	-	-

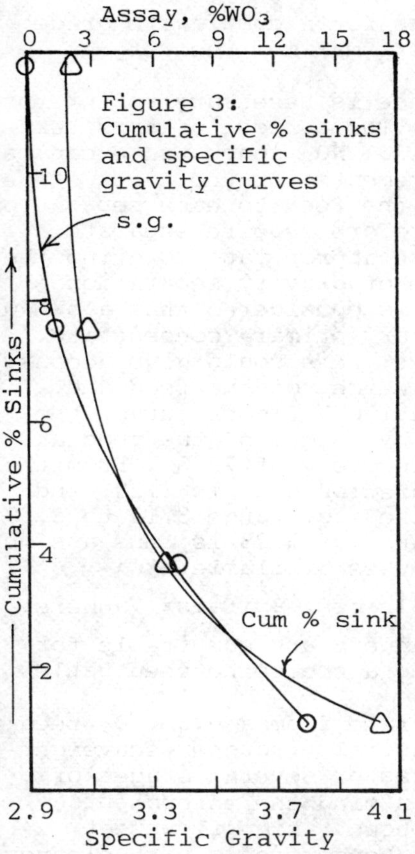

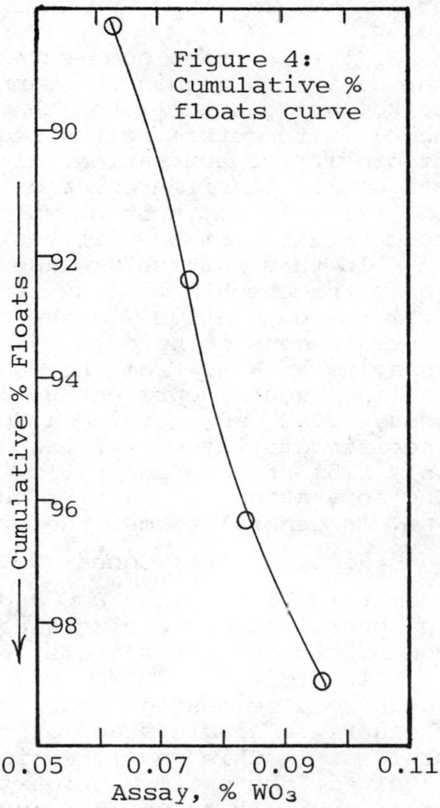

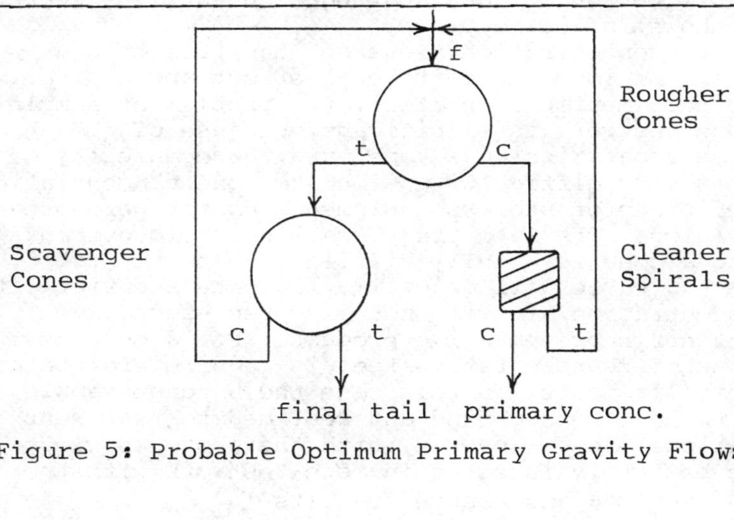

Figure 5: Probable Optimum Primary Gravity Flowsheet.

shows that the cumulative % sinks for a cumulative grade of 20% WO_3 is 1.05%, which gives a maximum recovery of 70.5%.

At this point some experience is necessary to interpret the data in a meaningful manner. The purpose of this testwork (having established that a 20% WO_3 concentrate can be made) is to determine the requirements for the first stage of gravity concentration, since the feed to each separation stage will be different and therefore require separate analysis. It may, for example, be appropriate to grind the concentrate from the first stage of gravity separation.

In this particular case it is considered that a 5% WO_3 concentrate would be a satisfactory primary concentrate. From the cumulative % sinks curve, this would give a concentrate consisting of 4.35% by weight of the feed and assaying 5% WO_3. From the cumulative % floats curve, the tailings would represent 95.65% by weight of the feed and assay 0.082% WO_3, giving a WO_3 recovery of 73.6%. From the specific gravity curve, the separation s.g. is 3.28, and only 1.5% of the feed falls in the s.g. range 3.18 to 3.38. The separation, if carried out at s.g. 3.28 is thus a simple one. In general terms, the processes available are:-

Jigs; Tables; Cones; DMS; Sluices; Spirals; Vanners.

From this list, jigs, DMS and vanners are unsuitable for reasons of particle size, leaving a choice between tables, cones, sluices and spirals.

To produce a 5% WO_3 concentrate from a 0.3% WO_3 feed would require 3-stage cones or spirals (rougher-scavenger-cleaner), a single stage of tables or several stages of sluices. In this case, bearing in mind the throughput involved, the optimum primary concentration flowsheet would probably be rougher and scavenger cones with cleaner spirals as shown in Figure 5.

A primary concentration flowsheet has thus been proposed with the minimum expenditure of effort and capital.

As has been mentioned earlier, the problem of sample size for cone testwork is significant. Because of the nature of the separation in a cone, and the complexity of stages, it is very difficult to simulate cone concentration on any other piece of process equipment, except perhaps a series of sluices. The laboratory scale unit closest in operation to a cone is a fully riffled wet shaking table. Although such a table will generally indicate the viability of cone concentration, it will not give representative performance, nor representative products. For a cone test about 5 tonne of representative feed is required for batch or closed circuit tests. In this case the 5 tonne sample would need to have been ground and deslimed to represent plant operational feed. However, with the above knowledge, capital can be safely invested in cone testwork with the certainty of success and assured results.

Chapter 19

EXAMPLES OF GRAVITY CONCENTRATION FLOWSHEETS

by Frank L. Bazzanella and Paul A. Weyler

Manager of Metallurgy
AMAX Arizona
Tucson, Arizona

Chief Metallurgist
AMAX INC.
Henderson Mill
Parshall, Colorado

ABSTRACT

A brief introduction to gravity concentration is presented, including some discussion on the more common minerals currently recovered. The important considerations in basic design of a gravity plant are mentioned and four currently operating flowsheets are presented. The plants discussed in detail include the Climax Molybdenum Co. By-Product Plant and the Canada Tungsten Ltd. Plant. Plants briefly highlighted include the Palabora Mining Co. Uranium Recovery Facility and the Bridge Hill Ridge Heavy Mineral Recovery Plant of Mineral Deposits Ltd.

INTRODUCTION

Gravity concentration is one of the oldest mineral dressing processes devised and remains an important method of physical concentration. Significant improvements have been made in the equipment over the past 25-30 years, and many modern high capacity plants have been brought on stream to recover such important commodities as tungsten, titanium, tin and diamonds, in addition to the more common iron and coal processing plants. In South Africa, Palabora Mining Company is recovering uranium as uranothorianite by gravity to provide feed for a chemical plant.

The primary purpose of this paper is to present some gravity flowsheets currently in use and to attempt to point out some of the thoughts and considerations that were involved in the design of these plants.

Minerals Recovered

As mentioned previously, many commodities are presently recovered by gravity concentration methods. The following is a brief discussion of the metals and mineral types recovered.

Tungsten. The major minerals of tungsten that are recovered by gravity or partially by gravity are:

Scheelite	$CaWO_4$	SG 5.9-6.1
Wolframite	$(FeMn)WO_3$	SG 7.2-7.5

Other minerals of tungsten include ferberite ($FeWO_4$) and huebnerite ($MnWO_4$), plus a host of minor minerals, including a tungsten sulfide called tungstenite, which resembles molybdenite in appearance.

Tin. The only commercially valuable tin mineral that is recovered at this time is cassiterite (SnO_2) with a specific gravity of 6.8-7.1. Only minor amounts have been produced in the United States at Climax Molybdenum Co., and in Canada, at Timmins, Ontario. Most of the world tin production comes from the Malaya States, Bolivia, East India, China, Thailand, Nigeria, Australia, South Africa and Cornwall England.

Titanium. Titanium is recovered in the form of two primary minerals:

Ilmenite	$FeTiO_2$	SG 4.5-5.0
Rutile	TiO_2	SG 4.2

A major portion of the titanium production comes from dredging and gravity concentration of beach sands. World production is found in the eastern United States, India, Norway, Brazil, Canada and Australia.

Iron. Where the mineralogy permits, iron ore is still concentrated by gravity methods, mainly spirals. Jigs and tables were popular in the past, but operating problems and limited capacity have tended to phase them out. Also, the larger tonnages of iron recovered from finely disseminated magnetite (taconite) ore bodies have placed gravity separation in the shadow of large magnetic recovery installations, but it remains an important method of recovery. Commercially important iron minerals include:

Hematite	Fe_2O_3	SG 4.9-5.3
Magnetite	$FeO-Fe_2O_3$	SG 5.2
Goethite	$Fe_2O_3.H_2O$	SG 4.3
Marlite	Fe_2O_3	SG 4.8-5.3 (minor)
Siderite	$FeCO_3$	SG 3.8-3.9 (minor)

Coal. At present, large tonnages of coal are processed by heavy media jigging operations to remove sulfur, usually as pyrite, plus other undesirable ash producing materials. This treatment is necessary to provide the low sulfur coals required by modern industry.

Gold. Historically, gravity concentration was the primary method of processing gold ore. However, in modern times, cyanidation has overshadowed gravity as a method of gold recovery.

Miscellaneous. In addition to the above, there are assorted minor minerals that are recovered by gravity methods. Columbium and tantalum are recovered in small quantities and, as mentioned earlier, uranium is being concentrated by gravity methods at Palabora Mining Co.

Types of Equipment

Over the past several hundred years numerous devices have been invented for the gravity concentration of minerals. Of the many devices introduced, a few have survived history while some new devices have been invented that have replaced some of the old. For the most part, the principles and basic equipment design of 100 years ago are still used today; only the application and particular equipment design have been modified. A previous chapter discusses specific equipment in detail.

DESIGN CONSIDERATIONS

When considering recovery of a mineral by gravity means, one of the most important considerations involves the design of feed preparation and handling facilities for the particular gravity equipment. The relative particle size distribution of the valuable constituent compared to the gangue constituent is very important as is the degree of liberation and liberation size of the valuable mineral. In many grinding circuits, for instance, overgrinding of the heavy mineral results from using hydraulic classifiers such as cyclones. In the case of by-product plants, the grinding circuit control is directed toward optimum liberation of the primary mineral and problems from overgrinding must be solved in other ways. In this instance, particular attention must be directed toward proper sizing of feed products to eliminate coarse gangue that will compete with finely ground values, to prevent interference from slimes, and to enhance the recovery of the by-product.

Other parameters, such as the per cent solids, must be closely examined. It is very simple for a plant water balance to be thrown completely out of control due to attempts to adjust per cent solids or utilize devices that require large volumes of dilution water.

The importance of sizing, particularly in cleaning operations such as tables, is pointed out by the fact that a quartz particle of 200 mesh will compete with a wolframite particle of 400 mesh or finer for the same site. When viscosity effects are taken into account, it is very possible the wolframite particle will report to tails and the quartz particle to concentrate, depending on the dynamics of the machine being utilized.

Most gravity recovery devices are very sensitive to the viscosity of the pulp; therefore, the per cent solids and the amount of viscosity producing slimes can become very important operating parameters. Proper adjustment of per cent solids, wash water addition on spirals and tables, and removal of slimes are mandatory. Related to the viscosity and flow consideration, of course, is feed rate. Higher viscosity pulps will necessitate a low feed rate in many operations because the higher viscosity hinders the transfer of the valuable mineral to the recovery area of the particular machine. Shaking tables, for example, are extremely feed rate sensitive, and interfering slimes, oversize and other heavy minerals can be very detrimental to the table operation.

In order to design a successful gravity flowsheet, close attention to the effects of other unit processes must be taken into account. These include thickening, flotation, heavy media, drying, magnetic separation, screening, grinding, and water conditioning. As with other milling processes, attention must be given to product handling and tailing disposal. One of the advantages of gravity operations is that, in most cases, very little or no reagents are added to the ore. Tailing disposal, at least as far as water pollution is concerned, is not too serious a problem.

Concluding this very brief and very general review of gravity concentration will permit dedicating the remainder of the paper to a more detailed discussion of currently operating gravity plants. These will include plants operated by Climax Molybdenum Co. in Colorado, U. S. A.; Canada Tungsten Ltd. of NW Territories, Canada; Palabora Mining Co., South Africa; and Mineral Deposits Ltd., Australia.

CLIMAX MOLYBDENUM COMPANY BY-PRODUCT PLANT

General

The Climax Mine in Colorado treats approximately 40,000 DSTD of molybdenum plant tailings through a gravity concentration by-product plant. This plant recovers tungsten as wolframite and tin as cassiterite. A schematic of the flowsheet is illustrated in Figure 1, and Table I contains a list of major equipment.

Briefly, the molybdenum plant rougher tails are pumped and distributed to 742 Humphrey spirals where a bulk gravity concentrate is produced. This bulk concentrate is pumped to dewatering cyclones, and the cyclone spigot is conditioned for pyrite flotation. After rougher and cleaner pyrite flotation, both non-float products are pumped to a cleaner spiral section consisting of 96 Humphrey spirals. The cleaner spiral concentrate is fed to the tabling circuit. All table middlings are recycled and table concentrates are advanced to dewatering and iron scavenging on permanent drum magnets. The non-magnetic fraction is dewatered in a screw classifier, then heated and conditioned with reagents for flotation of phosphate as monazite - $(Ce,La)PO_4$ - which is currently discarded to tails. The rougher and cleaner concentrates are scavenged on a shaking table for tungsten values, and the non-float is advanced to dewatering, drying and screening for preparation as feed to the cross-belt magnetic separators. Three banks of Wetherill type Dings magnetic separators produce a cleaned final tungsten product, a tin concentrate and a non-magnetic heavy mineral concentrate that is sold to third parties for further treatment.

Several products are directed to a clean-up ditch and are returned to the head of the circuit. The only materials of interest here are the tungsten and tin but the slurry contains varying amounts of pyrite, monazite, garnet and other minor minerals.

Equipment and Design Considerations

The plant can be divided into three main areas as far as equipment and design are concerned. The rougher-cleaner spiral area, the table area and final upgrading area.

Spiral Area. The spiral area consists of the 12 x 12 Hydroseal feed pumps, the pulp distributors, 742 rougher spirals, pyrite flotation and 96 cleaner spirals.

In the spiral area, the most important design considerations involved the distribution system to feed 838 spirals, and the launder system necessary to collect the products from each of these spirals. The distribution system consists of a series of compartment type distributors to feed each row of spirals. The rows consist of two rows each, back to back, with walkways between each double row. This permits fitting several spirals into a smaller area and consolidating the distribution and product launder systems.

The spiral rougher concentrate is pumped to four 10 inch cyclones with two 5 in. x 5 in. pumps. The cyclone underflow is conditioned with sulfuric acid for pH adjustment, and the pyrite is floated using xanthate and pine oil. The pyrite rougher and cleaner non-float fractions become the table feed. The plant is designed either to collect the pyrite concentrate and store it for local sale or discard it to tails.

Table Area. The table area consists of table feed pumps and twenty No. 6 Deister tables. Sixteen of these tables are rougher tables, three are middling tables and one is a cleaner middling table. Concentrate from all tables is advanced while middlings are recycled and tailings are discarded. Many modifications over the past several years have resulted in this present table circuit.

In addition to these tables, a one-half size table is used for scavenging tungsten from the phosphate concentrate. The amount of tungsten recovered is quite small, but over several years of observation, the operators have determined that significant amounts of tungsten can be lost by lack of attention to relatively minor details. Particularly important are losses that occur during and after the tabling of the material. At this point, the grade is quite high - plus 40 per cent tungsten.

All tables except the scavenger table are on one level of the plant. This makes routine inspection by the plant operator convenient and encourages more attention in this critical area of the plant. Because of their sensitivity to particle size and specific gravity, it is very important that regular inspections of table operations are carried out. Easy access and good visibility encourage regular inspection.

Final Upgrading Area. This small area of the plant contains a wide variety of equipment required to increase the grade of the table concentrate to a high grade saleable product with a minimum sacrifice in recovery. Equipment includes dewatering cyclones, wet drum magnetic separators, dewatering classifiers, flotation cells for monazite removal, drying, screening and crossbelt magnetic separators for final upgrading. Also included in this area are assorted small settling tanks and thickeners for recovering tungsten from the overflows of the various dewatering steps in this area. These are not shown on the flowsheet.

The most important design considerations in this area are:

1. Bin and feeder design for proper performance of the crossbelt magnetic separators. In order for the machines to do a satisfactory job, the material being treated must be evenly distributed on the feed belt in a very thin layer. Since the material is bone dry at this point, it flows easily providing for good control and uniform feeding to the belt. It must, however, be controlled or flooding of the belt can occur. A modified star feeder is used in one case and a vibrating feeder is used in another case at Climax to provide the desired feed rate and distribution to the belt. Both are satisfactory.

2. Careful design of belt and material handling systems to prevent losses of extremely valuable material is mandatory. In a by-product plant such as this, where tens of thousands of tons must be treated to recover a few thousand pounds of valuable product, belt design and speed, bin design and load-out facilities must be given particular attention.

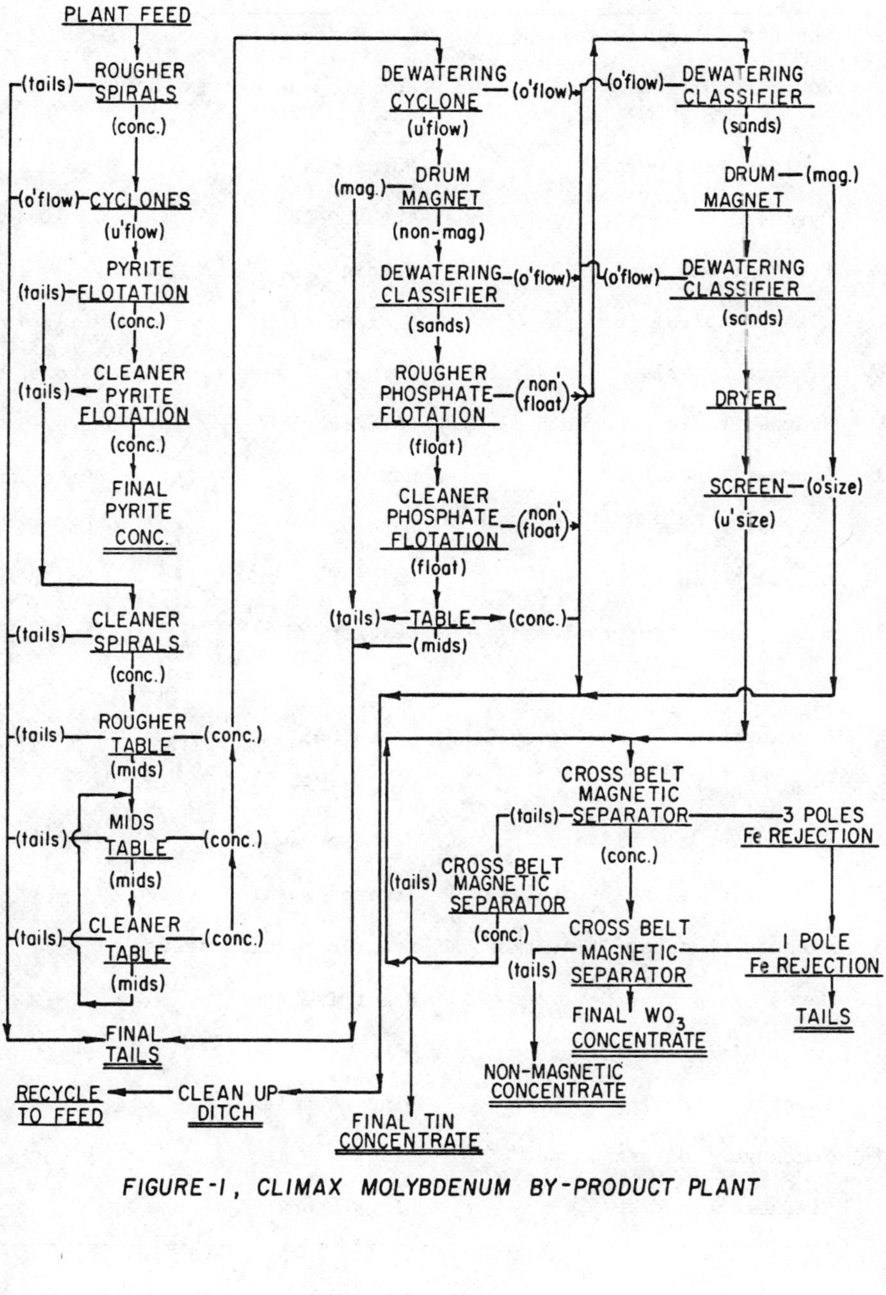

FIGURE-1, CLIMAX MOLYBDENUM BY-PRODUCT PLANT

TABLE I. Climax By-Product Plant - Table of Major Equipment

Qty.	Service	Type	Size
2	Rougher spiral feed pumps	ASH Hydroseal	12 in. x 12 in.
742	Rougher spirals	Humphrey 5-turn, single start	
2	Cyclone feed pumps	Horizontal	5 in. x 5 in.
4	Pyrite feed dewatering cyclone	Krebs D-10	10 in.
12	Pyrite flotation cells	Weinig	-
2	Cleaner spiral feed pump	Horizontal	5 in. x 5 in.
96	Cleaner spirals	Humphrey, 5-turn, single start	
1	Rougher table feed pump	Horizontal	5 in. x 5 in.
1	Rougher table feed pump	Horizontal	6 in. x 6 in.
16	Rougher tables	Deister	No. 6
3	Middling tables	Deister	No. 6
1	Middling cleaner table	Deister	No. 6
2	Cyclone feed pumps	-	-
2	Concentrate dewatering cyclones	Krebs D-4	4 in.
2	Drum magnet	Eriez permanent	
4	Dewatering classifiers	Wemco screw	12 in.
16	Rougher phosphate flotation	Stearns machine design	-
8	Cleaner phosphate flotation	Stearns machine design	-
2	Magnet feed dryer	Holoflite steam	-
1	Dryer discharge screen	Sweco	48 in.
1	Rougher magnet	Dings 8 pole	
1	Scavenger and tin magnet	Dings 4 pole	
1	Cleaner magnet	Dings 5 pole	

Summary

With the commissioning of the Climax By-Product Plant, Climax became the second largest producer of tungsten in the United States. Although the production volume is quite low, the attractive price of this by-product makes the recovery effort very economically attractive.

Recent modifications at Climax not shown in the flowsheet involve the installation of a module of Reichert Cone Concentrators. These devices have performed well, according to reports, and additional installations are a possibility in the future, particularly as replacements for spirals that are nearing the end of their useful life.

Four of the units are being used as rougher recovery devices, relieving some of the load on the original spiral circuit. In addition, another cone has been used to replace the cleaner spiral section. This results in having the original cleaner spirals available for use as roughers, further reducing the load on the spiral circuit. This reduction in load is expected to further improve recovery of valuable minerals.

CANADA TUNGSTEN MINING CORP., LTD.

General

The Canada Tungsten mining facilities are located in the Flat River Valley of the Mackenzie Mountains in the Northwest Territories of Canada.

The initial wooden structure of the concentrator, which was completed in October 1962, was totally destroyed by fire on December 26, 1966 and the new mill was built and put on stream on November 26, 1967. Since that time, various changes have been made in the flowsheet such as installing a cyclone in the grinding circuit, adding a regrind mill, and installing cyclones to provide sized material to the shaking tables. In addition a talc flotation step has been added and other modifications have been made to make the mill more efficient and to permit handling the changes in the ore characteristics.

The primary valuable mineral at Canada Tungsten is scheelite and the secondary valuable mineral is chalcopyrite. Impurities that must be removed include pyrrhotite, talc and calcite. The first two are removed by flotation and the latter by leaching. Two products are produced; Grade I which is produced in the gravity circuit and assays about 75 per cent WO_3 and Grade II, which is produced in the flotation circuit and assays about 25 per cent WO_3. This product is upgraded to 65 per cent WO_3 by leaching of the calcite.

Since scheelite is a friable mineral, considerable attention has been given to the grinding circuit to prevent overgrinding. Initially, the circuit consisted of 7 ft. x 10 ft. rod mill and 10.5 ft. x 6 ft. ball mill in closed circuit with Sala trommel screens. The cloths on the screen were 24 and 65 mesh, but various other sizes were tried in an effort to increase throughput and to remove the material at the liberation grind. At present, 24 and 50 mesh cloths are being used.

Also, a 15 in. cyclone was installed before the screens to remove the fine fractions. At that time, the overflow went to the 40 ft. thickener, but, with the increased talc content of the ore, a separate flotation circuit was installed to handle this product. The cyclone overflow amounts to 25 per cent to 30 per cent of the mill feed and contains approximately 80 per cent of the talc. The talc is detrimental to the scheelite flotation concentrate and cannot be leached by acid.

Equipment and Design Considerations

The plant is divided into different circuits, each interconnected through circulating loads. The flowsheet is shown in Fig. 2 and the major equipment is listed on Table II.

1. Grinding and classification circuit
2. Talc flotation circuit
3. Chalcopyrite flotation circuit
4. Pyrrhotite flotation circuit
5. Gravity circuit
6. Regrind circuit
7. Scheelite flotation circuit
8. Roasting and magnetic circuit

Briefly, the 15 in. cyclone overflow from the grinding circuit is fed to the talc flotation cells. The froth goes to the tailing pond and the non-float becomes part of the feed to the scheelite flotation section.

The Sala screen under size is thickened, conditioned and the chalcopyrite floated . In this circuit, a copper concentrate is produced, the cleaner tail is rejected to the final tails and the rougher tail is advanced to pyrrhotite flotation. At this point the pyrrhotite is floated and reports to the tailing pond. The non-float is advanced to the gravity circuit.

The tables in the gravity circuit are Concenco Deister "666" triple deck or No. 6 single deck. It should be noted that none of the table products are sent to the tailing pond but are sent to the regrind plant for further tabling. The cyclone overflow in this circuit is sent to flotation. There are three 6 in. cyclones prior to the rougher tables and one 6 in. and one 15 in. in the regrind circuit. The overflows from all these cyclones go to the 40 ft. "fines" thickener.

The overflow from the first 6 in. cyclones goes to the rougher tables which produce three products:

Rougher tables: Conc. to cleaner tables
Mids to 6 in. cyclone (regrind)
Tail to 15 in. cyclone (regrind)

Cleaner tables: Conc. to roaster
Mids to scavenger table
Tail to 15 in. cyclone (regrind)

Scavenger table: Conc. to roaster
Mid and tail to 6 in. cyclone (regrind)

The underflow from the 6 in. and 15 in. cyclones are fed to a 6 ft. x 8 ft., 125 HP ball mill, which is followed by scavenger pyrrhotite flotation. Any scheelite that was locked in the sulfides is thus liberated and any pyrrhotite not floated in the first flotation is removed.

The non-float is tabled as follows:

Rougher table: Conc. to cleaner tables
Mids to 6 in. cyclone (regrind)
Tail to 15 in. cyclone (regrind)

Cleaner table: Conc. to roaster
Mids to 6 in. cyclone (regrind)
Tail to 15 in. cyclone (regrind)

The mids and tail fine fractions are sent to scheelite flotation.

Feed to the scheelite flotation circuit consists then of the non-float from the talc flotation, 6 in. cyclone overflow from the gravity circuit and 6 in. and 15 in. cyclone overflows from the regrind circuit. The fine scheelite is floated and upgraded by two or three stages of cleaning and the rougher tails are discarded.

One additional gravity circuit is also employed. The concentrate from the tables is dewatered and roasted in a multiple-hearth furnace. This in turn passes through a 3 pole Carpco magnetic separator. The rejects from the first pole are screened at 80 mesh and the over size discarded. The under size and the magnetic fractions from the second and third poles are repulped and pumped to a cleaner table.

Cleaner table: Conc. to roaster
Mids to primary ball mill
Tail to Grade II product

General. The high specific gravity of scheelite makes it a natural mineral for gravity methods of concentration, but ore complexities

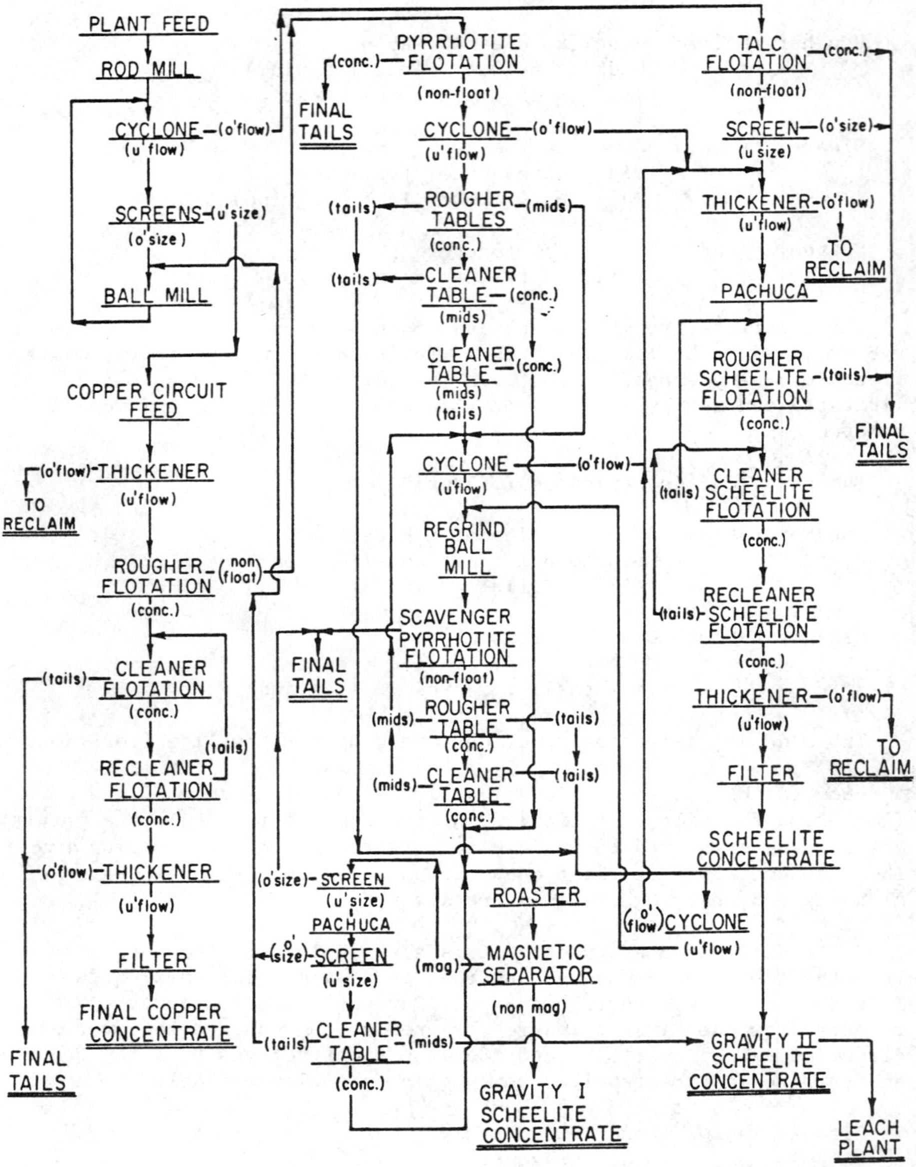

FIGURE 2, CANADA TUNGSTEN CONCENTRATOR FLOWSHEET

TABLE II. Canada Tungsten Ltd. - Table of Major Equipment

Qty.	Service	Type	Size
1	Primary rod mill	-	7 ft. x 10 ft. 200 HP
2	Cyclones	-	15 in.
4	Cyclones	-	6 in.
6	Wet screens	Sala	24 - 40 mesh
2	Thickeners	Tray	40 ft.
4	Copper rougher flotation cells	Denver	No. 24
4	Copper cleaner flotation cells	Denver	No. 24
2	Copper recleaner flotation cells	Denver	No. 24
10	Pyrrhotite flotation cells	Denver	No. 24
7	Rougher tables	Deister	No. 666
7	Cleaner tables	Deister	No. 6
1	Scavenger table	Deister	No. 6
1	Regrind ball mill	-	6 ft. x 8 ft. 125 HP
4	Pyrrhotite scavenger flotation	Denver	No. 21
6	Talc flotation cells	Denver	No. 24
8	Scheelite rougher flotation cells	Denver	No. 21
6	Scheelite cleaner and recleaner flotation cells	Denver	No. 12
1	6 Hearth roaster	Skinner	5 ft.
2	Magnetic separator	Carpco	3 pole

and the friable nature of scheelite can make upgrading quite difficult at times. This results in overgrinding interference from other high specific gravity minerals.

As with any other plant, ore changes can play havoc with an established flowsheet and this was experienced at Canada Tungsten when the open pit orebody was depleted and processing ore from the underground orebody began.

Although maybe not directly related to gravity methods of concentration, the following will show a sequence of events and how the present flowsheet evolved.

The most serious problem in the ore change was the increased amount of talc (serpentine) in the ore. The excessive talc resulted in the following difficulties:

1. Poor extraction of chalcopyrite and pyrrhotite, which raised the impurity level in the tungsten concentrate.

2. High SiO_2 levels in scheelite flotation concentrate which leaching could not upgrade.

3. Slime interference in the tables.

Laboratory tests to locate an effective depressant for talc in scheelite flotation were unsuccessful but did result in a talc depressant for the chalcopyrite and pyrrhotite circuits. Since the talc could not be depressed in the scheelite circuit, a flotation circuit was designed and put into operation to float the talc separately prior to scheelite flotation.

Nearly all of the design and flowsheet changes made in this circuit were to either achieve <u>closer sizing</u> of the products or to eliminate interfering minerals from a stream. The importance of closely sized products and the importance of removal of interfering minerals in gravity concentration is demonstrated in this plant.

PALABORA MINING COMPANY URANIUM RECOVERY PLANT

This flowsheet and the next are included without much detail only to emphasize the range of variations of application of gravity concentration methods. These two flowsheets were selected because one (Palabora) is a high tonnage by-product plant recovering a relatively small volume of a mineral of uranium in a low grade concentrate. This becomes feed for a chemical plant. The other flowsheet (Bridge Hill Ridge) is a relatively low tonnage primary beach sand operation recovering a large volume of lower value mineral. Both concentrates require additional chemical treatment. A brief description of each of the plants follows:

At Palabora, each primary copper circuit has an associated heavy mineral recovery section, as shown in Figure 3. Prior to feeding the copper plant tails to the gravity section, the tails are deslimed in hydro-separators, some of the magnetite is removed by rougher magnetic separation and additional slimes are removed by hydrocyclones. The hydro-separators and first stage magnetic separation are now shown. The cone section gives a 40-50 fold concentration of heavy minerals that include uranothorianite, baddeleyite (zirconium oxide) and magnetite. Additional magnetite is removed by magnetic drum separators following initial upgrading. The cleaned cone concentrate is treated on tables where three products are produced - uranothorianite concentrate containing 5 per cent U_3O_8 for chemical plant feed, baddeleyite concentrate, which is stockpiled, and tailings, which are retreated for recovery of copper. Research is ongoing for a method to recover the zirconium from the baddeleyite concentrate. The magnetite is presently stockpiled for possible future treatment.

It is important to include this type of flowsheet to point out some of the unexpected benefits that can result from gravity by-product plants. In addition to recovering a valuable commodity, such as uranium, the operators have realized an improvement in copper recovery. Also, the baddeleyite and magnetite may have commercial value in the future, depending on the results of the research programs being conducted by Palabora Mining Company.

MINERAL DEPOSITS LTD. BRIDGE HILL RIDGE HEAVY MINERALS PLANT

This flowsheet shown in Figure 4 is included primarily as an example of a total gravity flowsheet. This is, perhaps, one of the more simple and straightforward gravity flowsheets in existence. Mineral sands are dredged and pumped to a rougher cone circuit where they are upgraded about 6 to 1. Concentrates are advanced, middlings are scavenged on a separate cone circuit and tails from the rougher and scavenger are discarded. The scavenger concentrate is combined with the rougher concentrate, and this is cleaned and recleaned on cone concentrators. This concentrate is upgraded on spirals for a total upgrading ratio of 100 to 1 at a 92 per cent recovery which is impressive for such a simple circuit.

CONCLUSION

The main purpose of this presentation is to point out the types and variations of gravity concentration flowsheets that can be utilized. Of particular importance is the use of gravity in recovering by-product minerals from the tailings of large ore processing facilities, eg. Climax, Palabora and Ecstall. As operating costs increase, recovery of all important values in the ore becomes more of a concern year by year. Low cost, high throughput gravity operations will certainly become more important.

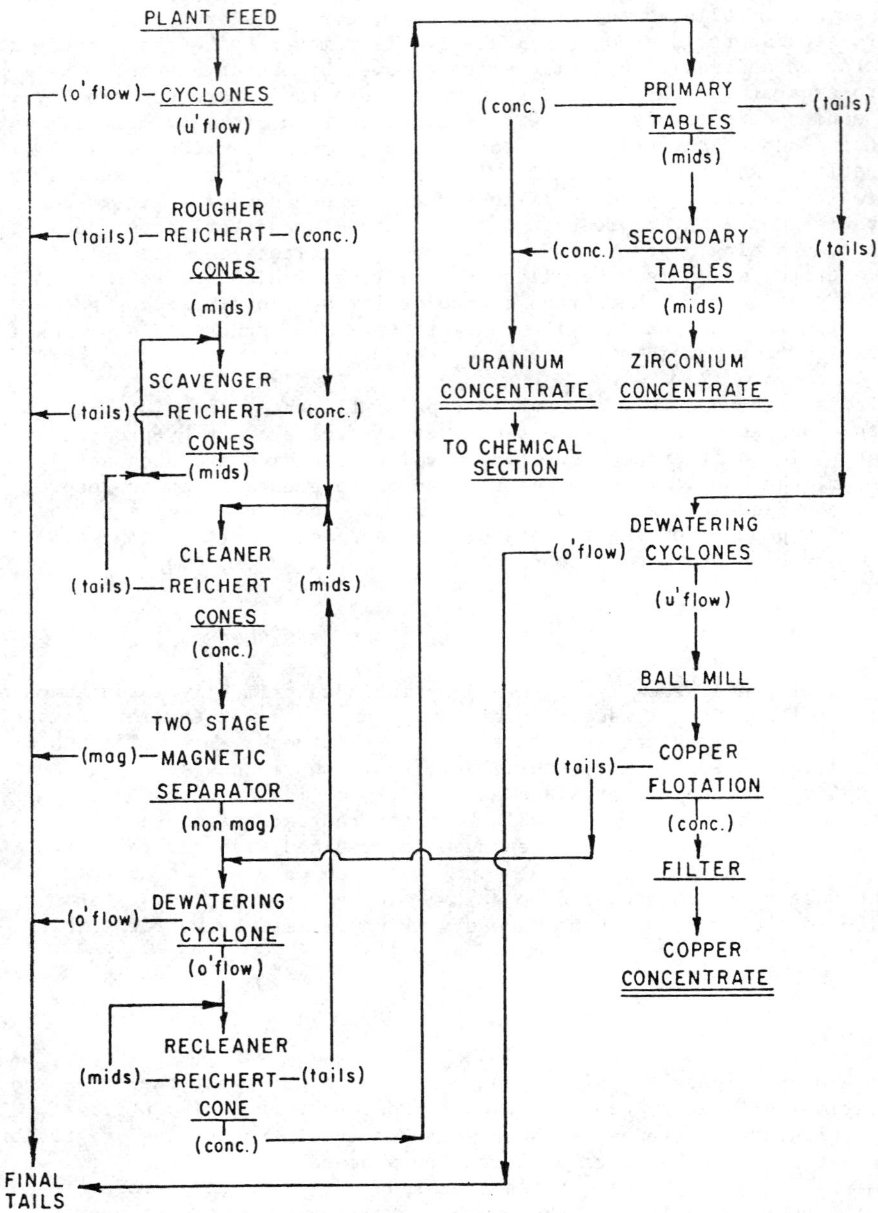

FIGURE 3, PALABORA HEAVY MINERAL PLANT

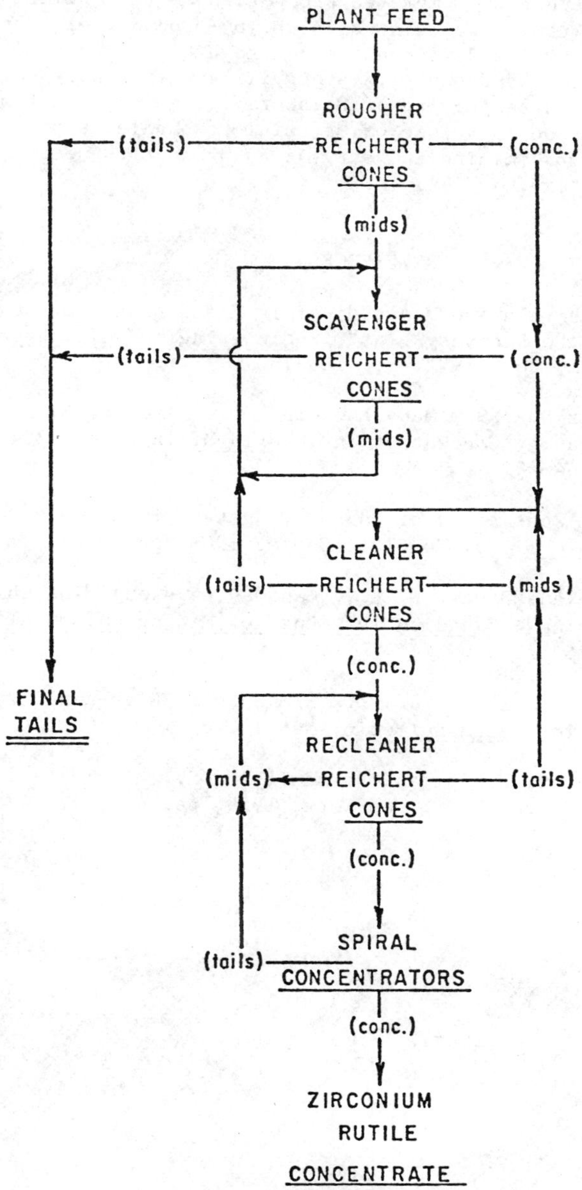

FIGURE 4, BRIDGE HILL RIDGE HEAVY MINERALS PLANT

As by-product gravity recovery circuits become more common, it will be important to consider the improved production of the primary mineral in the economic evaluation as well as in the by-product being studied. It will also be worthwhile during any study to at least consider the possibilities of other by-products. Stockpiling may be necessary as with the magnetite and baddelyite at Palabora. The low operating cost and the large reduction in volume of material to handle is proving to be very valuable to the mining industry as by-products become more economically attractive.

REFERENCES

1. Nel. Vic, "Palabora's New Heavy Minerals Plant Adds Uranium Concentrate to the Recovery List", Engineering and Mining Journal, Vol. 173, No. 11, Nov. 1972, pp. 180-187.

2. Peterson, Oliver D., "How Many Low Grade Ores Are Now Being Recovered by Gravity Concentration", World Mining, Vol. 30, No. 8, Jul. 1977, pp. 44-49.

3. Ellerman, Roy, "Operations of Canada Tungsten Mining Corporation Ltd.", Deco Trefoil, No. M4-B138, Summer 1969, pp. 9-24.

4. Cummings, W. W. and Bruce, D. E., "Canada Tungsten: The Change to Underground Mining", Canadian Institute of Mining and Metallurgy, Oct. 12-15, 1976.

5. Terrill, I. J. and Villan, J. B., "Elements of High-Capacity Gravity Separation", CIM Bulletin, May 1975.

6. Kraus, E. H., Hunt, W. F., and Ramsdell, L. S., Mineralogy, Fifth Edition, McGraw Hill, New York, 1959, pp. 455, 472-474.

5

Flotation-Magnetic-Electrostatic Separation

Chairmen: H. B. Treweek
J. H. Bassarear

Chapter 20

CONCEPTUAL DESIGN OF FLOTATION CIRCUITS

by N. Arbiter, P. B. Hobsbawn, J. F. Mahoney, and C. C. Harris

Professor Emeritus, Krumb School of Mines
New York, New York

Bechtel Corporation
San Francisco, California

Consultant
Tucson, Arizona

Professor of Mineral Engineering, Krumb School of Mines
New York, New York

ABSTRACT

Current trends in flotation applications are reviewed and effects of increased cell sizes and automation of circuits considered. The role of laboratory and pilot plant investigations in establishing design criteria and project feasibility are assessed and illustrated.

TRENDS IN APPLICATIONS OF FLOTATION

Flotation practice in the next decades will show a number of new trends in addition to those prevailing in the recent past. Lower grade feeds and more complex mineralogy will continue to pose their obvious problems. Major increases in iron ore flotation plants should develop as non-magnetic taconites eventually surpass magnetic taconites as source material. High intensity wet magnetic separation may offset this trend to some extent as witness its adoption for Brazilian hematite ores, but its relatively high capital costs may limit its use, particularly if flotation reagent costs can be reduced.

Flotation for phosphate production should increase. Changing mineralogy may eventually become important as simpler feeds in Florida give way to those containing higher proportions of shell and carbonate. Typical of this direction is IMC's new process[1] which involves not only conventional fatty acid and amine flotation steps, but a third dolomite flotation step after depression of phosphate minerals. But simpler flotation flowsheets are already in use commercially on igneous ores high in carbonate; typical are those at Phalaborwa (South Africa) and Jacupiranga (Sao Paulo, Brazil), which involve direct flotation of phosphate with depression of carbonates.

In the sulfide field no major changes in ore feed characteristics are anticipated. Circuits producing only copper concentrates, with molybdenum byproduct, will continue to dominate in tonnage. Duval's Sierrita mill in the U.S., and the Bougainville mill in New Guinea, are still the only plants near 100,000 tons per day, along with Kennecott's complex in Salt Lake City. No new mills of this size are in sight in the U.S. If work on the USSR's Udokan deposit is accelerated, this could easily support 100,000 tons per day, probably in two stages of construction. Other possibilities for large scale operations exist in Chile, if political stability continues and the world outlook for base metals improves. A number of non-Chilean companies are already active there.

At Udokan, in Chile, and also in two lower tonnage prospects in Arizona, copper flotation circuits may need to be integrated with leaching circuits or, less likely, sulfidization circuits for oxidized copper minerals, because of completely intermixed sulfide/oxide mineralogy. Another requirement for new sulfide flotation plants, particularly with low grade copper, may be the provision of more byproduct recovery circuits. Not only will molybdenum recovery continue to be commonplace, but so may circuits for tungsten recovery from porphyry coppers. Uranium too may be recovered, mainly from copper waste dump solutions, but also from primary streams where oxide copper is leached from mixed ores. The Twin Buttes (Arizona) project is now establishing practice for this, albeit on a selectively mined oxide portion of a sulfide/oxide ore body. Recovery of uranium from Florida phosphates may also develop.

Changes in requirements for design with lead/zinc plants will be less significant. However, if complex copper/lead/zinc massive sulfides are brought into production with attendant difficulties in flotation selectivity because of fine dissemination, there may be a trend to combinations of flotation with hydrometallurgy. This could involve bulk flotation of copper/lead/zinc sulfides for high recoveries, with iron sulfide rejection and leaching of the concentrates. Although the main emphasis so far in sulfide hydrometallurgy has been on oxidation processes with attendant corrosion or sulfate disposal problems, a shift to reduction processes or to selective metathesis may simplify the technology and make it more competitive with pyrometallurgy, particularly with complex concentrates of this type.

Another field for combined operations involving flotation and hydrometallurgy may be with low grade copper/nickel ores such as those in the Duluth gabbro in Minnesota, and the Heddleston complex in Montana. Bulk flotation of copper-nickel sulfides gives high recoveries with both types of ores, but the subsequent separations are more difficult. Another problem is that environmental protection requirements may either prevent smelting entirely at the mine site or make it very costly in those locations. Treatment of bulk concentrates to produce a higher grade copper sulfide product for further hydrometallurgy or for converting and a nickel/iron sulfate solution from which a nickel product could be obtained for conversion elsewhere, might alleviate the environmental problems.

Finally, if coal production is to be increased two or three-fold in the next decades, coal flotation may become substantial, with problems due more to large tonnage handling and dewatering rather than to flotation complexities. Table 1 shows U.S. flotation uses.

TRENDS IN EQUIPMENT AND CONTROL

In another direction, changes are taking place which will significantly affect the design problems for flotation circuits. These are first, the trend to still larger volume cells, and second, the rapid acceleration in the technology for and the use of automatic control of flotation circuits.

Larger Cells

The important considerations with the larger cells are, first, the assurance that metallurgical performances can be obtained at least equal to those with smaller cells; and second, the assessment of the direct and indirect savings which accrue by their use.

The ultimate proof of metallurgical performance unfortunately still requires operation of the larger cells at full scale because of absence of rigorously established scale-up criteria. However, there are enough of the sizes from 100 to 500 cubic feet now in operation to provide reasonable assurance of adequate performance. And with the general approach to rational scaleup principles already indicated,[2] manufacturers are adopting these principles in the development of still larger cells.[3][4]

With reference mainly to most widely used cells of U.S. manufacture, hydrodynamic aspects of large cell operation are remarkably similar (Table 2). All three (Agitair, Denver, and Wemco) operate within less than 100 ft/min of 1300 fpm. Air escape velocities ($ft^3/min/ft^2$ of cell area) average 4.3 ft/min. Power intensities for the larger sizes are between 0.06 and 0.09 hp/ft^3*, without allowance for blower air.

The Maxwell cell is claimed to operate at 0.02 hp/ft^3, with an air velocity lower by one-half[5] than for the U.S. cells. None of its reported dimensions or operating parameters explain this low power consumption so that in the absence of published operating data, as opposed to the manufacturers claims, judgment should be reserved.

In the light of these similarities, there is the reasonable expectation that performances will be similar on the same ore. Where large differences in capacity for a particular ore are found with cells operating conventionally, the differences are to be ascribed to differences in mineral flotation rates.

For experience teaches on the one hand that cell power intensity is a reasonably adequate measure of a cell's contribution to flotation kinetics; and on the other, that there may be as much as a five to ten-fold difference in flotation rates between fastest and slowest floating ores.

Statements of capacity per unit of cell volume, therefore, are meaningless unless qualified at least with respect to pulp percent solids and assumed flotation times. If they are so qualified, the only uncertainty becomes that in equating or extrapolating laboratory or pilot plant flotation times to those for commercial cell sizes. For a variety of reasons, this leads to the necessity of "safety factors" which are discussed below.

As a rough approximation in the absence of data, the summaries of manufacturers claims given by Harris,[6] after eliminating inconsistencies and absurdities, indicate a range of 1-2 dry tons

*Outokumpu 's OK 16 cell (565 ft^3) requires .05 hp/ft^3 for impeller and 0.02 additionally for blower air.[7]

per day per cubic foot of cell volumes. This is for roughing operation only, under "average" conditions.

Advantages In Larger Cells Although it is intuitively obvious that savings in purchase costs, costs of installation, and costs of operation should result from availability of larger cells, once it is established that results obtained with them are at least equal to those for smaller cells, there have been no data published to demonstrate this. To illustrate the thesis at least semi-quantitatively, Tables 3 and 4 have been prepared based on manufacturer's data.[8]

Table 3 takes the range of cell sizes 100 to 1000 ft^3, various daily operating rates, a fixed flotation time of eight minutes, and 20, 30 and 40 percent solids. Ore specific gravity which has a relatively minor effect is held constant at 3.0. The important aspects of these comparisons are:

1. The largest cell sizes, 1000 ft^3 and even moreso 2000 ft^2, would be useful in simple roughing only for the very largest plant sizes. Below 50,000 TPD, too few cells would be required to prevent short circuiting and to permit flexibility in operation. However, their use in rougher-scavenger circuits as rougher cells, followed by smaller cells for scavenging, merits consideration even for smaller capacity plants. In this way, part of the economies inherent with the larger cells could be utilized.

2. Cells of 500 ft^3 size appear useful for the 25,000 to 50,000 TPD range, but again are too large for 10,000 TPD or smaller plant sizes except for the rougher-scavenger role.

3. Cells of 100 ft^3 size or smaller are probably now limited to plants below 10,000 TPD capacity. They are definitely obsolete for larger plants except for cleaning circuits.

4. In general, the larger cells permit not a reduction in the number of cells in a row, which is controlled by short circuiting considerations to a probable minimum of eight, but in a drastic reduction in the number of rows, sections, pumps, distributors.

Table 4 illustrates this more specifically in terms of those factors contributory to purchase costs and to various installation and construction costs. Considering the effects of changing from the older 51 ft^3 Wemco cell to their more recent 425 ft^3 cell and their new 1000 ft^3 cell, now under plant testing at several locations, the following comparisons on a fixed treatment rate basis are obtained:

1. The number of cells is reduced to 12 percent and then 5 percent of the 563 units required with the 51 ft^3 cell.

2. The floor area - only that occupied by cells, feed boxes, junction boxes, etc., and not the aisles - is reduced to 44 and 29 percent of the original 16,800 ft^2 with the 51 ft^3 cell. This will have obvious effects on building costs.

3. The total length of cell rows is reduced to 23 and 9 percent of the original 3700 feet with the 51 ft^3 cell; this, together with the reduction in the number of cells, would have a major effect on installation costs and piping and wiring costs.

4. The cost of the cells themselves, based on extrapolation of relative figures quoted by Harris,(9) is reduced to 52 and 39 percent of that with the smallest cell. Inasmuch as overall constructed costs are generally from three to four times delivered equipment costs, it is not unreasonable to expect ratios of 1.0:0.33:0.25 for overall constructed costs for 51, 425, and 1000 ft^3 cell installations. This is on the unlikely assumption that inflation would be inoperative.

Trends in Measurement and Control Equipment A revolution has been taking place within the last decades in automatic control of flotation circuits. It is worth recalling that Taggart's 1945 Handbook does not even mention this subject; with respect to "Constant Conditions" and "Attendance," he said only that "quantified expression of the effects of a change in one variable on another is not possible in the present state of knowledge." To indicate fluctuating ore character, he described only the use of pilot tables on tailings, or "quick assays (20 to 30 minutes)." A review of the subject 15 years ago,(10) gave numerous details of level, flow, pulp density, and pH measurements, and mentioned x-ray fluorescence, but gave no references to integrated control of flotation circuits.

Today both hardware and software are available for full automation of flotation, with Australia and Finland playing leading roles in these developments and with Canada most active in the number of applications. Table 5 lists information on computer controlled flotation circuits in recent publications.(11)

The gains reported are in improved recovery, reductions in reagents, and possible decreases in manpower; less susceptible to quantitative measure are stabilization of operations by elimination of large fluctuations in throughput, and flexibility in adapting controls to maximize economic yields.

With respect to the impact of automation availability in principle to the design of flotation circuits, the following considerations are dominant.

1. For the near term, full scale automation will be applied to flotation circuits only after a new plant has been operated long enough to obtain the data accumulation necessary for control strategy development. This means that circuit design need incorporate only those control elements which would be difficult or costly to install after completion of construction. This includes equipment for XRF samples and to measure reagent flowrates, pulp levels, and air flowrates, etc. which should be installed initially where indicated. Computer hardware, and interfacing and automatic control equipment can probably be delayed until necessary for data accumulation and control strategy development.

2. The following preliminary guides for the eventuality of automation for a given plant size and product value can be utilized:[12]

a) With gross annual product values greater than $25 million (1976 prices), a detailed study of computer control feasibility is warranted. This means that installation of XRF equipment initially is also warranted with serious consideration of additional measuring equipment probably necessary.

b) In the $10-25 million gross value range, computer control will be warranted only where improvements at upper end of probable improvement[13] ranges are indicated. This in turn is to be expected only with complex ores or those with large variability in performance. In other cases, sophisticated measuring equipment installation is probably inadvisable initially.

c) With concentrator output values under $5 million, computer control is not likely to be feasible and sophisticated measuring equipment should not be installed initially, if at all.

3. With the larger plants for which automation is likely, flotation circuits design should be based on minimizing flotation sections and rows to save on installation costs for XRF hardware, sampling lines and other measuring and control equipment.

STAGES IN FLOTATION CIRCUIT DESIGN

The decision to build a plant including a flotation circuit will have been based on the demonstration that the mineral values in the ore can be mined, processed by flotation to produce concentrates which can be smelted to metal, or otherwise sold, with a net indicated financial yield attractive to those financing the project.

The decision to build will have been reached after a number of stages of activity with increasing intensity of effort, of work details, and of costs. With specific reference to the flotation component, these stages can be outlined as follows.

Stage I involves preliminary laboratory flotation testing with accompanying assay and mineralogical studies.

Its result is a preliminary report on amenability of ore samples to conventional flotation treatment. The report will give grades and recoveries, grind and reagent requirements, and note any unusual problems. Significant byproduct values will be shown together with recoveries or likelihood of recovery. Suggestions for a flotation circuit will be supplied.

The duration will be one to three months depending on availability of samples and ore complexity. A preliminary feasibility study will usually then provide order of magnitude cost of plant and operation.

A decision to go beyond this will depend on a) economic climate, b) where in the spectrum of possibilities indicated ore reserves lie; where amenability or refractoriness of ore response lies; where economic feasibility lies between low and highpprofitability. Decision then will be to continue to Stage II or to abandon effort.

Stage II involves detailed laboratory flotation testing, including:

a) Survey of reagents to establish optimum type and optimum levels.

b) Refining of grind limits.

c) Refining of circuitry; i.e., roughing vs. rougher-scavenging, establishing material to be reground, open circuit vs. closed circuit on intermediate products.

d) Often locked cycle testing.

e) Survey of ore types based on availability of samples.

Detail in this stage depends to some extent on size of ore body, and of potential plant profitability. Duration may be from six months to two years, depending on these same factors.

Stage III is often a more detailed feasibility study, in-house or by an engineering company. The results of this study and magnitude of project will determine whether a pilot plant is indicated and what size it should be, and whether preliminary engineering studies are to be undertaken along with pilot plant activities.

Stage IV will follow a decision to build or to use an existing pilot plant. The pilot plant program may then involve:

a) Intensive testing of major ore types.

b) Establishing reagent variability for ore types.

c) Establishing more definitive relationships among capacity, grind, and flotation results.

d) Establishing final details of circuitry.

e) Obtaining definitive circuitry for byproduct recovery and quantitative results.

f) Training foremen and operators.

g) Definitive estimation of flotation cell requirements and possibly of cell type.

h) Calculations of mass and flow balances from the final flowsheet.

i) A less obvious function of the pilot plant is to establish confidence of management and investors in the viability of the project.

Unrecognized dangers from the pilot plant program are the developing of over-confidence in the ability to handle variations in minerals and rock types, and to handle the effects of oxidation and weathering. This results from the tendency of the pilot plant to be overstaffed technically, and in operating manpower, so that operating problems are solved more readily than they will be in commercial plants and startup problems are under-emphasized. The complexity of complex processes may also be under-emphasized.

Stage V is then a pre-engineering estimate of plant costs and operating costs based on the pilot plant information. This will serve as the basis for management's decision to authorize the full scale plant, and to request bids for engineering and construction.

With the awarding of a contract for engineering and construction of the plant, the selected company, in collaboration with mining company representatives, will prepare the design criteria. Table 6 is a typical set for the flotation section of a hypothetical copper concentrator while Table 7 is the equipment list.

Also required is a detailed mass and flow balance, shown in the block diagram for the rougher circuit (Figure 1). Along with the retention times obtained in the pilot plant, this will serve as a basis for sizing all equipment in the circuit. Safety factors may be determined from the pilot plant test data on the softest and highest grade ore sample results(14) representing significant reserve tonnages; or else factors based on the collective experiences of the engineering and mining companies' staffs can be used.

Questions of layout of equipment on the flotation floor, of spatial arrangements among grinding, classifying, flotation and regrinding equipment, of the incorporation of spare pumps, of space for future expansion, and of extra equipment for flexibility, are matters for prolonged discussion and argument. Sometimes the decision is in favor of flexibility; more often, it is in favor of economy, particularly after the "final" pre-construction cost estimate is presented to a horrified management for appropriation of project funding.

This stage, the first months after the construction contract is signed, is the most critical in determining the final plant design; i.e., selection of type and size of equipment and flotation circuit details. This will produce the definitive cost estimate after requests for quotations are sent out and bids received, and then final selection of equipment to be purchased is made. Then, after all civil, mechanical, and electrical detailed drawings are produced, and on-site construction costs are estimated, the definitive cost is prepared.

While this last stage is critical for establishing the final form and costs of the flotation circuit, the preceding conceptual design stage will have determined, for better or worse, the configurations and the ultimate workability of the circuitry. It is in this penultimate stage that the maximum resources of both the mining company's metallurgists and the engineering company's staff engineers, should be brought to bear.

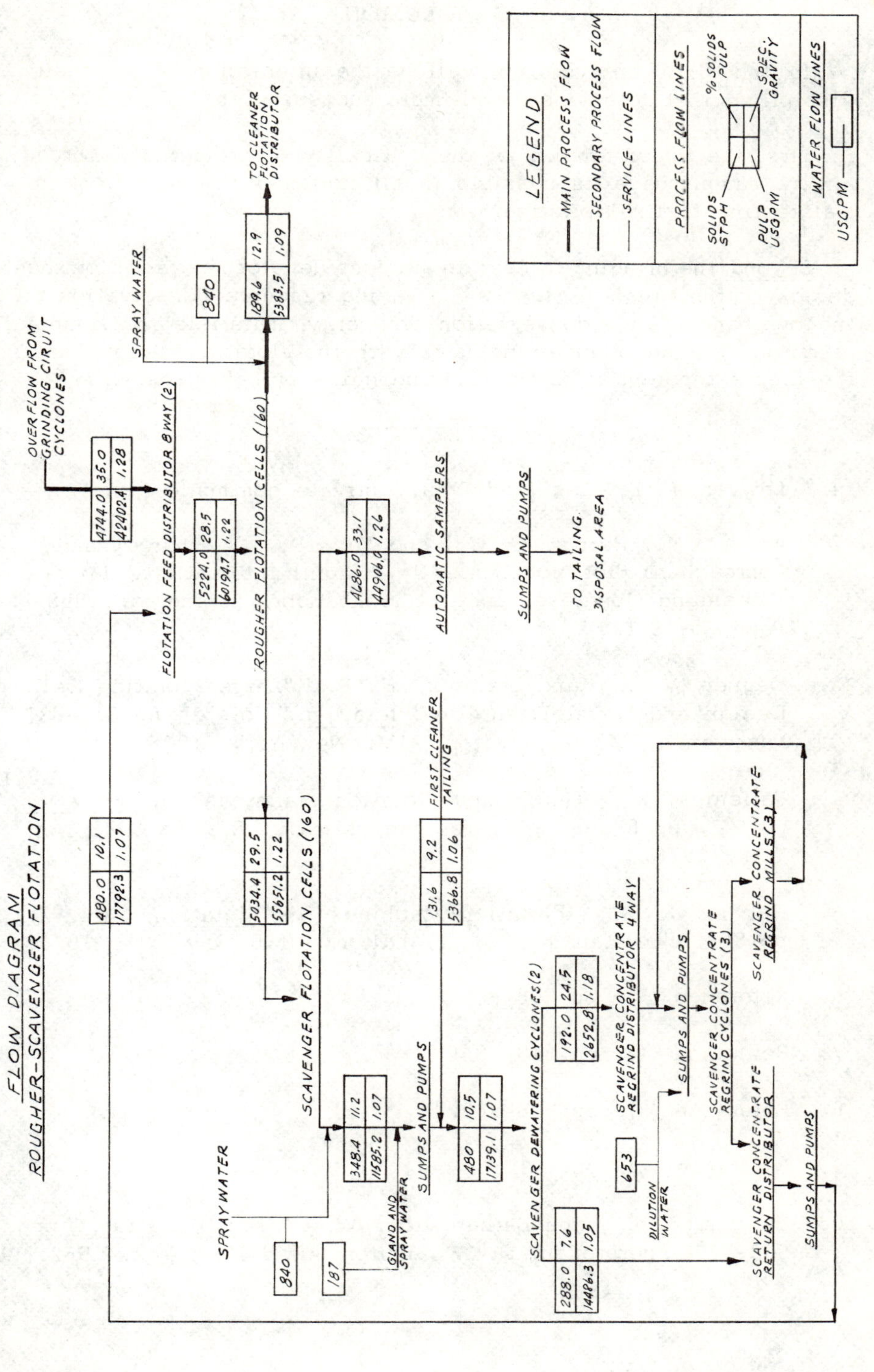
FLOW DIAGRAM
ROUGHER-SCAVENGER FLOTATION
OVERFLOW FROM GRINDING CIRUIT CYCLONES
FLOTATION FEED DISTRIBUTOR 8 WAY (2)
ROUGHER FLOTATION CELLS (160)
SPRAY WATER
TO CLEANER FLOTATION DISTRIBUTOR
SCAVENGER FLOTATION CELLS (160)
AUTOMATIC SAMPLERS
SUMPS AND PUMPS
TO TAILING DISPOSAL AREA
FIRST CLEANER TAILING
GLAND AND SPRAY WATER
SCAVENGER DEWATERING CYCLONES (2)
SCAVENGER CONCENTRATE REGRIND DISTRIBUTOR 4 WAY
SCAVENGER CONCENTRATE REGRIND CYCLONES (3)
SCAVENGER CONCENTRATE REGRIND MILLS (3)
DILUTION WATER
SCAVENGER CONCENTRATE RETURN DISTRIBUTOR
LEGEND
MAIN PROCESS FLOW
SECONDARY PROCESS FLOW
SERVICE LINES
PROCESS FLOW LINES
SOLIDS STPH
% SOLIDS PULP
PULP USGPM
SPEC. GRAVITY
WATER FLOW LINES
USGPM

CONCLUSION

In the final analysis, it will be the interaction of two main disciplines that determine the ultimate success of the entire project.

First the understanding of the criticality of providing as broad an ore sample base as possible in all stages of the work from initial testing to final piloting.

Second the bringing to bear in process development and flowsheet design of the greatest rigor in balancing resource conservation on the one hand against conservation of energy, materials and financial resources on the other. And it is with the flotation circuit that the greatest potential for maximizing gains probably rests.

REFERENCES

1. Lawver, J. E., Lakeland, Fla., private communication.
2. Arbiter, N., Harris, C. C., and Yap, R., "A Hydrodynamic Approach to Flotation Scale-Up," Eighth International Mineral Processing Congress, Institut Mekhanobr, Leningrad, 1968 Paper D-19:12.
3. Degner, V. R., and Treweek, H. B., "Large Flotation Cell Design and Development" in "Flotation," Vol. 2, pp 816-837, Fuerstenau, M. C., Editor, AIME New York, 1976.
4. Fallenius, K., "Outokumpu Flotation Machines" in "Flotation," Vol. 2, pp 838-861, Fuerstenau, M. C., Editor, AIME New York, 1976.
5. Harris, C. C., "Flotation Machines" in "Flotation," Vol. 2, pp 808, Fuerstenau, M. C., Editor, AIME New York, 1976.
6. Reference (5), pp 806-809.
7. Reference (4), pp 848, 850.
8. Reference (5), pp 806-809.
9. Reference (5), pp 771.
10. Lawver, J. E., and Barbarowicz, W., "Froth Flotation," pp 539-573, Fuerstenau, D. W., Editor, AIME New York, 1962.

11. Manlapig, E. V., Julius Kruttschnitt Mineral Research Centre, private communication.

12. Smith, H. W., "Computer Controls in Flotation Plants" in "Flotation," pp 916-993, Fuerstenau, M. C., Editor, AIME New York, 1976.

13. Reference (12), pp 983.

14. Arbiter, N., and Weiss, N., "Design of Flotation Cells and Circuits," Trans. AIME, Vol. 247, 1970, pp 340-347.

Table 1

Major Applications of Flotation in the United States*

Product	Number of Plants	Average Tons Per Day	Feed Assays	Ratio of Concentration
Copper	18	18,000	0.75	43
Copper-Mo	15	34,000	0.53	56
Copper-lead			0.29 Cu	118
Zinc	13	2,250	4.91 Pb	16
			0.99 Zn	78
Lead-zinc	21	1,500	4.78 Pb	16
			4.06 Zn	16
Zinc	10	2,000	4.38 Zn	15
Iron ore	5	17,000	36.8 Fe	2
Phosphate	19	19,000	11.6 P	4
Potash	8	6,000	16.4 K	4
Glass sand	22	1,800		1.2
Feldspar	13	800		5 Feldspar
				9 Quartz
				28 Mica

*Adopted from World Mining, January 1977, pages 44-46, Tables I, II and III. Tables are based on 1975 U.S. Bureau of Mines Minerals Yearbook (in press).

Table 2

Hydrodynamics of U.S. Flotation Cells

Volume ft^3	LxW	Air Flow/ Cell Area	Air Flow Number	Tip Speed ft/min	HP/ Volume
Galigher Agitair					
			$\times 10^3$		
60	5x5	5.0	4.2	1200	0.13
100	5x5	5.6	4.7	1200	0.09
150	6.5x6.5	4.3	3.5	1217	0.07
200	6.5x6.5	4.8	3.9	1217	0.06
300	10x7.5	3.3	2.7	1466	0.08
400	10x10	3.0	2.6	1151	0.08
500	10x10	3.5	3.0	1151	0.07
Denver D-R					
50	4x4	5.0	3.9	1300	0.12
100	5.2x5.2	5.1	4.1	1260	0.09
312	7.3x7.3	5.1	3.6	1410	0.06
507	8.8x8.8	5.2	3.5	1470	0.06
Wemco 2x2					
51	5.5x5.5	3.5	2.8	1268	0.16
85	5.5x5	4.1	3.6	1131	0.13
130	7x5.25	3.9	3.0	1299	0.09
255	10x7.5	3.6	2.8	1267	0.09
425	12x9	3.6	2.8	1293	0.08
Average (all cells)		4.3	3.4	1287	0.09

Table 3

Variations in Cells Required
With Cell Size at Different Tonnages
and Pulp Densities*

Dry Tons/Day % Solids	10,000	25,000	50,000	100,000
		100 ft^3 Cells		
20	77	192	383	767
30	49	122	245	489
40	32	81	161	322
		500 ft^3 Cells		
20	16	40	80	160
30	10	25	50	100
40	6	16	32	64
		1000 ft^3 Cells		
20	8	20	40	80
30	5	12	24	50
40	3	8	16	32

*Eight min flotation time; 3.0 specific gravity ore

Table 4

Comparison of Economic Factors
According to Cell Size*

Cell Size**		Number		Area		Total Length		Cost
ft^3	LxW(ft)	of Units	Ratio	ft^2	Ratio	ft	Ratio	Ratio
51	5.5x5	563	1.0	16,800	1.0	3700	1.0	1.0
425	12x9	68	0.12	7,350	0.44	840	0.23	0.52
1000	13.7x10	28	0.05	4,900	0.29	323	0.09	0.39

*Basis: 50,000 TPD, 20% solids six min rougher flotation or 28,750 ft^3 of cell volume
51 $\text{ft}^3 \approx$ 40 rows of 14 cells
425 $\text{ft}^3 \approx$ 5 rows of 14 cells
1000 $\text{ft}^3 \approx$ 2 rows of 14 cells

**Wemco 2x2 cells

Table 5

Computer Controlled Flotation Circuits

Plant	Computer	XRF Analyzer	Controlled Variables
Falconbridge-Lake Dufault	PDP-8	ARL	Collector, frother
Ecstall-Kidd Creek	Honeywell H 316	ARL	Collector
Inco-Clarabelle	IBM - 1800	ARL	Pulp level, collector
Hudson Bay	GE-PAC 4000	ARL	Frother, collector
Outokumpu-Pyhasalmi	Honeywell DD 516	Courier	NaCN-air
Outokumpu-Vuonos	IBM - 1800	Courier	Collector, air
Outokumpu-Kotalahti	PDP-11	Courier	Air, pulp level, pH, collector
Phelps Dodge-Morenci	N.A.*	N.A.	$(NH_4)_2S$, collector, water, frother
Mt. Isa Mines	HP 114B HP 2100A	Courier	Collector, air

*Not available

Table 6

Design Criteria

Nominal plant feed rate, TPD (calendar)	100,000
Grinding plant design rate, TPD (operating)	113,600
TPH (operating)	4,740
Cyclone overflow screen analysis,	
Cumulative % Retained	
+ 48 Mesh, %	8
+ 65 Mesh, %	20
+100 Mesh, %	30
+200 Mesh, %	50
-325 Mesh, %	38

Flotation Section

Copper content, %	
Feed to flotation	0.5
Rougher concentrate	12.3
1st cleaner concentrate	28.0
2nd cleaner concentrate	34.0
3rd cleaner concentrate	38.0
Retention time, minutes	
Roughers	12
Scavengers	12
Cleaners - 1st	9
- 2nd	10
- 3rd	15
Feed Solids, %	
To roughers	28.5
To scavengers	29.5
To 1st cleaners	11.8
To 2nd cleaners	14.8
To 3rd cleaners	18.7
Scavenger tails (final)	33.1
Reagent usage, lbs/ton plant feed	
Z-4	.05
Z-6	.05
MIBC	.0525
Cresylic acid	.0175
Lime	.20
1st cleaner concentrate regrind -	
Product size, minus 325-mesh, %	85
Tons per calendar day	2,160

Table 7

Mechanical Equipment Concentrator Complex

Flotation Area	Number of Units	Size	Rating	HP Each Unit
Flotation cells				
Roughers	160	No. 120	300 cf	40
Scavengers	160	No. 120	300 cf	40
No. 1 cleaners	27	No. 120	300 cf	40
No. 2 cleaners	24	No. 84	150 cf	15
No. 3 cleaners	18	No. 84	150 cf	15
Rougher concentrate pump	12	8" vert.	1800 gpm	75
Scavenger concentrate pumps	12	8" vert.	2000 gpm	100
1st cleaner concentrate pumps	2	8" vert.	2000 gpm	100
2nd cleaner concentrate pumps	2	8" vert.	2000 gpm	75
3rd cleaner concentrate pumps	2	8" vert.	1125 gpm	50
Clean-up sump pumps	6	3.5" vert.	250 gpm	50
Cranes flotation bays	5	60' span	3 ton	
Distributors				
8-way flotation feed	2	14'Øx5'		
Tailings launders	4			

Chapter 21

FLOTATION CIRCUIT DESIGN-CURRENT PRACTICE

by Harold B. Treweek

Chief Metallurgical Engineer: WEMCO Division of Envirotech Corporation, Sacramento, California

ABSTRACT

The development of full scale flotation circuits based on laboratory and/or pilot plant data is discussed. Auxilliary operations, equipment and layout essential for trouble-free start-up and operation are noted. The selection of flotation cell size is discussed as it affects plant capital and operation costs.

Chapter 22

MAGNETIC SEPARATION

by William J. Bronkala

President, MSI Industrial Products
Hudson, Ohio

ABSTRACT

Magnetic Separation is a proven means of effectively obtaining purification and/or concentration of mineral products. In addition, magnetic separators are widely used for tramp iron removal in the protection of mineral processing machinery such as crushers, conveyors, mixers, etc. This paper summarizes the various types of magnetic separators available and outlines their basic selection and usage. Current magnetic separator costing is provided for general guidance and estimating.

INTRODUCTION

General Description. Any solid placed in a magnetic field is affected by it in some way. Solids may be classified into two broad categories: (1) diamagnetic solids which are repelled, and (2) paramagnetic solids which are attracted by a magnetic field. For practical consideration, the more common paramagnetic solids are classified into three groups: namely, (1) strongly magnetic (ferromagnetic), (2) weakly magnetic, and (3) non-magnetic. The art of separating one solid from another by means of a magnetic field is called magnetic separation. The principal users of magnetic separators can be discussed in two broad categories: (1) tramp-iron removal and (2) concentration and purification.

Tramp-iron Magnetic Separators

Tramp-iron magnetic separators are used to protect handling and processing equipment such as crushers, pulverizers and material handling equipment. They are usually applied on dry material or on material which contains only surface moisture. Iron coarser than 1/8 in. is usually defined as tramp iron. Following are the most commonly applied separators.

Magnetic-head Pulleys. Magnetic pulleys are used to remove tramp iron from products handled on belt conveyors. Both electro-magnetic and permanent-magnetic pulleys up to 60 in. in diameter and 60 in. in width are available. Electromagnetic pulleys can be operated at belt speeds to 500 ft./min. Permanent magnetic pulleys can be operated at belt speeds to 435 ft./min. Table 21-18 shows typical information used in selecting magnetic pulleys. Enter the table with the belt width of the conveyor. Follow the belt-width column to the capacity equal or next larger to that being handled. Follow the appropriate horizontal column to the left side of the table to determine the diameter of magnetic pulley required. Check the right-hand column in this same horizontal column to determine normal recommended belt speed. If a higher speed is required, a larger-diameter pulley must be selected. In selecting magnetic pulleys for inclined-conveyor installation a correction factor must be applied to belt speed and capacity in cubic feet, as indicated in Table 21-19.

Table 21-18. Capacities of Magnetic Pulleys in Cubic Feet per Hour for Tramp-iron Removal on Normal Conveyor Operation

Pulley diam., in.	Width of belt, in.												Normal belt speed, ft./min.
	12	14	16	18	20	24	30	36	42	48	54	60	
12	790	1,080	1,460	1,880	2,330	3,500	...	...	...	...	...	...	175
15	900	1,240	1,670	2,150	2,660	4,000	6,300	...	...	...	...	...	200
18	1,010	1,390	1,880	2,420	3,000	4,500	7,100	10,500	...	...	...	...	225
20	...	1,550	2,090	2,690	3,330	5,000	7,900	11,700	16,600	...	...	...	250
24	...	...	2,380	3,060	3,800	5,700	9,000	13,300	19,000	26,200	...	...	285
30	...	...	...	3,540	4,400	6,600	10,400	15,400	22,000	30,400	...	...	330
36	...	...	...	...	4,860	7,300	11,500	17,000	24,300	33,600	43,300	...	365
42	...	...	...	...	...	8,000	12,600	18,700	26,700	36,800	47,500	59,000	400
48	...	...	...	...	...	8,700	13,600	20,300	29,000	40,000	51,600	64,200	435
54	...	...	...	...	...	...	...	22,200	31,700	43,700	56,400	70,100	475
60	...	...	...	...	...	...	...	...	33,300	46,000	59,300	73,800	500

Table 21-19. Correction Factors for Inclined Conveyors

Degrees incline	5	6	7	8	9	10	11	12
Correction factor	0.955	0.946	0.937	0.928	0.919	0.910	0.901	0.892

Degrees incline	13	14	15	16	17	18	19	20
Correction factor	0.883	0.874	0.865	0.856	0.847	0.839	0.829	0.820

Suspended Magnets. Suspended tramp-iron magnets are installed flat over a conveyor or at an angle over the conveyor head pulley. Height of suspension is determined by particle size and conveyor loading. Nominal burden is determined by the formula $D_e = 14.4\ CF/WV$ where D_e = center burden depth, in.; C = capacity, cu.ft./min.; W = width of belt, ft.; V = belt speed, ft./min. and F = factor for type of idler used and the point of magnet application. Table 21-20 indicates F values for normal operations. The size and shape of the tramp iron will influence magnet selection. The depth D_e plus a 2 to 4 in. clearance for collected magnetics determines suspension height. The ability of a rectangular magnet to attract tramp iron is dependent on the flux density (gauss) at the object and the rate of change of flux density at that point (gradient). Force index is a numerical

Table 21-20. F Values for Suspended-magnet Installations

Idler	Suspension position	
	Over pulley	Over belt
Flat	22.03	22.03
20°	18.72	22.32
35°	19.0	22.6
45°	20.16	23.9

expression of the ability of a magnet to attract a magnetic object. The static force index required to lift various types of tramp iron has been established. In addition, the additional force index required for the various depths of burdens has been established. The total force index required by the tramp iron is matched to the force-index rating of the rectangular magnet at the suspension height in making the selection of magnet size and series to be used. Force-index ratings for the various sizes and shapes of tramp iron at different burden depths are indicated in Table 21-21. To use this table, determine the size and shape of the minimum-size piece of tramp iron to be removed. Calculate the burden depth. Determine the suspension height required and the position of magnet installation. Enter the table at the desired tramp-iron size, note the static force index required, and add the force index required for burden depth and magnet position. This total force index is then matched to the magnet force-index curve(as supplied by the manufacturer) at the required suspension height. Cross-belt or in-line models are used when continuous removal of the tramp iron is required.

Magnetic Drums. Drum separators are used where insufficient space is available for other tramp-iron magnets or where chutes and ducts instead of belt conveyors are used to handle the flow of the feed material through the magnetic field. Feed is usually on the top, vertical center line, and the magnet arc covers approximately 165 deg.

Plate Magnets. Tramp iron is removed from bulk solids handled in chutes by trapping it against a magnetized plate as material flows down the chute. This type of magnet must be periodically cleaned. A chute angle of 45 deg. is recommended, and the plate magnet should be installed close to the feed point to eliminate velocity influences.

Grate Magnets. A grate magnet consists of a series of magnetized bars and is used to remove fine iron as well as tramp iron. This type of magnet must be periodically cleaned. Feed is usually vertical through the grate. Normally the grates are 1 in. in diameter and are spaced on 2-in. centers.

Metal Detectors. In those cases where the tramp metal is not magnetically responsive, an electronic detector is used to indicate its presence. Magnetic detectors are sometimes used to detect magnetic tramp iron when a suitable tramp-iron separator cannot be installed.

Table 21-21. Tramp-iron Force-index Ratings

Object size, in.	Static force index × 10^3	Angled over pulley: center belt depth, in.							Parallel over belt: center belt depth, in.						
		4	6	8	10	12	14	16	4	6	8	10	12	14	16
Bolts or Rods															
¼ × 1	13.5	22	27	31	35	39	43	47	100	171	240	325	420	525	640
¼ × 2	6.0	23	25	28	31	34	37	40	57	92	132	177	226	282	342
¼ × 4	2.8	26	28	29	30	31	32	33	44	67	93	122	155	191	230
½ × 2	13.5	23	25	28	30	32	34	36	57	92	132	177	227	282	342
½ × 4	6.0	20	22	23	24	25	26	27	33	50	69	90	113	140	168
½ × 8	2.8	16	17	17	18	18	19	19	22	31	41	52	64	78	93
¾ × 4	10.5	22	23	24	25	26	27	28	35	53	74	98	124	228	265
¾ × 8	4.0	14	15	15	16	16	17	17	20	27	36	46	57	70	83
¾ × 12	2.8	12	12	13	13	14	14	15	17	24	31	38	47	56	66
1 × 4	13.5	20	21	22	23	24	25	26	32	50	68	90	113	140	168
1 × 8	6.0	17	17	18	18	19	19	20	22	32	42	54	67	80	95
1 × 12	3.6	12	12	13	13	14	14	15	17	24	31	38	47	56	66
Hexagonal Nuts															
½	42	32	38	44	50	56	62	68	173	290	429	581	753	941	1148
⅝	42	26	31	36	41	46	51	56	131	223	321	448	588	743	936
¾	42	22	26	31	35	40	45	50	95	162	240	328	427	536	656
1	42	15	18	21	24	27	30	33	73	127	188	260	339	428	524
1½	42	11	13	15	17	19	21	23	48	84	127	176	231	293	360
Plates or Disks															
¼ × 1 × 1	36.5	30	36	41	47	53	59	65	70	116	170	228	294	368	448
¼ × 2 × 2	24.5	20	24	28	32	36	40	44	48	79	115	156	203	255	312
¼ × 3 × 3	16.5	14	16	19	21	24	26	28	33	53	77	104	133	158	192
¼ × 4 × 4	14.0	12	14	16	18	20	22	24	28	45	65	88	114	135	166
¼ × 5 × 5	12.0	10	12	14	16	18	20	22	24	39	56	76	97	121	146
½ × 2 × 2	36.5	15	18	21	24	27	30	33	36	58	85	114	147	184	224
½ × 4 × 4	24.5	10	12	14	16	18	20	22	24	39	57	77	100	125	152
½ × 6 × 6	16.5	7	8	9	11	12	13	14	16	26	38	52	67	84	102
½ × 8 × 8	14.0	6	7	8	9	10	11	12	14	23	33	44	57	71	86
¾ × 4 × 4	32.0	9	11	12	14	15	17	18	21	34	50	68	88	110	134
¾ × 6 × 6	24.5	7	8	9	11	12	13	14	16	26	38	51	66	82	99
¾ × 8 × 8	18.5	5	6	7	8	9	10	11	12	20	29	39	50	63	77
1 × 4 × 4	36.5	8	9	10	12	13	15	16	17	28	41	56	72	90	110
1 × 6 × 6	30.0	6	7	9	10	11	12	13	15	24	35	47	62	77	93
1 × 8 × 8	24.5	5	6	7	8	9	10	11	12	20	29	38	49	61	74
1 × 12 × 12	16.5	3	4	5	6	7	8	9	8	13	19	26	34	50	59
Steel Cubes															
½	66	33	39	45	51	57	63	69	213	358	528	715	924	1155	1408
¾	66	21	26	30	34	38	42	46	144	245	366	497	644	806	984
1	66	16	19	22	25	28	31	34	107	185	276	380	497	626	768
2	66	7	9	11	13	15	17	19	47	86	131	182	239	303	372

Magnetic Separators as Concentrators and Purifiers

Special magnetic separators have been designed for concentration and purification of solids. Generally, "concentration" involves magnetic separation of a large amount of magnetic-feed product, whereas "purification" involves the removal of small amounts of magnetic particles from a large amount of non-magnetic feed material. For practical application and selection, magnetic separators and purifiers are divided into (1) wet types and (2) dry types.

Wet Magnetic Separators. Three of the most frequently applied wet magnetic separators are (1) permanent and electromagnetic-drum separators, (2) permanent and electromagnetic filters, and (3) high-intensity wet magnetic separators. Wet drum separators are applied as medium recovery units in dense-media plants and in the concentration of ferromagnetic iron ores. Magnetic filters are applied to remove fine ferromagnetic particles from liquids or liquid suspensions. High-intensity wet separators are applied for the removal of fine weakly magnetic materials from solids carried in liquid suspension. Wet magnetic drums and wet high-intensity magnetic separators are designed for continuous magnetic discharge. Magnetic filters accumulate magnetics, and the filter element must be periodically removed and cleaned.

WET MAGNETIC-DRUM SEPARATORS. Magnetic-drum separators are applied to concentrate finely divided, strongly magnetic materials. Three basic types of drum concentrators are available: (1) concurrent drums for cobbing operations on minus 1/4-in. ore, (2) counterrotation drums for roughing operations on minus 10-mesh ore, and (3) countercurrent drums for finishing operations on minus 65-mesh and finer ore. Operations consist of introducing the feed into the separator feedbox which in turn introduces the pulp to the rotating drum shell where the magnetic particles are cleaned by the agitating action created by the alternating polarity of the magnet assembly. The basic principle of operation of the concurrent wet separator is shown in Fig. 21-84, and that of the counterrotation wet separator in Fig. 21-85.

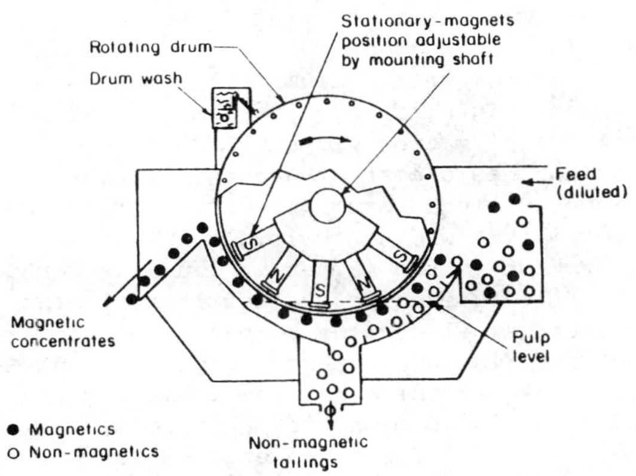

Fig. 21-84. Concurrent wet-drum separator.

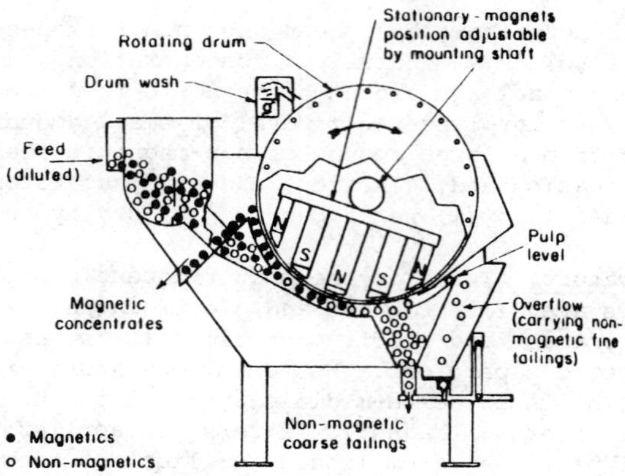

Fig. 21-85. Counterrotation wet-drum separator

WET-MAGNETIC-DRUM SPECIFICATIONS. Wet magnetic drums are constructed in both permanent and electromagnetic types. Permanent drums represent the large majority of installations because of their greater dependability and lower operating and maintenance costs. The electromagnetic wet drum is applied where extremely high magnetic fields are required or where a simple electric magnetic field control is desired. Most wet drums are 30, 36 or 48 in. in diameter with magnet widths to 10 ft. Multiple-drum separators are selected on the basis of the individual separation under consideration. Cobbing and finishing drums are commonly double or triple-drum separators. Satisfactory feed volume for cobbing separators is 75 to 125 gal./(min.)(ft. of magnet width); for roughers, 75 to 100 gal./(min.)(ft.); for finishers, 40 to 60 gal./(min.)(ft.).

WET-DRUM-SEPARATOR OPERATING COSTS. Wet-magnetic-drum separators are low in operating costs. Maintenance costs will vary widely but a figure of 3 to 5 per cent of initial cost per year is used in estimating maintenance costs. Initial cost of permanent drums will usually be lower than that of electromagnetic drums. A 30-in.-diameter single-drum wet permanent magnetic separator has a cost of about $1200 (1977 prices) per foot of magnet width. A 36-in.-diameter single-drum wet separator has a cost of about $2000 per foot of magnet width. A 48-in.-diameter single-drum wet separator has a cost of about $3000 per foot of magnet width. Multiple-drum costs increase in almost direct proportion to the number of drums required.

MAGNETIC FILTERS. Magnetic filters incorporate a filter element inductively magnetized by an external electro or permanent magnet source. The material to be cleaned is passed through this element where the magnetics are collected. Periodically, the element must be removed and cleaned of the accumulated magnetic particles. Capacities are relatively low, and the largest magnetic filter will handle about 2000 gal./ft. of a low-viscosity liquid. Multiple units in parallel with a suitable distribution system can be used for higher feed rates.

WET HIGH-INTENSITY MAGNETIC SEPARATORS. One of the newest types of magnetic separators is the high-intensity wet separator. This is designed to remove fine weakly-magnetic particles and minerals carried in liquid suspension. The application of a high-intensity wet separator eliminates the need for drying ahead of the previously used dry high-intensity separation and extends the range of high-intensity magnetic separation into recovery of particles finer than 200-mesh size. The principle of operation of a continuous wet high-intensity separator is shown in Fig. 21-86. The separator consists of an annular box rotating through one or more high-intensity magnetic

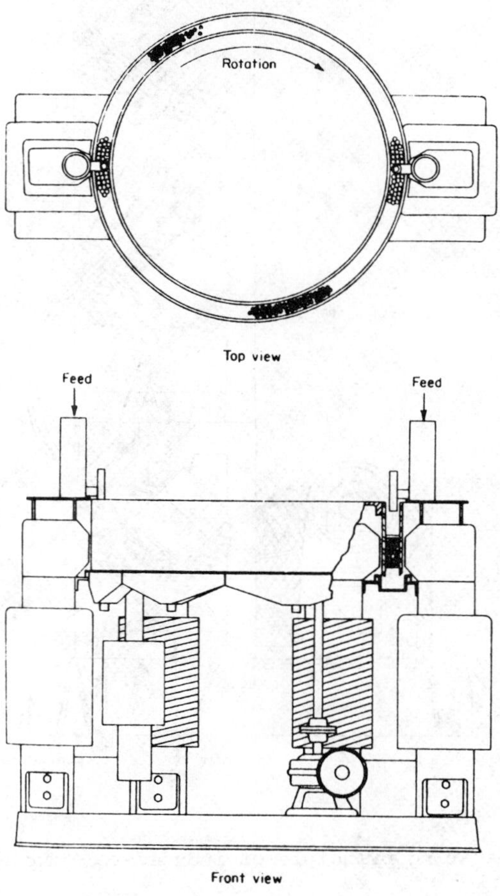

Fig. 21-86. Continuous high-intensity wet magnetic separator.

fields. The annular box is filled with shaped ferromagnetic material which is inductively magnetized to a high magnetic field intensity and which acts as a collecting medium for the weakly magnetic particles in the feed pulp. The separator is capable of producing a field in excess of 20,000 gauss with high field gradient between the induced poles. Non-magnetic particles pass directly through the annular ring, and the rotation of the annular box carries the magnetics out of the field where they are washed into a separate collection launder.

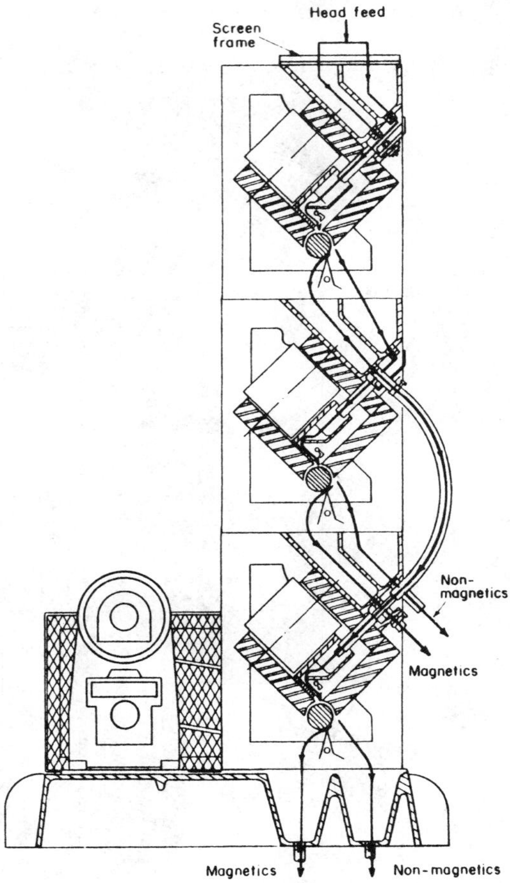

Fig. 21-87. Operating principle of induced-roll magnetic separator.

Dry Magnetic Concentrators and Purifiers. Dry magnetic separators can be classified on the basis of the magnetic field intensity that they develop. They include (1) high-intensity separators, (2) moderate-intensity separators, (3) low-intensity separators, and (4) high-speed low-intensity dry-drum separators. The individual applications are diverse, but generally high-intensity separators remove weakly magnetic particles, moderate-intensity separators remove moderately responsive magnetic particles, and low-intensity separators remove highly responsive magnetic particles. Physical conditions of the feed material also influence dry magnetic separator selection.

HIGH-INTENSITY DRY MAGNETIC SEPARATORS. Two types of high-intensity magnetic separators are most frequently applied: (1) the cross-belt and (2) the induced roll. The cross-belt separator is a selective concentrator. The cross belts pick the magnetics off the feed belt and discharge them to the side. Concentration is by direct lift, and the magnetic product is clean and free of entrapped nonmagnetic material. The cross belt has been used to concentrate wolframite, monazite, and other high-value mineral products. Cross-belt capacity is low and initial cost is high relative to feed capacity. An 18-in.-wide cross-belt separator with two magnet poles will handle up to 2 tons/hr. of feed and will cost about $27,000 (1977 prices).

The induced-roll separator is applied as a concentrator and a purifier on chemical and mineral products. Figure 21-87 illustrates the concentrating-type induced-roll arrangement. A lift-type induced-roll separator is also available and provides selective magnetic removal essentially equal to that of the cross-belt separator at about one-third of the cross-belt cost per unit of capacity. Induced-roll separators are manufactured in various roll combinations. The most frequently applied purifying type of induced-roll separator is the 30-in.-wide three-field twin-type separator having 60 in. of total feed width. Such a unit will clean 3 to 7-1/2 tons/hr. of product and will cost about $28,000 (1977 prices). The induced-roll separator has a low initial cost per unit of feed capacity.

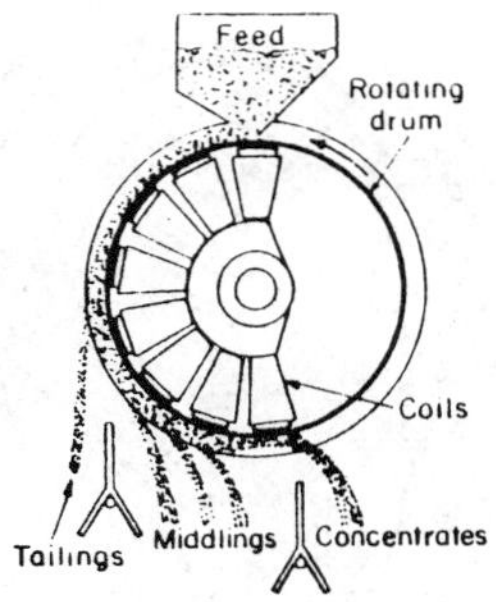

Fig. 21-88. Operating principle of alternating-polarity Ball-Norton separator.

MODERATE-INTENSITY DRY MAGNETIC SEPARATORS. The moderate-intensity magnetic separators include (1) the alternating-polarity drum separator and (2) the Uni-gap drum separator. The alternating-polarity drum was designed to treat a feed having a large amount of magnetics while obtaining a high-grade concentrate (see Fig. 21-88). This separator has been used to concentrate magnetite ores in the 1-1/2-in. by 100-mesh size range. The magnetic field of this drum will hold a large load of magnetics while providing sufficient agitation of the magnetics during treatment to shake out impurities entrapped in the initial magnetic pickup. The capacity of this alternating-polarity separator will vary inversely with the size of the feed. On coarser ores, capacities of 30 tons/(hr.)(ft. of magnet width) have been obtained. A 30-in.-diameter by 60-in.-wide magnetic drum will cost about $25,000 (1977 prices).

The Uni-gap drum is designed to produce a moderately high-intensity magnetic gap across the entire drum width at a single position on the drum circumference. This type of drum is useful in removing finely divided magnetic particles and is applied on material finer than 1/4 in. at feed rates of 3 tons/(hr.)(ft. of magnet width) or less.

LOW-INTENSITY DRY MAGNETIC SEPARATORS. Low-intensity magnetic separators are applied in concentration and purification where the magnetic particles to be removed are large in size and the magnetic responsiveness of the magnetics is high. Magnetic pulleys and low-intensity tramp-iron drum separators are used in these applications, although the magnetics to be removed may be lower in responsiveness than tramp iron. Modified versions of these tramp-iron separators are usually applied. Feed rates for these low-intensity separators are lower than those used in a tramp-iron application.

HIGH-SPEED LOW-INTENSITY DRY-DRUM MAGNETIC SEPARATORS. The high-speed alternating-polarity magnetic-drum separator is designed to treat very finely divided material - 100 mesh and finer - and to

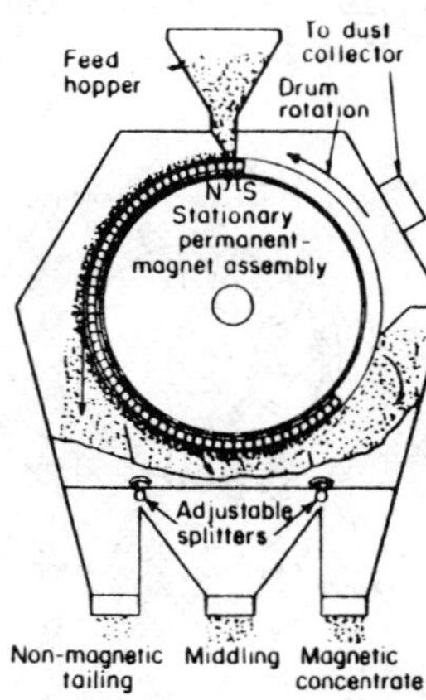

Fig. 21-89. Operating principle of dry-fines permanent magnet separator.

produce a high-grade magnetic concentrate product free of entrapped non-magnetics. It has been used to concentrate dry magnetite ores, fly ash, etc. Several models with different quantities of magnetic poles in the nominal 270-deg. magnet are available. A 17-pole unit permits the application of this high-speed drum separator on particles as coarse as one inch. Figure 21-89 shows the general principle of operation.

Types of Material Used in Magnet Manufacture. Magnetic separators are conventionally classified as (1) electro or (2) permanent. Electromagnets use insulated copper or aluminum-wire windings around a soft-iron core energized with direct current. Improved insulations have increased the temperature limits of electro coils to 220° C. Permanent magnets do not require external energization. The special permanent-magnet alloys continue to produce a magnetic field at a constant level indefinitely after initial charging unless they are exposed to demagnetizing influences. Alnico 5 continues to be used in special magnetic-separator applications, but the newer ceramic permanent magnets in the form of barium and strontium ferrites are most frequently employed. The higher coercive force, lower initial cost, and lighter weight of ceramic permanent magnets result in lower magnetic equipment costs without sacrifice in magnetic strength. New rare-earth-cobalt permanent-magnet alloys have been recently developed and have found application where the usage justifies the high cost of this new permanent magnet material.

BIBLIOGRAPHY

Bartnik and Stone, "Capital and Operating Costs of the Jones High Intensity Wet Magnetic Separator Plants for Iron Ore Concentration," Duluth, AIME Symposium, 1972.

Carpenter, "Electrical Methods for the Separation of Minerals," Minerals Sci. Eng., January, 1970.

Dean and Davis, "Magnetic Separation of Ores," U.S. Bur. Mines Bull. 425, 1941.

DeVaney, "New Developments in the Magnetic Concentration of Iron Ores," Proc. Intern. Mineral Process, Congr., Institution of Mining & Metallurgy, London, Paper 31, Group VI, April, 1960.

Gaudin, "Progress in Magnetic Separation Using High-Intensity, High-Gradient Separators," Denver, AMC Mining Convention, 1973.

"Mineral Processing Methods, Review and Forecast," Chemical Engng., June 20, 1977.

Pryor, Minerals Processing, Elsevier, New York, 1965.

Sherwood, "Cross-Belt and In-Line Primary Magnets," Salt Lake City, NSA Operators Meeting, February, 1976.

Taggart, Handbook of Mineral Dressing, Wiley, New York, 1953.

Chapter 23

ELECTROSTATIC SEPARATION

by William P. Dyrenforth

Director, Business Development
Ortloff Minerals Services Corporation
Golden, Colorado

ABSTRACT

The principle of electrostatic separation is discussed and elaborated on by dividing into the three main mechanisms of electrification as applied to separation of solid substances, i.e., (1) contact electrification, (2) electrification by conductive induction, (3) electrification by ion bombardment.

Types of equipment used for electrostatic processing are described and the effects of humidity mentioned. Processing of iron ore, titanium minerals, rare earths and cassiterite is discussed and some existing plants are described.

PRINCIPLE

The principle of electrostatic separation is based on the fact that, if one or more of the materials in a granular mixture can receive a surface charge on or before entering an electrostatic field, the grains of that material will be repelled from one of the electrodes and attracted toward the other, depending upon the sign of the charge on the grain. By causing such grains to fall into separate chutes, a separation or concentration results.

This "principle" is rather simply and yet broadly stated, and in view of the tremendous increase in the use of variations of it in industry today, it should be described more fully. Actually, the term electrostatic is a misnomer. In some processes the energy applied is almost entirely electrostatic; but in others, more commonly used, the energy is applied in the form of current flow and could more properly be called "electrodynamic".

The major electrification mechanisms of separation or beneficiation of solid substances can be divided into three groups:

1. Contact Electrification
2. Electrification By Conductive Induction
3. Electrification By Ion Bombardment

Each of the above mechanisms gives rise to a surface charge on the solid particles, and although each is a distinctly separate mechanism, two or more occur, more or less, in every separation, and thus, none can be neglected.

Charging By Contact Electrification

It has been observed that when the surfaces of two dissimilar materials are brought into contact, each surface will bear exchange charges the moment contact is broken. Contact electrification is also known as frictional electrification, although the role of rubbing is merely that of increasing the area of contact (provided temperature effects due to friction are negligible). In charging by particle-to-particle contact, the area of contact is usually quite small and it is necessary to provide some mechanical method of causing repeated contacts in order to build up an appreciable average surface charge on the particles. Any movement of a granular material causes repeated particle-to-particle contact. If the material is composed of poorly conducting particles, the resulting charge density often becomes high enough to use this mechanism as a means of electrical concentration. In fact, even if no effort is made to cause repeated particle-to-particle contact (as in the case of charging by either of the other two mechanisms), this mechanism cannot be disregarded. It was aptly said by J.E. Lawver (Ref. 2) that "the misleading terms reversible positive, reversible negative, and non-reversibility, commonly found in the literature on electrical concentration of minerals, were so coined because of the failure to recognize the importance of contact electrification".

Except at high temperature, the surface charge due to contact electrification of dry minerals can probably be attributed to the transfer of electrons. Coehn's rule states that "when two dielectric materials are contacted and separated, the material with the higher dielectric constant becomes positively charged". Although Coehn's rule is reported to have been verified for more than 400 substances, it is of limited value in the separation of solids because the contact potentials of the solids are related to the work functions of the solids, and impurities or lattice defects can reverse the predicted situation. The theory of contact electrification is complex, and certainly, a great deal more needs to be learned about the electrification properties of solids. No two triboelectrification series have been found to be identical, which can probably be attributed to the fact that the substances used to form the various series may have entirely different electrical properties because of previous temperature and chemical histories. The practical problem facing the engineer interested in controlling, and thus, using contact electrification is that of finding a means of altering (thermally, physically, or chemically) the surface of the materials so that a selective electrification, and hence a separation, can be made. In many simple cases, it is only necessary to have discrete surfaces in order to obtain suitable contact electrification. The treatment is then merely dedusting or desliming. (Removal of a common surface from materials to be separated is necessary, in any case, in the use of any of the three basic electrification mechanisms.)

Charging By Conductive Induction

This mechanism is simpler and more readily explained than contact electrification. If a solid particle is placed on a grounded conductor in the presence of an electric field, the particle will rapidly develop a surface charge by induction. Whether the particle is conductive or dielectric, it can be considered to have become more or less polarized. However, a conductive particle will, in an extremely brief interval of time, become an equipotential surface through its contact with the grounded conductor, and thus, become charged to the same potential as the grounded conductor. A dielectric particle is difficult to define in terms of its potential, because of the variation of charge on its surface. It is sufficient to say that the dielectric particle remains essentially polarized because of its inability to redistribute electrons, and thus, no net charge is generated (see Fig. 1). However, no substance is either a perfect conductor or a perfect dielectric. All real particles will require finite intervals of time to become polarized and conductive particles will also require finite time to transfer this charge stress to ground. One may assume the charging process to be equivalent to an RC circuit and a particle with finite conductivity will eventually obtain a total charge of $Q = C_pV$, where C_p is the capacitance of the particle and V is the voltage difference in the field. The mathematics of how the particle finally approaches this value are in Lawver (Ref. 2) and Mora (Ref. 3). It can be reasonably assumed that a non-conductive particle will have a high enough resistivity, and therefore, a low enough capacity to remain essentially uncharged under the RC circuit equation. Practically speaking, charging by conductive induction can be used to make a finite separation between relative conductors and non-conductors. If a grounded rotor is used to feed the material into the electric field, such as in Figure 2, the following results will more or less take place: The conductive particles will in a very short time assume the potential of the rotor, which is opposite to that of the active electrode, and will therefore, be attracted toward the active electrode. The non-conductive particles will be polarized, and will therefore, tend by virtue of their polar orientation, to be attracted to the rotor and repelled by the electrode. A properly placed cutter can transpose these reactions into a particle separation.

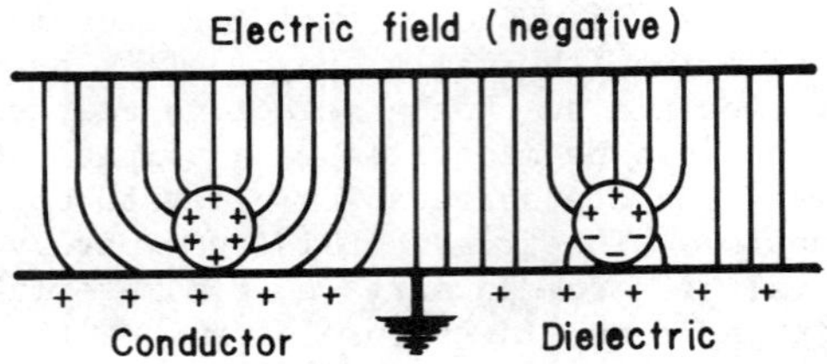

Fig. 1. Charging by conductive induction.

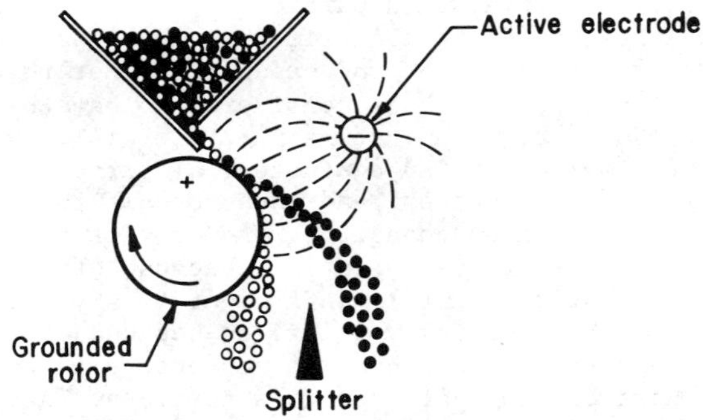

Fig. 2. Separation by conductive induction.

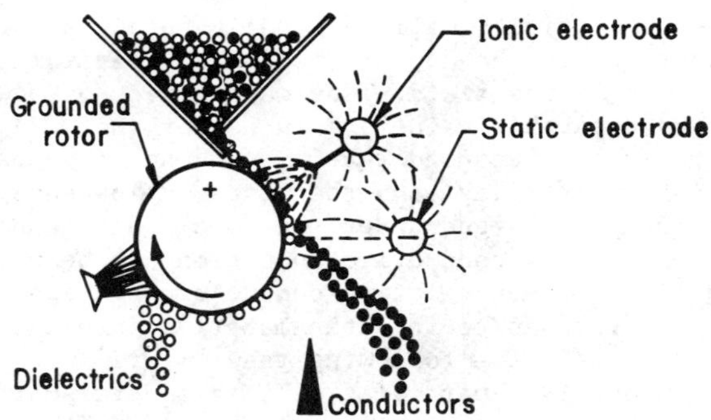

Fig. 3. Separation by ion bombardment.

Charging By Ion Bombardment

Ion bombardment is undoubtedly the strongest, and perhaps, the most effective mechanism for charging solid particles, but the selectivity of this mechanism is practically nil. However, when it is combined with the mechanism of conductive induction the selectivity of the compound process becomes very good. Charging by the use of mobile ions is definitely not an "electrostatic" process even though engineers generally use the term to describe any beneficiation process that uses a high-voltage electric field. In the past few years, the term "high tension" has, through constant usage, become the generally accepted name for the processes involving ionic bombardment. In the high-tension process, mobile ions are provided by a beamed electrode that produces a corona discharge while simultaneously concentrating this source of mobile ions in a given direction. If a dielectric particle and a conductive particle are placed in the path of these mobile ions, a portion of the surface of each particle will be given a strong electrical charge. On the conductor this charge is redistributed almost instantaneously, whereas on the non-conductor the redistribution of the same charge will be very slow. If a group of these particles is placed on a grounded surface in the path of these mobile ions, it will be found that, when the source of mobile ions is stopped, the conductors are free to leave the surface because they have shared their charge with the ground. On the other hand, the dielectric or non-conductor particles, which are not capable of readily losing their charge, are held to the surface by their own image force (Lawver, Ref. 2). The electrostatic-image theory is merely a method of solving Laplace's or Poisson's equations by inspection of symmetry conditions. In other words, these equations can be satisfied if a particle of equal and opposite charge is assumed to be placed at a mirror-image position with respect to the grounded surface and the particle in question. This image force is given as:

$$F = QQi/4\pi e_o(2s)^2$$

where $-Q = +Qi$ = total surface charge on the mineral and
s - the distance from the charge to the grounded surface
e_o - field strength of ionic field

For a detailed discussion of image theory, see Sommerfeld ("Electrodynamics", Academic Press, New York, 1952).

The mechanisms for producing a working corona discharge are complex. Loeb (Ref. 4), Bandel, and Amin have probably investigated these mechanisms more completely than any other group and have published some very informative papers on the subject. It has been established that the "point" in point-to-plane corona discharge should not actually be a point as such. According to Loeb's work, the optimum configuration is that of a small-diameter hemisphere, or for laterally extended corona, a fine wire gives a hemispherical cross section in the direction of ion flow. Both positive and negative corona discharges are useful and each has special application for certain separations. However, a negative electrode can produce a more intense corona in air before arcing because the flash-over voltage is higher.

Industrial high-tension separators make use of a grounded rotor for the "grounded surface" mentioned above in order to have a continuous surface that will introduce the particles to the source of mobile ions and then take them away. All particles are charged while under ion bombardment, but shortly after the particles are removed from this area, the conductors are free to leave the rotor and the non-conductors are pinned by their image force. This is the primary charging step and the initial separation step in this process. High-tension separation becomes extremely selective, however, when a static electrode (large enough in diameter to preclude corona discharge) is used following the ionic electrode (see Fig. 3). The dielectrics have been highly charged and strongly pinned to the rotor, and for the sake of simplicity, can be considered to be henceforth out of the system. The conductive particles, however, have shared their charge with the ground and are now being charged oppositely by conductive induction. This compound process produces very wide and distinct separation between discrete conductive and non-conductive particles and greatly increases the capacity over processes using conductive induction only.

TYPES OF EQUIPMENT

The two main types of equipment in use today are the plate type and the rotor type. The early plate separators developed at International Minerals and Chemical Corporation consisted essentially of two vertical, or near vertical, plates (with all edges sufficiently rounded to prevent corona discharge) suspended relatively close to each other, one charged positively and one negatively with a high voltage gradient between them. The materials to be separated must first be differentially charged by contact electrification (particle-to-particle charging) and then fed to the space between the electrodes. This type of separator was mainly useful in separating two non-conductive minerals from each other and is in sparse use today; in fact, to the author's knowledge, these separators are not manufactured for sale and the user is forced to design and build his own.

The modern plate separator is mainly used for cleaning a small quantity of conductive particles out of a large quantity of non-conductive particles. The charging mechanism here is conductive induction and these separators are quite efficient when used for such specific separations. A sloping, grounded plate is used under a large oval-shaped high voltage electrode and the minerals are allowed to slide down the plate into the electric field. The conductors, touching the plate, take on an opposite polarity to the electrode, and thus, tend to jump toward it. This, in effect, creates two streams of minerals at the lower end of the plate which are then separated by a cutter. The modern plate separator is manufactured by Mineral Deposits Ltd., of Southport and Readings of Lismore, both in Australia. Readings also produces them in Orange Park, Florida. The main use of these separators in industry is for cleaning small amounts of rutile and ilmenite out of a zircon concentrate.

The rotor type, or "high tension" separator, performs an extremely strong ionic charging step during its operation, and therefore, is not wholly dependent on charging mechanisms 1 or 2. Nevertheless, it uses mechanism 2 (conductive induction) in conjunction with ion bombardment to create a very selective separation. The original high tension separator manufacturer, and still the company with the most machines in the field is Carpco Research and Engineering, Inc. of Jacksonville, Florida. Mineral Deposits Ltd., and Readings are manufacturing these separators in Australia for the beach sand industry whereas Carpco has turned more toward iron ore and "recycle" high tension separators.

The Carpco machine is a very versatile separator of modular design wherein rotor units can be stacked vertically or horizontally in a single machine and any type of flow sheet can be performed in a single machine. Figure 4, shows a cross-section of a Carpco six-rotor vertically stacked machine, and it will be noted that the machine has been field-oriented on the right-hand side for parallel operation. In other words, each rotor is getting the same feed such as would be the case in an all "rougher" unit. On the left side, the machine is shown oriented for series operation such as would be the case with a small operation, wherein one rotor gives sufficient rougher capacity but the material needs more than one pass to make a final product. Although it is not shown, this high-tension separator can also be oriented to repass middlings or tailings products. Capacities of these machines are generally high and range from about 1500 to 3000 kilograms per hour per meter of rotor length. A Carpco 6 rotor machine (3 meter rotors) can, for example, handle up to 54 tonnes per hour of specular hematite ore in a one pass operation.

EFFECT OF HUMIDITY

It is imperative that all particles to be separated (under any one of the three basic electrification mechanisms) be completely surface-dry. This is not a serious problem in itself since drying of surface moisture is quite straightforward. However, a problem is encountered, at high humidities, of keeping the surfaces dry. Tests on specular hematite and silica have shown that a reasonable separation can be made on predried mineral, under high-tension conditions, at 20°C if the relative humidity is less than about 35 percent. These same tests have shown that at 60 percent relative humidity an ore temperature of 40°C is required to effect a separation, and at 90 percent relative humidity, a minimum temperature of 90°C is required. This is not a rule of thumb or a curve that can be applied to minerals in general, however. In the case of an ilmenite-rutile-zircon separation it will be found that the ilmenite and rutile (especially the rutile) must be

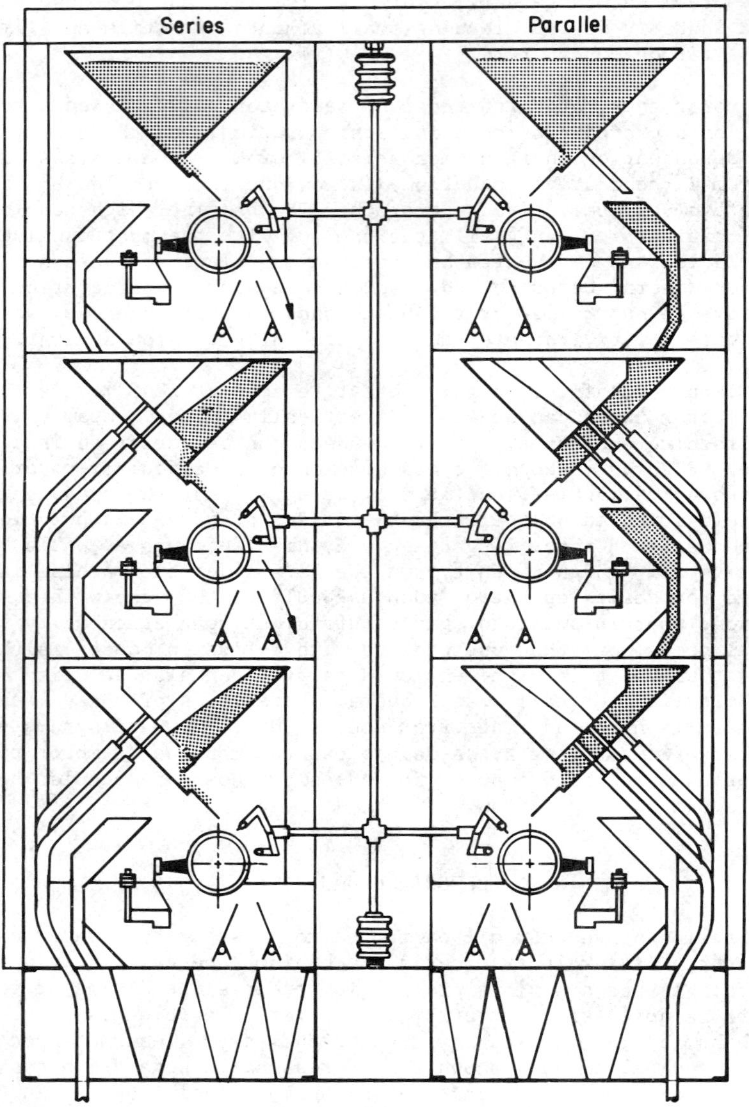

Fig. 4. Carpco high-tension separator.

kept at 90°C or above to perform an efficient separation, regardless of the relative humidity. In contact electrification, Florida phosphate rock contacted against Florida quartz will charge the quartz highly negatively and the phosphate equally positively over a temperature range of -20 to 300°C. Sylvite, however, requires a contact temperature of about 480°C to exchange charges with halite but will not begin to separate until cooled to about 150°C and discontinues separating below 90°C if the relative humidity is around 90 percent. It is sufficient to say that each case is individual and in each case ambient humidity plays a more or less significant part.

APPLICATION OF PROCESSES

Wabush Mines, Labrador, operated by Pickands Mather, is using the high-tension process to produce 1000 tonnes per hour of high-grade specular hematite concentrate. LKAB in its plant at Malmberget, Sweden, produces one million tonnes per year of high grade, low phosphorus specular hematite also using the high tension process. Almost every major tin-mining company in Malasia, Thailand and Nigeria uses the high-tension process to separate columbite, ilmenite, and cassiterite from gangue minerals. There are few titanium-mineral-producing plants in the world today that do not use the high-tension process for the separation of ilmenite and rutile from monazite and zircon and other non-conductors. E.I. du Pont de Nemours has two plants in Florida, each capable of processing up to 1000 tonnes per day of heavy mineral bulk concentrates and producing mainly ilmenite, rutile, leucoxene, and zircon by the combined use of high-tension separation and high-intensity magnetic separation. Titanium Enterprises, the newest heavy mineral plant in Florida, produces 100,000 tonnes per year of titanium minerals using the same processes as du Pont. Every beach sand mineral plant in Australia uses high-tension separation and a large number of them are now also using the modern plate separator for final zircon cleaning.

No typical flow sheets are given here because each case is quite individual. Some general conditions can be mentioned, however, to guide the engineer in the use of these processes: Generally speaking, most ores or substances must be ground to minus 8 mesh in order to have a sufficient charge-to-mass ratio to effect a separation. The size range of the particles has a definite bearing on the separation efficiency, and it may be said that the closer the size range, the more efficient the separation, particularly with respect to the separation of conductors from non-conductors. As stated previously, all particles must have discrete surfaces. This necessitates a relative degree of desliming. For example, a minus 8-mesh grind would probably need desliming at 200 mesh, a minus 20-mesh grind at 325 mesh, and a minus 35-mesh grind at 400 mesh, etc. Difficult ores may require classification at intermediate meshes in order to be successfully separated. According to Carpenter and Dyrenforth (Ref. 1), very fine particles such as minus 200 mesh plus 10 microns have been found to be amenable to electrodynamic separation using a direct current that pulsates at frequencies between about 300 and 700 cycles per second instead of the standard 120-pulse system. Specific gravity of the various minerals in an ore can greatly increase or decrease the efficiency of separation of conductors from non-conductors. For example, hematite (specular) has a much greater gravity than silica, and since it is the conductor and is to be thrown from the rotor, its greater mass aids the separation. On the other hand, in a separation of monazite from ilmenite, the non-conductive monazite has a greater gravity than the conductive ilmenite, which makes the separation more difficult.

SIZING OF EQUIPMENT

The high-tension separators (rotor type) have an inherently higher capacity than the modern plate separators and are generally used for all necessary "electrostatic" separation steps except final cleaning. A good rule of thumb for sizing high-tension equipment used to be about 100 lbs. of feed per hour per inch of rotor length but we will now call it about 18 Kilograms per hour per centimeter of rotor length. Closely sized coarse crushed ores, say minus 10 mesh plus 65 mesh, can often be treated at rates up to 30 Kilograms per centimeter, but the fines (65 mesh by 0) will probably require rates somewhat below the rule of thumb. It is best that the fine fraction of crushed ores (as above) be deslimed for high-tension separation, say minus 65 mesh plus 20 microns.

Beach sands have already been classified and deslimed by nature and usually have a mean mesh near 80. The rule of thumb of 18 Kilograms per centimeter holds well for almost all known beach sand operations.

If a new ore is being tested for high-tension separation characteristics, it is a simple matter on a laboratory separator to measure the width of the feed stream and the total feed rate for that stream width. Since the laboratory machines are indentical in cross-section to the industrial machines, scale up is simply a matter of industrial rotor length times the Kilograms per centimeter per hour rate established on the laboratory machine. Do not foget, however, that about 5 centimeters at each end of the industrial rotor are unusable (otherwise feed would spill off the ends) and should be subtracted from the actual rotor length for accurate capacity calculations.

REFERENCES

1. Carpenter and Dyrenforth, "Advances In Technique and Equipment For Iron Ore Separation", Annual Mining Symposium, University of Minnesota, 1965.

2. Lawver, "Fundamentals of Electrical Concentration of Minerals", AIME Tech. Paper, February, 1957.

3. Mora, "Study of Electrical Concentration of Minerals", MIT Thesis, 1958.

4. Loeb, "Recent Development in Analysis of Positive and Negative Coronas in Air", J. Appl. Phys., 19, 882, (1948).

5. Carpenter, "Technical Aspects of High Tension Separation", AIME, Tampa, Florida, December, 1949.

6. Dyrenforth, "Electrostatic Separation", Chemical Engineers' Handbook, Fifth Edition, McGraw Hill.

Chapter 24

PLANNING OF TURKISH DIVRIGI IRON ORE PROCESSING PLANT

by M. Sagheer and A. Yazan

Project Manager, Mineral Processing Division
Fried. Krupp GmbH, Krupp Industrie- und Stahlbau
Duisburg-Rheinhausen, West Germany

Deputy General Manager
Türkiye Demir ve Çelik Isletmeleri
Karabük, Turkey

ABSTRACT

To meet the demands of the planned development of Turkish steel industry an Iron ore processing plant is under construction at Divrigi, Anatolia and is planned to be commissioned mid of 1978. The plant has been designed for an annual capacity of 4 million tons of crushed magnetitic ore. About 1.9 million tons of coarse concentrate of 25 - 2 mm size and 1.3 million tons of pellet feed concentrate will be produced by wet screening, dry magnetic concentration of coarse size fractions and wet magnetic concentration of fine size fraction. The fine magnetic concentrate will be reground and filtered to prepare the feed for the pelletizing plant.

INTRODUCTION

The steel industry of Turkey is planned to be developed further in the next few years. The present yearly steel production of Turkey amounts to about 1.8 million tons which is intended to be increased by 150 % by 1978 and to be tripled by the end of this decade. There are two major steel works located at Karabük and Eregli in the northern part of Anatolia towards Black Sea, each producing 650 000 tons of steel annually.

The following table shows the present state of steel industry in Turkey and its planned development till the end of this decade:

FIG. 1: Planning of Steel Production of Turkey

Steel works	1975	1978	1980
	Production, tons		
Karabük	650 000	1 050 000	1 050 000
Eregli	650 000	1 500 000	1 500 000
Iskenderun	-	1 000 000	2 000 000
Metas	120 000	250 000	250 000
14 Mini-steel works Electric furnaces	350 000	650 000	900 000
Total steel production	1 770 000	4 450 000	5 700 000
Iron ore requirement	2 500 000	6 000 000	7 700 000

The extension at Eregli is underway. A part of the Iskenderun steel mill under construction with Soviet aid has already been commissioned.

The main requirement of the crude ore is met by Divrigi mine which is located about 700 km East of Ankara in the Province of Sivas in the Central Anatolia region of Turkey (Fig. 2). A negligible amount in addition is delivered by small mines. The open pit mine at Divrigi presently exploits only rich portion of the deposit. The ore is simply crushed, washed and screened and about 2.5 million tons of blast furnace ore and sinter feed are railed to the steel mills.

To satisfy the future demand for the steel mills the mining facilities at Divrigi are undergoing development. Besides development of the mine and infrastructure a new iron ore processing plant and a pelletizing plant are presently under construction.

The state-owned Turkish Iron and Steel works, Türkiye Demir ve Çelik Isletmeleri runs major steel works in the country. It contracted a German-Turkish consortium of Fried. KRUPP GmbH, Krupp Industrie- und Stahlbau, Duisburg-Rheinhausen, West Germany and Gama Endüstri Tesisleri Imalât ve Montaj A.Ş., Ankara, Turkey for design, manufacture and supply of equipment as well as construction of the processing plant involving an investment of over 60 million Deutsche Mark (DM). The processing plant includes the plant parts starting from bins for ore crushed to -25 mm and ending with the belt conveyor for the delivery of pellet feed concentrate.

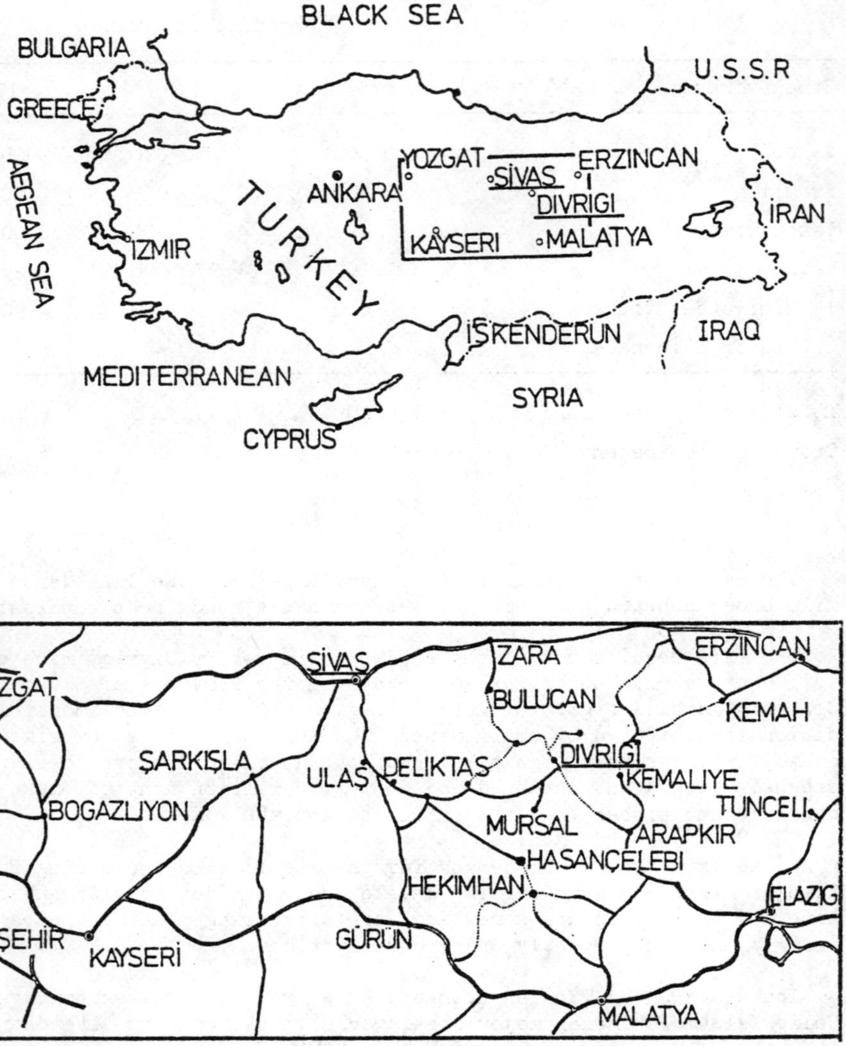

Fig. 2 LOCATION

DIVRIGI IRON ORE DEPOSIT

DEPOSIT, GEOLOGY AND MINERALOGY

The ore to be fed to the processing plant will be mined from A and B deposits. The formation of A-Body appears to be a typical pyrometasomatic type. The ore is found at the contact between the intrusive syenite and limestone with local serpentine inclusions in the contact line. The deposit lies along E-W direction dipping 35° south.

In the B-Body besides the pneumatolytic replacement origin a hydrothermal influence is probable. The ore body has an N-S strike direction dipping 45° to the west. The upper zone of B-Body is slightly martitized.

The proved mineable reserves of the A and B Bodies have been indicated at 83 million tons.

The predominant constituent of the ore is magnetite. The ore is of a distinctly irregular granular structure with the magnetite mainly occuring as irregularly shaped, comparatively pure aggregates. Magnetite impregnations in the country rock play a minor role. The ore grain or aggregate size varies within a wide range from 0.02 to 28 mm, the major part of grains fall in the size range of 0.2 - 0.6 mm.

The major part of the magnetite is entirely fresh and unweathered. Only sporadic samples exhibit martitization to different intensity.

The minor constituents of the ore are pyrite and its weathered product goethite; and traces of marcasite, chalcopyrite and limonite are also found. The intergrowth of pyrite with magnetic is comparatively loose. The limonite is in this case exclusively transformation product of pyrite.

The gangue minerals accompanying the ore include feldspar (plagioclase), biotite, augite, hornblende, serpentine, chlorite and preponderant in the contact lines skarn minerals.

ORE TYPES AND LABORATORY TESTS

The Divrigi iron ore from A and B Bodies is classified in five types according to the iron grade and magnetite content in the crude ore and weight recovery in the concentrate obtained after a predetermined scheme of laboratory tests. The grouping of the ore types is the result of serial laboratory tests carried out on drill cores and samples from galleries and shafts. The serial tests included size analysis, sink-float tests, Davis Tube tests and assay on different size fractions and test products.

Serial tests and laboratory semi-technical tests on the samples of different ore types gave the following conclusions as to the concentration of this ore.

a - It is amenable to concentration by dry magnetic separation at very coarse sizes and suitable coarse concentrates can be produced to meet the demands of Turkish blast furnaces and sinter plants.

b - It is possible to produce high-grade pellet feed concentrate by applying low-intensity magnetic separation on fine size fraction of the ore after grinding it to -0.1 mm or even coarser.

Laboratory and semi-technical tests on samples of individual ore types and on composite samples resulted into evolution of a preliminary process flowsheet.

PILOT PLANT TESTS

The final plant tests were carried out by the Studiengesellschaft für Eisenerzaufbereitung,Othfresen, West Germany, on a composite sample of different ore types. Following conclusions were drawn from the test results:

1 - The 25 - 10 mm and 10 - 2 mm size fractions screened wet from the crushed ore gave coarse concentrates grading about 62 - 63 % Fe by dry magnetic separation.

2 - The -2 mm size fraction of the ore as well as the coarse size middlings from the magnetic separation ground to -2 mm can be concentrated to about 63 % Fe by one stage of wet magnetic separation (cobbing).

3 - The -2 mm cobber concentrate requires an energy of 8 - 9 kWh/t for grinding to 0.1 mm prior to further concentration.

4 - The -2 mm cobber concentrate ground to -0.1 mm and treated in a cleaning magnetic separation stage produced a high-grade pellet feed concentrate with 67 - 68 % Fe content.

5 - The energy required for regrinding of the pellet feed concentrate to the fineness required for pelletizing will amount to 9 kWh/t.

6 - The filtration of the reground cleaner concentrate presents some difficulty. A moisture content of 9.5 - 10 % in the filter cake can only be achieved with a low filter capacity and by increasing the ratio of drying (filtration) time to the time for suction of the pulp on the filter.

7 - The results of the pilot plant tests are shown in Fig. 3 by a table of metallurgical balance.

FIG. 3: Metallurgical Balance, Processing of Divrigi Ore

Product	Weight %	% Fe	Fe-distribution %
Coarse concentrate 25-2 mm	46	62-63	51
Pellet feed concentrate	30	67-68	36
Total concentrate	76		87
Coarse tailing 25-2 mm	11	30	6
Fine tailing	13	28	7
Feed	100	56	100

PLANNING

To meet the demand of intermediate phase of increase of steel production for this decade about 4 million tons of crude ore is to be treated annually to give about 1.9 million tons of coarse concentrate and about 1.3 million tons of fine magnetic concentrate to be pelletized in a pellet plant located adjacent to the concentration plant.

The run-of-mine is crushed in an underground primary crushing plant and then in a surface secondary crushing plant to -25 mm. The crushed ore is conveyed to a stockpile where it is blended to prepare a uniform feed having a definite grade and magnetite content.

The feed to the processing plant amounts to 670 tons per hour of crushed ore on the basis of 6000 working hours per year.

FLOWSHEET

On the basis of the test results the process flowsheet (presented in Fig. 4) was worked out for Divrigi primary ore.

The pilot plant test results enabled to confirm the process and delivered data for preparation of equipment as well as for the selection of the equipment. The equipment flowsheet is presented in Fig. 5.

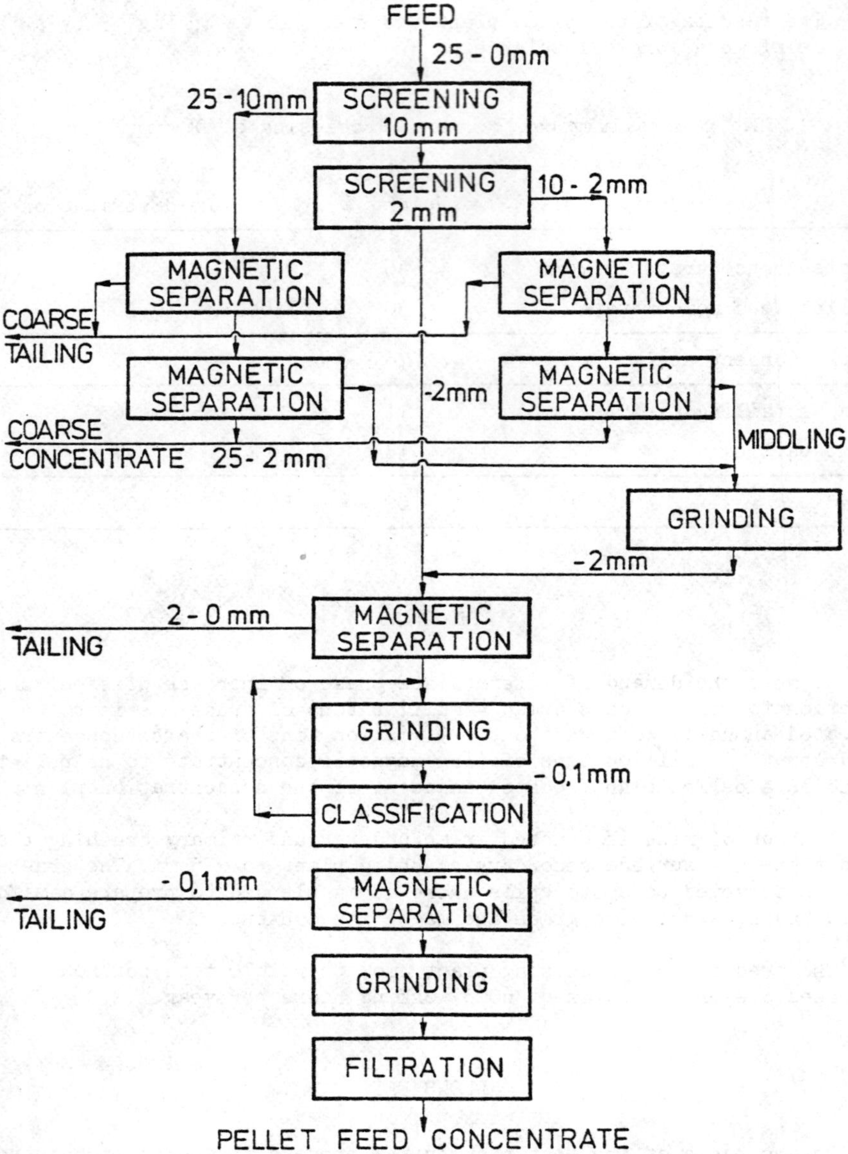

Fig. 4 PROCESS FLOWSHEET DIVRIGI IRON ORE, TURKEY

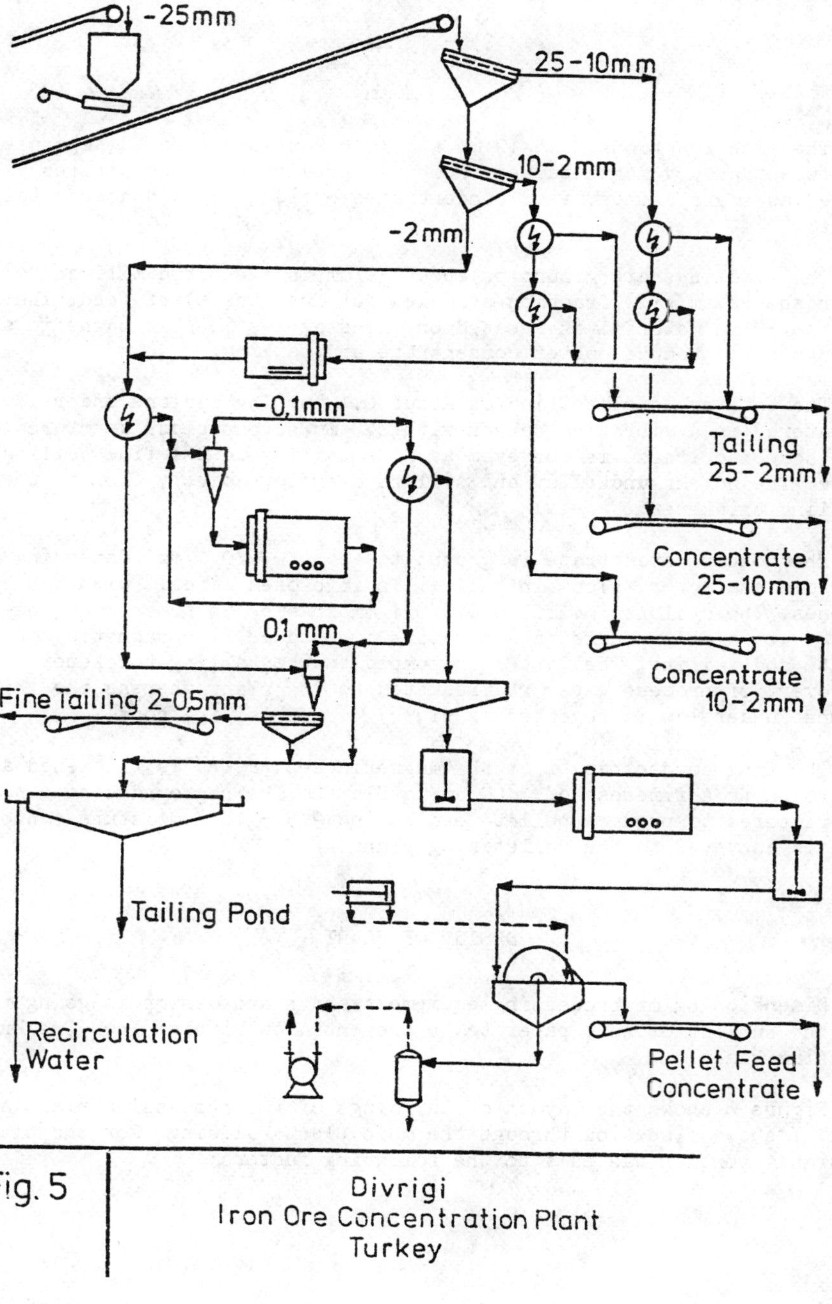

Fig. 5 Divrigi
Iron Ore Concentration Plant
Turkey

The process envisages two-stage wet screening of the crushed and blended -25 mm ore into three size fractions.

25 - 10 mm
10 - 2 mm
-2 mm

The size fractions 25 - 10 mm and 10 - 2 mm are treated separately on two-stage dry magnetic separators to produce the concentrates - named as coarse sinter feed concentrates - middlings and coarse tailings.

The middlings are ground to about -2 mm in a wet rod mill and mixed with the -2 mm size fraction screened out from the plant feed. The composite -2 mm material is treated on one-stage wet cobber magnetic separators to produce cobber concentrate and tailing.

The tailing is classified at about 0.1 mm. The cyclone underflow is dewatered on a vibrating screen with 0.5 mm screen cloth apertures. The screen oversize is conveyed as fine tailing to the fine tailing pile. The screen underflow and cyclone overflow jointly flow to the tailing thickener.

The cobber concentrate is ground to about 0.2/0.1 mm (depending on the ore characteristics) in ball mills in closed circuit with the cyclones. The cyclones overflow 0.2/0.1 mm in size is treated on one-stage wet cleaner magnetic separators to produce fine magnetic concentrate and tailing. The latter is pumped to the tailing thickener for recovery of process water recirculated to the plant whereas the thickener underflow is rejected as slurry to the tailing pond.

The fine concentrate from the magnetic separators is thickened and reground to a fineness of 1800 cm^2/g Blaine. The reground concentrate is filtered to produce pellet feed having 9.5 - 10 % moisture content and is conveyed to the pelletizing plant.

DESIGN OF PLANT

Dimensioning of processing equipment and plant design is going to be the subject of next paper to be presented on Divrigi Iron Ore Processing Plant.

Figure 6 shows the layout of buildings of the processing plant and Fig. 7 shows a section through the main plant building. For the plant design attention was paid to the following factors:

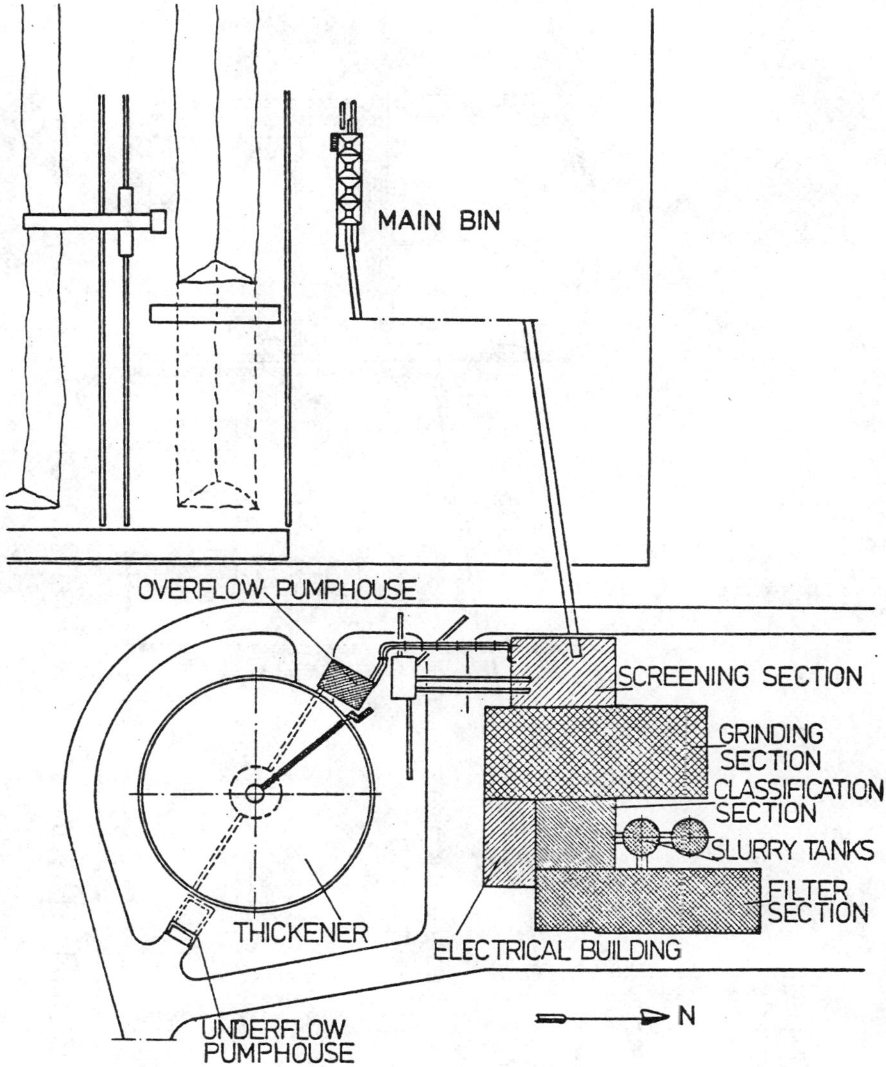

Fig. 6 LAYOUT OF CONCENTRATION PLANT DIVRIGI, TURKEY

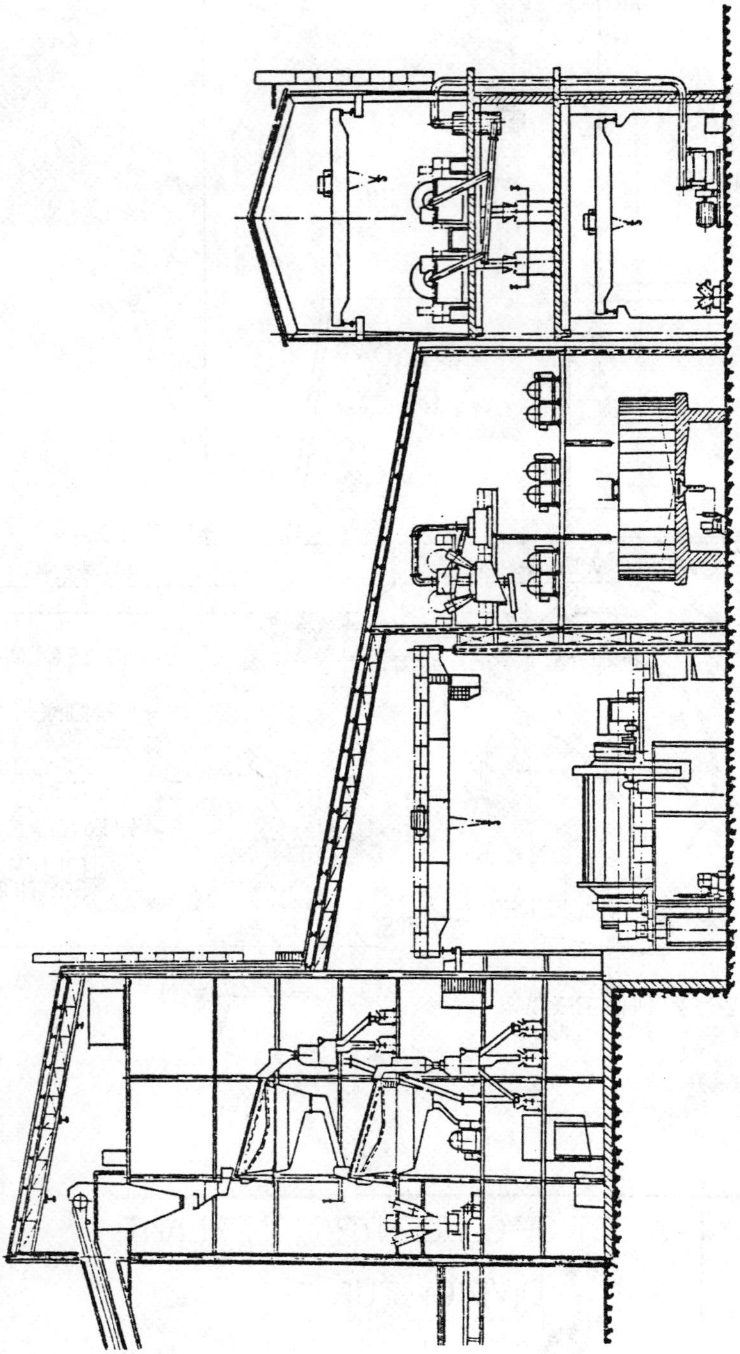

Fig. 7 SECTION THROUGH CONCENTRATION PLANT
DIVRIGI, TURKEY

a - To locate the equipment possibly to follow the flow of material in the process. The screening and cobbing magnetic separation section is located in the western building part, grinding section in the middle whereas cleaning magnetic separation section, concentrate thickener and filtration in the building part to the East.

b - To locate the heaviest machines in one building of the plant to facilitate the maintenance and repairs.

One rod mill 2.7 m diameter x 4.3 m long, 2 ball mills 4.4 m diameter x 7.0 m long and 1 regrind ball mill 4.4 m diameter x 8.8 m long, all of KRUPP design and manufacture are located in one building.

c - Extreme climatic conditions: The plant is located 1000 m above sea level in hilly ranges of Anatolia. The temperatures in summer rise to above 30 °C but the winter experiences frost period with temperature as low as -30 °C.

REFERENCES

1 - "Mineralogical Investigations on Iron Ore from Divrigi Deposits, Turkey", August 1975, unpublished work with Fried. KRUPP GmbH, Krupp Industrie- und Stahlbau, Duisburg-Rheinhausen, West Germany.

2 - "Iron Ore Deposits of Turkey", MTA-Institute Issue 118, Ankara 1964.

3 - "Tests Carried out on Semi-industrial Scale on Iron Ore Samples from the Divrigi Deposit, Turkey", December 1975, unpublished work with Fried. KRUPP GmbH, Krupp Industrie- und Stahlbau, Duisburg-Rheinhausen, West Germany.

Chapter 25 HEAVY MEDIA SEPARATION

METALLURGICAL, OPERATING, AND ECONOMIC CHARACTERISTICS OF THE DYNA WHIRLPOOL PROCESSOR

by Thomas J. Lien, and Roshan B. Bhappu

Senior Process Engineer
Mountain States Research and Development

Vice President and General Manager
Mountain States Research and Development

ABSTRACT

The DYNA WHIRLPOOL Processor (DWP) is a cylindrical dynamic heavy media vessel. It is used to beneficiate fine ore sizes from one and one-half inches in diameter down to 65-mesh. Process units available treat up to 80 tons per hour. Units can be manifolded for higher capacity plants. Finely ground magnetite and ferrosilicon are used for media.

Installed capital costs are from $4,000 to $6,000 per ton hour of feed. Operating costs are from $0.30 to $0.50 per ton of feed. The DWP is used to beneficiate metallic and non-metallic ores and coal.

The OCC Vessel is a static heavy media separator used to beneficiate coarse ore up to six inches in diameter and auto scrap. Process units available treat up to 300 tons per hour.

INTRODUCTION

Heavy media separation is generally divided into two basic separation methods: static and dynamic. Static separation is typified by the use of a trough, drum, or cone-type vessel where the separation is made in a relatively quiescent media under the influence of simple gravity forces. The ore can range from a maximum of six inches in diameter down to 3-mesh. Dynamic separation is typified by the use of centrifugal forces up to 20 times normal gravity on the media and ore inside a cylindrical-type vessel to effect a separation. The ore is normally less than two inches in diameter and may be as fine as 65-mesh.

The purpose of this paper is to review the operation, engineering design, capital cost, operating cost, and metallurgy of the DYNA WHIRLPOOL Processor as representative of dynamic heavy media separation. A brief description of the OCC Vessel, a static heavy media vessel used in the mineral and auto scrap industries, will be included.

DYNA WHIRLPOOL PROCESSOR

Description

The DYNA WHIRLPOOL Processor (DWP) is a cylindrical dynamic separation vessel which is used to beneficiate fine ores from about one and one-half inches in diameter down to 65-mesh. A cutaway section of the DWP is illustrated in Figure 1. The majority of the media (90 percent) is introduced tangentially into the DWP through the medium inlet (No. 5) at about 15 pounds pressure. The media separates into a sink and float fraction and exits the DWP at the sink discharge (No. 7) and the float discharge (No. 8). The specific gravity differential between the sink and float media is usually 0.2 to 0.5 gravity units. The remainder of the media (10 percent) enters the DWP at the raw feed inlet (No. 4) by gravity.

The circulating media stablizes very rapidly. The ore, normally not exceeding 10 percent moisture, is discharged into the feed hopper, mixes momentarily with the 10 percent media fraction, and descends with the media into the DWP. The heavier ore particles are forced to the wall of the DWP by the centrifugal forces and are carried upward by the sink media to the sink outlet and discharged. The lighter waste particles stay close to the internal vortex wall and descend downward with the float media exiting through the float discharge. The retention time of the ore in the DWP is on the order of a few seconds.

A unique feature of the DWP is that the ore is kept separate from the media until the moment of entry into the DWP. Most cyclonic heavy media separators mix the ore and media together in a sump and pump the mixture to the separatory vessel. The ore feeding the DWP is conveyed to the feed hopper significantly reducing pump wear and requirements over other systems. Because the media and ore feed systems to the DWP are independent they can be individually controlled and adjusted providing rapid response to changing system requirements. The fact that the ore is kept separate from the media until entry into the DWP permits the treatment of porous or vuggy ores. The short contact time between the ore and media limits the media penetration into the ore thereby reducing contamination of float with media and reducing the potential of float reporting to the sink product. Static head for the DWP is from two to six feet depending on the size of vessel compared to about 20 feet for cyclonic-type vessels.

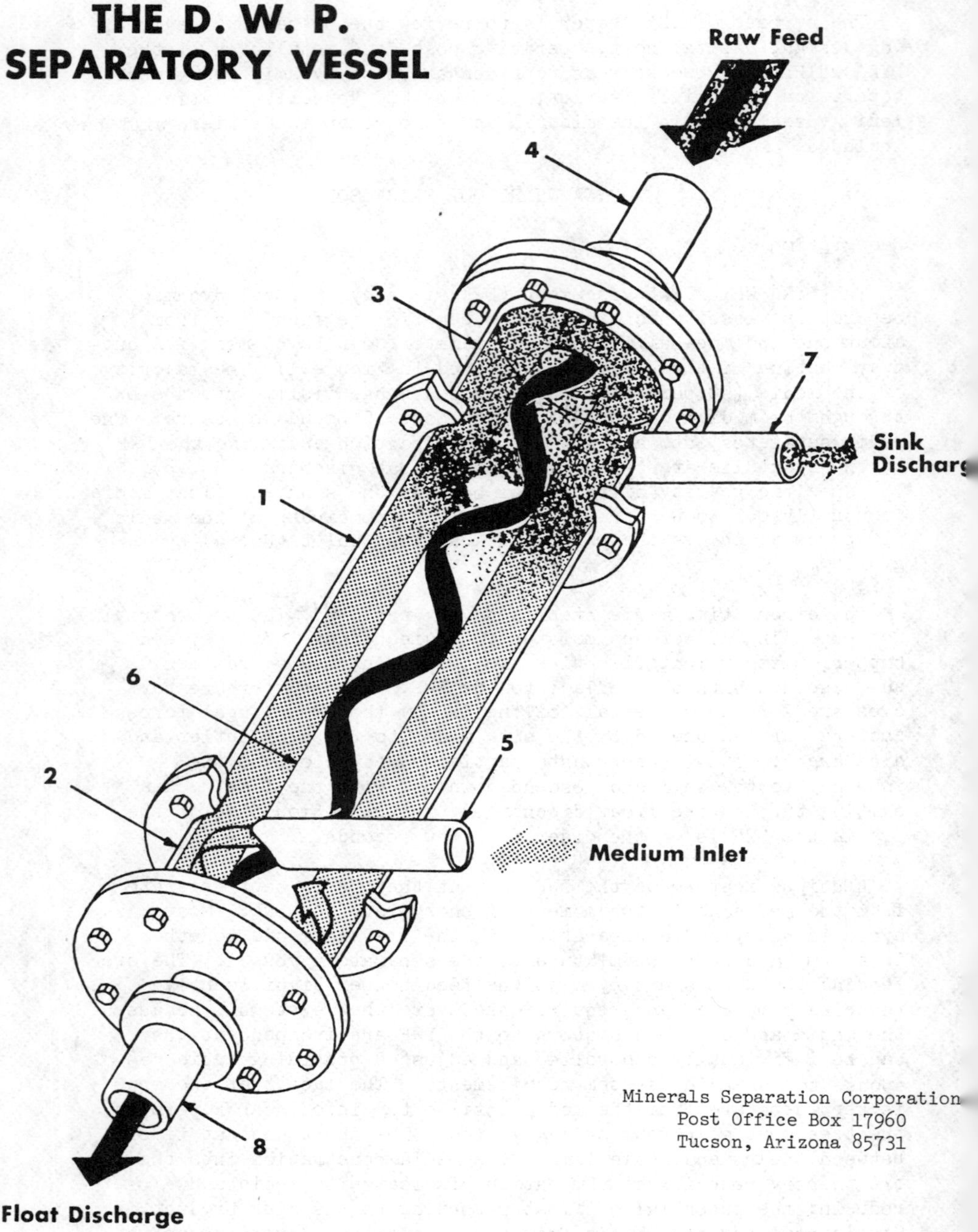

Minerals Separation Corporation
Post Office Box 17960
Tucson, Arizona 85731

Figure 1

DYNA WHIRLPOOL PROCESSOR

Testing

DYNA WHIRLPOOL Processor laboratory testing is similar to other heavy media testing procedures. The ore is visually examined, size analyzed, and assayed by size fraction to determine the valuable mineral content. The ore is then subjected to heavy liquid tests using several different gravities to determine the optimum separation gravity. The float and sink products are visually examined, weighed, and assayed. These laboratory tests are usually sufficient to determine the applicability of heavy media metallurgical techniques to the ore.

If the preliminary laboratory tests are favorable the ore is then processed through a DWP Pilot Plant. The operating parameters are established, process problems noted, and the metallurgical results analyzed. A preliminary feasibility study is made, based on the pilot plant testing, to determine the economic advantage, if any, for using heavy media separation to treat the ore. Dependent on the results, a recommendation is made for or against heavy media separation treatment.

DYNA WHIRLPOOL PROCESSOR

Equipment Sizes and Flowsheets

The available sizes, feed size limits, and tons per hour feed rates for the DWP are given in Table I. The lower feed size for all units is 65-mesh but the most common lower size treated is 20 to 28-mesh. Each ore must be individually tested to determine the optimum lower size of separation.

TABLE I

DYNA WHIRLPOOL PROCESSOR

Feed Size And Capacity

DWP		Feed Size		TPH Feed Rate	
I.D.(inches)	Length(ft.)	Max.	Min.	Max.	Min.
6	3.7	1/2"	65m	Pilot Plant Model	
9-1/2	5.1	3/4"	65m	20	10
12-1/2	5.9	1"	65m	40	20
15-1/2	6.4	1-1/4"	65m	60	40
18-1/2	6.8	1-1/2"	65m	80+	60

Equipment lists and general process flowsheets are illustrated in Figures 2 and 3. Figure 2 is the simplest DWP flowsheet and is commonly used in plants treating up to about 50 tons per hour. Figure 3 is the DWP flowsheet for plants treating in excess of 50 tons per hour. The difference between the flowsheets is in the increased number of sieve screens, washing screens, and magnetic separators required to recover and return the media for recycling in the system. Figure 3 also illustrates the fine sink product recovery circuit necessary when treating ores finer than 28-mesh.

The stationary media sieves are typically 60 degree sieve bend screens. The majority of the media is removed from the sink and float product on these screens and returned directly to the primary media sump. The sieve bend screen oversize with the remaining clinging media flows onto a single deck horizontal vibrating screen. Wedge wire screen cloth is commonly used for screen covers in this application. The size of opening for both the sieve screen cloth and the horizontal screen cloth is dependent upon the bottom size of the ore. The feed end of the horizontal screen is without water wash so that excess media can be shaken through and returned directly to the primary media sump. The center section of the screen has a spray wash system to remove media still clinging to the products. The washed media flows to a washing sump. The discharge end of the screen is without water wash, unless required, and excess water is removed from the products before discharging onto a conveyor.

The washed media is pumped to double drum magnetic separators for media recovery. The magnetic concentrate is densified in a spiral densifier. Densified media is demagnetized and recycled back to the media circuit. If the DWP separation gravity is 2.3 or less the spiral densifier can sometimes be eliminated from the circuit.

Approximately 900 to 1,000 gallons of media (about 10 tons) are required to process a ton of feed through the DWP. A small portion of the media is lost in the process and must be made-up at the primary media sump. If the ore is slimey, a portion of the primary media can be cycled through the magnetic separation circuit to remove slimes as required.

The DWP is inclined about 15 degrees from the horizontal for coal separation and about 25 degrees from the horizontal for ore separation. The unit can be adjusted, if required to improve the separation metallurgy for a specific ore.

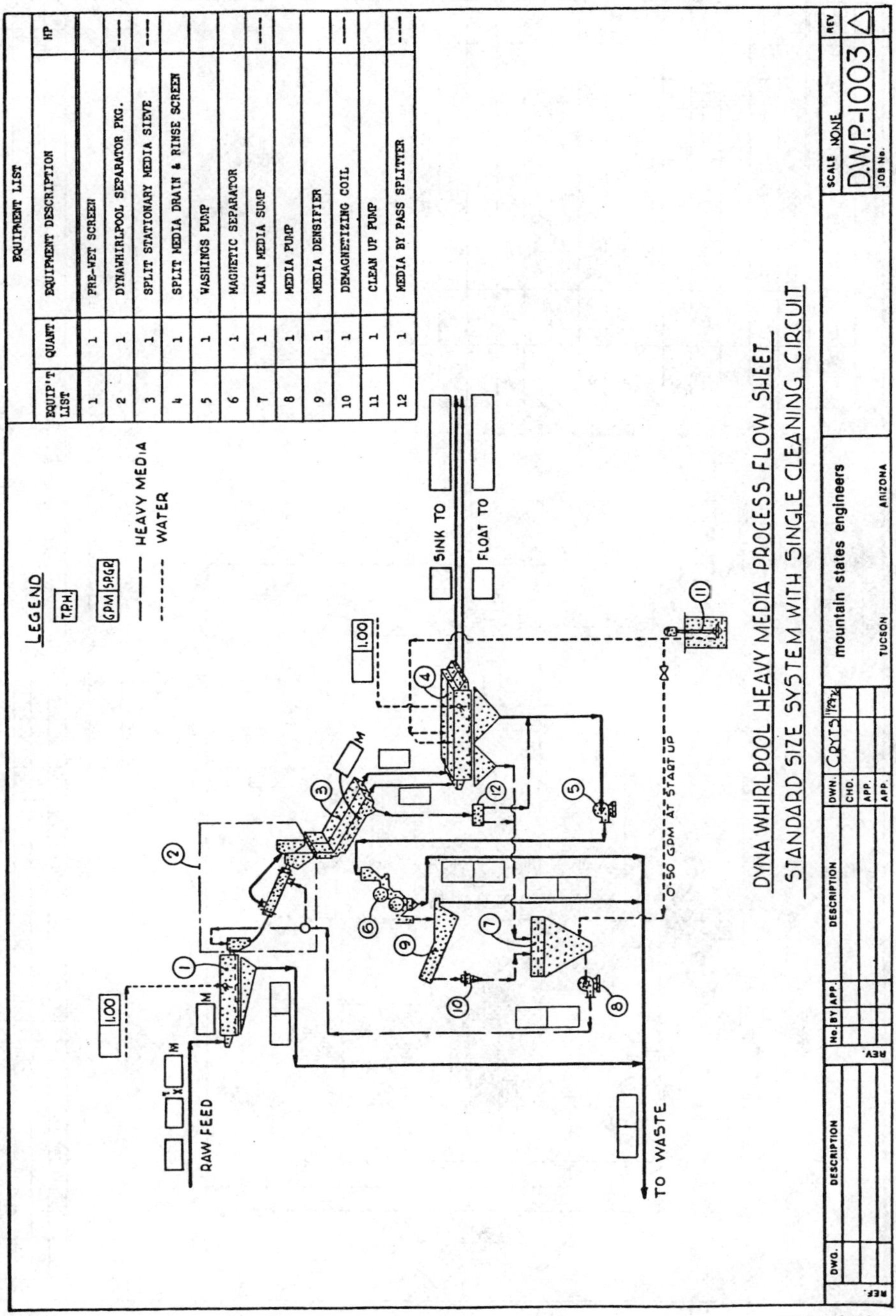
EQUIPMENT LIST
EQUIP'T LIST | QUANT. | EQUIPMENT DESCRIPTION | HP
1 | 1 | PRE-WET SCREEN
2 | 1 | DYNAWHIRLPOOL SEPARATOR PKG.
3 | 1 | SPLIT STATIONARY MEDIA SIEVE
4 | 1 | SPLIT MEDIA DRAIN & RINSE SCREEN
5 | 1 | WASHINGS PUMP
6 | 1 | MAGNETIC SEPARATOR
7 | 1 | MAIN MEDIA SUMP
8 | 1 | MEDIA PUMP
9 | 1 | MEDIA DENSIFIER
10 | 1 | DEMAGNETIZING COIL
11 | 1 | CLEAN UP PUMP
12 | 1 | MEDIA BY PASS SPLITTER
LEGEND
T.P.H.
GPM SP.GR
HEAVY MEDIA
WATER
RAW FEED
TO WASTE
1.00
SINK TO
FLOAT TO
0-50 GPM AT START UP
DYNA WHIRLPOOL HEAVY MEDIA PROCESS FLOW SHEET
STANDARD SIZE SYSTEM WITH SINGLE CLEANING CIRCUIT
mountain states engineers
TUCSON
ARIZONA
SCALE NONE
D.W.P.-1003
BRUNING 40-5333

Figure 2

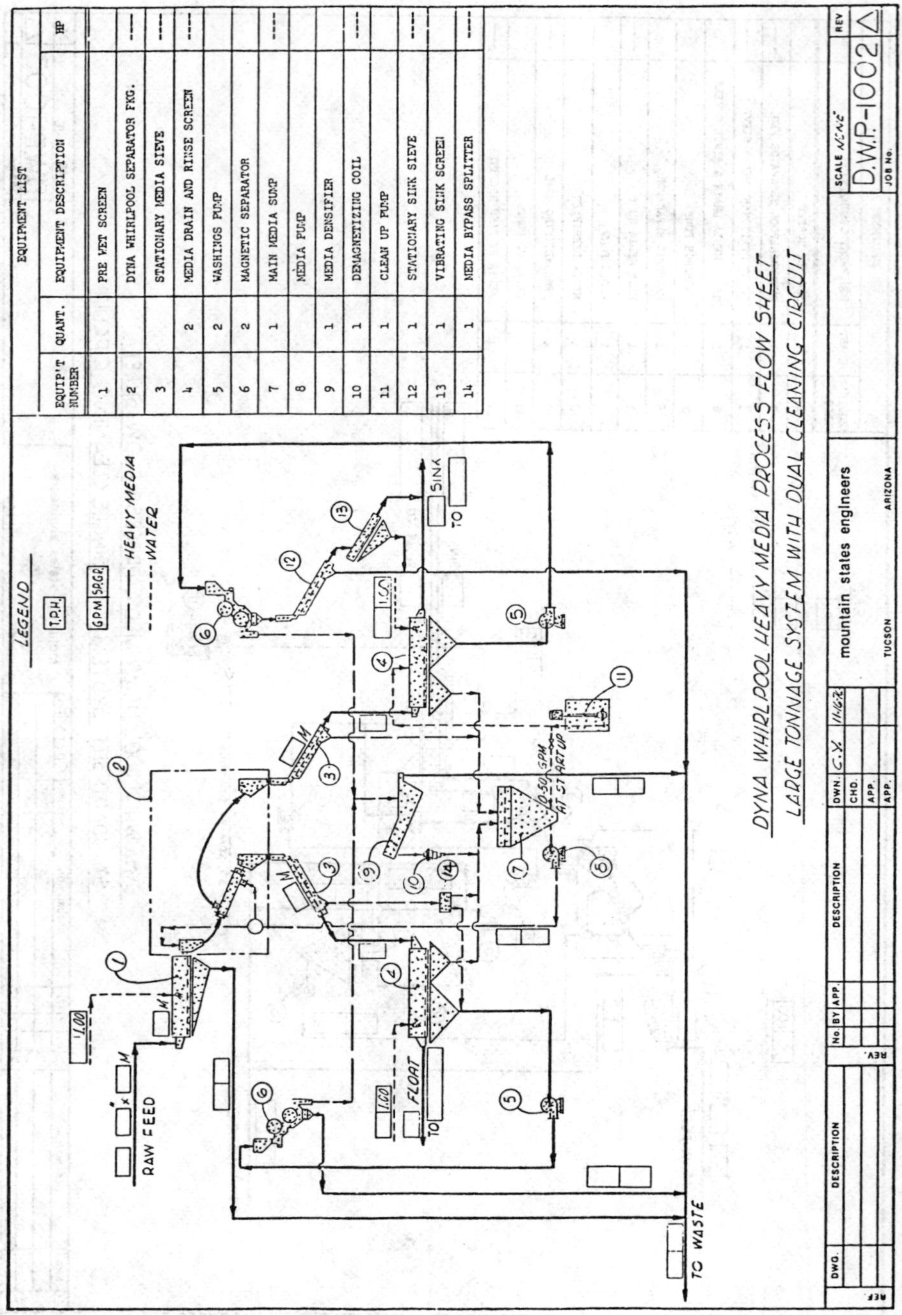
EQUIPMENT LIST
EQUIP'T NUMBER | QUANT. | EQUIPMENT DESCRIPTION | HP
1 | | PRE WET SCREEN |
2 | | DYNA WHIRLPOOL SEPARATOR PKG. |
3 | | STATIONARY MEDIA SIEVE |
4 | 2 | MEDIA DRAIN AND RINSE SCREEN |
5 | 2 | WASHINGS PUMP |
6 | 2 | MAGNETIC SEPARATOR |
7 | 1 | MAIN MEDIA SUMP |
8 | | MEDIA PUMP |
9 | 1 | MEDIA DENSIFIER |
10 | 1 | DEMAGNETIZING COIL |
11 | 1 | CLEAN UP PUMP |
12 | 1 | STATIONARY SINK SIEVE |
13 | 1 | VIBRATING SINK SCREEN |
14 | 1 | MEDIA BYPASS SPLITTER |
LEGEND
T.P.H.
GPM SP.GR.
HEAVY MEDIA
WATER
RAW FEED
TO WASTE
FLOAT
SINK
0-50 GPM AT START UP
DYNA WHIRLPOOL HEAVY MEDIA PROCESS FLOW SHEET
LARGE TONNAGE SYSTEM WITH DUAL CLEANING CIRCUIT
mountain states engineers
TUCSON ARIZONA
SCALE NONE
D.W.P.-1002
JOB No.
REV

Figure 3

DYNA WHIRLPOOL PROCESSOR

Feed Preparation

Successful application of heavy media separation in any size range and in any type of vessel is dependent upon adequate feed preparation. This critical fact is often overlooked or given insufficient attention in plant design.

DYNA WHIRLPOOL Processor feed top size is normally from one-half inch to one and one-half inches. Commonly, it is a secondary or tertiary crusher discharge product which is in closed circuit with a sizing screen. The bottom size is determined from testing. The feed preparation method should minimize the spread of the ore particle size distribution for optimum metallurgy and process weight throughput for a given size of DWP.

The crushed ore can be screened on a double deck vibrating screen with minimal water wash to remove the fines and slimes. The fines and slimes will bypass the DWP system and go directly to additional beneficiation if they contain significant values. The coarse ore is fed into a drainage bin to remove excess moisture prior to entry into the DWP.

DYNA WHIRLPOOL PROCESSOR

Media

Magnetite and ferrosilicon are the most common media choices in use today. They have a distinct advantage over other possible medias due to their magnetic property which permits a simple media recovery circuit for recycling.

The DWP media size can be from minus 48-mesh to minus 200-mesh and sometimes minus 325-mesh. Magnetite is limited to a maximum media specific gravity of about 2.6 to 2.8 while ferrosilicon can go as high as 3.3 to 3.5. Most heavy media separations in the mineral industry are in the range of 2.5 to 3.2. The coal industry uses lower separation gravities in the range of 1.3 to 1.8.

Testing of each ore is necessary to determine which media will yield the best metallurgy at the lowest cost. Coarser media minimizes problems encountered with slimes, which affect the media viscosity. Also reclamation, cleaning, and densifying of coarser media is more efficient than a finely ground media, thus minimizing media losses. The DWP has been successfully operated using magnetite, ferrosilicon, and a mixture of both.

DYNA WHIRLPOOL PROCESSOR

Capital And Operating Costs

The capital and operating cost ranges together with power and media consumption are given in Table II. The capital cost is the installed cost and includes feed preparation, the DWP system, and the media recovery system.

TABLE II

Installed Capital Cost Per/Ton Hr. of Feed		Operating Cost Per/Ton of Feed		Power	Media Loss Pounds/Ton	
Min.	Max.	Min.	Max.	KWH/Ton	Min.	Max.
$4000	$6000	$0.30	$0.50	3	1	2

The installed capital cost for DYNA WHIRLPOOL Processor plant is dependent upon such factors as plant size, feed preparation requirements, the size range of the feed, and auxillary facilities such as building requirements, power and water hookups, and similar costs. Generally, the larger the plant the lower the installed cost per ton. The size range of the feed, whether predominantly fine, or predominately coarse affects the maximum throughput tonnage of a DWP unit and consequently the number of units required.

The plant area for the DWP is small with respect to the tonnage processed. A 30 ton per hour plant will cover about 800 square feet and a 125 ton per hour plant (1,000,000 tons per year) will cover about 1,500 square feet. The equipment is located on three floors, including the basement, and is arranged to maximize gravity flow of ore and media.

The operating cost for the DWP plant is the sum of labor, power, maintenance supplies, media losses, water, and miscellaneous. The most significant cost saving in high tonnage plants is the unit cost of labor per ton of feed. Because the DWP circuit is relatively simple and can be automated, one man can operate several parallel circuits as readily as one circuit.

Costs such as power, maintenance supplies, media losses, and water requirements are typically more linear than the labor cost with respect to unit process costs based on feed tonnage. Approximately 60 to 70 percent of the power requirement will be consumed pumping the media. The remainder will be for conveying, magnetic separation, screening, and lighting. Maintenance costs will average from 15 to 25 percent of the total operating cost. Feed size and abrasiveness of the products are significant factors affecting maintenance costs. Makeup water requirements are about four gallons per minute per ton of feed.

DYNA WHIRLPOOL PROCESSOR

Metallurgy

In the mineral industry heavy media separation, as exemplified by the DWP, will be most commonly used as a pre-concentration tool. The purpose is to eliminate a significant portion of the feed, while maintaining high valuable mineral recovery, prior to expensive treatment techniques, particularly grinding. In certain cases, as with coal and iron ore, heavy media separation is used to make a final product. The following examples illustrate the versatility of the DWP as a metallurgical tool.

CASE 1 - POCAHONTAS SEAM COAL

Operating Criteria

1/2-inch x 60-mesh feed material,
13.20 percent ash in feed, and
Separation at 1.50 specific gravity.

Metallurgical Results

Product	Percent Weight	Assay (Percent) Coal	Ash	Percent Distribution Coal	Ash
Float	86.89	94.76	5.24	94.88	34.47
Sink	13.11	34.03	65.97	5.12	65.53
FEED	100.00	86.80	13.20	100.00	100.00

This particular West Virginia seam coal responded remarkably well to the DYNA WHIRLPOOL treatment. Ash content was reduced from 13.2 to 5.2 percent at a coal recovery approaching 95 percent. The BTU enhancement of the product more than pays for the 30 cents per ton treatment cost.

CASE 2 - MINNESOTA GOETHITE ORE (Pilot Test)

Operating Criteria

Minus 3-mesh x 28-mesh feed,
47.2 percent iron, 29.0 percent insoluble feed,
Media feed gravity: 2.71,
Media float gravity: 2.43,
Media sink gravity: 3.02, and
Minus 48-mesh magnetite and minus 200-mesh ferrosilicon media.

Metallurgical Results

Product	Percent Weight	Percent Assay Fe	Percent Assay Insol.	Percent Distribution Fe	Percent Distribution Insol.
Float	26.04	18.0	72.0	9.93	64.59
Sink	73.96	57.5	13.9	90.07	35.41
FEED	100.00	47.2	29.0	100.00	100.00

Checking the sink product in heavy liquids after processing the ore through the DWP indicated a high separation selectivity. Less than four percent of the sink product was lighter than the DWP separation gravity and should have reported in the float. Similarly only about six percent of the float product should have reported to the sink.

CASE 3 - TENNESSE ZINC ORE

Operating Criteria

5/16-inch x 35-mesh feed material,
3.77 percent zinc in feed,
Separation at 2.85 specific gravity,
500,000 tons per year throughput to DWP.

Metallurgical Results

Product	Percent Weight	Assay (Percent) Zinc	Percent Distribution Zinc
Sink	19.90	17.7	93.37
Float	80.10	0.308	6.63
FEED	100.00	3.77	100.00

In the original flowsheet of this installation static heavy media was used on the 2-inch by 5/16-inch fraction of the ore feed. The heavy media float was discarded and the heavy media sink was combined with the minus 5/16-inch undersize, ground to 65-mesh, and concentrated by flotation. The institution of the DWP system on the 5/16-inch by 35-mesh fraction resulted in a decrease in load to the grinding circuit of 13.5 percent. The total plant throughput was increased so that the grinding circuit could operate at maximum efficiency.

CASE 4 - BOLIVIAN TIN-TUNGSTEN ORE

Operating Criteria

3-mesh x 28-mesh feed (vein ore),
0.21 percent tin, 0.42 percent tungsten in feed,
Media feed gravity: 2.99,
Media float gravity: 2.88,
Media sink gravity: 3.14, and
Minus 200-mesh ferrosilicon media.

Metallurgical Results

Product	Percent Weight	Assay Percent		Percent Distribution	
		Sn	WO_3	Sn	WO_3
Float	74.82	0.062	0.058	22.11	10.36
Sink	25.18	0.64	1.48	77.89	89.64
FEED	100.00	0.21	0.42	100.00	100.00

The effect of installing the DYNA WHIRLPOOL Processor to preconcentrate this ore will be to eliminate 75 percent of the feed weight prior to conventional concentration. This will reduce the capital requirement for grinding, tabling, jigging, and flotation equipment including size of building and manpower requirements.

CASE 5 - WESTERN U.S. BARITE-FLUORITE ORE (Pilot Test)

Operating Criteria

Minus 1/2-inch x 20-mesh,
13.2 percent fluorite, 11.8 percent barite,
Media feed gravity: 2.67,
Media float gravity: 2.37,
Media sink gravity: 2.92, and
Minus 48-mesh magnetite and minus 200-mesh ferrosilicon media.

Metallurgical Results

Product	Percent Weight	Percent Assays CaF_2	Percent Assays $BaSO_4$	Percent Distribution CaF_2	Percent Distribution $BaSO_4$
Float	57.86	2.0	0.15	8.79	0.73
Sink	42.14	28.5	27.9	91.21	99.27
FEED	100.00	13.2	11.8	100.00	100.00

This ore responded exceedingly well to DWP treatment, resulting in a 58 percent weight rejection while maintaining very high fluorite and barite recovery.

The DWP has been employed in over 80 industrial applications, approximately half of which were coal, since the first unit was installed in 1963. A partial list of installations include separation of garnet from quartz (New York), chromite from serpentine (Turkey), spodumene from quartzite (North Carolina), fluorspar from gangue (several), iron ore from silicates (Africa), copper minerals from dolomite (Zaire), manganese minerals from amphibolites (Brazil), lead-zinc minerals from limestone (Iran), tin and tungsten minerals from shale (Bolivia), and Coquina shells from clam shells (Peru).

DYNA WHIRLPOOL Processor plants are currently operating with media specific gravities from 1.3 to 3.5. The float to sink product ratios are from five percent float and 95 percent sink to 95 percent float and five percent sink. Plant feed capacities are from 85,000 to 3,300,000 tons per year.

A list of DWP installations is available, upon request from:

Minerals Separation Corporation
Post Office Box 17960
Tucson, Arizona 85731

OCC VESSEL

Description

The OCC Vessel is an example of the static type of heavy media vessel. The static vessel features a relatively large quiescent pool permitting the float product to rise to the surface and the sink product to descend to the bottom of the vessel. Figure 3 illustrates the OCC Vessel. Feed material is introduced to the vessel at the surface level of the media slurry. Float material travels across the vessel and discharges over the float overflow weir. Sink material drops to the bottom of the vessel and is removed by an oscillating rake which alternately carries sink to the discharge lips at each end of vessel. Sink material overflows into launders and then to the sink discharge chute, which may be designed to discharge at either the front or rear of the vessel. Float barrier screens in front of each discharge lip effectively prevent the loss of float material with the sink at the instant of discharge to the sink launder.

The OCC Vessel features minimum headroom and space, thereby reducing building requirements. The only moving part in the media is the rake and the only friction in the vessel is between the rubber parts of the blade and the vessel bottom. The media pool is quiet, reducing entrapment problems due to eddy currents in the media. Power consumption is low.

The design of the OCC Vessel makes it particularly adaptable to high float ores or ores where a large quiet pool is critical to achieve satisfactory ore/waste separation.

OCC VESSEL

Equipment Sizes And Flowsheet

The available sizes, feed size limits, and tons per hour feed rates for the OCC Vessel are given in Table III. The longer units in each capacity range are to accommodate slower separating ores.

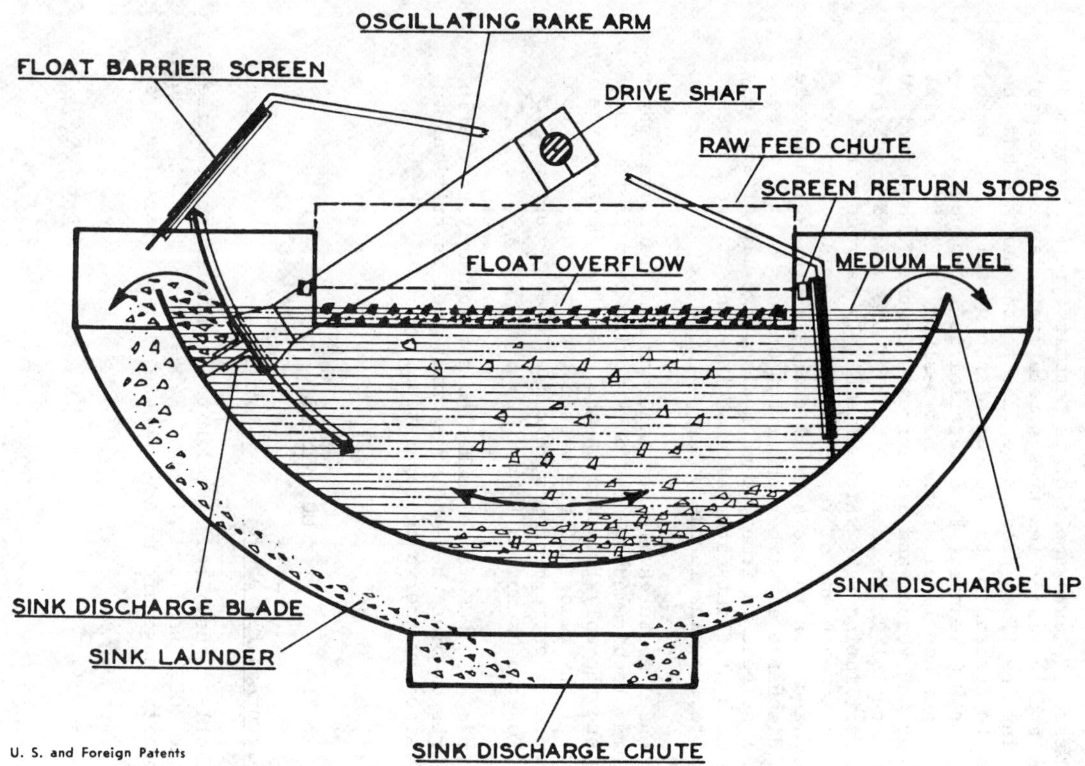

Figure 4

OCC VESSEL

Minerals Separation Corporation

TABLE III

OCC VESSEL

FEED SIZE AND CAPACITY

Model No.	Arm Radius (Inches)	Length (Feet)	Pool Area (Sq.Ft.)	Feed Size Max. (Inches)	Feed Size Min. (Inches)	TPH Feed Rate
405	40	5	18.5	3	1/4	25
525	52	5	27.5	6	1/4	100
526	52	6	33.0	6	1/4	100
528	52	8	44.0	6	1/4	100
606	60	6	36.0	6	1/4	200
608	60	8	48.0	6	1/4	200
686	68	6	42.5	6	1/4	300
687	68	7	50.0	6	1/4	300
688	68	8	57.0	6	1/4	300

The basic equipment list and process flowsheet are illustrated in Figure 5. The media recovery and densifying equipment is similar for all heavy media plants and operate the same way whether recovering media for a static or dynamic separation.

OCC VESSEL

CAPITAL AND OPERATING COST

The capital and operating cost ranges together with power and media consumption are given in Table IV. The capital cost is the installed cost and includes feed preparation, the OCC Vessel, and the media recovery system.

MINERALS SEPARATION CORP.

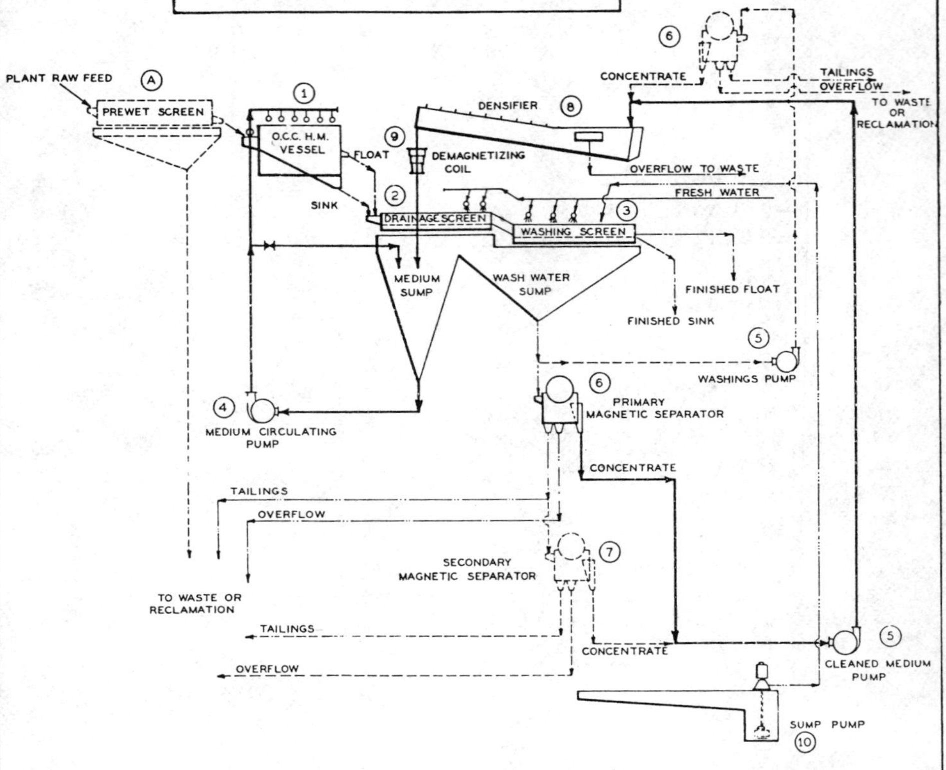

EQUIPMENT LIST

ITEM NO	NO REQ'D		SIZE	HP
A		PREWET SCREEN		
1		OCC H.M. VESSEL		
2		DRAINAGE SCREEN		
3		WASHING SCREEN		
4		MEDIUM CIRCULATING PUMP		
5		CLEANED MEDIUM PUMP		
		WASHINGS PUMP		
6		PRIMARY MAGNETIC SEPARATOR		
7		SECONDARY MAGNETIC SEPARATOR		
8		DENSIFIER		
9		DEMAGNETIZING COIL		
10		SUMP PUMP		

DRAWING NUMBER
O.C.C. O-179 - B

Figure 5

TABLE IV

GENERAL OCC VESSEL COSTS

Installed Capital Cost Per Ton Hr. of Feed		Operating Cost Per Ton of Feed		Power	Media Loss Pounds/Ton	
Min.	Max.	Min.	Max.	KWH/Ton	Min.	Max.
\$2000	\$4000	\$0.20	\$0.40	2	0.5	1

The capital and operating cost factors for the OCC Vessel are similar to the DWP. The large tonnage capacities of these units permits a sizable reduction in installation and operating costs per ton hour over lower tonnage plants.

OCC VESSEL

Metallurgy

The OCC Vessel is used in numerous mineral applications and is particularly well adapted to high float/low sink ore splits such as placers. A recent used has been in the auto scrap industry separating aluminum from die cast zinc alloy in several locations. The large, quiet pool of the OCC is particularly adapted to difficult separations such as irregular pieces and near gravity applications.

Chapter 26

HEAVY MEDIA SEPARATION FLOWSHEET DEVELOPMENT

by Colin G. Wilson

PRODUCT MANAGER

WEMCO DIVISION,ENVIROTECH CORPORATION
Sacramento, California

ABSTRACT

This paper considers the heavy liquid laboratory tests used to determine the amenability of a potential ore to the heavy media process. The results of the laboratory testing are correlated to actual plant performance and establish design parameters for the heavy media process. These parameters are used in the development of the metallurgical flowsheet and in the sizing of equipment in the metallurgical plant design.

INTRODUCTION

The metallurgical plant design program for a heavy media system can be divided into two stages. The first stage is that of laboratory testing of a representative sample of the potential ore to determine the amenability of the heavy media process. The laboratory testing also delineates certain parameters that are required in the selection of equipment within the final plant.

The second stage is that of developing the flowsheet and selecting the necessary equipment to perform the metallurgical function. Heavy media separation is somewhat distinct from other metallurgical concentration processes in that there is generally no necessity to include the pilot plant stage into the program. This is related to the exactness of the laboratory heavy liquid testing and the ability of the heavy media system to accurately reproduce the laboratory results.

LABORATORY TESTING

The collection and the size of the sample required for testing will not be included in the scope of this paper but the importance of obtaining a representative sample cannot be overstressed. Without a representative sample both the test work and resulting design consideration will be of no consequence.

The purpose of laboratory heavy liquid testing is to determine the washability characteristics of an ore. This establishes the specific gravity and particle size at which one of the components is liberated and concentrated.

Method of Analysis

The tests are performed by consecutively immersing the sample in a series of heavy liquids at different densities and by collection of the float and sink products at these densities. Depending on the specific gravity at which most of the component will be removed at an early stage, the test will either start at the highest density and work through the different densities to the lowest density or visa versa. For samples of one or two pounds, a 2 liter beaker is most suitable for the heavy liquid tests. A Buchner funnel and flask are useful in washing the liquids off the products. The products are washed with alcohol or methylated spirits to remove any remaining heavy liquids. These wash liquids evaporate rapidly leaving dry solids. The washings are collected for recovery of the heavy liquids.

Initially, visual inspection of the sample will in many cases permit an estimation of the appropriate size reduction required to obtain liberation of the components. A test sample is prepared by crushing to this size and screened to remove fine materials.

The samples obtained from each separation are then analyzed for the main components. Examination of the test results will give indications as to liberation of the components at that size range. Dependent on the degree of liberation, further tests will be carried out at a coarser or finer size reduction. Typical results of such test work are shown in Table I.

Heavy Liquids Used

The heavy liquids most commonly used are as follows:

1. Carbon Tetrachloride diluted with benzene used for densities up to 1.6.

2. Zinc Chloride in a water solution used for densities up to 1.8.

3. Tetrabromidoethane used for densities up to 3.3.

Either Dibutyl phthalate or carbon tetrachloride are used for dilution of tetrabromide or methylene iodide for intermediate densities.

When large quantities of samples are required to be tested, the heavy liquids are made with suspension of media (magnetite or ferrosilicon) in water.

TABLE I

RESULTS OF HEAVY LIQUID TESTS

	Sample I - 3-15mm			Sample II - 3-30mm		
Density	Wt.%	Pb %	Pb Distribution	Wt.%	Pb %	Pb Distribution
<2.65	9.30	0.20	0.4	4.40	0.31	0.3
2.65 - 2.70	22.10	0.29	1.3	16.70	0.35	1.2
2.70 - 2.75	24.40	0.30	1.6	21.50	0.37	1.7
2.75 - 2.80	6.90	0.35	0.5	10.80	0.39	0.9
<2.8	62.70	0.29	3.8	53.40	0.36	4.1
>2.8	37.30	12.36	96.2	46.60	9.83	95.9
2.80 - 2.85	5.40	0.70	0.8	7.10	0.68	1.0
2.85 - 2.90	3.70	1.20	0.9	6.90	1.28	1.8
- 2.90	28.20	16.06	94.5	32.60	13.63	93.1
	100	4.79	100.0	100	4.77	100.0

Separation Density - 2.8

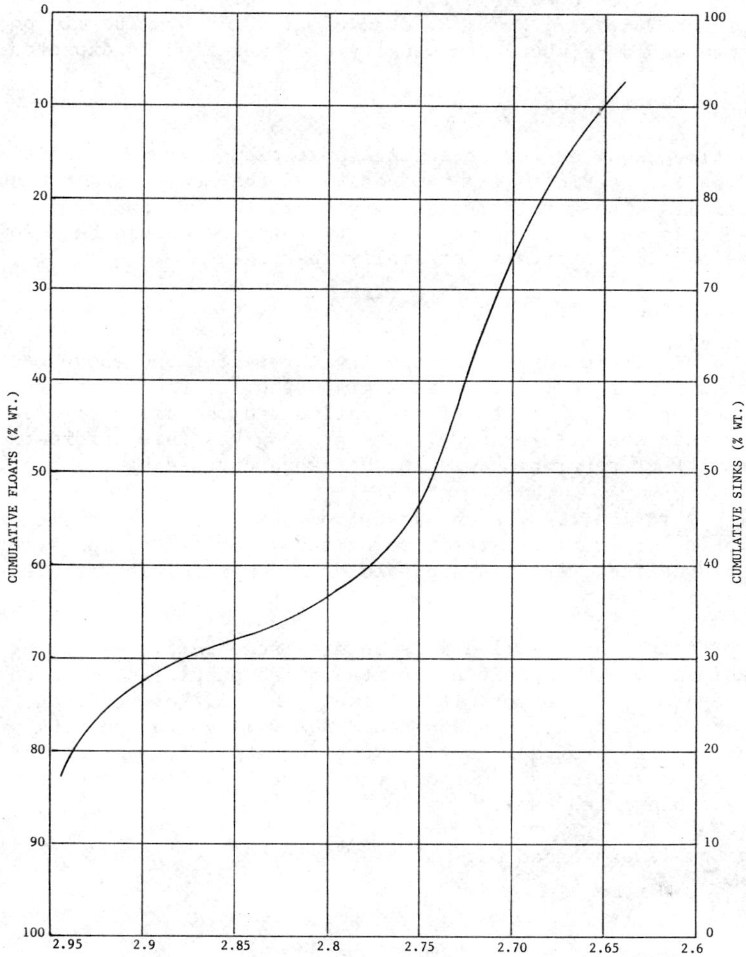

Figure 1. Densimetric Curve

The utmost care must be taken when using these hydrocarbons because of their volatility and the toxicity of their fumes. Testing should always be carried out in a down-draft fume cupboard.

Range of Densities

For the heavy liquid tests a range of densities are selected in progressing steps of 0.05 or 0.1. A typical suite of specific gravity would range 2.6, 2.7, 2.75, 2.80, 2.85, 2.90, 3.0. The mid point of the range would be where the density of separation is expected to fall.

Time Required for Separation

The time required for effecting separation of floats and sinks will depend on the particle size, viscosity of the media, and the quantity of particles with a relative density close to the density of separation. For large particles, at least two minutes should be allowed with increasing time intervals for smaller particles.

Interpretation of Test Results

The densimetric curve (washability curve) for the above test results are shown in Figure 1. This is a graph showing the relationship between the specific gravity of separation and the amount of float or sink at that specific gravity. The graph gives this information for any gravity of separation within the range covered by the test.

The separation results obtained from the test work are in affect a description of the ideal response of the material to separation at the various specific gravities and should be compared with figures obtained from practice.

In practice there will always be misplaced particles; it can be said that the closer the specific gravity of the particles is to the density of separation, the more it is likely that 50% of these particles will report to the float product and 50% will go to the sink product. The way to arrive at a measure of the deviation from a perfect separation is by a statistical method, the Ecart Probable Moyen (this is usually abbreviated Epm).

The Epm is expressed as:

$$Epm = \frac{d_{25} - d_{75}}{2}$$

where d_{25} is the specific gravity at the distribution number 25, and d_{75} is the specific gravity at the distribution number 75.

The Epm value is influenced by the type of vessel in which the separation takes place, the viscosity of the media, density at which the separation takes place, distance apart of the material particles in a separation pool of heavy media, time available for separation in the heavy media pool and particle size and shape of the materials. The Epm value is obtained from plant testing of a particular vessel in the H.M.S. circuit.

Figure 2 shows various values of Epm's at a separation density of 2.80. The deviation (partition coefficient), from this graph results in a tool for reconstituting the ideal test results to obtain a prediction of the plant performance.

Table II shows the influence of Epm's of 0.040 and 0.045 on the metal grade of the floats, on the metal recovery, and the amount of floats which can be expected from a heavy media separation plant when treating this material. These calculations also show the influence of liberation on the product weights and grade. At the finer feed condition a larger quantity of reject material is generated and a higher recovery of metal is achieved. In this example it would appear that there is an advantage to go to the 3-15mm feed sizing. However, in a total plant concept, consideration should also be given to the fact that liberation size has an economic limit. The amount of fines increase with reduced feed sizing and these fines effect the overall plant economics.

PROCESS FLOWSHEET DEVELOPMENT & EQUIPMENT SELECTION

From the results of heavy liquid tests the process flowsheet is then developed taking into consideration tonnage rate, liberation size required, and the equipment best suited for the specific application. A variety of heavy media separation vessels are used in heavy media plants and can be classified into two groups. One group can be defined as using a quasi-static pool of media for separation, and would include the Aitkens Classifier, Drag Tank, Drewboy, Drum and Teska Vessel. The top size of feed to these units varies from 150mm (6") to 600mm (24") and the minimum size of feed varies from 6mm (1/4") to 0.5mm (28 mesh). The second group can be defined as using a relatively fast moving stream of media for separation and would include the Dyna-Whirlpool and Heavy Media Cyclone. The top size of feed to these units varies from 50mm (2") to 15mm (3/4") and the minimum size of feed varies from 0.5mm (28 mesh) to 0.2mm (65 mesh). For the purpose of this paper remarks will be limited to a heavy media system incorporating a WEMCO Heavy Media Separation Drum. From a flowsheet point of view, only slight variations to the flowsheet would be required when considering other separatory vessels. Figure 3 shows a typical Process Flowsheet for the Heavy Media System. The design considerations for each unit of equipment in the flowsheet will now be considered.

Feed Preparation Section

The importance of the feed preparation section prior to heavy media

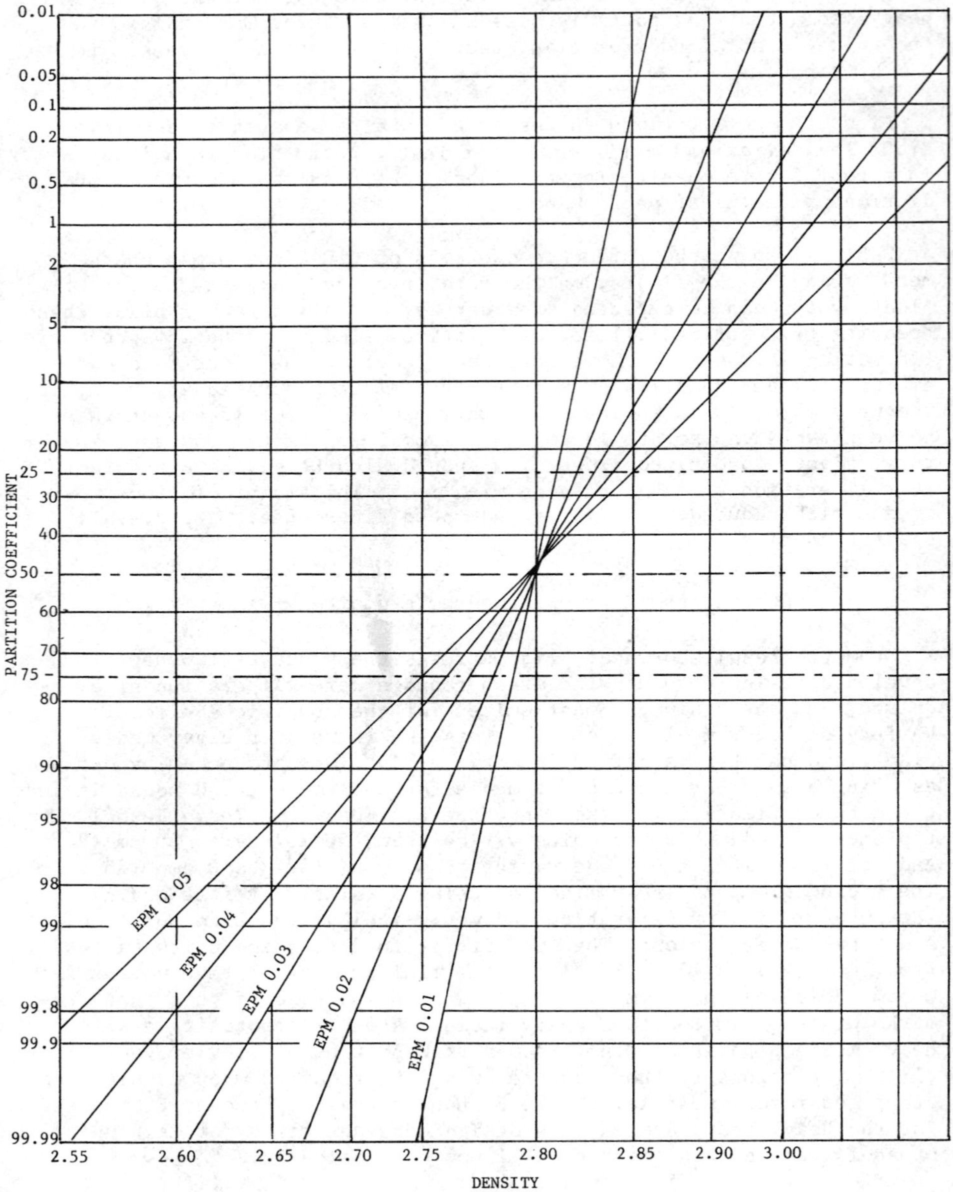

Figure 2. EPM Curve

TABLE II

CORRECTED RESULTS OF HEAVY LIQUID TESTS

Size 3-15mm

Density	Wt.%	Partition Coeff.	Epm 0.040				Partition Coeff.	Epm 0.045			
			Float		Sink			Float		Sink	
			Wt.%	Dist. of Pb	Wt.%	Dist. of Pb		Wt.%	Dist. of Pb	Wt.%	Dist. of Pb
<2.65	9.30	99.85	9.29	0.4	0.01	0	99.55	9.26	0.4	0.04	0.0
2.65 - 2.70	22.10	98.2	21.70	1.3	0.40	0	97.0	21.44	1.3	0.66	0.0
2.70 - 2.75	24.40	89.5	21.84	1.4	2.56	0.2	87.0	21.23	1.3	3.17	0.3
2.75 - 2.80	6.90	66.0	4.55	0.3	2.35	0.2	64.0	4.42	0.3	2.48	0.2
Σ	62.70		57.38	3.4	5.32	0.4		56.35	3.3	6.35	0.5
Σ	37.30		2.24	0.6	35.06	95.6		2.39	0.9	34.91	95.3
2.80 - 2.85	5.40	33.5	1.81	0.3	3.59	0.5	35.0	1.74	0.3	3.66	0.5
2.85 - 2.90	3.70	10.0	0.57	0.1	2.33	0.8	13.0	0.48	0.1	3.22	0.8
>2.90	28.20	0.2	0.06	0.2	28.14	94.3	0.6	0.17	0.5	28.63	94.0
Σ	100.00		59.62	4.0	40.38	96.0		58.74	4.3	41.26	95.7
Assay % Pb			0.32		11.4			0.35		11.1	

Size 3-30mm

Density	Wt.%	Partition Coeff.	Epm 0.040				Partition Coeff.	Epm 0.045			
			Float		Sink			Float		Sink	
			Wt.%	Dist. of Pb	Wt.%	Dist. of Pb		Wt.%	Dist. of Pb	Wt.%	Dist. of Pb
<2.65	4.40	99.85	4.39	0.3	0.01	0.0	99.55	4.38	0.3	0.02	0.0
2.65 - 2.70	16.70	98.2	16.40	1.2	0.30	0.0	97.0	16.20	1.2	0.05	0.0
2.70 - 2.75	21.50	89.5	19.24	1.5	2.26	0.2	87.0	18.71	1.4	2.79	0.3
2.75 - 2.80	10.80	66.0	7.13	0.6	3.67	0.3	64.0	6.91	0.6	3.89	0.3
Σ	53.40		47.16	3.6	6.24	0.5		46.20	3.5	7.20	0.6
Σ	46.60		3.13	0.7	43.47	95.2		3.50	1.1	43.01	94.8
2.80 - 2.85	7.10	33.5	2.38	0.3	4.72	0.7	35.0	2.49	0.3	4.61	0.7
2.85 - 2.90	6.90	10.0	0.69	0.2	6.21	1.6	13.0	0.90	0.2	6.00	1.6
>2.90	32.60	0.2	0.06	0.2	32.54	92.9	0.6	0.20	0.6	32.40	92.5
Σ	100.00		50.29	4.3	49.71	95.7		49.79	4.6	50.21	95.6
Assay % Pb			0.41		9.2			0.44		9.1	

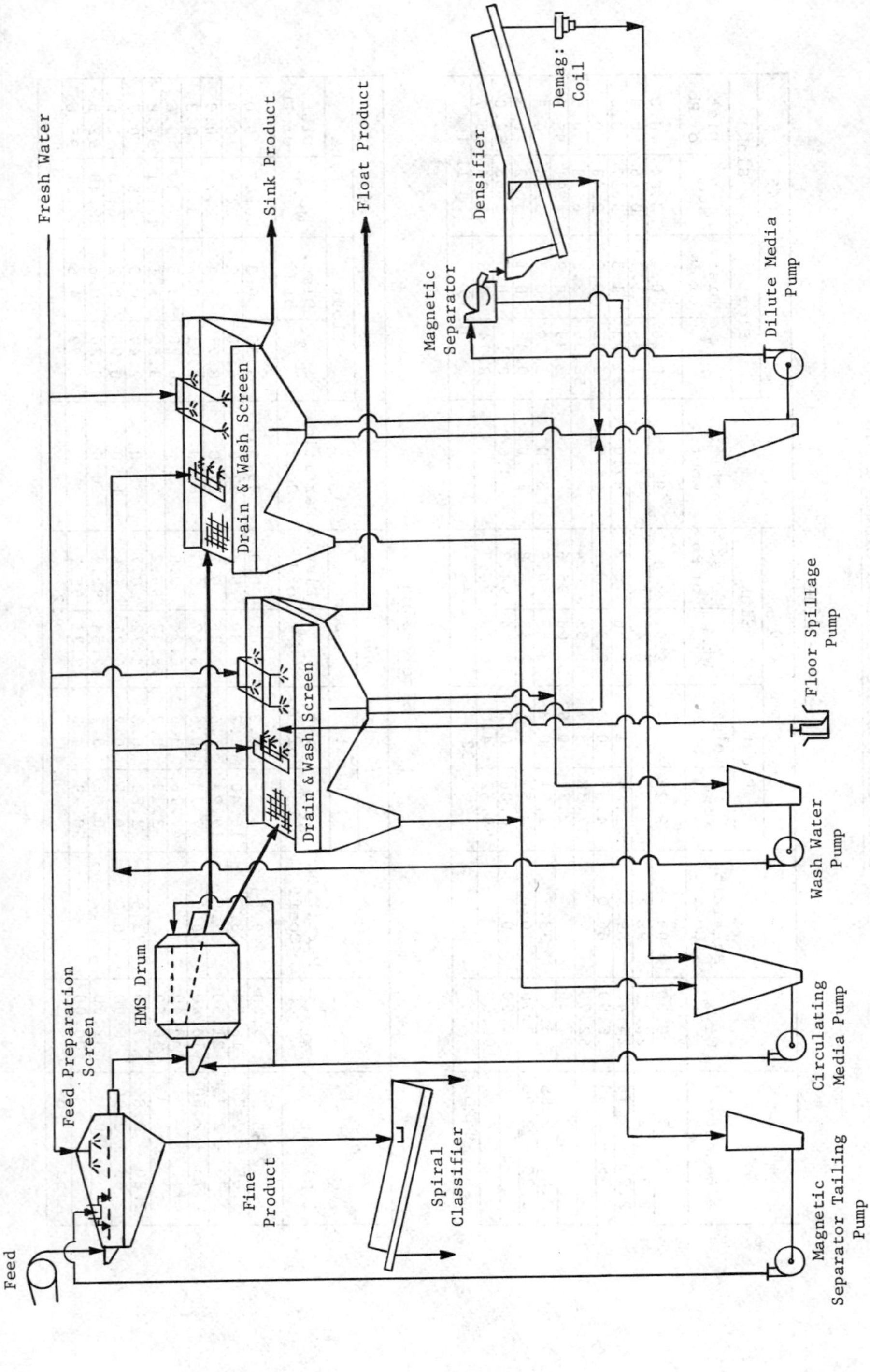

Figure 3. Typical HMS Flowsheet

separation cannot be over stressed. It is essential to present the feed material to the separatory vessel with a limited amount of fine or clay material. The addition of excessive amounts of undersized materials to the media circuit will increase the viscosity of the media with resultant reduced separation efficiency.

When the feed material comprises excessive clay materials adhering to the particles, it would be necessary to initially scrub the material in a scrubber which would then be followed by a screening operation. A feed condition with only a small amount of fines present would only necessitate a screening operation. When the feed material comprises excessive clay materials adhering to the particles, it would be necessary to initially scrub the material in a scrubber which would then be followed by a screening operation.

Horizontal vibrating screens are preferred as feed preparation screens because they permit more efficient dewatering of the feed to the H.M.S. vessel. Water does not run along the screen frame into the H.M.S. vessel, as is the case with inclined screens. The introduction of water at this point will dilute the media in the H.M.S. vessel and reduce the specific gravity of the separation.

Single deck screens have the advantage of easy access to the deck for repairs and the water-sprays are easier to maintain. On the other hand, if the bottom size feed for the H.M.S. requires a relatively fine deck and the top size is large, the capacity and life of a single deck would be too short and benefit can be obtained by going to a double deck screen.

Generally speaking, for quite low tonnage rates, 3 048mm (10') long screens should be considered as the minimum. 3 656mm (12') and 4 267 mm (14') long screens are more often used, but 4 877mm (16') long screens are recommended for all cases where the quantity of fines in the feed is high, where the material layer on the screen is several particles thick, and where there are difficult surface cleaning problems.

If the deck opening is relatively fine, the screen decking should be wedgewire. Woven wire, with the high percentage of open surface, is an advantage, but the life of woven wire screen is relatively short. Perforated steel plate lasts a long time, but the relatively low open surface has to be compensated by a larger screen size.

A good guidance figure for feed preparation screen capacity is 74.41 t/m (25 t/ft) width of screen, with deduction of 76mm (3") for the wall effect on both the screen sides. This figure will vary with the size composition, the bottom and top size and the specific gravity of the feed material. Screen manufacturers will always assist in the final selection of the screen.

The amount of spray water used on the feed preparation screens depends largely on the nature of the material, a good guidance figure is

16.7 liters per second, per ton per hour (4.4 US GPM per ton per hour) of feed material. On coal feed preparation screens, the spray water requirement can be 8-10 liters per second per ton per hour (2.0-2.6 US GPM per ton per hour) if there is a dry screening ahead of the feed preparation screen. Applications such as fluorspar, calcareous minerals, some siderite ores and similar, with nearly no clay or clayish materials in the feed require minimum water while siderite ores with limonite are, on the contrary, generally very clayish and need ample spraying.

A very efficient means of utilizing return water or magnetic separator underflow is the use of flood boxes at the entrance of the feed to the screen. These boxes can have a single or double overflow and build a water curtain which efficiently flushes the fines through the deck. Fresh water sprays should be used for the first one third to two thirds of the deck length while the last third of the deck length, on the discharge end, is used as a dewatering deck for drainage of water from the materials. The treatment of feed preparation screen undersize fraction will not be covered in the scope of this paper. Generally, this fraction requires dewatering prior to further treatment. This dewatering stage can include classification to remove the coarser fraction followed by thickening of the classifier overflow product.

The Heavy Media Separation Vessel

The H.M.S. Drum is one of the versatile and popular vessels used for heavy media separation on feed materials within the range of 6mm to 200mm (1/4" to 8"). The H.M.S. Drum is used for either two, three or four product separations and is illustrated in Figure 4. The single compartment drum is used for the two product separation.

The two compartment drum is used for the three or four product separation and is illustrated in Figure 5. A radial partition divides the drum into two separate compartments with each compartment functioning individually. This unit is adapted to three feeding arrangements, namely:

1. On dual-gravity operation: new feed material is fed to the first compartment for low gravity separation. The sink product from this compartment is elevated by the lifters and feeds into the second compartment for high gravity separation.

2. On dual-gravity operation: different feed material is fed into either compartment with compartments operating at different densities.

3. On single gravity operation: either the same feed or a different sized feed is fed into each compartment operating at the same density.

The design conditions will dictate which of these systems are applicable. The size of H.M.S. drum is determined by taking into consider-

Figure 4. WEMCO H. M. S. Drum
(WEMCO DIVISION, ENVIROTECH CORPORATION)

General specification data

Single-gravity systems

Single drum
Sizes: 6′ to 12′ diameter
Capacities: to 400 tph
Power requirements: 5 to 20 hp

Single drum, dual overflow
Sizes: 6′ x 11′ to 15′ x 23′
Capacities: to 900 tph
Power requirements: 10 to 30 hp

Dual-gravity systems

Dual-gravity, three-product drum
Sizes: 6′ x 10′ to 12′ x 20′
Capacities: to 325 tph
Power requirements: 10 to 20 hp

Dual-gravity, four-product drum
Sizes: 6′ x 10′ to 12′ x 20′
Capacities: to 650 tph
Power requirements: 10 to 20 hp

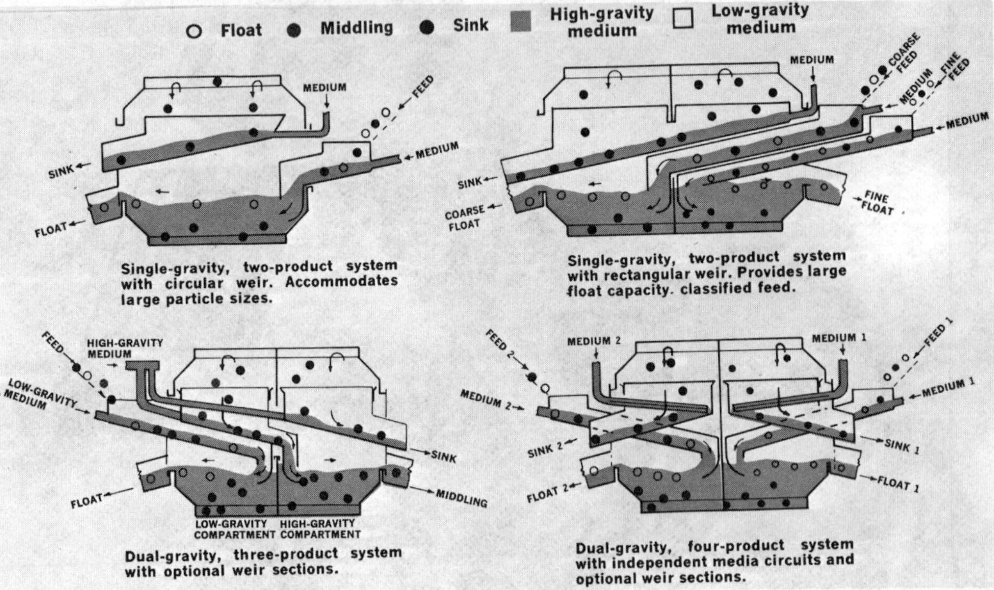

Figure 5. Diagram of Drum Product Systems
(WEMCO DIVISION, ENVIROTECH CORPORATION)

ation the feed tonnage rate, expected tonnage rate of sink and float products, feed particle sizing, density of separation and specific gravity of the materials. These functions effect requirement of effective pool area, retention time in the drum, lifting capacity and depth of media over the drum discharge lip. Table III illustrates some of these dimensions for various sizes of H.M.S. Drums.

The H.M.S. Drum is selected from the parameter giving the largest diameter of the drum.

Pool Area

The effective pool area requirement is dependent on the mean feed particle size and the quantity of float/sink material. The unit capacity rate will vary from 13.6 Kg per second per M^2 (5 tph per ft.^{2}) of pool area, for an average feed particle size of 100mm (4") down to 2.7 kg per sec per M^2 (1.0 tph per ft.^{2}) of pool area for an average feed particle size of 10mm (1/2"). A certain amount of experience factor is used in the establishment of this parameter.

Thus knowing the feed tonnage rate and the unit capacity for the feed sizing, the size of drum can be established for this condition.

Lifter Capacity

The lifting capacity of a drum is determined by the number of lifters, bulk density of the material, the breakage shape, the volume of the lifter, and the speed of the drum. The maximum particle size limits the spacing of the lifters and determines the maximum number of lifters that can be used in a particular size drum. The lifter spacing should be at least 2.5 cm (1") greater than the largest particle size in the feed material. An exception to this would be on slabby materials. Table III illustrates the lifting capacity of various drums. Thus knowing the above parameters the size of drum can be established for these conditions.

Drum Weir Overflow

The required depth of overflow material from the drum has to be greater than the maximum particle size in the feed as the particle has to be carried over the weir. This condition will determine the quantity of media that is required. When there is a considerable tonnage of float material, a rectangular discharge is used. For a particular top size of feed material less media is required for the rectangular weir than for a circular weir.

Retention Time

The retention time required is a function of the tonnage of float material, the volume of media required and the settling rate of the smallest particle in the sink material. The horizontal velocity of the floating layer in the pool will, for a particular length of drum,

TABLE III

WEMCO DRUMS LIFTING CAPACITY AND MEDIA FLOW

	1 219mm (4') Dia. Drum			1 824mm (6') Dia. Drum			2 433mm (8') Dia. Drum			3 048mm (10') Dia.Drum			3 658mm (12') Dia.Drum		
	O'Flow			O'Flow			O'Flow			O'Flow			O'Flow		
No.of Lift-ers	Lifter Spacing in mm	Liters per Second	Tons per Hour	Lifter Spacing in mm	Liters per Second	Tons per Hour	Lifter Spacing in mm	Liters per Second	Tons per Hour	Lifter Spacing in mm	Liters per Second	Tons per Hour	Lifter Spacing in mm	Liters per Second	Tons per Hour
22	112	3.86	5.9	175	4.55	13.6	237	5.31	25.4						
26	93	7.76	6.9	150	10.20	16.3	196	11.67	30.9	239	13.25	63.0	292	14.25	117.0
30	79	15.37	8.2	123	17.97	18.7	167	20.83	35.4	203	23.33	74.0	240	26.11	135.0
34	69	22.70	9.1	107	27.74	21.3	145	32.50	40.0	177	35.97	84.0	217	40.56	153.0
38	61		10.4	87	40.36	23.6	128	46.1	44.5	156	51.75	94.0	192	56.94	121.0
42	54		11.4	84		26.4	114	63.06	49.0	140	70.69	104.0	172	77.64	188.0
46	49			76		28.8	103		54.5	126	41.53	113.0	155	102.78	206.0
50	45			69		31.4	94		61.0	115	113.61	124.0	142	128.06	224.0
54				64			86		63.5	106	135.56	133.0	130	156.94	243.0
58							80		68.0	98		143.0	120	189.44	270.0
For 1 RPM. Sp.Gr. 2.7															
Recommended Maximum Particle - mm	89			125			150			200			250		
RPM	2.25 - 3			1.5 - 2.2			1.10 - 1.5			0.9 - 1.2			0.75 - 1		
Lifter Volume (liters) for 305mm length				2.0			2.8			5.0			7.5		
Pool Area, m^3	0.836			1.951			3.283			5.110			7.525		

Note: The above table is illustrative only as these values will vary significantly with the material being treated.

determine the retention time. The retention time has to be of long enough duration to prevent the smallest sink particle being carried over the weir overflow.

Drain and Wash Screen

The purpose of the drain and wash screen is three fold, drainage of media from the float/sink products, washing of media from the float/sink products and drainage of water from the float/sink products. These functions are carried out in three distinct areas along the length of the screen.

A horizontal vibrating screen is preferred as drain and wash screens for similar reasons as explained for the feed preparation screen. The minimum length of this screen is considered to be 4 877mm (16 ft.). The screen width required is determined by the bed depth. For efficient operation this will vary from 8.0 times the smallest particles size to 0.15 times the largest particle size and is also related to the particle distribution of the feed material. The width of drain and wash screens will depend on the tonnage rate of the float and sink product. When a single screen will handle both products, this is achieved by using a divider along the length of the screen to keep the products separated.

The screen deck preferred is the wedge wire screen. The deck openings are kept to 1.0 to 2.0mm. The opening required is dependant on the amount of degradation of the material as it passes through the system as small particles have to be restricted from diluting the media circuit.

The hopper under the drain and wash screen is divided into two compartments to keep the drained media and the dilute washed media separated. A splitting device is incorporated in the hoppers so as to deflect a required amount of drained media into the dilute media section to allow cleaning of the media and to achieve a water balance within the media circuit.

The Media Circuit

To allow separation to take place within the drum vessel a certain amount of media is required and is related to the tonnage of feed material. To satisfy this condition a ratio of 1 volume of material to 2.5 to 3.0 volumes of media is required. The media at the correct density is added with the feed to the drum. Under certain conditions, when the percentage of float material is high, a portion of this feed media is directed behind the curtain on the down-going lifter side of the drum. This media stream behind the curtain prevents the lifter from trapping any float material on its downward path and the subsequent inclusion in with the sink fraction.

Media at the correct density is also used to transport the sink product in the drum sink launder. The quantity of this media will vary from 1.0 to 1.5 volumes of media to 1.0 volume of material.

The media drainage from the drain section of the drain and wash screen is returned to the current media sump for recirculation in the media circuit.

Dilute Media Circuit

The dilute media circuit is used to recover and dewater the media. A typical circuit is illustrated in the Flow Sheet (Figure 3). This circuit will vary under certain conditions; a thickener may replace the densifier.

The dilute media from the wash section of the drain and wash screen is fed to wet magnetic separation for recovery of the media. The concurrent single drum wet type Magnetic Separator with permanent magnets is the preferred and most commonly used unit in heavy media plants. The double drum concurrent wet type magnetic separator is also used and will permit higher feed volumes while obtaining the same recovery as can be reached at lower rates on single drum separators.

The efficiency of the Magnetic Separator will depend on concentration of magnetics and the rate at which the unit is fed. For a 30" diameter single drum wet type magnetic separator a typical capacity for this application would be $70 m^3/h/m$.

The media recovered by the Magnetic Separator is fed to a densifier. The densifier is used in H.M.S. installations to dewater the cleaned media from the magnetic separator and for storage of media. The media from the Magnetic Separator is in a flocculated condition which aids the settling rate in the densifier.

In coal preparation plants using magnetite media, the discharge from the Magnetic Separators can be returned directly without using a densifier. It is good practice to propose a densifier as a magnetite storage vessel, even though the media from the Magnetic Separator by-passes the densifier or can be by-passed partially. On installations using ferrosilicon as media, the densifier is a necessity, particularly in cases where the density of separation is above 2.8.

The densifier overflow can eliminate some slimes from the media circuit, but care should be taken that this overflow does not carry fine media. The densifier overflow is generally returned to the diluted media pump sump. Table IV illustrates the raking and storage capacities of densifiers.

The media from the densifier discharge is returned to the correct media sump via a demagnetizing coil. The purpose of the coil is to demagnetize the media to prevent floculation of the media. The demagnetizing coil is sized according to the tonnage rate of the media.

TABLE IV

CAPACITY AND STORAGE CAPACITY OF DENSIFIERS

Size	Raking Capacity TPH		Maximum Storage Capacity	
	Magnetite	Ferrosilicon	Magnetite	Ferrosilicon
24	4	5	7	9
36	9	12	15	19
48	13	16	45	59
54	15	20	64	83
66	23	29	122	158

The Wash Water Circuit

The wash water on the drainage and wash screen comprises three to four rows of fish tail sprays using fresh water. The amount of water used will be dependent on the particle size of the material and will range from 0.21 to 0.35 liters per sec per tph (3 to 5 gal/min/tph). Counter-current washing is sometimes used to reduce the amount of fresh water required. In this case, the number of fresh water sprays is reduced to two rows placed towards the back part of the wash section of the screen. This wash water is collected and pumped to a wash box situated at the front part of the wash section of the screen. The collected wash water (dilute media) is pumped to the magnetic separator for recovery of the media.

Floor Spillage

Spillage from the H.M.S. circuit is collected in a floor spillage sump and is pumped to the dilute media circuit via the wash section of the drain and wash screen. The floor spillage system can also be used to add make-up media to the circuit.

Density Measurement and Control

Density measurement of the correct media density is obtained by either bubble type, magnetic or nuclear methods. The details of this equipment are not included in this paper. The measurement is made on circulating media from the circulating media pump. Control of the media density can be obtained by various methods, one of which is to add water to the circulating media pump suction. In this instance, the circulating media sump is used to store over-dense media which allows the controlled amount of dilution water to be added at the pump suction to achieve the operating density.

Conclusion

Heavy liquid testing determines the feasibility of the heavy media separation process. The derived test data is used for determining parameters in the sizing of the process equipment. The flowsheet incorporates the feed and product materials, circulating media, wash water and media recovery circuits. Present and future projects for heavy media separation for minerals have and will continue to contribute to low cost separation or preconcentration techniques for the mineral processing industry.

ACKNOWLEDGEMENTS

The permission of the management of WEMCO Division, Envirotech Corporation to publish this paper is gratefully acknowledged.

6

Solid-Liquid Separation and Instrumentation

Chairmen: S. M. Moos
C. A. Richardson

Chapter 27

THICKENERS

by Donald L. King

Eimco Process Machinery Division
Envirotech Corporation
Salt Lake City, Utah

Manager, Development Engineering -
Sedimentation Business Center

INTRODUCTION

This paper has been prepared with the objective of providing basic information on thickening equipment as applied to mineral processing plant design. Thus, by design, the content is elementary, yet sufficiently informative to advise the reader concerning the thickening and clarifying process, equipment configurations, installation and operation. Practical considerations are provided for slurry samples for testing, evaluation of testing, preliminary sizing, duty classification, and thickener selection to enhance operation, servicing and control. Illustrations are provided to both inform and demonstrate present state of the art features and capability.

THICKENING DEFINITION

Continuous thickening and clarification - by an operation called Sedimentation - is the separation of suspended solid particles from a liquid stream by gravity settling. The primary purpose of thickening is to increase the solids concentration of the feed stream, while that of clarification is to remove solids from the feed stream. Thus, there is no precise distinction between thickening and clarification other than the principal result desired.

The inlet stream going to a thickener generally is called "feed" or "influent". Overflow from the unit may be called "overflow", "effluent" or "supernatant". Underflow may be called "pulp", "sludge", "slurry", "mud", etc.. The terminology will depend upon the industry and application.

HISTORY OF THICKENER DEVELOPMENT

Modern Sedimentation technology was developed in the mining industry. Prior to development of the continuous thickener, batch gravitational settling was employed. In such batch operations, dilute feed was pumped into a tank continuously until a clear overflow was no longer obtained. The feed was then discontinued and the tank left undisturbed until the solids had settled. After a suitable retention time, clear supernatant liquor was decanted and the thickened sludge removed.

Following batch settling operations, cone settlers were developed. These could be operated continuously by delivering a continuous feed stream to the settling cone; removing the settled solids underflow continuously from the bottom of the cone; and the overflow continuously over weirs into launders at the surface and on the periphery of the cone. To maintain a uniform underflow from the cone, and to insure proper removal of the solids, the slope of the cone often had to be fairly steep. A steep cone slope limited the size of the cone settler.

Since the effectiveness of gravitational settling is largely a function of area, it soon was recognized that the basic cone design had serious limitations. As the technology developed it became obvious that continuous movement of the solids deposited over a large area toward a common withdrawal point was necessary. Adopting this general principle, the continuous thickener - with its many modifications - was developed about 1905.

From the mining industry, the continuous thickener was applied to the chemical, water, and waste water treatment industries. However, with minor exceptions the heaviest duty machines are required in the metallurgical industry, while machines of lighter duty are used for water and waste water treatment in other process industries. It is obvious that a higher specific gravity of the solids, and/or larger amounts of solids to be handled by a thickener of given size, requires heavier construction of the thickener mechanism. Many metallurgical pulps have specific gravities of 2.6 or more, and settle to concentrations up to 60-75%, and these machines must be much heavier in design than for water and waste water treatment, where the specific gravity of the solids often are 1.1-1.3 and rarely settle to a concentration of greater than 10%. In some applications, even though the solids in the feed are finely divided and the unit acts as a clarifier, underflow concentrations can be relatively high, thereby requiring a semi heavy duty clarifier mechanism.

HOW A CONTINUOUS THICKENER FUNCTIONS

As a feed stream enters the thickener, the solids settle to the bottom. Clarified liquor overflows the top and the settled solids underflow is removed from the bottom.

Figure 1 shows a cross-section, schematically illustrating the operation of a continuous thickener. Zone A, which is the clear overflow liquor, is essentially free of solids in most applications. Zone B consists of a pulp of fairly uniform consistency which is nearly the same solids concentration as the feed stream. Zone C is an intermediate state in which the pulp is in a condition of transition between free unhindered settling and compression. Zone D shows the pulp in compression, where dewatering occurs by compression of the solids forcing the liquids out of the interstices.

In actual practice, special characteristics which distinguish zones B, C, and D are not readily discernible, except for an increasing solids concentration, and the description is more academic than realistic.

Figure 2 presents an illustration of that which actually occurs in a continuous thickener. Feed slurry becomes very diluted on entering the feedwell (unless the feedwell is submerged in a moderately concentrated pulp zone as illustrated in Figure 1), and leaves the feedwell as a dilute suspension from which particulate settling, rather than "zone" or "line" settling must occur. Considerable lateral movement occurs in the floc bed zone as liquid into the feedwell to sustain this dilution action. The floc particles agglomerate and settle to the surface of the thickening pulp, and continue to concentrate in this zone until underflow density is reached.

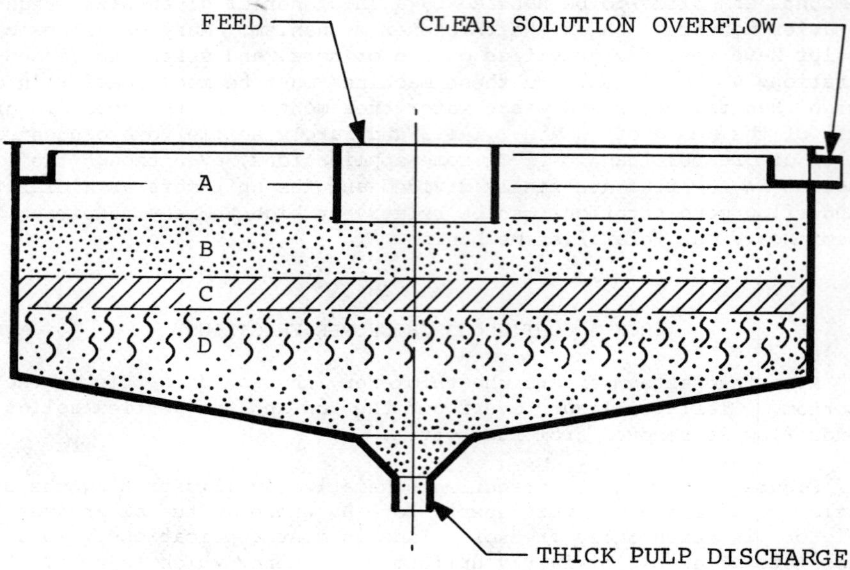

SECTION THROUGH A CONTINUOUS THICKENER ILLUSTRATING POSITION OF FOUR ZONES OF SETTLING PULP

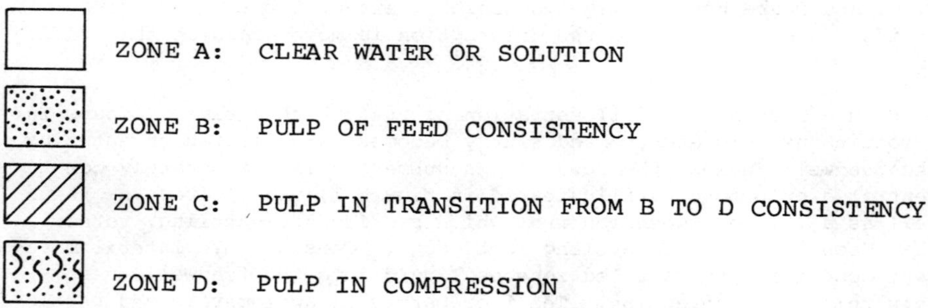

FOUR ZONES OF SETTLING PULP, ILLUSTRATING CONTINUOUS THICKENING

FIGURE 1

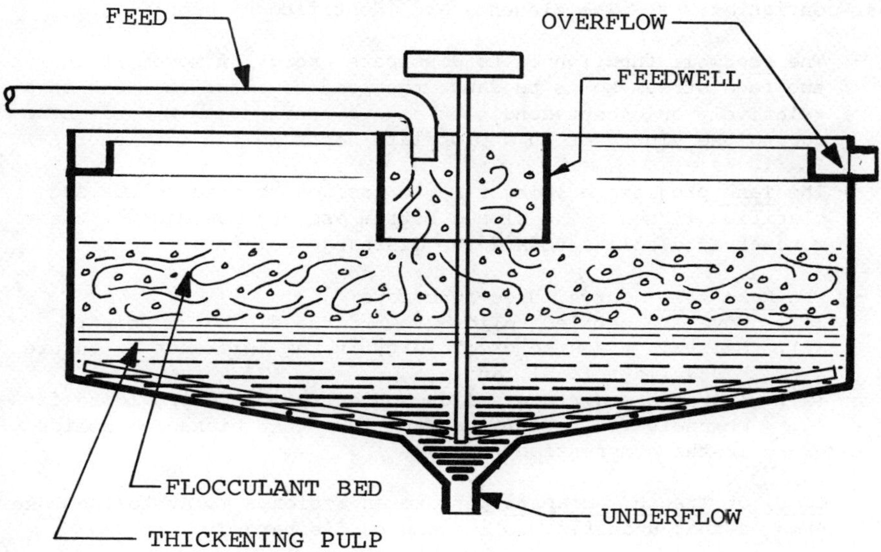

SECTION THROUGH A CONTINUOUS THICKENER
ILLUSTRATING THICKENING ACTION

FIGURE 2

ELEMENTS OF A CONTINUOUS THICKENER AND THEIR FUNCTION

Refer to figure 3 for schematic representations of two basic thickener configurations. The elements are identified by name:

- The Feedwell function is to dissipate energy of movement in the the feed stream so as to cause the feed to enter the tank in a relatively quiescent condition and to provide a means of introducing the slurry at an appropriate depth in the thickener.

- The Tank provides holding time to produce settled solids and clarified liquor. The sloped bottom assists movement of the concentrated solids toward the discharge point.

- The Rake Arms serve three functions:
  (1). Move the settled solids toward the discharge point;
  (2). Maintain a degree of fluidity in the thickener to ensure hydraulic removal, and
  (3). Increase underflow solids concentration by providing for channels for water to escape from the thickening solids in the compression zone.

- Cone, or Trench, Scrapers perform an action similar to the Rake Arms so that underflow solids can be discharged.

- The Overflow Launder collects clarified liquor for transport to an outlet. Best design practice is for a uniform rate of overflow around the tank periphery.

- The Rake Drive provides driving force (torque) to move the Rake Arms and Blades against resistance of the thickened solids.

- The Rake Lift provides a means of lifting the Rake Arms out of contact with the more concentrated solids so as to reduce the driving force demand from the rake drive. The lift will operate while the rake arms are rotating.

SOME FACTORS THAT SIZE CONTINUOUS THICKENER BASINS

Sizing the Thickener

Detailed sizing techniques are not discussed in this article, but certain basic considerations are covered.

Area: The area of a thickener must provide sufficient detention time to allow the slowest settling particle to reach the bottom of the unit. Thickener size is often expressed in area per unit weight of dry solids per day. (Meter squared/ton of dry solids/day).

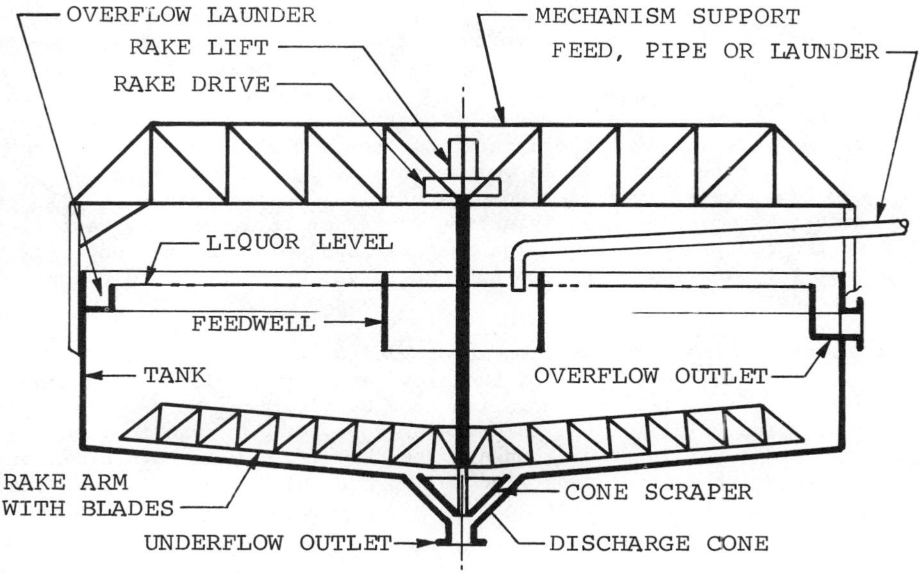

BRIDGE, OR BEAM SUPPORT TYPE

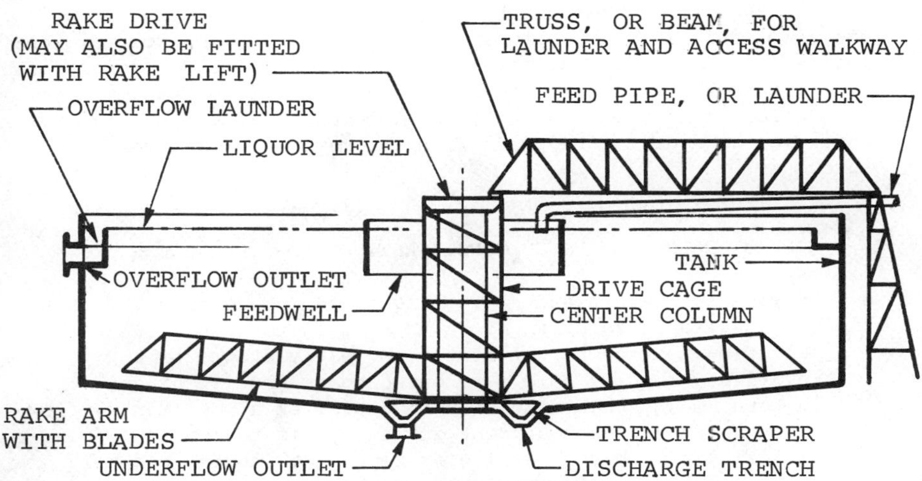

COLUMN SUPPORT TYPE

ELEMENTS OF A CONTINUOUS THICKENER

FIGURE 3

Overflow Rate: The overflow, or upflow rate, of the unit must be low enough so that excessive turbulence does not prevent separation of the solids from the liquid and not greater than the settling rate of the slowest settling particle or flocculi. Thus, in thickeners which are essentially clarifiers (that is, are used to remove very finely divided material from a dilute suspension) the overflow rate expressed as volume per day per unit area (meter cubed/day/meter squared) becomes a critical design factor. For materials which apparently settle very fast in laboratory tests, it is good practice to design the area and volume of the unit so that the overflow rate does not exceed values established by experience.

Detention Time: In addition to providing sufficient detention time to allow separation of the slow settling particles from the overflow, special consideration must be given to the compression zone. The volume of the compression zone, or for practical purposes the entire volume beneath the pulp level, will have a direct bearing on the final underflow solids concentration as long as the particles within this zone are in a state of subsidence, that is, not at rest.

In an ideal thickener operation the solids in the feed to be withdrawn in the underflow must move continuously toward the withdrawal point at the same rate as they enter the compression zone. The solids do not merely settle on the bottom and then are pushed by the rakes toward the discharge point. The pulp is retained in the compression zone, which must provide adequate time for thickening to final density, but too long a retention time can result in overload to the thickener mechanism. The underflow rate must be controlled so that pulp density, as withdrawn, will be the maximum practical at a given feed rate with the minimum pulp level in the thickener.

PRACTICAL MILL DESIGN CONSIDERATIONS FOR THICKENERS

A. Utilize the expertise of established equipment manufacturers or consultants experienced in this field to insure the best selection for the requirement. The early stages of a project invariably require some form of economic justification. This, in turn, means that someone must establish a preliminary flowsheet and size the equipment involved. If the Design Engineer is familiar with the particular flowsheet, he can probably complete this preliminary phase without outside help. However, if the flowsheet, or parts of it, are outside the realm of his experience, he must look to others for assistance. Suitable help may be found within his company's files or the general technical literature, or it may be necessary to contact equipment manufacturers or consultants. Equipment manufacturers have dealt with a wide range of flowsheets and applications and, with a reasonable description of the materials involved and process requirements, can quickly estimate equipment size and performance with sufficient accuracy for most preliminary feasibility studies. Many engineers make use of this source of knowledge. Others may spend an unnecessarily large amount of time searching the literature without developing accurate information that could be obtained by consulting a reputable equipment supplier

B. Define the requirements for the thickener. In addition to process requirements, upset, or unusual operational conditions, must be anticipated.

C. Verify the thickener requirements. Pilot plant, or laboratory cylinder tests, are very desirable, but the slurry sample must be representative of the actual slurry. Many fail to recognize the importance of a truly representative slurry sample. The emphasis here is in reference to simulated slurries or slurries produced in small quantities in a bench scale apparatus. These samples must be carefully characterized and their properties, such as size distribution, suspended solids concentration, pH and temperature, can have very pronounced effects on equipment capacity and product quality.

Once it has been determined that the slurry in question is reasonably representative, it is necessary to perform meaningful sizing tests. Although required test procedures and equipment have been prescribed in literature in some detail, if experience is lacking in running these tests, it would be well to consult a recognized authority in this field. There remains a significant degree of "art" in performing solid-liquid separation tests and interpreting the results. Proper thickener selection depends as much upon the recording of significant test observations as upon the routine recording of required test data.

D. Pilot Plant design and operation is important. As the plant design project progresses, it is frequently necessary to confirm the equipment sizing and performance. This may mean simulating the complete operation including continuous pilot plant equipment. It is at this point that many miss one of the prime opportunities to obtain meaningful design data. When a continuous pilot plant is designed, it is important that all of the various operating units have sufficient capacity to enable the process to operate at or near capacity. It is unfortunate, from equipment sizing basis, that the liquid-solid steps usually have the greatest degree of uncertainty and that the pilot plant equipment is almost always oversized, so that the process may operate continuously regardless of the characteristics of the slurries.

When pilot plant equipment is oversized, it is extremely difficult to get meaningful sizing information. The data obtained is important, but for the results to be interpreted correctly, this must usually be done in conjunction with a good bench scale testing program. Operating conditions in a bench scale test can be varied in a controlled manner that enables one to determine the effect of many variables and to arrive at a reasonable design operating condition. Since conditions in a pilot plant operation are a great deal more difficult to control, pilot plant operation should be used only to confirm design operating conditions, such as how the feed quality varies with time, and to obtain confidence that the type of unit chosen is capable of giving consistent long term results. The observation of feed variations and the confirmation of long term operating reliability are the most important functions of a solid-liquid pilot plant operation.

E. The thickener feed velocity should be as low as practical. Thickener operation starts with a system that introduces the feed slurry. The feed velocity must be high enough to prevent solid material "sand out" in the feed launder, but not so high as to cause excess turbulence in the thickener feedwell and tank. Most feed arrangements are such that the feed launder is above liquor level, and for those slurries, that contain solids of specific gravity 2, a feed launder slurry velocity of 2.5 to 3 meter/second will usually be sufficient. As a general rule a launder slope of 1 to 1 1/2% will provide the necessary velocity.

F. Thickener feed entrance into the feedwell should be such as to provide feed stream energy dissipation. Some ways to accomplish this are:

- Minimize elevation difference between feed launder and liquor level. Often the available elevation difference is much greater than a minimum required, and the feedwell of the thickener is used to dissipate the kinetic energy resulting from this elevation difference. This practice can, and usually does, have detrimental effects on thickener operation. Most feedwells are not designed to accept excessive entrance velocities, and there is always a danger that the slurry will enter the feedwell with a significant downward component of velocity that will cause it to stir up the settling pulp. The result can be both decreased underflow solids concentration and increased turbidity in the clarified overflow. Also, there is always a certain amount of air entrainment in any overhead feeding arrangement that requires a vertical drop. If the thickener feed has come from a flotation step, or if it has natural foaming tendencies, excessive feed entrance velocities will aggravate already undesirable conditions by producing froth which floats on the liquor surface causing loss of solids in the overflow.

- The launder should terminate its horizontal run as near the liquor level as possible, and feed should be introduced below liquor level. A good scheme is to introduce feed such that it exits the launder with flow divided into two equal, opposite directions, and nearly horizontal. The opposite, horizontal feed streams are usually introduced at slightly different elevations to create feed stream shear rather than impingement. Feedwells usually have a bottom shelf to help prevent feed stream "short circuit" to the overflow. If the launder is enclosed, provide air release at its feedwell termination.

G. Make the provision to remove oversize material from the feed. Remember, a thickener is designed to operate within specified limits, and exceeding those limits can cause operation problems. As a general rule the feed to a thickener should contain very little material larger than 250 micron (+60 mesh). There are a number of applications where this rule is violated, but special features are designed into such applications to accommodate larger material.

Oversize tramp material cannot be tolerated, and it must be removed by screening. It is preferable to remove such prior to the feed entering the launder.

H. Flocculation of the feed must be controlled and must be complete. There are many instances where flocculation control has been incomplete, and in nearly all such instances, thickener results were less than expected. It must be remembered that a sedimentation device is usually designed to provide both clarification and solids concentration. Emphasis is often on one or the other only, and there is a tendency to consider that clarification depends only upon providing an area equivalent to an upflow rate less than the settling rate of the finest particle removed. While this is certainly one of the mechanisms involved, it is seldom that this is the rate controlling mechanism. Most clarification problems involve time dependent functions of coagulation, or flocculation, or a combination of both. Unless the material is naturally flocculent, some type of chemical must be added to force or promote formation of settleable flocs from the dispersed fine particles. The chemicals which are added must be uniformly dispersed if they are to function properly. When the feed slurry requires only the addition of multi-valent cations, such as alum, the time required to accomplish the dispersion is of a minor consequence, so long as it is possible to provide the necessary mixing in a conveniently sized vessel. However, when synthetic polymers are used, it is usually necessary to provide an efficient and rapid mix. In this case, the main problem is that of getting the polymers dispersed among the suspended solids before their active sites are all used up by particles in their immediate vicinity. Inefficient mixing means that unnecessarily high polymer dosages will be required. Once the required chemicals have been added to the feed slurry, it is frequently necessary to provide a period of mechanical contact time to allow the flocs to grow to a size that will settle rapidly. If the suspended solids concentration in the original feed is too dilute, it may not be possible to grow flocs of useful size and good floc growth could require that additional solids be recirculated to the feed stream to increase the solids concentration. This may be done either by external recirculation or by internal recirculation of thickened solids in a suitably designed unit. Regardless of the design of the particular unit, it is essential that the engineer make certain that the unit does provide that environment that is necessary to produce an agglomeration with a usable settling rate. Since final clarification is almost always a time function, the clarification zone of the unit must not only provide an equivalent upflow rate low enough to prevent particles from being swept out in the effluent, but enough retention time to allow the additional flocculation which does occur within the clarification zone.

I. Underflow handling. When designing thickener underflow piping for metallurgical and other similar slurries, it is essential to remember that the piping must be designed so that operators can, if necessary, purge or "blow back" with either high pressure water or air. The question is not "What to do *if* the underflow plugs?" but rather, "What must be possible to do *when* the underflow plugs?" There are a number of suitable underflow piping and pumping arrangements. The correct one for a particular installation is a function of the solid-liquid system to be handled, tank size, pumped or gravity flow, economics of tunnel versus center pump room, etc.. These considerations may, in turn, be modified by local site conditions such as high ground water, hard rock formation, etc..

Some underflow handling schemes that have been used are discussed here with comment as to desirable and undesirable features of each.

Underflow Handling Scheme	Comment
1. Tank elevated. Underflow valves, piping and pumps at ground level.	1. Usually the most expensive tank design as tank size increases. Excellent accessibility to the underflow system. Tank drain system through underflow outlet.

Underflow Handling Scheme	Comment
2. Tank at ground level. Underflow valves, piping and pumps in a tunnel under the tank.	2. By far the most common design for medium to large size tanks. Usually attractive for reasons of installed cost. OSHA rules must be followed for access. Second best for accessibility to the underflow system. Tank drain system thru underflow outlet.
3. Tank at ground level. Submerged underflow pumps in center column near bottom, with pump motor above liquor level and usually above and centered over the thickener drive.	3. The least installed cost system. Disadvantages are a spare pump is not in place, and separate tank drainage features are required. Pump servicing requires the thickener to be out of service. Use of such a system influences thickener mechanism design.
4. Tank at ground level. Underflow pumps on bridge with underflow inlet in center column near bottom.	4. Installed cost is slightly more than item 3 above. A spare pump can be included, however disadvantages are limited service room, limited center depth for suction lift, and separate tank drainage is required. Use of such a system influences thickener mechanism design.
5. Tank at ground level. Underflow valves, piping and pumps in a room at the center bottom of the thickener tank. Underflow pumped through pipes up through a large center opening. A thickener of this design is called a Caisson Thickener. A Caisson Thickener requires a drive with large center opening. Two types are available. One utilizes a large diameter hydrostatic type bearing	5. Very large tanks may be less expensive when constructed in this manner. This construction eliminates the need for an access tunnel, but requires a more costly thickener mechanism. The center drive has to be large enough to allow worker access via stairways and to permit removal of pump equipment without affecting the access stairways. Installed cost of such a design is usually dependent on soil conditions, tank size, application, and type of support bearing chosen.

(fluid film). Another type utilizes a large diameter mechanical rolling element type of bearing.	Two designs for such are illustrated on figure 5. Of the two designs, the one using the mechanical rolling element bearing (tapered column) is less expensive both in equipment cost and installed cost. In addition, the hydrostatic bearing requires continuous power input. At 1977 construction costs, this design begins to receive serious consideration at tank sizes above 122 meter (appoximately 400 ft.).
6. Tank at ground level and underflow line buried with pumps at tank periphery.	6. For most mineral processing industries, this scheme is not practical, however, when it is practical, it is obviously an economical scheme.
7. Tank at ground level. Underflow at tank periphery and flat bottom tank.	7. There are a few mineral processing applications where this scheme is used. Red mud thickeners and washers in alumina plants have used this scheme.

Tank arrangements illustrating the foregoing schemes, except for item 7, are in the part of this article titled "Thickener Tanks and Mechanism Selection." Refer to figures 4, 5, and 6.

J. Servicing. Oil change, condensate drain, location of lubrication fittings, walkway access, enclosures, location of instrumentation, night lighting, local electrical plugs for tools, etc., all need to be considered for ease of routine maintenance and for more extensive maintenance.

K. Operator control requires certain observations and measurements. Many are now made by instrumentation. The need exists for periodic verification, and this requires design to minimize the chance of foreign objects falling into the tank. Hard hats for example, make excellent underflow outlet plugs. Tools and other objects, which can get into the underflow outlet, cause pump damage or stoppage. Some installations are provided with protection nets to catch falling objects.

L. Emergency bypass, or shut-off, of feed and underflow recirculation, are must features. Standby underflow pumps are common.

MAJOR FACTORS INFLUENCING THICKENER DESIGN

A. The quantity of solids to be handled. Usually expressed as area per unit weight of dry solids per day. The smaller the number the greater the chance of an upset. This usually requires a combination of a stronger thickener mechanism and a lifting device.

B. The amount of material larger than 250 micron (+60 mesh) in the feed. This affects tank bottom slope, drive and strength of mechanism. It may also require a mechanism lifting device.

C. Specific gravity of the solids. The greater the specific gravity the more likely a stronger drive and mechanism will be required.

D. Overflow launders and feedwell capable of handling additional material when other thickeners are out of service.

E. Feed and underflow material settling characteristics that may require special rake construction such as blades located a distance below the rake arms on posts or spikes on the blades to cut into packed solids.

F. Scale build up tendency of feed slurry may require special arms and drive.

G. An operating requirement to accumulate solids for defined periods of time will require a special mechanism design, as it is not a normal operating procedure.

H. Froth control or removal.

I. Slurry temperature, vapors, gases, etc. may require covered and/or insulated tanks with attendant seals.

J. Soil conditions and ground water elevation affect foundation design and may determine mechanism type.

K. Climatic conditions may require special considerations, such as enclosures around the drive and instrumentation.

L. Measurement and control methods and control room location.

M. An operating requirement to use a powered lift for purposes other than occasional corrective action. Refer to the section on "Thickener Operation and Operator Control."

THICKENER TANKS AND MECHANISM SELECTION

A. Tanks

Tank construction features to be selected are determined mainly by size, feed characteristics, underflow handling requirements, and topography. Tanks may be constructed of steel, concrete, combinations of steel and concrete, wood, steel, or concrete wall and lined earth bottom, or all lined earth. Corrosion protection for such may be paint, elastomer or plastic.

As described in a preceding section, underflow handling may require tanks:

- Elevated above ground, (or)
- With access tunnel under a tank on the ground, (or)
- A large center pump room wherein pumps may be located and serviced without access tunnels under the tank.

Bridge supported mechanisms are generally furnished with tanks up to 37 meter (approximately 120 ft) diameter. Some have been furnished to 45.7 meter diameter (150 ft.)

Center column supported mechanisms are generally economical starting at about 27.4 meter (90 ft.) diameter and are presently available for tanks up to 183 meter (600 ft.) diameter.

Some tank illustrations are shown on Figures 4, 5 and 6. Mechanisms are regularly being installed in tanks of 122, 137 and 152 meter diameter (400, 450 and 500 ft.). Large diameter center column (Caisson) to 7.3 meter (24 ft) are available for center pump rooms with walk-thru center drive openings of 3.9 meter (12 ft 10 inch) diameter, for mechanical rolling element bearings, and 5.8 meter (19 ft.) diameter for hydrostatic (fluid film) bearings. Some photographic representations of these examples are included.

Thus it is obvious that the choices of thickening equipment designs to suit specific purposes and economic construction are many. No one article can acquaint a person responsible for mill design with the optimum choices. Familiarization with a few basics is all that can be hoped for.

With the above reference as to what has been done and can be done, presentation of the total sedimentation problem to an experienced manufacturer of thickeners will result in offerings which will suit the process, installation, operation and economic needs.

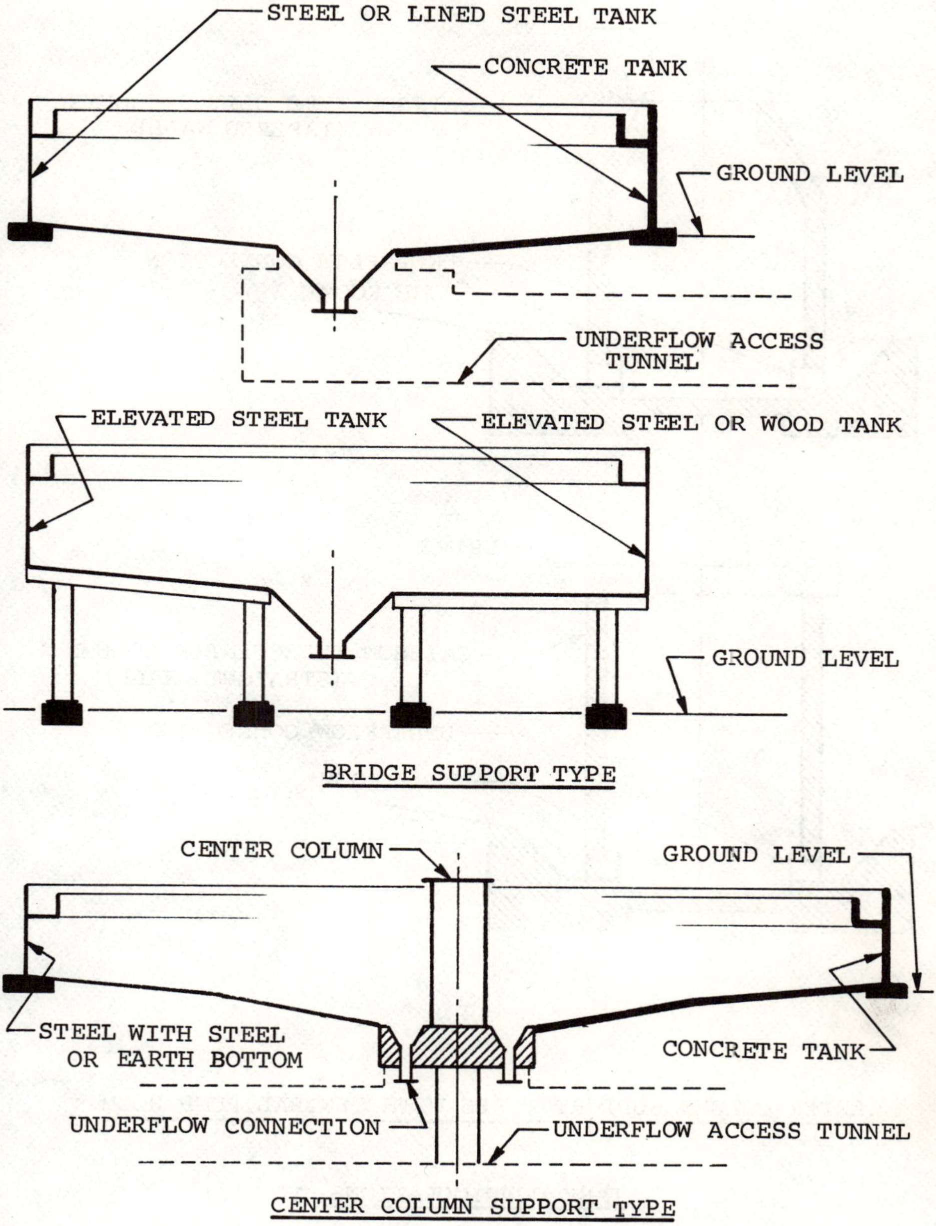

TANK CONFIGURATIONS

FIGURE 4

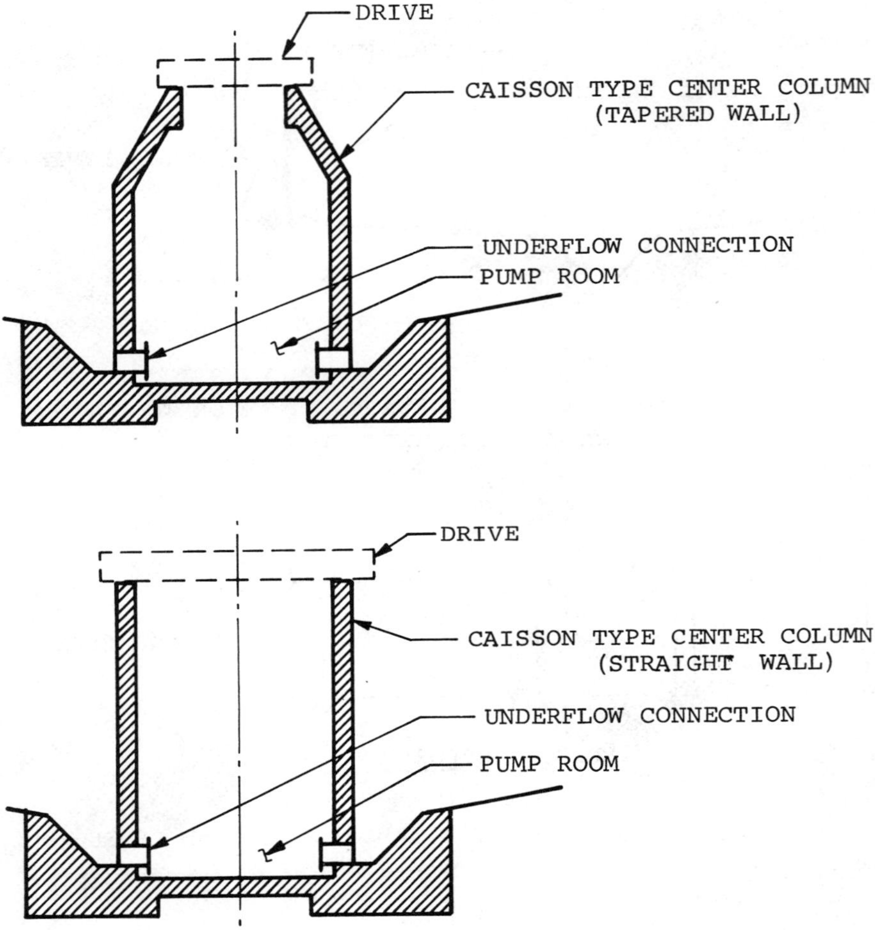

CENTER COLUMN SUPPORT TYPES WITH CENTRAL PUMP ROOM

TANK CONFIGURATIONS

FIGURE 5

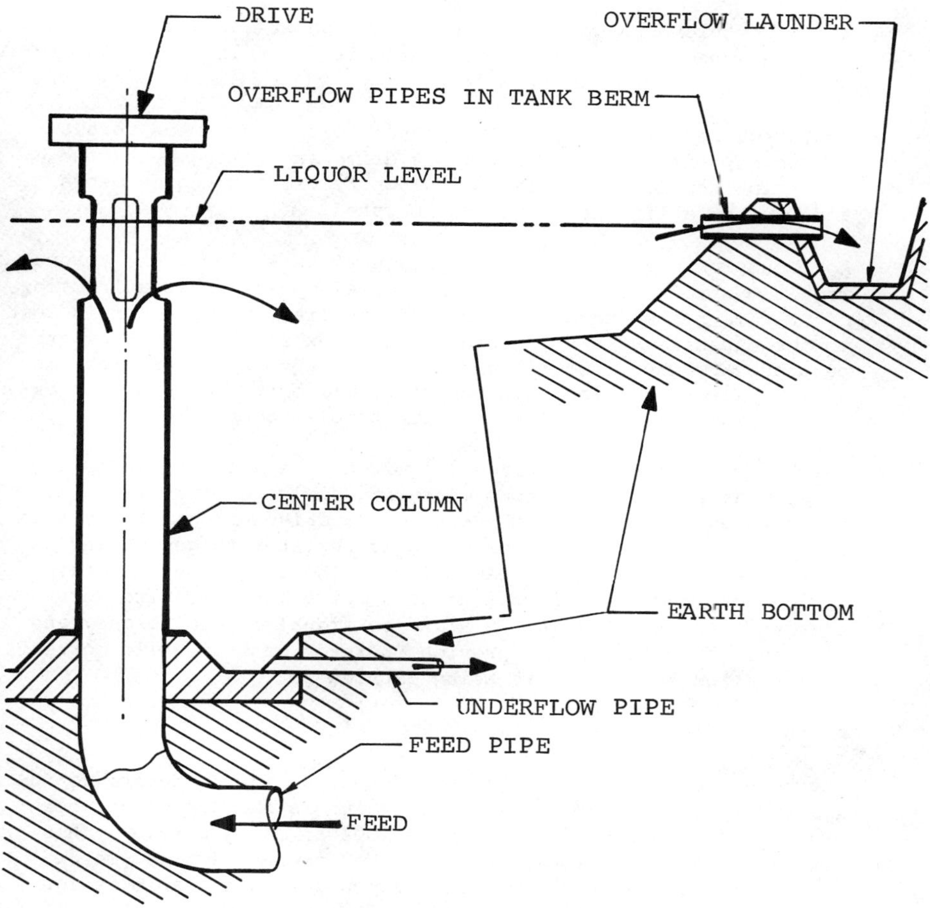

Tank illustration where solids in feed will permit under the tank feed introduction and underflow pipes may be buried without tunnel access. Tank is all earth construction and center column sits on foundation in center. This has been applied to phosphate slimes.

FIGURE 6

B. Mechanisms

When the thickener tank size, type, and underflow scheme have been selected, the required mechanism is determined by an analysis of duty requirements.

Duty requirements are usually related to torque demand versus torque capability of the thickener drive and raking mechanism which in turn determine the main support bearing requirements and whether a lift is to be incorporated.

A discussion of torque is appropriate at this time. The torque requirement of a thickener during normal operation is a fraction of what is recommended, selected and applied. The reason is that torque capability is a function of what experience has shown to be necessary when (not if) upsets or other operational problems occur. Torque availability beyond normal operation provides extra driving force to allow time for problem correction.

Thickener manufacturers place a limit on the deliverable torque by providing a torque output measuring device on the thickener drive which will interrupt power to the drive when that limit is reached. The measuring device should register torque demand as a percent of rated torque capability. The torque output measuring device, usually referred to as a Drive Control, also serves the function of actuating an alarm to signal rising torque, and lift and lower actions to provide operator control while corrections to the operation are being carried out.

<u>As an example</u>:

Torque selection for a mechanism of 500,000 Newton Meters (approximately 360,000 pound feet). This value is the torque at which the measuring device will register 100% of rated torque. This torque, and the rotational speed of the rakes, determines motor power applied to the drive. This torque, with suitable design safety factor, is the value used to design the steel raking mechanism.

DRIVE CONTROL AND TORQUE INDICATOR

FIGURE 7

There is not a precise method of selecting torque to be applied to a thickener mechanism. Some generalities have been tabulated to use as a guide for preliminary selection for inward raking mechanisms. As shown on Table I, these provide a torque determination as a function of tank diameter to the second power (area) by employing a K factor, ie KD^2. Final selection should be made in conjunction with an experienced thickener manufacturer, or recognized consultant.

C. Lifting Device actuated by power separate from the Drive unit. Should a rake lift mechanism be specified?

Use of a powered and controlled lift is an operator assist to provide time for corrective action and is usually dependent on the following:

a. What is the worth of "down time?"

Often, the initial cost of a lift device is far less than the cost of one shut down.

TABLE I

Preliminary torque selection chart for Inward Raking Thickener Mechanisms.

ITEM	DUTY CLASSIFICATION			
	LIGHT	STANDARD	HEAVY	EXTRA HEAVY
Solids Loading Meter Squared Per Ton Per Day	Above 4.7	1.4 to 4.7	0.5 to 1.4	Less Than 0.5
(Ft^2/Ton/Day)	(Above 50)	(15 to 50)	(5 to 15)	Less Than (5)
Underflow Concentration % Dry Solids	5 or Less	5-30	30-50	More than 50
% of Total Solids smaller than 74 Micron (% minus 200 mesh)	100	85 to 100	50 to 85	Less than 50
% of Total Solids Larger than 210 Micron (% +65 mesh)	0	0 to 5	5 to 15	More than 15
Specific Gravity of Dry Solids	1.0 to 1.25	1.25 to 3.0	3.0 to 4.0	Above 4.0
Torque Determination where torque = KD^2				
K for Newton Meter torque where D is tank diameter in meters.	15 to 58	73 to 131	146 to 292	Above 292
K for Pound Feet torque where D is tank diameter in ft.	1 to 4	5 to 9	10 to 20	Above 20

Some examples of duty classification as applied to Sedimentation equipment.

Light Duty - River, or lake water clarification, Metallic oxides, Brine clarification.

Standard Duty - Magnesium oxide, Lime softening, Brine softening

Heavy Duty - Copper Tails, Iron Tails, Coal refuse tank, Coal, Zinc or lead concentrates, Clay, Titanium oxide, and phosphate tails.

Extra Heavy Duty - Uranium Counter Current Decantation (CCD), Iron Ore concentrate, Iron Pellet feed, Titanium Ilmenite.

b. How much material larger than 250 micron (+60 mesh) may be fed to the thickener?

A lift will furnish operational correction time, thus minimizing shut down possibility.

c. Lift operation may be an operational necessity to routinely disperse solids accumulation detrimental to thickener operation. Refer to section "Thickener Operation and Operator Control."

Some typical drive and lift configurations are shown in figures 8, 9 and 10. Powered and controlled lifts have been furnished on mechanisms to 104 meter (340 ft.) diameter, and are available for mechanisms 122 meter (400 ft) diameter.

D. Lifting Device not operated by power separate from the Drive. Certain applications will allow use of a lift design not operated separately from the drive. Such a lift is incorporated in a thickener wherein the rake arms are both pulled through, and suspended in, the slurry by means of cables attached to torque arm and to the center shaft, or cage. The torque at which lift occurs is determined during the thickener mechanism design. Some field adjustment of the lift torque is possible through the use of ballast on the arms.

This design is a distinct improvement over earlier non powered lift schemes known as "hinged arms". The basic improvement is the small rake arm structure to be moved thru the pulp. Figure 13 illustrates this type mechanism as an Eimco Swing Lift Thickener.

The lift action is accomplished by the arms moving backward with respect to rotation direction and upward. Lift is greatest at the ends of the arms and proportionately less toward the center. An operating feature of this type mechanism is that the center scrapers and the center of the arms remain in the thickened pulp at all times. This in contrast with the powered lift which usually raises the arms and scrapers equally across the tank diameter.

It is possible that in some applications, the use of a Swing Lift Thickener may reduce the need for very heavy lift capability as demonstrated in Figure 9, because scale formation (of the type shown in Figure 11. Figure 12 shows the arm prior to scale formation) may be substantially reduced with a smaller rake arm section in the pulp.

DRIVE AND TORQUE TUBE LIFT FOR
BRIDGE SUPPORTED MECHANISM

FIGURE 8

DRIVE AND CAGE TYPE LIFT FOR CENTER
CENTER SUPPORT MECHANISM

FIGURE 9

DRIVE AND PLATFORM LIFT FOR BRIDGE SUPPORTED MECHANISM

Application was for Alumina service where heavy scale loads occur.

FIGURE 10

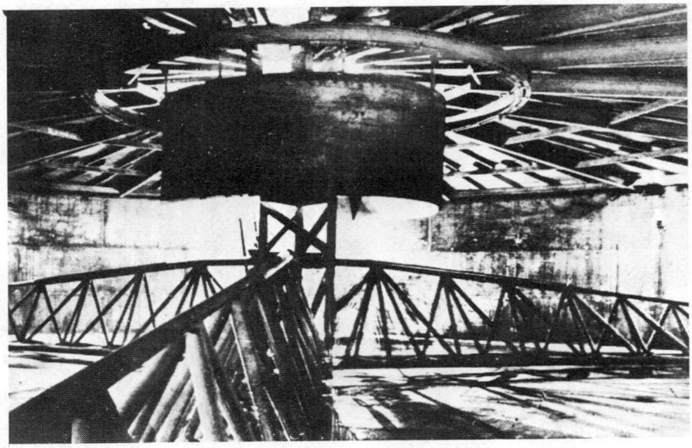

THICKENER ARM WITH HEAVY SCALE LOAD

FIGURE 11

30.5 METER (100 FT) DIAMETER THICKENER ARMS

PRIOR TO SCALE FORMATION
SHOWN IN FIGURE 11.

FIGURE 12

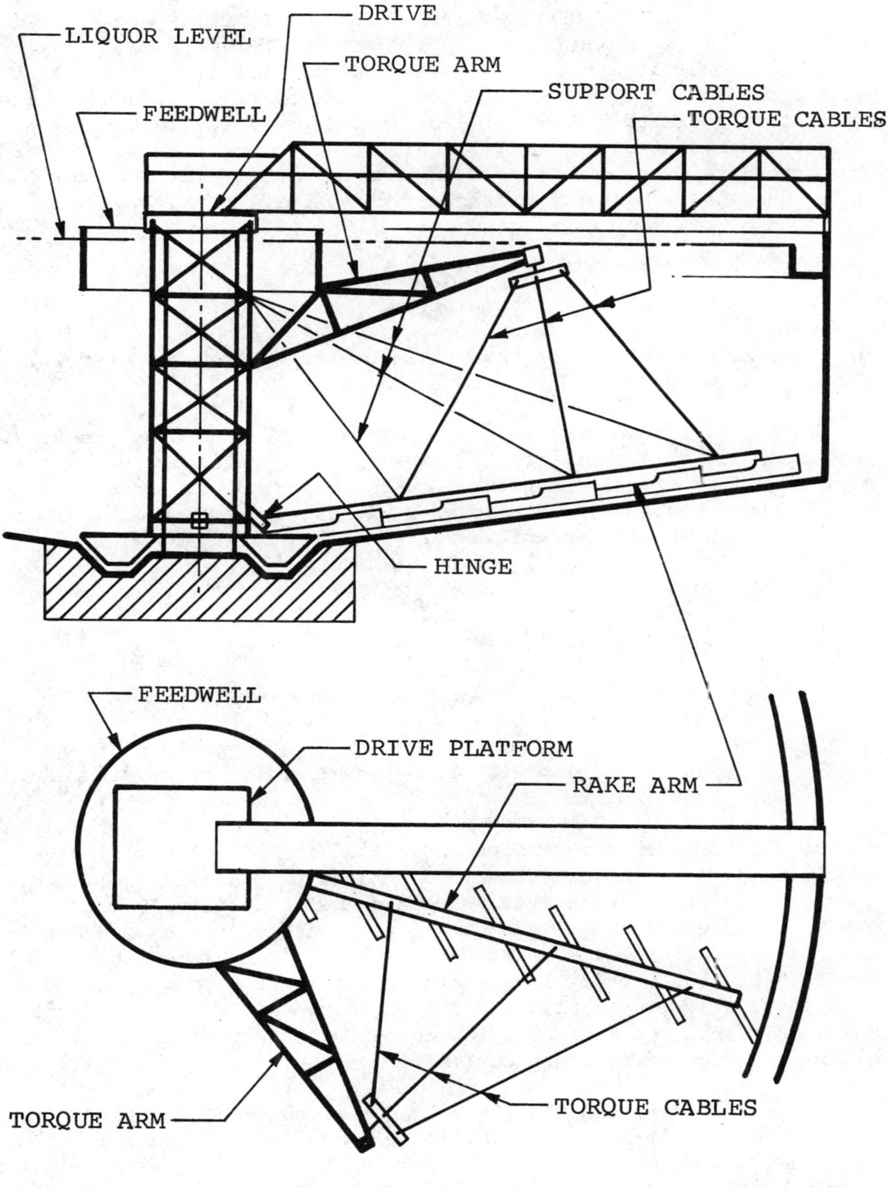

EIMCO SWING LIFT THICKENER

CENTER COLUMN SUPPORT TYPE SHOWN,
BRIDGE SUPPORT TYPE IS SIMILAR.

FIGURE 13

THICKENER OPERATION AND OPERATOR CONTROL

A thickener is a machine with a limited function. It is important to realize that a thickener is not just a "pass through" device. It must be operated and controlled within specified limits if the required end result is to be obtained. Selection of the proper operating procedure or control method requires an understanding of how the unit functions and the importance of the variables involved.

Thickener Operation Rules

A. Input minus output = Accumulation.

B. Excessive accumulation results in operating problems, usually requiring shut down and cleanout = disaster.

These two statements are so simple that they should not require emphasis, yet a majority of thickener operation problems are a result of a lack of understanding of these basic rules. If solids are allowed to accumulate in a thickener without corrective action, one or more of the following will occur:

a). The pulp will begin to exit the tank with the overflow.

b). The underflow will become too thick to pump.

c). An "island" will form in the thickener and the underflow density will start to approach the feed density.

d). The rake mechanism will become overloaded and be stopped by the Drive Control.

The seriousness of the above is related to the characteristics of the solids which are to be handled. Magnesium hydroxide solids, for instance, usually can be left in a thickener indefinitely without any problems developing with respect to underflow solids concentration or island formation as long as the rake mechanism continues to rotate. Of course there is a limit to the storage volume, and the pulp level will eventually rise and begin to exit with the overflow. Tailings, (iron, copper, etc.) or concentrate accumulation will not allow very much time before causing problems.

Coarse solids will usually overload a mechanism. Coarse solids, 840 to 250 micron (20 x 65 mesh), should generally be kept out of a thickener. However, there are circumstances under which these solids can be handled. Solids in this size range produce a very high rake arm load and require significant torque to transmit them to the discharge point. As the inventory of coarse solids increases the torque requirement rises rapidly.

Operator Control

That which occurs inside a thickener is not obvious to visual observation. The operator must rely upon representative (continuous is best) measurements of:

- Pulp level. Increasing level is indicative of accumulation, possibly due to flocculation or insufficient underflow withdrawal.
- Feed. Density, size distribution, and quantity of solids.
- Underflow. Density, solids size distribution, and quantity. Solids in the feed are expected to report to the underflow.
- Torque. Increasing torque indicates overload may be occurring.
- Flocculant dosage and mix.
- Position of rake arms with respect to normal elevation, if a lift device is employed.
- pH of liquor, when such is required.

Interpretation of the meaning of these measurements provides the operator with actions required to stay out of trouble. There is no universal set of rules as to what changes in the above measurements mean. There are only generalities for similar installations. Each installation has its own operating peculiarities.

Some of the generalities for Mineral Processing are:

a). An increase in torque is usually the result of an increasing quantity of coarse material in the slurry and on the tank bottom, and may be accompanied by a rise in pulp level and increased underflow density requiring a bulldozing action by the rake arms. Corrective action is to get the coarse material out, increase underflow withdrawal rate, and, if possible, increase mechanism speed. If corrective action appears to require more time to prevent rake arm stoppage, raise the mechanism until torque decreases - provided a lift is incorporated.

b). An increase in torque associated with a decrease in underflow density may be the result of "island" formation of the solids.

The general tendency when operating a thickener is to decrease the rate of underflow withdrawal as the density decreases, but this move may work in favor of increased island formation. As an island forms there is often an increase in drive torque due to viscous drag through the slurry and the friction between the island and the bed beneath the rakes.

An increase in torque indication, together with decreasing underflow density is usually a sure sign of island formation. The problem then becomes how to break up the island before it builds up to the point where the machine will shut down due to excessive torque.

If the island has developed to the point that thickener performance is severely handicapped, feed to the unit should be stopped, since it will probably require considerable time to clear out the island. Continued addition of solids will only aggravate the problem. If the mechanism is fitted with a lifting device, raising the rakes will frequently cause the island to slough off and flow, or slide, into the discharge outlet. It should be remembered, however, that raising the rakes removes the main support of the island, and a greater weight of solids will have to be pushed along the thickener floor temporarily. This will cause an immediate increase in torque. Therefore, the extent to which the rakes should be raised at any given time must be conditioned by the percentage of torque already being used.

Some plants have employed the powered lift device to prevent island formation by routinely raising and lowering the thickener arms. Such raising and lowering may be as often as once per operating shift.

If the mechanism does not have a lifting device, the operator must use high pressure water or air lances to try to break up the island and move it to the discharge point. If this fails, the only alternative may be to shut down and dig out the unit.

Island formation is not always predictable, but some generalizations about it are discussed:

A large number of thickener applications make use of synthetic polymers to increase performance. As polymer dosage is increased, there is frequently an increase in underflow viscosity. There is a point at which the thickened solids lose their fluidity and the mechanism rake arms may no longer cause flow toward the discharge point. Instead, a viscous, or gelatinous, mass tends to travel along in front of the rake arms, and this mass will slowly build up and start to accumulate in the rake arms. As this happens, those solids which should move to the center of the tank will be blocked by the "stationary" mass. Additional residence time tends to consolidate the solids in this mass and make them more immovable. The net result is the formation of a fairly solid accumulation which slides along the floor of the thickener and eventually fills the rake truss itself. If allowed to continue long enough, additional solids accumulate in front of the mass contained in the rake and the total accumulation can eventually grow to form a complete ring. This formation is that commonly referred to as an "island". The island effectively blocks settled solids from the central discharge outlet, and any solids which do reach the outlet must pass up and over this island or short-circuit directly from the feed inlet. With insufficient detention time in the thickener, a much lower solids concentration can result. Thus, the underflow becomes a diluted slurry.

Polymers are not the only contributing cause of island formation. It may also be a result of allowing the solids to stay in the thickener as long as possible to increase underflow density. However, this must be done with discretion, and successful application of this principle is dependent upon the type of solids being handled and the type and quantity of polymers being used.

Finally, a very important factor is to understand the thickener operation and to have the patience to allow changes to take place once corrective action has been taken.

THE HIGH CAPACITY THICKENER

The availability of new and powerful flocculents make possible thickener mechanisms in relatively small tanks and having a very small area per unit weight, of solids per day, ratio. Some are installed and operating with ratios as small as .014 meter squared (.15 square foot)/ton/day. Such units require very thorough mixing and dispersing of the flocculent and slurry. The Eimco High Capacity Thickener, wherein mixing, flocculation and solids contact are combined in one vessel, is represented in Figure 14.

ILLUSTRATIONS

Some selected photographs of unique thickener installations are included to illustrate that which has been done. They may be of value as reference when a mill design is being considered.

The illustrations are by no means a complete listing and represent only a few of the thickener designs with which the author is intimately familar.

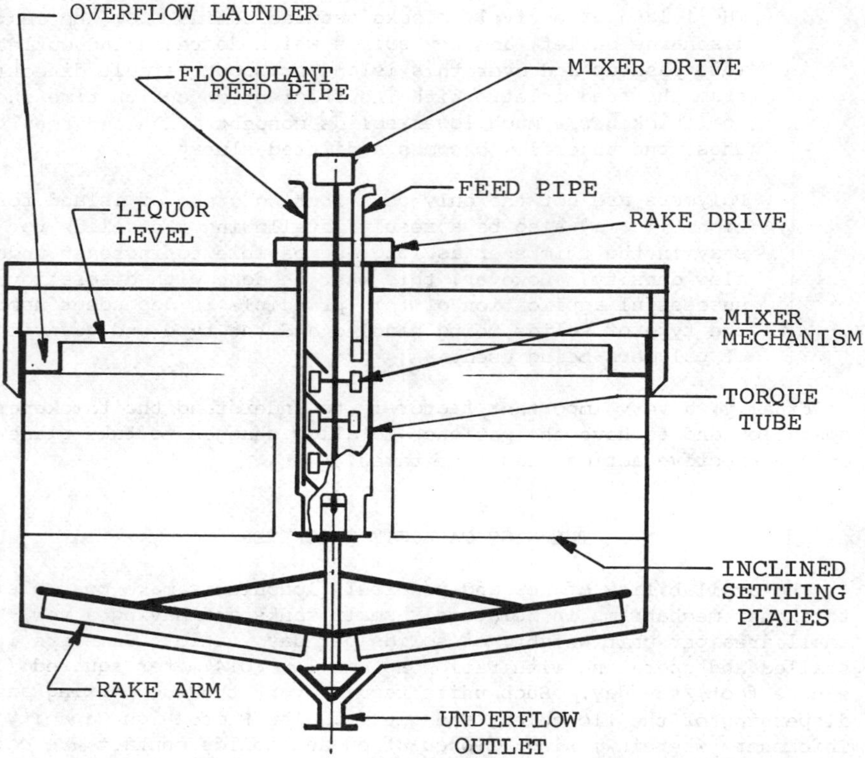

EIMCO HIGH CAPACITY THICKENER

FIGURE 14

Nine 38.1 Meter (125 ft.) diameter thickeners in elevated steel tanks and with Lift Device. Note state of construction and scalloped tank bottom.

FIGURE 15

Four 42.7 meter (140 ft.) diameter thickeners. Elevated steel tanks and with Lift Device

FIGURE 16

91.4 Meter (300 ft.) diameter thickener with Center Lift and drive enclosure for cold climate operation. Tank is steel wall with earth bottom.

FIGURE 17

View of overflow launder during construction of thickener shown in Figure 16. Note uplift supports and uplift relief holes.

FIGURE 18

121.9 meter (400 ft.) diameter thickeners with steel tank wall and lined earth bottom.

FIGURE 19

130.8 meter (429 ft.) diameter Caisson Thickener with concrete tank wall and earth bottom. The earth bottom had a chemically treated clay surface for liquid seal.

FIGURE 20

153.6 meter (504 ft) diameter Thickener with earth tank, under the tank center feed, and buried underflow pipe. Tank is the type illustrated in Figure 6 with overflow through pipes in the earth wall. Service access to the drive in the center is by boat, as an access bridge to the center was not furnished.

FIGURE 21

Peripheral Traction Drive Thickeners. Tank is concrete wall and lined earth bottom. Tractor rail shows on the tank wall. For this type thickener, the access bridge and feed launder must be elevated to clear the drive tractor on the wall. Servicing of the drive tractor requires access around the entire tank periphery.

FIGURE 22

Chapter 28

HOW TO SELECT AND SIZE FILTERS

by Donald A. Dahlstrom

Vice President & Director, R&D

Envirotech Corporation
Salt Lake City, Utah

ABSTRACT

The selection and sizing of filters depends on the characteristics of the type of filter and the feed slurry. In addition, the rate functions of cake formation, cake washing and cake dewatering must be determined in order to design the filter station on the basis of desired performance. This chapter discusses the applied theory used to determine the rate functions and illustrates how to incorporate these into the filter cycle to obtain the required objectives. This also permits the sizing of the filter and auxiliaries as well as the selection of the proper type of filter.

INTRODUCTION

With the growth in technology and size of the mineral processing industry, together with a greater emphasis on hydrometallurgy, most filter applications now employ continuous filtration. This chapter will develop the methods utilized to select and size these units.

TYPES OF CONTINUOUS FILTERS

Types of continuous filters must first be considered as their characteristics will influence data interpretation and sizing. The percent of the filter cycle that can be devoted to cake formation, cake washing and cake dewatering will vary with the basic units. Cake discharge methods can vary within a basic type, whether cake washing can be practiced and even if two or more counter-

TABLE I

CHARACTERISTICS OF BASIC TYPES OF CONTINUOUS FILTERS

Basic Type of Filter With Options Indicated	Range of Filter Cycle Percent Devoted to *			Number of Washing Stages Possible	Minimum Cake Thickness mm
	Cake Formation	Cake Washing	Cake Dewatering		
		Cake Formation Against Gravity			
Disc Filter With and Without Agitation	5 - 40%	Not Recommended	5 - 45%	--	8
Drum Filter	5 - 50%	0 - 30%**	0 - 60%	1	
Scraper Discharge					6
Roller Discharge					1
String Discharge					6
Continuous Belt Discharge					3
Precoat					~0
		Cake Formation With Gravity			
Top Feed Drum Filter	5 - 15	0 - 20	25 - 70	1	12
Horizontal Belt Filter	5 - 90	0 - 90	5 - 90	As Many as Desired	3
Horizontal Table or Tilting Pan	5 - 70	0 - 70	5 - 75	2-3	20

* With Horizontal Belt Filter, only active area is considered in cycle. All others include total area including dead time in cycle.

** Can go to 50% if drum is flood washed -- not common.

current washes can be employed.

Table I lists the ranges of the filter cycles that can be devoted to cake formation, cake dewatering and cake washing, as well as the minimum cake thickness that is required for efficient discharge. The Table has been divided into two groups -- those forming their cake against gravity (bottom feed) and those forming cake with gravity (top feed).

The factors given in Table I definitely influence the filter selection as well as optimization. For example, as will be shown later, if a filter can discharge a thinner cake, productive capacity per unit area can be greater because of the shorter cycle time. In addition, with apparent submergences greater than 40%, stuffing boxes on the filter trunnions will be required for disc and drum type filters which may not be desirable for the type of feed slurry involved.

The filters forming their cake with gravity are generaly applied to relatively fast filtering material and particularly if the solids cannot be left in suspension by reasonable agitation of the type applied in a filter tank. The exception to this is the horizontal belt filter where very thin cakes (as thin as 3 mm) can be discharged and advantage can be taken of counter-current washing while minimizing wash fluid addition. Installation of 5 stages of counter-current wash have been used on the horizontal belt filter.

APPLIED THEORY OF CONTINUOUS FILTRATION

The continuous filter cycle can be divided into three rate functions which can be analyzed separately. These are (1) cake formation rate (2) cake dewatering rate and (3) cake washing rate.[2] While one of these will usually be controlling, all of them must be considered in order to incorporate them into the filter cycle. This is another reason why the percentage of filter cycle time that can be devoted to the various three functions for each type filter must be considered.

Cake Formation Rate

As indicated in earlier publications[1&2] this rate function can be defined by the following equation:

Form Filtration Rate = Z = (1)

$$= \left[\frac{Kw\Delta P}{\mu\alpha\theta_f}\right]^{1/2}$$

Where Z = form filtration rate expressed as weight of dry solids per unit area per unit time of cake formation

K = proportionality constant

w = weight of dry cake solids per unit volume of filtrate

ΔP = pressure drop across the cake

μ = viscosity of liquid, centipoises

θ_f = cake formation time per cycle

α = specific resistance of the cake

Z is normally expressed either as kilograms of dry solids per square meter per hour of form time or pounds of dry solids per square foot per hour of form time.

It is stressed that the cake formation rate Z does not take into account the remainder of the filter cycle (cake wash time, cake dewatering time, dead time for discharge, and percentage of apparent submergence required by the specific type of filter). However, once the other rate functions are known, this rate formation can be worked into the total cycle by multiplying by the fraction of the cycle that will be employed for cake formation.

The form filtration rate can be converted to the volumetric rate of filtrate by dividing both sides of the equation by w to obtain:

$$Y = \left[\frac{K\,\Delta P}{\mu\alpha w\,\theta_f}\right]^{1/2} \qquad (2)$$

Where Y = form filtration rate, volume of filtrate per unit area per unit time of cake formation

Equations 1 and 2 are ideal equations and do not consider such things as the resistance of the filter cloth or the internal drainage network of the filter. The former is generally negligible in comparison to the specific resistance of the cake. However, it will be noticed that varying form filtration rates will be experienced with different filter media. This is not generally due to the resistance of the cloth but instead the nature of the cake formed next to the filter cloth. As a tighter filter cloth is employed, the pores of the media bridge over with finer particles yielding a higher specific resistance at this critical layer. After that layer of formation, the cake itself becomes the filter media and the filtrate becomes practically clear in most cases. Therefore, the most open filter media that gives reasonable filtrate clarity for process requirements consistent with good cake discharge and life should be employed.

Specific resistance α is influenced by particle size, particle size distribution and shape. If colloidal solids are present, flocculation may significantly reduce specific resistance. In addition, feed solids concentration can also influence specific resistance. As it will generally decrease to some extent as solids concentration is increased, this further emphasizes the importance of maintaining high filter feed solids concentrations.

Cake Dewatering Rate

The cake dewatering rate is a very complex phenomenon which can be reduced to its simplest form by the following equation:[3]

$$\%M = f(Fa, d, \%M_r) \quad (3)$$

Where $\%M$ = weight % moisture of the discharged cake

$\%M_r$ = Wt.% residual moisture which is the moisture at equilibrium if 100% saturated gas is pulled through the cake at the pressure drop ΔP

Fa = approach factor which depicts the approach to $\%M_r$.

d = particle size, particle size distribution and shape factor

The term d is normally employed as a parameter as specific surface area alone does not give the total influence, although it is undoubtedly a major factor. All of the surface area is wetted by a film of liquid and as this increases, moisture will increase. However, a particle size distribution parameter is normally used such as % -200 mesh, % -10 microns, etc.

$\%M_r$ very seldom needs to be measured because of the correlation methods that will be indicated. However, it should be realized that it is influenced not only by d, but also by ΔP, viscosity of the liquid, sometimes by cake thickness, and can be a function of the filter cloth employed primarily by the phenomenon indicated earlier. Surfactants can also influence $\%M_r$ by decreasing surface tension or interfacial tension. Flocculants generally have little effect on $\%M_r$ because the specific surface area is not changed.

The term Fa is the major factor that is employed to determine optimum moisture content that can be achieved and the requirements to obtain it for a specific feed. From basic theory, it can be shown that Fa can be described by the following equation:[4]

$$a = Fa \left[\frac{\Delta P}{W}\right]\left[\frac{CFM}{Ft^2}\right]_d \Theta_d \qquad (4)$$

Where W = cake weight of dry solids per unit area per cycle

$\left[\frac{CFM}{Ft^2}\right]_d$ = gas pulled through the cake measured at down stream pressure as cubic feet per minute per square foot during the dewatering portion of the cycle

Θ_d = dewatering time during the cycle normally measured in minutes

A plot of %M versus Fa will yield a descending curve becoming asymptotic to some minimum value ($\%M_r$) as Fa is increased. It should be noted that the term ΔP may still have to be used as a parameter as it can affect $\%M_r$.

The term Fa is valuable from two standpoints. Not only does it predict moisture content as a function of primary variables which can be optimized, but it also yields energy requirements. The two terms ΔP and $(CFM/Ft^2)_d$ will give the requirements for the vacuum pump or compressor which is used as the driving force for filtration.

From experience, it has been observed that the term Fa can be simplified to the following equation as long as $(CFM/Ft^2)_d$ is less than 20 CFM/Ft^2:[5]

$$Fa = f\left[\frac{\Theta_d}{W}\right] \tag{5}$$

The same type of curve is experienced but parameters of ΔP must be employed. The $(CFM/Ft^2)_d$ is still measured in order to determine energy requirements.

It should also be noted that as the correlating factor is increased either by increasing Θ_d or decreasing W, productivity per unit area of filtration will also decrease. Thus, it is obvious that economics of capital and operating costs are involved to optimize moisture content.

Cake Washing Rate

This rate function has two rate equations that must be considered. These are the rate of flow of the wash fluid through the cake and the efficiency of displacement of the mother liquor by the washing fluid. The first has been shown to follow the expression: (6)

$$\Theta_w = K' \Theta_f n \tag{6}$$

Where Θ_w = wash time during the filter cycle, normally measured in minutes

K' = proportionality constant

n = volume of wash fluid/volume of cake liquor

Thus, a coordinate plot of wash time Θ_w as a function of wash ratio n should yield a family of straight lines whose separation is proportional to Θ_f and will pass through the origin. Another way of plotting to obtain a single line is to plot Θ_w/Θ_f versus n.

It should be noted that Equation 6 assumes that the mother liquor and wash fluid are miscible and of relatively the same viscosity. If the viscosity of the wash fluid is less than the mother liquor the plot will show a faster penetration rate (less slope) after about a 0.7 wash ratio is achieved.

While the wash flow rate can be determined, this does not yield the efficiency of removal of the soluble solids by the washing fluid. This removal can occur both by displacement and diffusion. The ideal would be plug flow whereby one wash ratio applied would remove all the mother liquor from the cake. However, because of the zero velocity of the liquid film next to the particle surface, some of the mother liquor can only be removed by diffusion. The removal efficiency has been shown to best agree with the following semi-log function:(6)

$$\frac{R}{100} = \left[1 - \frac{E}{100}\right]^{n} \tag{7}$$

Where R = % of soluble solids remaining in the cake after washing with 100% being that in the cake prior to wash

E = wash efficiency % or 100-R at n=1.0

Thus, a plot of the log of the % remaining R as a function of the wash ratio n should yield a straight line. Experience has shown that the agreement is good until a wash ratio of 1.5 to 2.5 is reached where a curved relationship begins and becomes asymptotic to some minimum value. This is believed due to blocked capillaries which can only be removed by diffusion or reorientation of the cake.

Wash efficiency E has ranged from a low of about 45% to a high of 86% for various applications investigated. Most values will range between 60 and 80%. However, some reduction in wash efficiency is almost always experienced in going from the first stage of wash to the second and so on. This will usually range from 3 to 5 percentage points per stage. Wash efficiencies will, in practically all cases, be higher than simply mixing the unwashed cake and refiltering and at the same time will require less wash fluid.

APPLIED THEORY USE IN PREDICTING FULL SCALE RESULTS

The applied theory is very useful in prediction of full scale results from representative samples. In most cases, only a small quantity of feed slurry is available and performance requirements of the filtration step must be determined. Such things as obtainable removal of soluble solids, final moisture content, or dilution of the mother liquor by washing fluid are important economic factors and must be predictable. In addition, filtration rate, and thus capital costs, will be influenced by these same factors.

Normally a 0.1 square foot filter leaf is employed for developing data due to the small amount of sample available. At the same time, a much broader investigation can be made with the test filter leaf if proper correlation methods are used. In this way, all rate functions can be determined, the optimum type of filter can be selected, the rate functions can be combined into the filter cycle to obtain desired performance and the resultant filtration rate can be calculated on the basis of performance.

In performing the filtration step, various items are relatively easily measured.

- Feed solids concentration
- Solids size distribution
- Vacuum level or pressure drop
- Cake weight (wet) and thickness
- Cake moisture content
- Volume of filtrate
- Filtrate solids concentration
- Volume of wash fluid
- Soluble solids content of the cake and mother liquor
- Cake formation time
- Cake washing time
- Cake dewatering time
- Rate and volume of gas pulled through the cake
- Observations are also made on ease of cake discharge and required type of discharge.

From these data, all the rate functions can be determined. It is stressed that every effort should be made to simulate probable full scale conditions such as feeding method, agitation, temperature, etc.

Sizing Based on Cake Formation Rate

To obtain the cake formation rate, form filtration rates are first determined by dividing the dry weight of the cake collected by the area employed (usually 0.1 ft = $0.0093M^2$ and by the time of cake formation in minutes and multiplying by 60 to obtain kilograms per square meter per hour of cake formation time.

According to Equation 1, a log-log plot of form filtration rate as a function of cake formation time should yield a straight line with parameters of feed solids concentration, pressure drop and temperature. In addition, because of the importance of cake thickness and its influence on cake discharge for the various types of filters, it is desirable to indicate cake thickness.

Ideally, the line should have a slope of -0.50. However, if there is migration of solids after cake deposition or there is a shape factor effect, the slope can be more negative. With one exception, the slope should always lie between the limits of -1.0 and -0.50. Any other values indicate erroneous data. The one exception is the precoat filter where the resistance of the precoat bed can be greater than the deposited cake which can occur generally with feed solids concentration of 300 ppm or less. In these instances, the slope can be between zero and -0.50.

As the slope becomes more negative, this will generally result in shorter full scale filter cycles to obtain economic filtration rates. However, sufficient cake thickness must be obtained for proper discharge with the type of filter employed.

Figure 1 is a typical plot of the log of form filtration rate as a function of the log of form time. Parameters of feed solids concentration are indicated. Also, the influence of flocculation is also indicated for the particular material. When applicable, a sizeable increase in filtration rate is experienced with flocculation.

Normally, about 5 tests at different form times are sufficient to determine the straight line relationship spread over a 5:1 to 10:1 cake formation time range. Validity of the data can be further checked by back-calculating the feed solids concentration and comparing with the measured value. The back-calculation will normally be slightly lower because the solids must be accelerated to the cake; whereas, the liquid does not. If the back-calculated values are less than about 95% of the measured, solids must be stratifying or settling such that the coarser ones are not being pulled to the cake. Thus, erroneous results will be predicted and corrective action should be taken.

To obtain a full scale rate, the form filtration rate at the desired cake thickness is multiplied by the fraction of the filter cycle time that will be devoted to cake formation and by a scale-up factor. The latter is normally 0.8, but will depend on the type of filter.[7] The scale-up factor should also take into account such things as reasonable fluctuations in feed quality, filter media blinding with time and other operating factors.

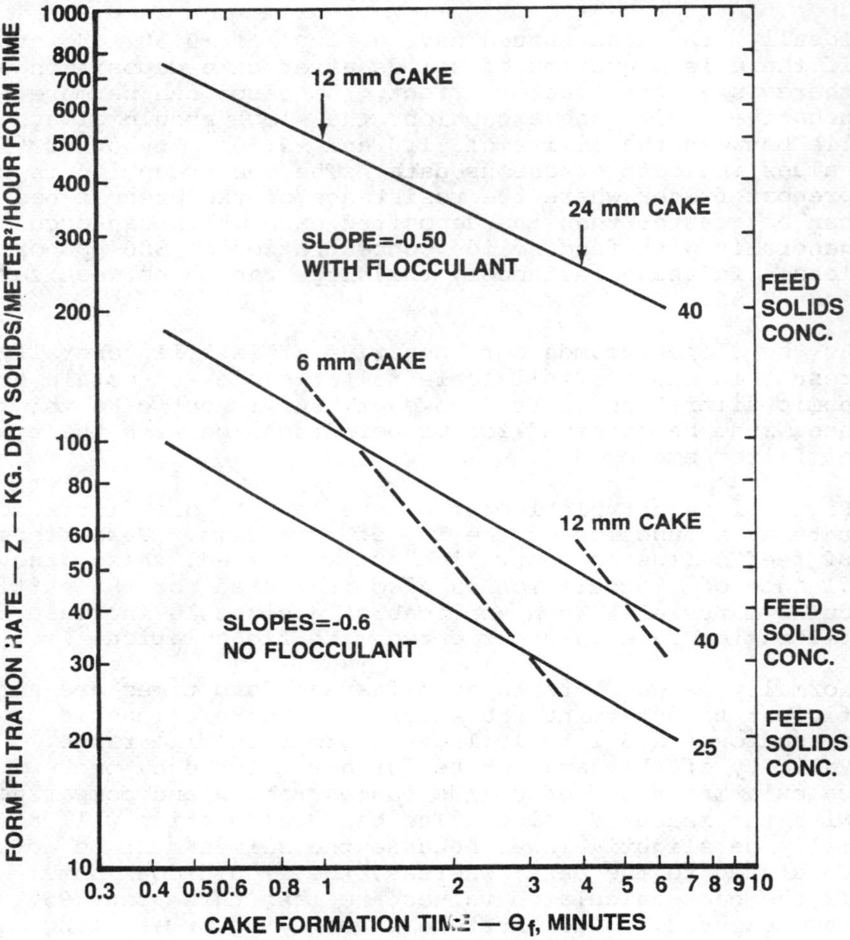

TYPICAL PLOT OF FORM FILTRATION RATE VS. FORM TIME

FIGURE 1

As an example of scale-up based on cake formation rate only, Figure 1 can be employed. It will be assumed that this is of disc type feed and flocculation is required. Furthermore, the feed solids size distribution contains some coarse material that must be kept in suspension so that the Agidisc Filter must be employed. A 35% effective submergence (% of cycle exposed to vacuum under slurry) will be used. Also, a 12 mm cake thickness for good discharge is selected. From Figure 1 at 40% feed solids concentration and with flocculation, the form filtration rate is 500 Kg dry solids/M^2/hour of cake formation time. With the disc filter, an 0.8 scale-up factor should be used. Accordingly,

$$\text{Full scale design rate} = 500 \times \frac{35}{100} \times 0.8$$

$$= 140 \text{ Kg dry solids}/M^2/\text{hour}$$

$$= 28.7 \text{ lbs.dry solids}/F^2/\text{hour}$$

It should be emphasized that the scale-up factor takes into account % of cake discharge, and reasonable changes in filterability and solids concentration of the feed. A higher value of cake thickness greater than minimum (8 mm) has been selected so that increased capacity is available for higher production rates when needed.

Scale-up Based on Cake Dewatering Rate

Cake dewatering rate is determined by a coordinate plot of discharged moisture content as a function of the correlating factor Θ_d/W. Parameters of vacuum level may also be included. Figure 2 illustrates a typical plot with two vacuum levels. It should be noted that if the $(cfm/ft^2)_d$ is above 20 during the dewatering step, the correlating factor $\left(\frac{\Delta P}{W}\right)\left(\frac{CFM}{Ft^2}\right)_d \Theta_d$ should be employed.

A sufficient spread of the correlating factor should be tested to determine the knee of the curve and about 100% beyond that point. This will usually entail a 10:1 range in Θ_d/W.

To convert this to a production rate, the desired moisture content should be selected and accordingly a value of Θ_d/W. As a scale-up factor this value should normally be multiplied by 1.2 to take into account off-quality feeds and other factors. The cake weight per unit area per cycle is selected that yields a dischargeable cake so that the required cake dewatering time can be calculated. If dewatering time is measured in minutes, the filter cycle time can be calculated in minutes per revolution (MPR) by dividing the cake dewatering time by the

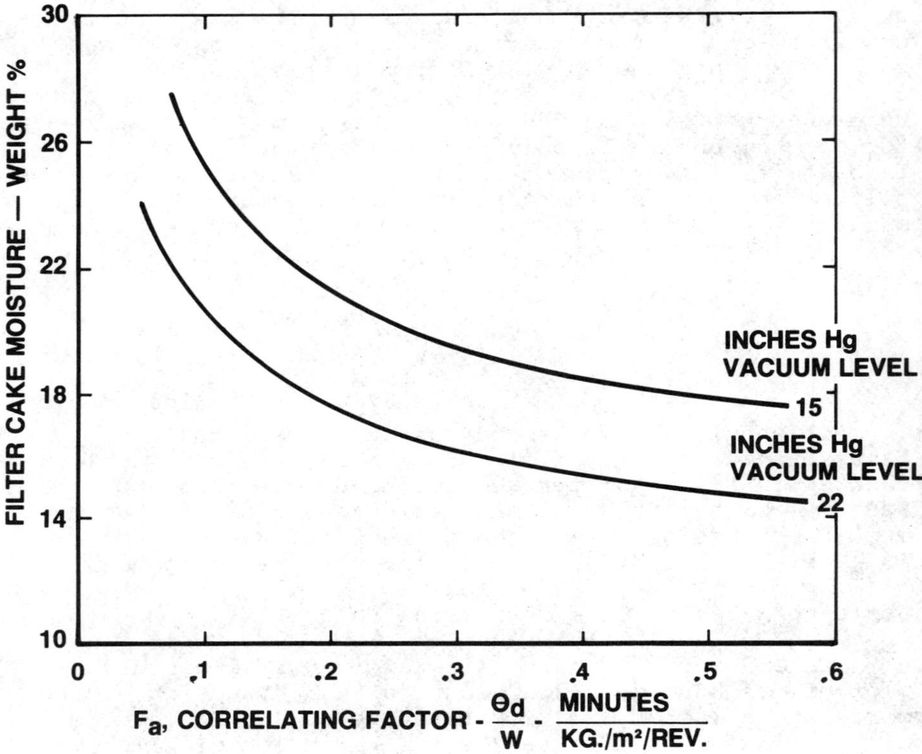

TYPICAL PLOT OF MOISTURE CONTENT VS. THE CORRELATING FACTOR, F_a

FIGURE 2

fraction of the filter cycle that can be devoted to cake dewatering. Finally, the filtration rate is determined by multiplying W by 60/MPR. Thus

$$\text{Filtration Rate} = \frac{60x}{1.2\ (\Theta_d/W)_1} \tag{8}$$

Where Filtration Rate = weight of dry solids per unit area per hour

x = Fraction of cycle time devoted to dewatering

$\frac{\Theta_d}{W}_1$ = required value to achieve desired moisture

$$MPR = \frac{1.2(\Theta_d)_1}{x} \tag{9}$$

Where $(\Theta_d)_1$ = cake dewatering time obtained from $(\Theta_d/W/)_1$ at proper cake thickness

It should be noted that the value of Θ_d/W should normally be selected at or beyond the lower portion of the "knee" of the curve as the sharply descending portion is relatively unstable. Furthermore, if moisture content is critical to later processing, it may be desirable to increase the scale-up factor.

As an illustration, if 18% moisture were desired for the material of Figure 2, undoubtedly 22 inches of mercury vacuum would be employed as it could easily be justified. It will also be assumed that a drum filter should be employed as thin cakes are involved. Accordingly, from Figure 2, Θ_d/W at 18% moisture = 0.18 minutes per Kg dry solids per meter^2 per revolution. The solids involved have a specific gravity of 2.7 and a dry solids filter cake density of 101.1 pounds per cubic foot (1609 Kg/M^3). With 60% dewatering time on the drum filter and from Equation 8,

$$\text{Filtration rate} = \frac{60(0.6)}{1.2(0.18} = 166.7\ \text{Kg. dry solids/hr/M}^2$$
$$= 34.2\ \text{lbs.dry solids/hr/ft}^2$$

The filter cycle time can also be determined assuming

an 8 mm cake thickness.

$$\frac{\theta_d}{W} = 0.18(1.2) = \frac{\theta_d}{1609}(6/1000)$$

$$\theta_d = 2.09 \text{ minutes}$$

$$MPR = \frac{2.09}{0.6} = 3.48 \text{ minutes/revolution}$$

By the same method, it is possible to determine moisture content as a function of filtration rate by selecting various θ_d/W values from a plot similar to Figure 2. There will be an upper limit at which cake thickness is too thin to permit effective discharge. Such a plot is shown in Figure 3 which would be typical for some iron ore concentrates. This permits selection of the filter station on the basis of desired performance.

A plot can now be made of moisture content as a function of predicted filtration rate. There will be an upper limit at which cake thickness is too thin to permit effective discharge. A typical plot is shown in Figure 3.

To determine energy requirements (vacuum pump or compressor), a measurement of gas rate through the cake as a function of dewatering time is made. This should be taken at the desired cake thickness. A typical plot is shown on Figure 4. It will be noted that the instantaneous gas rate increased rapidly and levels off to some maximum value as the capillaries become dewatered. Generally, the maximum rate is employed as the desired value to ensure proper vacuum operation. As the filter internal drainage network must also be evacuated prior to filtration, a scale-up factor of 1.1 is usually employed. As vacuum or compressor requirements are usually based on total area, the following equation permits calculation:

Required compressor or vacuum pump capacity =

$$= 1.1 \left(\frac{CFM}{Ft^2}\right)_{d2} \times \qquad (10)$$

Where $(CFM/Ft^2)_{d2}$ = maximum gas rate during dewatering expressed as cubic feet per minute per square foot measured at downstream pressure

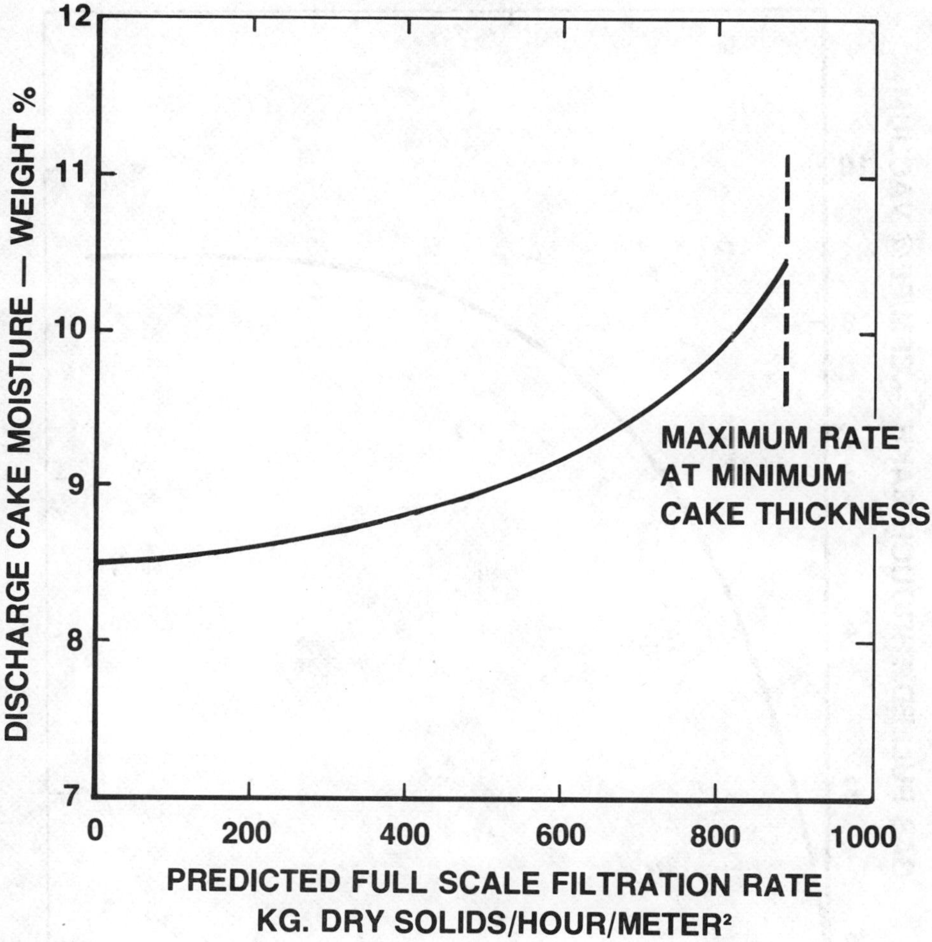

TYPICAL CURVE OF CAKE MOISTURE CONTENT AS A FUNCTION OF FILTRATION RATE

FIGURE 3

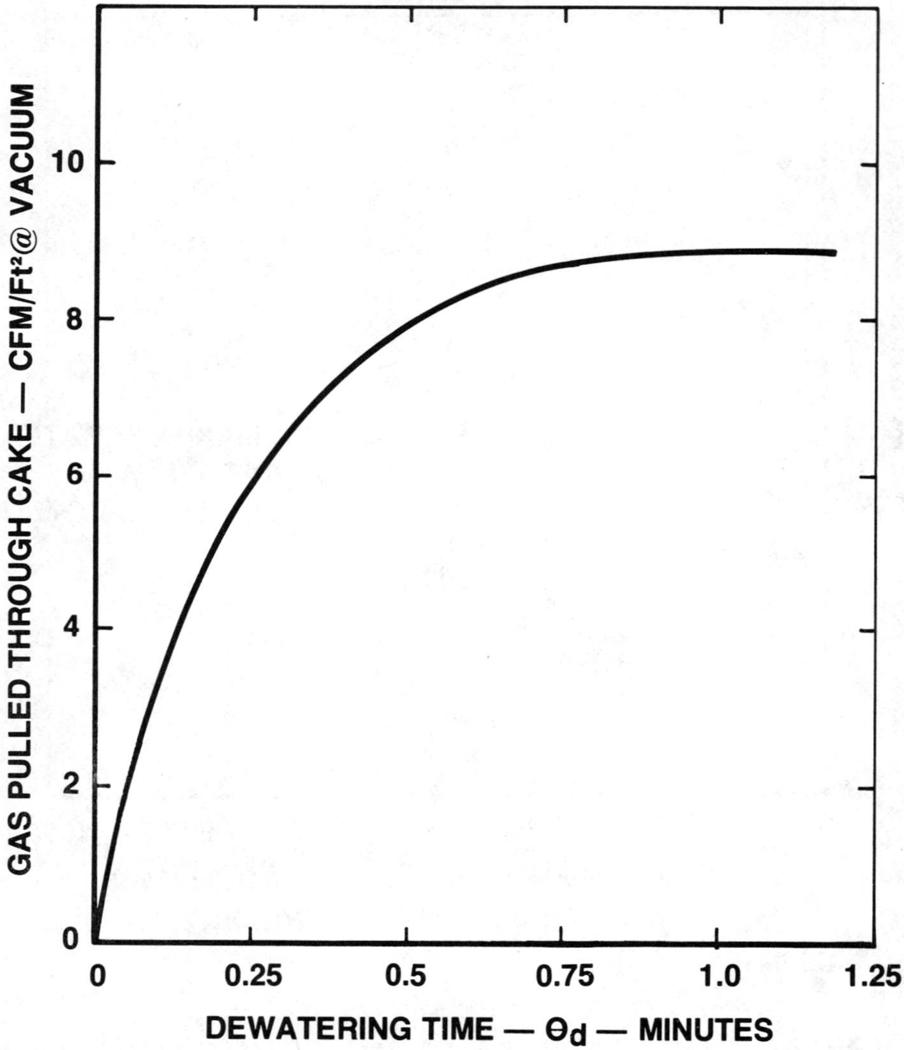

TYPICAL PLOT OF GAS RATE THROUGH THE CAKE AS A FUNCTION OF DEWATERING TIME

FIGURE 4

Scale-up Based on Cake Washing Rate

To determine the cake washing rate functions two plots are required. The first is a plot of Θ_W as a function of wash ratio n with parameters of Θ_f. A typical plot is shown in Figure 5. To check validity, the ratio of cake washing time at any one wash ratio should be approximately equal to the cake formation time. However, this will generally "fan out" slightly more because of certain phenomenon.

The second is a plot of R as a function of wash ratio n on semi-log paper. A typical plot is shown in Figure 6. Another convenient plot is % soluble solids in the washed cake as a function of the wash ratio on semi-log paper. Figure 6 permits the calculation of the wash efficiency term E at a wash ratio of 1.0. As a scale-up factor, the calculated value of E is usually lowered by 5 percentage points to account for uneven distribution of wash fluid and reasonable off-quality of feed.

Full scale filtration rate for cake washing can now be calculated. From Figure 5, a wash time is selected at a form time that yields a sufficient cake thickness for discharge and at a specific wash ratio. This value divided by the fraction of time during the filter cycle devoted to cake washing yields the cycle time. A scale-up factor of 1.2 is normally employed. Multiplying the cake weight selected by 60/MPR yields the full scale filtration rate. Thus

$$\text{Filtration Rate} - W \ \frac{60 \quad S}{1.2\ \Theta_W} \qquad (11)$$

Where filtration rate is measured in weight of dry solids per hour per total unit area W and Θ_W are selected values with Θ_W measured in minutes

S = Fraction of filter cycle devoted to cake washing

From the full scale filtration rates calculated, it is now possible to draw plots of rates as a function of filter cycle time with parameters of wash ratio n. A drum filter selection will be assumed for the data of Figure 5. Cake weight at 8 mm cake thickness obtained at a 1.0 cake formation time will also be assumed. Cake weight is 2.48 pounds dry solids/Ft^2/Rev. (12.08 Kg. dry solids/M^2/Rev.)

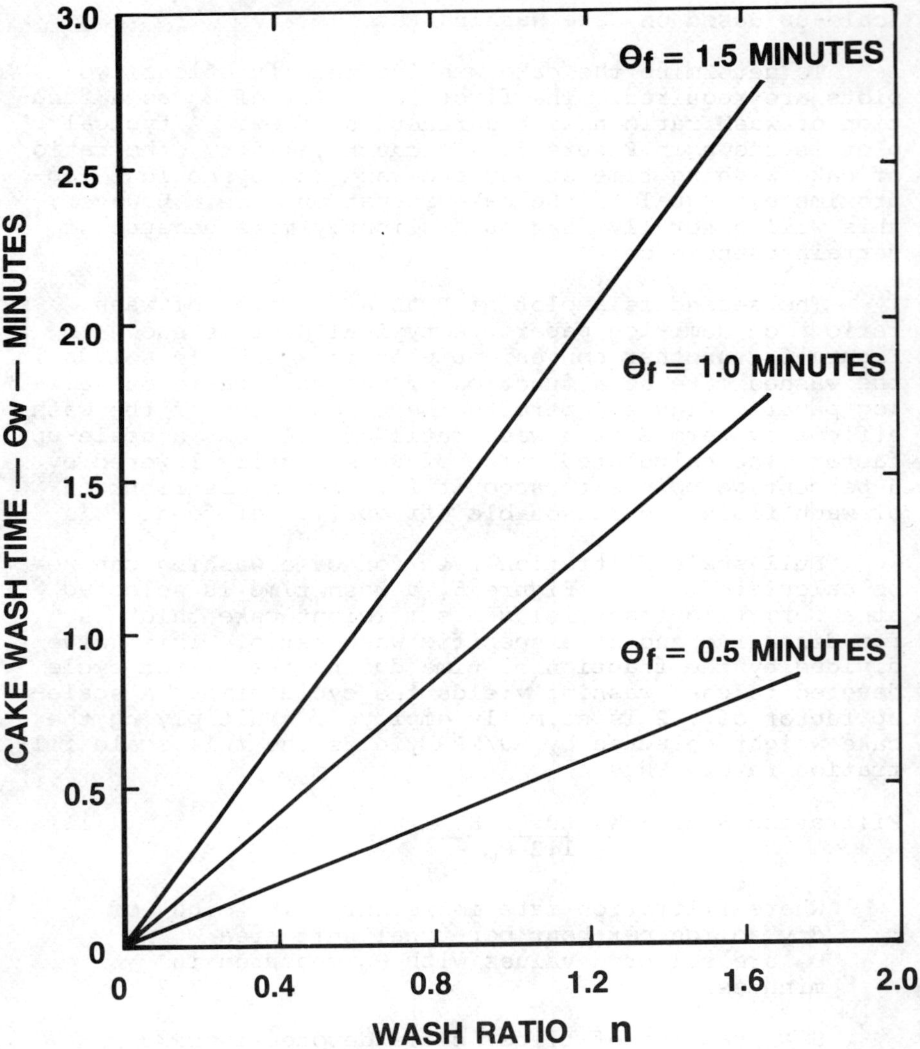

TYPICAL PLOT OF CAKE WASH TIME AS A FUNCTION OF WASH RATIO PARAMETERS OF CAKE FORMATION TIME

FIGURE 5

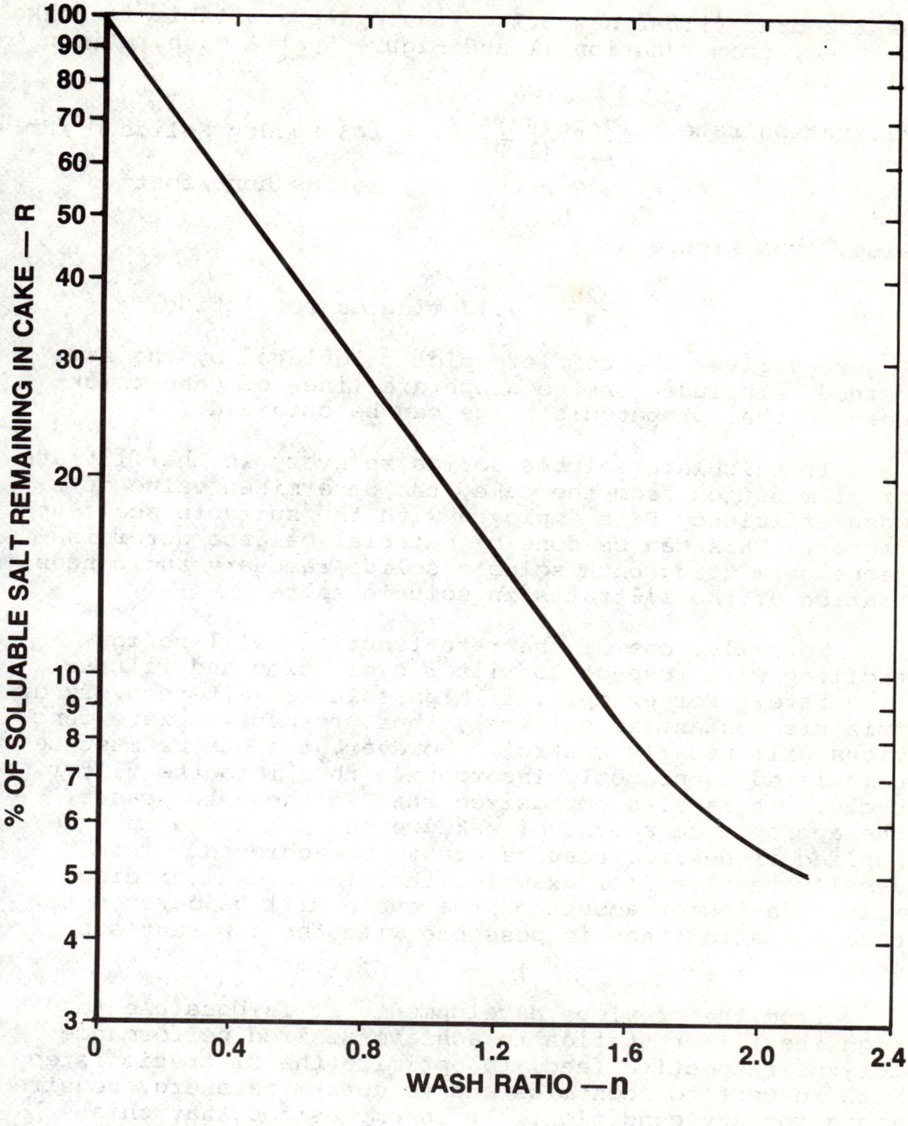

TYPICAL PLOT OF % REMAINING OF SOLUABLE SALT IN THE FILTER CAKE AS A FUNCTION OF WASH RATIO

FIGURE 6

As the drum filter has 30% of the cycle available for cake washing, from Equation 11 and Figure 5 at a wash ratio n of 1.2,

$$\text{Filtration rate} = \frac{12.08(60)0.30}{1.2\ (1.25)} = 145 \text{ Kg.dry Solids/M}^2\text{/Hr.}$$

$$= 29.7 \text{ pounds dry solids/hour/foot}^2$$

Also, from Figure 5

$$\text{MPR} = \frac{1.25}{0.3} = 4.17 \text{ minutes/revolution}$$

Figure 7 gives the complete plot calculated by the same method. Included on the graph are lines of cake thickness so that proper discharge can be obtained.

To calculate soluble solids recovery in the filtrate or elimination from the cake, the determined value of wash efficiency E is employed with the suitable scale-up factor. This can be done by material balance determinations permitting both soluble solids recovery and concentration of the filtrates in soluble salts.

Normally, one of the rate functions will be controlling with respect to filter cycle time and filtration rate. For example, if high soluble salt recovery or moisture content is critical, these respective rate functions will usually control. However, the others must be calculated to properly incorporate them into the filter cycle. It is also emphasized that in these two cases, the appropriate amount of cake weight per cycle must be applied if desired results are to be achieved. This usually results, for example, that for a drum or disc filter, a lesser amount of the cycle will be devoted to cake formation than is possible with the apparent submergence.

From the previous development, it is possible to design the filter station to achieve desired performance for the respective feed, to optimize the filtration step with respect to economics and to determine energy requirements for any condition. It is emphasized that this stems from applied theory, all of which was developed from the fundamental theory.

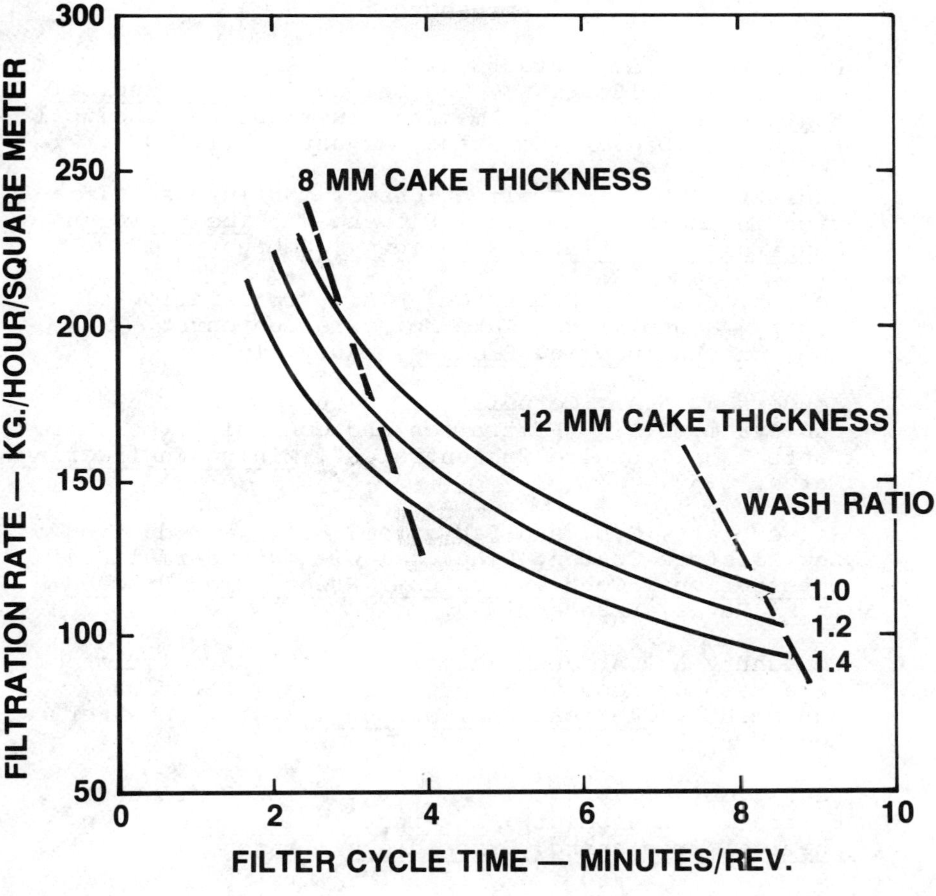

TYPICAL PLOT OF FULL SCALE FILTRATION RATE AS A FUNCTION OF FILTER CYCLE TIME PARAMETERS OF WASH RATIO AND CAKE THICKNESS

FIGURE 7

REFERENCES

1. Dahlstrom, D.A., "Practical Use of Applied Theory of Continuous Filtration", The Second Pacific Chemical Engineering Congress, American Institute of Chemical Engineers, Denver, Colorado, August 28-31, 1977.

2. Dahlstrom, D.A., "Applications of Continuous Filtration in the Metallurgical Flowsheet", The Canadian Institute of Mining & Metallurgy, February, 1959.

3. Nelson, P.A. and Dahlstrom, D.A., "Correlation of Rotary Vacuum Filter Cake Moisture Contents", Chemical Engineering Progress, July, 1957.

4. Henderson, A.A., Cornell, C.F., Dunyon, A.F. and Dahlstrom, D.A., "Filtration and Control of Moisture Content on Taconite Concentrates", Mining Engineering, March, 1957.

5. Silverblatt, C.E. and Dahlstrom, D.A., "Production of Low Moisture Content Fine Coal Without Thermal Drying", International Congress on Coal Preparation Proceedings, Paris, France, 1973.

6. Choudhury,A.P.A. and Dahlstrom, D.A., "Prediction of Cake Washing Results With Continuous Filtration Equipment", Chemical Engineering Journal, December, December, 1957.

7. Dahlstrom, D.A., "Scale-up Methods for Continuous Filtration Equipment", International Symposium on Chemical Processing, Proceedings, London, May, 1959.

Chapter 29

CONCENTRATE DRYING, HANDLING AND STORAGE

Kenneth W. Kolthammer

Metallurgical Engineer
Wright Engineers Limited
Vancouver, British Columbia
CANADA

ABSTRACT

Fundamentals of drying and the equipment used are reviewed. Dryers employed in the mineral industry include rotary dryers, fluid bed dryers, spray dryers, hearth dryers and rotary tray dryers.

Concentrate characteristics, corresponding problems and handling methods are likewise discussed. Important factors include weighing and sampling, moisture content, market timetables, transportation availability and climatic conditions.

INTRODUCTION

Drying systems are differentiated from other evaporative processes by being defined as those which thermally remove moisture from solids where such moisture is considered to be relatively small. In general terms, dryers are sized and selected on the basis of evaporative load, drying gas temperature drop across the system and the drying rate required. The commonly recognized reasons for drying materials are related to convenience and cost.

In contrast, the handling and storage of concentrate is predicated on many criteria besides the obvious one of material movement at minimum cost. The type of concentrate, its unit value, location, distance from the market, the climatic conditions at the plant site, and the type of transportation available give rise to specific problems which must be solved for each individual case. As mineral properties are developed in still more remote areas and in more extreme environments, changes in technology have been forced on the industry. In many cases the exploitation of difficult ores requires the handling of finer, stickier and wetter concentrates.

CONCENTRATE DRYING

In-Plant considerations for drying include the requirement to:

- Facilitate the handling of the material in subsequent processes,
- Increase the capacity or decrease the sizing of related equipment,
- Preserve the characteristics of a product during storage and shipment.

Cost considerations for drying include the need to:

- Reduce shipping costs per unit by minimizing moisture content,
- Obtain the best market by providing material to suit a customer's particular requirements,
- Economically balance the cost of an optimum dryness against the cost of drying,
- Simplify packaging of a marketable product,
- Minimize customer and carrier handling charges relating to shipping and storage.

Drying Fundamentals

A number of systems have been developed to describe and categorize dryers so that they can be selected from a table of functions. Most of them are based initially on the physical characteristics of the material such as its consistency, uniformity and handling characteristics. In the mineral processing industry such definitions may be unnecessary and dryers may be classified according to a combination of the heating method employed (direct, indirect, or combined) and whether batch or continuous. "Direct" is understood to be an intimate contact between hot drying gases and wet material with the moisture leaving the vessel in the out-going gases. "Indirect" uses a wall between these two streams through which the heat is conducted. The moisture which is driven off by the contact of the wet material with the hot surfaces is removed separately from the stream of gases used for heating. A "combined", or "direct-indirect" system separates the heated gas stream initially for conductive heating and then joins the gas stream and its remaining heat with the material being dried. "Batch" or "continuous" refers to the handling method for the material. Combining these options with many types of mechanical and manual handling allows a wide range of choices to solve any drying problem.

Direct, continuous dryers are most commonly utilized in the mineral processing industries, i.e., direct contact between the wet solid and hot gases with some means of throughput of the mineral being dried, heat transfer by convection, conduction and radiation, and the mass transfer of the liquid vapour from the drying chamber.

Drying process calculations and design are affected by moisture which is free and that which is held by the particle surface, interstices and absorption. The migration of moisture from the center of the material can have a considerable effect on the time required for drying and equipment sizing. However, the theories involved are not usually a part of mineral processing plant calculations.

The final choice of dryer is based on a number of variables relating to the material to be dried and the heated gas stream. Among the most important solid characteristics are sizing, flowability, and contact with the hot equipment surfaces. The range of sizes of particles is usually quite broad and while the larger particles may move uniformly through the reaction area, smaller particles may present no resistance to being swept completely from the chamber requiring large collector and cleaning devices. Particles of some materials may present difficulties by breaking down or, conversely, by agglomerating. Flowability of the material will affect retention time and spacial separation in the drying stream, and must be considered in the design of the dryer.

The handling properties of the feed and discharge materials will usually be quite different and in the movement of the solids through the drying process numerous variations of these properties may be experienced. Equipment surfaces must be below any temperature that might affect the solid, but should provide as much help in the drying procedure as possible. Process temperatures are limited by the materials being handled and the equipment being used, and the common maximum drying temperature is about 750 °C. The time available for drying within the framework of the plant operating time-table also affects the sizing and selection of the equipment required.

As to the gas stream; temperature, flow rate, and inlet/outlet humidity are the controlling factors. These must be compatible with the equipment sizing, sequence of operation, and the effect of the gas medium upon the material being dried. Initial selection of dryer possibilities is made from these known or calculated details, and a cost analysis for capital and operating expenses.

Pre-treatment of the material may also enter into the study since it usually costs less to remove as much moisture as possible by other than thermal means, e.g., a new or improved filtering step may be indicated which would allow a less expensive drying stage.

Other influences to be considered are the physical and chemical characteristics of the materials being dried. Those which are corrosive or abrasive in the media involved will affect the materials of construction and these must be selected carefully. On the other hand, analysis or past experience may indicate that special materials are necessary through only a part of the paths which the solids and gases travel, and cheaper materials can be used for the remainder of the equipment.

The cost and installation of the auxiliary equipment required by specific dryers will also affect final dryer preference. As mentioned above, pre-treatment may lower the total installation cost and size. Product must also be collected and handled and waste gases which contain a large percentage of the final product must be cleaned. Environmental reasons may require more efficient cleaning and a larger installation than would otherwise be the case.

Dryer Selection

After preliminary calculations have been completed, the choice of possible dryers has usually narrowed down to two or three possibilities and if past experience does not provide a suitable answer, practical testing may be carried out.

Commercial suppliers have drawn up their own data collection sheets, programmes and graphs which relate as many of the above known and required details to each other as possible. They may include:

Feed characteristics	- rate, weights, moisture, specific heats, physical/chemical properties
Product characteristics	- moisture required
Gas stream	- velocity, humidity in and out
Liquid removal	- calculated
Heating medium	- fuel analysis, moisture content
Heat balance	- calculated
Materials of construction	- corrosion, erosion
Space requirements	- capital, material flow
Known related data	- present practice, known practices, design restraints

Drying tests, with representative equipment based on this data may be pilot scale or full scale. The results, coupled with historical information collated by the manufacturers from past trials and operations, will provide good back-up for final selection. Since the cost of testing is small compared to the cost of the equipment, and dryers once selected are relatively inflexible to change, the investment is well worthwhile in testing a selection of the different classifications under consideration.

Drying Equipment

Drying equipment employed in the mineral industries include rotary dryers, fluid bed dryers, spray dryers, hearth dryers and rotary tray dryers.

Rotary Dryers. Rotary dryers can be used to meet most drying requirements and are by far the most commonly employed in the mineral process industries. A typical rotary dryer is shown in Figure 1.

The dryer is comprised of a nearly horizontal revolving cylinder through which the material, in contact with the hot drying gases, passes toward the discharge end which is located slightly lower than the feed end.

The shell is usually fitted with lifting flights which raise the material to allow it to rain through the passing hot gases, thus providing more intimate contact between the two media. The design and materials of construction are based on the nature of the feed, the amount of build-up which may occur on the walls, and any material physical changes which occur during the drying cycle.

The flight patterns have a large effect on the efficiency of a given dryer. Inlet sections are usually spiralled to provide quick entry of the solid material into the cylinder. Flight heights are about 20% of the radius and are spaced to provide full lift of the contained solids related to residence time, although they may be deleted in the last 10% to 20% of the dryer length when dust carry-over is a problem. Sticky materials are usually handled with radial flights with no lips and some help may be obtained in keeping them clean by the use of external knockers or internal chains, although these will increase the maintenance costs incurred. Free flowing materials, or those developing this characteristic during passage, may have lips added to the flights allowing a full width distribution of the cascade of material, which may permit a reduction in length and rotational speed of the dryer.

Standard dryers come in sizes up to 30 m in length, with slopes of up to 8%. The diameter is not directly related, usually being in the range of 7% to 10% of the length, with cleanout and maintenance space requirements also considered. Peripheral speed may vary from 0.1 to 1 m/s. Air volumes or mass velocities are the maximum that can be tolerated without excessive carry-over of the product into the exit gas stream, and cannot be calculated accurately, so this is best left to practical testwork or past operating experience.

Heated gases may be supplied in direct, indirect, combined or louvered configurations and in concurrent or countercurrent flow. They may be either under pressure, suction or neutral. In the former case, sealing of the rotating cylinder may be necessary which is difficult; under suction it is minimal. The third method, a balanced

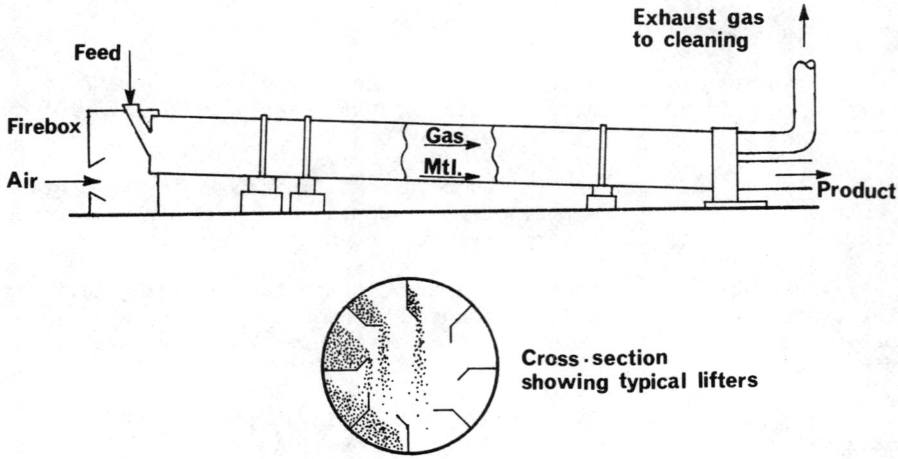

Figure 1 **Rotary Dryer**
Direct-heat, Co-current Flows

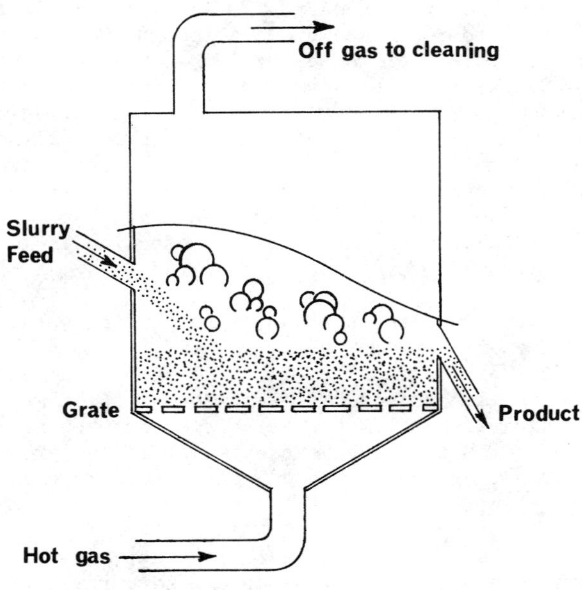

Figure 2 **Fluid Bed Dryer**

blower-exhauster system, which equalizes the inside pressure with that on the outside of the cylinder, is the best control for leakage and heat loss problems.

Major auxiliary equipment includes that for feed preparation and mixing, recycle product handling, and exit gas cyclones, scrubbers, etc. so that cyclone products are either fed directly back into the dryer discharge, or slurried for discharge to the dewatering system which precedes the dryer.

Corrosion/erosion can be a significant cost factor in the operation of a rotary dryer but operating costs are relatively low, with maintenance about 6% of the original cost.

Product dryness can be maintained within quite close limits and concentrates may be dried to a controlled moisture content commensurate with other factors.

Fluid Bed Dryers. Figure 2 shows a fluid bed dryer. These can be used to produce a very homogeneous product or else to divide the material into separate products of the coarser and finer fractions in the drying zone product and the dust collection system respectively.

Fluid bed dryers are vertical cylinders with grate hearth bottoms through which the drying gases are blown from beneath. The exit gases leave from near the top of the dryer, usually through a plenum or restricted area to minimize the amount of dust leaving in the gas stream. In the reaction area of the chamber the gas stream space velocity is such that the materials appear to boil like a fluid, thus the name. In the simplest operation, the slurry or wet solids are directed into the reaction area from above and are collected as an overflow on the circumference which opens from the top of the fluid body drying space across from the entry port. Due to the relationship between the space velocity required to keep the solid particles moving and drying, a large proportion of the material may be carried out with the gas stream from which it is collected.

Design of this type uses common features escept that the grate area may be divided up in various configurations to give a homogeneous fluid bed or a split activity by channelling or spouting the air stream through the body of the material (grate areas may exceed 15 m^2 with height equal to 1-1/2 to 3 times the diameter). Volume-heat transfer ratio is smaller than in a rotary dryer, and location is simplified since moving parts, footing allowance for mechanical impact, weight and floor space are all less. However, sizing can be excessive and costly compared to a rotary. If a low air velocity is used due to the fineness of the particles, a large area will be required, or if a higher air velocity is used, larger and more expensive dust collection systems may be required.

Drying can be carried out to well-defined limits and fuel consumption is comparatively low. The chamber can be used as a means of separating the fines from the body of the material which can lend itself to improved methods of handling the product and changing its characteristics, e.g., very fine particles in the exit gas can be agglomerated and remixed with the coarser material which has been produced as the main product. The fine particles are collected in expansion chambers, cyclones, baghouses, or precipitators and if not otherwise treated, made to rejoin the circuit or product at any point. Other auxiliary equipment is minimal.

Corrosion/erosion of the chamber is not very severe, although some special construction materials may be necessary at solid feed and exit gas areas. Maintenance costs are reasonable, 2 or 3% of the capital cost.

Spray Dryers. Figure 3 represents a typical spray dryer. This type of dryer is particularly applicable in the drying of materials such as slurries high in moisture content, or where a bone dry product is required.

The dryer consists of a vertical drying chamber which may be one of various shapes but is most commonly cyclindrical with a conical bottom section. The slurry to be dried enters the top of the chamber as droplets which may be formed by spray head or high pressure nozzle or by directing a stream of slurry onto the face of a disc which is rotating at high speed. The hot gas stream enters the chamber concurrently and is directed to contact the droplets as rapidly as they are formed, and drying is virtually instantaneous. Particle size of product, while related to the sizes in the feed, is also a function of the spray characteristics which are affected by the media viscosity and density and solids content. Particles are usually fairly large, uniform and hollow or have many voids.

The most popular spray forming and dissemination method is by use of a whirling disc and its velocity and shape must be defined by experience or experimentation. Because of the umbrella pattern in which the particles are thrown, the chamber diameter is normally very large, being more or less equal to the height, with the bottom section in the shape of an inverted cone to simplify the gathering of the product. Dryers have been designed with diameters exceeding 14 m.

Most of the product may be swept out of the chamber in a few seconds, particularly if it is in the "bone dry" state (0.5% moisture, maximum) and this is collected in the usual cyclones, hoppers, baghouses, etc. Since drying is almost instantaneous, the temperature of the product is seldom as high as the outlet gas stream temperature and pyrophoric concentrates can be dried to completion with no problem of spontaneous combustion.

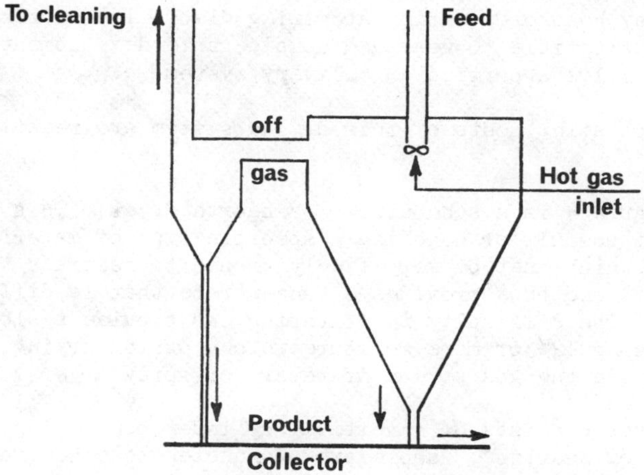

Figure 3 Spray Dryer

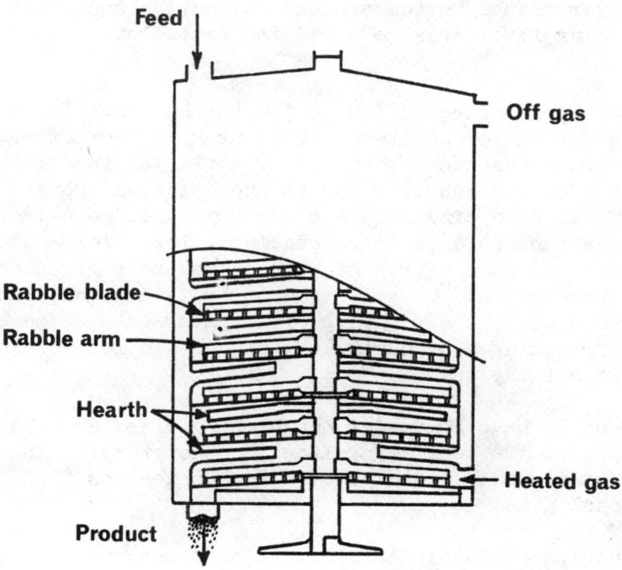

Figure 4 Stationary Hearth Dryer

Corrosion/erosion of most of the equipment is not excessive except for the spray head or rotating atomizing disc. The latter in particular is susceptible to wear and considerable development may be required to solve a particular delivery system.

Capital and operating costs of this drying system are relatively high.

Hearth Dryers. Figure 4 is a schematic of a hearth dryer. This dryer has been found most useful for handling a specific type of material, namely, for an ore which must be very finely ground to separate the mineral constituents and thus provides a concentrate that is difficult to filter and dry. The difficulty in filtering can provide a sloppy concentrate containing 12% or more moisture which, during drying, will all be carried away in the gas stream of rotary or spray dryers.

The hearth dryer consists of a vertical cylinder containing stacked circular floors or hearths. Rabble arms connected to a central rotating shaft extend across each floor and the solids are raked middle to edge and edge to middle on succeeding floors to provide contact with the hot gases. Each floor has openings alternating between rim and center which permit the material to proceed down the height of the dryer to the bottom product collection and discharge floor. The building floor area required for installation is quite small.

The material to be dried is fed to the top hearth. The gas stream may be concurrent or countercurrent, usually the latter. Since contact between the two streams is primarily at the fresh surfaces presented by the rabbling and as the material showers to the next lower floor, the time required for drying is relatively long, and the treatment rate is correspondingly low. Temperature control is precise and some of the gas stream may be recirculated. With no violent movement of the solids, dusting is minimal, indeed, under some circumstances, the rabbling will agglomerate a concentrate slightly, providing a product which can be handled more easily than if this were not the case.

Hearth dryers usually have a greater height than diameter. The process is related to the total travel and contact time in the dryer and it is easier and less expensive to add hearths than to add cross-sectional area.

Auxiliary equipment is similar to that of other dryers but gas cleaning and dust collection systems are smaller than most systems since gas rate and solid content are minimal.

Capital cost is higher than rotary dryers of the same capacity but operating and maintenance costs are considerably less.

Rotary Tray Dryers. These dryers are depicted in Figure 5. Such devices are useful, particularly in smaller sizes, for small throughputs of high value materials. Manufacturers have recently designed a single small unit for drying, followed by calcining, yellow cake (ammonium diuranate) to produce uranium oxide.

The dryer is a vertical cylinder containing superimposed circular floors or trays which rotate about a central axis. This axis contains fans which circulate heated air across the surface of the material which is fed onto the top tray and passed consecutively down the height of the dryer. Height of the unit is usually 1-1/2 to 3 times its diameter. It is easier to add trays than area and design changes tend to add height rather than diameter. They require less room than rotary dryers for installation.

In operation, the material is spread across the tray with a level bar, travels nearly one revolution and then is swept by a second bar onto a slotted tray section through which it falls to the tray beneath. Travel distance is about 80% to 85% of a circumference per level and vertical drop is minimal. Collection is from the bottom floor.

Gas flow may be concurrent or countercurrent at any or all levels with the most common configuration being primarily the former with sections of the latter regularly interspersed.

Dryer sizes range up to 10 m in diameter by 20 m high with evaporative capacity of 1,100 kg/hr. Feed preparation is normal and minimal dusting occurs so that the gas cleaning system can be small.

Other Dryer Systems. Other dryer systems are available to the mineral processing industries such as various configurations of drum, belt, flash, infrared, etc., but the types which have been described cover most of the qualities which must be considered in the design of a drying system.

CONCENTRATE HANDLING AND STORAGE

Concentrates produced in a mill usually go directly from the last process, which may be either filtering or drying, into storage. Since there is no differentiation in transportation rates between the concentrate and the water contained, it is normal practice to minimize the moisture content as far as possible commensurate with the cost of its removal. Such factors as climate, flow characteristics, intermodal transport and customer requirements also play a part. Newly produced concentrate can seldom be loaded immediately from the mill into carriers unless they are held solely for this

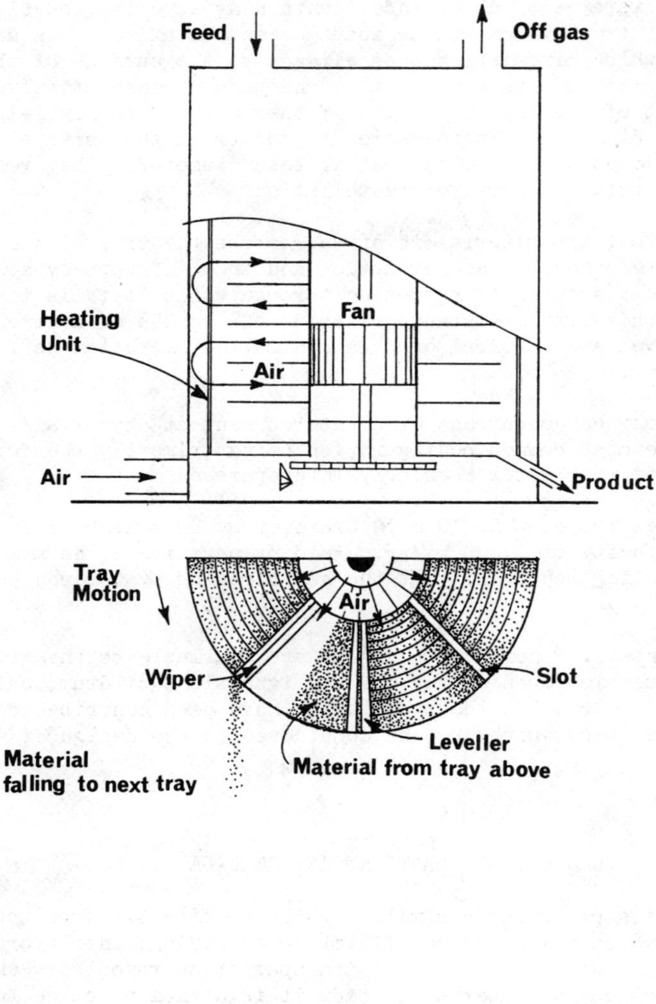

Figure 5 Rotary Tray Dryer

service as captive transport. More normally, rental and demurrage charges dictate that a minimum stockpile will be maintained which will be loaded out to suit the requirements at some other location, such as an interim stockpile at a wharf or smelter.

Loading to stockpile or storage bin from the end of the concentrator line is usually as automatic as possible, either by conveyor belt, dront end loader or by feeding directly from the last filter or dryer. Most concentrates compact too readily and flow with difficulty which tends to restrict the delivery, storage and reclaiming methods employed. Equipment must be selected carefully for the change in these characteristics of the material during handling. Bulk density has a wide range depending on whether it is on a conveyor belt, bucket, or in stockpile and design must make allowance for this, e.g., wet concentrates may be carried on steeper inclines than dry concentrates.

Weighing and Sampling

Control of plant metallurgy, inventory and accountability make the weighing and sampling of concentrate mandatory at the plant and design must include facilities for this. Normally, this is done after the material has been loaded and trimmed as necessary. Since it is usually moved by container, truck or train, weigh bridges are installed for regular and accurate accounting of the amounts shipped. Samples are taken primarily to establish the moisture content and the quantities of contracted minerals that are contained. It is particularly difficult to get a representative sample of high grade concentrate containing elemental metals, and for these it is usually necessary to develop special sampling methods agreeable to producer and client. Sampling can also be done continuously or by compositing batch grab samples while loading.

Moisture Content

While the effect of contained moisture in concentrate affects the flow characteristics, the handling methods will not be expected to change greatly once the plant has been designed for a certain operation. However, these, along with subsequent handling, storage and transportation, must be considered and the consequences of various moisture levels are interesting in their effect.

Granular concentrates usually remain cohesive and homogeneous with moisture levels up to about 14%. However, very fine concentrate which contains over 10% moisture can be fluidized by vibrational forces since they may be thixotropic in nature. On land this may lead to losses through holes, hopper door cracks, etc., while on the water there is the added safety hazard of shifting cargo, mixing cargoes or damaging bulkheads. Consequently, as far as water transport is concerned, various regulatory bodies have adopted standards similar to that laid down by the Inter-Governmental Maritime

Consultative Organization who have promulgated transportable moisture limits which define the maximum moisture level permitted in various concentrates and ores. In addition, testing methods to establish flow moisture points are recognized as useful guides to safe loading.

On the other hand, some concentrates may be highly fluid going into storage but if they settle over an extended period of time, with or without being subject to vibrations from nearby mechanical equipment, become a concrete-like block with expulsion of most of the contained water. Still others react to moisture movement through the mass by becoming more sticky in numerous pockets.

A contained moisture between 4% and 8% is most common since it is usually within these limits that the cost of reducing moisture becomes greater than the cost of shipping water. However, in this range, metal sulphide concentrates may be pyrophoric, i.e., ignite spontaneously, during storage. If air can permeate the pile or move past the loosened surface providing a continuous supply of oxygen, heat will be produced at an increasing rate and sintering or combustion will occur. In this range, if the other factors are present, the moisture will enhance the intimacy between the reactive particles. At levels below 4%, this contact is not so readily available, and at levels above 8% the cooling effect of the moisture and evaporation help to offset these reaction factors, as well as minimizing the number of void spaces in the pile through which air can pass.

At moisture contents below 4%, the problem of dust losses becomes very important. Containment is necessary so that environmental restrictions are met. Very dry concentrates which contain less than 0.5% moisture must be stored in enclosed areas and are usually moved pneumatically or by screw conveyor.

Storage

The storage of concentrate is related to a marketing timetable, transportation availability, and climatic conditions.

Marketing schedules are preferably based on long term contracts and once a pattern has been established relate more to size of storage required than methods of placement.

Transportation availability is similar and is involved with the problems of vehicle rental and demurrage, shipping priorities, interim storage points, and possible wharfage and shiploading facilities to be utilized.

Climatic conditions affect both handling and storage, where concentrates may be subject to extremes of moisture, temperature and winds.

Heavy precipitation may cause material losses as well as difficulty in handling and sampling. Covered storage areas will be necessary for normal concentrates and the use of protected equipment, area and carrier will be necessary. In areas where torrential rainfalls occur it may be better to store the concentrate as a slurry in a covered area. It can then be pumped and transferred via pipeline, and filtered and dried at a more convenient location.

On the other hand, in areas of particularly low humidity it may be necessary to prevent extreme dusting of the concentrate by spraying.

Concentrates located in frigid or temperate zones at altitude have to cope with freezing temperatures for a part of the year and storage and retrieval systems must be designed for these periods as well as for normal times. Minimizing moisture content to less than 5% helps to limit the effects of freezing granular concentrate, and briquetting or agglomeration of the concentrates may also be possible. Facilities which can be warmed if necessary can be used to control the temperature and keep out snow and other moisture.

Shipping delays from such areas should be minimized and use made of interim or customer storage in warmer climes where practical.

Winds affect stockpiles by first drying out the surfaces of the pile and then carrying away valuable dust. This is costly in loss of material and is not appreciated environmentally. The problem may be solved by neatly trimming the piles in open storage, by covered storage, and by location of the stockpile remote from structures, etc., which may funnel winds past at high velocity and in vagrant paths.

Specifically, concentrate storage design takes all of these potential problems into consideration and a choice is made of compatible storage and reclaim methods or the latter may be interspersed with direct loading to the transportation equipment. Covered versus uncovered stockpiles usually relate to precipitation and wind losses. Open storage facilities are the least expensive, and a level concrete base pad may be employed. Different concentrates may be segregated by bulkheads or by using separate storage areas with materials usually served by an overhead tripper conveyor belt system. Ready access to store and recover at maximum efficiency and minimum cost must be allowed for. Bin storage may be required, the initial capital cost being chosen as the preferred alternative to reclaim labour and inconvenience. Here, design for reclaim becomes more involved and the effects of moisture, internal and kinematic angles of friction, flow factors, and arching must be taken into account. Withdrawal may be by hopper or belt conveyor and the retrieval methods employed for these devices is also a part of the design.

Where loading directly to transport is invariably the rule, it may be found useful to replace an interim emergency stockpile system with a ponding system, recovering the slurry through the thickener circuit later.

SUMMARY AND CONCLUSIONS

Dryers are sized on the basis of evaporative load, temperature drop across the system and drying rate required. Both cost and in-plant considerations are involved in the decision to dry concentrates. Direct continuous dryers are most common in the mineral industries. Factors to consider in making a choice include solid characteristics such as flowability in contact with hot surfaces, particle size, chemical characteristics and cost of installation. Data required are: feed and product characteristics, gas stream specifications, estimated liquid removal, heating medium available, heat balances, material of construction, space requirements and past experience with drying. Equipment most often chosen include rotary dryers, fluid bed dryers, spray dryers, hearth dryers and rotary tray dryers.

Moisture content plays a role in decisions on the design of concentrate handling and storage facilities. Contained moisture between 4% and 8% is common, where the cost of reducing moisture becomes greater than the cost of shipping water. Concentrate loading to a stockpile or storage bin is usually as automatic as possible, either by conveyor belt, by front end loader or by feeding directly from the last filter or dryer.

Concentrate storage design must consider such factors as marketing schedules, transportation availability, area precipitation, wind, and local temperatures. Cost considerations are important.

BIBLIOGRAPHY

1. Katta, S., and Gauvin, W. H., "Some Fundamental Aspects of Spray Drying", J. AIChE, 21, 1975, pp 143-152.
2. Katta, S., and Gauvin, W. H., "Basic Concepts of Spray Drying Design", J. AIChE, 22, 1976, pp 713-724.
3. Keey, R. B., Drying Principles and Practice, Pergamon Press, Oxford 1972.
4. Lee, D. A., "Rotating Shelf Dryers", Amer. Ceram. Soc. Bulletin, 55, 1976, pp 498-499.
5. Leonard, J. W., and Mitchell, D. R., "Thermal Drying", Coal Prep. AIME, New York, 1968.
6. Masters, K. A., Spray Drying, Leonard Hill, London, 1972.
7. Perry, J. H., et. al., Chemical Engineers Handbook, 4th Ed., McGraw-Hill, New York, 1963.
8. Sloan, C. E. and Wheelock, T. D., and Tsao, G. T., "Drying", Chemical Engineering, 74, 1967, pp 167-214.

BIBLIOGRAPHY CONTINUED

9. Tait, R. J. C. et. al., "Process Control at the Ecstall Concentrator", Lead and Zinc Symposium, 1, AIME, New York, pp 571-595.
10. Walker, W. H., et. al., Principles of Chemical Engineering, McGraw Hill, New York, 1937.
11. Williams-Gardner, A., Industrial Drying, Leonard Hill, London,1971
12. Fagerberg, B., "Hazards of Shipping Granular Ore Concentrates", Canadian Mining J., July, 1965, pp 53-57; Aug 1965, pp 81-86.
13. Kirshenbaum, N. W., and Argall, O., Minerals Transportation, Miller Freeman Publications, San Francisco, vol. 1 and 2, 1974.
14. Simpson, R. N. F., "Spontaneous Combustion in Sulphide Concentrate" Trans. IMM, London, 80, pp C240-249, 81, 1972, pp C31-35
15. Wasp, E. J., "State of the Art of Slurry Pipelining" Pipeline Engineer, 41, 1969, pp 56-63; staff interview, pp 31-35.

Chapter 30

BELT CONVEYOR SELECTION AND STOCKPILING AND RECLAIMING APPLICATIONS

by F. D. Dietiker

Chief Estimator, Mintec International
Division of Barber-Greene Co.
Aurora, Illinois

ABSTRACT

After determination of the requirements for a given application, proper selection of a belt conveyor is made from tables and formulas commonly used by conveyor designers. Belt conveyors, which are universally accepted as a means of transporting bulk materials, also perform the important functions of stockpiling and reclaiming.

BELT CONVEYOR DESIGN

A belt conveyor, as defined in the Conveyor Equipment Manufacturer's publication Belt Conveyors for Bulk Materials, is an arrangement of mechanical components which support and propel a conveyor belt, which in turn carries the bulk material to be conveyed.

It is a relatively simple machine (see figure 1) comprising five principal components:

1. The belt, which forms the moving and supporting surface on which the material rides.

2. The idlers, which support the carrying and return strands of belt.

3. The pulleys, which support and move the belt and controls its tension.

4. The drive, which powers one or more pulleys to move the belt.

5. The structure, which supports the idlers, pulleys, and drive machinery.

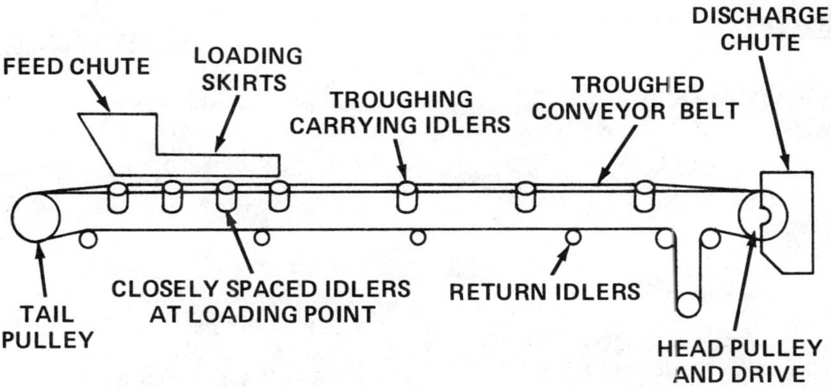

FIGURE 1. TYPCIAL CONVEYOR COMPONENTS

Belt conveyors can be designed for an almost unlimited number of profiles or paths of travel including horizontal, inclined, declined, convex or concave vertical curve or any combination of these. (See fig. 2)

In order to design a belt conveyor for a specific duty it is essential to know the following:

1. Material characteristics

   a. Density in lbs. per cubic feet

   b. Size and gradation

   c. Abrasiveness

   d. Physical condition. Is it wet, dry, sticky, dusty, oily, etc?

2. Capacity required in tons per hour. It is essential to know the peak rate at which material will be fed to the conveyor and not just the average rate or total tonnage for a day.

3. The configuration of the conveyor so that the horizontal and vertical centers of the head and tail pulleys can be established.

4. Type of power available. This is almost always electric.

5. Climatic conditions including temperature range, rainfall, snowfall, wind speeds and direction and any unusual conditions which would affect the operation of the machinery.

Only after these facts have been established, can the conveyor be designed.

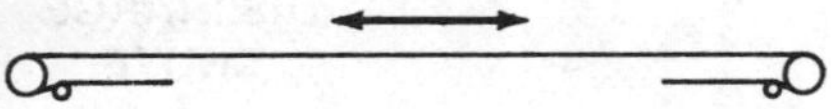

HORIZONTAL BELT.

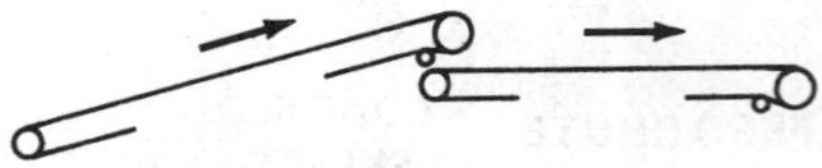

ASCENDING AND HORIZONTAL PATH, WHEN ADVISABLE TO USE TWO CONVEYOR BELTS.

HORIZONTAL AND ASCENDING PATH, WHEN SPACE WILL PERMIT VERTICAL CURVE AND BELT STRENGTH WILL PERMIT ONE BELT.

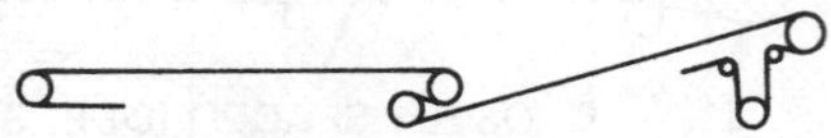

POSSIBLE HORIZONTAL AND ASCENDING PATH, WHEN SPACE WILL NOT PERMIT VERTICAL CURVE BUT BELT STRENGTH WILL PERMIT ONE BELT.

ASCENDING AND HORIZONTAL PATH, WHEN BELT TENSIONS WILL PERMIT ONE BELT AND SPACE WILL PERMIT VERTICAL CURVE.

COMPOUND PATH WITH DECLINES, HORIZONTAL PORTIONS, VERTICAL CURVES, AND INCLINE.

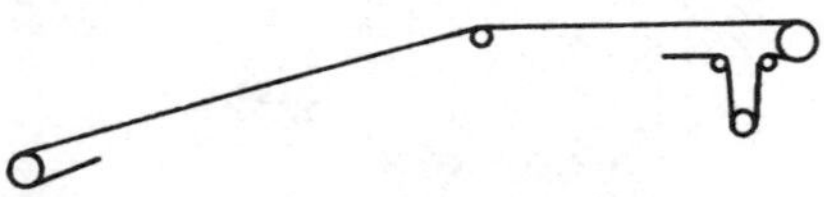

ASCENDING AND HORIZONTAL PATH; OR HORIZONTAL AND DESCENDING PATH, WHEN SPACE WILL NOT PERMIT VERTICAL CURVE BUT ONE BELT CAN BE USED.

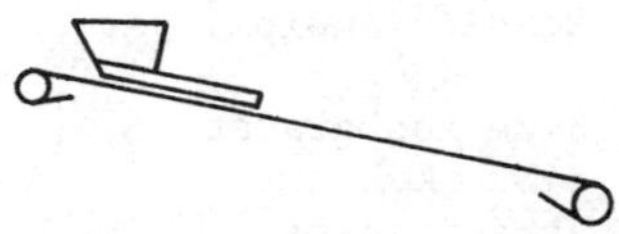

LOADING CAN BE ACCOMPLISHED, AS SHOWN, ON MINOR INCLINES OR DECLINES.

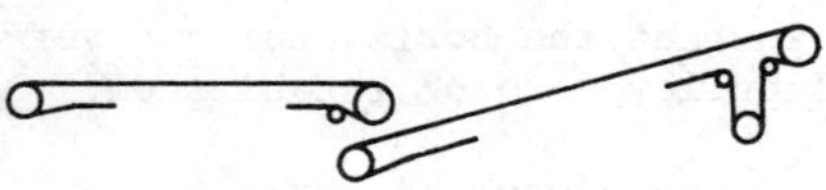

POSSIBLE HORIZONTAL AND ASCENDING PATH, WHEN SPACE WILL NOT PERMIT A VERTICAL CURVE OR WHEN THE CONVEYOR BELT STRENGTH REQUIRES TWO BELTS.

FIGURE 2. TYPCIAL BELT CONVEYOR TRAVEL PATHS

Belt width is normally determined by consideration of material density, lump size, idler configuration, the speed at which the belt will be operated, and the required capacity.

TABLE 1

CAPACITY IN TPH OF EQUAL ROLL TROUGH BELT CONVEYORS (1) FOR 100 LB./CU. FT. MATERIAL AT 100 FPM BELT SPEED

(4) MAXIMUM SIZE LUMPS FOR 20° SURCHARGE

BELT WIDTH	(3) ANGLE OF SURCHARGE												UNIFORM SIZE	(2) MIXED WITH 50% FINES
	20° TROUGH IDLER			35° TROUGH IDLER				45° TROUGH IDLER						
	20°	25°	30°	10°	20°	25°	30°	5°	10°	20°	25°	30°		
18"	50	56	63	53	65	70	75	55	60	70	75	80	4"	4"
24"	96	108	120	102	122	132	142	106	115	132	140	150	5"	7"
30"	157	175	195	167	200	215	232	175	187	215	230	244	6"	10"
36"	230	260	290	248	295	318	343	258	278	318	340	360	7"	12"
42"	320	360	400	344	408	442	475	358	386	440	470	500	8"	14"
48"	430	480	530	457	540	585	630	475	510	584	623	660	10"	16"
54"	547	612	678	585	693	750	806	608	655	748	797	845	11"	18"
60"	680	762	844	730	863	933	1000	758	815	930	992	1050	12"	20"

1. FOR MATERIAL OTHER THAN 100 LBS./CU. FT. – CAPACITY = TABLE CAPACITY X $\frac{\text{WT/CU.FT.}}{100}$
2. 50% FINES IS DEFINED AS 50% SHALL BE LESS THAN 1/2 THE MAXIMUM.
3. ANGLE OF SURCHARGE IS ABOUT 15° LESS THAN ANGLE OF REPOSE.
4. FOR 30° SURCHARGE, USE 1/2 LUMP SIZE OF 20° SURCHARGE.

TABLE 2

MAXIMUM RECOMMENDED BELT SPEEDS IN FPM

MATERIAL			BELT WIDTH							
	CHARACTERISTICS	EXAMPLE	18"	24"	30"	36"	42"	48"	54"	60"
Maximum Size Lumps	Non-Abrasive	Coal, Earth	350	400	450	500	550	600	600	600
	Smooth - Abrasive	Gravel	300	350	400	450	500	550	550	550
	Sharp and Jagged - Abrasive	Stone, Ore	250	300	350	400	450	500	500	500
Half-Max. Size Lumps	Non-Abrasive	Coal, Earth	400	450	500	550	600	650	700	750
	Smooth - Abrasive	Gravel	350	400	450	500	550	600	650	700
	Sharp and Jagged - Abrasive	Stone, Ore	300	350	400	450	500	550	600	650
Granular 1/8" to 1/2"		Sand, Grain Wood Chips	400	500	600	700	800	900	900	900
Aerating Powders		Cement, Flue Dust	200 to 300							
Conveyors With Plow Discharge			200							

Table 1 is a convenient method of determining the belt width since it provides for all the considerations that affect proper selection. The narrowest belt that can be used is that which will handle the design capacity and lump size at the maximum recommended belt speed. However, other considerations including equipment life usually dictate a speed lower than the maximum. Table 2 can be used to determine the maximum recommended speed.

TABLE 3 – IDLER SPACING

BELT WIDTH IN INCHES	SUGGESTED NORMAL SPACING						
	TROUGHING IDLERS						RETURN IDLERS
	WEIGHT OF MATERIAL IN LBS./CU. FT.						
	30	50	75	100	150	200	
18	5'-6"	5'-0"	5'-0"	5'-0"	4'-6"	4'-6"	10'-0"
24	5'-0"	4'-6"	4'-6"	4'-0"	4'-0"	4'-0"	10'-0"
30	5'-0"	4'-6"	4'-6"	4'-0"	4'-0"	4'-0"	10'-0"
36	5'-0"	4'-6"	4'-0"	4'-0"	3'-6"	3'-6"	10'-0"
42	4'-6"	4'-6"	4'-0"	3'-6"	3'-0"	3'-0"	10'-0"
48	4'-6"	4'-0"	4'-0"	3'-6"	3'-0"	3'-0"	10'-0"
54	4'-6"	4'-0"	3'-6"	3'-6"	3'-0"	3'-0"	10'-0"
60	4'-0"	4'-0"	3'-6"	3'-0"	3'-0"	3'-0"	10'-0"

Idler spacing can be determined from Table 3 which gives the normal spacing for various belt widths and material densities.

Horsepower requirements at the headshaft of a belt conveyor is the sum of the following three factors:

x) Horsepower to overcome friction of all moving parts of the conveyor and to drive the empty belt at the specified speed.

y) Horsepower to move load horizontally at the specified tonnage rate.

z) Horsepower to lift or lower material vertically at the specified tonnage rate.

Tables 4, 5, and 6 tabulate these factors and provide a convenient and simple method of determining the requirements for conveyors in widths from 18" through 48" and lengths up to 500 feet.

TABLE 4 – FACTOR (x)

HORSEPOWER AT HEADSHAFT FOR EMPTY CONVEYOR AT 100 FPM BELT SPEED

(For other speeds use direct proportion)

CONVEYOR CENTERS FEET	BELT WIDTH					
	18"	24"	30"	36"	42"	48"
25	.44	.53	.62	.72	.82	.98
50	.47	.57	.67	.77	.89	1.06
100	.52	.63	.76	.87	1.02	1.21
150	.57	.69	.85	.97	1.15	1.36
200	.62	.76	.93	1.08	1.28	1.50
250	.67	.82	1.02	1.18	1.41	1.65
300	.72	.89	1.11	1.29	1.54	1.80
350	.77	.95	1.20	1.39	1.67	1.95
400	.82	1.02	1.28	1.50	1.80	2.40

TABLE 5 – FACTOR (y)
HORSPOWER AT HEADSHAFT TO MOVE LOAD HORIZONTALLY
ANY BELT SPEED – ANY MATERIAL

CONVEYOR CENTERS	CAPACITY - TONS PER HOUR									
FEET	50	100	150	200	250	300	350	400	500	600
25	.25	.50	.76	1.01	1.26	1.51	1.77	2.02	2.52	3.03
50	.28	.57	.85	1.14	1.42	1.70	1.99	2.27	2.84	3.41
100	.35	.69	1.04	1.39	1.74	2.08	2.43	2.78	3.47	4.17
150	.41	.82	1.23	1.64	2.05	2.46	2.87	3.28	4.10	4.92
200	.47	.95	1.42	1.89	2.37	2.84	3.31	3.79	4.73	5.68
250	.54	1.07	1.61	2.15	2.68	3.22	3.75	4.29	5.36	6.44
300	.60	1.20	1.80	2.40	3.00	3.60	4.20	4.80	6.00	7.20
350	.66	1.32	1.98	2.65	3.31	3.97	4.64	5.30	6.63	7.95
400	.72	1.45	2.17	2.90	3.63	4.35	5.08	5.81	7.26	8.71
450	.79	1.58	2.36	3.16	3.94	4.73	5.52	6.31	7.89	9.47
500	.85	1.70	2.55	3.41	4.26	5.11	5.96	6.82	8.52	10.23

NOTE: The above HP values are for idlers with anti-friction bearings.

TABLE 6 – FACTOR (z)
HORSEPOWER AT HEADSHAFT TO LIFT LOAD VERTICALLY
ANY BELT SPEED – ANY MATERIAL

VERTICAL LIFT	CAPACITY - TONS PER HOUR									
FEET	50	100	150	200	250	300	350	400	500	600
5	.25	.51	.76	1.01	1.26	1.51	1.76	2.02	2.52	3.03
10	.51	1.01	1.52	2.02	2.52	3.03	3.53	4.04	5.05	6.06
20	1.01	2.02	3.03	4.04	5.05	6.06	7.07	8.08	10.10	12.12
30	1.52	3.03	4.55	6.06	7.57	909	10.60	12.12	15.15	18.18
40	2.02	4.04	6.06	8.08	10.10	12.12	14.14	16.16	20.20	24.24
50	2.53	5.05	7.58	10.10	12.62	15.15	17.67	20.20	25.25	30.30
60	3.03	6.06	9.09	12.12	15.15	18.18	21.21	24.24	30.30	36.36
70	3.54	7.07	10.60	14.14	17.67	21.21	24.74	28.28	35.35	42.42
80	4.04	8.08	12.12	16.16	20.20	24.24	28.28	32.32	40.40	48.48

THE TOTAL HP AT THE HEADSHAFT IS THE TOTAL OF FACTORS (x) + (y) + (z)

NOTE: If factor (z) exceeds 1/2 the sum of (x+y), backstop is necessary.

For conveyors over 500 feet long, decline conveyors, irregular belt lines, and wider widths the CEMA (Conveyor Equipment Manufacturers Association) formula is widely accepted. This formula states that:

$$H.P. = \frac{T_E \times \text{belt speed}}{33000}$$

$$\text{when } T_E = L\left[K_t(K_x + K_y W_b + 0.015\, W_b)\right] + K_y L W_m \pm H W_m + \text{Accessories}$$

when

L = length of the conveyor along the belt line, or centers, ft

H = height, ft, that the material is lifted, including the lift at any tripper; or the height the material is lowered.

$0.015W_b$ = return belt friction, lbs per ft of conveyor length.

K_t = ambient temperature factor.

K_x = force, in lbs per ft of conveyor length, to rotate the idler rolls, carrying and return, and to cover the sliding resistance of the belt on the idler rolls.

K_y = force, lbs per foot of length to flex the conveyor belt and load over the idlers.

W_b = weight of the belt, in lbs per ft of length.

W_m = weight of material, lbs per ft of belt length.

Accessories. In applying the CEMA horsepower formula for effective belt tension, Te, the following six items must be accounted for; belt tension to cover pulley friction from non-driving pulleys, including any tripper pulleys, belt tension for belt propelled trippers, including chute seals if any; belt tension for plows; belt tension to cover skirtboard friction; belt tension to cover material acceleration; belt tension to cover any type of conveyor belt (or pulley) cleaning device.

It is beyond the scope of this paper to discuss the application of this formula. However, its use gives accurate results for any configuration of the conveyor and it is the basis for almost all the sophisticated computer programs which have been developed by the manufacturers of conveyor systems.

Even though a belt conveyor is comprised of several important parts, none is more economically important than the belting itself, which, therefore, must be selected with great care. It consists of three elements; top cover, carcass, and bottom cover. The primary purpose of the covers is to protect the carcass against damage from the material handled, from wear, and from the environment. The carcass provides the strength to carry the tension forces, to absorb the impact energy of loading, and to give the belt stability for belt alignment and load support.

In recent years the development of new fibers and weaves has produced a multitude of carcass constructions. It is not practical to discuss them here; however, the most widely used are classified as:

1. Multiple ply
2. Reduced ply
3. Steel cable
4. Solid woven

Selection of the belt is predicated on the maximum anticipated belt tension, on the physical characteristics of the material handled, and on the type of idler used. Only with this information can the best selection be made.

There are other important considerations in the design of conveyors. However, it is the purpose of this discussion only to point out the data required before proper design can be established.

STOCKPILING AND RECLAIMING

Although the most common application for belt conveyors is the transportation of bulk materials, they also perform other functions of great importance to you who are responsible for the economical and efficient operation of a mill, primarily stockpiling and reclaiming bulk materials.

There are two important stockpiling applications at a mill. One is to provide a surge pile of the coarse ore between the mine and the mill. The other is to provide for storage at some intermediate point in the processing of the material.

FIGURE 3. 54" X 250' STACKER PROVIDES SURGE PILE BETWEEN MINE AND MILL

Optimum operating efficiency is obtained when the feed of coarse ore is uninterrupted, with no peaks or lows, and when the ore is consistent in quality. To assure these conditions most mills now operate with a surge pile of coarse ore between the mining operation and the mill. The surge pile is applicable whether the primary crusher is located at the mine and the ore conveyed to the mill or whether the crusher is located at the mill and the ore trucked to the crushers. See figure 3.

The surge pile not only assures a uniforn flow of material to the mill, but it also has two other advantages. It provides a means of blending different ores so as to provide a consistent feed to the mill. It also allows operation of the mine independently of the mill so that each can be operated the optimum hours necessary to reach desired production levels. Frequently the mining operation is reduced from three shifts to two shifts or from two shifts to one shift.

The economical justification for the surge pile is apparent since it is estimated that the efficiency of a mill is increased from 10% to 25% when the surge pile is incorporated in the system. This increased efficiency lowers production costs by increasing capacity without adding equipment or manpower and by significantly reducing energy requirements.

There are several conveyor arrangements used to develop stockpiles including the fixed stacker, tripper, shuttle, traveling winged stacker, and the radial stacker. Each has its advantages.

FIGURE 4. FIXED STACKER FORMS CONICAL PILE

The fixed stacker is a conveyor, usually inclined, cantilevered at the head to eliminate the need for supports in the pile or in the live area of the pile. See figure 4. The material discharges over the head pulley to form a conical pile. Piles formed by fixed stackers are usually under 100 ft. high. However, there are instances where a

FIGURE 5. FIXED STACKER FORMS 180 FT. HIGH STOCKPILE WITH NO SUPPORTING STRUCTURE IN PILE

FIGURE 6. FIXED STACKER DISCHARGES INTO WEATHER PROTECTED STOCKPILE

higher pile is justified as shown in figure 5, which shows a fixed stacker discharging into a 180 ft. high pile. The stacker is the simplest and most economical of all conveyor arrangements for moderate sized piles. However, it provides no means for blending. It can be equipped with a telescoping chute at the head end to minimize dusting and can be used to form an open pile or it can be used to form a pile protected by a hodd or building as shown in figure 6.

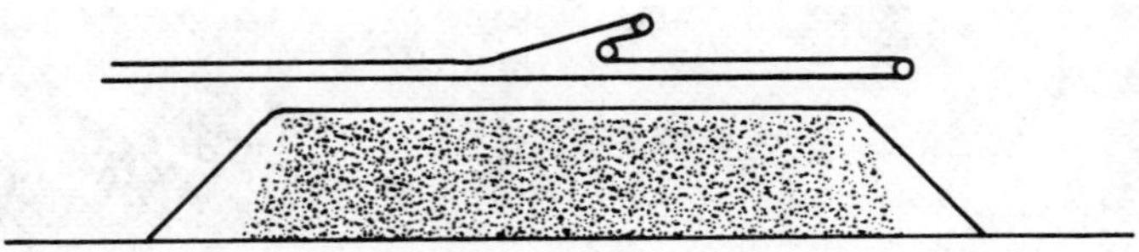

FIGURE 7. TRIPPER FORMS LONG EXTENDED PILE

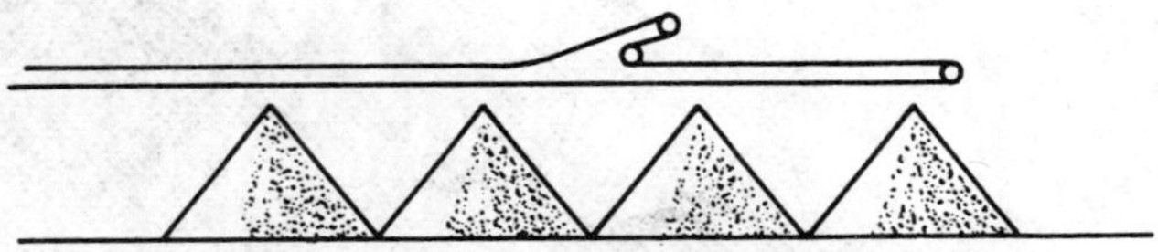

FIGURE 8. TRIPPER FORMS SERIES OF CONICAL PILES

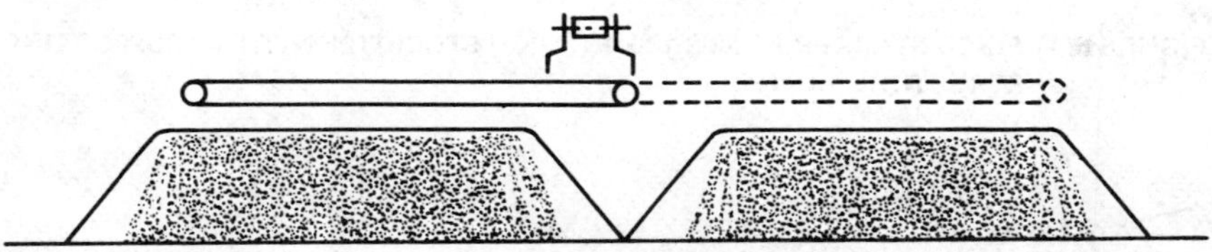

FIGURE 9. REVERSING SHUTTLE LENGTH IS ONE-HALF OF PILE LENGTH

The tripper conveyor is a horizontal belt with a movable tripper which lifts the belt and its load off the horizontal run and discharges the load at any intermediate point along its path. It is commonly used in storage buildings supported from the roof members. It also is used for outdoor ground storage and frequently to fill surge bins or silos. It can form a long, extended pile or a series of conical piles, and can be equipped to provide for automatic positioning. See figures 7 and 8. As an alternative to the tripper, a reversing shuttle conveyor is commonly used to stockpile. A reversing shuttle is a horizontal reversing conveyor mounted on a traveling carriage. Movement of the carriage along its runway and reversal of the direction of belt travel allow the conveyor to discharge at any point along the full length of the carriage travel. Since the feed to the shuttle is located at the midpoint of the shuttle run, the shuttle length is just a few feet longer than one half the total length of stockpile. Usually the shuttle is less costly than the tripper when the feed conveyor is at right angles to the shuttle, since the tripper conveyor must be as long as the pile length while the shuttle length is one-half as long as the pile. See figure 9.

Both the tripper and shuttle can provide for blending. If the units shuttle back and forth along the pile, the materials are layered and mix when reclaimed. If the units form separate piles for each quality of ore, a blend can be achieved by combining the flow from selected feeders onto a reclaim conveyor. See figure 10.

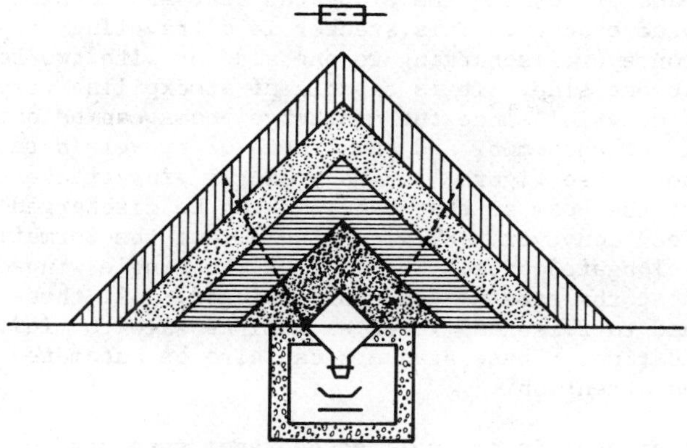

FIGURE 10. BLENDING FROM LAYERED PILE

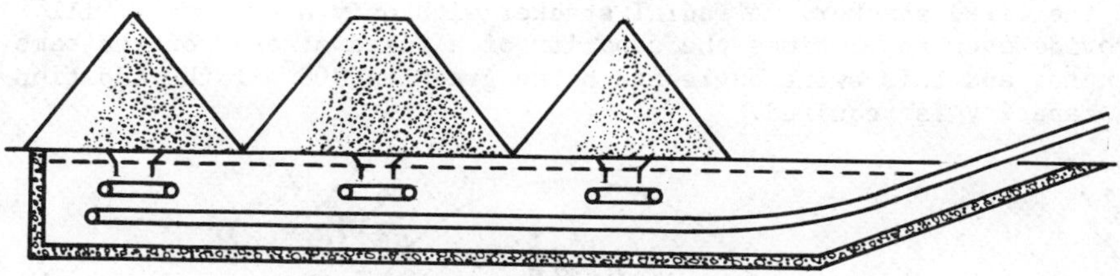

BLENDING BY SELECTION FROM INDIVIDUAL PILES

FIGURE 11. TRAVELING WINGED STACKER

Another means of forming the elongated tent shaped pile is by a traveling winged stacker. This stacker is a traveling tripper with a single boom conveyor discharging to one side or with two booms each discharging to one side. It is capable of stockpiling very large quantities of material since the discharge booms can be quite long, 150'-200' are not uncommon, and the length of travel is restricted only by economy. See figure 11. Many single wing stackers provide for slewing of the boom so that material can be discharged to either side of the feed conveyor. Slewing also allows the formation of a flat topped, elongated pile. The cost of the single winged slewing stacker is about the same as the double winged. All these stackers can be designed to raise and lower as well as slew to minimize dusting and degradation. These stackers can also be automated to operate with almost no attention.

The radial stacker is an inclined conveyor supported on a structure which permits radial movement of the unit on pneumatic tires or rail mounted steel wheels around a pivot at the tail end to form a kidney shaped stockpile. The radial provides the greatest amount of stockpile for the lowest cost for capacities beyond the economical limits of the fixed stacker. A radial stacker with only a 60° swing will provide over three times the capacity of a fixed stacker of the same length, and this swing angle can be as great as 300° if the additional capacity is required.

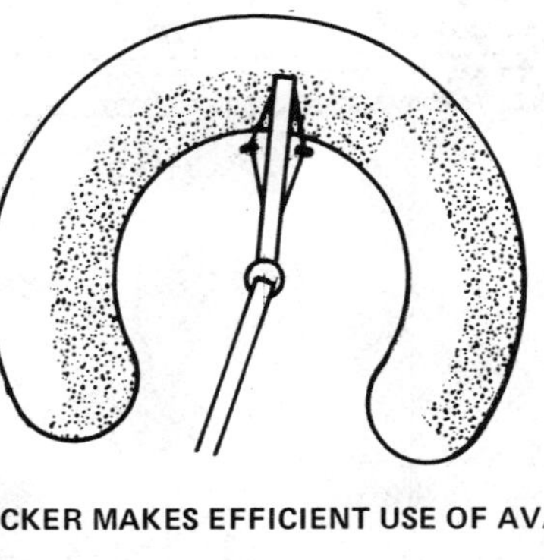

FIGURE 12. RADIAL STACKER MAKES EFFICIENT USE OF AVAILABLE GROUND AREA

The radial is also efficient in its use of ground area since the space requirement is almost square eliminating the need for a long, relatively narrow area required by the traveling winged stacker for an equivalent stockpile volume. See figure 12.

Blending by the radial stacker arrangement is accomplished by programming the radial to continuously traverse the pile and to reverse when it reaches the ends of the run. This method creates a layered pile, the depth of the layers determined by the speed of the radial movement. When ore is reclaimed by tunnel, the layers mix as they flow into the feeder opening, thus providing a relatively homogeneous feed of ore to the mill.

Another and more exact method of blending is to reclaim from two or more separated piles of material by means of feeders which can be regulated to discharge simultaneously on the collecting reclaim belt at any predetermined rate thus providing a very precise blend of ore. See figure 10.

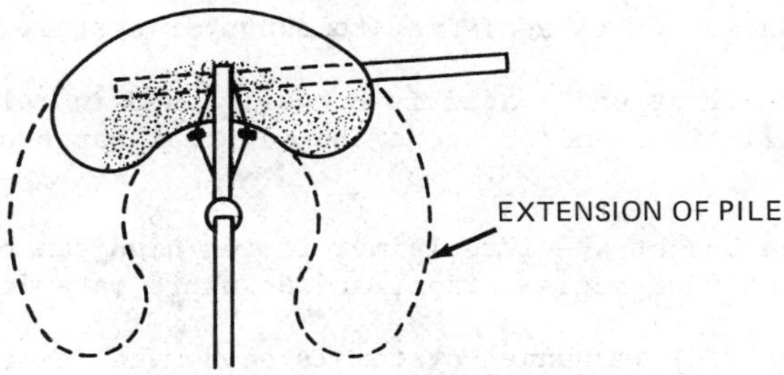

FIGURE 13. RADIAL STACKER PROVIDES RESERVE STORAGE WHEN REQUIRED

Another advantage of the radial is its ability to discharge over a reclaim tunnel to provide live storage and to also provide reserve storage by slewing beyond the range of the tunnel. When the live storage is depleted, the reserves can be moved by other equipment to the reclaim tunnel. See figure 13.

The radial can be supplied with the boom fixed at one elevation or with a boom which raises and lowers to minimize dusting and degradation.

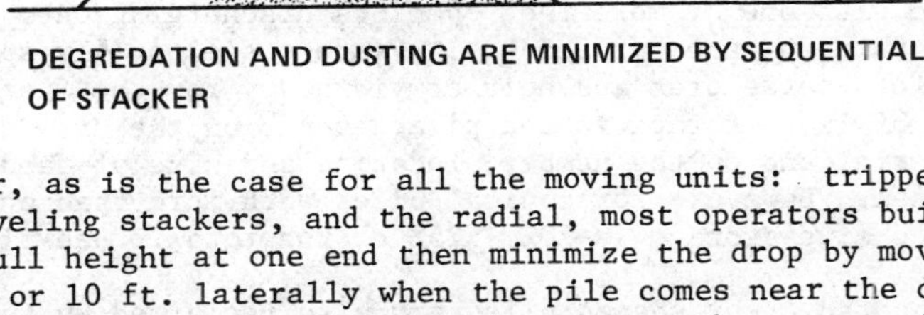

FIGURE 14. DEGREDATION AND DUSTING ARE MINIMIZED BY SEQUENTIAL MOVEMENT OF STACKER

However, as is the case for all the moving units: trippers, shuttles, traveling stackers, and the radial, most operators build up the pile to full height at one end then minimize the drop by moving the stacker 5 or 10 ft. laterally when the pile comes near the discharge pulley of the stacker. A tilt switch or some other sensor indicates the location of the pile when it nears the discharge pulley and signals the stacker to move. See figure 14.

Stockpiled materials are reclaimed from stockpiles by several different systems among the most common of which are:

1. Front end loaders to conveyor system.

2. Reclaim tunnel system.

3. Bucket wheel reclaimer to conveyor system.

The front end loader is normally used in relatively low capacity installations and/or where the material hangs up and will not flow by gravity.

The bucket wheel reclaimer is for high capacity installations requiring 100% reclaimation, and/or where material hangs up.

The reclaim tunnel system is economical since it requires a minimum amount of manpower to operate. It is practical for both high and low capacity requirements and does not interfere with stockpiling operations. Since it utilizes gravity flow to place the material onto the reclaim belt, it can be used only where the material reclaimed is free flowing. It is especially suited for blending by feeding from any combination of openings.

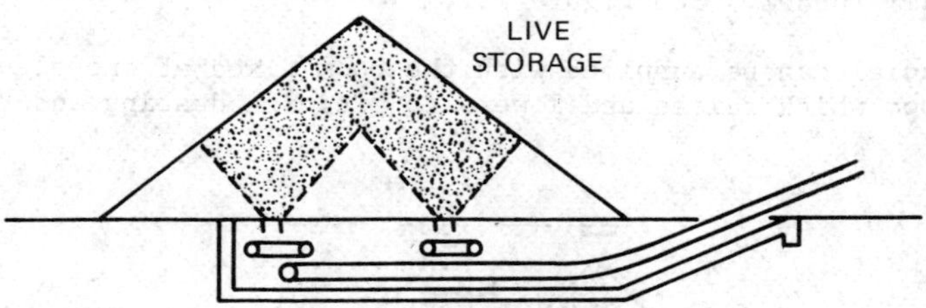

FIGURE 15. LIVE STORAGE IN CONICAL PILE APPROXIMATES 25% OF TOTAL VOLUME STORED

Conical piles are reclaimed by a tunnel running through the center of the pile with one or more feed openings discharging onto the reclaim belt through gates or feeders. Feeders are usually apron or vibratory for coarse ores and belt or vibratory for the finer ores. The amount of live storage in the pile depends on the draw down angle of the material and on the number, location and size of feed openings in the tunnel. However, for conical piles with more than one opening the expected live storage is 20% - 25% of the total. See figure 15.

The elongated, tent shaped piles are also reclaimed by tunnels located at or near the center line of the pile or piles. See figure 16. The anticipated live storage is about 30% - 35% of the total.

Even though there is no technical limitation on the length of a tunnel conveyor, the cost for tunnels under exceedingly long piles must be analyzed to justify their installation. However, it is not unusual to provide tunnels 1000' long.

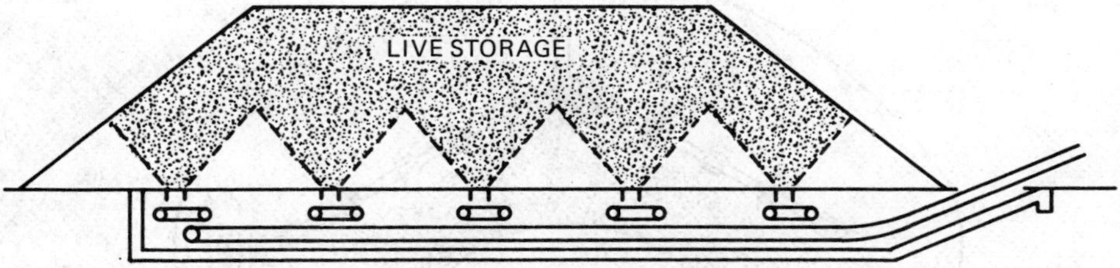

FIGURE 16. ELONGATED PILE LIVE STORAGE APPROXIMATES 35% OF TOTAL STORED VOLUME

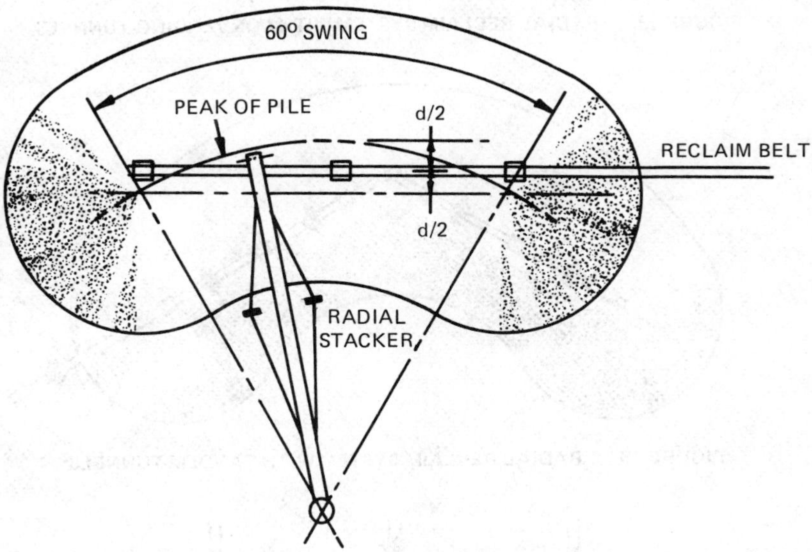

FIGURE 17. TYPICAL RECLAIM TUNNEL ARRANGEMENT UNDER RADIAL STOCKPILE

Tunnel reclaiming from a radial pile is most efficiently accomplished by a single conveyor whose center line corresponds to the chord of an arc scribed by a 60° swing of the stacker. See figure 17. Additional capacity can be realized, of course, by providing two tunnels which converge at a common discharge point or by providing a series of tunnels in tandem so that the reclaim belt in one tunnel discharges onto the tail end of the next in line. See figures 18 and 19. In all cases the belt conveyor lifting the reclaimed material to the discharge point above the stockpile floor will require a vertical curve to accomplish the transition from horizontal to inclined and therefore will be provided with a gravity take-up and soft start to prevent lifting of belt from the conveyor hardware.

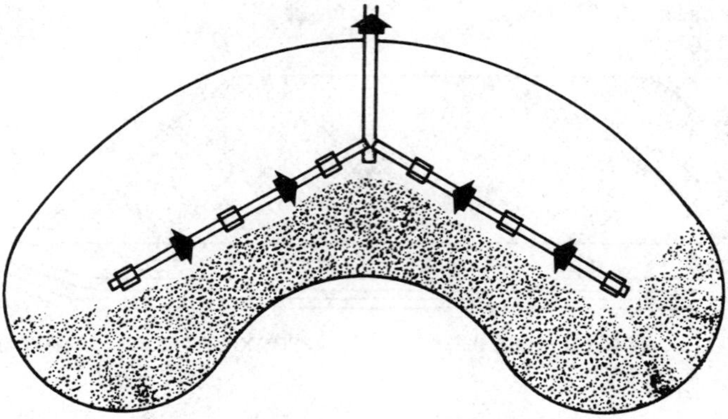

FIGURE 18. RADIAL RECLAIM SYSTEM WITH CONVERGING TUNNELS

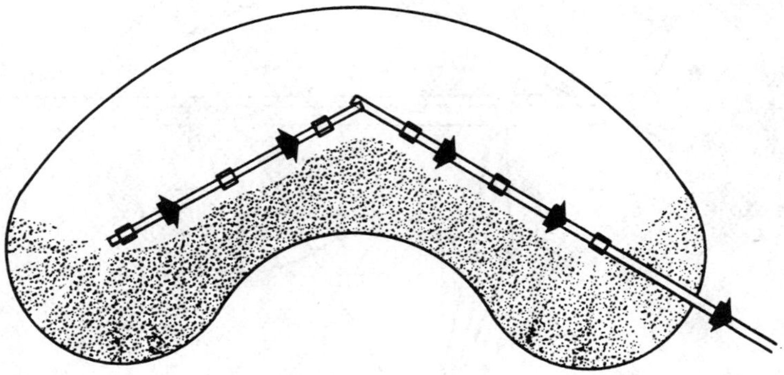

FIGURE 19. RADIAL RECLAIM SYSTEM WITH TANDEM TUNNELS

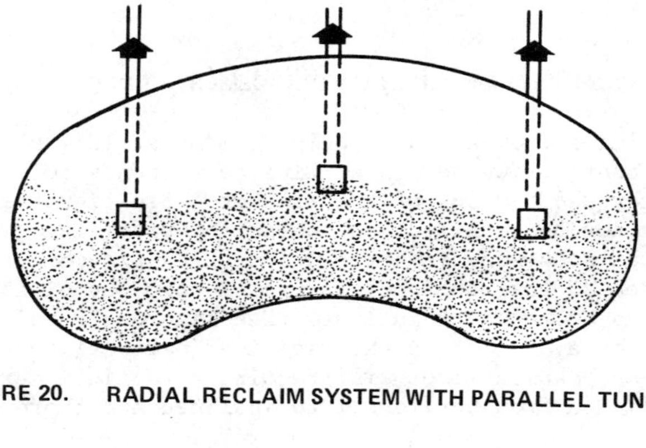

FIGURE 20. RADIAL RECLAIM SYSTEM WITH PARALLEL TUNNELS

Another system of reclaim from the radial piles is by means of two or more parallel tunnels whose direction is approximately perpendicular to a chord of the arc scribed by the stacker discharge. See figure 20. This arrangement provides for simultaneous feed to two or more systems such as crushing/screening circuits in the mill.

The first part of this presentation was concerned with the basics of conveyor design showing tables and formulas commonly used by conveyor engineers. In addition, it pointed out essential information required by the conveyor engineer to establish the proper design of the conveyor. The complete design, however, covering mechanical, electrical, and structural determinations, is beyond the scope of this paper so those interested must obtain information from other sources.

We have also attempted to emphasize the application of conveyors in mill design. Knowing that most operators acknowledge the universal acceptance of the conveyor as a transporter of bulk materials, we have emphasized the use of conveyors as a means of stockpiling and reclaiming. We hope that the remarks and illustrations contained herein will be of some assistance in your future planning.

Chapter 31

AUTOMATION/INSTRUMENTATION

by Larry S. Diaz and Om P. Pande

Consulting Engineer
Granada Hills, California

Instrument Engineer, The Ralph M. Parsons Company
Pasadena, California

INTRODUCTION

Modern mineral processing plants could not operate efficiently, or even exist, without instruments which indicate, record, control, and in some cases anticipate the many changes that occur in a process. Even in plants which are operated manually, instruments are used to tell the operator when he should turn a valve or press a handswitch. Accordingly, all mineral processing plant engineering and construction projects must give careful consideration to this area of design activity.

Instrument Engineering Group

It follows that an engineering company performing a mineral processing plant design and construction project needs the services of a capable, experienced, and well-informed instrument engineering group, the organization of which varies in each engineering company. A typical project-oriented organization, as used by The Ralph M. Parsons Company, is shown in Figure 1. In general, one lead instrument engineer (or job engineer) is assigned responsibility for the instrumentation aspect of the project. Depending on the size of the project, subjob engineers are sometimes assigned responsibilities for individual sections of the plant, under the direction of the lead instrument engineer.

The lead instrument engineer begins his work early in the development of the plant design and is backed up by engineering and drafting groups within the instrument engineering division of his company. He becomes involved in the process design stage and works closely with the chief process engineer and, at times, with the owner's engineers. During this stage, process and instrument engineers locate required instruments at process points best suited

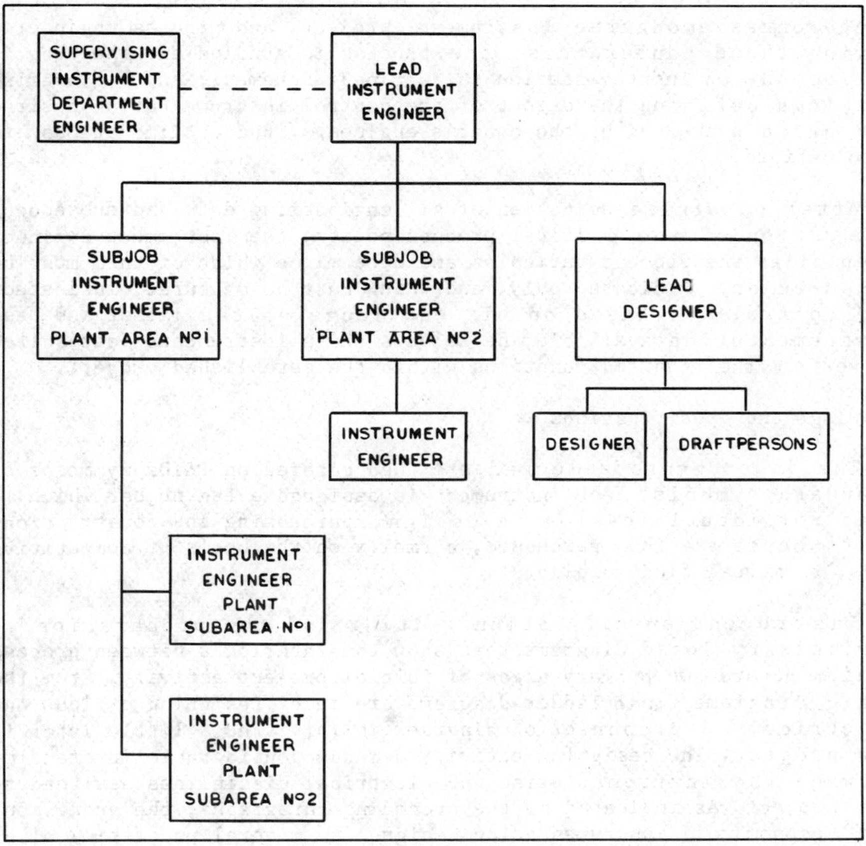

Figure 1. Organization of project instrument engineering group.

to the characteristics of both the instruments and the process stream.

Another very important responsibility of the lead instrument engineer is to plan a budget for his area of work. He must estimate the manhours needed to perform the required work and develop a schedule. Later, when vendor quotations are received, the lead instrument engineer will price out the hardware.

Selection of Instruments

When process design has been completed and the piping and instrumentation diagrams (P&IDs) are being developed, all instrumentation, including measurement devices, transmitters, indicating alarms, controllers, etc., is determined and interlocks between operating equipment are specified. Typically, this is accomplished by work conferences among the instrument, project, and process engineers. During these conferences, the process is studied and the owner's philosophy on instrumentation is examined. Schematic and engineering drawings defining the extent of the control instrumentation desired are often provided by the owner's engineers. Budget limitations are also defined.

After careful examination of all engineering data and subsequent evaluation of the critical process points, the instrument engineer identifies the process variables and determines which of them must be measured and indicated only, and which must be measured, indicated, and controlled. Based on his knowledge of the latest and best instrumentation available, he selects those instruments best suited to perform the required functions within the established budget.

Drawings and Specifications

The instruments selected are then entered on P&IDs by means of standard symbols. Each instrument is assigned a tag number which is used for identification in design, purchasing and construction. Instruments are then purchased, normally on the basis of competitive bids, from qualified suppliers.

Instrument specification writing and the preparation of preliminary logic diagrams that show the interlocks between process equipment are two primary areas of initial project activity. From the logic diagrams final ladder diagrams are made. Designing of loop and electrical interconnection diagrams is initiated a little later in the project. The designing activity demands considerable interaction between the instrumentation and electrical disciplines assigned to the project. As indicated by the preceding paragraphs, the successful (and economical) instrumentation design of a mineral processing plant demands communication and cooperation among process, project, electrical, and instrument engineers.

Plant Control Centers

One clear trend in the design of mineral processing plants is the provision for centralized control rooms (control centers). The measured process variables are transmitted from their respective locations in the process plant stream to an environment-controlled control center that contains the display and control instruments. In this way, a number of large capacity units of process equipment, such as crushers, grinding mills, flotation cells, etc., can be operated by only a few experienced operators.

Miniature instruments mounted in a high density pattern are normally used to reduce the size of the control panels that house them. Indicating controllers which require panel cutouts smaller than 76 mm by 152 mm (3 in. by 6 in.) and illuminated start/stop handswitches with drop-out alarms that measure only 25 mm by 51 mm (1 in. by 2 in.) make possible the design of small panels. Typically, control panels contain five categories of instruments:

- Indicators and indicating alarms
- Indicating controllers and recorders
- Alarm annunciators
- Illuminated start/stop handswitches
- Time and tonnage integrating counters

The design and arrangement of panels vary considerably. Semigraphic-type panels provide a simplified flowsheet of the process. Their usefulness is more directed to new in-training operators during initial plant operations than to experienced operators.

Computer-Based Control

In recent years, the use of digital process computers in mineral processing plant control rooms as an alternative to conventional analog hardware has increased steadily. There are several reasons for this:

(1) Digital computer control can handle slow, interacting processes such as many of those found in mineral processing plants in a manner which is superior to conventional analog control.

(2) Reliable direct digital control can be implemented at a lower cost than analog beyond a certain number of control loops.

(3) The use for control of process variables that cannot be measured directly but that can be calculated from direct measurements of other variables is made possible.

(4) More complex and integrated control strategies for crushing, grinding, and flotation processes, aimed at operating the mineral processing plant closer to process equipment maximum capacities, can be implemented.

(5) Simplified operator interfaces, designed around cathode ray tube (CRT) displays with functional keyboards, allow the measured data and controls to come to the operator and not the reverse.

Distributed Control

Very recently, the concept of distributed control has been introduced. This concept, made possible by modern, low cost microprocessor technology, enables direct digital control with CRT display with or without a linked digital computer. Data acquisition and control functions may be physically distributed to remote locations of the plant and linked to centralized operator display-and-control equipment. One outstanding feature of distributed controllers is their ability to change control functions (ratio, cascade, PID*, etc.) by simple push button configuration procedures.

Remote Multiplexing of Signals

Direct-connected measurement and control devices require each field sensor and final control element to have its own connecting wires to transmit noise-sensitive signals to the central control room. This necessitates large quantities of expensive cabling. In mineral processing plants, the wiring problem is magnified by the fact that large numbers of sensor and control valves are spread over wide areas that are distant from the central control room.

Remote multiplexing stations solve this problem by acquiring or receiving instrument signals right at each process area, and digitizing and transmitting them over a 2-wire link (data bus) to the control room. These stations permit adding more field instruments without adding more wiring. At the control room, multiplexers route the digital signals to process control computers or convert them to analog signals that are used in the normal way. Signals originating from controllers in the control room are handled similarly but in reverse direction.

Frequently, microprocessor-based multiplexing units that can perform scanning, limit-checking, output functions, event sequencing, timekeeping, and automatic annunciation of process changes and alarms are used with computer-based and distributed control systems.

Special Characteristics of Mineral Processing Plant Controls

As an ending to this introduction, it is most appropriate to comment on the special characteristics of mineral processing operations that have significant influence over the design of plant instrumentation. For one, the measurement of process variables is frequently done on slurry streams containing coarse and abrasive solid particles. This adverse physical condition requires sensing devices that not only can measure a given process variable but that can also withstand excessive wear and tear for reasonably long periods of

*proportional-integral-derivative

time. In many instances, the adverse conditions cause the failure of process measuring devices that would otherwise perform adequately.

A second characteristic is that crushing, grinding, and flotation processes frequently take place under closed-circuit conditions. The presence of recycling loads inevitably leads to highly interactive control loops. Finally, there is the ore-type change problem. In many plants, there is a high frequency of change in the hardness and size distribution of the ore to the grinding circuit. These characteristics require that the design of automation/instrumentation for mineral processing plants must be engineered with care and the instruments chosen correctly to ensure success.

PROJECT DESIGN GUIDELINES

Before any of the more basic activities of proper instrumentation design can begin, it is necessary for the lead instrument engineer to establish design guidelines for the project. These guidelines, which are essentially general in nature, define the extent of instrumentation and automation for the project and are based on the following activities:

- Process analysis
- Formulation of instrumentation concepts
- Definition of extent of instrumentation
- Selection of Control Strategies

The design guidelines are used in the preparation of the project instrument design criteria and P&ID drawings.

Process Analysis

Each mineral processing area of the proposed plant is critically analyzed from the view point of its instrumentation requirements. Basic data used here are process flowsheets and general arrangement/plot plan drawings, and of course, the instrument engineer's own experience. The lead instrument engineer, with the support of the process engineers, determines on a preliminary basis the process variables to be measured and related final control elements, if any. In addition, all reasonable equipment interlocks are identified. Typical design areas of a copper mineral processing plant project are:

- Primary crushing and coarse ore conveying/stockpiling
- Fine crushing and screening
- Concentrator (grinding, flotation, regrind)
- Reagent handling
- Concentrate thickening and filtering
- Tailings thickening and disposal

- Water recovery
- Molybdenum plant

Formulation of Instrumentation Concepts

Based on the data gathered through process analysis, preliminary instrumentation concepts are formulated. Typically, the concept formulation activity proceeds concurrently with process analysis. The resulting concepts are rendered into logic diagrams and instrument sketches and are used in planning the extent of instrumentation.

Definition of Extent of Plant Instrumentation

Identifying the extent of the instrumentation to be designed into the plant is probably the key design activity of the entire project. The extent of the instrumentation will be strongly influenced by the control strategies adopted. This activity is normally performed in close contact with the owner's engineers. In defining the extent of instrumentation, the planners also define the scope of the instrumentation design criteria.

The lead instrument engineer and the chief process engineer carefully assess the preliminary instrumentation concepts in the light of expected operating conditions. The following important considerations must be included in the assessment.

<u>Geographical Location of the Proposed Plant.</u> In a plant to be located in a remote, underdeveloped area where qualified operators are not readily available, it is advisable to plan for the use of a wide range of instrumentation. However, only relatively few automatic control loops should be designed so that the plant would not become "over-instrumented". Suitable instrumentation under remote location conditions would include centralized control rooms equipped with indicating alarms, field instruments to transmit data, and automatic interlocks to protect process equipment. The judicious application of instrumentation minimizes the number of qualified operators required for efficient operation and the instrumentation can be maintained by relatively few trained electronic technicians.

In plants to be located in more favorable geographic locations where qualified personnel are readily available, the use of sophisticated control instrumentation might be advisable, if suitable. Experience and the proven reliability of modern control instruments recommend extensive instrumentation with the maximum number of control loops for a fully automated plant, when practical. Such a plant can achieve the highest possible operating efficiency with the minimum number of operating personnel.

<u>Budget Allocated for Plant Instrumentation.</u> The extent of instrumentation will be limited by the amount of money economically justified for the purpose.

Specific Control Objectives for Each Plant Design Area. The degree of automatic control and the degree of accuracy of measured data is determined by the specific control objectives. For example, is it required to control ore weight rate to the grinding circuits? If so, to what degree of accuracy? Or, is it required to control the rate of water flow to the cyclone pump feed box? If so, to what degree of flow rate accuracy?

Specific Requirements of the Plant Owner. Normally, the owner specifies preferred control practices based on experience in other plants. For example, the owner might propose to use onstream X-ray fluorescence analysis. X-ray systems require the installation of sampling equipment, which will affect plant design.

The Design Philosophy for the Project. The extent of instrumentation is guided by the design or control philosophy of the planners. It may be a philosophy of maximum control through extensive instrumentation tending toward full automation or one of limited instrumentation in support of many manual procedures.

In today's large scale mineral processing operations, extensive instrumentation incorporating centralized control rooms is a universal practice. Additionally, electronic control hardware is preferred over the pneumatic type in general design philosophy.

Another control philosophy-type consideration concerns alarms. Are control panel alarms to be mounted in a matrix, on the graphics, or both? Is audible annunciation of alarms required, and if so, at how many levels? Is an alarm scanner and printer required?

Selection of Control Strategies

Having determined the extent of instrumentation for the project, it is now necessary to specifically define a control strategy for each design area of the project. Like the previous activity, it is performed in close contact with plant owner's engineers.

On the basis of information generated by the previous activities, the lead instrument engineer selects for each design area (or interrelated design areas) the degree of automatic control required to implement the specific control objectives.

One question which must be resolved early is that related to analog or computer-based control - does the extent of instrumentation determined for the project warrant the use of process control computers as a more flexible and lower cost alternative to analog hardware? In this day and age, the answer is frequently yes, even when a relatively narrow extent of instrumentation is specified. In fact, the control strategy set for any one of the plant design areas may require a computer for its implementation.

An example of such a situation is the indirect approach to particle size-based control of a grinding circuit. A simplified block diagram of a typical circuit is shown in Figure 2. As indicated on the diagram, particle size of the cyclone overflow is inferred from mass flow rate to cyclone feed and other process variables.

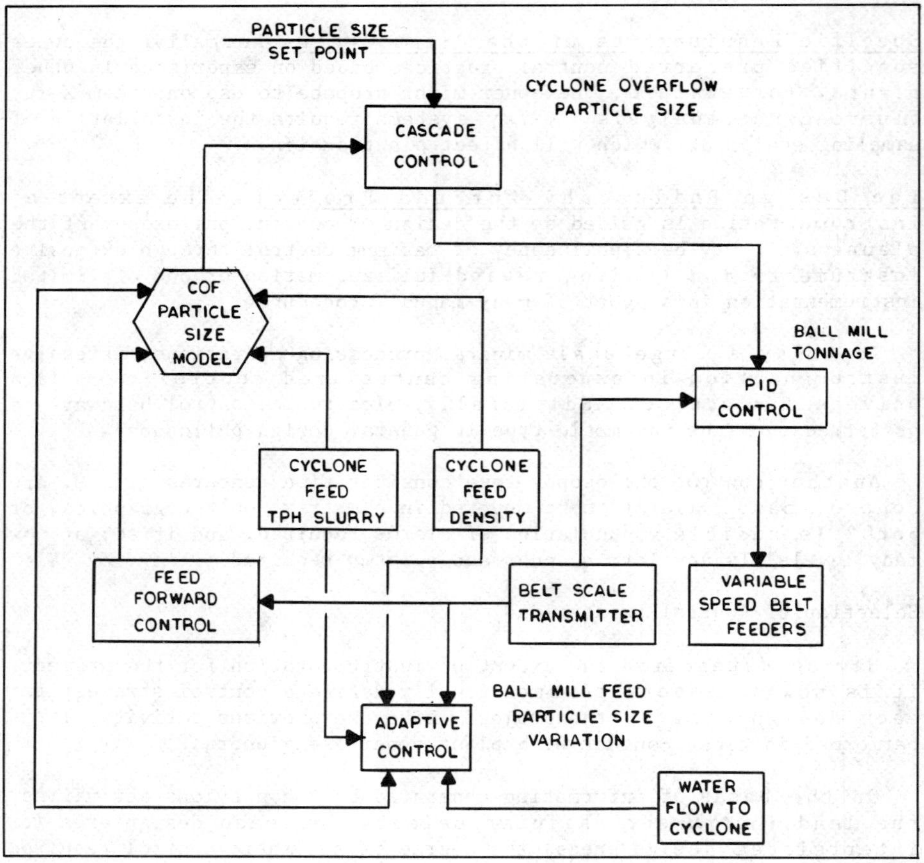

Fig. 2. Indirect size-based control of a single stage ball mill grinding circuit.

Three important questions come up immediately in the event that computer based control is selected.

- Is a stabilizing control strategy to be implemented or are there design areas that will require optimizing control?

- Is direct digital control (DDC) or supervisory control, or both, to be used?

- Is analog back-up required? The lead instrument engineer and owner's engineer must decide which computer based controls require analog back-up and its type: manual back-up or automatic analog control back-up.

At this point it must also be decided how computer software is to be developed. Will a subcontractor be selected, or will the instrument engineering group do the work? This decision obviously influences the size of the group.

Analog control strategies can vary all the way from simple indication and alarming to complex multiloop feedforward/cascade control loops. For example, the temperature of the inlet water to a crusher lubricating oil cooling unit need only be indicated, either on the control panel or locally. Only if the temperature of the water increases to a value above which proper oil cooling will not occur is an alarm annunciation necessary. On the other hand, the control strategy for ball mill feed weight rate may require the measurement of density and volumetric flow rate of the cyclone feed slurry plus flow of water to the cyclone for its implementation.

Most strategies used in mineral processing plant control systems can be implemented with one or more of the following methods:

(1) PID Control. The PID feedback controller applies proportional, integral, and derivative control actions to the deviation of the input signal from the set point to determine the output to the final control device, usually a valve or a variable speed drive.

(2) Cascade Control. In this method, two PID controllers are linked together. The output of the first controller in the chain supplies the set point to the second PID controller which in turn controls the final control device.

(3) Ratio Control. The input to the ratio controller is multiplied by a selectable constant and its output supplies a set point to a PID controller.

(4) Feedforward Control. The output of the feedforward controller to a final control device or PID controller is calculated by manipulation of its input by means of predefined functions. It produces a control action before the effect of a change in process conditions is reflected in a measurement change downstream. Ratio control is a simple form of feedforward control.

(5) Adaptive Control. The mathematical function which relates output to input of a control loop usually has constant parameters. With adaptive control these parameters are continuously updated on the basis of operating data.

Figure 2 illustrates some of the control methods described. Obviously, the control strategies selected will have an enormous influence on the extent of plant instrumentation.

DESIGN PROCEDURES

When the design guidelines have been established, instrumentation engineering and design work is carried out in a sequential manner. The procedural steps include:

(1) Obtaining project data such as contract agreement, basic design criteria, and the client's special instructions for specification and vendor selection
(2) Preparation of instrument design criteria
(3) Development of P&IDs and instrument summary
(4) Estimation of manhours
(5) Preparation of vessel level sketches
(6) Sizing of panels and control valves
(7) Writing specifications for measurement and control instruments
(8) Development of logic diagrams, ladder diagrams, and loop diagrams
(9) Procurement of instruments
(10) Design and review of installation details
(11) Panel inspection
(12) Field installation, calibration, and start-up

Figure 3 depicts a typical instrumentation section project work flow.

Instrumentation Design Criteria

The instrumentation design criteria describe the types of instruments that must be installed and establish minimum standards of quality and performance. The criteria maintain consistency of instrumentation schemes and consistency in the selection of instruments for similar application in the various design areas. Mineral processing plant projects frequently involve complex operations and rugged environments. Accordingly, instrumentation design criteria require considerable plant-related expertise for their preparation and must take into careful consideration process details that can influence the selection of instruments. As a result

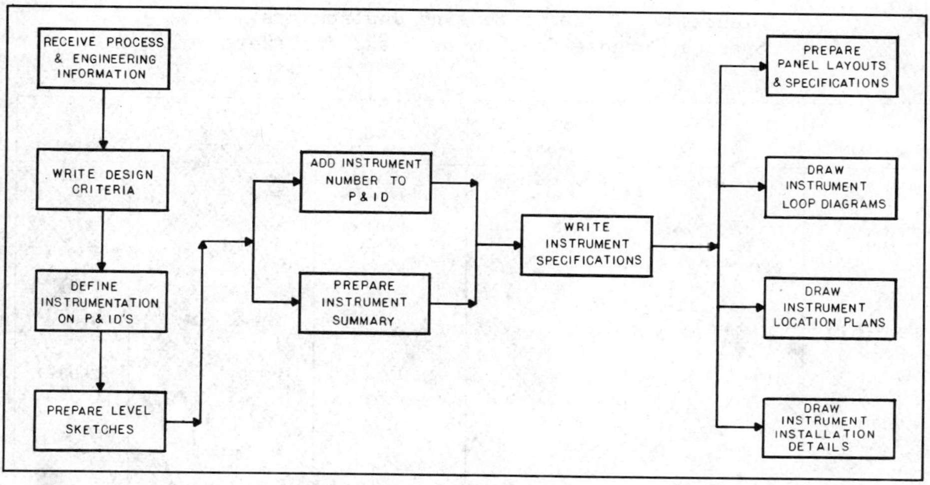

Fig. 3. Instrumentation section work flow.

criteria vary from one design area to another because of varying process details. Typical instrument design criteria include the following items:

- Scope
- General
- Control Centers
- Flow Instruments
- Level Instruments (Dry Materials and Liquids)
- Pressure Instruments
- Temperature Instruments
- Special Purpose Instruments
- Control Valves
- Design (Control) Philosophy

Following are lists of the various types of instruments commonly used in mineral processing plants for sensing level, flow, pressure, and temperature variables, plus special purpose instruments and analyzers. (See Table 1.) The criteria specify which type of instrument is to be selected for a specific function.

P&IDs and Instrument Summary

Following completion of the instrumentation design criteria, development of P&IDs becomes a key task. These drawings map out all of the instrumentation and control strategies and show the relationship of the instruments to the process equipment and material flow.

Table 1. Types of Sensing Instruments, Special Purpose Instruments, and Analyzers

Level	Flow	Pressure	Temperature	Special Purpose Instruments	Analyzers
Ultrasonic	Electro-Magnetic	Diaphragm Type Transmitter	Thermocouple w/Well	Nuclear Density Gage	Particle Size
D/P Cell (Purged)	D/P Cell (Purged) w/Orifice Plate or Annubar		Bi-Metallic Thermometer w/Well	Metal Detectors	X-Ray
Nuclear					pH/ORP
Conductivity	Area Flow (Rotameter)			Tilt Switch	Moisture
Tape	Vortex			Conveyor Belt Scale	Turbidity
Bubble Tube	Turbine				
Displacement				Automatic Sampler	
				Thermal Converter (Power)	
				Tachometer	

A typical P&ID for a single stage ball mill grinding circuit is shown in Figure 4. Interlocks between the process equipment are indicated by the symbol ⟨I⟩ adjacent to the equipment. Occasionally, special purpose P&IDs are generated to detail segments of control loops or such safety features as the interlocking between equipment. Control strategies and instruments are identified on the P&IDs by standard symbols and tag numbers. Tag numbers, along with other information, are logged on instrument summary forms.

Standard Symbols. A widely used system for indicating instruments on P&IDs is described in "Instrumentation Symbols and Identification," ANSI Y32.20-1975, originated by the Instrument Society of America (ISA). Some ANSI standard symbols and their use on typical control loops are shown on Figure 5.

Tag Numbers. After in-house and owner reviews of the P&IDs, every instrument is assigned a tag number, which is entered on the drawings on the instrument symbol. The tag number is constructed in accordance with ANSI Y32.20 and identifies the function of the instrument in addition to providing it with a plant unit or area number for reference. Figure 6 explains the components of a tag number.

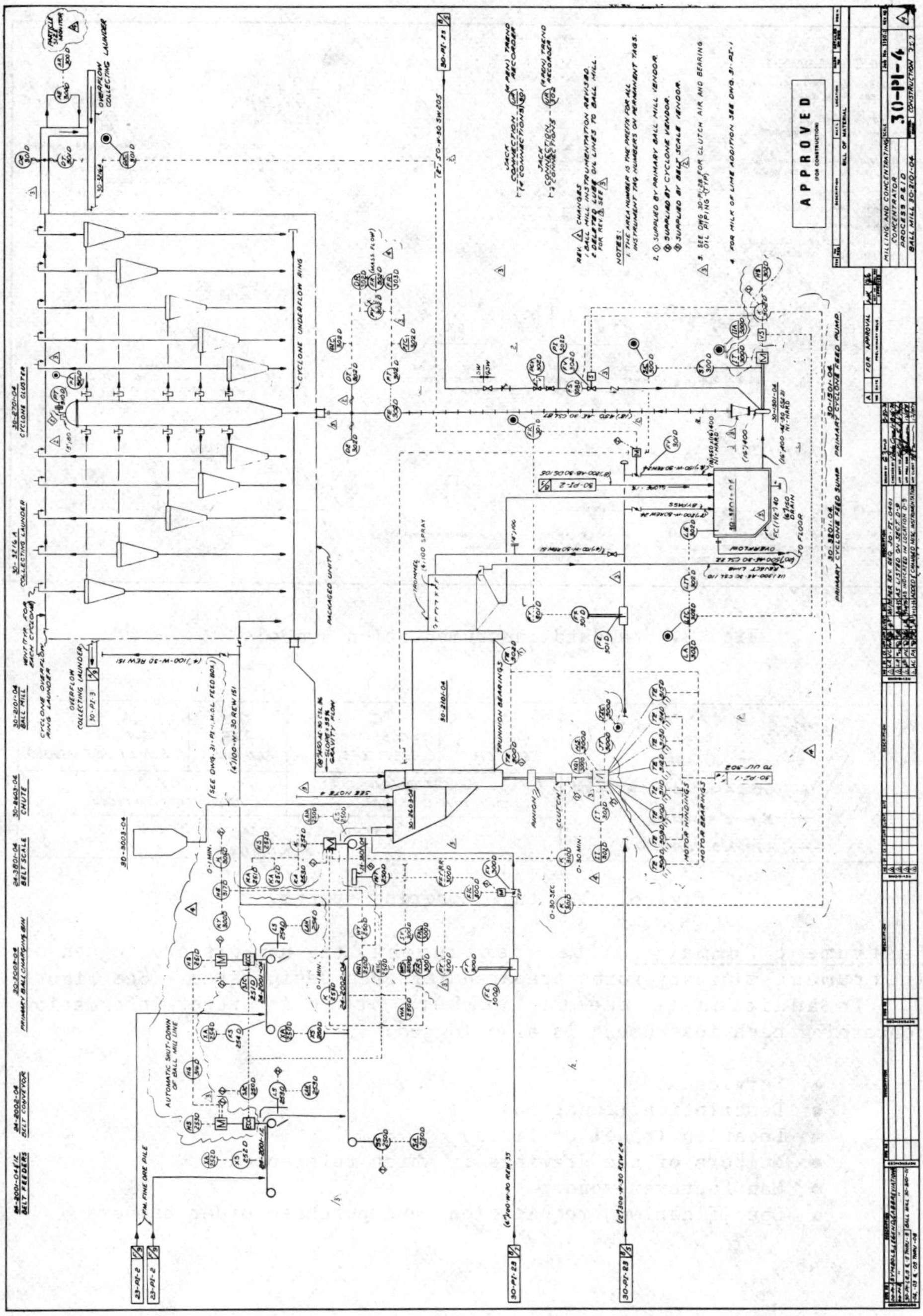

Fig. 4. P&ID for single stage ball mill grinding circuit.

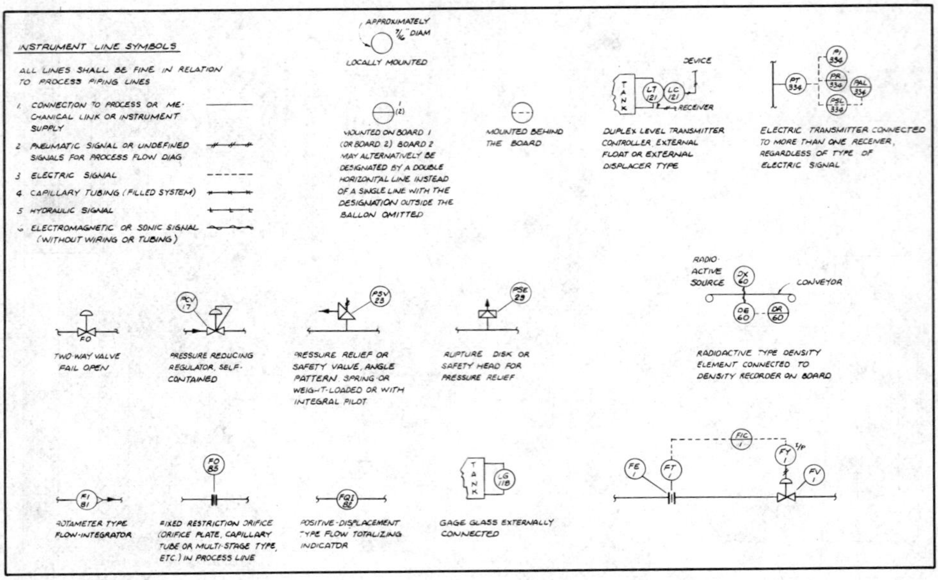

Fig. 5. Standard instrumentation symbols.

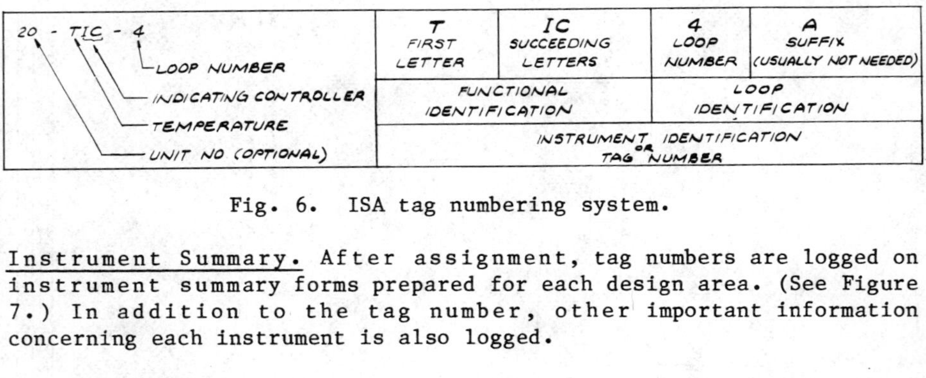

Fig. 6. ISA tag numbering system.

Instrument Summary. After assignment, tag numbers are logged on instrument summary forms prepared for each design area. (See Figure 7.) In addition to the tag number, other important information concerning each instrument is also logged.

- Service
- Description (function)
- Location (panel or field)
- Numbers of the drawings in which referenced
- Manufacturer/vendor
- Specification, requisition, and purchase order numbers

Table 2. Definitions of Instrument Functional Identification Letters

	FIRST LETTER		SUCCEEDING LETTERS		
	MEASURED OR INITIATING VARIABLE	MODIFIER	READOUT OR PASSIVE FUNCTION	OUTPUT FUNCTION	MODIFIER
A	Analysis		Alarm		
B	Burner Flame		User's Choice	User's Choice	User's Choice
C	Conductivity (Electrical)			Control	
D	Density (Mass) or Specific Gravity	Differential			
E	Voltage (EMF)		Primary Element		
F	Flow Rate	Ratio (Fraction)			
G	Gaging (Dimensional)		Glass		
H	Hand (Manually) Initiated)				High
I	Current (Electrical)		Indicate		
J	Power	Scan			
K	Time or Time-Schedule			Control Station	
L	Level		Light (Pilot)		Low
M	Moisture or Humidity				Middle or Intermediate
N	User's Choice		User's Choice	User's Choice	User's Choice
O	User's Choice		Orifice (Restriction)		
P	Pressure or Vacuum		Point (Test Connection)		
Q	Quantity or Event	Integrate or Totalize			
R	Radioactivity		Record or Print		
S	Speed or Frequency	Safety		Switch	
T	Temperature			Transmit	
U	Multivariable		Multifunction	Multifunction	Multifunction
V	Viscosity			Valve, Damper, or Louver	
W	Weight or Force		Well		
X	Unclassified		Unclassified	Unclassified	Unclassified
Y	User's Choice			Relay or Compute	
Z	Position			Drive, Actuate, or Unclassified Final Control Element	

Because of the comprehensive information it contains, the instrument summary is the main control document for instrumentation and is essential to quickly identify the location and function of each and every instrument in the process flow throughout the project.

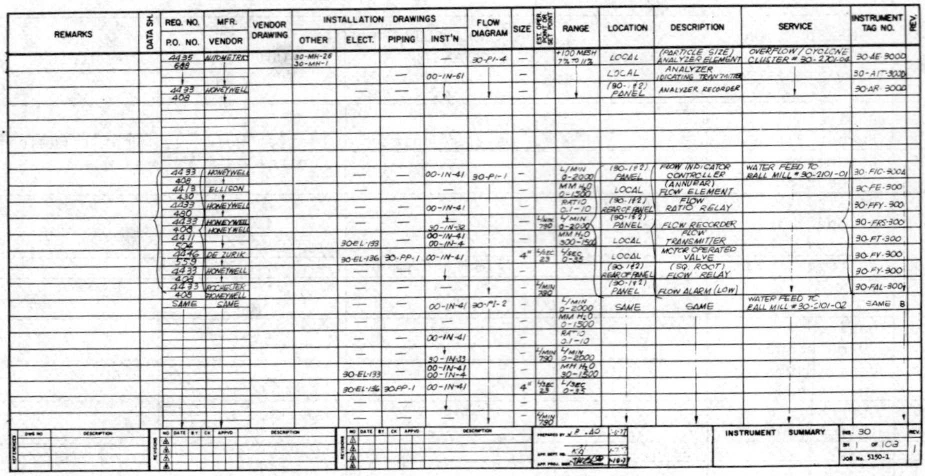

Fig. 7. Instrument summary form.

Manhour Requirements

The instrumentation manhour requirements for the project are estimated on the basis of the activities to be completed. An instrumentation activity list itemizing each instrument shown on the P&ID is prepared. The manhours for every activity required to be done relative to each instrument are estimated, entered on the list, and totaled. (See Table 3.)

Table 3. Instrumentation Activity List

Instrument	Hours per Activity					Total Hours
	Specification Writing	Bid Requisition	Bid Analysis	Purchase Requisition	Vendor Drawings Review	
Control Valve (Ball, Globe, Butterfly, or Pinch)	3	1/4	1/2	1/4	1/2	4-1/2
D/P Transmitter (Pneumatic or Electronic)	2	1/4	1/2	1/4	1/2	3-1/2
Electro-Magnetic Flowmeter	3	1/4	1/2	1/4	1/2	4-1/2
Nuclear Level Transmitter	4	1/4	1	1/4	1	6-1/2

Vessel Level Sketches

Following the development of preliminary P&ID drawings, level sketches are prepared for each vessel that requires a level instrument. These sketches help to establish the range and elevation data for level controllers (if any) and vessel nozzle locations.

Control Panels and Control Valves

Because of the long delivery periods required, control panels and control valves must be sized and purchase orders issued soon after the development of preliminary P&ID drawings. Panel sizes are also required to develop control room layout.

Control Panels. As stated earlier, the plant is divided into design areas. Separate control panels, installed in centrally located environment-controlled control rooms, are usually provided for the primary crushing, fine crushing and screening, concentrator, and molybdenum plant areas. These central control panels frequently are of the semigraphic type. The upper section of the panel, sloped in a reverse angular configuration, displays graphic representations of all the major equipment to illustrate the process flow in its respective area. Status and alarm condition lights are located on the graphic symbols of the equipment, arranged in matrices, or both.

A panel designed to control a copper ore leaching operation is shown in Figure 8. The total number of panel-mounted instruments, annunciators, handswitches, and push buttons determines the size of the control panel required. The project architect plans control room space requirements based on the size of the panel.

Numerous drawings are prepared in the design of control panels. These drawings show the arrangement, number, and kinds of instruments to be mounted and direct the vendor in fabricating the control panel. When fabrication has been completed, each central control panel is thoroughly inspected at vendor's shop to ensure the quality of work and the correct wiring of each loop.

Control instruments are also located on local panels mounted adjacent to the process equipment. The local panels are usually self supporting, vertical-face designed, and totally enclosed with rear doors. They are commonly specified as part of the equipment vendor package.

Control Valves. Many control circuits require the adjustment of flows of liquids and slurries. Control valves are required for this purpose. As soon as hydraulic data is available, specifications and purchase orders must be completed for their procurement.

Fig. 8. Semigraphic control panel.

Because of abrasion and corrosion problems, rubber-sleeve pinch valves are advisable for on-off and modulated slurry service in mineral processing operations. Despite poor flow characteristics and limited allowable pressure drop, the performance of this type of valve is satisfactory. Concentric-type rubber-sleeve valves operate like the shutter of a camera and give good throttling control. This feature also contributes to long life. Butterfly valves are very suitable for water and flotation air control.

Proper sizing of control valves is critical and is based on the control requirements. Proper sizing permits operation over as wide a range as needed without generating instability. Vendor supplied data advise the engineer as to the correct sizing procedure. The sizing procedure for a 102 mm (4-in.) butterfly valve controlling the water flow rate to a ball mill is presented below.

Sizing Formulas:

$$P_a = K_m(P_i - \gamma_c P_v)$$

$$C_v = Q\sqrt{\frac{G}{\Delta P}}$$

Where:
C_v = liquid sizing coefficient
G = specific gravity of fluid
ΔP_a = maximum allowable ΔP
P_i = valve inlet pressure, kg/cm^2(psig)
P_o = valve outlet pressure, kg/cm^2(psig)
P_v = vapor pressure of fluid, kg/cm^2(psia)
Q = flow rate, liters/min (gpm)
ΔP = differential pressure across valve, kg/cm^2(psi)
K_m = valve recovery coefficient
γ_c = critical pressure ratio

Hydraulic Data

Line size	152 mm (6 in.)
Q max.	2082 liters/min (550 gpm)
Q normal	1514 liters/min (400 gpm)
P_i (maximum)	1.82 kg/cm^2(26 psig)
P_o (normal)	0.14 kg/cm^2(2 psig)
ΔP sizing	1.54 kb/cm^2(22 psi)
ΔP maximum (shutoff)	1.82 kg/cm^2(26 psi)

Calculated Data

P_a = 1.58 kg/cm^2(22.58 psi) - This value is greater than ΔP sizing, therefore, no cavitation or flashing will occur.
C_v (Q maximum, ΔP sizing) = 117.26 (valve open 50 deg)
C_v (Q normal, ΔP sizing) = 85.28 (valve open 45 deg)
C_v (1/3 Q normal, ΔP maximum) = 27.22 (valve open 30 deg)

From the manufacturer-supplied C_v/valve size tables, select a 102-mm (4-in) butterfly valve on the 152-mm (6-in.) line. When the valve size is selected, use a swaging correction factor to correct the calculated $C_v$values. This factor is a function of the line diameter to valve size ratio.

Specifications

After owner's approval of the P&IDs, specification sheets are prepared for every instrument on the drawings and used for their procurement. Typical specification sheets for a control valve, orifice plate, level instrument, and relief valve are shown in Figure 9. Material of construction limitations for all components of a particular instrument should be clearly stated in the specifications. Specifications for indicating controllers require the definition of control modes. (See Table 4.)

Table 4. Definition of Indicating Controller Modes

Mode of Control	Controlled Process Reaction Rate	Transfer Lag or Dead Time	Process Load Changes
On/Off	Slow	Slight	Small and Slow
Proporational	Slow or Moderate	Small or Moderate	Small
Proportional Plus-Derivative (PD)	Slow or Moderate	Moderate	Small
Proportional Plus Integral (PI)	Fast	Small or Moderate	Slow, but any amount
Proportional Plus Integral Plus Derivative (PID)	Fast	Moderate	Fast

Note: Integral action is frequently referred to as reset action. Derivative action is frequently referred to as rate action.

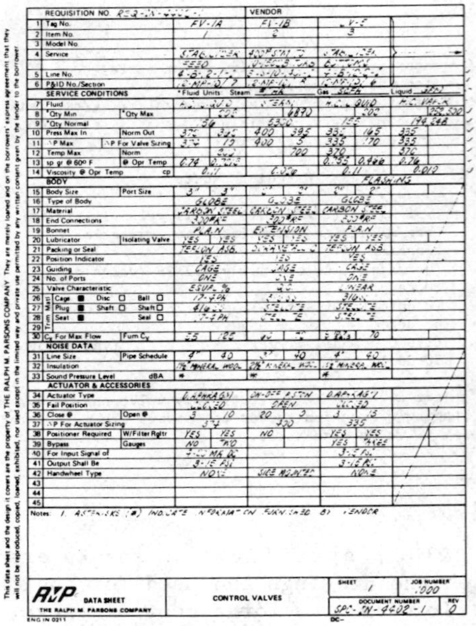

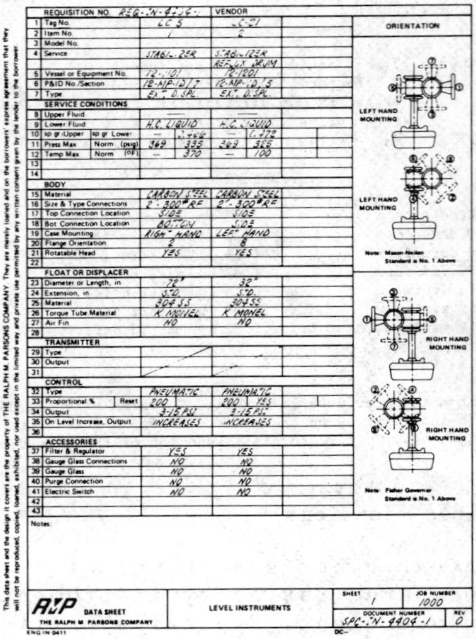

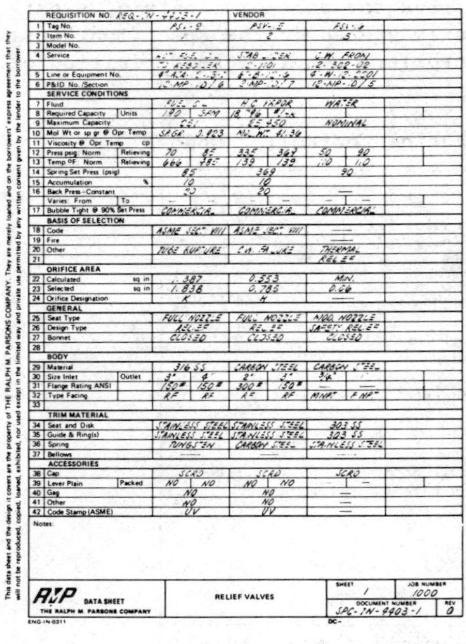

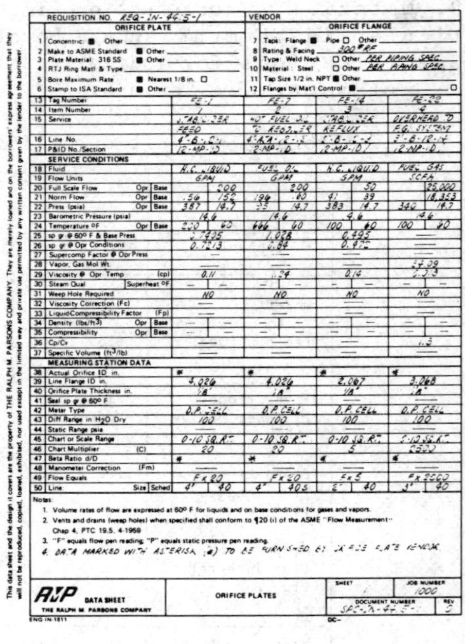

Fig. 9. Data sheet forms used in preparing specifications.

Logic, Ladder, and Loop Diagrams

The frequency and sequence of signals to initiate or prevent events, details of electrical circuits and components, and details of the interconnection of instruments and control devices are graphically presented by means of special purpose drawings: logic diagrams, ladder diagrams, and loop diagrams. These drawings are often produced by computer based systems.

Logic Diagrams. Logic diagrams are simple, compact, and symbolic representations of the number and sequence of events (signals) required to initiate or to prevent another event. They are utilized in the design of electrical and hydraulic control circuits and frequently serve only as preliminaries for ladder diagrams.

The symbols used in logic diagrams are defined in ANSI Y32.14-1973, "Graphic Symbols for Logic Diagrams (Two-State Devices)". Figure 10 demonstrates the use of standard symbols in a logic diagram for a single stage ball mill shutdown sequence.

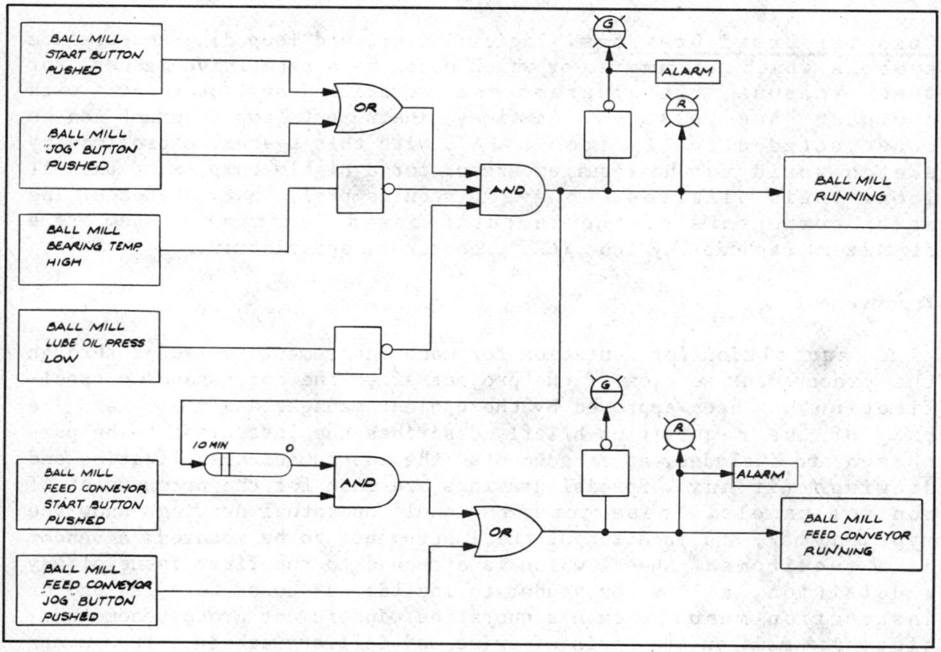

Fig. 10. Single stage ball mill shutdown sequence.

All interlocks between units of process equipment are detailed on the logic diagram. One diagram is produced for each interlock system in a design area.

Ladder Diagrams. Ladder diagrams are detailed representations of logic diagrams. They show the relays, relay contacts, handswitches, shutoff switches, timers, etc., that are required to meet the logic functions in the electrical circuits of each unit of process equipment and are used in developing electrical interconnection drawings (loop diagrams). Ladder diagrams are an important tool in troubleshooting during plant start-up and operation. Figure 11 is a computer-generated ladder diagram for a grinding ball mill.

Loop Diagrams. The interconnection of the various components of each instrumentation loop on the P&IDs are detailed in loop diagrams. (See Figure 12.) These diagrams show the interconnection wiring between field devices and panel-mounted instruments and tell the electrical discipline in the project what field wiring is required for each loop. In addition, loop diagrams show the control panel fabricator the interconnections internal to the panel.

Computer Based Drawings. Logic, ladder, and loop diagrams utilize symbols which are common, or which occur in a repetitive manner. For these reasons, the diagrams can be drafted at faster rates with computer based systems. At times, instrument loop diagrams can be constructed directly from a P&ID with this system. A preliminary sketch would not be required except for a highly complex or unusual loop. Logic diagrams, however, require preliminary sketches. The major components of the computer based drafting system are a digitizer, cathode ray tube (CRT), computer, and plotter.

Procurement

A requisition for quotation for each instrument is issued through the procurement section of the project after the corresponding specification has been approved by the project manager and the owner. The body of the requisition briefly describes the instrument to be purchased and includes, as attachments, the instrument specification and drawings (if any). Special drawings are made for the procurement of control panels. These control panel conceptual drawings show the type, number, and location of the instruments to be mounted. A vendor data requirement sheet, which is attached to the first issue of any requisition, allows the vendor to include any costs for drawings or instruction manuals in his quotation. Instrument group recommendations are made on the basis of price and delivery schedule from among those bids which comply with the specification. After the bid analysis has been approved and the specification revised to conform to the material being purchased, a purchase order is issued. The vendor later submits data such as drawings and instruction/maintenance manuals, which are reviewed and approved by the instrument group.

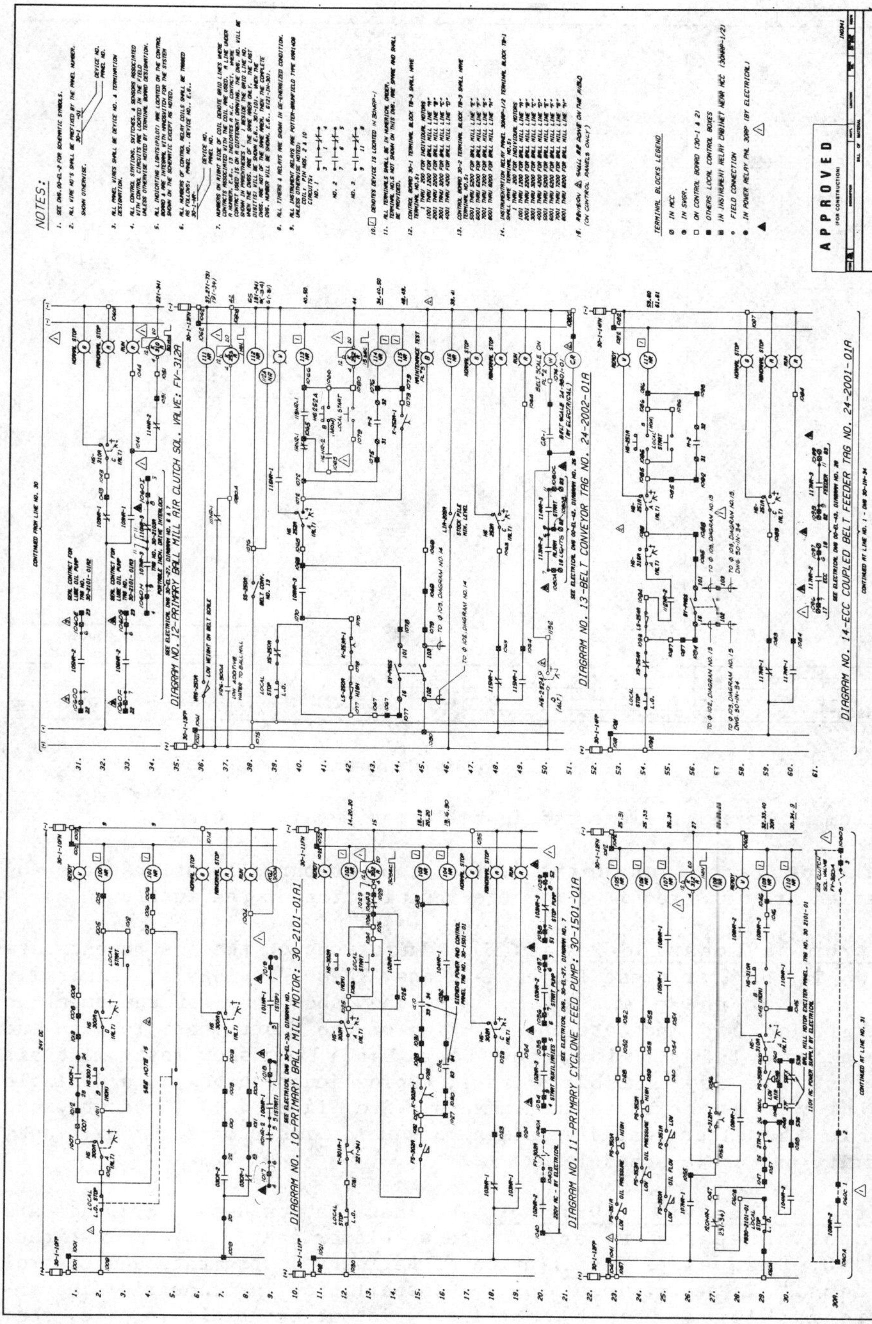

Figure 11. Computer-generated ladder diagram for grinding ball mill.

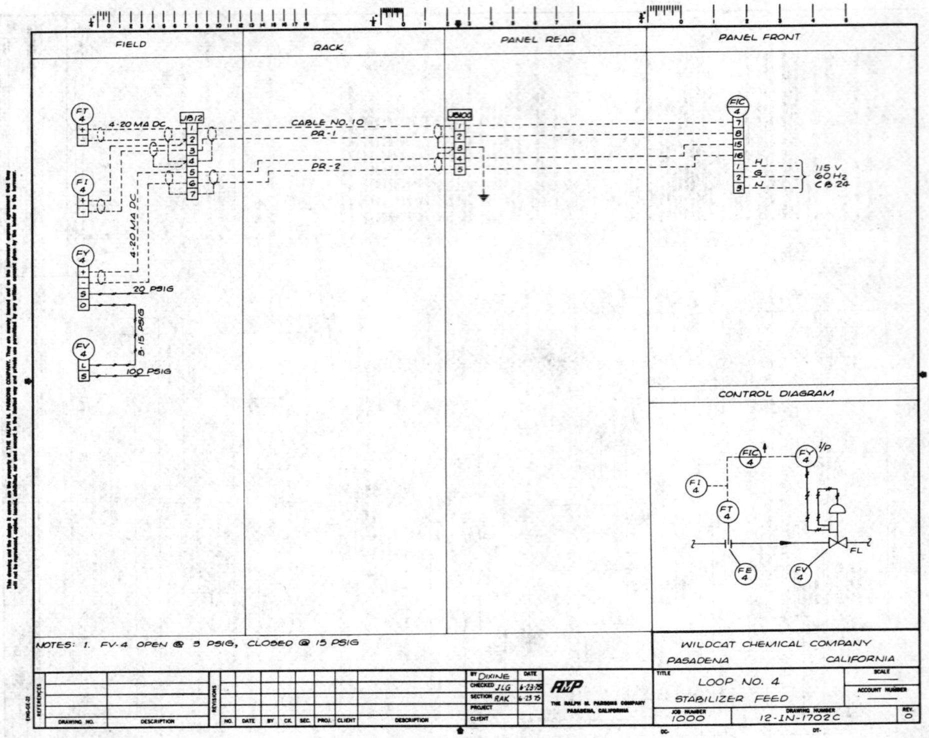

Figure 12. Loop Diagram

Instrument Location Plans and Installation Detail Drawings

The project instrument design group prepares location plans and detailed drawings required for the installation of the instruments.

Instrument Location Plans. The locations of the instruments are shown in the instrument locations plans. These drawings are basically overlays of the plant plot plans. They show the main equipment as background and indicate the position of the instruments in plan and elevation. In addition, the location plans show instrument air headers (if any), tubing trays, tubing junction boxes, and single tubing (if present) to instruments. (See Figure 13.) These drawings enable the electrical discipline to develop instrument wiring plans, conduit, and cable schedules.

Installation Detail Drawings. The installation detail drawings are commonly prepared in accordance with American Petroleum Institute RP-550, "Manual of Installation of Refinery Instruments and Control Systems." These drawings show the installation in isometric/diagrammatic fashion and each contains a list of materials (nipple, tee, plug, tubing, etc.) for the particular installation. (See Figure 14.)

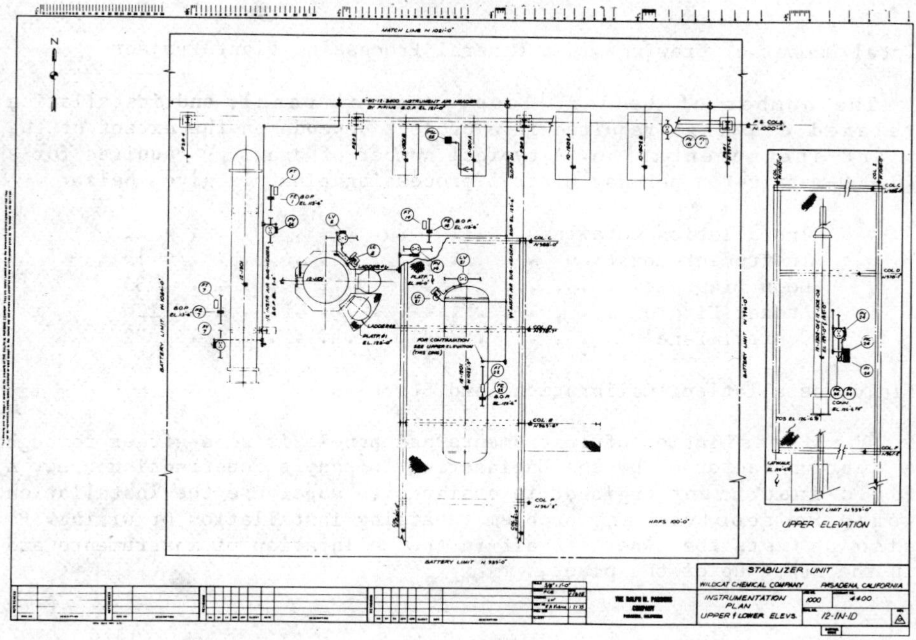

Figure 13. Instrument location plan.

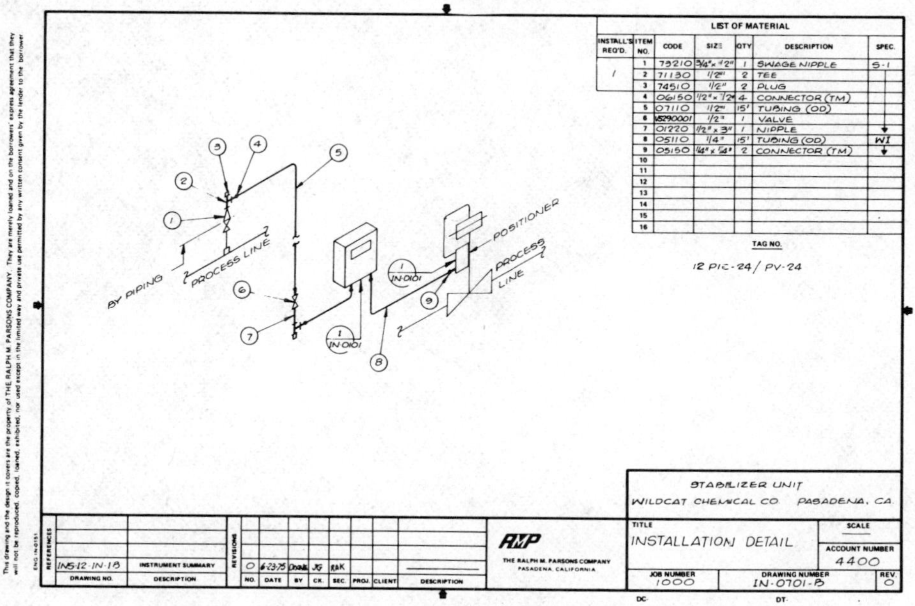

Figure 14. Instrument installation detail.

Total Number of Drawings in a Mineral Processing Plant Project

The number of ladder, loop, control panel, and installation related drawings required in a project depends on the extent of the plant instrumentation. A typical number of drawings required for a 40,000-metric-ton-per-day mineral processing plant is given below:

Installation Details	11
Instrument Location	15
Loop Diagram	33
Ladder Diagram	126
Control Panel	22

Field Installation, Calibration, and Start-up

The installation of instruments and panels is done either through a subcontractor or by the Engineering Company's construction crew. A field instrument engineer is assigned to supervise the installation work. He resolves any problem regarding installation or wiring. He also assists the owner's staff in the calibration of instruments and in the start-up of the plant.

7

Auxiliary Operations

Chairmen: I. L. Sandberg
J. Miller

Chapter 32

THE DESIGN AND APPLICATION OF A CENTRIFUGAL SLURRY PUMP

by Jon E. Fatzinger

Assistant Sales Manager
A-S-H PUMP
Division of Envirotech Corporation
Paoli, Pennsylvania

ABSTRACT

The design of a centrifugal slurry pump is approached as a compromise between efficiency and wear life of the wetted parts. This compromise is effected by the choice of proper materials of construction and geometry. However even the most optimally designed pump will fail if due consideration is not given to its installation and operation. System head curves and hydraulic profiles are discussed as useful tools to be used in the operational analysis of a slurry system.

DESIGN OF A CENTRIFUGAL SLURRY PUMP

The design of a centrifugal slurry pump is not an easy task. Not only must the designer be concerned with the efficient conversion of mechanical shaft energy into pressure energy or head, but also with the attainment of an economic wear life for each pump element which contacts slurry. There are design decisions where an increase in efficiency also means an increase in wear life; but, there are also many more decisions which involve a trade-off between efficiency and wear life. These decisions must be evaluated in terms of the application(s) for which the pump is to be designed. Obviously, in designing a pump for mill circuit service, wear considerations outweigh efficiency considerations, while in designing a scrubber pump for a flue gas desulfurization system, the converse is true. However, in the design of high volume tailings pumps, the desired balance between wear and efficiency is less tenable and the design choices are more difficult to resolve.

Typical Slurry Pump Construction

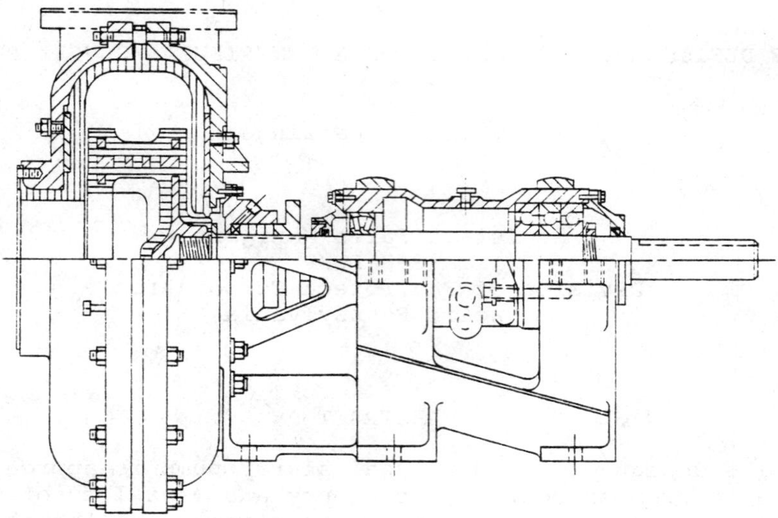

Fig. 1. Typical Slurry Pump

Figure 1 shows the cross-section of a typical slurry pump commonly employed in the minerals industry. Characteristically, a slurry pump is larger in size than its conventional water pump counterpart over a comparable range of capacity and head.

Wet End Design   Flow passages through the impeller as well as between the periphery of the impeller and volute of a slurry pump must be sufficiently large to pass the largest solid particle and prevent excessive internal velocities which will cause rapid wear. Usually, the volute tongue or cutwater of a slurry pump is somewhat rounded in an attempt to spread wear over as broad an area as possible. These geometric concessions are usually made at the expense of hydraulic performance.

Slurry pump impellers have relatively large diameters as compared with other types of centrifugal pumps in order that a given capacity and head can be achieved at a minimum rotational speed. It is an accepted axiom that wear in a slurry pump increases in proportion to the third power of impeller speed increases. Hence, maintaining low impeller speeds in a slurry pump is an important factor as related to wear. Unfortunately, hydraulic efficiency cannot be enhanced by using larger diameter impellers, and pump construction costs rise rapidly with an increase in impeller diameter.

Impeller Design This is the most critical factor in slurry pump construction and such details as the number of vanes, vane contour, shape of the leading vane edge, and entrance angle to the vanes are very important with respect to wear and hydraulic performance. Available open area at the impeller eye is also a concern as it directly determines entering slurry velocities. Practically speaking, suction nozzle velocities for slurry pumps lie in the range of 1.5 m/s to 5.5 m/s maximum. The velocity must be high enough to insure that the slurry solids reach the impeller eye, yet not be so high that wear is rapid.

Closed type impellers are generally preferred in a slurry pump where higher efficiencies are required. It is important with this design to maintain a tight clearance (1 mm) between the impeller and suction-side pump liner in order to minimize internal slurry circulation between the high pressure region at the impeller periphery and low pressure region at the impeller eye. Maintaining a tight clearance, therefore, not only results in power savings but also extends wear life, in that large particles cannot be trapped in this clearance and lead to "record grooving" of the suction liner. Some manufacturers put expeller type vanes on the impeller shroud to prevent circulation leakage by pumping away slurry from this area. This technique is useful in extending wear life but is intrinsically a power consumer.

An open impeller (no suction shroud) is basically an inefficient design but does have utility in pumping materials which can clog a closed impeller and in pumping frothy slurries where separated air bubbles cannot accumulate and choke-off the entrance to the impeller eye as in the case of a closed impeller.

The Casing As can be noted from Figure 1, the design of a slurry pump's wet-end often features a vertically split casing with replaceable liners in the areas of critical wear. Normally, the casing is constructed of cast iron, ductile iron, or cast steel depending on the pressure rating of the pump or the possibility of shock loading. Most pump manufactures prefer that suction and discharge piping be supported in order that nozzle forces and moments transmitted to the pump casings are minimal so that casing distortion with subsequent disruption of vital internal clearances is avoided.

Shaft and Bearing Assembly Because the specific gravity of slurries can be as high as 2.5 - 3.0, as for example, in pumping 70% by weight of mineral concentrates, the shaft and bearing assembly of the pump must be sufficiently rugged to withstand the imposed hydraulic and external loads.

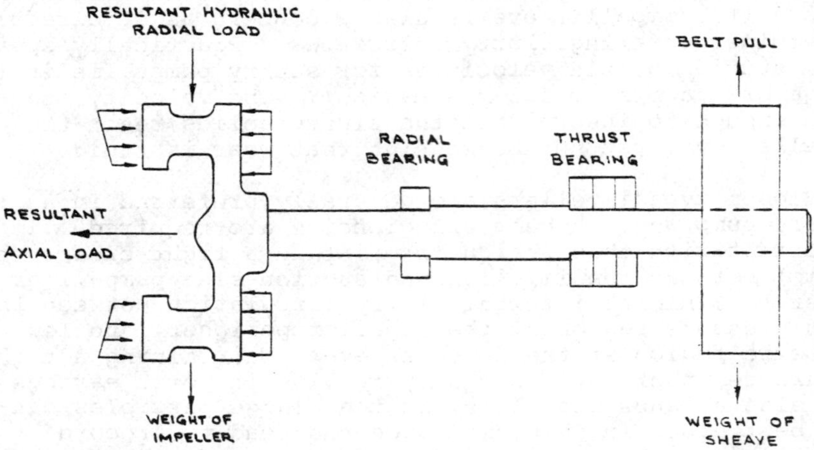

Fig. 2 Typical Bearing Loads

Figure 2 shows the types of loads which the bearings must carry. The axial thrust load primarily results from the resolution of the differential pressure on the impeller shrouds and vanes. Typical radial loads result from hydraulic sources, impeller assembly weight, sheave weights, and belt pull.

Pump Sealing This is generally accomplished by introducing seal water into a packed stuffing box and is the most positive method of sealing the pump. Centrifugal sealing also utilizes a packed stuffing box but also relies on auxiliary expeller vanes on the engine side of the impeller or within the packing box to prevent solids from entering the packing area. To work efficiently, this type of seal usually depends on operating the pump at its higher range of speed. Centrifugal sealing is not usually recommended for applications with high suction heads, which can essentially negate the head created by the expeller vanes.

Materials of Construction Wear performance and efficiency are, of course, not the only major concerns of the pump designer. He must also keep a rein on the capital cost of the pump as built with an eye to minimizing the replacement cost of the wearing components. This again is not an easy task, as a slurry pump by virtue of its size and thick-walled wear components implies inherent high cost compared with other types of centrifugal pumps.

The wetted parts of a slurry pump are normally constructed from a variety of materials ranging from tight grained irons such as manganese steel, 28% chrome, and nihard to elastomeric materials, such as natural rubber, neoprene, butyl rubber or a variety of other synthetics. The pump designer must make his selection of materials based on the type of mechanical and corrosive wear to which the pump will be subjected.

Traditionally, hard metals have been used where the solids particles to be pumped exceed 1/4" in size and elastomers for slurries which contain fine particles. The usefulness of metal resides in their hardness as related to the mineral hardness. Direct impact wear is the most destructive type of mechanical wear to metals and depending on the size and gravity of the particle, this type of wear causes permanent deformation. The elastomers, particularly natural rubber, owe their utility to the ability to flex on impact and absorb the particles' kinetic energy.

The angle of impact also has a great deal to do with rate of wear on hard alloys and rubber. Figure 3 illustrates this point. Because the pump volute is exposed to a wide angle range of impact, the designer's treatment of volute geometry and material of construction is of significant importance.

Recently, the rubber technology gained in applying heavy, dense, molded rubber in grinding mills has been put to use in the design of pumps for coarse duty service, such as primary mill to cyclone feed applications. It has been shown that shell-liners of densely molded thick rubber sections so that the "cut-through" value of the rubber is not exceeded, is a viable alternate to the higher initial costs and replacement values of hard alloy shell liners and casings.

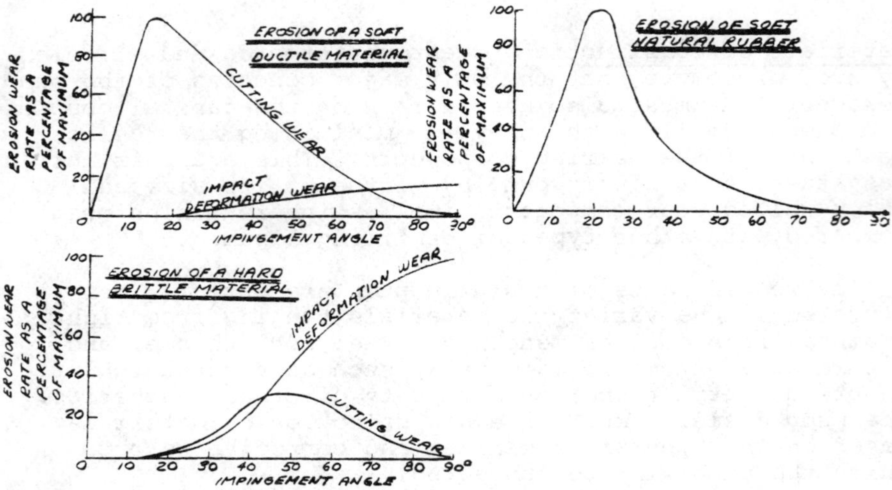

Fig. 3. Erosion/Impingement Angle
Various Materials

SLURRY PUMP APPLICATION AND INSTALLATION

Even the most optimally designed slurry pump cannot operate successfully unless it is properly selected to suit the application and then properly installed.

Slurry Pump Application

System Head Curve From a hydraulic viewpoint, the best way to select a pump is to match the system head curve against the proposed pump characteristic head-capacity curves. The system head curve, which is a dynamic profile of the operating system, is plotted by summing the system friction loss at each flow rate to the system static head, which is constant. The intersection of the system head curve and pump curve at the design flow and head determines pump speed. See Figure 4.

Ideally, it would be best to have the design point located as nearly as possible to the best efficiency point (B.E.P.) on the pump's characteristic curve (Figure 5).

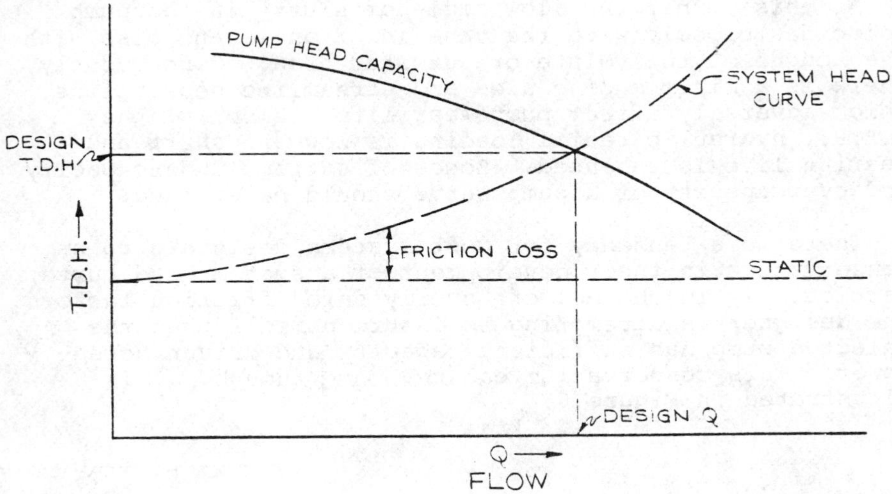

Fig. 4. System Head Curve

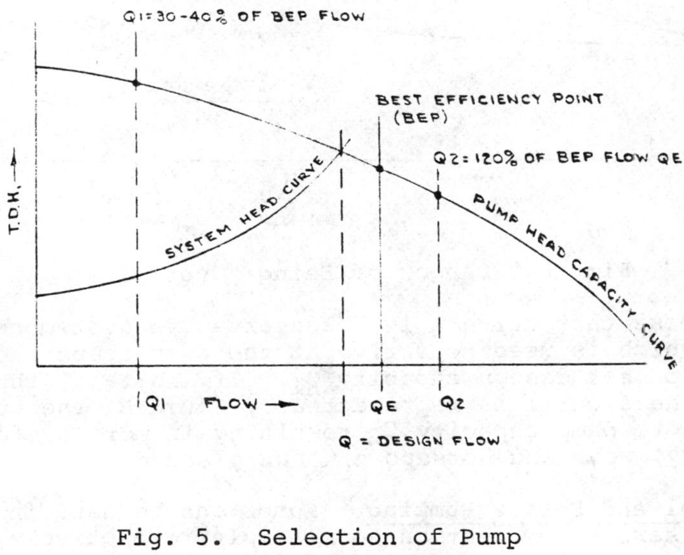

Fig. 5. Selection of Pump

At this point, the flow angle of slurry in the pump coincides closely with the vane inlet angle and also with the tongue of the volute or cutwater area. Consequently, there is a minimum of eddies and streamline separations which adversely affect pump wear life. Also, at the B.E.P., hydraulic radial loading is low and shaft and bearing life is enhanced. Zones of extreme undercapacity and overcapacity of a pump curve should be avoided.

There is a tendency for most systems designers to be conservative in their development of a system head curve, particularly in the use of "overly safe" friction factors. The designer is attempting to insure himself that the selected pump has sufficient capacity and driven horsepower. This conservatism can backfire, though, as is illustrated in Figure 6.

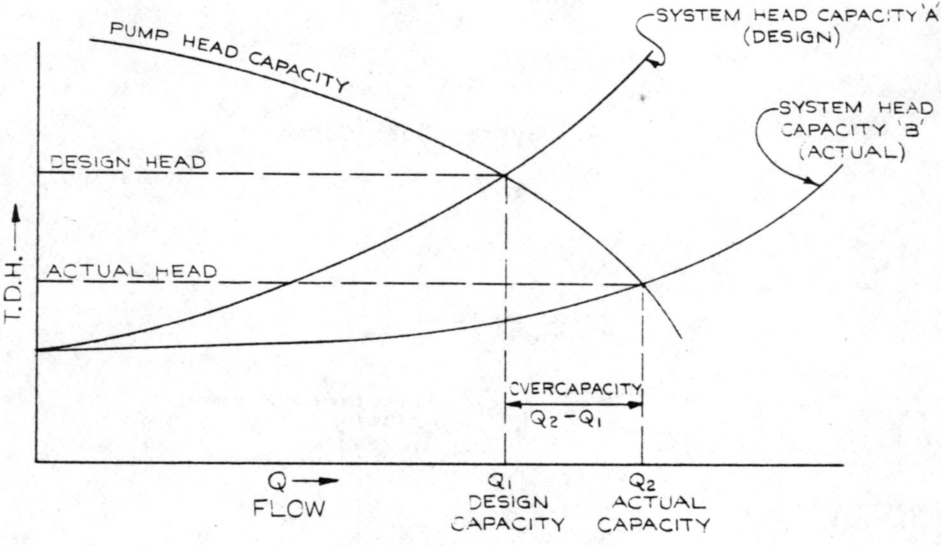

Fig. 6. Danger of Being "Over-Safe"

Suppose that Curve A is a conservative system head curve which is used to arrive at the pump speed shown in order to meet design capacity Q_1. If, in fact, the actual system head curve is represented by Curve B, the pump will attempt to pump capacity Q_2 resulting in more system flow, velocity, wear and horsepower than planned.

<u>Parallel and Series Pumping</u> Pumps can be used in parallel or series combination to multiply respectively, the

capacity or head capability of a single pump unit.

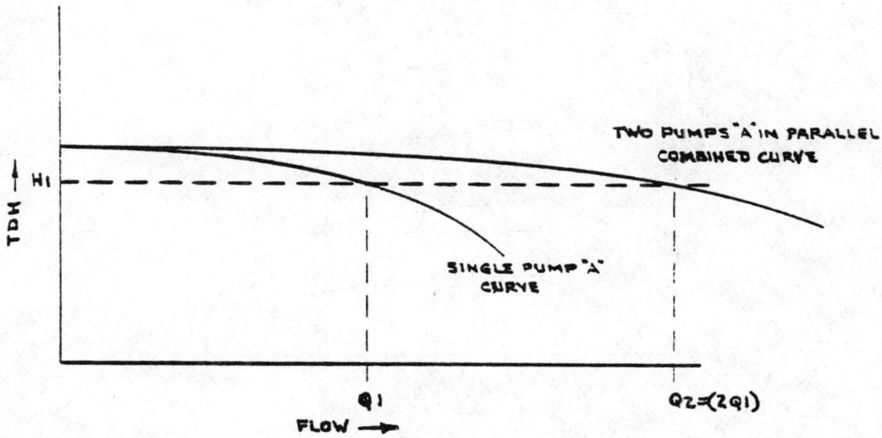

Fig. 7. Parallel Pumping

<u>Parallel Operation</u> Figure 7 illustrates the combined head-capacity curve for two identical pumps in parallel combination, identically speeded. Because of the relative flatness of slurry pump curves, it is often difficult to balance the hydraulic duty equally between the two pumps as even small variations in the driving motor speeds, wear, and clearances of the individual pump units can cause major imbalances. It is best in such systems to pick the pump with the steepest head-capacity characteristic available at the design point and to install the pumps with as nearly symmetrical pipe and fittings as possible from the sump to the point of juncture into a common discharge line. Control valves or variable speed drives are normally employed to help fine tune the system.

<u>Series Operation</u> Series pumping is normally encountered in the design of long distance slurry transportation systems such as the pumping of mill tailings. See next page for Figures 8 and 9.

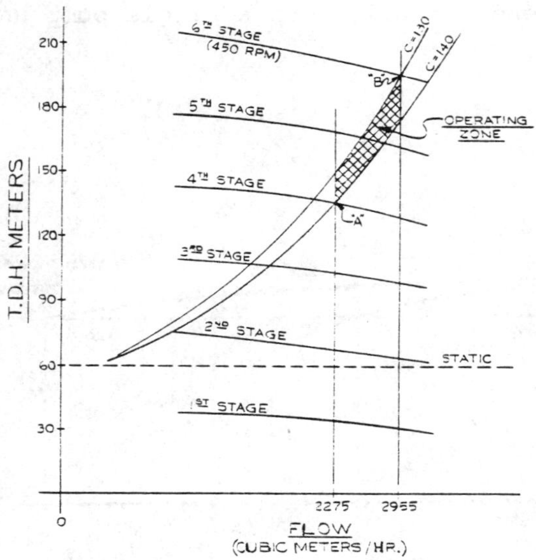

Fig. 8. System-Head Curve of 330 m Length Tailings Pipeline

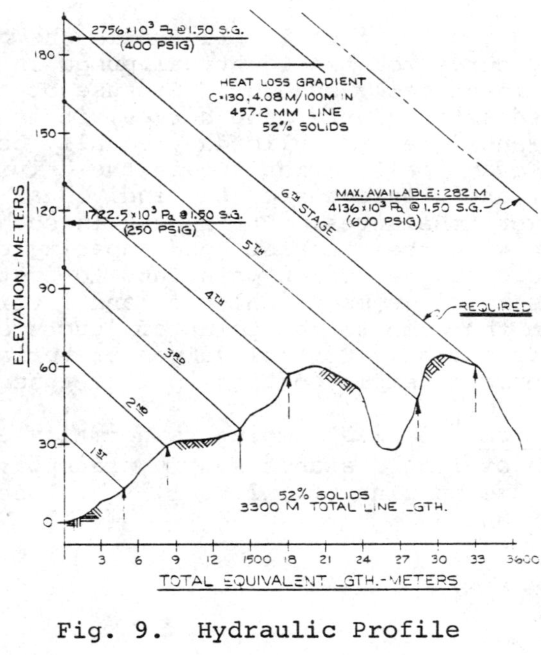

Fig. 9. Hydraulic Profile

Figure 8 shows the use of 6 stages of pumps in series in order to transport mill tailings over a 3300 meter length of 460 mm I.D. pipe. The curve for each stage represents the additive effects of that stage and all preceding stages. Superimposed on the pump curves are two system head curves. One curve (C=130) may represent the design losses for the normal flow conditions and the other curve (C=120) may represent maximum anticipated losses under conditions such as coarse grind or partially sanded line. Vertical lines are drawn at volumes representing the minimum safe carrying velocity for the slurry and at the maximum volume, perhaps corresponding to peak mill output. The intersection of these variables defines the operating zone of interest.

Figure 8 demonstrates the useful application of system head curves matched against a series sequence of pump curves. The system designer can see at a glance the number of stages required to perform a particular hydraulic duty as well as establish the number of variable speed units required to cover the anticipated operating zone.

Hydraulic Profiles The hydraulic profile is another very useful tool which is used to analyze long-distance series pumping applications. This profile is a comparison of the pipeline contour and the hydraulic gradient for the system.

Figure 9 shows the hydraulic profile for the series pumping application presented in Figure 8. As can be readily noted, the hydraulic profile quickly shows the necessary pressure ratings for the pump casings as well as for the pipeline at any point along the pipeline. In the example given, all pumps are located at the beginning of the line and have a case working pressure rating of 2756×10^3 Pa (600 psig). However, if the maximum working pressure rating of the pump casing were only 1722.5×10^3 Pa (250 psig), only 3 pumps could be located at the beginning of the line. The logical placement of a booster or remote station would then be approximated by the intersection of the third stage hydraulic gradient at the pipeline contour.

NPSH Perhaps 90% of operating pump difficulties which occur can be traced to the suction side of the pump. Perhaps the most common design abuse is the lack of sufficient NPSH or net positive suction head. Simply put, NPSH is the net absolute pressure of the slurry at the suction eye of the impeller over and above the vapor pressure of the slurry liquid. The greater the pump speed and pump capacity, the more NPSH is required by the pump to perform its job. Therefore, for any application, the system engineer must determine the NPSH available to the pump (NPSHa) and compare this to NPSH required (NPSHr) by the pump at the design point. As an example of calculating NPSHa refer to Figure 10 which shows a bridge mounted pump over a thickener. The pump is attempting to lift the thickened material, with slurry specific gravity of 1.5, from the bottom of the thickener.

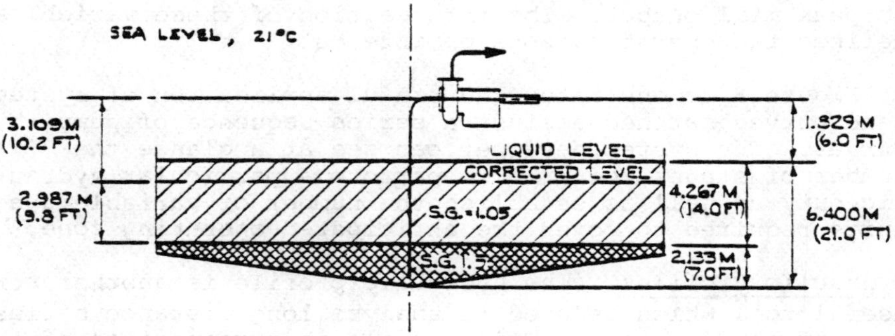

Fig. 10. Bridge Mounted Thickener Underflow Pump.

The familiar NPSHa formula is:

NPSHa = Ha ± Hz - Hf - Hvp

1. Ha is the atmospheric force trying to push the slurry into the impeller. In the example, Ha in terms of meters of slurry is $\frac{10.36\text{m}}{1.5} = 6.91\text{m}$. That is, while atmospheric pressure could normally support a column of water 10.36m, it can only support a column of slurry 6.91 m.

2. Hz is the net static height between the slurry liquid level and the center-line of the pump. In the case of flooded suction, this is a positive contribution which, with the available atmospheric pressure,pushes the slurry into the pump. In the example shown, however, Hz is negative as there is a lift condition. Note that Hz is not -1.83m but actually -3.11m. This is because 4.27m of the upper layer of liquid at a specific gravity of 1.05m is only equivalent to 2.99m of the slurry being pumped, which is at a specific gravity of 1.5.

3. Hf is the friction loss in the suction pipe and fittings and for the example shown, is given as -0.91m for the capacity being pumped.

4. Hvp is the vapor pressure of the slurry at the given temperature of the slurry, which is 100°F. Usually, the vapor pressure of the slurry carrier fluid, in this case, water, is taken and converted to equivalent meters of slurry. Therefore, in this case:

$$Hvp = \frac{0.67m}{1.5} = .45m$$

5. Now, the net absolute pressure of the slurry at the suction flange above the vapor pressure of the slurry or NPSHa is:

$$NPSHa = 6.91 \ -3.11 - .91 - .45 = 2.44m$$

Consequently, the pump selected must theoretically require a minimum of 2.44m of NPSH to operate successfully. If the pump's NPSH requirement is beyond what is provided by the system, loss of pump performance due to cavitation can occur. If there is a large disparity of NPSHa to NPSHr, cavitation can be severe and pump damage might be the result.

Pump Installation

In addition to insuring that pump NPSH requirements are met, most other suction side pump problems can be eliminated by careful attention to sump design and pump installation. Reference 1 gives some excellent suggestions in this regard, which are enumerated below.

<u>Sump Design</u> Some good points to remember in the design of a sump are as follows:

1. Provide adequate sump volume to accommodate surges.
2. Sloped sides must have an angle in excess of the angle of repose of the solids.
3. The entire sump should be screen covered for tramp protection and/or...
4. Provide a pocket below the suction pipe to collect tramp.
5. Provide maximum sump height which will provide maximum NPSH available to the pump.
6. Install baffles as required to prevent vortexes.
7. Slurry feed and possible air entrainment should not be in a direct path to the suction line.

Suction Piping Suction piping can often make or break a pump's operation. Some common sense rules to follow are as follows:

1. Keep lines short and straight as possible.
2. Use only non-restricting type isolation valves, if required. Pumps should not be throttled from suction side as friction is increased and NPSH to the pump is reduced.
3. Avoid excessive fittings and packed-stem valves which can leak air.
4. Line size should provide adequate carrying velocities to prevent sanding; however, excessive velocities, like sanding, will spoil NPSH available to the pump and increase wear.
5. Anti-vortex vanes may be necessary to prevent prerotation of the liquid in the suction line. Vortexing will destroy NPSH characteristics of a pump.

CONCLUSION

The successful application of a slurry pump depends not only on a good pump design but also on correctly matching the pump with the anticipated hydraulic duty. The use of system-head curves and hydraulic profiles are useful tools in aiding with this selection and total system design. The physical pump installation is also important and the practice of employing time-proven techniques in this regard will go a long way toward avoiding trouble.

BIBLIOGRAPHY

1. Horne, C.A.,"A Guide to Good Slurry Pumping Practice", Canadian Mining Journal November, 1976, December, 1976 issues.

2. Dalstad, J. Ingemar, "Slurry Pump Selection and Application", Chemical Engineering April 25, 1977, pp. 101-106.

3. Wilson, G. "The Design Aspects of Centrifugal Pumps for Abrasive Slurries", The Canadian Mining and Metallurgical Bulletin, May, 1973 issue,pp. 91-102.

Chapter 33

PROCESS PIPING AND SLURRY TRANSPORTATION

By John C. Loretto and E.T. Laker

Manager and Assistant Manager, Mining & Metallurgical Division
H.A. Simons (International) Ltd., Vancouver, Canada

ABSTRACT

This chapter deals with the process piping aspects of moving slurries in mineral processing plants and the transportation of slurries such as tailings and concentrates over longer distances. The basic data required to select the process piping is given together with an example of line sizing and material selection. Alternative piping materials are discussed together with pumps and pumpbox design and the types of valves and other fittings available to the process piping designer. The long distance transportation of slurries is briefly discussed and a number of references given on this, including papers on actual installations.

INTRODUCTION

The design of the process piping in a mineral processing plant is normally evolved from the following steps:

- Finalizing the process flowsheet.
- Identifying the various process fluids.
- Establishing critical and minimum transport velocities of fluids.
- Selection of pipe diameter and determining friction factors.
- Finalizing the process piping layout.
- Selection of piping materials, fittings and supports.
- Listing of pipe and fittings including specifications.

BASIC DATA REQUIRED AND SOURCES OF DATA

To carry out the design of the process piping, the following basic data is required:

- Flow or volume of fluid.
- Characteristics of fluid.
- Configuration of piping layout.
- Operating conditions.

SOURCE OF BASIC DATA

Flow

The flowsheet generally shows an average flow figure but the "normal" maximum and minimum flows must also be determined and are usually extrapolated from the average flow based on the plant operating schedule together with the experience and judgement of the process engineer.

Fluid Characteristics

Characteristics	Source
Number and type of phases	Flowsheet.
Grain size of solids	Flowsheet or testwork.
Percent solids	Flowsheet based on testwork.
Critical velocities	Data tables based on testwork and operating experience.
Friction factors	Data tables based on testwork and experience.
Abrasiveness	Experience or Miller Number.
Temperature, pH or corrosiveness	Testwork.

Configuration

This is determined from the plant general arrangement drawings and is frequently clarified by means of an isometric drawing of each individual pipe run. This drawing will include static head, total pipe length, bends and other fittings such as elbows and valves.

Operating Conditions

The main operating conditions to be considered include:

- Continuous or intermittent operation.
- Ambient temperature.
- Elevation.
- Geographic location; whether remote or close to population centers.

- Continuous or Intermittent Operation - Where the operation is continuous, greater attention than usual must be paid to installing process piping that requires a minimum of downtime for maintenance and in some cases, a duplicate line is installed. Where the operation is intermittent, the design might well be less stringent, but attention must be paid to the startup and shutdown capability of the system. In this case the lines would be capable of being conveniently drained or flushed at shutdown.

- Ambient Temperature - This can be of considerable significance in cold areas where insulated or heated lines may be required to avoid freezing, particularly where the process piping runs outside of buildings between sections of the plant. In extreme cases such as in exposed tailings lines, an automatic drainage system may be necessary to avoid a long line frozen for the rest of the winter.

- Elevation - Although this usually has little or no direct effect on the piping itself, it can drastically alter the pump suction conditions, thereby affecting pump operation and eventually the piping.

- Geographic Location - If the plant is situated in a remote area where there is likely to be a minimum of experienced operating labour, then the process piping design must take this into account. For example, the design should ensure ease of startup and shutdown to allow for the more frequent than usual stoppages that could be expected under these circumstances.

EXAMPLES AND USE OF BASIC DATA

The following typical examples of basic data are included in the Appendix at the end of the chapter:

- Flowsheet (Figure 1).
- Velocity versus friction factor tabulation (Table 1).
- Tabulation of particle size versus critical velocity (Table 2).
- Table of Miller Numbers - abrasiveness (Table 3).
- Isometric of piping layout (Figure 2).
- Mill condition details (Table 4).
- Piping specification sheet (Table 5).

The following calculation is typical of how the process piping design for mineral processing plants makes use of the basic data to determine the line diameter and recommended piping material. It is also prudent to calculate the total dynamic head of the system so as to ensure that the type of pump envisaged can handle the duty satisfactorily.

EXAMPLE - Cyclone Feed Piping Selection

Flow - m^3/second:	Average	0.175	
	Minimum	0.158	
	Maximum	0.211	
Grain size of solids:	80% passing size		1,000μ
% solids by weight:	Average	52%	
	Minimum	46%	
	Maximum	59%	
Critical velocity at 46% solids:	From Table 2		
Friction head loss:	From Table 1		
Abrasiveness of slurry:	Moderately high - Miller No. 140		
Slurry temperature and pH:	25° C. and pH 9.0		
Configuration:	From relevant isometric drawing		
Operating conditions:	Continuous operation. Ambient temperature 4 - 15° C. Elevation 4,100 meters. Remote location.		

Calculation

- From Table 2, Appendix, it can be seen that the critical velocity of an 80% passing 1,000μ slurry at 46% solids is 1.9 m/sec. Adding 0.5 m/sec. as an experience safety factor gives a minimum line velocity of 1.9 + 0.5 = 2.4 m/sec.

- From standard pipe diameter/line velocity tables, the line velocities for the various flows through alternative pipe sizes are:

	Inside Pipe Diameter	
	0.305 m (12 inch)	0.356 m (14 inch)
Average flow	3.21 m/sec.	2.36 m/sec.
Minimum flow	2.92 m/sec.	2.15 m/sec.
Maximum flow	3.51 m/sec.	2.57 m/sec.

- From these figures it can be seen that the velocity of the minimum flow of 0.158 m^3/sec. in a 0.305 m inside diameter pipe is 2.92 m/sec. which is safely above the minimum line velocity of 2.4 m/sec. Also the line velocity of the maximum flow in this size of pipe is 3.51 m/sec., which as Table 1 shows, does not give excessively high friction head losses. A 0.305 m (12 inch) inside diameter pipe is therefore selected.

- Since the abrasion is only moderate, steel pipe with rubber bends will be satisfactory. Using this material, the friction head loss per 100 m of 0.305 m (12 inch) pipe for the varying flows will be:

Average flow of	0.175 m^3/sec.	head loss is	4.62 m
Minimum flow of	0.158 m^3/sec.	head loss is	3.88 m
Maximum flow of	0.211 m^3/sec.	head loss is	5.39 m

- From the appropriate isometric drawing, the total pipe length equivalent taking into account the bends and fittings is 34 meters, the static head is 18 meters and the pressure head required at the cyclone is equivalent to 5 meters.

- From Table 2, Appendix, the friction head losses for the various flows through 34 meters of 0.305 m pipe are as follows:

Average flow of	0.175 m^3/sec.	4.62 x 0.34 = 1.57 m
Minimum flow of	0.158 m^3/sec.	3.88 x 0.34 = 1.32 m
Maximum flow of	0.211 m^3/sec.	5.39 x 0.34 = 1.83 m

- The total dynamic head at each of the various flows is the sum of friction + static + pressure, as follows:

Average flow	1.57 + 18 + 5 = 24.57 m total dynamic head
Minimum flow	1.32 + 18 + 5 = 24.32 m total dynamic head
Maximum flow	1.83 + 18 + 4 = 24.83 m total dynamic head

Using these total dynamic heads the pumping design engineer decides the type of pump and what pump speed and horsepower should be used to ensure that these duties can consistently be handled. Although a variable speed slurry pump could be used, virtually all modern horizontal rubberlined slurry pumps could handle this range of duties at a fixed speed. Lower than anticipated flows could be compensated for by adding water through a float valve. Higher flows would cause the level in the pumpbox to rise, decreasing the net static head, enabling the fixed speed pump to pump the increased volume until the flow returns to normal.

PROCESS PIPING MATERIALS

The principal materials used in process piping are:

- Steel
- Rubberlined steel
- Rubber hose
- Plastic

Plain Steel Pipe

This is the most common piping in use in mineral processing plants as it is easy to install, has reasonable wear resistance and its rigidity makes it relatively easy to support. Where abrasive wear is significant, thick wall steel pipe can be used, but in extreme cases, special wear-resisting material will be required.

To facilitate installation, maintenance and replacement, a complete system of convenient pipe couplings and fittings of the Victaulic, Couplox, Gustin-Bacon or other similar styles have been developed. This type of coupling is relatively inexpensive, needs a minimum of skill to install and is also suitable for rubberlined pipe. One of the limitations of this type of coupling is that fittings such as elbows are usually of short radius, which together with the gap at each coupling can lead to turbulent flow and encourage erosion at these locations.

Rubberlined Steel Pipe

Rubberlined steel pipe usually consists of a 50 - 100 mm rubber lining glued inside standard or thin-wall steel pipe which is joined with Victaulic or flanged coupling. Wear resistance of rubberlined pipe is many times greater than plain steel pipe, which gives it a substantial advantage when handling highly abrasive and coarse slurries.

Complex fittings of tees, elbows, etc. can be fabricated in steel and then rubberlined, allowing the piping designer great flexibility in arranging interconnecting piping. However, once set in position in the plant, rubberlined pipe is difficult to modify which may well give maintenance and modification problems during plant operation.

Rubberlined Hoses

The rubberlined hose was developed to provide greater wear resistance together with flexibility in slurry piping arrangements. Mineral processing plants have adopted it for many purposes such as abrasion-resisting bends and piping from concentrate launders to pumpboxes.

Rubberlined hose is available from 0.05 m to 0.75 m diameter and is designed for use with pressures up to 750 Kg/cm^2. It consists of a lining, a carcas and a cover. The lining is usually of rubber between 25 and 50 mm thick to resist the abrasion and corrosion. The carcas consists of several layers of a woven terylene or nylon fabric and is designed to resist the internal pressure and the longitudinal pullout. A steel wire helix is often embedded in the carcas to resist vacuum pressure or to avoid kinking of the hose under excessive bending. The cover is a 10 - 20 mm rubber or synthetic formulated to resist weathering and abrasion from rough handling and often includes corrugated covers to provide better impact resistance. One of the disadvantages of rubberlined hose is that long lengths must be supported with wide pipe clamps spaced along its length and these supports must be able to swing free to allow the normal stress expansion to occur.

Plastic or Plastic Lined Pipe

Certain areas of mineral processing plants have adopted these materials quite widely where there are problems of erosion and corrosion. Flexible plastic pipes of the polyethylene type exhibit excellent friction factors in the region of C = 150 and can resist coating by gypsum and lime and their flexibility greatly reduces water hammer problems.

The types of plastics available include:

- Polyurethane - Usually used to line steel pipe and is comparable to rubberlined steel pipe. A 1973 report by a clay company in Britain noted an installation of 12 miles of this type of pipe handling sand quartz/mica slurry that had a life of between 12 and 20 times that of steel pipe on the same duty.

- High Density Polyethylene - Since the early 1970's pipe made entirely of this material has been used extensively for tailings and sand lines throughout North America in diameters up to one meter. Tests on high density polyethylene have demonstrated a 5:1 wear ratio over steel for a factory price per meter slightly above that of steel pipe.

  For example, in 1973 a major mine in British Columbia, Canada installed 8 km of 0.9 m high density polyethylene pipe to transport copper tailings with normal line velocities of 2 - 4 m/sec., but with velocities as high as 6 m/sec. at steep gradient locations. Wear losses were measured at 3 mm/year for normal flows whereas asbestos cement and woodstave in the same installation had wear rates several times this figure giving line life of one year and two years respectively.

  Although polyethylene has a limited high temperature range of approximately 50° C., it has a low thermal loss which is beneficial on outside lines in cold climates. However, it has a high thermal expansion so where the line is installed on steep slopes, care must be taken to clamp the pipe adequately to prevent it creeping downhill due to the daily temperature changes.

For in-plant use, polyethylene must be fully supported throughout most of its length and different pipe supports and layout techniques from that used in steel must be developed. For tailings lines and water lines, polyethylene can be placed on a graded dirt or gravel right-of-way without any supports or guides. To reduce heat loss in cold climates, 0.2 - 0.6 m of earth has been heaped over the pipeline to provide a cover.

Other plastics such as ultra high-density polyethylene, polypropylene, polybutylene, polytetraethylene (Teflon) are all being used but their relatively high cost, limited availability and more critical installation and operating procedures suggest their use for special applications only.

In addition to its use for handling slurries, plastic piping has been used for water piping and has the advantage of lower corrosion, less scaling, lower friction factors, no contamination and its low total cost. Against these advantages, are the disadvantages of the additional support needed, the potential fire hazard from welding and its vulnerability to physical damage.

Polyethylene has been widely accepted for use as reagent piping as it does not react to most reagents and is easily handled and relocated. The fittings, however, are usually PVC or nylon but these must be checked for each reagent and where there is a reaction between plastic fittings and the reagent, stainless steel fittings may be required.

VALVES AND OTHER FITTINGS

The number and type of valves used in a typical mineral processing plant is relatively small when compared with the range used in other industries and the main types available are:

- Pinch valves
- Diaphragm valves
- Plug valves
- Gate valves
- Butterfly valves

Pinch Valves

The pinch valve has a minimum number of wetted parts and can be manually operated or used as a process control valve. It consists of a length of rubberlined hose with an external clamp and is used extensively for slurries in pipe sizes from 2.5 cm to 50 cm and for pressures from 25 to 250 Kg/sq. cm. They are also available in small diameters down to 30 mm for special applications such as on reagents and similar application. The hose is available in materials such as rubber, neoprene and urethane to allow the selection of a material to resist abrasion or reaction between slurry and the valve hose material.

Diaphragm Valves

These operate somewhat similarly to pinch valves except that the moving diaphragm closes against the fixed valve body. The closing is generally by hand wheel and screw, air or hydraulic and they are frequently used as control valves. Although rubber diaphragms are standard, they are available in a wide range of synthetic materials.

Plug Valves

There are many industrial uses for the common plug valve but for slurry services, the full port type, together with pressure lubrication must be used to ensure positive operation.

This type of valve is a medium price unit suitable for relatively high pressures and fine slurries or water services with the advantage that it can be clearly seen whether it is open or closed. It has less bulk than most other valves and can be easily fitted for mechanical operation. However, its rapid rate of closure can cause severe shock in fluid systems and in the event that this can cause serious damage, a gear fitting can be used to decrease the rate of closure and consequent hydraulic shock.

Ball valves are a type of plug valve but are not as resistant to slurries and generally more expensive. Small valves of the order of 50 mm are often used for service water or purge connections. Vee-ported ball valves, or eccentric plug valves are sometimes used in slurry services particularly as control valves at higher pressure with wear being a function of the slurry's abrasiveness.

Gate Valves

These are of limited use on abrasive slurries but are often used on plant water lines and non-abrasive slurries, the rising stem type has the advantage that it can be clearly seen by how much it is open.

Butterfly Valves

These are more usually used to control the flow of pure liquids and are frequently used in flow control systems, particularly where a significant pressure drop is required across the restricting device.

Special Valves

Tech-Taylor ball valves - These are often used when two pumps deliver into the same line, particularly where both pumps might operate together or the standby pump might be cut in automatically.

Other Fittings

Cleanouts - In slurry transfer, cleanout connections are often required to avoid dismantling sections of the line when a choke occurs. These should be accessible and are usually tees or elbows with removable plugs. In cases of frequent use, valves can be fitted together with high pressure water or air connections for flushing out the blockage. The underside of a concentrate thickener is an example of a location where a sophisticated cleanout system is recommended.

Cleanouts must be located where operators can get at them and still use their tools and equipment and where the release of a plugged line will not be dangerous to the operator or cause the uncontrolled draining of upstream equipment with damaging results.

PUMPS AND PUMPBOXES

The process piping should not be considered in isolation from the pumping unit and for this reason pumpboxes and the principal types of pumps presently used in mineral processing plants are briefly discussed.

Horizontal Rubberlined Sand Pumps

Since their development in the late 1930's, the design of these pumps has progressed considerably, particularly in the areas of abrasion-resisting materials. Modern pressure moulded rubber pump parts often last in excess of 2 years pumping slurries and rubber impeller lifes of over 6 months are common even on duties such as pumping cyclone feed.

Vertical Pumps

One of the limitations of most types of horizontal pumps is that aerated froth can cause air locking and consequent spillage from the pumpbox. A sloping suction can do much to minimize this but with many froths a vertical pump is the only sure way to eliminate air locking. Due to their inherent design, these pumps have not developed hydraulically to the same extent that horizontal rubberlined sand pumps have progressed, and they are limited to heads of about 30 meters. Nevertheless, vertical pumps have the advantage of being able to handle varying flows of highly aerated material without air locking. In addition to this, they can be mounted in flotation launders thereby eliminating the need for a sub-basement containing pumpboxes and pumps. If this layout is used, the vertical requirements of the flotation section can be significantly reduced.

Diaphragm Pumps

This type of pump is normally limited to light suction lift duties and as metering devices on applications such as concentrate thickener underflows. Air-operated diaphragm pumps have overcome some of the duty limitations of the mechanically-operated diaphragm pump and in addition to this, they can pump against quite high delivery heads.

Plunger Pumps

Since centrifugal slurry pumps are at present limited to total dynamic heads of 40 or so meters per pump, the positive displacement pump is often used where very high heads are encountered or where a single pumping unit is preferred over a series of centrifugal pumps. These duties are rare in mineral processing plants themselves and although they are sometimes encountered in pumping tailings, they are more likely to be found when pumping concentrate or mine backfill and this is considered under the section on Slurry Transportation.

Pumpboxes

Although pumpboxes are not the prime subject of this chapter, they are an integral part of the process piping and slurry transfer system and some discussion on them is appropriate.

For instance, if a badly designed pumpbox causes periodic sliding of coarse material into the pump, this can lead to choking of the process piping. Thus, the pumpbox design has to be taken into account when considering the overall process piping and slurry transportation system. When varying flows are anticipated and vari-speed pumps are to be avoided, a fixed speed pump with a tall pumpbox is one method of overcoming the flow variations.

Another point to avoid is delivering slurries to the pumpbox too close to the suction of the operating pump, as this can result in air entrapment and inefficient operation of the pump.

Pumpboxes should generally be fitted with an overflow that directs any surplus slurry to where it is least inconvenient and also fitted with a quick release drain plug or valve for draining and flushing.

PROCESS PIPING LAYOUT AND DESIGN

In mineral processing plants the varying properties of the different slurries do not allow application of the horizontal rack principle as in oil refineries and many chemical plants. Nevertheless, every attempt is made to organize the process piping as follows:

- Minimum length run and number of bends and fittings.
- Good accessibility for maintenance and replacement.
- Supported from existing steelwork with a minimum of special support steel.
- Easily drained and vented.
- Routed clear of equipment requiring regular removal or maintenance such as welding.
- Arranged so that control valves, etc. can be easily operated or maintained with a minimum of special platforms and ladders.

Process piping layouts must be considered in the early stages of design in order to:

- Ensure sufficient vertical and horizontal clearances.
- Provide reasonably accurate data for the process pump calculations.

- Allow adequate time for preparation of the piping and installation in conjunction with the rest of the construction.

Starting at the feed end of the line, the first point is that the pump suction lines should cause minimum restriction to the flow and should therefore be short with a minimum of bends. Valves on suction lines can often be a significant restriction and should be avoided unless they are required for operation or maintenance reasons.

Generally the most demanding duty in mineral processing plants is the line between the cyclone feed pumps and the cyclones and if the problems of layout and abrasion for this piping system can be solved, then the layout of the other process piping is usually relatively simple. Experience has strongly suggested that in such a process line it is better to use only vertical and horizontal runs of pipe if at all possible. It is believed that sloping lines have the disadvantage of both vertical and horizontal lines in that settling can occur followed by sliding back of the settled material. This sliding often results in heavy abrasion and also a potential for choking the pipeline where the sloping line joins the horizontal line.

Thus, a typical cyclone feed pump could discharge horizontally using large sweep bends of abrasion-resisting material such as rubber hose to turn in the direction of the cyclones. Similarly the transition from the horizontal to the vertical should be made with abrasion-resisting sweep bends and also any transition that might be necessary from the vertical pipeline into the actual cyclone or cyclopack.

Where each ball mill uses a nest of cyclones, then the operating and standby cyclone feed pumps would both feed into a single cyclopack feed line using a valve of the Tech-Taylor type to bring the two pump lines together. However, where each ball mill uses only one or two cyclones, then it is often preferable to have a separate line between each cyclone feed pump and its cyclone or pair of cyclones.

Launders have been frequently used instead of piping for froth transfer, and they have the advantage of being able to handle varying flows of highly aerated material. However, the disadvantage of the vertical space requirements for launders often more than offsets their advantages as this can result in a taller and more expensive building.

Piping Engineering and Design

It has been often assumed in the past that it is sufficient to show simple lines on an engineering drawing to indicating the piping runs, leaving the field construction forces to decide the actual details of the piping run itself. However, it is now generally realized that it is more economic to detail the piping design and prefabricate the piping rather than field run it, particularly as the drawing office and pipe shop almost invariably have better working conditions than the construction site. This has led to the increased use of models of the mineral processing plant to show the piping arrangements and facilitate the detailed design of the pipe runs and location of valves and other fittings.

Warrantees, Performance and Specifications Required

Normally, warrantees and performance guarantees are not specifically applicable to process piping providing the material conforms to the particular ASTM standard or equivalent. However, where a new type of piping is being introduced for an unusual duty, then performance guarantees are often called for by the user who wishes to be covered against the early failure of the experimental material.

There are many variations of specifications, but those shown on Figure 7 are typical of the process piping specifications required in the design of mineral processing plants.

SLURRY TRANSPORTATION

Pipelines have been used as far back as the early 1900's and slurry transportation through a pipeline has been developed both in the mining industry and also in non-mining industries. The mining industry initially used slurry transportation to move tailings and concentrates. However, as tailings areas became further away from concentrators, longer tailings lines were required so that the tailings line became a slurry transportation system rather than an appendage on the end of the concentrator.

The use of pipelines to transport concentrate slurries has undergone development since the 1950's with installations such as Savage River in Tasmania being typical early examples.

Slurry transportation has developed from necessity when it was the only way of doing the job to the point where it has been selected for transporting materials over several hundred miles, even where alternative methods were available.

Apart from the fact it may be the only way, one of the main advantages of slurry transportation is that it has minimum labour requirements once it is constructed and in operation, whereas alternative forms of transportation are often labour intensive. In addition to this, a pipeline can traverse poor terrain, whereas a conventional transportation system may have extreme difficulty in doing so and have a significantly higher capital cost than a pipeline.

The operating costs of a pipeline are usually low and after construction is finished, the environmental impact of a pipeline is generally minimal. Also, the only escalation to which a pipeline is subject is usually that of the power required to propel the material through the pipeline. Some limits to the use of pipelining slurries are the inflexibility of transporting varying tonnages, the possible effect on the migration of wildlife and susceptibility to intentional damage.

During the analysis of alternative methods of slurry transportation, consideration will often be given to the implications of having many pumphouses versus a lesser number pumping at higher pressures. Frequently this is a matter of economics. With the use of multiple pumphouses, instrumentation and communications become important as during startup and shutdown or draining of the line, it is very necessary to have the proper sequencing of events.

The design of a long distance slurry transportation system requires basically the same data as process piping design in a mineral processing plant, except that the long distance slurry line requires the data to be known to a much greater degree of certainty. For this reason pipeline tests are often run, particularly where it is believed the operating limits of the pipeline will be quite narrow.

During the pipeline investigation consideration is often given to changing some of the base data, particularly the particle feed size. For example, if a coarse material has to be transported a long distance and requires a relatively high velocity, it can be economic to crush or grind the material to a smaller size, thus enabling the line velocity to be decreased with consequent savings in power, wear and general maintenance. Although fine grinding can sometimes lead to dewatering problems at the delivery end of the pipeline, it is particularly worth considering if, in any event, the material has to be crushed or ground eventually.

One of the limitations of pipelines is that it is very difficult to pump small tonnages over anything other than short distances, particularly if the materials are fairly coarse. With coarse material, high velocities are necessary which with small diameter pipes will lead to extremely high friction factors and line pressures. This will need either extremely high pressure piping and/or intermediate pump stations, both of these tending to be uneconomic when moving small tonnages.

Alternative Pumping Systems

The choice between centrifugal pumps and positive displacement pumps in slurry transportation is often determined by straight economics. Centrifugal pumps are favoured for high volumes and where abrasion is a significant factor and positive displacement pumps are preferred where high pressures are required. However, if abrasion is a serious factor, there are available positive displacement pumps that have greatly offset the problem of abrasion by keeping the slurry away from the plungers and cylinders by use of an oil slurry interface and an external valve system.

The alternative pumping arrangements generally considered for slurry transportation are:

- A single pumping station of centrifugal pumps in series - generally limited to 250 m total dynamic head.
- Single pumping stations of a positive displacement pump - up to 750 m total dynamic head.
- Several pumping stations of centrifugal pumps in series - each station limited to 250 m total dynamic head.
- Multiple pumping stations of positive displacement pumps - each station limited to 750 m total dynamic head.

REFERENCES

The subject of long distance slurry transportation and slurry transportation in general has been extremely well covered in the literature since the mid 1960's and a list of references is given. It is considered that this is of more benefit than attempting to work through a typical example, when in most cases the base data can only be realistically generated from a pipelining test.

- Transportation of Solids in Steel Pipelines; Colorado School of Mines, 1963.
- Link, J.M. "Slurry Transport of Industrial Minerals", 72H75 Society of Mining Engineers, AIME, February 1972.
- Job, A.L., "Transport of Solids in Pipelines with Special Reference to Mineral Ores, Concentrates & Unconsolidated Deposits". MR 68/115-LD, Dep't. of Energy, Mines & Resources, Canada, June 1968.
- Horne, C.A., "A Guide to Good Slurry Pumping Practice", Canadian Mining Journal, 2 parts November and December 1976.
- Volcado, J.J. and Charles, M.E., "Transportation of Slurries", (CIM) Bulletin, October 1972.
- McElvain, R.E. and Cave, I., "Transportation of Tailings", Tailings Disposal Today, Miller Freeman Publications, 1973.
- W.F. McDermott, "Savage River Mines - the World's First Long Distance Iron Ore Slurry Pipeline", (CIM) Bulletin, December 1970.
- Flintkote Co. Ltd., "Pipelining Proves Cheapest Way for Calaveras to Move Cement Slurry 17.6 Miles", World Mining, January 1972.
- Bougainville Copper, "Pumping Concentrate - 17 Miles", World Mining, October 1972.

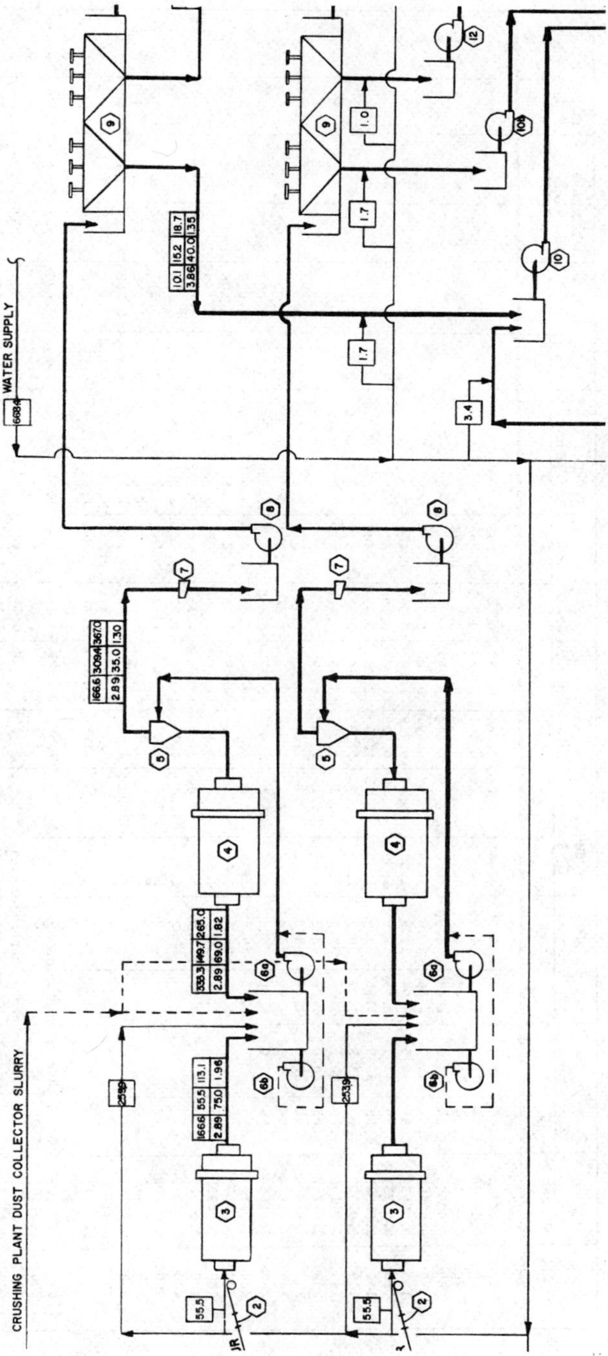

FIGURE 1

FLOWSHEET: GRINDING AND FLOTATION

TABLE 1:

LINE VELOCITY VS FRICTION HEAD LOSS - TYPICAL

USGPM	M^3/Sec.	2.067" I.D. 52.50 mm		2.469" I.D. 62.71 mm		3.068" I.D. 77.93 mm		4.026" I.D. 102.26 mm		6.065" I.D. 154.05 mm		7.981" I.D. 202.72 mm		10.02" I.D. 254.51 mm		12.00" I.D. 304.8 mm		13.25" I.D. 336.55 mm	
		V	h	V	h	V	h	V	h	V	h	V	h	V	h	V	h	V	h
50	0.00315	1.46	8.46	1.02	3.56	0.661	1.24	0.384	0.330	0.169	0.045								
60	0.00379	1.75	11.9	1.23	4.99	0.794	1.74	0.461	0.463	0.203	0.063								
70	0.00442	2.04	15.8	1.43	6.64	0.926	2.31	0.538	0.615	0.237	0.084								
80	0.00505	2.33	20.2	1.63	8.50	1.06	2.96	0.615	0.788	0.271	0.107								
90	0.00568	2.62	25.1	1.84	10.6	1.19	3.67	0.691	0.980	0.305	0.133								
100	0.00631	2.91	30.5	2.04	12.8	1.32	4.47	0.768	1.19	0.338	0.162								
120	0.00757	3.50	42.7	2.45	18.0	1.59	6.26	0.922	1.67	0.406	0.227								
140	0.00883	4.08	56.9	2.86	23.9	1.85	8.32	1.07	2.22	0.474	0.302	0.274	0.079						
160	0.01009	4.66	72.8	3.27	30.7	2.12	10.7	1.23	2.84	0.542	0.387	0.313	0.102						
180	0.01136	5.25	90.5	3.68	38.1	2.38	13.2	.38	3.53	0.609	0.481	0.352	0.126						
200	0.01262	5.83	110.0	4.09	46.3	2.65	16.1	1.54	4.29	0.677	0.584	0.391	0.154						
220	0.01388	6.41	131.0	4.49	55.3	2.91	19.2	1.69	5.12	0.745	0.697	0.430	0.183						
240	0.01514	6.99	154.0	4.90	66.4	3.17	22.6	1.84	6.01	0.812	0.819	0.469	0.215	0.298	0.071				
260	0.01640	7.58	179.0	5.31	75.3	3.44	26.2	2.00	6.97	0.880	0.950	0.508	0.250	0.322	0.083				
280	0.01767			5.72	86.3	3.70	30.0	2.15	8.00	0.948	1.09	0.547	0.286	0.347	0.095				
300	0.01893			6.13	98.1	3.97	34.1	2.30	9.09	1.02	1.24	0.586	0.325	0.372	0.108				
350	0.02208			7.15	130.0	4.63	45.4	2.69	12.1	1.18	1.65	0.684	0.433	0.434	0.143				
400	0.02524			8.17	167.0	5.29	58.0	3.07	15.5	1.35	2.11	0.782	0.554	0.496	0.183	0.346	0.076		
450	0.02839					5.95	72.3	3.46	19.3	1.52	2.62	0.880	0.689	0.558	0.228	0.389	0.095		
500	0.03155					6.61	87.7	3.84	23.4	1.69	3.19	0.977	0.838	0.620	0.277	0.432	0.115	0.355	0.071
550	0.03470					7.28	105.0	4.22	27.9	1.86	3.80	1.08	0.999	0.682	0.330	0.476	0.137	0.390	0.087
600	0.03785					7.94	123.0	4.61	32.8	2.03	4.46	1.17	1.17	0.744	0.388	0.519	0.161	0.426	0.100
700	0.04416							5.38	43.6	2.37	5.93	1.36	1.56	0.868	0.516	0.605	0.214	0.496	0.132
800	0.05047							6.15	55.8	2.71	7.60	1.56	1.99	0.992	0.660	0.692	0.275	0.567	0.170
900	0.05678							6.91	69.3	3.05	9.44	1.76	2.48	1.12	0.821	0.778	0.341	0.638	0.211
1,000	0.06309							7.68	84.3	3.38	11.5	1.95	3.02	1.24	0.998	0.865	0.415	0.709	0.256
1,500	0.09464									5.08	24.3	2.93	6.39	1.86	2.11	1.30	0.878	1.06	0.542
2,000	0.1262									6.77	41.4	3.91	10.9	2.48	3.60	1.73	1.50	1.42	0.924
2,500	0.1577									8.46	62.1	4.89	16.5	3.10	5.44	2.16	2.27	1.77	1.40
3,000	0.1893											5.86	23.0	3.72	7.61	2.59	3.17	2.13	1.96
4,000	0.2524											7.82	39.2	4.96	13.0	3.46	5.39	2.84	3.32
5,000	0.3155											9.77	59.3	6.20	19.6	4.32	8.15	3.55	5.03
6,000	0.3785													7.44	27.5	5.19	11.4	4.26	7.05
8,000	0.5047													9.92	46.7	6.92	19.4	5.67	12.0

V = Line velocity in m/s

h = Loss of head in meters for clear water per 100 m of standard weight straight steel pipe

TABLE 2:

PARTICLE SIZE VS CRITICAL VELOCITY
(TYPICAL ONLY)

	Particle Size (80% passing)	% Solids (by weight)	Critical Velocity m/sec.
Cyclone feed	1,000μ	46	1.9
Flotation feed	100	30	1.0
Rougher concentrate	80	25	0.8
Tailings	100	35	0.9
Final concentrate	50	60	0.6

Critical velocity - pipeline velocity at which settling occurs.

TABLE 3:

MILLER ABRASIVITY NUMBERS

Material	Miller Abrasivity Number
Oil well mud	10
Potash - Brine	11
Limestone	27
Magnetite	67
Copper concentrate	128
Backfill - minus 100 mesh	180
Pyrite	194
Corundum 400 mesh	241
Corundum 200 mesh	1,040

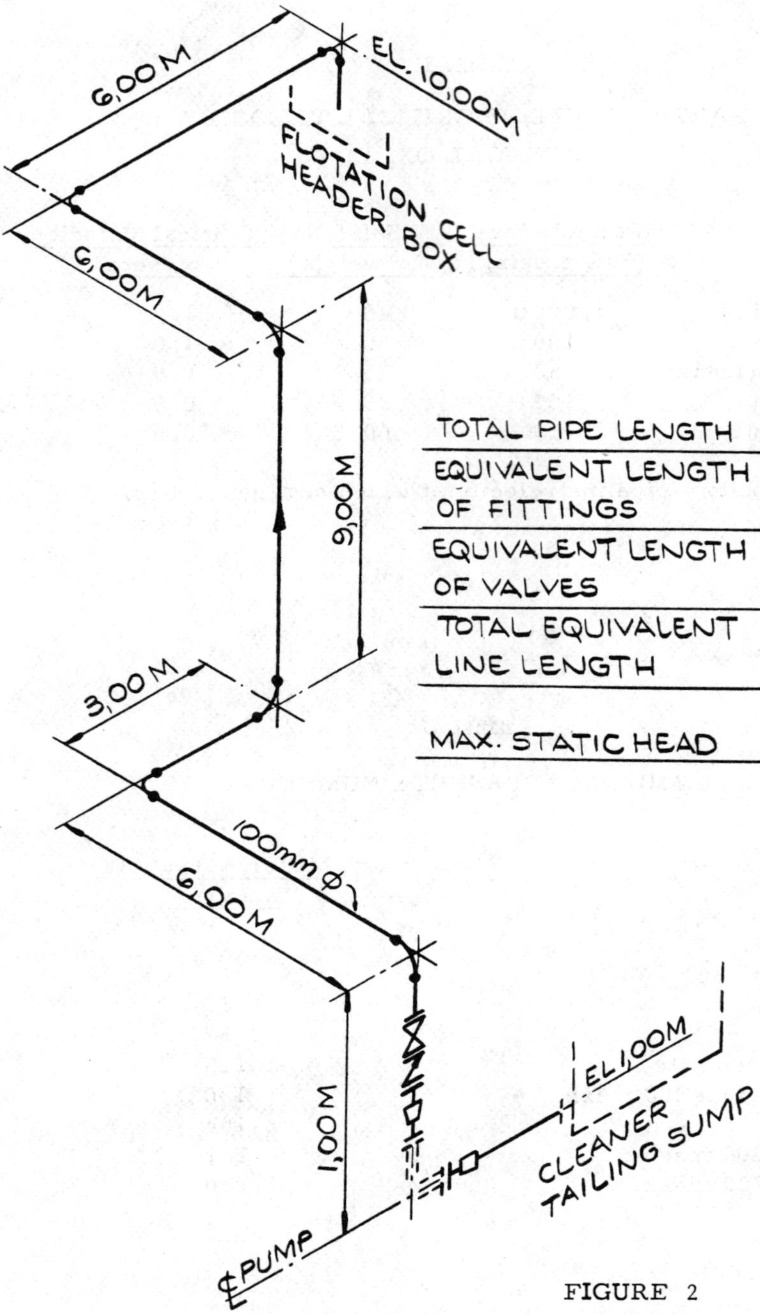

TOTAL PIPE LENGTH	30,00M
EQUIVALENT LENGTH OF FITTINGS	13,50M
EQUIVALENT LENGTH OF VALVES	15,00M
TOTAL EQUIVALENT LINE LENGTH	58,50M
MAX. STATIC HEAD	9,00M

FIGURE 2
ISOMETRIC PIPING LAYOUT

TABLE 4:
(Excerpts from Internal Report)

PLANT AND SITE CONDITIONS

This section describes conditions imposed by either the site or proposed plant designs as the Purchaser's preferred standards. The site information is freely offered to the best of our present knowledge. In any case where the Bidder's guarantee is dependent on the accuracy of this information, it is to be clearly understood that the Bidder is solely responsible for independently investigating and satisfying the Bidder's requirements for such basic design information.

The Bidder shall make note of the Purchaser's preference where specifically related to the materials or equipment offered and the tender shall confirm acceptance of these requirements or clearly state otherwise.

Where a specific instruction is given in the Requirements and Specification section, it shall take precedence over the following conditions, standards and codes.

DESCRIPTION AND LOCATION OF OPERATION

A copper mine located high in the Peruvian Altiplano in the Department of Cuzco, approximately 1,600 km south of the equator.

Conditions and Utilities

Plant ground floor elevation above sea level approximately 4,100 meters.

Outdoor temperature:

Summer:	extreme high:	18° C. D.B.	
	design:	___° C. D.B.	___° C. W.B.
Winter:	extreme low:	-14° C.	
	design:	-__° C.	

Temperature range of water supplied to mill: summer; winter 1° C.
Pressure of water supply: 4.21 kg/cm^2
Electric power for motors:
- 200 HP and less (including fractional HP) 240 volts, 3 phase, 60 Hz.
- 250 HP and up 4,160 volts, 3 phase, 60 Hz.

Electric power for control circuits ___ volts from individual controlled transformers.

TABLE 5:

PIPING SPECIFICATION SHEET - CYCLONE FEED

Item	Nom. Size Inches	Description	Specification
Pipe and Fittings	0.305 m (12 inches)	Std. wt. carbon steel, E.R.W.	ASTM A53. A/B
		Alternative 1 Rubberlined Terylene reinforced rubber hoses. Neoprene cover flanged ends.	See H.A.S. 4S-80.1
		Alternative 2 Rubberlined carbon steel pipe bend. Bend radius 5 x dia. flanged ends.	See H.A.S. 4S-80.2
Flanges	"	70 Kg forged steel. Slip-on drilled, flat face	ANSI B16.5 ASTM A181
Gaskets	"	8 mm thick red rubber	ANSI B16.21
Couplings	"	Split quick connect, self gripping coupling	Victaulic Style "99"
Bolts/Nuts	All Sizes	Carbon steel, American standard machine Class 2 fit	ASTM A307.B ANSI B18.2 ANSI B1.1

Instructions:

- Material and fabrication specification for rubberlined steel pipe and fittings. 4S-80.2
- Material and fabrication specification for rubberlined material conducting hose. 4S-80.1
- Erection specification for rubberlined pipelines. 4S-70.1

Chapter 34

DESIGN AND CONSTRUCTION OF TAILINGS PONDS AND RECLAMATION FACILITIES - CASE HISTORIES

J.D. Jones
Manager of Environmental Engineering

KILBORN LIMITED
Toronto, Ontario

ABSTRACT

Unique aspects of tailing pond design and reclamation facilities for three separate projects are described.

INTRODUCTION

Historically, man's approach to waste has been to dispose of it as effortlessly as possible with the absolute minimum of associated cost. Today the same criteria exist, waste disposal must be achieved at minimum costs but taking into consideration a multiplicity of constraints. There was a time, and not too long ago, when tailings would be discharged into the nearest lake or river; alternatively, if there were none convenient, an impounding area would be developed and the waste piled up behind dykes constructed from the tailings, to be abandoned when the ore ran out.

Today pollution in every form is the major concern of the public in general, and consequently, our governments'. The disposal of any type of waste is now given the utmost scrutiny and each day there seems to be new regulations controlling disposal. Since the mining industry produces such vast quantities of highly visible waste, much of the legislation is directed toward it and the disposal of tailings is being more and more rigidly controlled. Failure of embankments with the ensuing uncontrolled discharge of tailings is no longer tolerated. Consequently, the dams must be competently designed and properly constructed of sound materials. In most areas it is now necessary to collect the effluent from the tailings containment area for recirculation and to treat the excess before discharge to a receiving body. What was once a simple procedure, and the last to be considered in constructing a new facility, has become a major engineering problem requiring extensive studies and investigations.

The tailings disposal system for a modern development can represent a very substantial proportion of the capital cost as well as contributing very significantly to the long term operating costs; consequently a variety of inter-related investigations must be undertaken during the preliminary phases of the development to evaluate practical alternatives to arrive at the one representing the least cost over the life of the mine.

In locating the mill its proximity to the mine is of great importance the intent being to minimize the distance the ore must be transported. Similarly, the tailings impoundment area must be selected as close to the mill as practical and it is this selection process which requires the utmost consideration. Once the site has been identified and the operating procedures established, the facilities for its development can be designed, bearing in mind that ultimately it will be abandoned.

The design requirements for all the various components of a tailings disposal system have been adequately covered in innumerable articles, papers and text books (1) (2) and will not be discussed again in any detail. The intent of this paper is to outline some recent projects in which Kilborn has been involved where environmental constraints have dictated the need for special consideration in the design of the tailings containment area, the reclamation of process water, the treatment of tailings effluents and the stabilization of embankments in preparation of abandonment.

UNION MINIERE EXPLORATIONS AND MINING CORPORATION LIMITED (UMEX) Thierry Project, Pickle Lake, Ontario

The UMEX Thierry project is a 4000 TPD copper mine and concentrator located at Pickle Lake in north-western Ontario. It is a combined open pit and underground operation with one half the ore coming from the open pit during the initial three years of operation and all of the ore being obtained from underground operations over the remaining 12 years of the mine's expected life. Concentrate is produced by the standard flotation process requiring approximately 12 million litres of water per day of which 75% is reclaimed.

Environmental legislation in Ontario for base metal sulphide ores in force when the project was implemented required:

(a) the collection and treatment of all runoff from the plant site and waste rock pile before discharge to any watercourse

(b) the impoundment of all tailings

(c) adjustment of pH and precipitation of heavy metals prior to the discharge of tailings effluents to any watercourse

(d) a minimum 40% water reclamation for recirculation through the concentrator.

As an initial step in the implementation of this project, environmental studies were undertaken to establish the baseline conditions and from these investigations recommendations were made on how to minimize the impact of the development on the various ecosystems. Obviously then, they weighed heavily in the selection of the ultimate tailings impoundment area and determined to a large extent the treatment which would be necessary for the tailings effluent. The site selected was a swampy area at the top of a watershed where, by constructing the necessary embankments in such a way as to retain some of the swamp between the tailings area and the lake, a natural buffer would be provided for the treated effluent. The studies also identified that the aquatic ecosystems in the swamp and the lake were highly sensitive to fluctuations in pH levels with the natural pH ranging between 7.5 and 8.5. This then dictated that the tailings effluent with an induced pH in the range of 3.5 due to sulphide oxidation would have to be treated with lime.

Once the environmental studies had confirmed the suitability of the tailings containment area, extensive geotechnical investigations were initiated to confirm that the glacial till encountered at the mine site extended into this area; to determine foundation conditions for the design of the embankments and to identify sources of material for embankment construction. The investigations substantiated the existence of the till overburden which with a permeability in the order of 10^{-7} cm/sec would be ideally suited for tailings containment and also the construction of the water retaining structures.

The facilities required at the UMEX development for tailings disposal were quite extensive. To develop approximately 7.3 million cubic meters of storage in a swamp which by its very nature is quite flat required an embankment close to 800 meters long and about 10 meters high at the centre. A water reclaim system was required as well as treatment facilities for the effluent. The scheme that was devised is comprised of a main dam which forms the tailings containment area, a shorter and lower dam creating the reclaim pond, and a third and still smaller dam to form a retention and sedimentation pond for the treated effluent.

In the design of the main dam it was recognized that the primary purpose of the structure is to retain the solids and not the liquids. Consequently, a pervious structure was designed which could be constructed from the readily available sand in the area and waste rock from the mining operations. Further economies were realized with this type of structure in that it could consolidate after construction to a very considerable degree with minimal effect on its structural stability. This then permitted the dam to be constructed on top of the muskeg which eliminated the need for extensive stripping and site dewatering. It also required careful design though, which identified the need for berms on each side of the embankment to maintain the stability of the foundation during construction and which also identified the procedures to be used during construction. A typical cross-section of the main dam is shown in Fig. 1.

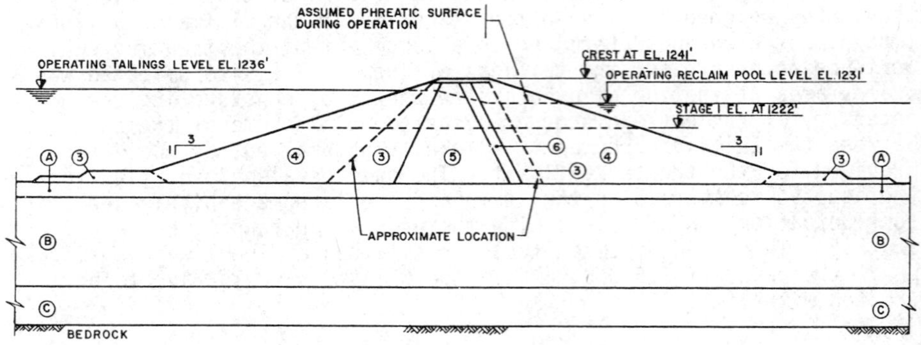

TYPICAL SECTION
TAILINGS DAM

LEGEND:

CONSTRUCTION MATERIALS

(3) FINE SELECT MINE WASTE
(4) COARSE SELECT MINE WASTE
(5) FINE SAND
(6) MINUS 1 INCH MINE WASTE

FOUNDATION SOILS

(A) MUSKEG
(B) SILTY FINE TO MEDIUM SANDS
(C) GLACIAL TILL

FIGURE 1

The secondary and tertiary embankments are water retaining structures which required a different approach in their design but the same care was taken to utilize the materials available at the site. A similar cross-section for the two structures was used and this is shown in Fig. 2.

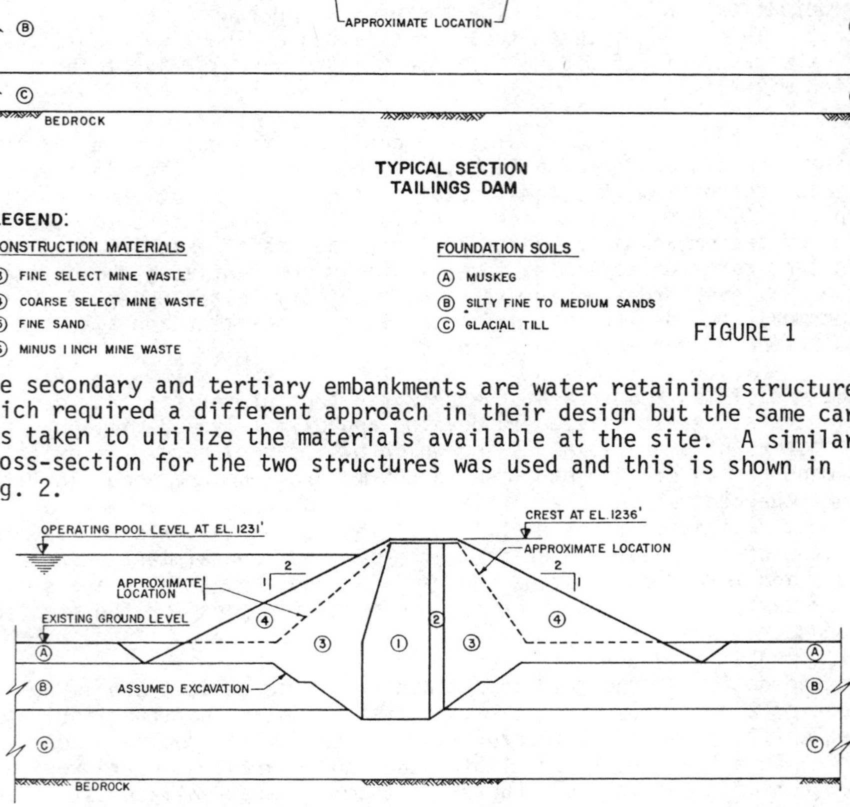

TYPICAL SECTION
RECLAIM DAM AND
RETENTION DAM

LEGEND:

CONSTRUCTION MATERIALS

(1) CORE: GLACIAL TILL FROM OPEN PIT
(2) FILTER: PROCESSED GRAVELLY SAND
(3) TRANSITION: FINE SELECT MINE WASTE
(4) ROCK SHELL: COARSE SELECT MINE WASTE

FOUNDATION SOILS

(A) MUSKEG
(B) SILTY FINE TO MEDIUM SANDS
(C) GLACIAL TILL

FIGURE 2

The obvious difference between the main dam and these water retaining structures is that the organic overburden beneath the latter has been stripped and the impervious till cores are keyed into the underlying till stratum.

In the operation of the system all of the drainage from the mine and the plantsite as well as the tailings is directed to the top end of the tailings containment area (3) where most of the solids settle before the effluent is decanted to the reclaim pond. From here the water is either reclaimed for recirculation to the mill or treated and discharged to the retention pond. The outflow is measured by means of a venturi meter in the spillway and slurried lime is automatically added to raise the pH of the effluent to 8.5 to provide for a rebound to 7.5, the natural pH of the receiving waters. The retention time in the final pond is 5 days to permit the precipitation and collection of heavy metals prior to discharge into the swamp and the watercourse.

The total capital cost for the tailings disposal system on this project was $6,000,000.

Comments

In the foregoing, the environmental constraints were known at the outset and although they added materially to the cost of the tailings disposal system, they could be accommodated, and with imaginative design and the utilization of locally available material some economies were achieved. Major problems arise though when existing systems must be altered to satisfy modifications to plant operations or to conform to changing regulations.

MADAWASKA MINES LIMITED
Bancroft, Ontario

Madawaska Mines is a 1500 TPD uranium mine and concentrator located near Bancroft in southern Ontario - the center of one of the most popular summer recreation areas in the province. The property was initially developed during the 1950's and operated until 1964 when the market for uranium diminished. The recent resurgence in uranium prices has made it economically feasible for the mine to be put back into production with operations starting again in 1976.

Subsequent to the closure of the mine, environmental legislation in respect to the disposal of radioactive wastes has changed considerably. The one significant change in respect to tailings disposal is that the desirable objective for radium 226 in the effluent from the tailings is now 3 pCi/l compared to 10 pCi/l initially. This can be reasonably attained by recirculating a substantial portion of the effluent from the tailings containment and treating that which must be discharged with barium chloride. The problem the reduced limit imposes though, is in the use of the old tailings impoundment areas.

When the plant was reactivated the choice for tailings disposal was between Pond No. 1 and Pond No. 2 as shown in Fig. 3. Pond No. 2 required extensive dykes be constructed adjacent to Bentley Creek with

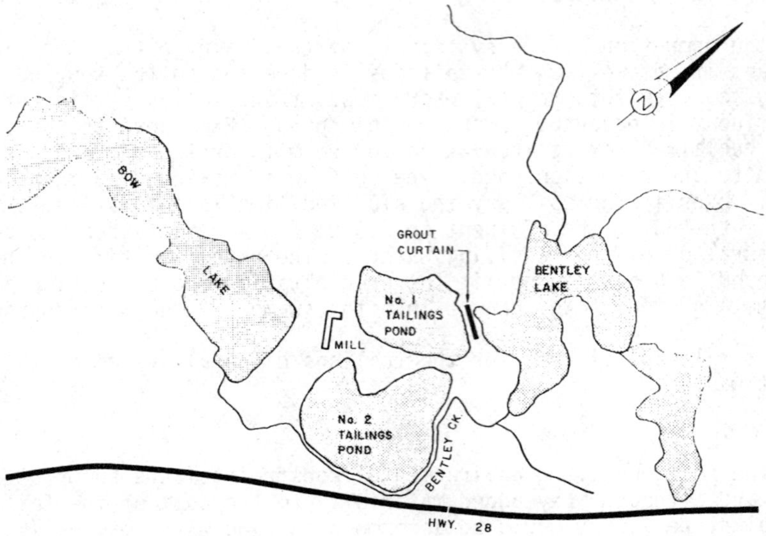

FIGURE 3

considerable reclamation facilities to prevent any additional seepage into the creek. Pond No. 1 is remote from the creek, but it had been developed on a deposit of sand and gravel which forms a direct connection to Bentley Lake where the levels of radioactivity exceed the maximum permissible as a result of the leachate from the tailings. To develop this area for tailings disposal required some form of cutoff be constructed to intercept any contaminated groundwater flow. For obvious economic reasons it was decided that Pond No.1 would be improved and the necessary remedial actions undertaken.

The Pond is located in a natural depression with rock out-cropping close to both the east and the west ends and along both the north and the south sides. At the west end the overburden is quite shallow and here it has been possible to intercept the groundwater flow by constructing a concrete cutoff wall to bedrock. The west end has presented much greater problems for here the overburden ranges up to approximately 70 meters in depth and is comprised of sands and gravels with large boulders. All of the known techniques for constructing an impervious cutoff, from sheet piles to slurry trenches were considered but in the final analysis overburden grouting seemed to be as good a solution as any of the other alternatives and could be achieved much more economically.

The operating procedure that was finally adopted was to drill and grout three rows of holes across the valley. The upstream and downstream rows were to be grouted first, using a clay/cement mixture followed by chemical grout in the middle row. The curtain would extend from bedrock up to a depth where the grouting pressures produced excessive ground heave. To complete the cutoff, a trench was excavated to expose the top of the grout curtain and backfilled with concrete. To date, the upstream and downstream rows have been grouted and the concrete cap installed. Pumping tests have also been carried out to evaluate the effectiveness of the curtain and groundwater samples have been taken to determine if there has been any reduction in radioactivity.

The results of these tests are quite interesting. The pumping tests show conclusively that the curtain is not continuous. Drilling and grouting operations commenced on the west side of the valley with all three rows of holes being grouted at the overburden/ bedrock interface and the two outer rows of holes were grouted to the surface. During this period there was much experimenting with mixes, pressures and procedures so that by the time the work had progressed toward the centre of the valley both crews and equipment were working proficiently, and as the tests prove, effectively. There appears to be little seepage through the easterly 2/3 of the curtain, but there is definitely seepage through the westerly 1/3, the area of experimentation. But the work has been successful to some degree in that the groundwater level downstream of the curtain is lower than on the upstream side. An ancillary benefit of the curtain, of which we were not aware when the program was initiated, is that radium 226 has an ion exchange affinity for some types of soils and the clay in the grout seems to be reacting to reduce the amount of the element being conveyed past or through the curtain. Analysis of water samples obtained above and below the curtain clearly show the radioactivity to be less on the downstream side.

Comments

When the Madawaska Mine was shutdown, the tailings were covered with gravel to prevent wind erosion but no particular precautions were taken to safeguard the environment from any harmful affects of the effluent eminating from the containment area. Throughout the north, and the world for that matter, there are abandoned tailings disposal areas which continue to give off unacceptable levels of polluting agents. With the public generally having easier access to remote areas these abandoned disposals sites are becoming more widely known and are generating considerable concern, to the point, in fact, where some jurisdictions are now insisting upon reclamation prior to abandonment and probably all will in the future. In Ontario for instance the gravel producers are required to submit a final grading plan before a permit is issued to open a pit.

The catchment area tributary to Pond No. 2 is approximately 130 ha which can generate substantial inflows. This was recognized and a spillway to provide discharge capacity in addition to the decant system was an integral part of the embankment from its inception. But the meteorological conditions preceding the failure were extreme and certainly unusual for the area in the middle of winter. By mid-December the Gaspe had been exposed to a normal winter. There was about 200 cm of snow on the ground and the ponds were covered with about 25 cm of ice. On the 13th the temperature began to rise to well above freezing with rain saturating the snow pack. These conditions prevailed until the 21st when the temperature exceeded 10° C and was accompanied by a very heavy rainfall. The resulting runoff was extreme and the inflow to the pond exceeded the spillway capacity. As a result the water in the pond continued to rise until it flowed over the lowest point on the crest of the dam and erosion commenced. The breach in the dam was about 50 meters wide and extended to bedrock, through about 10 meters of overburden. The onrush of water and tailings followed the route of the watercourse as shown in Fig. 4 washing out a considerable portion of the mine access road, power lines, pipe lines and the fresh water pumphouse. The breach was plugged immediately to contain the tailings remaining in the pond and permanent repairs were initiated and completed early in 1974.

As a result of the failure many studies were undertaken and one for the Province culminated in a number of recommendations along with the following relating to the ultimate abandonment of the tailings disposal areas;

1. The abandonment of the tailings disposal sites to be planned to guard against any ponding of water on the tailings.

2. A decant system or spillway to be installed with sufficient capacity to discharge any runoff resulting from rainfall or snow melt.

3. The embankments to have a minimum of 5 feet of freeboard above the calculated water level.

4. The embankments be modified to ensure their long term stability.

5. The sites to be covered with soil to encourage re-vegetation.

A schematic representation of the details developed for the abandonment of Pond No. 1 are given in Fig. 5 which are also representative of the work to be undertaken for Pond No. 2.

With adequate pre-planning of the costs associated with abandonment can be minimized but restoration after the fact can be very expensive.

MADELEINE MINES LTD.
Gaspe Nord, Quebec

Madeleine Mines is a copper mine and concentrator located in the Gaspe Peninsula about 30 kilometers south of St. Anne-de-Mont in the Gaspesie Provincial Park. The mine went into production in 1969 with ore reserves which, when processed at the mill capacity of 2500 TPD, would last until about 1980. In 1976 with the decline in the price of copper, the mill was temporarily closed with the intention of starting up when the operations again become economical.

In December of 1973 attention was drawn to the mine when an embankment containing the tailings failed with the subsequent contamination of the north-east branch of the Ste. Anne River, one of the better salmon streams in the area. Investigations at the time indicated that the dam, constructed of locally available till and mine waste, was structurally sound as was its foundation and that the failure was the result of extremely high water level in the pond and the subsequent over-topping of the embankment.

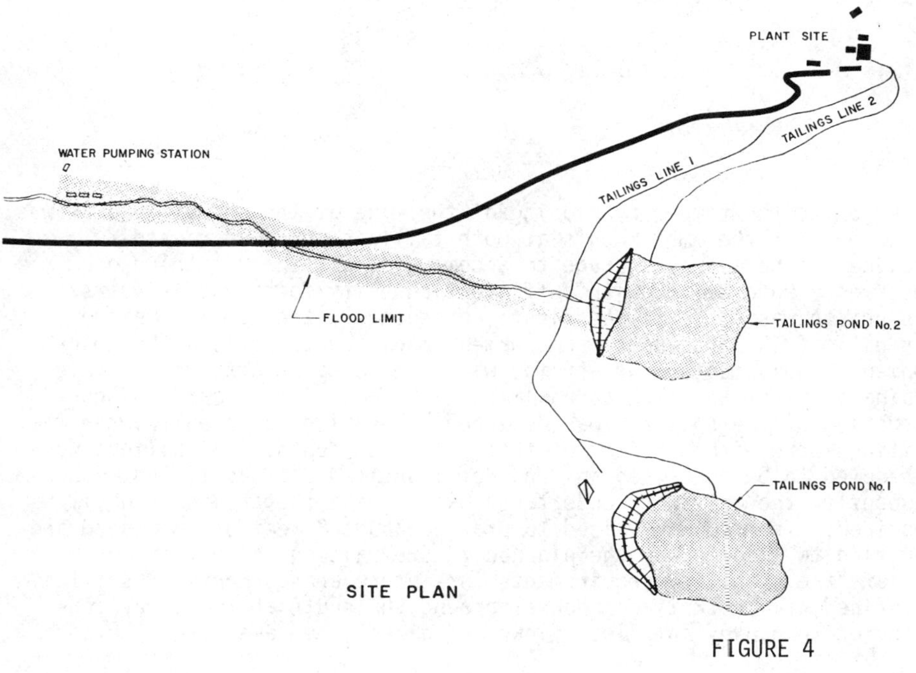

FIGURE 4

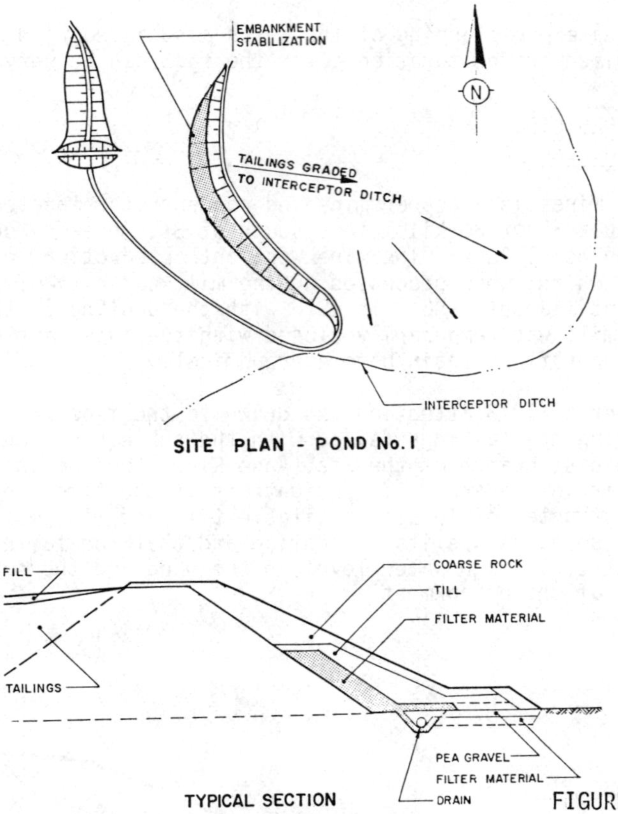

FIGURE 5

Prior to the temporary closing of the mine at the end of 1976 it was intended that the embankments at both tailings ponds be raised to provide the necessary storage to accommodate all the remaining ore reserves. Pond No. 2, with sufficient capacity for about 1½ years production was to be filled first; consequently the dam was raised to its ultimate height and the embankment modified to satisfy the abandonment requirements. No attempt was made to construct the ultimate drainage system and with approximately 25 feet of freeboard the construction of the spillway was deferred, scheduled to be built when the tailings were within 5 feet of their ultimate depth. No tailings were scheduled to be deposited in Pond No. 1 until No. 2 was filled. Consequently, the only work undertaken was to make the area safe until required. The dam was raised to provide about 8 feet of freeboard and modified to accommodate the planned future raising as well as to satisfy the stability requirements for future abandonment. A spillway comprised of a rock lined channel around the south abutment was constructed to convey any flood flows and direct them away from the toe of the embankment. Nothing was done to incept any runoff, this also being deferred until ultimate abandonment.

The material required to improve the stability of the embankments was placed on a contract basis at considerable cost. Had the abandonment requirements been known when the project was initiated the work could have been phased over the life of the project and constructed by mine personnel, as the original embankments were. The more efficient use of material, men and equipment would most certainly have resulted in more economical structures and reduced costs.

CONCLUSIONS

Tailings disposal is no longer a simple matter requiring a minimum of consideration. The necessary facilities must be given a great deal of thought and their design integrated with that of the process and the mill. The long-term operation of tailings impoundments requires due consideration at the outset and long range plans developed for its operation and expansion in order to minimize the costs of abandonment.

To arrive at that point will require the input of a multi-discipline design team capable of recognizing all the various constraints and dealing sensibly with them to produce an economical and practical disposal system.

Proper planning at the outset can minimize capital costs for the construction and operation of the tailings disposal system and eliminate the need for extensive and expensive remedial measures later on. There is nothing new in the approach to be taken in designing from the outset to satisfy all government constraints but if any is ignored it can prove to be an expensive afterthought.

REFERENCES

1. Aplin, C.L., Argall, G.O. (1972) "Tailing Disposal Today", First International Symposium on Tailings, Tucson, Arizona.

2. Klohn, E.J. (1972) "Design on Construction of Tailings Dam" Canadian Mining and Metallurgical Bulletin, April 1972.

3. Robinsky, E.I. (1975) "Thickened Discharge - A New Approach to Tailings Disposal". Canadian Mining and Metallurgical Bulletin, December 1975.

Chapter 35

ENVIRONMENTAL CONSIDERATIONS IN MILL SITING

by Ronald E. Versaw

Senior Engineer
Dames & Moore, Denver

ABSTRACT

Environmental legislation, both existing and proposed, requires that mills, tailings dams, evaporation ponds, and related processing facilities be located, constructed, and operated within stringent guidelines. An early step in planning a new processing facility is performing a site selection study in which topographic, hydrologic, geotechnical, and environmental parameters are evaluated for several alternate sites. Once a site is selected, then an environmental report, including a description of the proposed action, alternatives, baseline surveys, and discussions of potential environmental impacts, mitigation measures and monitoring programs is required. The federal legislation, siting studies, and environmental studies (along with the implications of these as they affect design) are discussed. In addition, costs for performing these studies and implementing control of effluents are discussed.

INTRODUCTION

Since passage of the National Environmental Policy Act (NEPA) in 1969 and the disastrous failure of the solid waste disposal facility at Buffalo Creek, West Virginia, 1972, the Federal Government has become involved in regulating the environmental and safety aspects of the mineral industry. The concern for public safety and environmental issues has resulted in the creation of several federal laws, regulations and agencies to address these concerns. As a result a mining company planning a new mill and waste disposal facilities will have to satisfy a host of governmental agencies and have plans that comply with all applicable environmental and safety regulations prior to obtaining permits to project construction and operation.

This discussion reviews some of the regulations which a mining company must be aware of when planning a new project and the actions the actions which must be taken by that company in order to comply with the regulations. Specifically, this discussion summarizes existing legislation and describes what is required for site selection studies, preparation of environmental assessment reports and compliance with safety guidelines. In addition, the cost for performing these types of studies and the costs of implementing environmental and safety controls at waste disposal facilities are presented.

It is beyond the scope of this paper to discuss environmental and safety regulations which control the design and operation of the mill itself. Instead, the main emphasis will be on waste disposal facilities. Also, the scope of this paper includes only federal legislation regarding mill and waste disposal facilities. In many locations, state or local environmental laws may be more stringent than federal laws and compliance with regulations from local level through federal will be required. However, this subject is far too complex and dynamic to discuss herein.

LEGISLATIVE BACKGROUND

NATIONAL ENVIRONMENTAL POLICY ACT

The federal government's role in environmental matters was established by passage of the National Environmental Policy Act in 1969. As stated in Section 2 of this Act, "The purposes of this Act are: To declare a national policy which will encourage productive and enjoyable harmony between man and his environment; to promote efforts which will prevent or eliminate damage to the environment in biosphere and stimulate the health and welfare of man; to enrich the understanding of the ecological systems and natural resources important to the Nation; and to establish a Council on Environmental Quality (1)."

The part of this Act which has the most direct effect on the mining company is Section 102, where it is stated that any action which is deemed a major federal action will require the preparation of an environmental impact statement (EIS). The responsibility for preparing an EIS rests on whichever governmental agency is denoted as the "lead agency" for a particular project. In most cases, the lead agency requires the applicant (in this case the mining company) to provide an environmental assessment report (EAR) which the lead agency, at its option, may use in part or in whole as the basis of its own EIS (2). Although a controversy can occur over whether or not a project will be deemed a major federal action under the guidelines of NEPA it is safe to assume that virtually any new mining project involving federal leases or surface will be considered a major federal action. The operator can expedite both his internal planning and development and the regulatory processes by commencing the site selection and environmental studies as soon as feasible. Performing the necessary siting studies, preparation of the EAR and EIS and obtaining final approvals commonly takes at least 2 to 3 years. All of this has to be completed prior to beginning construction.

ENVIRONMENTAL LEGISLATION

To provide legislative guidelines for enforcement of environmental policy, several acts of Congress have been passed which regulate and control mineral processing and waste disposal. The four major Public Laws (PL) which regulate milling and waste disposal are:

PL 92-500, Federal Water Pollution Control Act
PL 92-253, Safe Drinking Water Act
PL 94-580, Resource Conservation and Recovery Act
PL 93-319, Clean Air Act

From each of these laws, rules and regulations are developed to implement the laws and to define those steps or measures necessary to comply with these laws. All of these laws require a permit for discharges into the environment.

Federal Water Pollution Control Act (PL 92-500)

The Federal Water Pollution Control Act (3) regulates effluent discharge into surface waters. The law states that existing mine and mill discharges must be neutral in pH and contain no toxic or hazardous substances. Future mines and mills will be required to use the best practicable control technology available to minimize discharge of pollutants (4). Ideally, the intent of this law is to require a closed system with no discharges into the environment by 1985. Also, the Environmental Protection Agency (EPA) desires to expand enforcement of this law to include requirement for a contingency plan for possible spills from waste disposal structures (5).

Safe Drinking Water Act (PL 92-253)

The Safe Drinking Water Act (6) protects public drinking water supplies. Part C, "Protection of Underground Sources of Drinking Water," is pertinent in its application to contamination of underground aquifers. EPA proposed regulations (7) state that underground injection includes any contaminant which may be put below ground level and moves from the point of introduction. Seepage from waste disposal facilities and mills therefore would fall under this category, and this type of seepage must not pollute sources of underground water supply. That is, the operator will be required to prove that any seepage or spills from his processing facilities will not pollute sources of underground water supply.

Resource Conservation and Recovery Act (PL 94-580)

The Resource Conservation and Recovery Act was passed in October 1976 (8). Regulations should appear within 18 months of that date.

The law will apply to all existing and proposed waste disposal facilities. The law is comprehensive and essentially covers all environmental concerns relating to solid waste disposal and land use that are not covered by existing laws. The law is commonly referred to as the Solid Waste Disposal Act.

The law states that hazardous wastes must be kept together whenever possible; inventoried; monitored; not allowed to escape into the air, water or environment and be placed in a form easily retrievable for reprocessing to recover additional minerals. Definitions of hazardous wastes will be published in the regulations but mine and mill wastes would, in all likelihood, be considered hazardous. Clearly this act has far reaching consequences for the mining industry.

Clean Air Act (PL 93-319)

The Clean Air Act (9) can also apply to process waste disposal. Ambient air quality standards for particulates have to be maintained and, in the case of a waste disposal facility, dust cannot be allowed to cause a significant degradation of air quality. In the past, this law has not been a crucial problem to operating waste disposal facilities. However, the fine character of the slime portion of most tailings can cause dust problems, and it is likely that tighter controls and enforcement of regulations will be applied to this aspect of processing in the future.

REGULATORY AGENCIES

Three governmental bodies have been assigned the primary responsibility for regulating and monitoring mill and waste disposal operations. These are the Environmental Protection Agency (EPA), the Mining Enforcement and Safety Administration, Department of the Labor (MESA), and the Nuclear Regulatory Commission (NRC). Each of these agencies is responsible for different aspects of the milling and waste disposal as outlined below.

EPA

The EPA is primarily concerned with air and water quality regulations. However, it is likely that their jurisdiction will be extended to regulation of solid waste disposal when the regulations for the Resource Conservation and Recovery Act take effect. Therefore, the EPA is involved in the siting, operation and reclamation of processing facilities.

The intent of the EPA is to assure that the best reasonable technology, given the specific site and environmental conditions, is applied to lessen or alleviate environmental problems associated with the mill and waste disposal facility. From an operator's point of view, application of the best possible technology presents a problem, because this technology is constantly changing. Therefore, the guidelines and regulations are subject to continual change.

MESA

MESA, which is currently a part of the Department of Interior, is responsible for mining safety. Following the disastrous Buffalo Creek failure in West Virginia in 1972, Congress extended MESA's responsibility to refuse piles and impounding structures associated with coal mines (10). At the present time, legislation pending before Congress would place MESA in the Department of Labor and extend its jurisdiction to all metal and non-metal refuse piles and impounding structures.

NRC

The NRC's jurisdiction is limited to uranium processing facilities. At the present time with the surge of activity in the uranium mining industry, NRC is the primary regulatory agency for a large percentage of the mills presently in planning stages. The NRC's guidelines which primarily affect mills are summarized in Regulatory Guides 3.5, 3.8, 3.11, 3.23 and 4.14. These guides include all of the EPA guidelines plus specific environmental regulations for radioactive materials, operations monitoring regulations and design regulations. In addition to the safety requirements of MESA, which also apply to uranium processing facilities, the NRC also has safety regulations specially for containment of radioactive tailings. The environmental and safety regulations of these agencies as they affect the siting of mill facilities are discussed in subsequent sections of this report.

SITE SELECTION

Once a go ahead decision has been made on a project, the first step should be a site selection study. The objective of the site selection study is to select the most advantageous site or sites from the standpoint of technical, economic and environmental factors. Although the site selection process is predominantly an engineering study, environmental concerns must be included in the study in order to select the most appropriate site. The environmental, topographic, hydrologic, and geotechnical studies which need to be completed in a proper site selection process are discussed in the following paragraphs. Essentially site selection studies are desk studies aided by reconnaissance level site surveys. Undesirable sites are eliminated from the start so that monies and time are not wasted doing environmental assessment and engineering studies at unlikely sites.

In addition to selecting the best site for a tailings dam, regulatory agencies (especially the NRC) are now requiring that alternative methods of disposal be considered. Such methods might include backfilling a mine, excavating a special pit, and transporting off-site to a more suitable site. An adequate site selection study must include making engineering, environmental and economic comparisons of alternatives to conventional tailings dams.

ENVIRONMENTAL FACTORS

The initial site selection studies should consider environmental factors an integral part of the site selection process. A site which may be considered ideal from an engineering point of view may not be satisfactory from an environmental point of view, with the result that necessary construction and operating permits may not be granted. Types of environmental factors which should be considered in the site selection process include vegetation patterns, wildlife abundance and habitats, potential impacts on hydrology and water quality aesthetics, proximity to residences or recreation, special use areas, historically or archaeologically significant sites, and social economics. Closely allied to environmental issues are zoning regulations and land use planning, which would normally have to be considered at local governmental levels but nevertheless have to be evaluated as part of the project planning.

In general, environmental concerns are greatest for those sites where a unique environmental situation occurs. If a given waste disposal site would destroy a large portion of the habitat of certain plants or animals then that site would be less environmentally favorable than an alternate site which was covered by plant or animal life abundant at other locations in the general area. In extreme cases, entire projects have been blocked because of one single species' habitat which would be either eliminated or partially destroyed by the construction of a facility (12).

TOPOGRAPHIC FACTORS

Probably the easiest and least expensive way to rank sites according to their relative desirability is according to topographic factors. The types of topographic information which should be obtained at potential tailings disposal sites include comparisons of the volume of materials required to construct the tailings embankment for a given volume of tailings storage capacity, comparisons of the surface area versus volume of tailings storage, the amount of grading required for road construction and site preparation, the area of drainage basin above disposal site, required diversion of surface runoff to protect facilities and prevent downstream contamination, and transport of tailings from the mill to the tailings disposal area.

The essential item required to perform this phase of a site selection study is detailed topographic mapping. Although U.S.Geological Survey 7 1/2 minute and 15 minute quadrangle maps may be useful for making broad appraisals of potential sites and determining approximate distances from mine to mill sites, they are of little value in performing the more detailed aspects of site selection based on topographic factors. For detailed calculations, topographic mapping at a scale of approximately 1 inch = 200 to 500 feet is required.

HYDROLOGIC FACTORS

The major hydrologic factors which should be considered in the site selection process include potential surface water flows into the tailings disposal area; flooding hazard at the site; access to a supply of water for use in the milling operations; potential construction problems for the mill and tailings disposal areas; precipitation data including at least 100-year and probable maximum precipitation; evaporation data coupled with water balance data for mill; and proximity, depth and character of underground aquifers.

Every mill site situation is different. However, the safety of the system is of certain common concern to all disposal facilities. Therefore, flooding potential is always a primary consideration. Normally the smaller the drainage area above a site the better the site, since the flood size would be less.

GEOTECHNICAL FACTORS

The principal geotechnical factors used in a site selection analysis are availability of materials for site grading, embankment construction and reclamation; containment or control of seepage from mill and tailings disposal areas; permeabilities of in situ soil/bedrock and planned borrow materials; foundation support for mill facilities and the tailings embankment; required site preparation for construction of the tailings disposal facility; site seismicity; and locations of major geologic structures such as faults, joints, solution cavities and the like which could lead to instability or excess seepage.

An obvious aid to geotechnical evaluation of the site is good geologic mapping. The geotechnical site investigation may also require excavation of test pits, drilling holes and/or geologic mapping. The geotechnical field work done during this phase does not have to be as thorough or as complete as is required for final design. But, the work has to be done with sufficient detail to allow reliable review of alternative sites and selection of best sites.

SUMMARY

Careful consideration of the environmental, topographic, hydrologic and geotechnical factors of alternative sites is necessary to make a thorough evaluation of the most favorable sites. Much of the information collected during this phase of the investigation can be used in the more detailed design studies which will be required prior to construction and operation. The site selection information will also provide necessary input to the environmental assessment report discussed in the next section of this paper. For a typical project, the site selection study might require from 4 to 8 man weeks of effort and cost in the range of $10,000 to $20,000.

ENVIRONMENTAL ASSESSMENT REPORT

Once a site has been selected and plans are being finalized, the next logical step is preparation of the environmental assessment report. As previously indicated, the EAR has to be prepared in accordance with the requirements of the lead agency responsible for the project. The EAR may then be used in part or whole to enable the lead agency to prepare an environmental impact statement required by NEPA.

The environmental assessment report is typically a several-hundred page document which thoroughly discusses all aspects of the proposed project, the environment in which the project will be constructed, and the effects that the project will have on its environment. Although each department and agency has somewhat different requirements for the environmental assessment report, the reports are generally similar (13,14,15,16). The following topics are usually required in an EAR.

DESCRIPTION OF PROPOSED PROJECT

This section would normally include a description of the mine and orebody; discussion of land ownership and lease agreements; descriptions of project facilities such as water supply, power supply, roads, sanitary facilities, buildings and structures; descriptions of mining operations including equipment (if applicable), processing operations, and tailings disposal system; description of packaging and transportation; and discussions of plans for post-operational disposition of facilities and reclamation.

Detailed project planning and site selection studies have to be completed prior to completing this section of the environmental report. Besides just meeting the requirements of preparing an environmental report, this type of planning is useful to the mine/mill operator and probably would be completed even if not required for the environmental assessment.

DESCRIPTION OF EXISTING ENVIRONMENT

This section of the EAR describes in detail the physical, biological and social economic environment into which the proposed project will be placed. These site specific descriptions of baseline conditions commonly require substantial amounts of field time to complete.

Under the description of the physical environment, the following items should be discussed: climatology, including regional and site climatology; geology, including site topography, stratigraphy, geological structure, soils, seismicity, description of mineral deposits to be mined and description of other mineral resources; surface water hydrology including regional surface water drainage, local surface water drainage, quality of surface water and surface water impoundments; ground water hydrology, including the occurrence of aquifers, characteristics of local aquifers, ground water conditions at the site, and quality of ground water.

A description of the biological environment necessitates inventories of the terrestrial and aquatic biota and analysis of ecosystems, production and structure. These baseline studies normally require at least a year of field work, so that studies and sampling of the flora and fauna can be made during all seasons of the year. The format of these descriptions includes lists of all the plant and animals present at a site and discussions of the interrelationships of these flora and fauna to one another and to their environment.

The social and economic environment section of the report necessitates a thorough discussion of the regional and local history of the site; a description of the demography of the site including population density, population distribution, and population projections; land use descriptions, both regional and local; water use descriptions, both regional and local; water use descriptions including surface water and ground water; economic bases including the regional and local level and descriptions of housing, services, culture and aesthetics.

ENVIRONMENTAL IMPACTS OF PROPOSED PROJECT

This part of the report describes the impacts which a milling operation would have on the environment described in the previous section of the report. In effect, each of the subsections which were used to define the baseline environmental conditions are again addressed in terms of impact the project will have on that particular part of the environment. In this section of the report impacts on such things as threatened and endangered or protected species, aesthetics and the like become of particular significance; and a tie-in to the site selection studies is clear. That is, during the site selection study such things as a site where a unique environment or threatened, endangered and protected species are located would have been eliminated.

MITIGATING MEASURES, MONITORING AND RECLAMATION

This section of the report describes what measures will be taken during the construction and operation of the project to lessen the environmental impacts of the project. For instance, a possible mitigating measure might be paving access roads or watering haul roads to prevent dust; or erosion may be controlled by revegetating slopes; or water consumption might be reduced by recycling more water from the tailings dam.

Monitoring programs also have to be described. A sampling program has to be completed prior to project authorization as part of the baseline studies. The monitoring has to be continued on a periodic basis throughout the life of the project to measure and control effects the project may be having on the air or water quality.

Finally, a description has to be given of the stabilization and reclamation plans following milling and waste disposal operations. Details have to be given regarding how the reclamation will be accomplished, what methods and plant species will be used to revegetate and stabilize areas and what amounts of capital and long-term maintenance costs will be associated with the stabilization and reclamation. In most cases bonding will be required to insure that adequate funds are available to complete the stabilization and reclamation, if the operator fails to do so.

UNAVOIDABLE ADVERSE IMPACTS

Unavoidable adverse impacts include those items which cannot be mitigated or eliminated by careful project planning and construction practices. Typical examples of unavoidable adverse impacts are disruption of vegetation at the tailings disposal site, increases in dust and surface water erosion from bare land areas, and an increase in the ambient sound levels.

SHORT-TERM USE VERSUS LONG-TERM PRODUCTIVITY OF ENVIRONMENT AND IRREVERSIBLE AND IRRETRIEVABLE COMMITMENTS OF RESOURCES

These two sections of the EAR are normally relatively short sections which describe how the environment will be used during the milling operations and how the productivity of the environment will be reestablished following mining. The irreversible and irretrievable commitments of resources include discussions of the depletion of the mineral resources, fuel consumption, water consumption, discussions of the changes in or loss of biotic habitats and range resources that are destroyed, and also descriptions of the cultural resources which are changed by the project. The latter includes changes in land use, aesthetics, and archaeological and historical sites which may be affected.

BENEFIT COST ANALYSIS

This section describes the quantifiable external benefits of the proposed action and the quantifiable external costs. Generally the benefits of a project include increased employment, higher standard of living, increased tax base and the benefits that society at large gains from production of the product of the mill. The cost of the project includes both actual costs of developing the project as well as less quantifiable costs such as disruption of the biotic habitats, changes in land use, or interference with archaeological and historical sites.

ALTERNATIVES TO PROPOSED ACTIONS

This section directly ties in with the site selection studies previously discussed. Alternatives that have to be considered in this section are no action, alternative processing methods and alternative locations of processing facilities, that is, locating the tailing disposal system in a different area or such things as placing tailings back in the mine or even transporting tailings off site to another location. The purpose of this report section is to demonstrate that the proposed action is the best action from technical, economic and environmental points of view.

ENVIRONMENTAL APPROVALS AND PERMIT REQUIREMENTS

This last section of the EAR includes a listing of all licenses, permits and other approvals for construction and operation required by federal, state, and local authorities. Along with the listing of the permits and licenses, it is usually necessary to provide an indication of the status of these permits; if they have already been granted or if they have yet to be granted.

SPECIAL REQUIREMENTS FOR URANIUM PROCESSING FACILITIES

The Nuclear Regulatory Commission has published a series of regulatory guides and Regulatory Guide 3.8 of that guide is entitled "Standard Format and Content of Environmental Reports for Uranium Mills." The environmental report (ER) required by the Nuclear Regulatory Commission is essentially similar to the EAR required by most other governmental agencies. The major difference is that careful consideration has to be given to radiological impacts of the proposed processing facility. In particular, information has to be obtained on background radiation and positive action plans have to be developed for monitoring and controlling the spread of radiation into the physical and biological environments surrounding the mill and waste disposal facilities.

SUMMARY

The description of tasks required to complete an environmental assessment report indicates that a thorough multi-disciplined study is required. For this reason and others, such as expediency and benefit of third party approach, environmental assessment reports are typically done by outside consulting firms such as Dames & Moore. An environmental assessment report will typically require on the order of 18 months to complete with a cost in the range of $300,000 to $500,000 for most typical mine/mill projects. However, special site problems and necessity for public hearings and reviews can escalate the environmental report preparation cost and extend the time for at least an additional year.

SAFETY CONSIDERATIONS

As previously indicated, MESA is the governmental agency which probably will have been granted by 1978 the authority to regulate the design and construction of waste disposal facilities from a safety standpoint. In the case of uranium milling operations, the Nuclear Regulatory Commission also has authority over the safety of the tailings disposal facility. In addition, many states through the state engineer's office or similar agencies have jurisdiction over impounding structures and state level requirements would have to be complied with also. In general, however, most state requirements are less restrictive than MESA or NRC requirements. Therefore, if a given project complies with federal safety guidelines, it probably will comply with the state safety requirements. The highlights of the MESA and NRC requirements as they relate to waste disposal structures are outlined in the following paragraphs.

MESA REQUIREMENTS

MESA's requirements for waste disposal structures are outlined in their publication entitled "Design Guidelines for Mine Waste Piles and Tailings Dams" (16). This seven-page document outlines the specific requirements which must be met by any impounding structure prior to its approval. The guidelines are subdivided into hydraulic and hydrologic considerations, geotechnical considerations, and a plan review check list.

Hydrologic and Hydraulic Considerations

MESA expects that the waste disposal structure will use a conservative approach with respect to flood protection. Therefore, the probable maximum precipitation must be determined for the drainage area behind the site along with the amount of runoff which would flow into the tailings disposal facility. MESA requires minimum design freeboard of at least three feet for those structures whose capacity is less than 100 percent of the probable maximum precipitation. Diversion ditches around embankments are required when the structure cannot impound the runoff from a 100-year frequency storm of a 6 hour duration or greater storm. Spillways have to be properly designed and spillway or decant capacity such that at least 90 percent of the probable maximum precipitation can be evacuated within 10 days or the facility will have to be designed to store an antecedent storm of similar volume.

References are provided in the text of this guideline for use in evaluating the hydrologic and hydraulic considerations which are required for the waste disposal structure. The references are all readily available government publications and engineering texts.

Geotechnical Considerations

The geotechnical considerations which MESA requires for waste disposal structures are commonly accepted factors which would be used in the design of most conventional water storage dams. The stability the impounding structure should have is a minimum static and dynamic factors of safety of 1.5 and 1.2 respectively under full anticipated design capacity. Foundations for the embankments and impoundments should be properly prepared, and filters, drainage blankets, and so forth should be designed in accordance with U.S. Bureau of Reclamation standards (17).

The upstream, center line, or downstream method of increasing the height of the tailings impoundment (18) can be utilized as long as the operator can demonstrate that the embankment is stable and meets all hydrologic and hydraulic design considerations. However, the dumping of the material over the freeboard area of the dam crest to extend and raise the embankment is not allowed unless it is in accordance with an approved plan. The construction of the dam addition must be engineer controlled with compaction of materials as in conventional water storage dams. MESA's authority extends to existing as well as planned structures. If, in the process of reviewing existing structures, MESA determines that there is sufficient reason to suspect the hydraulic, hydrologic or geotechnical integrity of existing structures, then a safety evaluation will have to be performed for those structures also. Those embankments will be required to meet the same standards as new structures. This may require extensive renovation of unsafe structures, particularly in hydrologic design and slope stability.

Plan Review Checklist

MESA requires that detailed plans and drawings of the proposed tailings disposal structures be submitted to their office for review prior to MESA's approval. Plans and drawings could also be required for any existing structure which is shown to be potentially unsafe and in need of remedial work. The detailed plans and drawings which MESA requires include details of the site description, drawings of the site, future expansion plans, description of the waste embankment and tailings impoundment, abandonment plans, and design calculations.

NRC GUIDELINES

The Nuclear Regulatory Commission's Regulatory Guide 3.11 provides their criteria for design, construction, and inspection of embankment retention systems for uranium mills. NRC's guidelines are generally similar to MESA's but include more specific methods for hydraulic, hydrologic and geotechnical analyses. A thorough review of the NRC's requirements is beyond the scope of this paper; however, some highlights of their requirements, particularly those that differ from MESA, are discussed.

Since obvious contamination problems would result from their use, spillways are not permitted in a uranium tailings disposal facility. Surface flows which cannot be diverted have to be stored in the tailings disposal facility. The NRC requires that the storage capacity of the retention system be designed to store 1.4 times the probable maximum precipitation plus a 100-year storm. At most sites, this will require several feet of additional storage capacity in addition to the normal three feet of freeboard required for wave action.

The NRC has also taken the position that the downstream method of construction is the most satisfactory method. It is unlikely, based on our recent experience, that they would license anything but embankments constructed and/or extended by the downstream method.

The NRC gives specific requirements for the type of laboratory testing and analyses which will be performed in order for them to approve a tailings disposal design. Unlike MESA, which has in-house technical capability to review dam design projects, the NRC relies heavily on outside government agencies such as the Bureau of Reclamation, the Corps of Engineers, and private consultants to review proposed tailings embankment designs. Effectively, the position adopted by the NRC is that an acceptable uranium tailings disposal embankment will be designed in a manner essentially identical with the design criteria and specifications used for conventional water storage embankments, only with stringent seepage control criteria as discussed later.

SUMMARY

This brief summary of the federal regulations which control the safety of tailings embankments clearly shows that the government is emphasizing the safety of tailings structures. Typically, in the past, relatively little time and engineering effort has been spent in the design and construction of tailings structures. This will not be the case in the future, since tailings structures will have to be constructed to essentially the same criteria as has been the practice for water storage embankments.

Because of the time required to complete the field, laboratory and engineering evaluations, preparation of plans and specifications and then submittal of these to the appropriate governmental agencies, it is unlikely that approval could be granted for a tailings project in less than a year from the initiation of engineering studies. This time may be greatly extended if a uranium tailings disposal site is involved. Costs for completing the engineering studies and preparation of the plans and specifications* probably would range between $50,000

*NOTE: Costs are for engineering costs associated with the hydrologic, geotechnical and seepage control aspects of the design. Detailed engineering of decant systems, pumps slurry lines and related non-earth structures are not included in the estimate.

and $100,000 for non-uranium projects and would be somewhat higher for projects which have to be reviewed by the NRC. In addition to the engineering costs associated with tailings embankments, the construction costs for tailings disposal systems which comply with federal regulations would, in most cases, be substantially increased over the cost of systems constructed prior to these regulations. That is, the added costs of flattening slopes, constructing by the downstream method, allowing storage for probable maximum flood in-flows, putting filters and cores in embankments as in water dams and the like will increase costs. Even for small mills, on the order of 500 to 1000 tons-per-day capacity, initial construction costs in the range of $2,000,000, not including lining costs for sealing the reservoir which are described in the next section of the report, could become typical (19).

IMPACT OF ENVIRONMENTAL REGULATIONS ON THE DESIGN AND OPERATION OF TAILINGS DISPOSAL SYSTEMS

The regulations which presently have greatest impact on waste disposal are air quality and water quality standards. The federal laws which have been responsible for the stringent seepage and air pollution standards were discussed in a previous section in this report. Seepage control is probably the biggest problem and the most expensive to control.

If federal regulations were controlled to the letter of the law, and zero discharge were, in fact, required, this would be virtually impossible to achieve at any cost. Even artificial membranes such as PVC, Hypalon, butyl rubber and the like are not totally effective in stopping seepage and certainly their service over more than a few decades is questionable. Probably the most effective sealants are natural clays which may or may not be available at a given site but even clays will allow some seepage.

The answers to the questions of acceptable seepage control methods and seepage rates are uncertain at the present time. However, it is safe to say that positive control methods such as clay linings or artificial linings or location of sites in naturally impervious basins will be required in order to comply with environmental regulations in the future. Since even the cost of clay linings would range up to several thousand dollars per acre (20), lining costs for small mills would probably be in the range of several million dollars. In many cases, lining costs would exceed the cost of constructing the embankment itself.

As the cost of seepage control to comply with regulations becomes higher, alternative processing techniques or waste disposal treatment methods will probably be incorporated into the mill design. For instance, a relatively simple step technologically, and often very beneficial step, is neutralization of the waste effluent. Other possible steps might include precipitation of toxic or hazardous materials prior to disposal of tailings. The toxic or hazardous materials could be disposed of in a well-sealed, but relatively small area. The bulk of the tailings, which are relatively inert sand and silt-size rock fragments, could be disposed in a more conventional tailings disposal facility from which some seepage could be tolerated because of the innocuous character of the effluent. Whichever methods are implemented, waste disposal is certainly going to cost more than it has in the past. These costs must be included early into the planning and development stages of a mine/mill operation.

Normally air pollution problems from tailings disposal facilities can be controlled by keeping the surface wet by either submerging or sprinkling the tailings. Tailings at completed disposal facilities can be stabilized chemically or by covering with topsoil and planting, which is preferable.

NRC GUIDELINES

In general, the guidelines imposed by the Nuclear Regulatory Commission for uranium tailings disposal facilities are more stringent and more demanding than those which would be applied to other facilities by the EPA where the tailings are not radioactive. In a position paper (21) the NRC has indicated at least some of the criteria it will use to determine the adequacy of proposed tailings disposal facilities.

MONITORING SYSTEMS

Typically, monitoring systems for water quality include cased wells for ground water and sampling stations on surface water courses. The number of monitoring stations and specific type of installations will vary according to the specific site conditions, but an adequate water monitoring system could be constructed for less than $100,000 in most cases. Periodic monitoring of these installations may require as much as one man-week of effort per month for the life of the project. With proper planning based on site selection studies, the monitoring wells and stations established to perform the necessary baseline surveys may also be used for monitoring effluent seepage from the tailings disposal area.

Air quality monitoring normally is concerned with particulates; high or low volume samples are used for this purpose. An exception is for uranium facilities where radon gas also has to be monitored. Normally the sampling station set up to perform the baseline surveys may also be used during the life of the tailings disposal facility to monitor airborne emissions from the reservoir. Similar construction costs and operation costs would be incurred for air quality sampling as for water quality sampling.

CONCLUSIONS

The result of the existing and proposed environmental and safety regulations will be that the milling industry would come under much tighter scrutiny from federal and state regulatory agencies than in the past. Therefore, substantially more time and money will be required to design and construct systems which will comply with the numerous regulations.

The major conclusions which should be derived from this paper are:

1. The milling industry has to be ready to adopt "state-of-the-art" methods of tailings disposal (not necessarily conventional tailings embankments) and pollution (principally seepage) control;

2. The industry has to be prepared in many cases to upgrade existing waste disposal facilities so that they will comply with existing and proposed safety and environmental regulations;

3. The industry should become more active in research and innovation of techniques to comply with regulations. Often the industry has taken the attitude that it will not do anything until forced. However, rules and regulations are far more likely to be reasonable if the industry actively participates in their implementation than if solutions and guidelines are left strictly up to the regulatory agencies such as the EPA and the NRC; and

4. The industry should allow at least two to three years of lead time to perform siting and environmental studies and to obtain necessary permits and licenses prior to constructing a new mill and waste disposal facility.

ACKNOWLEDGEMENT

The author wishes to thank Dr. K.R. Porter, Senior Ecologist for Dames & Moore, who made substantial contributions to the sections of this paper dealing with environmental reports and environmental regulations.

REFERENCES

1. The Bureau of National Affairs, Inc., 1975, The National Environmental Policy Act, in Environment Reporter, Federal Laws, p. 71:0101-71:0103, Bureau of National Affairs, Washington, D.C.

2. United States Department of Agriculture, Rural Electrification Administration, May 20, 1974, Revision of REA Bulletin 20-21: 320-21, p. 7.

3. The Bureau of National Affairs, 1977, Federal Water Pollution Control Act, in Environment Reporter, Federal Laws, p. 71:5101-71:5131, Bureau of National Affairs, Washington, D.C.

4. The Bureau of National Affairs, Inc., 1977, Environmental Protection Agenccy, Effluent Guidelines and Standards for Ore Mining and Dressing, in Environment Reporter, Federal Regulations p. 135:0881-135:0888, Bureau of National Affairs, Washington, D.C.

5. Rouse, Jim, 1977, Environmental Protection Agency, Denver, Colorado, personnel communication, January 1977.

6. The Bureau of National Affairs, 1975, Safe Drinking Water Act, in Environment Reporter, Federal Laws, p. 71:6041-71:6051, Bureau of National Affairs, Washington, D.C.

7. Environmental Protection Agency, 1976, Grants for State Underground Water Source Protection Programs; State Underground Injection Control Program, Proposed Regulations. Federal Register, v. 41, n. 170, p. 36726-36745.

8. The Bureau of National Affairs, 1976, Resource Conservation and Recovery Act, in Environment Reporter, Federal Laws, p. 71:3101-71:3116, Bureau of National Affairs, Washington, D.C.

9. The Bureau of National Affairs, 1974, Clean Air Act, in Environment Reporter, Federal Laws, p. 71:1101-71:1127. Bureau of National Affairs, Washington, D.C.

10. Mining Enforcement and Safety Administration, Department of the Interior, Part 77-Mandatory Safety Standards, Surface Coal Mines and Surface Work Areas of Underground Coal Mines, September 9, 1975, Federal Register v. 40, no. 175, p. 41775-41778.

11. Fujimoto, Robert, MESA, Denver Technical Support Center, Denver, Colorado, personnel communication, August 1977.

12. Skillings' Mining Review, "Kennecott Plans Court Action Against EPA," October 16, 1976, p. 23.

13. United States Department of Agriculture, Rural Electrification Administration, May 20, 1974, Revision of REA Bulletin 20-21: 320-21.

14. United States Department of Agriculture, Soil Conservation Service, Compliance with the National Environmental Pllicy Act, 1969, Federal Register, August 8, 1977, Part IV, p. 4011-40188.

15. United States Department of the Interior, National Park Service, Environmental Assessments and Statements Guideline NPS-12, November 1976.

16. United States Department of the Interior, Bureau of Reclamation, Design of Small Dams, Second Edition, 1973.

17. Swaisgood, J.R. and Toland, G.C., "The Control of Water in Tailings Structures" Tailings Disposal Today. Miller Freeman Publications, San Francisco, 1973, pp. 138-163.

18. Dames & Moore, unpublished report, March 5, 1976.

19. Toland G.C. and Versaw, R.E., "Design of Impoundment and Evaporation Ponds and Embankments for Cyanide and Other Toxic Effluents." Society of Mining Engineers, AIME, Transactions Vol. 260, March 1976, p. 71-74.

20. Dames & Moore, unpublished report, 1977.

21. United States Nuclear Regulatory Commission, "Branch Position - Uranium Mill Tailings Management," Fuel Processing and Fabrication Branch, 1977.

Chapter 36

DUST COLLECTION SYSTEM -- GENERAL GUIDELINES FOR THE MINERAL PROCESSING INDUSTRY

by L. M. Higgins

V.P., Hennesy Equipment Sales Company
Phoenix, Arizona

Dust or fume control systems are devices for capturing contaminants at their source and preventing their discharge to atmosphere.

The Primary Components Of Any Dust Control System Include:

An enclosure around any potential dust source such as conveyor belt transfer, crusher, screening machine, bin loading and loadout.

Ductwork including hood takeoffs (transitions) to convey the collected dust laden air away from the source of dust generation to a convenient location.

A dust collector for separation of particulate matter from the air stream prior to its discharge to the atmosphere.

A prime mover generally a fan, to draw air into an enclosure and to overcome resistance to flow in the ductwork, dust collector and stack.

Plant operators prefer to see the primary ore processing machinery in operation such as feeders, screening machines and crushers. They do realize that it is necessary to enclose this equipment to prevent dust escapement, however, the primary objective of plant operation is to efficiently process a given tonnage of ore. Dust control is a secondary consideration, useful for extending the life of mechanical equipment and creating a better working environment, healthier climate and worker morale, however salaries are paid by monies generated through ore processed, a product produced and sold. Therefore it is very important that the enclosures, duct routing and collector location result in the least amount of interference to plant operation. Hinged quick acting inspection door, preferably self closing, and peep holes should be located to allow easy inspection of process equipment. Enclosures should be designed to allow quick access and, ease of maintenance and adjustment to equipment. Enclosure panels, hoods and duct sections may be constructed for fast, complete removal. Should a dust control system be too confining, difficult to work around, or high maintenance, plant workers will invariably cut holes, leave off sections of enclosures, doors open, dampers misadjusted and numerous other things detrimental to dust collection system operation.

In The Design Of Any Successful Dust Control System, Several Factors Must Be Taken Into Consideration At The Outset. These Include:

A. Volume of air withdrawn from a given potential dust source adequate to prevent localized discharge.

B. Velocity of air. This includes velocities at slot openings in enclosures and equipment high enough to prevent localized discharge, and duct transport velocities adequate to convey all collected particles without settling (saltation) nor high abrasion.

C. Enclosures. Designed to contain the dust at its source, allow settling of larger material and air withdrawal commensurate with the process contained therein. Three distinct velocities are encountered within enclosures. Slot velocity at openings (the higher the better) cross sectional velocity (the lower the better to allow settling) withdrawal velocity, through the discharge opening allowing a smooth transition to duct transport velocity, resulting in a minimum of large particle pick up. Generally the best controlled source will have the lowest dust loading.

D. Ductwork transitions, branch entries, routing, method of balance (either balanced draft or damper control) and materials of construction.

E. Dust collector type commensurate with plant operation, sizing equipment and specifying operating conditions. You should determine early in the design phase what types are acceptable, since there are appreciable differences in physical size, electrical requirements and collected dust handling that can seriously effect other areas of plant design.

F. Fan operating conditions, size and type, and method of control. This item is frequently included with the collector bid package. In the case of dryers and other processes requiring specific conditions for process reasons the fan and control system should be carefully investigated to be sure of volume and pressure operation over the desired range.

Dust control system design begins whenever primary process equipment type and general location have been determined. Simplified general arrangement plan and elevation drawings are useful for locating arecs to be controlled, number of systems desired, general dust collector locations and air volumes. Single line schematic drawings may be prepared to determine preliminary information required for duct routing, energy losses, collector specifications and collected dust return or disposal. Collector type should be determined if possible. Dry or wet collector, fabric filter, electrostatic precipitator, granular bed filter and cyclonic are dry choices. Fabric filters may be shaker type, reverse air,pulse jet, or combinations. Recommendations between manufacturers vary widely, however we attach Table 1, showing the results of a recent survey of manufacturers reporting air to cloth ratios for pulse jet collectors. Shaker type and reverse air ratios are significantly lower, thereby requiring larger equipment. The pulse jet has become the most popular filter for general dust control, where a dry type system is required. Electrostatic precipitators and granular bed filters are high in initial cost and used only for specific operations within the mineral processing industry. These include gypsum, lime and similar calcination operations. Cyclone type collectors are sometimes used as pre-cleaners to reduce inlet dust loadings to final control equipment, however their use as a final method of dust collection is very limited. They are extremely useful tools for process purposes such as in fine material grinding and classification.

Wet collectors include scrubbers and irrigated precipitators. Many different scrubbers are manufactured, however, they can be categorized into cyclonic spray chambers, liquid level, tower type, dynamic and venturi types. Generally the higher the energy input the higher the collection efficiency.

Cyclonic or spin chamber types generally are constructed using a vertical tank with drain bottom, tangential, lower side gas inlet with or without a spin damper to induce higher velocity at the tank wall, internal spray nozzles and a top vertical gas outlet. Pressure drop is 2" - 4" in W.G., however liquid spray pressure can be quite high.

Liquid level scrubbers depend on a fixed height water level in a hopper with the gas being forced to pass under a baffle or the inlet duct discharging below liquid level. Make up water must compensate for slurry drain to assure a constant liquid level and therefore differential pressure across the scrubbing zone. Pressure drops are generally 4" - 6" W.G.

Tower type scrubbers are constructed similar to cyclonic spin chamber types except that the scrubbing zone is a horizontal perforated plate, or vane-like affair, with uprising gases and decending liquid. The gas velocity passing through the holes or slots, causes a liquid buildup and highly agitated scrubbing

TABLE 1. AIR-TO-CLOTH RATIOS FOR FABRIC FILTERS USED FOR EXHAUST EMISSION CONTROL

Industrial segment	Air-to-cloth ratio[a] ACFM/Ft.^{2}
Sand and gravel	7.
Clay	6.
Gypsum	6.
Lightweight aggregate	7.5
Pumice	4.5
Feldspar	4.
Borate	5.
Talc and soapstone	5.
Barite	5.
Diatomite	6.
Potash	6.
Rock salt	4.5
Fluorspar	6.
Mica	6.
Kyanite	4.5
Sodium compounds	6.
Asphat and related bitumens	N.R.[b]
Crushed and broken stone	7.

[a] Ratio is based on operating surface required to obtain a particulate concentration of 0.02 grain per standard cubic foot in the outlet stream from the filter. In all cases, the filter is a pulse-jet type operating at 6 in. W.G. differential pressure. The filtering medium is felted polypropylene or polyester.

[b] No recommendation for this segment.

CONVERSION FACTORS

TO CONVERT FROM	TO	MULTIPLY BY
Feet per minute (FPM)	Meters/Second (M/S)	5.08×10^{-3}
Feet, cubed per minute	Meters cubed/Second (M^3/S)	4.72×10^{-4}
Inches of water	Pascal (Pa)	2.49×10^{2}
Foot, squared	Meter squared (M^2)	9.29×10^{-2}

zone above the plate, or vane. As liquid build up increases due to the addition of scrubbing water, a balance is established dependent on fan pressure capability. Upon establishment of pressure balance, the continually added scrubber water displaces liquid and collected dust which descends to the slurry hopper or drain at the bottom of the tower. Pressure drops are generally 4" - 8" W.G.

Dynamic or wet fan type utilize the rotational motion of the fan wheel to force a wetted surface through the dust laden gases, thereby increasing the probability of collision between particulate and a wet surface and as an agglomeration device for an airborne water droplet - dust mixture. Special blade shapes are sometimes used to aid in liquid gas separation (eliminator stage). The wet fan also serves as prime mover for the dust control system. Preconditioning of the gases by upstream spray and scalping sections play a large part in the success of a dynamic scrubber both in terms of dust collection and maintenance. Energy useage is similar to the liquid level or tower types, but dust collection efficiency is significantly higher.

Venturi scrubbers can operate at very high pressure drop levels and therefore, offer the highest degree of dust collection efficiency. They are sometimes used at lower pressure drops such as 10" - 12" W.G. in a fixed throat or picture frame affair, with vertical down commer gas inlet. Horizontal gas inlet versions are available with liquid introduced via a manifold into a vertical slotted throat. The vertical down commer type is most popular with liquid introduced via a weir or large pipe upstream of the venturi throat. An elbow connects the verturi section to a vertical tower mist eliminator. Variable and fixed throat designs are available for operations requiring 90"+ W.G. Δ P. for submicron fume. Water and power consumption are high. Abrasion can be very significant due to the extremely high velocity of gases through the throat section. 35" - 40" W.G. differential pressure requires approximately 20,000 F.P.M. velocity, wherever a large quantity of submicron particulate is anticipated, a venturi should be considered. Applications requiring this degree of collection efficiency within the mineral processing industries are limited. Fine submicron particles are more commonly generated through thermal or chemical degredation, sublimation or condensation. The venturi can be considered the wet counterpart of the dry electrostatic precipitator in terms of dust collection efficiency.

Irrigated or wet precipitators have been utilized for acid mist collection for many years. They are essentially an electrostatic precipitator that uses water to wash away the collected dust. Applications within the minerals processing industry are very limited.

The venturi's offer the widest range of dust collection efficiency. They must be matched with corresponding capability in the fan system in order to take advantage of this feature. The combination venturi scrubber - fan system can be very costly to operate, due to high horsepower, water consumption and abrasion. The simpler liquid level and tower types can be utilized for many applications within the mineral processing industry and still meet stringent air pollution codes. Applications include coarse ore crushers and reclaim systems, fine ore bin ventilation and any application having a low grain loading and/or coarse particles. Fans preceeding or following their scrubbers are subjected to abrasion or build up resulting in inbalance. The dynamic scrubbers have received wide acceptance for many mineral processing applications. They are utilized for crushers, material handling, screening,dryers and calciners. Dust collection efficiency is very good down to 1 micron and fan wear and balance problems are minimal due to continual water flushing.

A-a. VOLUME REQUIREMENTS

Wherever possible, volume requirements should be based upon previous experience on the same, or similar material and application. Tight fittings, enclosures and hoods require the least amount of air. The volume of air must be adequate to withdraw aspirated air caused by material falling or machinery movement such as chutes, conveyors and crushers, as well as providing an in draft velocity adequate to prevent dust escaping from openings in the enclosure. The higher the velocity of particles within an enclosure, the higher the slot velocity requirement to prevent their escapement. Drop boxes and dead beds with V shaped plates should be used for high drops to break the fall of material and reduce the velocity of material impacting near an opening such as conveyor skirt board slot and to minimize aspiration effect. Slotted rubber curtains are useful where conveyor belts exit from enclosures, oversize from screens and for bucket elevators. Tight fitting adjustable skirt boarding should be used for conveyor sides. See sketch #1, 2 and 3 for a typical design.

B-a. TYPICAL CONTROL VELOCITIES

Operation	Indraft Velocity FPM	Comments
Barrel Filling	100	Tight fitting hood
Bins and Hoppers	200	Add for conveyor transfer
Conveyor transfer and conveyor loading	200 - 300	See sketch
Crushers	200 - 300	Jaw, gyratory and cone
Bucket elevators	200 - 300	Up to 30' high
Bucket elevators	300 - 500	Greater than 30'

TYPICAL CONVEYOR TO CONVEYOR TRANSFER WITH GREATER THAN 3 FOOT DROP.

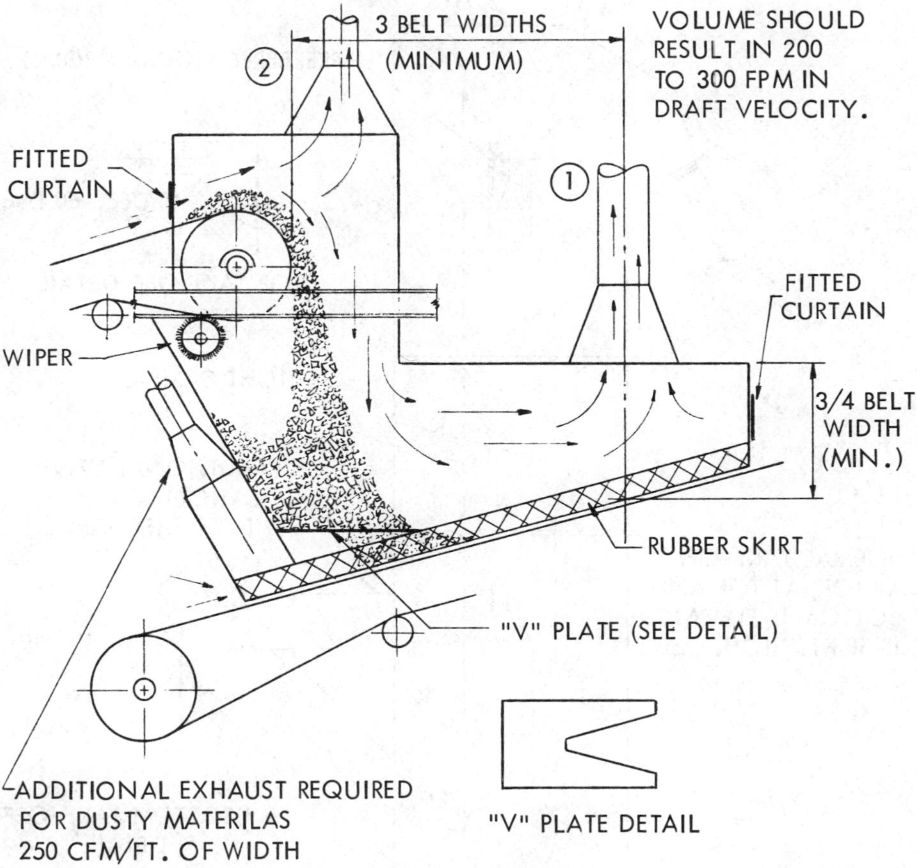

"V" PLATE DETAIL

FITTED CURTAIN DETAIL

	CFM/FT. OF BELT WIDTH	
BELT SPEED FPM	LESS THAN 250	250 TO 500
(1) LOAD-OUT POINT	500 W	750 W
(2) TRANSFER POINT	250 W	350 W

W=BELT WIDTH (FEET).

SKETCH 1

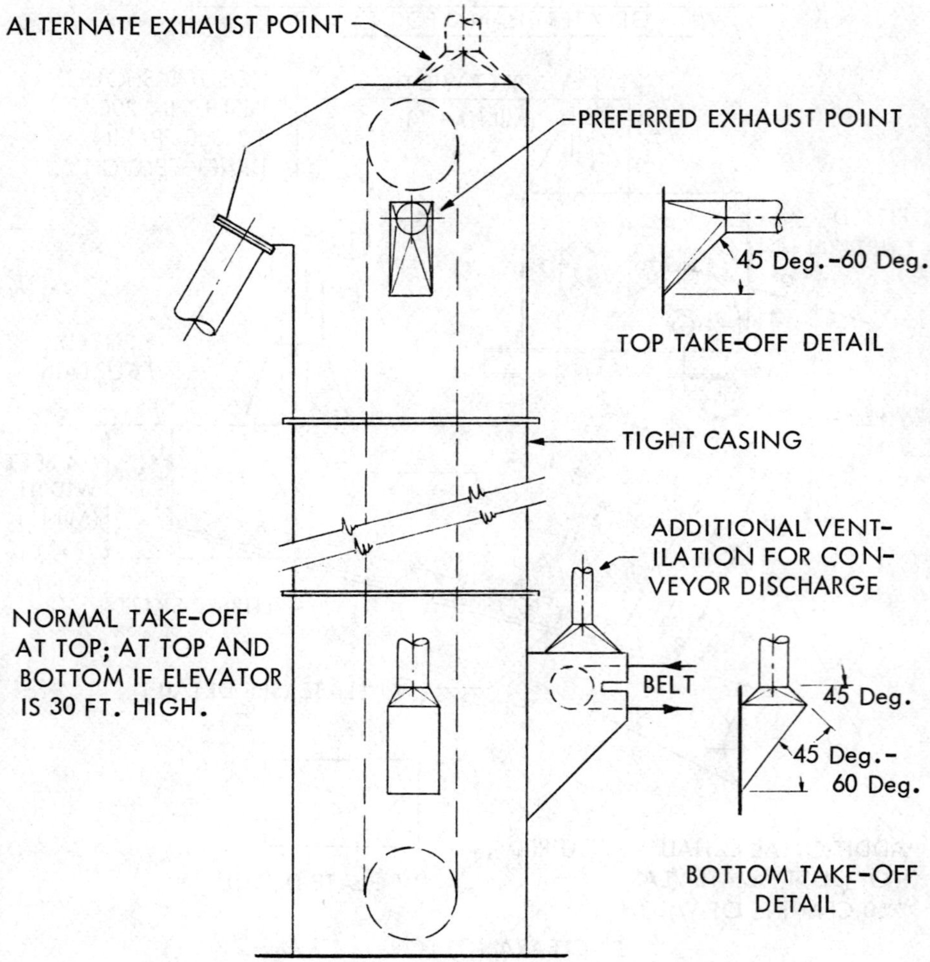

BUCKET ELEVATOR VENTILATION

SKETCH 2

FLAT DECK SCREEN

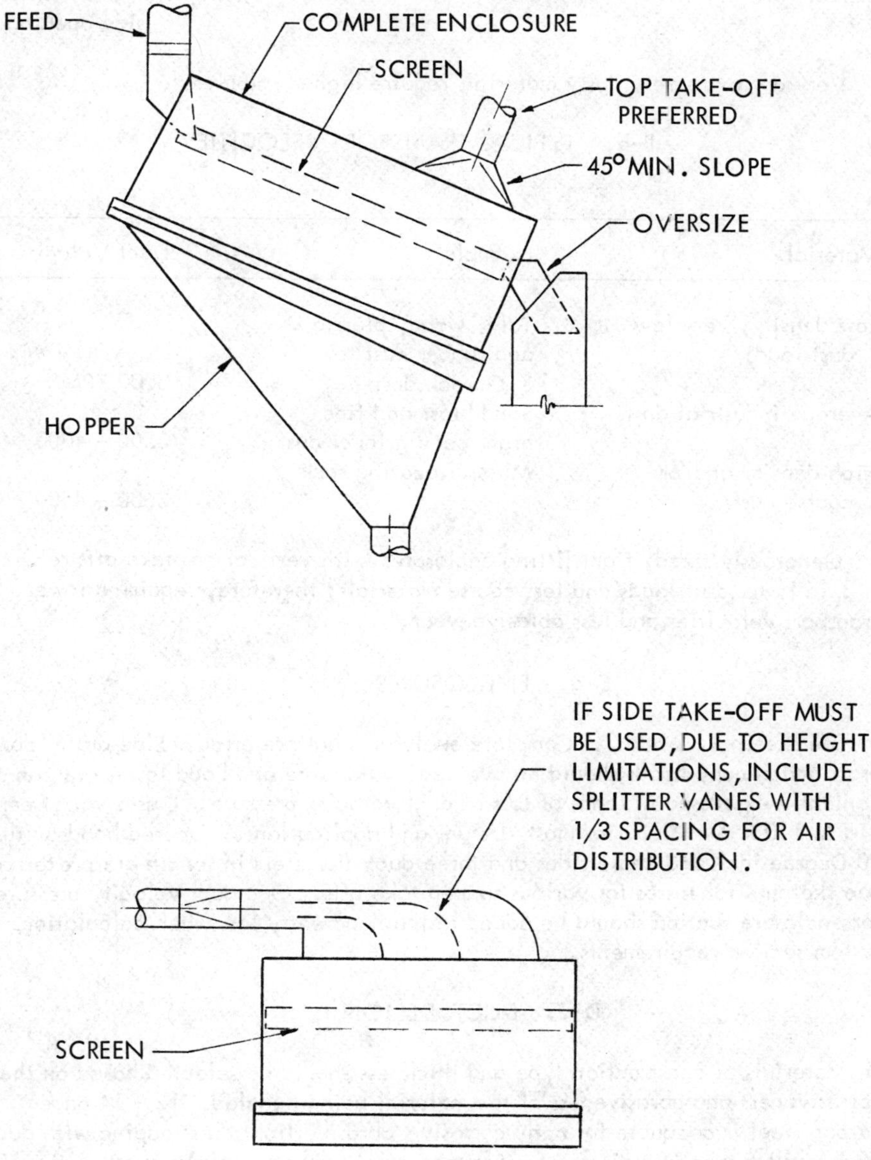

SKETCH 3

Vibrating screens	300 - 500	Not less than 75 CFM/Ft.^{2} of deck
Impact mills	500 - 1500	Minimize inlet side opening

Very dry extremely dusty materials require higher volumes.

B-b. TYPICAL TRANSPORT VELOCITIES

Material	Example	Duct Velocity
Low density, very low dust loads	Talc, wood, plastic and rubber dust low S.G. rock dust	3000 FPM
Average industrial dusts	Sand blast and fine high speed grinder dust	3500 - 4000
High density and/or coarse dusts	Mineral bearing rock	3600 - 4500

Generously sized, tight fitting enclosures with vertical gas take offs result in lower dust loads and less coarse materials, therefore, requiring lower transport velocities and less abrasive wear.

C-a. ENCLOSURES

See sketch 1, 2 and 3. Complete enclosures are preferred. Side draft, down draft and canopy hoods should be avoided. Enclosure and hood losses are conveniently expressed in terms of loss of duct velocity pressure. Losses vary from .15 to 1.0 x duct V.P. for most designs and applications. Tapered hoods with 90 Degree included angle sides and three duct diameters in length are preferred. See sketch 4 for losses for various shaped take offs. One duct velocity pressure for enclosure suction should be added to the hood entry loss when calculating system suction requirements.

D-a. DUCT DESIGN

Materials of construction type and thickness should be selected based on the corrosiveness and abrasiveness of the material being handled. 12 - 14 gage carbon steel is adequate for non corrosive/abrasive light dust loading with ductwork smaller then 30" diameter. For moderately abrasive applications, 10 - 12 gage duct work with elbow backs 2 gage sizes heavier is preferred. For severe applications 3/16" thick steel is preferred. Extreme abrasive materials and high

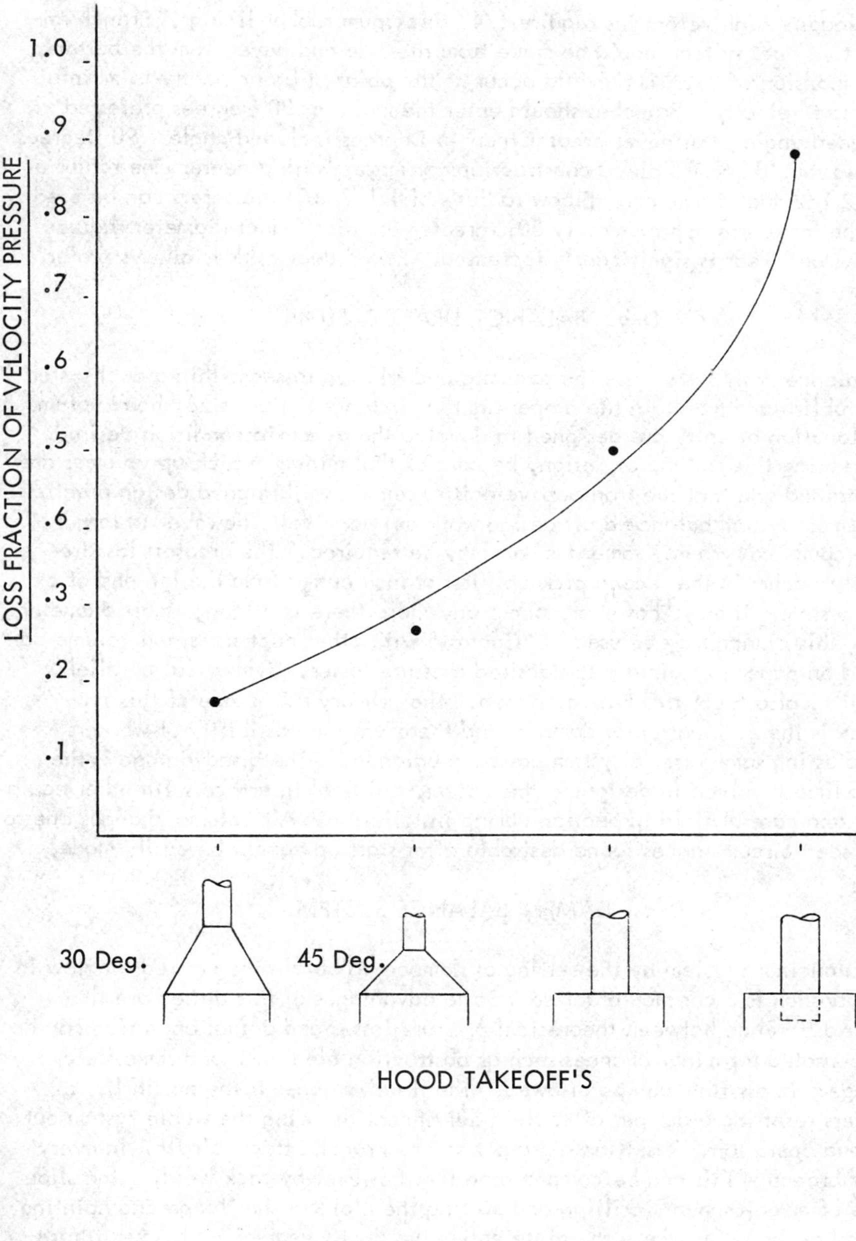
LOSS FRACTION OF VELOCITY PRESSURE
1.0
.9
.8
.7
.6
.5
.6
.3
.2
.1
30 Deg.
45 Deg.
HOOD TAKEOFF'S

SKETCH 4

dust loads - high velocities require 1/4" thick gum rubber linings. Branch entries to a duct system should be made from the side and never from the bottom. An expansion of duct size should occur at the point of branch entry to maintain transport velocity. Branches should enter the mains at 30 Degrees preferred included angle, but never greater than 45 Degrees included angle. 90 Degrees elbows should be of 5 piece construction, or more, with a center line radius of 2 to 2 1/2 duct diameters. Elbow radius's of 1 1/2 duct diameters can be used, but the losses are approximately 50% greater than for 2 duct diameter radius elbows and wear is significantly increased. Round duct work is always preferred.

D-b. BALANCE DRAFT DESIGN

Balance draft system use the dynamic and friction losses within a duct system to establish and maintain the proper air flow balance. Duct size, hood volume and location of entry are designed to develop the dynamic condition desired. When using this method of design, be careful that minimum pick up volumes are maintained and that the transport velocities remain within good design practice. In a small system balance draft design work out very well, however in larger more complex systems, some dampers may be required. The greatest loss frequently occurs in the second pick up (first branch entry) from the fan end of a duct system. It may, however, occur anywhere there is a long, small diameter run. This branch may be used as "Control" with other duct sizes and volume adjusted on paper to maintain the desired pressure losses. Symetrical parallel circuits, also favor this type of system. The primary advantage of this type system is the elimination of dampers and therefore the possibility of workers misadjusting same after a system has been balanced. The disadvantage is the extra time involved in designing the system, crictical interface with other equipment and careful field inspection during installation. Air volume changes due to unforseen circumstances found desirable after start up cannot be easily made.

D-c. DAMPER BALANCE SYSTEM

Balancing a system by the setting of dampers to obtain the desired air flow in each branch is a common practice. Some advantages of this design are that minor difference between theoretical pressure losses and actual operation can be compensated for, minor changes such as obstruction clearance and some field changes in air flow can be allowed. The disadvantages is the possibility of workers resetting a damper after field adjustment throwing the whole system out of good operation. Resetting a damper in one branch affects air flow in every other branch. This can be compensated for, however by tack welding the slide plate after adjustment, drilling and bolting the plate to the flange and painting a band at the point where the plate enters the duct, becoming obvious to foreman that the damper has been moved. Abrasion and build up can be a problem with dampers poorly located or designed. Installation in vertical runs is desireable, but not too close to a hood, or poor air flow distribution will occur.

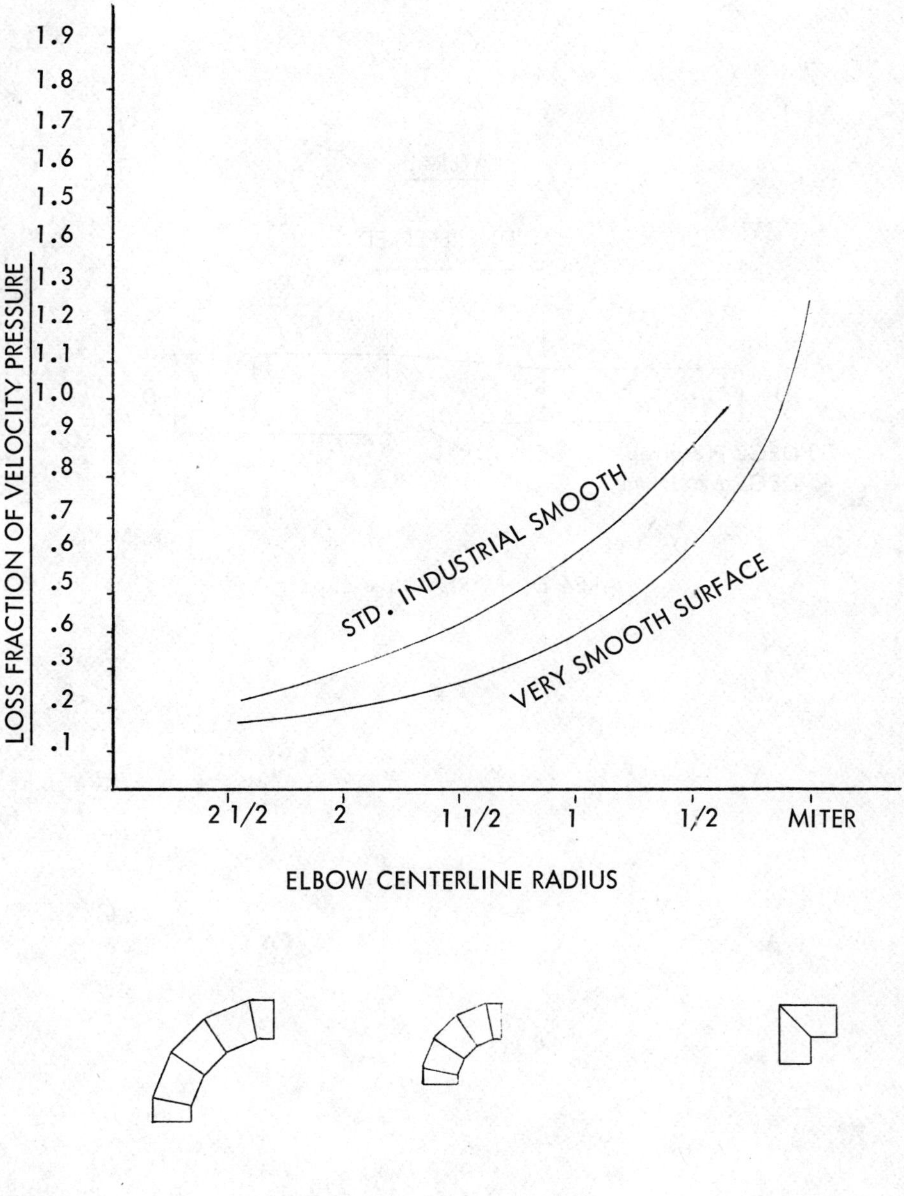
LOSS FRACTION OF VELOCITY PRESSURE
1.9
1.8
1.7
1.6
1.5
1.6
1.3
1.2
1.1
1.0
.9
.8
.7
.6
.5
.6
.3
.2
.1
STD. INDUSTRIAL SMOOTH
VERY SMOOTH SURFACE
2 1/2
2
1 1/2
1
1/2
MITER
ELBOW CENTERLINE RADIUS

SKETCH 5

LATERAL

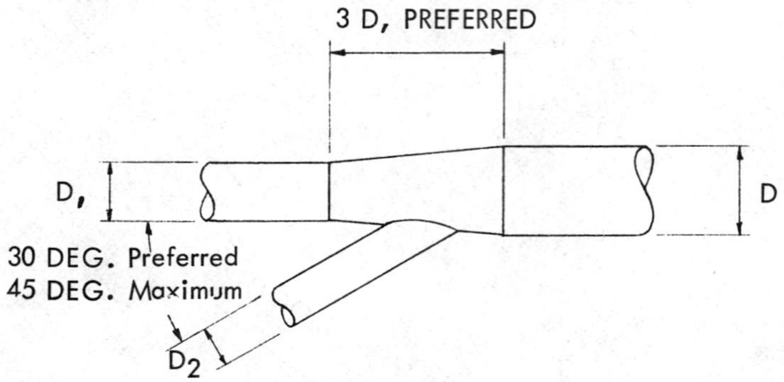

AREA D, + AREA D_2 = AREA D

SKETCH 6

Branch location should be arranged to result in the least pressure differential at branch to main junction without dampers if possible. Occasionally high differential pressures are unavoidable, requiring a more closed damper setting to obtain air flow balance. A spacer plate extending into the duct and rounded edge slide damper results in an adjustable orfice affair minimizing wear by creating the high velocity in the center of the duct as opposed to one side wall as with a simple slide damper.

Specific procedures regarding pressure losses within the dust control system may be found from several sources. The most popular reference is "Industrial Ventilation" by the American Conference of Government Industrial Hygenists and printed by Edwards Brothers, Ann Arbor, Michigan. The American Society of Heating, Refrigerating and Air Conditioning Engineers and several equipment manufacturers, also offer information and tables. Most all tables including fan performance charts are based on standard air at .075 Lbs/CF. It is convenient to make system loss calculations using standard air data however, care should be taken to correct for actual conditions when making final fan speed determinations. Altitude corrections are sometimes overlooked resulting in excess fan speed, air flow and/or dampers closed more than normal to balance the system causing accelerated wear from abrasion.

E-a. DUST COLLECTOR TYPE

Dust collector type and approximate size should have been determined previously with recommendations received from manufacturers. With completion of duct design air flow and system resistance calculations specific sizing, speed and horsepower information may now be obtained from the dust collector and fan manufacturers. If the fan is not selected as part of the collector package, then the fan manufacturer will require the dust collector pressure drop as well as duct system pressure losses and temperature and pressure of the gases. Disposal of the collected solids with a dry type system generally involves rotary air locks and screw conveyors, items familiar to most engineers. The only precaution to take is to assure the material does not become airborne, creating another problem. A dust pickup may be required as at any other potential dust source.

Wet collector slurry disposal presents different problems such as settling and pipe build up. Slurry discharge drains from scrubbers require an air seal. Since it is desirable to have as large an opening as practical in the scrubber to allow the unrestricted discharge of collected material, local sealing under the scrubber is common. Sketch number 7 shows 2 typical methods of sealing. Method A shows one common arrangement. Method B shows a type where the slurry pipe is submerged at the far end, yet allowing for liquid overflow near the scrubber, should the pipe become clogged.

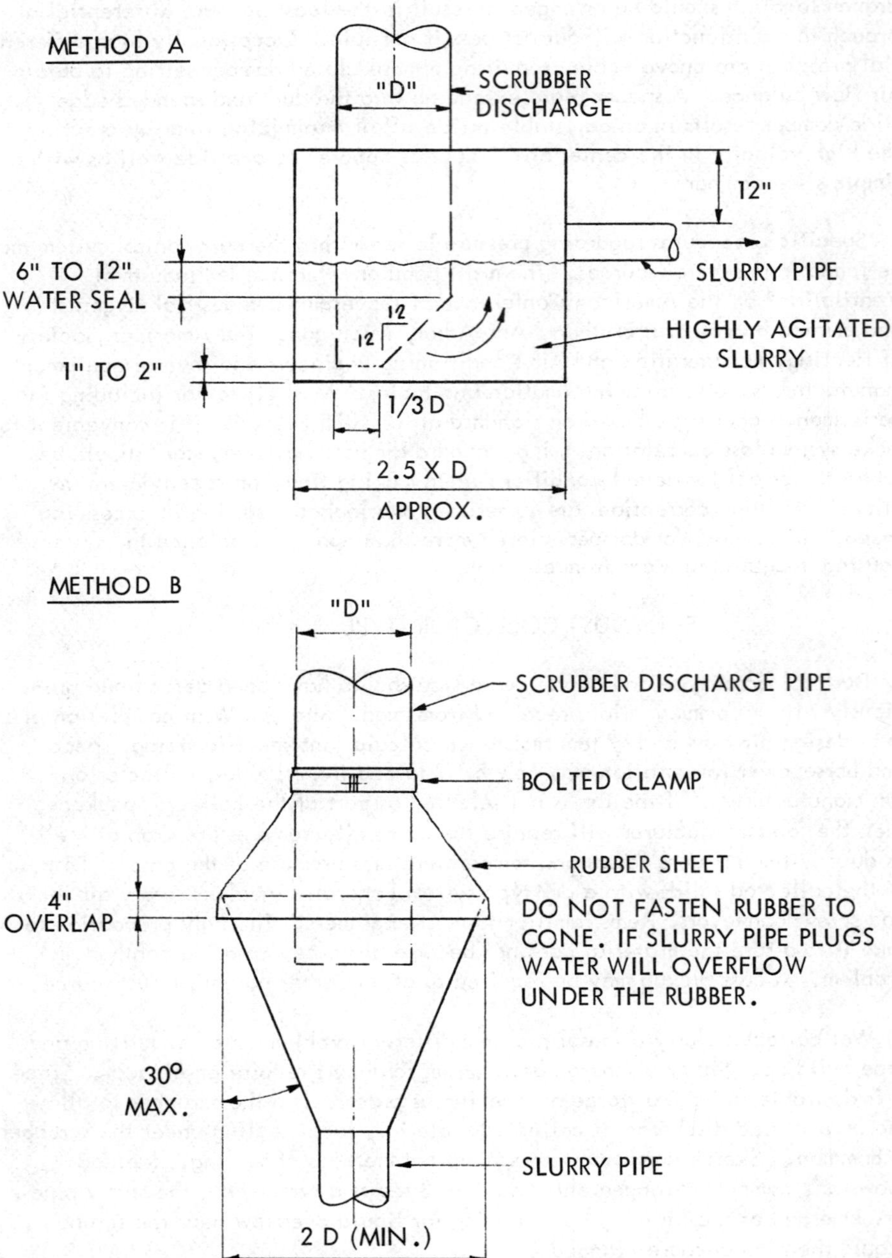

TYPICAL METHODS OF SEALING SCRUBBER SLURRY DISCHARGE PIPE

SKETCH 7

F-a. FAN OPERATION

Fan operating conditions supplied to the dust collector or fan manufacture should be checked against information received back from those manufacturers and overall system requirements. Be sure enough flexibility exists within a particular fan design to allow safe operation over the range of conditions anticipated. For instance, if the collector is purchased anticipating future addition to the dust control system, then the fan should be selected accordingly. Likewise, very large oversized equipment may be inefficient. Control for critical air flow system may require special inlet or outlet dampers, and variable speed devices. Plot constant speed curves if necessary, from manufacturers tables to illustrate volume and pressure characteristics throughout the operating range. Use simple paddle wheel fan blades of heavy gage construction, where possible. Avoid hollow constructed, intricate shaped blades following scrubbers, since moisture can enter the blade, throwing it out of balance. Clean out doors should be large enough for a man to work through.

FIELD CHARGES AFTER DESIGN

The operation of any system is dependent on the dynamic balance of all its component parts. Dust collection systems commonly incorporate both series and parallel circuits. Not only must component relationship be considered during the design phaze; but careful system balance must be accomplished after installation and adequately maintained for long system life.

The addition of pick ups and the change in shape of components have a profound effect on the entire system and should not be made in the field without careful review of the entire dust control system. What may appear to be a rather insignificant addition to a large system could start a chain reaction resulting in serious deterioration that may not show up for several months.

As an example, lets assume a new pick up is added in the field; the new duct would probably be cut into the main near the collector end. A 12" diameter duct to handle 3000 CFM introduced into a 40" diameter main would not appear to upset a system too badly. (See Sketch 8 and 9). Sketch 10 shows a typical fan performance curve and system operating points. Duct resistance varies approximately as the square of the volume. Scrubber Δ P variations with air flow is dependent on the type of scrubber and % of operating capacity. A typical volume Vs Δ P curve is shown in Sketch 11. Sketch 10, Point B shows the new system operating point after the introduction of the new 12" diameter duct. The new duct would most probably offer less resistance to air flow than the main, so a damper would be adjusted in the branch to maintain 3000 CFM.

The new total system volume will not be 33,000, but something less than that

SKETCH 8

TYPICAL SYSTEM

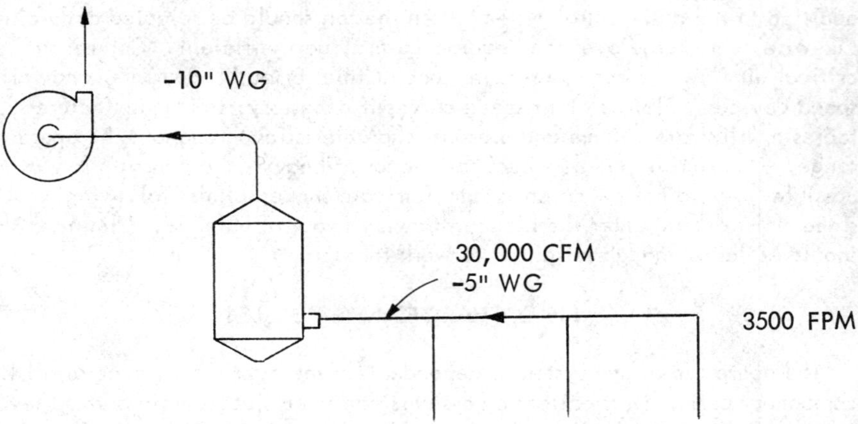

SKETCH 9

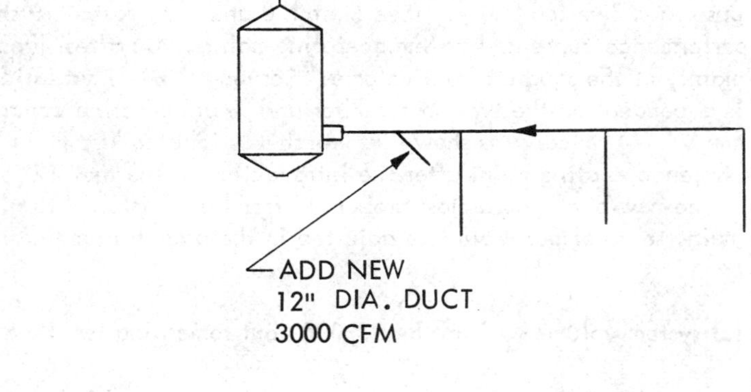

SKETCH 10

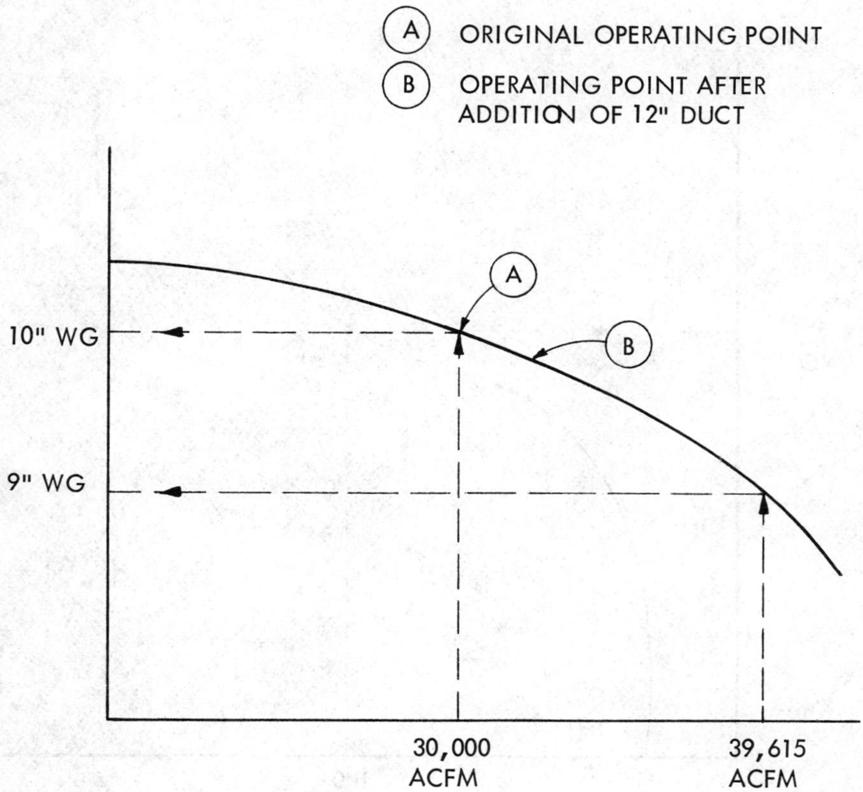

TYPICAL FAN PERFORMANCE CURVE

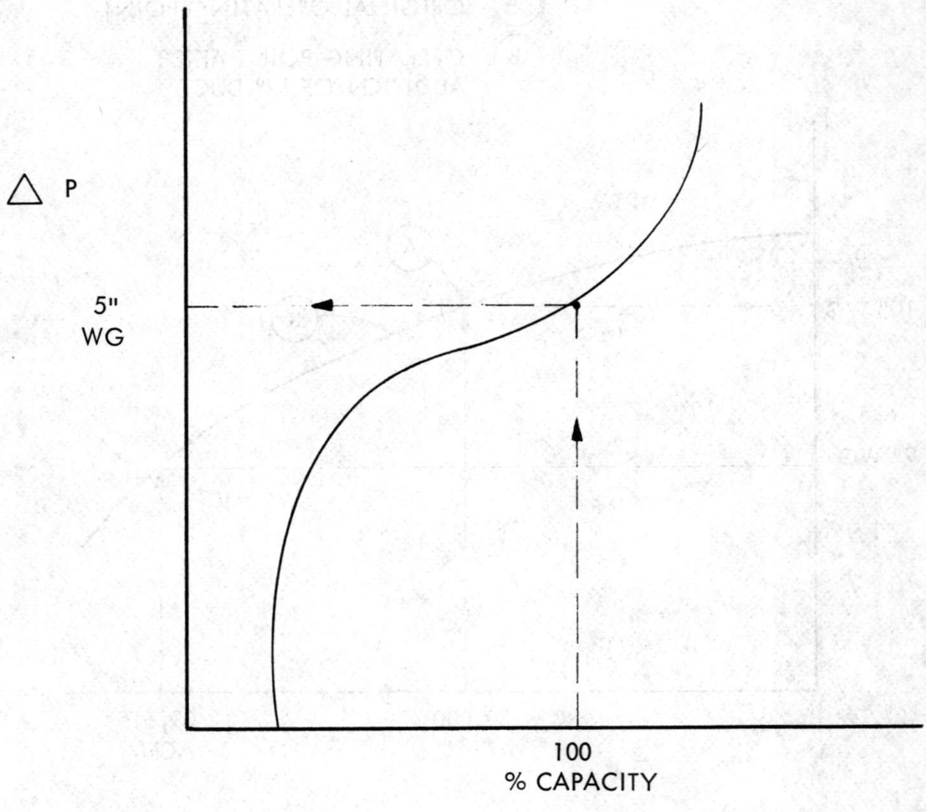

TYPICAL CURVE SCRUBBER △P
VS
% CAPACITY

SKETCH 11

depending on the fan curve, scrubber curve, and system resistance. In this case31,150 ACFM satisfies the new operating points. (See Sketch 10, Point B).

Fan Static Available	9.75" WG
Scrubber Δ P	5.35"
Available Suction	4.40" WG

Main Volume $\sqrt{\frac{4.40}{5.0}} \times 30,000 = 28,150$ ACFM

Branch Volume	3,000
	31,150 ACFM

Duct velocity in the main and other branches is reduced in direct proportion to air flow. The new velocity is now 3,280 FPM.

Although, the changes seem small, this velocity is now too low for the transport of larger particles and drop-out will occur. Drop-out of dust in one part of the system will cause an increase in air flow at some other branch, eventually resulting in near complete blockage at one point and excessive velocities at other locations. Localized dusting will occur at the plugged branch, while severe abrasive cutting of the ducting will occur at the high velocity zone.

This condition could probably have been avoided by a careful study of the system design and operation. Most systems are of the blast gate balance type, which allows changes to be accomplished more easily than with the balance duct method. A change in fan speed and dam per adjustment could probably have allowed the duct addition, while still conforming with the required transport velocities.

The above exam ple is one of the least severe modifications encountered and the numbers shown are realistic.

Chapter 37

ELECTRICAL SYSTEM DESIGN AND EQUIPMENT APPLICATIONS

By W. Macdonald, M.J. Sheriff and D.H. Smith

H.A. Simons (International) Ltd., Consulting Engineers
Vancouver, British Columbia
Canada

ABSTRACT

The paper outlines the basic procedure used in electrical system planning and design. Consideration is given to the use of applicable codes and standards, development and use of a system One Line Diagram, power availability, load requirements and power factor.

Also discussed are the factors influencing the selection of power supply and distribution arrangement, voltage levels, system grounding, distribution and utilization equipment such as switchgear, transformers, motor control, cable systems, motors.

Short circuit calculation methods and system protection are discussed.

INTRODUCTION

This paper will provide guidelines for the basic electrical planning and design of a mineral processing plant. The discussion is generally centered around the development of a typical Plant One Line Diagram, and the selection of specific system components.

The presentation is non-technical. It describes the power system design method, and factors affecting electrical equipment selection. References are provided as a guide towards more technically detailed sources of information.

BASIC SYSTEM PLANNING

Use of Codes and Standards

Initial planning of the electrical system requires that the designer be familiar with electrical and safety codes applicable to the area where the plant will be located. In North America, the applicable codes are:

U.S.A.:	Federal Mines Public Law No. 89577; 30CFR, Chap. 1, Parts 55 - 58; National Electrical Code Sponsored by National Fire Protection Association (N.E.C.)
Canada:	Canadian Electrical Code, Part I, and Canadian Electrical Code Part V (Use of Electricity in Mines) (C.S.A. 22.1 & 22.5) Published by the Canadian Standards Association

In addition, many standards and guides are available through the IEEE and other organizations, which provide recommended practices for the various aspects of system design. References 1 to 7 provide a sampling of this resource material.

The One Line Diagram

General The key control document used in the development of the entire electrical system is the "One Line Diagram"; see Figure 1. This diagram shows in simple schematic form the design philosophy of the power supply and distribution systems. From this diagram, a designer will:

a. identify all loads;
b. determine the capacity of the power source(s);
c. determine the capacity, connection, impedance, and grounding connection of all transformers;
d. identify special equipment such as lightning arresters, surge capacitors, protective devices, metering transformers, etc.

The stages of development of this diagram will be discussed in the following sections:

Stage 1 - System Loading A load evaluation must be made to determine the required capacity of the equipment used in the power supply and distribution system. This is done in coordination with the process and mechanical designers. Two loading conditions must be considered; steady state loading and transient or short time loading.

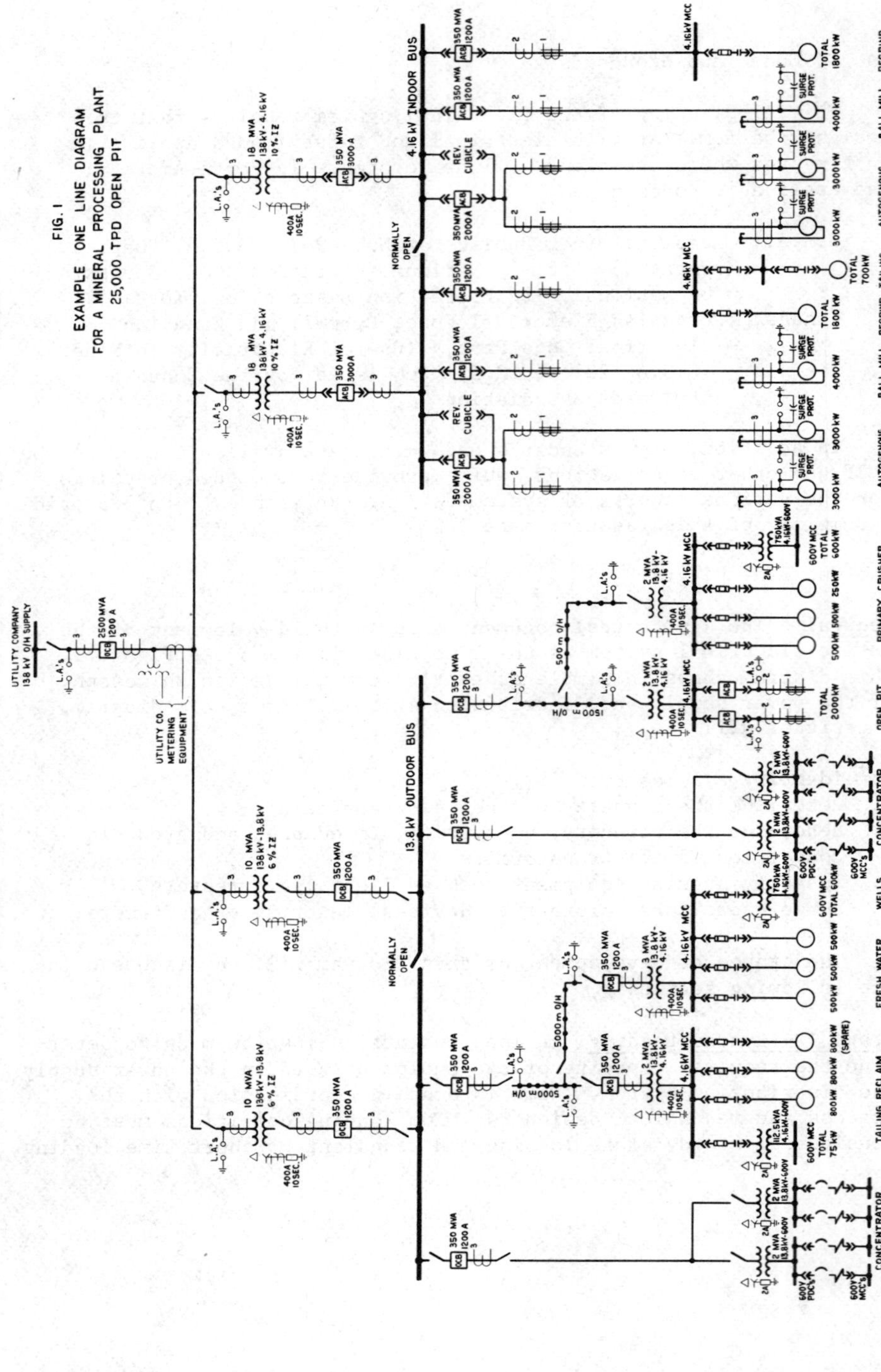

FIG. 1
EXAMPLE ONE LINE DIAGRAM
FOR A MINERAL PROCESSING PLANT
25,000 TPD OPEN PIT

STEADY STATE LOADING: An evaluation of steady state loading will determine:

- the power supply capacity required from utility and/or in-plant generation;
- capacity of system components such as transformers, feeders, cable circuits, switchgear, motor control;
- physical sizing of electrical rooms and other specific areas set aside for electrical equipment;
- selection of suitable supply and utilization voltage levels.

It is necessary to obtain from the mechanical designers an accurate assessment of the connected loads in the areas forming the total plant complex.

Special attention must be given to the rating of motors above 150 kW in order to select the most satisfactory utilization voltages.

A commonly accepted voltage division in North America for motors is 0-150 kW motors 480 volt (USA) or 600 volt (Canada); motors larger than 150 kW, 2.4 kV or 4.16 kV or higher, depending on motor rating, economic, and system considerations. No general rule can be followed; each case must be evaluated on its own, and special studies may be necessary (see Stage 6).

When system loading information is obtained, appropriate multipliers can be applied to calculate average and peak demands on both power sources and individual circuits. Table 1 provides typical steady state loading values derived from Fig. 1.

TRANSIENT LOADING: An evaluation is necessary to:

- advise the utility supply company of the magnitude and frequency of transient power swings that will arise in the plant;
- establish the capacity of in-plant generation;
- establish special requirements for power system components;
- examine alternative system designs which absorb or reduce the effects of transient loads.

The main sources of significant transient load fluctuations are:

- sudden dropping of a block of load as a result of system disturbances;
- starting of large motors;
- rapid, heavy power swings resulting from the normal operating cycle of specific drives, such as mine pit shovels.

In most cases, a detailed evaluation of the effect of these load variations requires special analysis (see Stage 6).

TABLE 1 PLANT LOAD PREDICTION AND MOTOR LIST

COLUMN: 1	2	3	4	3 x 4 5	6	5 x 6 7	8		7 x 8b 9
MOTOR DRIVE NO.	NAME	SHAFT KW	EFFICIENCY	INPUT KW	AV. DRIVE LOAD %	AV. MOTOR LOAD KW	a POWER FACTOR	b	TOTAL MOTOR KVAR
							COS Ø	TAN Ø	
4.16 KV BUS 'A'									
32-001-1 32-001-2	AUTOGENOUS MILL NO.1	6000	.96	6250	80	5000	1.0	0	0
32-003	BALL MILL NO.1	4000	.95	4238	85	3602	1.0	0	0
FEEDER TO REGRIND & TAILINGS									
33-006	CONE CRUSHER NO.1	400	.86	465	70	326	.88	.54	176
33-007	CONE CRUSHER NO.2	400	.86	465	70	326	.88	.54	176
33-008	CONE CRUSHER NO.3	400	.86	465	70	326	.88	.54	176
33-009	CONE CRUSHER NO.4	400	.86	465	NIL	----	.88	.54	---
33-010/ 33-070	CONVEYING AND SCREENING TOTAL	200	.92	217	80	174	.84	.65	113
35-009	TAILINGS PUMP NO.1	200	.88	160	75	120	.87	.57	68
35-010	TAILINGS PUMP NO.2	200	.88	160	75	120	.87	.57	68
35-011	TAILINGS PUMP NO.3	200	.88	160	NIL	----	.87	.57	---
35-067/081	TAILINGS AUX.	100	.92	92	60	55	.84	.65	36
TOTAL ESTIMATED RUNNING LOAD ON 18 MVA TRANSFORMER NO. 1:						10,049 KW			813 KVAR

Stage 2 - Power Supply

UTILITY SUPPLY: Early discussions with the supplying utility company are essential. The plant distribution system adopted depends heavily on the nature of this power supply.

Ref. 1, pp. 35 proposes a "check list" of information to be discussed with the utility critical to the plant power system design. The following is a part of this list:

a. Information to give the Utility Company:

- plant site plan;
- plant loading in connected kVA and maximum demand kVA;
- anticipated plant operating power factor;
- details of the nature of the load and identification of large motor starting requirements;
- preferred point of delivery;
- preferred supply transformer connections;
- construction and start-up schedule;
- unusual requirements, such as unacceptable high speed reclosing.

B. Information to obtain from the Utility Company:

- choice of supply voltage (nominal and range of change);
- point of delivery;
- billing rate schedule;
- options available on supply substation ownership;
- supply substation space requirements, if utility owned;
- three phase and phase to ground short circuit capacity at the supply point, and system X/R ratio;
- special requirements for metering and protective relaying;
- type of grounding used in the supply system;
- supply system reliability data.

Referring to Fig. 1, the following information was received from the Utility Company:

- supply voltage 138 kV $\pm$ 10%;
- short circuit capacity at 138 kV bus: maximum future three phase and phase to ground = 2000 MVA; X/R ratio 10;
- supply point - line side of 138 kV oil circuit breaker;
- metering point - load side of 138 kV oil circuit breaker;
- the complete substation including the 2-10 MVA and 2-18 MVA transformers and incoming 138 kV oil circuit breaker to be purchased by the customer and approved by the utility company;
- supply system - capable of supplying plant demand and starting the Autogenous Mill (2-3000 kVA synchronous motors);
- plant power factor must not be less than 0.9;
- billing rate - \$3.95 per kVA of demand per month plus \$0.003 per kwhr per month.

IN-PLANT GENERATION; This must be considered when:

- the proposed plant is in a location where it is not possible for the utility company to provide a suitable source of power;
- the utility supply is incapable of supporting the total plant load;
- it is necessary or desirable to have a source of "emergency power" during loss of the main power supply;
- use of plant fossil fuel fired or hydro-electric generating units is more economical than utility supplied power.

Reference 8 discusses the items to be considered when planning for in-plant generation. The key items are:

- overall economic evaluation;
- total steady state and transient loading requirements;
- selection of individual generator unit size;
- fuel selection;
- voltage;
- protective relaying and metering;
- voltage regulator and governor requirements;
- grounding method;
- synchronizing requirements;
- maintenance requirements;
- need for parallel operation with the utility.

<u>Stage 3 - Power Supply and Distribution System Arrangement</u> Several factors should be considered in selecting the power system which best meets the needs of the plant:

- the potential economic loss and/or safety hazard if service to an area is interrupted, i.e. is single radial, duplicate radial or ring feeder justified?
- how will the utility supply govern the system arrangement?
- how much redundancy is justified in order to perform maintenance?
- the necessity of electrically isolating loads having fluctuating power requirements from critical process loads;
- how can the in-plant emergency generator be used to the best advantage?
- the process plant production life (complex distribution systems are seldom justified for short term operations.)

Fig. 1 shows a modified radial system which is one of the simplest system arrangements. It has enough flexibility for most process plants. The simple radial system is modified in that each main supply transformer is rated to supply the total load on the two buses separated by the normally open disconnecting switches. This will allow one 138-13.8 kV and one 138 kV - 4.16 kV transformer to be out of service.

There are many variations from this basic arrangement which will meet specific process requirements. Ref. 1 illustrates several alternatives such as Primary Selective, Primary Loop, and Secondary Selective.

Stage 4 - Voltage Selection

GENERAL: The subject of voltage consideration is treated in detail in Ref. 1, Chap. 3. This or an equivalent source will give guidance in the selection of system voltages. The following is a brief summary of the major considerations:

MAIN SUPPLY VOLTAGE: If the plant power is supplied by a utility company, the main transmission line voltage will be predetermined. This voltage may be too high for distribution within the plant so a step-down transformer substation will be required; see Fig. 1.

In plant generators usually operate at the plant distribution voltage. Economics may dictate using a lower generator voltage with transformers to step-up to distribution level.

DISTRIBUTION VOLTAGE: If not predetermined by the utility, the selection of a distribution voltage to supply power to the plant load centers is generally governed by plant load. For plants with a capacity of less than 10,000 kVA, 4.16 kV, or less, is usually the most economical system voltage. Above 10,000 kVA, system voltages of 6.9 kV or 13.8 kV are required.

In open pits and where remote loads exist, special loading conditions and long distances between load centers may require a higher distribution voltage than is used in the process plant.

Fig. 1 provides an example of two distribution voltages; 13.8 kV is used for supply to remote areas such as Tailing Reclaim, Fresh Water Supply, Open Pit and Crusher; 4.16 kV is used for the large mill drives which are close to the center of power supply.

UTILIZATION VOLTAGE: The most economic plant voltage is generally dictated by motor rating. Table 2 provides a guide to standard North American utilization voltages.

Except in very small plants, step-down transformers are required to reduce the distribution voltage to the utilization voltages in the plant process areas. See Fig. 1.

TABLE 2
VOLTAGE RATINGS FOR STANDARD
NORTH AMERICAN INDUCTION MOTORS

Nominal Utilization Voltage	Motor Nameplate Voltage
Single Phase Motors	
120	115
240	230
Three Phase Motors	
208	200
240	230
480	460
600	575
2400	2300
4160	4000
4800	4600
6900	6600
13800	13200

Stage 5 - System and Equipment Grounding

GENERAL: Reference 3 treats the subject of system and equipment grounding in detail. This and the applicable safety codes should be consulted for assistance in the design of grounding systems. The following is a brief summary of items to consider:

SYSTEM GROUNDING: System grounding refers to the intentional connection between system conductors and the earth. The connection is normally made at the supply transformer secondary bushings (Ref. 19), by specifying the winding configuration. This is most commonly delta-wye. The main objectives are to reduce transient overvoltages, and to minimize fault damage by limiting the magnitude of the fault current. Faulted circuits are isolated by the use of ground current monitoring devices. The methods shown on Fig. 1 illustrate recommended practices in North America. These are:

1. 138 kV supply system - generally dictated by the utility supply company and usually solidly grounded. Ground relays connected to the 138 kV oil circuit breaker will isolate ground faults in the 138 kV section of the substation.
2. Plant Distribution Systems (in this case 13.8 kV) - low resistance (400A) grounding of the neutral point of the wye connected transformer winding. Ground relays in the 13.8 kV feeder circuit breaker units and in the transformer neutral ground connection will isolate ground faults in the 13.8 kV system.

3. Plant Utilization Voltages Above 600 Volts (in this case 4.16 kV) Low resistance (400A) grounding of the neutral point of the wye connected transformer winding, with the same relaying as the 13.8 kV system, except in the mine pit area where special requirements are necessary because of the use of trailing cables and the potential hazard to personnel; see Ref. 3, Chap. 2.
4. Plant Utilization Voltages 600 Volts and Below. For motors use high resistance grounding with ground current monitoring and first fault alarm rather than trip; for lighting - solidly grounded.

NOTE: Historically, the mineral processing industry has followed mining practice in using solidly grounded wye systems at 600 volts and below. Recent recognition of the severe damage which can result from arcing ground faults on such systems (Ref. 3 and 10) has led to a code revision in North America which requires special ground fault relaying on most solidly grounded systems. This and other factors have had the effect of changing grounding practice so that modern practice favours either solidly grounded systems with special ground relaying systems to isolate faults quickly, or alternatively, high resistance grounded neutral wye systems which limit the ground fault current to a value slightly in excess of the system capacitive charging current. This low magnitude current can usually be allowed to flow safely so that removal of the fault can be scheduled.

The main advantages of high resistance grounding are:

- a firm limit on the ground fault arc energy available;
- predictable transient or maximum imposed system fault voltages applied to the insulation systems;
- permissive operation of the power system during fault conditions, by monitoring and alarm devices.

EQUIPMENT GROUNDING: Equipment grounding refers to the system which interconnects all metallic structures which carry energized conductors. The main objectives of equipment grounding are:

- to reduce the potential shock hazard to personnel. The approximate limits of electrical discharge via humans which may cause fatality can be as low as 30 mA A.C., 60 HZ and about 10-12 volts (see Ref. 11). Actual values depend on resistivity of human body contact and ground conditions underfoot.
- to present a continuous low impedance path for the safe dissipation of ground fault current, to reduce the possibility of sparking and subsequent equipment damage, loss of production, and fire hazard.
- to protect buildings and equipment from lightning strikes.

For further details regarding preferred methods of equipment grounding and for details of special requirements for static and lightning protection, see Ref. 3, Chap. 3.

INSTALLATION GUIDELINES: To ensure that all the grounding requirements are met, a power system ground grid must be designed that will, under worst possible system ground fault conditions, eliminate dangerous levels of potential where men normally walk (step or mesh potential) or come into contact with equipment (touch potential), (Ref. 12).

The practical restraints imposed by thermoweld jointing of copper ground grids have given impetus to the development of a compression-type jointing system, which is not so demanding of operator expertise or cable and joint preparation.

Problems of broken wires for equipment grounding are relieved by using ground wire enclosed in conduit or integral ground wires in Teck cable.

Stage 6 - Special Studies As an integral part of power system design, it is often necessary to perform special studies to ensure satisfactory plant operation. The digital computer can be a helpful tool in performing such studies. It is recommended that a specialist be employed to perform this work and present the results in report form. The studies which are usually necessary are:

- LOAD FLOW ANALYSIS. (See Stage 1) Used to determine power supply requirements and system arrangement for steady state and transient loading conditions.
- SHORT CIRCUIT ANALYSIS. Used to determine system short circuit (fault) current levels, and to aid in the selection of protective device settings. For details on the most up-to-date calculating techniques, see Ref. 1, Chap. 4.
- PROTECTIVE DEVICE COORDINATION. Used to select coordinated protective device settings which will provide maximum protection and minimize system outages during fault conditions. Ref. 2 provides a detailed coverage of this subject.
- VOLTAGE DROP ANALYSIS. Used to determine voltage conditions when starting large motors and when supplying remote loads.
- POWER FACTOR ANALYSIS. Used to establish system operating power factor and to determine the best method of improvement. See Ref. 1, Chap. 7.

Some additional special studies which may be necessary are:

- Transient Stability. Used to check the ability of synchronous generators and motors to remain in synchronism during short circuits and other system disturbances; See Ref. 13.

- Voltage Regulation Analysis. Used to determine the need for special voltage regulating devices for maintaining a constant system voltage; see Ref. 14.
- Reliability Analysis. Used to evaluate the degree of reliability which exists in various system arrangements; see Ref. 15.
- Voltage Surge Analysis. Used to provide a system of protection for lightning and switching surges; see Ref. 1, Chap. 3.

EQUIPMENT DESIGN AND APPLICATIONS

Overhead Lines, Cables and Supports

Overhead Lines Within a mineral processing plant, fume or gaseous atmosphere pollution may require additional insulators to be added to increase creepage distances.

Where occasional traffic access may be required (for cranes, shovels), the use of preformed suspended aerial cable may be more appropriate for 25 kV (or below) lines, rather than bare conductors. All routing must be examined to avoid hazards and relocation in future.

Cables

INSULATION SYSTEMS P.V.C. and E.P.R. are well established as insulating materials. The manufacturing development of cross-linked polyethylene (X.L.P.E.) seems to indicate that this will be dominant as an insulant for cables up to 69 kV in future (see Ref. 16).

TABLE 3
TEMPERATURE OF CABLE INSULATION SYSTEM

Insulation	Maximum Operating Temp. °C Continuous	Short Circuit
Impregnated paper	65 - 80	160 - 250
Polyethylene (P.E.)	70	130
Polyvinylchloride (P.V.C.)	70	150 - 160
Cross-linked P.E. (X.L.P.E.)	90	250
Ethylene propylene rubber (E.P.R.)	90	250

The major advantages of X.L.P.E. are thinner insulation, (lighter cable) higher operating temperature (thermo-setting material as opposed to thermo-plastic P.V.C.), and excellent mechanical properties. Major disadvantages are that it is more difficult to manufacture, and has a more rapid flame propagation than P.V.C. Table 3 shows various cable temperature ratings, and Fig. 2 compares thermal performance.

For flexible cables, E.P.R. has now replaced other materials, principally Butyl Rubber, for similar reasons as X.L.P.E. is replacing P.V.C. E.P.R. still requires an external jacket as additional protection.

ALUMINUM VERSUS COPPER CONDUCTORS: Generalized comparisons between aluminum and copper as conductors are difficult to make. As the application of aluminum conductors in industrial situations is now well understood, a pricing and engineering study should determine which is suitable. Cable costs can be as high as 20% of all electrical material costs, so some savings may well be available by using aluminum conductors.

CABLE PROTECTION: External cable jacket materials can be made to give whatever physical properties are important to the application. The most commonly used materials in jackets are rubber, neoprene and P.V.C. (Experiments with polyurethene currently in progress do not yet lend themselves to general statements regarding applications.)

Mineral insulated metal sheathed cables should be used in specialized applications where cables are at risk of mechanical damage.

Recent industrial and utility fires have given impetus to the development of flame or smoke reducing cables which may be necessary in essential service areas. Consult with the cable manufacturer to determine what is available. As examples, the trade names "X-Flam", "Lo-Tox" and "Lo-Smoke" are self-explanatory in describing some designed properties of the cable jacket material.

Cable Supports The use of ladder type cable trays is well established (Ref. 17). This is available in up to six meter standard lengths, and need be supported at no more than this interval with reasonable deflection. Precise calculations and guidance on cable derating for stacked or closely packed cables should be obtained by reference to the N.E.C., Article 318, and the C.S.A. 22.1, Para. 12-1212, and prospective vendor data. Users should note also that restrictions exist for tray system grounding requirements. The use of both conduit and tray systems may give the best combination of material and installation costs. These may also be minimized for either system by recognizing major electrical cable runs at the design stage, and keeping changes in direction to a minimum.

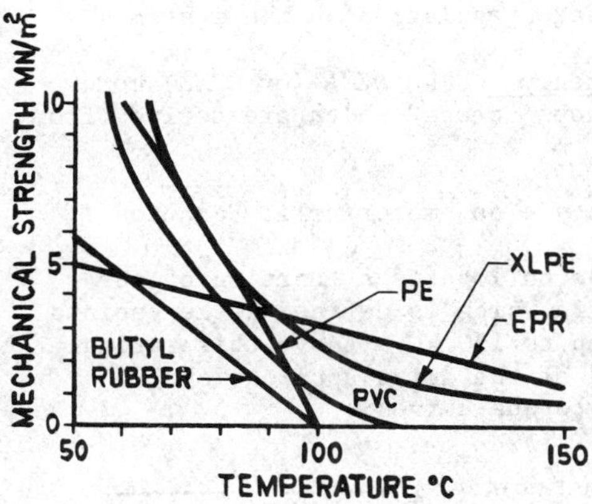

FIG. 2 RELATIVE THERMAL PERFORMANCE OF CABLE INSULATION MATERIALS.

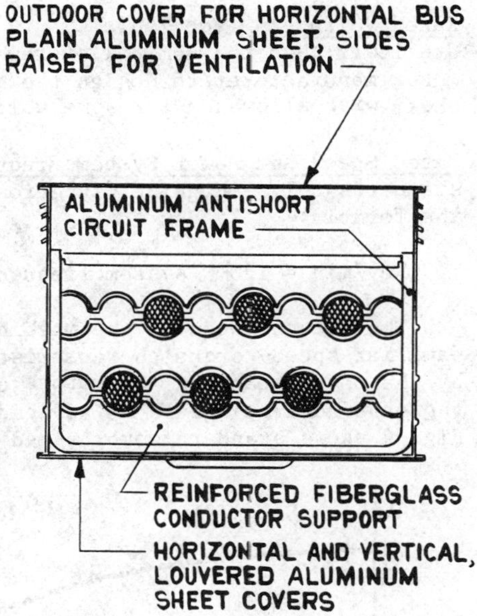

FIG. 3 TYPICAL OUTDOOR CABLE DUCT ASSEMBLY, WITH 2 CONDUCTORS PER PHASE.

Cable Duct and Bus Duct Large currents are handled using either cable or bus duct systems. Cable duct is normally easier to install under field conditions found in plants under discussion, and will probably prove more reliable in the long term due to a greater tolerance to pollution; see Fig. 3. Bus duct requires careful joint preparation and maintenance for reliable service.

Motors

General The following notes are designed to avoid the more obvious pitfalls in application design:

Electric Motor Enclosure Types The definitions of enclosure type are a guide to the ability of the electrical machine to withstand physical conditions which may be encountered in service (see Ref. 18, MG1-1.25/26). Historically, the totally enclosed fan cooled (T.E.F.C.) squirrel cage induction motor (S.C.I.M.) has been applied to most industrial applications where either permanent or occasional pollution by the process is a feature of the service. Other enclosure types may be more suitable in some applications.

Rated Voltage; (see Table 2) Note that motor terminal voltages are lower than the nominal system voltage in order to allow the motor manufacturer to design the machine to give rated shaft output power with allowance for some voltage regulation on the system.

Rated Speed and Power System Frequency The NEMA (or CEMA) motor standards only recognize "synchronous speeds" which are derived from the formula:

Rev/min = 120 x system frequency ÷ no. motor poles (Equation 1.)

The actual speed of the machine on load is a function of the familiar speed-torque characteristic which is defined at key points by the motor standard for motors up to 150 kW. Above this rating, the manufacturer should be advised of the drive application. Fig. 4 shows standard motor speed torque curves.

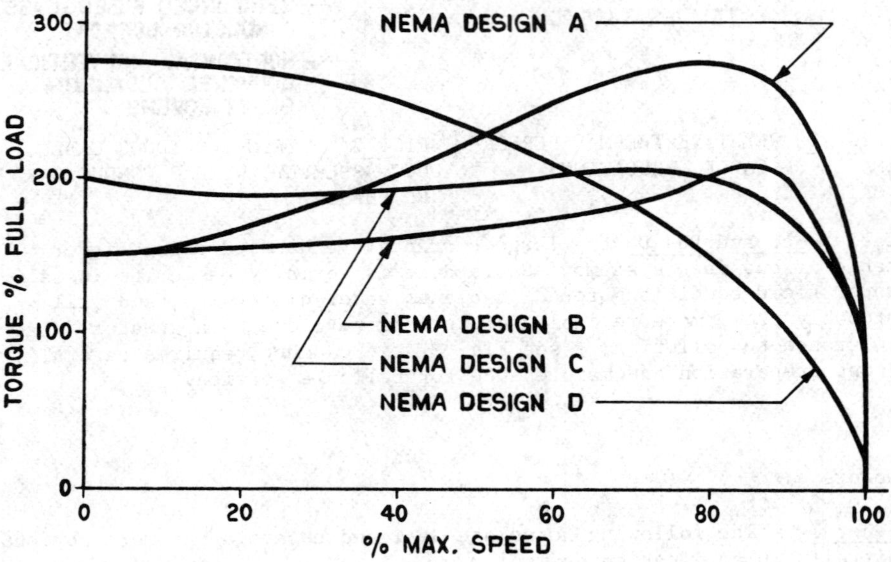

FIG. 4 NEMA MOTOR STANDARD SPEED TORQUE CURVES FOR MOTORS TO 150 KW RATING.

Industrial T.E.F.C. motors are normally designed to NEMA 'B' torque characteristics, to give the best combination of starting current and starting torque. Special drive applications may require a different speed-torque design. Note that the curves are very steep at normal operating speeds, and each manufacturer may have a slightly different speed for a given full load torque.

Direction of Rotation A.C. motors normally have shaft mounted cooling fans. In the lower kW ranges these are usually bi-directional. In larger kW ranges, (above about 37 kW) the fans may be directional, and therefore, rotation should be specified. (Consult with the motor manufacturer to confirm.) "Normal" rotation is clockwise facing the non-drive end (NEMA MG1-14.06).

Type of Duty For larger machines, it may be economical to have a motor "time rated" rather than "continuously" rated. This is defined (MG1-10.35) as being a ratio of loaded to unloaded time of 5, 15, 30 or 60 minutes based on a two-hour repetition time.

For other than "on-off" duty cycles, a reasonable method of calculating a derating factor is to use an R.M.S. method based on a two-hour (or less) time interval where:

$$Pe = \sqrt{\frac{P_1^2 t_1 + P_2^2 t_2 + P_3^2 t_3 + \ldots\ldots P_n^2 t_n}{(t_1 + t_2 + t_3 \ \ldots\ldots \ t_n)}} \ \text{kW} \qquad \text{Equation 2}$$

Pe = equivalent continuous rating.

Pi = actual rating for time ti.

The application of a "service factor" (normally 1.15 for open, drip-proof motors) allows an additional temperature margin in the winding of 10°C under normal conditions. This should be regarded as a safety margin for drives subject to occasional overload.

Method of Starting If the power system can deliver sufficient starting kVA and if the drive can tolerate the loads imposed, direct on-line full voltage starting is to be preferred above all other starting methods because of the simplicity of the starter and the development of maximum starting and accelerating torque for the drive. Any other starting method is more expensive and requires more maintenance; see section on Applications of Motor Control.

Mounting Arrangements Standard motors are designed to drive horizontally mounted in-line coupled loads. Other arrangements may require modifications to either bearing designs or to ventilation and condensate drain arrangements. T.E.F.C. motors generate moisture, due to the winding "breathing" between successive load cycles. The manufacturer should be advised of any other arrangement that may be necessary.

Winding Insulation System The basic material costs of Class 'F' or 'H' insulation materials are only 3-5% higher than those of Class 'B' materials. The higher temperature class of material is normally present in all modern low voltage, standard motor designs.

Increases in winding temperature will result in a deterioration of the insulation system. A good rule of thumb is the "half life per 10°C" rule. A normal motor insulation system which is not abused by temperature, dust, power system transients, etc., will normally last longer than the mechanical components. If it is consistently overloaded beyond the maximum design winding temperature eventual burnout is inevitable. Fig. 5 shows motor insulation temperature limits. Because of the high winding temperature design, typically within 5°C of maximum, motor casings are too hot to touch - typically 65-70°C for about 15°C ambient conditions. However, a "hot" motor is not necessarily overloaded.

Operating Characteristics Fig. 6 illustrates the penalty paid for running at less than rated output in terms of falling power factor and non-linear kVA damand. (Ref. 19).

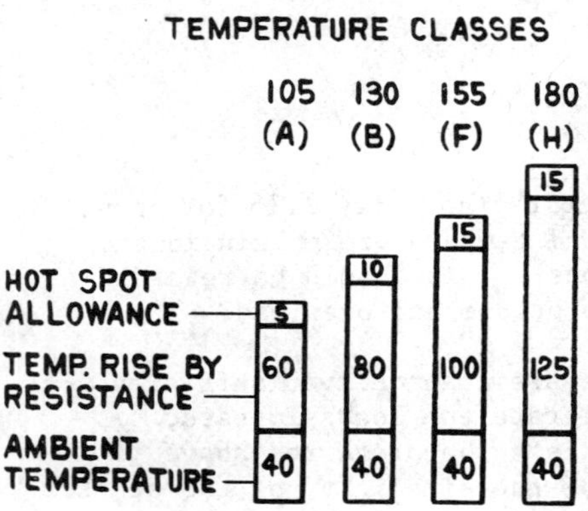

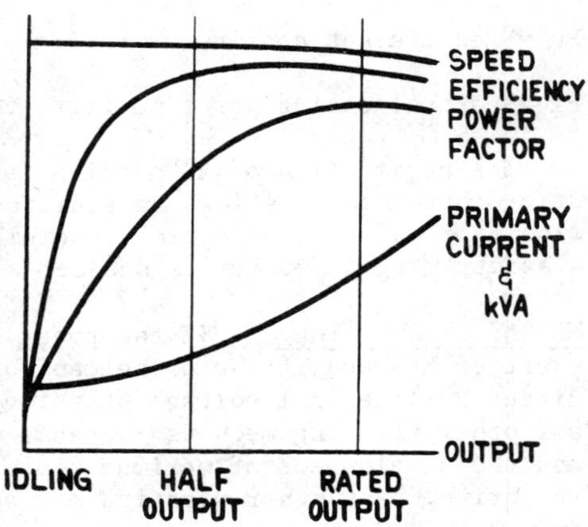

FIG. 5 TEMPERATURE RISE LIMITS FOR THE TEMPERATURE CLASSES SPECIFIED IN NEMA MG1-12.42 INTEGRAL H.P. TEFC MOTORS 1.0 S.F.

FIG. 6 TYPICAL MOTOR SPEED, EFFICIENCY, POWER FACTOR AND PRIMARY CURRENT GRAPHS SHOWN AS AS FUNCTION OF THE SHAFT OUTPUT.

Other Considerations The NEMA standard includes a list of recommended data to be advised to the motor manufacturer in sections MG1-14.02, and .03. This deals with unusual ambient conditions, altitude, shock loading, deviations from normal power system arrangements, etc.

Comments

The following notes are pointers to the more obvious design pitfalls resulting in motor failure:

Ventilation With precise motor designs and small temperature safety margins, the importance of adequate cooling is essential to maintain a reasonable machine life. All mineral processing plants produce both dust and moisture or slurry at different points in the process, and material should be prevented from building up on the cooling fins.

Duty Cycles Short time rated motors must never be used elsewhere where continuous output is required, or the duty cycle specified no longer meets the design.

Starting Frequency "Jog" starting of heavily loaded conveyors or plugged slurry pumps creates overheating problems because of the repeated inrush kVA in a short time period, which in turn eventually damages the insulation system in the motor. The NEMA and CEMA standard does not recognize more than two consecutive starts per hour for any integral H.P. machine. For machines up to 150 kW, a manufacturer may allow repeated starts up to a certain limit, depending on the frequency of starting and the load driven. The maximum starting repetition frequency allowed will be a function of accelerating time and motor thermal capacity, and therefore, will be unique to each drive situation. One manufacturer suggests the following method of derating a drive to allow repetitive starting or jogging, but the formula includes design constants applicable to that manufacturer only. Fig. 7 shows graphically how rapidly the derating factor decreases with increasing drive starting frequency.

$$\text{Permitted output H.P.} = \text{Continuous rated H.P.}\sqrt{1 - \frac{m}{mo}} \qquad \text{Equation 3}$$

$$\text{Where } m = \text{No. starts per hour} \; \frac{\text{X (motor inertia + load inertia)}}{\text{motor inertia}}$$

and mo - maximum permitted starts per hour at no load, from manufacturers data.

Example: 18.5 kW 4 pole drive, indirectly coupled, with

$\frac{\text{load inertia}}{\text{motor inertia}}$

ratio of 14

mo = 640 (from manufacturer's data)

Vibration Machine transmitted vibration will cause damage to windings, rotors, bearings and sometimes cast frames. Small air gaps will result in destroyed motor assemblies after bearing collapse, due to frictional contact of rotor and stator.

Acceleration Time If insufficient accelerating torque is developed by the motor, to allow run up times within normal design margins, the motor components will experience additional thermal stress. Normal design margins are about 10-15 seconds acceleration (locked rotor) time for occasional starting for motors to about 7.5 kW, and up to 20 seconds for motors to about 150 kW. The precise shape of the speed torque characteristic will determine how rapidly the drive will accelerate a given load. A design error that may occur is to rate the drive system for process starting (empty) under "normal" load conditions. It should be noted, that most motor designs beyond about 100 kW rating are rotor limited thermally, and therefore, cannot be easily protected from overheating under excess load during acceleration. For an excellent practical example of how to calculate accelerating time for a given load torque from a given motor design speed torque characteristic, see Ref. 21.

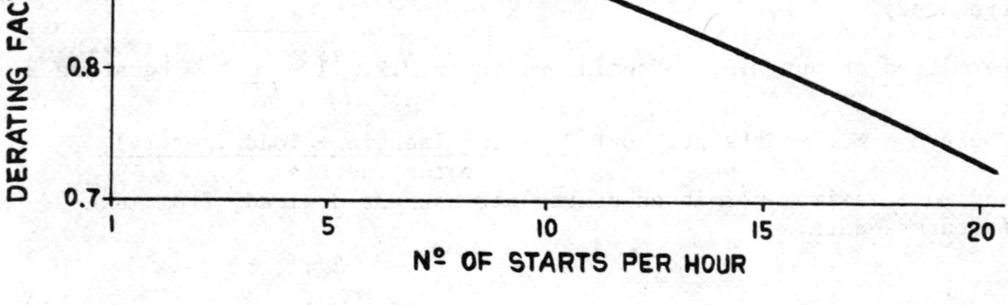

FIG. 7 TYPICAL DERATING CURVE FOR AN 18.5 KW 4 POLE MOTOR SUBJECT TO FREQUENT STARTING DUTY

Motor Control and Protection

Motor Starters The power system one line diagram, Fig. 1, shows that three kinds of motor starters are applied on the system. A full explanation of the details of the selection of devices for motor control can be found in Ref 22, Tables 1-5. This reference also gives guidelines for motor protection.

The first type of starter used in the 4.16 kV system is an air circuit breaker. For reasons relating to transient currents (both inrush and fault), contact ratings, equipment short circuit interrupting capability and motor circuit protection, this type of starter is selected as most suitable for the tandem Autogenous Mill drives (2 x 3000 kW). Fig. 1 shows that the synchronous motor windings are protected against internal faults by means of a differential current scheme and against system transient voltages by means of voltage surge arrestors. The motor protection package would be specified to also protect against circuit overcurrent and ground faults, negative phase sequence, undervoltage and excitation faults (out of step, field failure, etc.). The reversing of the mills is achieved in this case by removing the circuit breaker carriage, and plugging it back into an adjacent cell, energizing the motor leads with the phase leads connected for opposite rotation. Note that the operation of tandem synchronous drives can be electrically complex (see Ref. 23).

The second type of starter used in the 4.16 kV system shown is defined as a "Class E2 controller". Fuses are used in combination with the contactor in order to protect the motor circuit from high magnitude fault currents. Most motor protection devices can be designed to operate in conjunction with a Class E2 medium voltage controller. Contactors used as shown are normally energized continuously, and are controlled using conventional 110 volt or 220 volt 2 wire or 3 wire circuits in the same way as low voltage motors.

The third type of motor controller used is the low voltage combination starter. Features designed into these starters are: interchangeability of units of a given rating as defined by the NEMA size code (e.g. a 460 volt 18.5 kW motor requires a size 2 starter), protection for motor short circuits and overloaded conditions, and control normally remote from the motor control center.

REDUCED VOLTAGE STARTERS The previous section dealt briefly with the description of full voltage starters. Reduced voltage starters are still used in certain applications, for example, where the power system voltage regulation is poor, or if the application of full starting torque to the load creates problems on the driven machinery. The traditional resistance, liquid or transformer type is giving way to the phase controlled triac or thyristor starter. The effect of using reduced voltage starting can be explained by reference to Fig.8.

Since for S.C.I.M. machines, torque developed is proportional to the square of the voltage, the application of less than full starting torque to the drive (typically a thickener rake which has a slow moving but large mass) allows the drive train to take up and accelerate smoothly. The problem in commissioning this type of drive is to determine at which point the transition steps should be made. The solid state starter has a smooth and continuously increasing torque/speed capability from about 25% of normal starting torque for three phase combination diode-thyristor starters or from some other preset torque level for triac starters. This allows the drive train to pick up smoothly, and has been used with success on long conveyor systems (see Ref. 25).

WOUND ROTOR MOTOR STARTERS The addition of resistance (either metallic or liquid) in the rotor circuit also allows less than full load torque to be developed at the drive. The electrical method is somewhat different, as the torque speed relationship is modified by stepping from discreet characteristics at different points in time for stepped resistance starters, and is continuously variable for liquid resistance starters; see Fig. 9.

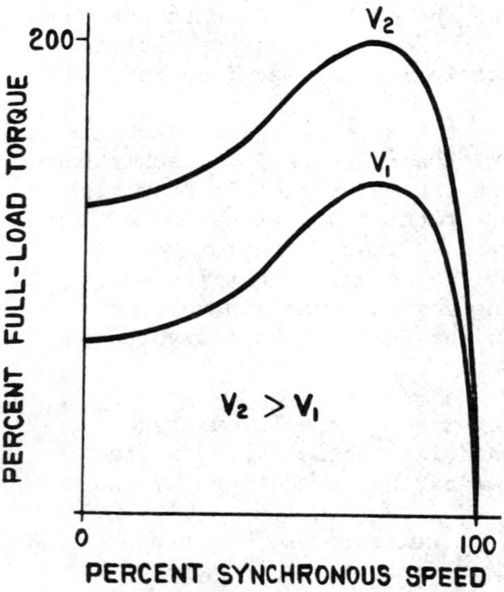

FIG. 8 BASIC SPEED-TORQUE RELATIONSHIPS FOR DIFFERENT TERMINAL VOLTAGES.

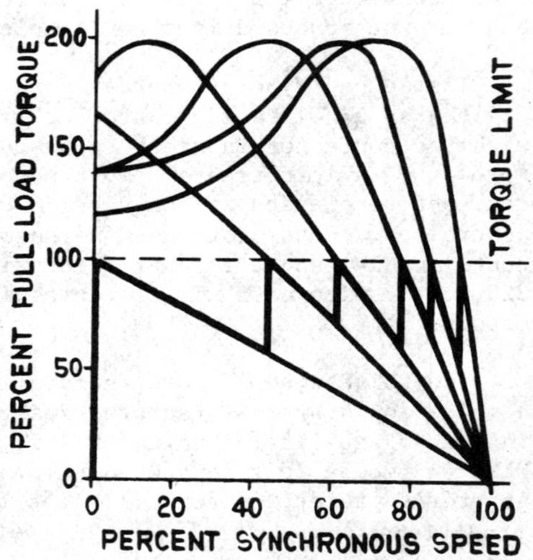

FIG. 9 SIX STEP WOUND ROTOR MOTOR CONTROL SPEED TORQUE RELATIONSHIP.

Also, in order to obtain smoother transition and starting performance, equipment is available incorporating thyristors used to modify the resistance in the rotor circuit during starting periods in order to allow a smoothly decreasing average value of resistance (Ref. 26). An added benefit of this is that with continuously rated current carrying components, some measure of coarse speed control is also available.

Motor Starter Protection For plants of the type under discussion here, the starters will be grouped in motor control centers (MCC's) and located in electrical rooms adjacent to the plant area being served. Manufacturers offer a variety of wiring design combinations, defined typically as NEMA Class 1 Type A or Class 2 Type B wiring for the individual combination starters. Class 1B or 1C wiring usually offers the best combination of initial cost versus flexibility of control design. MCC's are specified in terms of required continuous bus rating and material, short circuit current bracing, protection requirements, etc.

The location of the motor control gear in closed, ventilated electrical rooms avoids the expense of specifying enclosures for the starters capable of withstanding the physical conditions experienced in the plant operating areas, and removes the equipment from the risk of interference by untrained personnel.

The most commonly specified electrical equipment enclosures within the plant areas would be at least NEMA Type 4, which will provide protection against hosing down; Ref. 24. Gaseous or acidic atmospheres may require either stainless steel or moulded synthetic enclosure materials to a more rigorous and expensive specification.

Electrical Control Concepts

In order to satisfy the electrical control system designer, a close rapport with the potential plant operators must be established. Control concepts vary widely, depending both on the degree of automation required and expense of designing, purchasing, and installing monitoring and protective devices.

Discussions over expanded flow sheets (Process, Instrumentation and Control Diagrams) usually result in a clear statement of owner/operator requirements. Each sub-system of a process should be examined in detail, especially where obvious process areas overlap. It is essential to emphasize what is important to equipment, and the process, in order to determine the operator's function, for each event and device.

The complexity of remote control from a single plant control room is often not necessary and is itself the cause of operating problems. For example, valve operation may be automatic, but then sensing of valve position is necessary to prove operation, resulting in a two-fold increase in complexity. Physical inspection of machinery during a "local" electrical operation may prevent a more serious mechanical problem developing.

The design and positioning of consoles is important to ensure adequate visual control, light, etc., but most mineral process functions are slow and do not require extremely rapid operator corrective action. Examples of justifiable automation without operator participation are thickener rake operations and mill feed/mill load control.

The design tendency of modern plants to large scale single line flow operations should not necessarily involve a totally electrically interlocked system. Advantage should be taken of natural or increased storage from process to process, so that electrical isolation of operations between processes should be possible.

Control Equipment

This should be selected strictly on the basis of "suitability for the job". Severe mechanical and physical duty may be involved, and reliability is paramount. If a history of failure develops around a protective or monitoring device, this should be treated as a minor disaster, as the <u>necessity</u> of the device is already established from the process control concepts developed with the plant operator.

The reliability of modern components is generally very good, but problems may occur because of incorrect applications; for example, a pressure/flow monitoring device operating at the marginal end of its range. Also, physical problems may not be foreseen; for example, bin level probes, installed and not protected from rocks and falling ore. Also, the routing of the final cable to the device may give problems; for example, if used as a convenient foot rest for a greaser on a maintenance routine.

<u>Variable Speed Applications</u> The increasing use of feedback and feed-forward process control loops within modern plants (Ref. 27) means that electrical variable speed systems are becoming more common. A link between some measured variable (flow, pressure, level) and the drive system (pump, feeder) allows either an operator selected or an automatic preset feed condition to be maintained, regardless of the material density or rate of flow of material passing through the system.

Feedback signals for speed control may be derived from any variable or combinations of variables by the use of interposing and summation electronic convertors, which reduce the instrument signal input to the desired control input level. The most common control signal inputs acceptable by the drive control module are the 4-20 mA input signals derived from process instrument monitors. Any variable may be used with suitable convertors, in isolation from process instrumentation if necessary.

The reasons for having speed control on a drive are to develop a more efficient process system, by eliminating partially checked flow points, and to develop the maximum utilization of the process equipment actually being supplied.

TABLE 4
SUMMARY OF ELECTRICAL VARIABLE SPEED SYSTEMS

Type	Principle Power Source	Drive Motor
Wound rotor induction motor	A.C.	W.R.I.M.
Eddy Current Coupling	A.C./D.C.	S.C.I.M./ D.C. Coupling
Solid state variable voltage (V.V.) fixed frequency	A.C.	S.C.I.M.
Solid state V.V. & variable freq.	A.C.	S.C.I.M.
Solid state V.V. D.C.	D.C.	D.C.

Ref. 26 discusses each of these systems in terms of their torque/speed relationships.

Switchgear

General Circuit breakers are required to energize the major system components, and to isolate these components during normal or transient fault conditions. They should be selected to provide full system operation and protection without major modification for both the present system and any foreseeable future system expansion. The medium in which the interrupting contacts reside normally defines the circuit breaker type, e.g. air, vacuum, bulk oil, etc. Vacuum circuit breakers are now used where space, safety, and maintenance requirements dictate. Applications and ratings for circuit breakers are available in ANSI and IEEE standards, published by ANSI Committee C37, (Ref. 23).

Assemblies and Enclosures (Metal-Clad, Metal-Enclosed) The metal-clad breaker assembly, although it has a high first cost, can, in the long term, be the more economical enclosure.

In general, metal-clad switchgear consists of the bus work, CT's, control circuits, etc., all enclosed in a steel enclosure. In addition, it includes:

- bus insulated for its full length;
- a draw-out type circuit breaker, i.e., it is mounted on wheels or some connections are stab-in, allowing complete withdrawal;
- bus section is isolated from the breaker section with shutters that close off any openings to the bus section when the breaker is removed;
- line and load bus sections are isolated from each other;
- control wiring completely isolated from the H.V. section.

Metal-enclosed breaker assemblies, on the other hand, consist basically of a steel enclosure assembled around the breaker bus work and control compartments:

- circuit breaker is not necessarily draw-out, i.e., it may have bolted connections to the bus;
- bus work is not isolated from the breaker section; when the breaker is removed the bus is exposed;
- the bus is not insulated.

The net result is that maintenance of a breaker in metal-clad switchgear is easier and safer than in metal-enclosed switchgear. Corrosion is less likely to cause problems in metal-clad switchgear. A bus fault in metal-clad switchgear is contained within a separate enclosure and therefore is less likely to damage other parts of the switchgear assembly.

Transformers

General

Only the conventional types of installation are discussed. It should be recognized that there are many special applications or combinations of factors outside the scope of this discussion that may affect the decision as to the type of transformer to be used. Engineering judgment coupled with economic evaluation should be used in reaching a decision. Additional data may be obtained in ANSI and IEEE Standards (Ref. 29), and from manufacturers.

TABLE 5 TRANSFORMER COMPARISON

Type of Transformer	LIQUID FILLED Oil	Flame Retardant	DRY TYPE Open Ventilated	Sealed Class B	Sealed Class H
Impulse strength	100%	100%	50%*	50%*	50%*
Total loss at 75° C.	100%	100%	100%	100%	100%
Insulation	Class A	Class A	Class B	Class B	Class H
Temperature ratings:					
Average rise	55° C.	55° C.	80° C.	120° C.	120° C.*
Hottest spot rise	65° C.	65° C.	110° C.	140° C.	140° C.*
Audio sound level	X db	X db	(X+10 db*)	(X+10 db*)	(X+10 db*)
Weights	100%	125%	80%	125%	125%
Dimensions:					
Floor space	100%	100%	100%	120%	120%
Height	100%	100%	90%	110%	110%
Normally available for application:					
Indoor or outdoor	Outdoor**	All	Indoor only	All***	All***
Submersible	Submersible				
Fire and explosion resistant	No	Yes	Yes	Yes (Plus)	Yes (Plus)
Maintenance required:					
Liquid	Normal	Infrequent	None	None	None
Internal cleaning	None	None	Frequent	None	None
External cleaning and painting expense	Normal	Normal	Subnormal	Minimum	Minimum
Special precautions before energizing either initially or after shutdown	None	None	Yes	None	None

* NOT YET COVERED BY INDUSTRY STANDARDS.

** MAY BE INSTALLED INDOORS IN A FIRE RATED ROOM.

*** APPLICABLE FOR ALL TYPES OF INSTALLATION ASSUMING NO EXPENSE TO LIGHTNING OR ASSUMING ADEQUATE PROTECTION AGAINST IMPULSE VOLTAGES CAN BE PROVIDED.

TABLE 6 TRANSFORMER OPERATING CONDITIONS

		TYPE OF TRANSFORMER RECOMMENDED		
		Open Dry	Flame Retardant	Oil Filled
Exposure to Lightning	Where transformers are connected to circuits exposed to lightning and the usual protection is provided.	No	Yes	Yes
	Where study has determined the amount of lightning exposure is negligible, or the possible resultant voltage stresses can be adequately taken care of by lightning protection.	Yes	Yes	Yes
Atmospheric Conditions	Where atmospheric conditions are clean, such as in clean, dry vaults or electrical rooms.	Yes	Yes	Yes
	Where dirt conditions are severe, such as in milling halls or other similar dusty and dirty locations.	No	Yes	Yes
	Where moisture conditions are severe, such as in geographic locations of high humidity and where the transformers may be subjected to partial or total submersion by water.	No	No	No
	Where units are subjected to acid, oil, or corrosive vapors.	No	Yes	Yes
	Where units are subjected to oil or inflammable vapors and location is classified as hazardous or semi-hazardous. (See Article 500 of National Electrical Code for guidance.)	No	Yes	No
	Where transformers are located overhead on platforms or in roof trusses.	Yes	Yes	No
Future	Where possible rearrangement may cause transformers to be moved outdoors at a later date.	No	Yes	Yes

Classes

Liquid Filled:	- Oil Filled - Flame Retardant Liquid Filled
Dry Type:	- Open Ventilated - Sealed

Cost Comparison The oil filled transformer has the lowest first cost. However, if the transformer is to be installed indoors, the cost of a fire rated room may negate any price advantage.

The following is a comparison of relative first costs of various transformers:

Type	First Cost
Oil	100%
Flame Retardant	125%
Dry Core Open Ventilated	125%

Although both liquid filled and dry type transformers have been mentioned above, liquid filled units are generally preferred to dry type in industrial application because of their reduced susceptability to damage from lightning and other system transients.

Overload Capacity When specifying transformers, it is wise to consider liquid filled units with premium insulation and provision for future fan cooling. The premium insulation provides a dual rating, i.e. kVA at $55^{\circ}C$ and $65^{\circ}C$ rise. The difference in kVA rating from $55^{\circ}C$ rise to $65^{\circ}C$ rise is approximately 12%. The addition of fans at some future date will further increase the capacity by 15% to 20%. This provides a built-in load increase factor of up to 40%. In addition, for guidance on short time overloading factors, see Ref. 30 which is currently being revised.

The use of 'askeral' in flame retardant liquid filled transformers has been abandoned because of possible environmental damage, and new products, such as silicone based liquids are being developed. Since final acceptance of these new fluids is not complete, it would be wise to consider only oil filled transformers for outdoor applications.

Check List The following is a list of basic items to be considered at the time of ordering a transformer:

- kVA requirements;
- need for future fan cooling;
- primary and secondary voltage;
- winding connections (most common and problem free is delta-wye; Ref. 9);
- insulating/cooling systems, i.e. paper/oil, epoxy-mica/air, etc.
- impedance (determined from short circuit study);
- auxiliary equipment:
  - (a) tap changers, type and range;
  - (b) relays - over-pressure, sudden pressure, over-temperature, etc.;
  - (c) lightning arresters;
  - (d) CT's - if built-in, type and range.
- location, including elevation, if over 3000 ft.;
- environmental conditions.

Lighting

In today's energy conscious society, lighting is one area in which energy conservation can be achieved while providing an improved system. The use of gaseous discharge lamps in place of incandescent lamps can provide energy savings of up to 50%.

Lamps In production areas, maintenance areas, and outdoor lighting, the most efficient and controllable lighting is produced by high intensity discharge, H.I.D., mercury, metal halide, or high pressure sodium lamps. In office areas and other low or relatively low ceiling rooms, lighting produced by fluorescent lamps is generally more efficient.

Luminaire In high and medium ceiling areas the hi-bay or lo-bay luminaire with HID lamps is an efficient means of providing the required illumination. Since these lamps are in effect high intensity point sources, consideration should be given to proper shielding to reduce the brightness to reduce fatigue.

In offices, etc., fluorescent luminaires with good quality shielding can create an atmosphere conducive to efficient work.

TABLE 7
ILLUMINATION LEVELS

Based on I.E.S. standards, the following is a list of areas and the recommended illumination:

AREA	LUX
General production and maintenance (specific areas may require greater illumination)	300
Control rooms with vertical face control consoles	540
Laboratories	540
Offices	750-1100
Drafting	1600
Washrooms	300

Quick Estimating Data Figure 10 provides a guide to power consumption to achieve a given illumination with various types of lamps. For a recent cost comparison of lamps of different types, see Ref. 31.

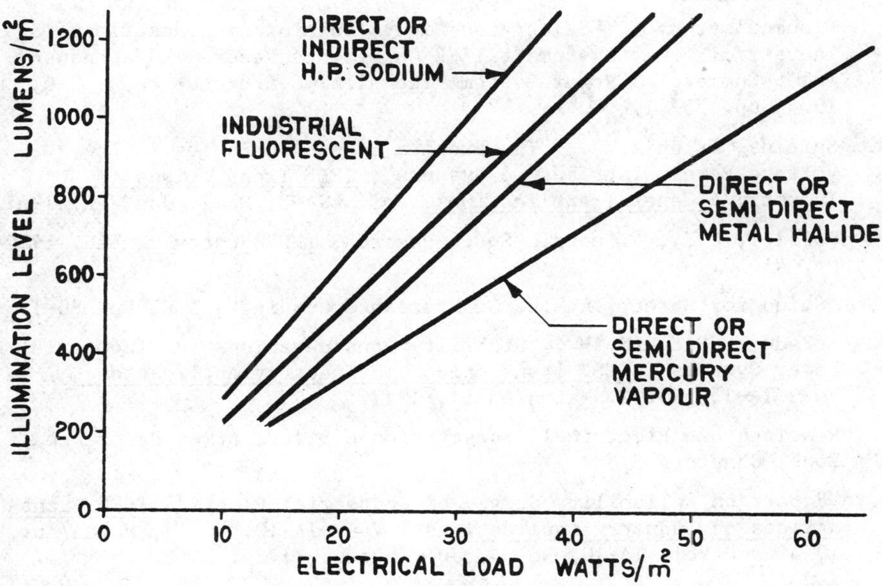

FIG. 10 SKETCH SHOWING RELATIVE LOADING DENSITY FOR TYPICAL INDUSTRIAL LIGHTING

REFERENCES

1. "Recommended Practice for Electric Power Distribution for Industrial Plants", IEEE Standard 141 - 1976

2. "Recommended Practice for Protection and Coordination of Industrial & Commercial Power Systems", IEEE Standard 242 - 1975

3. "Recommended Practice for Grounding of Industrial & Commercial Power Systems", IEEE Standard 142 - 1975

4. "Recommended Practice for Emergency and Stand-by Power Systems", IEEE Standard 446 - 1974.

5.a. McPartland, J., & editors of Electrical Construction & Maintenance Magazine "How to Design Electrical Systems", (McGraw-Hill, 1968)

b. McPartland, J., & Novak, W. "Electrical Design Details", (McGraw-Hill, 1960)

6. Fink, D.G. & Carroll, J.M. EDS "Standard Handbook for Electrical Engineers" (McGraw-Hill, 1968)

7. Beeman, D., "Industrial Power Systems Handbook", (McGraw-Hill, Lib. Congress Cat. No. 54-10629)

8. Thomas, M.A., "Power Plant Design For the Mining Industry", CIM Bulletin, January, 1976

9. Bloomquist, W.C., "Select the Right Transformer Connections for Industrial Power Systems", IEEE Conference Record, 1974 Industrial & Commercial Power Systems Technical Conference, June 2-6, 1974, pp. 35-41.

10. Shields, Francis, J., "The Problem of Arcing Faults in Low Voltage Power Distribution Systems", IEEE Transactions on Industry & General Applications, Vol. LGA-3, No.1, Jan/Feb. 1967.

11. Dalziel, C.F., "Electric Shock Hazard", IEEE Spectrum, Feb. 1972, pp. 41-50.

12. "Guide for Safety in A.C. Substation Grounding", IEEE No. 80-1976

13. McFadden, Richard, H., "Stability Considerations for Industrial Power Systems", IEEE Transactions on Industry Applications, Vol. 1A-13, No. 2, March/April, 1977.

14. Westinghouse Electrical Transmission & Distribution Reference Book, Chapters 5 & 7.

15. "Report on Reliability Survey of Industrial Plants", IEEE Transactions on Industry Applications, Vol. 1A-10, No. 2, March/Apr. 1974, and Vol. 1A-10, No. 4, July/August, 1974.

16. King. A., "Mine Electrical Cables - 1975", Canadian Institute of Mining & Metallurgy Journal, Vol. 69, No. 771, July, 1976 pp. 72-82.

17. Chamlee, R.D., Swanson, G.W., "Cable Tray Installation for Process Plants", IEEE Transactions of Industry Applications, Vol. 1A-13, No. 2 March/April 1977, pp. 115-119.

18. "Motors & Generators", NEMA Standards Pub.No. MG1-1973.

19. Knowlton, A.E., "Standard Handbook for Electrical Engineers, 8th Ed., McGraw-Hill, Section 7-207, pp. 711-712.

20. "Totally Enclosed Three Phase Squirrel Cage Motors", ASEA Ltd., Cat. No. MK20E, Part B1, Ed. 7, Feb. 1977, File M, pp. 2.

21. Roberts, O.S. "Conveyor Belt Drive Selection", Society of Mining Engineers Proceedings, Sept. 22-24,1971, Seattle, Washington, (Available as reprint No. 71-B-301)

22. "Guide for A.C. Motor Protection", IEEE Std. 588-1976, ANSI C37.96 - 1975.

23. Valentine,R.D., Trasky, J.G., Rippin D.R., "Load Sharing of Dual Motor Grinding Mill Drives". IEEE Trans. on Industry Applications, Vol. 1A-13, No. 2, March/April, 1977, pp. 161-167.

24. "Industrial Controls & Systems", NEMA Standards Pub. 1CS-1970

25. Dunmire, H.T., "Solid State Motor Starting for the Mining Industry" (Abstract). Proceedings of 3rd WVU Conf. on Coal Mine Electrotechnology, Morgantown, W.Va. 4-6 Aug. 1976, (IEEE) pp. 7/1-6.

26. Hollis, C.R., "A Review of Five Types of Electrical Adjustable Speed Drives & Their Application to a Centrifugal Pump", Paper Ref. PS-MON-5, Emerson Electric Co. Internal Conf. 1975.

27. LeGuen, F., "The Control of a Flotation Process", Canadian Institute of Mining & Metallurgy Bulletin, Vol. 68, No. 756, April 1975, pp. 113-128.

28. "Application Guide for A.C. H.V. Circuit Breakers," ANSI C37.010, IEEE Std. 320.

29. "Standard Requirements for Distribution, Power & Regulating Transformers", ANSI C57.12.00-1973, IEEE Std. 462-1973.

30. "Guide for Loading Oil Immersed Distribution & Power Transformers," ANSI C57.92.

31. Chen, K., "The Energy Orientated Economics of Lighting Systems", IEEE Transactions of Industry Applications, Vol. 1A-13, No. 1, January/February, 1977, pp. 62-68.

8

Engineering-Construction and Future Plant Design

Chairmen: T. Henderson
J. M. Keane

Chapter 38

THE ROLE OF THE CONSULTING ENGINEERING FIRM IN MINERAL PROCESS PLANT DESIGN

By J. C. Bowling

Vice President, Engineering
Wright Engineers Limited
Vancouver, B.C.
Canada

ABSTRACT

The services provided by consulting engineers to the mineral process industry are briefly described. The organization of a multi-discipline consulting firm is outlined. Finally, the role of a consulting engineering firm is amplified by a description of its activities in the design and construction of a mineral process plant from inception to start-up.

INTRODUCTION

World-wide development of the mineral processing industry has been accompanied by a parallel growth in the number and size of consulting engineering firms providing services to the industry. Some of these firms have developed to where they can provide a full range of services from initial studies to start-up of a plant. Many others provide specialist advice and design services in such areas as mine evaluation, soils investigations, the environment, water supply, tailing dams and transportation, which are included in the development of most new projects.

The need for consulting engineering services by a mine operating company arises when the company is:-

1. Developing a new project.
2. Expanding or modifying an existing plant.
3. Requiring specialist advice.

Generally, a mine operating company (the client) will not have on staff the specialists required for all its activities. Furthermore, major new projects or plant expansions do not occur with sufficient frequency to justify a mine operating company carrying a staff large enough to undertake them. Consequently, a consulting engineering firm is retained to provide the required services.

In addition to providing the necessary professional and support staff, a consulting engineering firm brings to a new project a diversity of experience and information gained from previous projects.

In some instances, particularly for a large project when a wide range of services is required, a number of firms will combine to form a "Joint-Venture" group to carry out the project.

Many of the topics discussed in this chapter are covered in greater detail in other chapters. They are included here to illustrate the scope of services and range of activities of a consulting engineering firm in the field of mineral process plant design.

SCOPE OF SERVICES

The consulting services required for a particular project vary with the size and scope of the project. The overall development for a typical project consists of the mine, the process plant, contained within well-defined limits (sometimes called "battery limits"), together with essential support facilities such as access road, water supply, power supply, housing, communications and a terminal for shipping the final product. These support facilities are often termed "offsites" inasmuch as they are outside the main plant site.

The engineering disciplines of mining, metallurgical process, mechanical, civil/structural, electrical and instrumentation are commonly represented on the staff of a consulting firm engaged in the design of mineral process plants. Specialized services for geotechnical (soils), environmental, water supply, tailing disposal, architectural and other work may also exist in-house, or may be carried out by independent consultants practicing in these fields.

The role of a specialist consultant is limited to investigations, reports, design and possibly construction supervision of the specific task for which the firm is retained.

The services provided by a large multi-discipline consulting engineering firm may range from feasibility studies to start-up of a plant, and can be considered in six major categories:-

1. Feasibility Studies.
2. Project Management.
3. Engineering Design.
4. Scheduling and Estimating.
5. Purchasing (Equipment) and Construction Contracts.
6. Field Activities.

Feasibility Studies

Technical studies, cost estimates and cash flows are required before large expenditures for engineering and construction are committed.

Technical studies involve the assessment and correlation of project information, including mine and metallurgical data and the results of preliminary site investigations. These studies are carried out to the extent necessary to satisfy the consulting engineer and the client that each area of the project is technically feasible.

Estimates of capital and operating costs are then prepared from the technical studies. Finally, these costs, together with other cost data such as taxes and finance charges are incorporated into cash flows for the projected life of the plant. From the cash flows the client can assess whether the development will provide an acceptable financial return for the investment involved.

It is on the basis of a feasibility study that a client normally decides to proceed with a project and make the necessary financial arrangements. Major lending institutions such as the World Bank and private banks customarily require a feasibility study before committing funds for the project.

Sometimes, before a full feasibility study is carried out, a consulting engineer is retained to prepare a pre-feasibility study consisting of a general assessment of the technical aspects of the project together with a preliminary cost estimate and cash flow.

Project Management

Project management is the co-ordination of all the activities on a project from the time the decision is made to proceed with design and construction until the start-up of the plant and its final acceptance by the client.

The overall objectives in the managment of the project are:-

1. An operating plant that meets the performance requirements and product quality established by the client.
2. Total costs within approved budget.
3. Completion on schedule.

Engineering Design

Engineering design commences after a feasibility study has established the economic viability of the project and the decision to proceed has been made.

The first phase of the design activity is variously termed conceptual engineering, basic engineering or preliminary engineering. It consists of:-

- Preparation of final process flow sheets.
- Major equipment selection.
- Thorough site investigations.
- Preparation of site layouts and plant arrangement drawings.
- Technical and cost optimization of each area of the project.

Final design and the preparation of construction drawings and specifications are carried out after approval by the client of the conceptual engineering for each area of the project.

Scheduling and Estimating

At the beginning of a project an overall plan, frequently a precedence network, is prepared. The network relates the various project activities one to another, and incorporates a duration estimate for each activity. From this plan the overall time required for the project is established and schedules are prepared for the design and construction of the project.

Cost estimates, with improving degrees of accuracy, are required as a project progresses. After the feasibility study, an early requirement is a budget capital cost estimate, usually prepared at the end of the conceptual engineering phase to serve as a basis for cost control. Later on, as the project is defined in greater detail, a more definitive estimate may be required for more accurate cost control.

The information in the overall schedule and the budget capital cost estimate is integrated in the preparation of the cash flow for the design and construction phases of the project.

Purchasing and Construction Contracts

Information on the equipment for a new process plant, such as dimensions, weights and horse power is required for final layout and design of the plant. Hence, it is common practice for the consulting engineering firm to select suitable equipment, obtain quotations, analyze bids and recommend a supplier to the client. The consulting engineer may then either requisition the equipment for purchase by the client or may place the order himself as agent for the client.

When construction drawings and specifications for items such as excavation, concrete, structural steel and equipment installation are complete, they are combined with the other documents required for contract purposes and then bids are solicited from suitable contractors. The bids are analyzed and a construction contract is awarded to the selected contractor.

Field Activities

The services provided by a consulting engineer extend into the construction and start-up phases of a project and may include:-

- Field Engineering.
- Construction Supervision.
- Contract Administration.
- Cost Control.
- Administration of the Construction Camp, Warehouse, Field Office and Related Items.
- Start-up Services.

Terminology varies for the field activities carried out by consulting engineering firms, but for the purpose of describing the nature of the work involved in each activity the following definitions are used:-

Field engineering covers distinct engineering activities at the site such as:-

- Surveying.
- Design for field changes.
- Measurement of quantities.
- Preparation of as-built drawings.

Construction supervision is chiefly supervision and inspection of construction for compliance with the drawings and specifications.

Contract administration is the administration of construction contracts which have been awarded either in the engineering office or the field.

Cost control is the process of continuous comparison of actual and budgeted costs, with the objective of maintaining the project cost within budget.

Construction management is the overall co-ordination of the foregoing activities together with administration of the construction camp, warehouse for construction supplies, field office and related items. The scope of these activities will vary, depending upon whether there is a general contractor with subcontractors for the trades or there are several contractors co-ordinated by the construction manager.

Start-up services consist of assisting construction crews and future plant operators in the start-up and run-in of the process equipment, and in obtaining the specified performance requirements of the plant.

ORGANIZATION

The personnel of most consulting firms, particularly the larger ones, engaged in the design of mines and mineral process plants, consist of specialists, project personnel (mainly professional engineers and support staff) and administrative staff.

The overall administrative organization of a firm will depend upon the size, the range of services offered and its historical development. An effective organization is required for carrying out concurrently a number of projects within the framework of the administrative organization. Today, this is usually achieved by a matrix-type organization where the requisite personnel from the various engineering and other departments (e.g. estimating and contracts) are assigned to each project to form a project team. This type of organization achieves the two objectives of providing project teams to carry out each project effectively, and at the same time ensuring that the overall expertise of the company is available to each project. Figure 1 illustrates this matrix concept diagramatically.

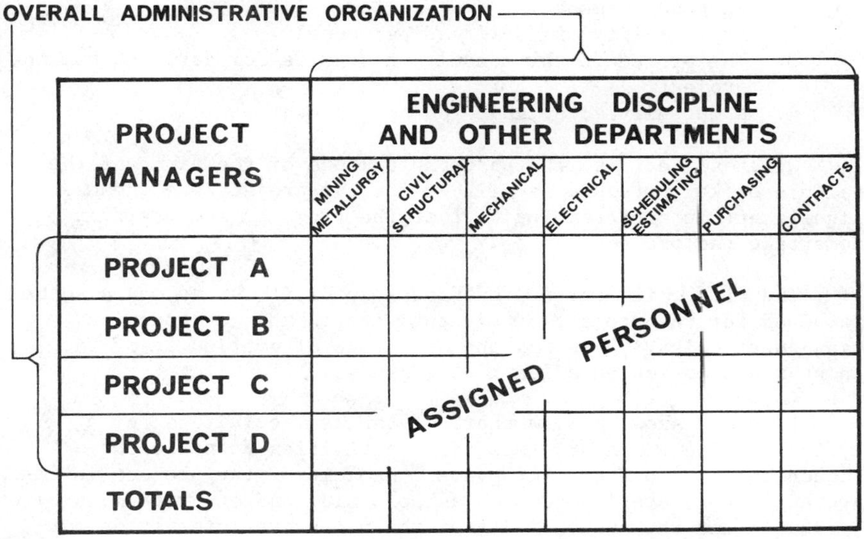

Figure 1: Matrix organization for a multi-discipline consulting engineering firm.

Each project team is assigned a project manager with overall management responsibility for the project. The technical and administrative personnel assigned to the project report to the project manager for the project requirements and to their appropriate department head (discipline design chief) for technical supervision. The project managers and department heads report to the overall administrative organization of the firm.

Staff specialists within the organization are available for consulting on all projects.

TYPICAL PROJECT

The start of a project usually occurs when its economic viability has been established, suitable financing has been arranged, and sales contracts for the final product, e.g. concentrate, have been arranged.

In selecting a consulting engineering firm for the project the client may invite proposals from qualified firms and then select the most suitable on the basis of:-

1. Past performance.
2. Availability of qualified personnel.
3. The procedure the consultant submits for carrying out the project.
4. A mutually acceptable fee arrangement.

Of these factors, the past performance of the firm and the calibre of key personnel available for the project are the most significant in ascertaining whether the firm can successfully undertake the project.

The firm which carried out the feasibility study may also be retained for the project management, design and construction management roles. This has the advantage of continuity of the thought and action entailed in the study.

Figure 2 shows a flow chart of the key activities for a typical project. In practice these activities do not occur sequentially. In fact, the overall economy of the project is improved if they overlap as much as possible, commensurate with client approvals and the availability of the necessary information for each activity.

The periodic review and approval by the client of the ongoing activities of the project team is also illustrated in Figure 2.

In order to demonstrate more fully the role of the consulting engineering firm in a typical project each of these activities is now described in greater detail.

Project Inception

After a consulting firm has been selected for a project, the first step in the formation of the project team is the appointment of a project manager to be responsible for the overall management of the project. The client appoints his own representative for continuous liaison with the project manager.

The project manager and the client's representative jointly review the Terms of Reference for the project and establish the scope of services to be provided. Usually the contract for engineering services is executed at this time.

The number of key personnel, and their organization, depends upon the size and scope of the project and the extent of services to be provided. Figure 3 shows the organization for a typical large project where the services extend from design to plant start-up. Engineering and administrative personnel are assigned to perform the various functions and to constitute the project team.

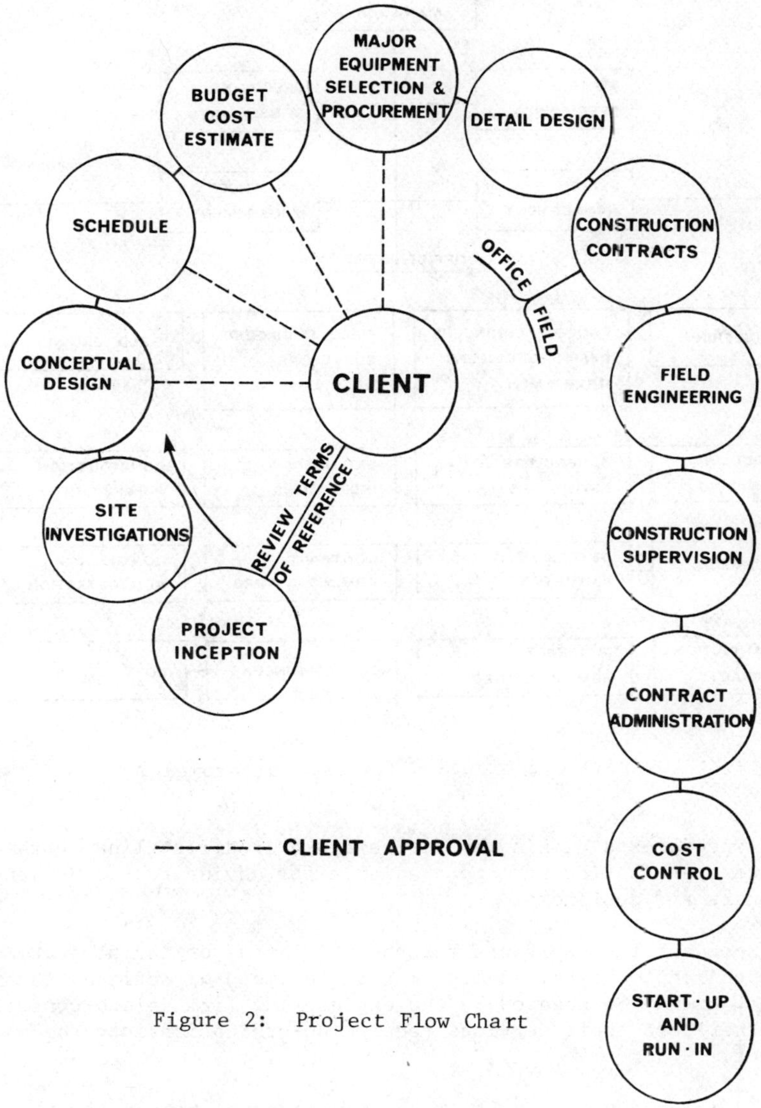

Figure 2: Project Flow Chart

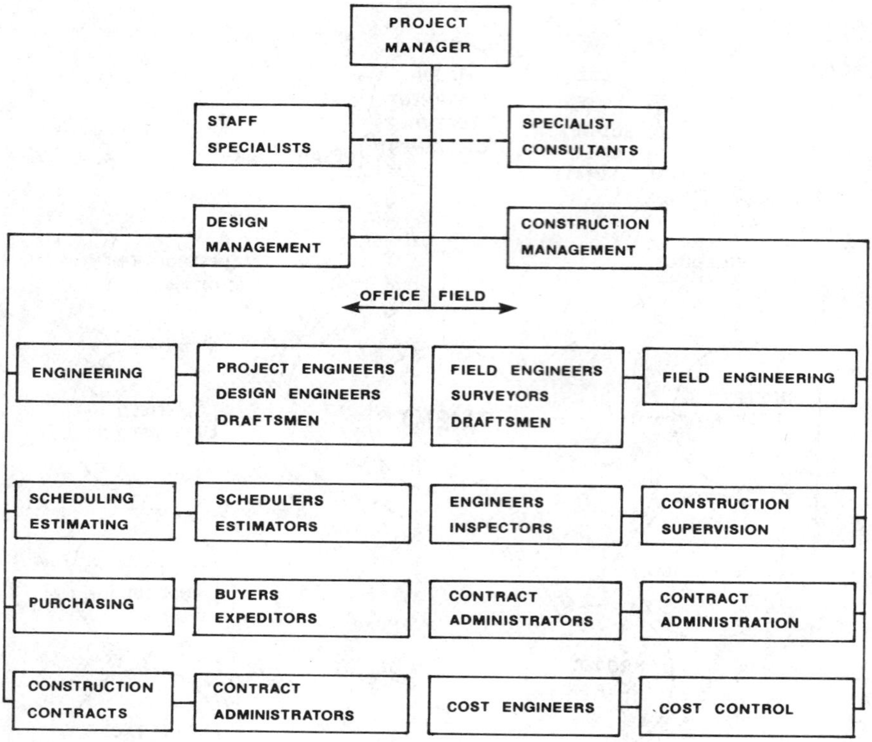

Figure 3: Overall organization for typical project.

Staff specialists within the firm are not in direct line in the project organization but are available for advice to the project engineers and designers.

Specialist consultants for specific areas of the project, or for specific technical advice, may be retained directly by the client, or may be engaged by the engineering firm as sub-consultants. In either case their services require co-ordination into the project.

It is good practice at the beginning of a project for the consulting engineer to prepare a manual of project procedures for approval by the client. Some of the more important items to include in this manual are:-

- Defining responsibility and time limits for approvals.
- Distribution of correspondence.
- Numbering system to be used.
- Purchasing and expediting procedures.
- Drawing issue procedure.
- Circulation and review of vendors' drawings.

Site Investigations

The collection of information from the field is an essential part of the assembly of data for engineering purposes. This information is required for studies, design, construction and operation, and for assessment of the impact on the environment. The extent of necessary information depends upon the size of the project and the nature of the site. In the case of a complex development in difficult terrain or close to large centres of population, data collection can be an expensive and lengthly process. An indication of the field information required is contained in the following broad classification:-

(a) Physical and chemical data relating to:
- earth (soils, topography);
- water (groundwater, water quality, permafrost);
- atmosphere (climate records, air quality);
- natural processes (erosion, slope stability).

(b) Cultural factors such as:
- land use (agricultural, residential, recreation);
- aesthetics and human interests (parks, archeology);
- cultural factors (employment and population densities);
- costs and availability of labour, materials, equipment and power

(c) Man-made facilities such as:
- communications;
- transportation;
- utilities;
- housing, hotels, schools and medical facilities;
- local consultants and construction companies;
- industries;
- waste treatment.

(d) Administration:
- government and regulatory authorities;
- legislation and permits;
- taxation;
- labour unions.

Some of the information listed above will have been accumulated in the study phase of the project and would be augmented by more detailed investigations for design purposes.

Conceptual Design

In this, the conceptual phase of the project, the assurance of a successful operation is established, and the most significant cost savings can be made. The most important activities in the design of a process plant are: the selection of the major equipment; the preparation of the final flow sheet; the optimization of the general arrangements of each of the unit processes and off-site facilities, and their overall layout on the site.

Alternative arrangements and cost estimates are prepared with due consideration of:-

- Site conditions: topography, foundation soils, climate.
- Access to the overall site and to each unit process or facility.
- Proximity to the underground mine or open pit.
- Physical and metallurgical properties of the ore and beneficiated products.
- Operating and maintenance features.

The best of the alternative schemes is selected, taking into account the technical, operational and cost aspects. The client's approval is then obtained before proceeding with final design.

Schedule

A preliminary schedule is prepared as soon as the work can be identified. Before detail design commences, the schedule must, of necessity, be broad in scope and may specify durations but not identify actual dates when work is to be performed. Such schedules often take the form of bar charts but there is an increasing trend to the use of precedence networks for scheduling.

Schedules are guides to assist management in processing the work efficiently. They are continually up-dated to reflect the latest information available. Up-dating consists of shortening durations where practical, as well as lengthening them, or changing starting dates. The schedule becomes more accurate as design is completed, orders for equipment are placed and contracts are let.

It is important, from a management point of view, to ensure that the schedule does not become an end in itself. It provides useful information which is of value only if it is available and used by those who need it. Durations and start dates should not be arbitrarily imposed but should be decided in co-operation with contractors and equipment suppliers.

As a project develops the number of activities often becomes unwieldy to handle on one precedence network and it becomes necessary to create a number of convenient sub-networks. A scheduler can handle about 200 activities manually; however, if the number of activities exceeds 200 a computer program should be considered.

Budget Cost Estimate

The budget cost estimate is prepared to guide the client and engineer on the capital cost aspects of the plant and to serve as a tool for cost control. The estimate is prepared from general arrangement drawings after major equipment has been sized.

The budget cost estimate provides a good check for project management on unacceptable increases in the scope of the work, e.g., if quantities of materials or costs of equipment for the completed design are shown, by quantity take-offs and suppliers' quotations, to substantially exceed the comparable figures in the budget cost estimate, then a check can be made to ensure that the authorized design has not been changed.

It is necessary to consider all costs when preparing the budget cost estimate. Taxes, freight, import duties, camp costs and client's administration costs all have to be paid, and adequate allowances must be made for them in the budget. It is also important to make allowance for unforeseeable events: foundation rock may not be as sound as anticipated; adverse weather may impede construction; strikes may delay equipment deliveries. A contingency allowance must be included in the cost estimate to cover such events.

Equipment Selection and Procurement

Technical specifications are prepared for each type of equipment. These are basically "performance type" specifications incorporating the following information:-

(a) Particulars of the environment such as: geographic location; whether housing in a heated building is necessary.
(b) Operation and duty.
(c) Any limitation of size; power availability; water and air supply.
(d) General design requirements or preferences such as adherence to national standards for gearing; raw materials; welding.
(e) The scope of the specification, e.g. with or without driving motor, foundation bolts, etc.

(f) A questionnaire covering design details, to be completed by the bidder.

Commercial documents are also prepared. These documents vary from project to project to incorporate clients' terms and conditions, and include instructions for such items as pricing structure, method of packing, shipping and payment. The commercial conditions are combined with the technical specifications to create the Bid Documents.

The consultant prepares a list of bidders for each type of equipment. This list is influenced by his direct experience of past performances; by geographic location; and by the terms of the contract. Client approval of the bidders' list is obtained before quotations are solicited.

The specifications are sent to the selected bidders with a clear statement of the delivery required for the equipment, and a specified closing date for the submission of the bid.

The bids are evaluated from both technical and commercial aspects:-

Technical
(a) General conformity to the specification.
(b) Design features; operating cost; durability; maintenance.
(c) The merit of any alternatives that may be suggested by a bidder.
(d) Ease of shipping and assembly.
(e) Bidder's plant facilities and capabilities.

Commercial
(a) Total cost delivered to site.
(b) Delivery schedule.
(c) Cost and availability of spare parts.
(d) Compliance with requested terms and conditions.
(e) Bidder's performance record.

The evaluations are usually prepared in a standardized, tabular form, with the final choice covered in a short memorandum to the client, high-lighting the basic reasons for the choice and indicating the order of preference for the remaining bidders. After approval by the client the purchase order is placed with the successful bidder.

When the consultant is performing the total purchasing function, he handles all the associated paperwork, such as licences, invoices, bills of lading and export documentation.

Placing orders, particularly for major equipment, as soon as possible in the project program, results in early receipt of the equipment information necessary for final design and preparation of construction drawings.

Inspection and expediting of equipment during manufacture and delivery is frequently carried out by the consulting engineer as part of his purchasing services.

Detail Design

The final design and preparation of construction drawings and specifications constitute the largest phase, in terms of engineering man hours for the design of a mineral process plant.

A manual of design criteria is prepared for use in this phase which incorporates:-

General Information:
- Location; climate; earthquake probability.

Operating Data:
- Plant life, capacity, manpower.
- Operating times.
- Metallurgical data.
- Characteristics of the ore and beneficiated products.

Design:
- Applicable codes.
- Design and drafting requirements and standards for specific items in each engineering discipline.

An engineering budget for the detail design phase is prepared from the information established in the conceptual design phase. The budget consists of a tabulation of all the detail activities involved, including supervision, design, specifications and each individual drawing. Start and finish dates, and the estimated engineering man hours for each of these activities, are incorporated. The final document serves as a tool for controlling the progress and cost of the engineering work.

During the detail design phase the consulting engineer uses his previously established design and drafting standards to speed the progress of the work and to incorporate workable details of proven quality. Check lists for final checking of the drawings in each of the engineering disciplines are also utilized.

Continuous liaison and interfacing between the various engineering disciplines and specialist consultants are necessary to ensure integration of all the design and details. Costly field changes can result if this liaison is not rigorously performed. A client is not particularly impressed by a steel column through the middle of a conveyor belt.

Any significant changes to the conceptual design, which may evolve during the detail design, are estimated and compared with the budget capital cost estimate previously prepared. Approval by the client is obtained before these changes are incorporated into the final design. The capital cost estimate is then appropriately adjusted.

Towards the end of the detail design and drawing phase, a definitive capital cost estimate may be prepared to serve as a more accurate cost control tool during construction.

Technical specifications covering the quality of materials and workmanship are prepared and are combined with the other contract documents for construction contract purposes.

Periodically (usually at the end of each month) a progress report on the detail design is prepared and submitted to the client.

Construction Contracts

The procedure employed for soliciting and placing construction contracts is similar to the procedure previously described for the purchasing of equipment. Construction contracts are generally of three types, lump sum, unit rate and cost plus.

The lump sum contract is the most satisfactory from an engineer's and client's point of view. If, however, the work cannot be clearly defined, the design is incomplete, or is subject to change, the cost of the resulting extra work can make a lump sum contract less desirable.

Unit rate contracts are similar to lump sum contracts in so far as they define the price for a unit of work and allow the quantity of units to be determined at a later date. Excavation, concrete and steelwork are particularly suitable to unit rate contracts.

Cost plus contracts are used only when time limitations, or the type or work, prohibit the use of the other two types. If it becomes necessary to use a cost plus contract there must be adequate allowance for its supervision.

Field Engineering

The engineering for a process plant does not end in the design office. There are key field activities which are an essential part of the overall engineering of the plant. Some of them are strictly field engineering while others are an extension of the office engineering.

A field survey group is normally required for establishing baselines and co-ordinates for setting out the work, and for checking the contractor's layout in the various areas of the plant.

Design changes,necessitated for example, by over-excavation for foundations, are carried out in the field. However, it is good practice to refer these changes back to the design office so that their significance can be checked.

The field engineering office is responsible for checking quantities for approval of contractors' claims. This is particularly important for unit price contracts, but is also required for measurement of additions and deletions to lump sum contracts. Thorough documentation of field changes is necessary so that contractors' claims can be verified.

Other field engineering work, related to the mechanical and electrical trades, includes such items as pressure testing of pipelines and the checking of electrical circuits.

The preparation of as-built drawings is an on-going field activity. These drawings may be completed in the field or the information may be sent back to the design office to be incorporated into the record drawings for the project.

Construction Supervision

Supervision and inspection of construction is required to ensure that the materials and workmanship comply with the intention of the drawings and specifications.

Typical inspection activities in the various engineering disciplines are:-

Civil:
- Foundation conditions.
- Compaction and density tests.

Structural:
- Concrete inspection and testing.
- Installation of bolts in steelwork.
- X-ray of welds.

Mechanical:
- Setting and alignment of equipment.

Contract Administration

Regardless of the type of contract, provision must be made for its administration. With lump sum contracts, procedures must be made and followed for handling extra work; fair prices must be negotiated for changes in the scope of the work and assessment must be made of the amount of the work completed before the submission of a progress claim.

Unit rate contracts, in addition to the administration required for lump sum contracts, require that the quantities for each unit are taken off the drawings, or, if required by the contract, measured in the field.

More work is involved in cost plus contracts than in the other two types. Procedures for control need to be stringent and enforced. The purchasing of materials must be competitive; the use of labour must be efficient; equipment must be worked productively; each purchase order, delivery slip and invoice must be approved and checked.

Cost Control

Cost control begins, in practice, with the design of the work. There is very little that can be done to reduce the cost of overdesigned or "gold plated" work once it is issued for construction.

The process of cost control consists essentially of a continuous comparison of the budget cost estimate with the actual quantities and costs established as the project proceeds.

As the work progresses it is constantly checked against the budget cost estimate. Quantities of work are measured from the completed drawings and compared with quantities from the estimate. Prices for construction work and equipment purchases, solicited from bidders, are compared with the estimate.

In the event that there are significant increases in the quantities and prices for the completed work compared to those in the estimate, then decisions must be made on what action to take. Sometimes the design can be modified; sometimes the scope can be curtailed; sometimes equipment or contract specifications can be altered. In any case, it is imperative that cost control information be provided in sufficient time to permit remedial action to be taken.

Start-up

Towards the end of the design phase of the project the consulting engineer may be requested to prepare equipment and operation manuals for use by the plant operators. He is also in a position to recommend to the client the timing for the hiring of staff, engineers, supervisors and operator-maintenance personnel as construction nears completion. The equipment and operation manuals can then be used by these future operators during the run-in and start-up of the plant.

The consulting engineer's metallurgical, mechanical, electrical and instrumentation engineers, who have been involved in the design and construction of the plant, are ideally suited to assist the future operating personnel in the sequential start-up of the plant until the performance levels required by the client are achieved.

CONCLUSIONS

To provide engineering services for project management, studies, design, purchasing, scheduling, estimating, construction management and start-up supervision for a new mineral process plant requires a multi-discipline organization with a wide spectrum of engineering and administrative personnel.

A matrix organization of engineering disciplines and project teams is the most suitable for effectively carrying out a number of projects concurrently, and at the same time ensuring that the expertise of the entire organization is available to each project.

A project benefits from the background and experience which a consulting engineering firm has gained on previous projects. Past performance and the calibre of key personnel are the most important considerations in selecting a consulting engineer.

Close liaison with the client, particularly during the conceptual design and equipment selection phases of a project, is highly desirable. It is in these phases that the assurance of a successful operation is established, and the most significant capital and operating cost savings can be made.

The extent and variety of activities required for the design and construction of a mineral process plant illustrates the need for overall project management to ensure that the quality and performance requirements are met, the project is within budget, and is completed on schedule.

Chapter 39

CONSTRUCTION OF A MODERN MINERAL PROCESSING PLANT

G. M. Ellis

Vice President, Technical Services
Commonwealth Construction Company Limited
Vancouver, B. C.
Canada

When developing a mineral processing facility, the mining company management and engineering staff should have some knowledge of the construction industry in order to work with the contractors involved as a team. Of course, the situation varies with the nature of the project and its size, but relations among owners, engineers, and contractors always will be enhanced if the mining company has a knowledge of three construction industry facets: (1) the makeup of a construction company, (2) the functions performed by its staff, and (3) the various types of contracts and tendering procedures available.

There are several options open to a company wishing to bring a new mine into production from the grass roots, depending upon corporate structure, capabilities, and financial strength. A company that already employs a substantial engineering staff may elect to manage and perform the design requirements itself, and may handle some or even all of the actual construction work. But in cases where bank financing is required, the loan is often dependent upon adequate professional construction help being used.

THE MAKEUP OF A CONSTRUCTION COMPANY

Construction companies can vary from a one-man housebuilder to the national contractor and can include specialties of every type imaginable: painting, plastering, pile driving, drilling, tunneling, etc. An individual, such as a carpenter or plumber, who hires out his services, may form a company to limit his liability to the extent of the business' net worth. In larger organizations, limited or incorporated companies are formed to limit the liability of its members.

Financing of a Construction Company

Banks are very conservative in most countries, but even so, have suffered some very heavy losses due to construction company failures. As a result, they have developed a set of rules and procedures for the contracting fraternity.

To start a business, working capital or a bank guarantee must be put forward, after which the bank will extend a line of credit from one to three times the working capital at any one time. The lower limit applies to a new company and the upper limit to a company which has three years of successful operation. For computation of the credit line, banks require a general assignment of book accounts, defining working capital as the difference between current assets (any cash receivable within one year) and current liabilities (accounts payable within one year). Items such as equipment and buildings are not considered current assets.

Banks can move into a contractor's office and take over his receivables at any time, except in certain places such as British Columbia where other lien holders have first right. Under the trust provisions of the Mechanic's Lien Act, wages and materials used on a project must first be paid, provided, of course, the claimants file notice.

Banks do not actually loan the full amount of working capital plus the line of credit. Instead, they withhold 10-20% in a special noninterest-bearing account known as the compensating balance. For example, if $50,000 cash were placed in a bank for working capital, $5000-10,000 of it would be set aside as the compensating balance, giving the construction company an effective $40,000-45,000 balance, plus the line of credit. In effect, the construction company is paying the bank interest to borrow some of the company's own money.

Cash Flow in Construction

A contractor needs working capital to pay ownership expenses on equipment, land, and buildings; to pay his bills within the month they are received, or before the discount date in some cases; to meet payrolls, WCB, UIC, and income tax payments when they are due. All of these payments must be met whether or not the contractor has any work going, and whether or not he is getting paid or making a profit. Consider the normal circumstances where a contractor has several projects in progress, from which he gets paid on a monthly certification of work completed, less a normal 15% holdback.

There are two financial problems that result from this system. One is that the claim is usually presented for payment by the eighth of the month following the time the work was done, but is not paid to the contractor for four to eight weeks thereafter. This slow payment means that the contractor is always one to two months behind in receivables and must finance this lag. The second problem is the 15% holdback, which creates an additional amount of money to be financed by the contractor.

This lag in payment is not entirely financed by the general contractor, as the system permits him to pass the burden on to his subcontractors. He must, however, pay his payrolls, assessments, rentals, and material invoices as they are presented. In fact, if he does not pay his suppliers before the invoice discount date, he may become known as a bad risk and be put on a COD basis. Many suppliers, such as cement manufacturers and lumber mills, offer a 1 or 2% discount on bills paid in a given length of time, usually 8-15 days.

For the contractor, cash flow backed by adequate financing is a critical part of business.

Various Types of Construction Companies

The basic function of a contracting business is to provide service to the community and industry through the abilities of its people and the use of its equipment. Contracting companies vary both in size and capabilities, the most basic being the one-man plumber or painter service to house owners. On a larger scale, there are numerous types of specialist contractors--highway, paving, dam, electrical, mechanical, structural steel, etc.

General contractors carry out the basic agreement with a client to build a complete function, be it a house, an office building, or a mine concentrator facility. In order to perform the contract he may do some of the work himself and subcontract the balance of the project functions. General contractors are usually civil contractors, doing the excavation and concrete work together with any carpentry work required, but some specialist general contractors have gone a step further with the capability of doing additional services such as steel erection, equipment setting, piping, and electrical work. The more a contractor can do himself, the more direct control he has on the project, plus the owner benefits by lower costs and fewer subcontractors to deal with.

FUNCTIONS WITHIN THE CONSTRUCTION COMPANY

A construction company sells a service to the public through the abilities of its people and the use of its equipment. Regardless of size, the basic organization is similar in all contracting companies, although smaller firms may employ only two or three persons to perform all the different activities involved. Basic functions are: administration, purchasing, accounting, estimating, personnel, contract management, planning and scheduling, and construction management.

Administration is performed by the senior management of the company and covers the basic melding of the business, making policy decisions, and supervising department heads and intermediate management. Administration also includes contract negotiations, proposals, and handling disputes with clients.

Accounting includes payroll and timekeeping specialists, as well as invoice clerks, cost accountants, machine programmers and operators, field accountants, and numerous other specialists. Major projects, such as the Gibraltar open pit mine/mill in British Columbia, have a self-contained field staff who perform all accounting functions, with the possible exception of making out checks to suppliers and subcontractors. Smaller projects may only have a one-man field accounting office which handles all details. Where feasible, it is preferable to use the facilities and staff at the contractor's home base for all accounting functions due to the reduced costs, centralized staff, and better facilities.

The accounting staff performs an important control function in any project by checking delivery slips against materials received. The signature, made by someone authorized to receive goods, signifies acceptance of the condition of the shipment and the quantity listed.

Cost Control

The accounting department also keeps a cost control record and therefore controls the construction budget. Project estimates are usually broken down into (1) areas of work and (2) functions, based on a standard cost code index used for every project. For example, a new mineral concentrating facility may have its estimates and costs divided into areas such as site development, primary crushing, secondary crushing, coarse ore storage, fine ore storage, concentrate grinding, concentrate flotation, concentrate thickening, concentrate storage, tailings disposal, water supply, power supply, shops and stores, assay lab, general office, changehouse, and others as needed. Each area has a different code number, and within each, work functions would be separated into items such as excavation, backfill, foundation concrete, superstructure concrete, foundation forms, superstructure forms, rebar, structural steel, cladding and roofing, piping, equipment setting, lighting, power wiring, instrumentation, and others as needed. The same code number is used for each work function in all areas. A more detailed cost breakdown is often used in the category "Equipment Setting," to further identify the specific equipment being set, such as a ball mill or a filter.

This information can be fed into a computer, resulting in a printed cost report for labor after every pay period and for materials and subcontracts on a monthly basis. The prerequisite for this system is that the time sheets be coded daily by the foreman in charge of each crew in order to properly allocate costs.

The cost report performs several functions. First, as a project control during work progress, the construction management staff can determine how each activity and work area is progressing in comparison with the budget and time schedule. To do this, someone has to make an accurate estimate of the percentage completion of each activity for the work period. This report allows construction management to review methods and crews on the activity before the phase is completed and to make adjustments to improve costs and efficiency if possible. The cost report also allows management to check once a month on the position of the budget for the project and to predict cash flows for the next few months. The final cost report is of value to the owner in sorting out the different capital cost items for tax and assessment purposes, and for the contractor, it can be used for estimating purposes for other projects.

Other Accounting Functions

It is the accounting department's function to keep employee time records and to meet payrolls when due, as well as to allocate correctly and on time all compulsory payments such as taxes, pension funds, and union dues. There should be some kind of check system to see that everyone on the payroll is working and that time is being allocated to the correct cost function. Company accountants also must keep an internal record of business for the tax department, auditors, and company owners. Many contractors' projects are performed on a cost plus or a management contract basis, in which case the contractor's records are open to audit by the client if he so desires.

It is also the accountant's function to pay material and subcontract bills on time. On the other hand, the accounting department must send the contractor's claims out on time for payment by clients. In many cases it is also necessary to follow these billings through all the channels to ensure they do not get lost or laid aside on someone's desk.

Estimating

The estimating department is important to company success since all prospective work flows through this group. The people themselves are specially trained to read plans from an analytical viewpoint and to take off the quantities of work functions required in project construction. They come from technical institutes, universities, and from overseas where quantity surveying is a profession, although in some cases, young people without previous knowledge are trained in contractor offices. There are two classifications used: (1) quantity surveyors, who take off quantities of work from plans or actual field measurements; and (2) estimators, who can price the work as well as take off quantities. This knowledge of pricing must come from cost records or actual field experience, and in some specific types of construction, such as that involved in mineral processing facilities, the estimator must use past experience records to develop costs for the new work.

Most mining facilities are similar in construction and cost from a contractor's point of view; the flowsheet for metal mines requiring a flotation process is made up of the same basic components, varying only in size and quantity. Consequently, a construction company can develop an expertise in estimating the cost of mineral processing facilities after constructing a few and keeping good records. Many companies can compile a detailed construction cost estimate of a new plant from preliminary information such as the flowsheet, general arrangement drawings, and process equipment list.

Personnel Department

The personnel people in a contractor's office usually hire the staff and hourly workers, and they keep records of all tradesmen hired to determine the ones most suitable for reemployment.

At the start of every industrial project, a prejob conference should be held with the business agents who represent the trade unions that may be involved in the proposed work. The project is reviewed in detail at this meeting and the assignment of work by trades is agreed to by all and then is recorded. The hours of work, travel time, living-out allowance, and camp conditions are also reviewed. All of this results in a more harmonious project, with less possibility of work stoppages due to jurisdictional disputes or working conditions. It requires a person trained or experienced in labor relations to chair the prejob conference and to handle work grievances during the course of the project. Negotiations with the unions at the expiration of each agreement are also handled by these people.

Personnel functions are usually handled by the head office staff, but on occasion a project may be large enough to warrant a man on site performing this function. It is common to give this field personnel officer other allied duties to fill his time, duties such as safety control and camp management.

Contract Management

The contract management staff comprises people who do what the name implies - they look after the business aspect of project construction, beginning their jobs as soon as a contract is signed or authorized. The staff makes sure that all materials, equipment, and subcontracts are ordered; that deliveries are scheduled when required; that all necessary correspondence is sent to clients, engineers, architects, suppliers, subcontractors, municipal authorities, and others. They make up claims or charges for contract extras and credits and process all shop drawings, spool sheets, etc., when required. The contract management staff, usually known as coordinators, divide their time between site visits and the home office, handling several different construction projects at one time, depending on the size and complexity. On large projects, this function is often performed by a project engineer who spends most of his time at the site.

Planning and Scheduling

Planning and scheduling, a very important aspect of contracting, is done as a joint effort of a scheduling staff and the operational staff. Before any project is started, no matter how small, the scheduler meets with the project supervisor to develop a schedule that usually takes the form of a critical path method (CPM) precedence network for computer input.

At major mining projects, for example, this schedule should start with the design and supply of working drawings, and include the soliciting of tenders and awarding of contracts, or issuing of orders for all components of the work as well as orders for production equipment. The actual construction phase can then be included in its proper perspective and the schedule used to plan the project from an equipment and manpower point of view.

An initial schedule often will be shown to require adjustment because of uneven manpower peaking or uneconomical use of construction equipment or camp accomodations. Frequent updating is required, but this new information can be used in a more precise estimation of budget expenditure.

Purchasing

The purchasing department of a company does the actual soliciting of prices and the consequent buying. The overall purchasing requirements of a major project dictate that either the owner, his consultants, or his contractor be familiar with the details of soliciting quotations for all of the elements required in a plant. This includes crushers, milling equipment, conveyors, piping, electrical switchgear, transformers, and instrumentation, as well as concrete materials, structural steel, chutes, hoppers, and various construction subcontract components. To effectively do their jobs, purchasing agents must be aware of all possible sources of supply for each item required and, if possible, have some idea of how busy they are, or their ability to meet required delivery dates. This includes transportation problems, if any, of the product to the millsite.

A detailed and accurate assessment of each bidder's ability to perform or deliver is of utmost importance to the success of the project and, in many cases, it means a visit to the shop or manufacturing plant of the supplier to verify the statements he makes. The final award of a contract or a purchase order should be made after consideration of maintenance, spare parts, or durability rather than initial cost. It is also common practice to standardize on certain manufacturers throughout a company organization for such things as motors, crushers, conveyor components, mills, flotation cells, etc., for ease in transferring parts in emergencies or shutdowns.

Subcontract Documents and Material Orders

The construction industry has evolved a set of standard documents dealing with subcontractors to cover most situations which occur on construction sites. The use of these documents, particularly the subcontract order form, makes dealing with subcontractors easier, since there is wording to cover most eventualities. It is of utmost importance in soliciting a material supply order or a subcontract that the request for tenders contains all of the information necessary for the work to be done or material to be supplied. This must include wording regarding quantity, quality, performance, shop drawings, delivery dates and method, inspection required, erection supervision required, bolt settings, and payment terms. It is equally important in evaluating the offers received that all of the requirements are met and there are no exceptions or counter offers.

Expediting

The purchasing department should be responsible for the expediting of materials and equipment after the order is placed. It is essential to get correct information on deliveries in order to plan the work intelligently and, in this regard, there should be a monthly expediting report listing all items required for the project. This includes all process equipment as well as construction materials. The list should be made from the flowsheet and feasibility estimate as early as possible, and should include columns with information such as date of orders; name of supplier; date of shop drawings received, approved, and returned; delivery requirement dates; and latest information on actual delivery dates as reported on a monthly basis. This report serves two important functions, one of which is to ensure that someone has ordered all required items, the other is to allow accurate scheduling of work.

Construction Management

Construction management people are the technical supervisors, superintendents, project engineers, project managers, and general superintendents - they are the ones who see that the actual physical work is done at the construction site. Major companies choose field supervisors to suit each project, basing their selection on past experience and capability in that type of work. A top supervisor, based at the home office, is in charge of this management team and takes the responsibility of allocating men to a project and assisting in construction start-up. He must also visit each project frequently to give assistance in construction methods, problems, and equipment use, plus remain on call in the home office, should there be problems.

At a major project, all of the detailed supervision must be carried out by site-based people. Quality control is often handled by separately hired supervisors or consultants who test materials such as soil, aggregates, cement, and steel. There is usually someone at the site whose duty is to interpret the engineers' drawings and to act as the owner's representative in site work inspection.

In some countries, the construction management group must be capable of acting as teachers as well as supervisors for the site labor, especially in the case of mechanical, piping, instrumentation, and electrical work. There are competent civil contractors in most countries, but the setting of electronically controlled precision equipment such as autogenous mills requires detailed supervision. On occasion, contractors have had to go as far as to set up pipe and steel fabrication plants, training schools for welders and equipment operators, and other job training programs to integrate the local work force into the project.

Contractors usually take responsibility for plant start-up after construction is completed by training a local operating staff several months before the scheduled turnover of the plant. They also provide operating manuals and instructions, as well as "as built" drawings, to aid in plant maintenance and operation. It is usual to have at least a month trial period when the plant equipment is adjusted to suit the material being processed. The major pieces of equipment, such as the mills, have a manufacturer's representative on site during erection to certify construction within the required tolerance, thus preserving the warranty. The contractor's trial-run supervisors should not leave the project until the owner is satisfied that his staff can handle the plant.

CONSTRUCTION TENDERING PROCEDURES AND CONTRACTS

Tendering procedures vary considerably, depending upon the nature of the work, the owner requirements, the financing, and the urgency of the project. For projects involving public funds, such as hospitals or bridges, the standard contracting procedure is to have a consulting engineer or an architect prepare a complete set of working plans and specifications, then publicly advertise a request for contractors to prepare tenders for submission at a given time. These tenders, or offers to do work, are usually in the form of a lump sum or a series of unit prices made firm for acceptance within a specified period, usually 30 or 60 days. These are often accompanied by a bid bond to ensure acceptance by the contractor if he should be chosen. The professional consultant normally supervises the construction and certifies the payments to the contractor.

In these cases, the contractor is given from three weeks to three months to prepare a tender, depending upon project size. With public tenders it is common to ask the successful contractor to post a bond for performance of his contract and/or labor and materials payments. This protects an owner from unscrupulous, unsuccessful, or incapable contractors. It should be understood that public tender calls are nonrestrictive - they are open to any individual or company, regardless of size or capabilities, who can meet the tender bonding requirements. The fact that a company can hold a bond does not indicate its construction capability, only its financial capability. The very nature of the public tendering system results in public money being spent in the least efficient way, since there is no choice among company capabilities. In some parts of the world this problem is controlled by licensing and classification of contractors, limiting the value and type of work they are permitted to quote upon, according to capability, both financial and technical.

Contract Payment Terms and Contract Changes

Lump sum projects are paid in monthly progress installments evaluated by the portion of the work completed. Unit prices are often submitted with these contracts in order to evaluate changes in foundation quantities from the working drawings. Any changes from the requirements of the contract plans and specifications are subject to price revisions, the terms of which are usually covered in the contract. The most common arrangement is a cost plus 15% for overhead and 10% for profit. Changes should be certified and authorized by a representative of the owner.

There is usually a total of at least 5%, and often 10%, of the contract value resulting from changes or extra work orders during the course of a project. This can be kept to a minimum by a vigilant consulting group which makes few mistakes, and a thorough understanding by the owners of all that is being provided in the contract documents. Once construction begins, changes are very expensive; it is wise to have department heads review the plans and specifications before calling tenders.

The man has never lived who never made a mistake; however, consultant errors can be minimized by allowing time for a thorough review of plans in order to catch one of the most common mistakes - lack of coordination among different work functions or different consultants. Examples include heating ducts run through steel beams not designed to be cut, ceiling lighting fixtures and sprinkler heads located in the same place, mechanical equipment without an electrical connection. Most contractural difficulties or disputes result from items of additional work involved in rectifying errors and omission of consultants who are not strong enough to admit their mistakes. Many consultants attempt to protect their image by hiding behind "weasel" clauses in the contract.

Firm Price Contracts

Firm price contracts are used when the owners must know how much a project will cost before building begins, although unforeseen extras must always be anticipated and for which provision must be made. Firm unit price quotations are also given and can be used to check the owner's budget allowance, in which case the preliminary quantities will determine the amount of the bids. Actual quantities determine payment to the contractor, so the budget must be assessed to determine the probable accuracy of the preliminary quantities. Additional work or extras can still result from changes to the contract, errors, or omissions, and must be allowed for in the overall budget. A common situation that occurs in this type of contract is additional work that is required for which no price units are quoted.

The firm price contract often is used for bridges, roads, and industrial plants such as pulp mills and mine concentrators. Actual working drawings are not needed to call tenders as long as the scope of work is clearly defined and drawings of similar installations are shown.

Tender Information and Selection

Public tender calls are generally restricted to lump sum or firm unit price contracts, in which case the owner must wait several months for complete working drawings before tenders are called. In private industry, the individuals and companies are not tied to specific tendering and contract procedures and will usually specifically select one or more contractors to submit tenders or proposals. Final selection generally is based on capability and past performance.

An owner will sometimes solicit and negotiate a proposal from one contractor only; in other cases, an owner may solicit five or six contractors, although he has preselected one or two of them as favorites. The other contractors, in essence, provide a free estimating service and as this may cost many hundreds of dollars to provide such estimates, the practice is certainly unfair. There is no requirement that the lowest tender should get the work, but it is expected by the contractors that all bidders can perform the job equally. Tenders are sometimes requested to give other information which may be used in assessing bids, such as naming project supervisors or providing project organization charts and construction schedules.

Some projects include a bonus/penalty clause at completion. This is usually a daily amount, perhaps $1000 for every day over or under the completion date stated in the contract. It is also quite common for the tender to name all intended subcontractors. This serves two purposes: (1) to give the owner and his consultants a better idea of who will be working on the project, and (2) to prevent a contractor from shopping subcontractor prices after he is awarded the work. In many instances of public or open tender calls, the subtrade bids are covered by a depository bid system in which all subtrade bids are placed in a central agency 48 hr before the main tender closes. The general contractors who are bidding then pick up all quotations addressed to them and use the lowest or best of each trade in making up the tender. This system eliminates subtrade shopping and irresponsible bidding.

Cost Plus Contracts

There are a number of ways in which construction work can be done on a reimbursable cost basis, including cost plus a percentage, or cost plus a fixed fee with variations covering end costs and target dates. Major industrial plants are generally built, using this type of contract, mainly because of expedience. Construction can proceed while working drawings are still being made and main equipment components are undelivered.

Financing of major expansions is an important consideration for any company and, consequently, it is sometimes more economical to accept additional construction costs to save financing expense, since charges begin when money is committed and do not end until all borrowed capital is repaid. From this point of view, time of construction is measured from the day the first order is placed or the first consultant is hired. It is essential that the contractor work as part of the management team from the very beginning of the project, providing his knowledge of planning and scheduling.

Sometimes it is difficult to determine if a contractor has made his best effort in a simple cost plus or percentage fee contract, and for that reason some type of incentive preferably is added. This arrangement involves a target price compiled by the contractor's estimating staff, who use plans from similar projects and a set of assumed quantities and agreed units for labor, material, and construction equipment costs. The units become the basis for computing a final price after the project is finished and are agreed upon by the owner or his consultant before work begins. Items which cannot be expressed in units are broken into parts which can be given lump sum estimates. At the end of the project, the actual number of as-built items is evaluated at the previously agreed upon unit price in order to compute the final price.

The contractor and owner share, in a previously agreed proportion, any savings or overrun of costs. This forms the contractor's incentive, which is added to a guaranteed fixed fee. Major projects which use this type of contract are generally self-contained and owner-financed, meaning that all personnel and facilities required by the contractor are site-based. In practice, this is impossible since some scheduling, purchasing, spooling, pipe fabrication, accounting, and personnel facilities must be used at the home office, or in the nearest town. These are often supplied at an agreed cost or for a lump sum, which is added to the fee. Owner financing of a project requires that an impressed account be set up in a local bank from which payrolls, materials, and subtrade accounts are paid. With this arrangement, the contractor has a totally reimbursable contract with all costs paid out of this account.

Management Contracts and Project Management

Project managers or construction management contractors generally restrict themselves to major construction projects, such as a downtown skyscraper or a turnkey mine and mill. In this arrangement, the owner contracts with one group or company to handle the complete project, from design through construction, equipment setting, and start-up. This can be done on a fixed fee basis and may include a target price for design work as well as for construction. It varies considerably, from the case where the prime management contractor is not permitted to do any construction work himself to the case where he does all of the work.

The project management contract has been very successful in overseas projects. Most countries will not permit a foreign contractor to build a mill or smelter, yet the local contractors are incapable of performing the detailed work. The answer to this impasse is a project manager who will hire local contractors and supervise the work with his own key personnel, as well as teach the local work force when necessary. It is important that the project manager be employed by the contractor, rather than by the consultant or owners. There is usually a management committee appointed, with representatives from all three parties involved - owners, managers, and contractors.

Construction Management

In the construction management contract, the project owner enlists a firm to perform all contracting services with the exception of design control. This could mean that the contracted firm performs a portion of the work, or, as is often the case, does no construction work with his work force but merely acts as a supervisor and controller. Construction management usually makes up tender calls and analyzes bids; plans and schedules the sequence of work; supervises construction; administers progress claims; and gives approvals of extras, credits, and work acceptance. These services generally are performed for a fixed fee.

The tender call usually will require a list of personnel assigned to the project and its expected duration. Sketch drawings and a basic outline of the scope of the work are required in the tender call for the contractor to properly assess the work.

Construction Equipment

Arrangements often can be made for mine/mill equipment to be used during the construction phase. Machinery such as dozers, road graders, truck cranes, drills, compressors, welders, etc., can be supplied by the owner to the contractor, thus saving a significant expense in mine facility construction. It would be advisable to have a list of such equipment available at the time the proposals for construction are solicited. In existing plants where additions are being made, common use might be made of the owner's fuel and maintenance facilities. In new plant construction, the permanent equipment required by the mine and mill should be reviewed; items that can be used in construction should then be ordered early enough to be assigned to the contractor.

The most expensive machinery requirement in the construction of mining facilities is heavy earth- and rock-moving equipment for road construction, plant site preparation, tailings and water supply dams, pipelines, power lines, and mine stripping. In cases where the mine is not able to provide any equipment or to do this work itself, it is sometimes cheaper to consider buying some of the required equipment and selling it when the work is done instead of paying equipment rental rates. This is particularly true when the equipment can be used for a year or more.

Construction for the Mining Industry

In the category of heavy construction, the mining industry makes use of the following types of work: road building, bridges, power, water, tailings dam construction, site clearing and general excavation, pit stripping, and overburden removal. Townsite development and housing are items of a different specialty, and normal building construction of facilities such as offices, assay labs, shops and stores, and warehouses come under the category of general contracting. Industrial and specialist contractors construct such items as power lines, water lines, tailings lines, crushing and concentrating facilities, and smelters. There are specialist contractors who can do a single type of work, and there are general and management contractors who can supervise all of these classes of work while performing some of them.

Critical Path Scheduling

Whenever possible, every project with a duration of three months or longer should be scheduled according to the critical path method (CPM). In the first step, a scheduler and the project supervisor meet to divide the work into properly sequenced events. These events are selected to encompass time periods that can readily be controlled, usually two to three weeks. The number of units of work to be done in each event is estimated by man-hours in each worker category and the necessary construction equipment is allocated to the event for the duration needed.

For example, one event could be construction of ball mill foundations, which may require 550 man-hr of laborers, 1000 man-hr of carpenters, 600 man-hr of reinforcing ironworkers, 100 man-hr of machine operators, and 50 man-hr of cement finishers to build three bases. If the logical duration is three weeks, then crew sizes of each trade are determined. In the same manner, a mechanical event such as setting a ball mill or an electrical event such as wiring from mill motors to the control panel is also estimated according to manpower and equipment needs.

The principle of CPM scheduling is to resolve the interdependence of one upon another - certain events must be completed before another particular event can be started. When the complete information is fed into the computer, the resulting schedule print-out shows the critical path and duration of the project. Critical path is the sequence of events which takes the longest time to complete, including commissioning and trial run of equipment. This schedule, set up in calendar form, totals all manpower requirements for each category during every day of work. Construction equipment requirements are also placed on the calendar. The result is a manpower and equipment use curve which may have to be adjusted if too erratic. Items or events of work may have to be shifted, or crew sizes and durations might be changed so that manpower utilization is constant and suits the availability of men, equipment, and camp accomodations. CPM scheduling in major projects will also include the design and production of working drawings and the ordering and delivery of materials and process equipment. Maintaining the schedule on a current status, i.e., correct to within the last month, is essential to project management since it allows them to forecast costs, expenditures, manpower and equipment requirements, and completion dates. This is where the expediting report is an essential tool.

PROJECT REVIEW, GIBRALTAR MINE, B.C.

The Gibraltar open-pit copper mine produces about 40 000 t/d (44,000 stpd) from a location 19 km (12 miles) off the Cariboo Highway and about 61 km (38 miles) north of Williams Lake, B.C. The mine and mill are on a ridge over 900 m (3000 ft) in elevation, and water had to be tapped from the Fraser River, a distance of 13 700 m (45,000 ft) with a rise of about 600 m (2000 ft). Power was brought in 19 km (12 miles) from B.C. Hydro's main line.

A townsite was not needed at Gibraltar since Williams Lake was near enough to accomodate employees. Other engineering considerations included tailings, which were to be taken to a nearby valley impoundment, and the climate, which is such that outside concrete work would be impractical from October 15 through April 15.

Project planning and construction was a team effort that combined the design services of Wright Engineers Ltd. and the construction management services of Commonwealth Construction Co. Ltd., both under the direction of Placer Development Ltd.'s engineering department. The team began working together at the feasibility stage when Commonwealth was called in to put construction cost estimates together with input from the others. At this early stage, some delivery and construction problems were recognized. For example, most Gibraltar construction was at elevations over 900 m (3000 ft), and since it was in central British Columbia, climate was a very important consideration. Concrete work had to be completed early enough in the year for steel frames and building cladding to be erected before winter. On this project, the site needed to be cleared and leveled before any of this could start. There was not a sufficient labor force of the skilled trades in the area, so camp construction was also necessary before work began.

Early Construction Start

After it was decided to proceed with mine development, the first consideration was preparation of a mill site and camp facilities so construction could begin as soon as possible the next spring. To do this, site clearing began in October, followed closely by development of camp services. Orders were placed for camp buildings and key mill equipment, and engineering design was commenced.

Project management decided there would be no major premium to pay for having the site excavation proceed through December, January, and February. As the frost was known to penetrate the ground for 1.2-1.8 m (4-6 ft), construction work on the foundations normally could not begin until May. But by preparing the entire site in February, foundations started into the ground in March. Frost was controlled by leaving a meter of soil unexcavated on the building sites until immediately before work began. By doing this, the work areas were never exposed to the cold for a sufficient amount of time to cause problems.

The construction team engaged a firm to supply concrete aggregate and, to get work started at the site, hired some local workers who commuted from Williams Lake. The 600-man camp (size was determined from CPM scheduling) was erected during April and May. Site offices were also set up and manned with sufficient staff for the project to be self-contained except for some purchasing, expediting, scheduling, and personnel services carried out in Vancouver.

Construction Supervision by Areas

Construction equipment requirements were reviewed with the owner to coordinate the end requirements for mine operation. For administration purposes, the mine was divided into five main areas and a senior supervisor placed in charge of each. A senior supervisor was responsible for the complete construction - including all civil, mechanical, and electrical work to make it operational - in one of the following areas: (1) service buildings, including the office complex, assay lab, changehouse, shops and warehouse, carpenter shop, concentrator storage, and guardhouse; (2) crusher buildings, including primary and secondary crushers, coarse ore stockpile, and conveyors; (3) crushing services, fine ore storage, electrical substation, thickener tanks, tailings tunnel, and electrical tunnel; (4) concentrator and dryer buildings; and (5) services - freshwater line, tailings line, tailings reclaim water line, water storage tanks, and site services.

Construction Pace

Construction work proceeded full pace throughout the late spring and early summer with crews peaking at 800 men. Some 33 000 m^3 (43,000 cu yd) were formed and poured from mid-April to November using a site-based concrete plant and ancillary equipment. As working drawings were completed, tender calls were solicited for vital building components such as structural steel, cladding, and roofing, with deliveries and erection made to suit the schedule and to allow closing in of the buildings by late fall or early winter. Deliveries of grinding and milling equipment were such that the mechanical and electrical installations and the process piping were done in the fall and winter, inside enclosed and heated buildings. The major pipeline and outside electrical line work was done throughout the summer with maximum crews working.

Labor Relations

An important feature of the project was the handling of labor relations. The project began with a prejob conference involving all the union business agents and served to allocate most of the possible jurisdictional problems and review the camp facilities. On-site meetings were held periodically to review current problems and camp conditions. This attention resulted in a nearly trouble-free project.

Actual trial run of the mill was started in March, one year after site work began on major construction. Technical supervisors aided in machine setting, instrumentation, and power wiring, and worked with the owner's staff and the vendor's supervisors for at least a month after initial start-up to iron out the bugs in the flow process. This was made possible by preplanning, early ordering of principal components, and early start of camp and site development.

Contract Data

The project was run on a cost plus a fee basis with target unit prices for most items to gage the savings resulting from target underrun. The owner set up an impressed bank account which was kept current by budget analysis from cost report data. The entire project was cost coded with a monthly computer print-out of costs. The final total project cost was on the order of $70 million, which was under the project budget.

The Mine Construction Team

We have worked on many different types of construction projects, from minor alterations to major plants, and in the course of this work, have had the opportunity of working with various types of consultants and owners, from architects and engineers to public bodies and major corporations. We can safely say that the best results are achieved when everyone works as a team.

After all, the owner pays for the project whether it is done on a firm price basis or a negotiated basis. Unfortunately, there are too many projects where the contractor and the professional consultant and/or owner are adversaries from the beginning. This type of project does not receive the full benefit of the contractor's experience and expertise, and often there is someone who is always looking for ways to take advantage of the other party.

We strongly recommend the total team concept in mine plant construction.

Chapter 40

MODERN MILL DESIGN

by Robert S. Shoemaker and Wayne D. Gould

Manager of Metallurgy
Bechtel Corporation
San Francisco, California

Metallurgical Consultant
Tucson, Arizona

ABSTRACT

Modern mill design is a combination of many factors which permit the efficient production of a mineral concentrate from operating, maintenance, and capital requirements. In general, mills should be designed to use the largest possible equipment compatible with the requirements of the ore and the owner's operating philosophy. The paper discusses the major parts of concentrating mills, from primary crushing through comminution, both conventional and autogenous, and the production of a final concentrate. Features of various plants are compared and discussed in the light of operating practice and maintenance requirements.

INTRODUCTION

In the minds of many engineers mill design means capital cost. Unfortunately, there have been some mills constructed, even in the last few years, where it has seemed that conservation of capital was of no object; yet, these mills have no better (but sometimes worse) operating costs than mills designed on strict budgets. It is nice to go first class with unlimited equipment and spacious buildings, but in the final analysis, the profitability of a project must be a proper balance between capital and operating costs under the business conditions in which the plant must operate.

Good mill design must consist of a composite of three different phases:

1. Process design
2. Equipment selection
3. Plant layout

The most economic design may not be easily recognizable and can only result from a cooperative effort between the owner's most experienced plant operators and maintenance personnel and the engineer/constructor. It must be recognized, however, that bias from past practice can enter into new designs and can have a considerable overall cost effect. It is relatively easy to analyze the effects of invested capital, but it is not so simple to analyze the savings or loss to plant layout or equipment selection. Mining company managements can easily calculate accountable items, but it is not well known in the area of plant design when, where, how, and what to look for in order to end up with a minimum capital cost as well as a minimum operating cost.

While there are many facets to good mill design, a summary of the most important considerations might read as follows:

1. Use of the largest possible equipment consistent with the conditions under which it must operate.
2. Minimum storage.
3. Provision for reasonable plant expansion.
4. Provision for easy maintenance access to all equipment.
5. Use of reliable and easily maintained equipment.
6. Simple and effective instrumentation.
7. Operating flexibility.
8. Ease of general housekeeping.
9. Safety.
10. Prevention of excess noise.

In regard to large equipment, the statement is still heard that multiple processing lines permit a higher operating availability; but the fact is that operating availability is a function of the design of the processing lines and the ease and type of their maintenance, not of their number. Operating availability is in turn related to the amount of storage both up- and down-stream of the processing section; and tying two major pieces of equipment together reduces the availability of both. In-process storage is necessary in many flowsheets; but it must be realized that while any storage involves both capital and operating costs, storage does nothing to an ore, except in the case of sulfides where storage permits oxidation. Unless the storage pile is replaced occasionally, it will become nearly worthless as a source of mill feed. Money spent for excess storage could be better put to use in purchasing more reliable processing equipment or in plant design which will permit higher operating availability.

Provision for plant expansion is a necessary but not always expensive item in the design of any plant. Often an ore body will change in character as mining progresses, and a capacity imbalance will develop between crushing and grinding or grinding and concentrating sections. Rounding out plant capacity by the easy addition of a crusher, a grinding mill, or some flotation cells will often result in a sharply increased capacity divisor that will reduce operating costs per ton. Then, too, increased production of the final product is often more easily and economically attained through expansion than by increased recovery.

In these days of high-cost money, prompt, planned, and inexpensive maintenance is essential in obtaining the highest operating availability possible. Equipment which sits idle a portion of the available time can only mean the original capital investment had to be increased by that same amount. This means, of course, the use of a modern preventive-maintenance program; although it is not always safe to assume during design that because such a program works well in Company A it will work equally well for Company B. Just as important, however, is plant design for easy and rapid maintenance. A pump that can only be removed and replaced by the crane which may be involved in a mill relining job is a liability, and availability of equipment exists only if accessibility to that equipment is built into the plant.

Overall plant availability or operating time also depends on the use of well-designed and well-built equipment that does not necessarily have the lowest first cost. Unfortunately, equipment selection is often based on price rather than on quality; and the old adage which says, "You get what you pay for" has not been proved wrong even in mill design. Although it is difficult to single out equipment items which are the worst offenders in this category, conveyors, trippers, and vibrating screens seem to be at fault most of the time.

Plant operators generally expect a mysterious quality called flexibility built into their plants. This quality can be reduced to one of the plant being able to handle ores which range from hard to soft and dusty to sticky. Not only do these last two qualities give the operator the most trouble, but designing for them can be expensive and sometimes difficult to justify on the basis of relatively few tons processed through a pilot plant or laboratory under conditions closer to ideal than realistic.

One of the most easily overlooked necessities of good plant design is the ability to process the ore while maintaining a clean working environment. Even with the best of designs, there will always be spills; and with today's large equipment and high treatment capacities, accidental spills are larger than ever. These spills must be removed and other dust and dirt must be cleaned up to permit both efficient operation and maintenance. Design for rapid and easy clean-up, if strictly adhered to, will add to capital cost but will more than repay the investment in lowered operating and maintenance costs. In all cases, however, a clean plant will be a safe plant; but the dollar value of an easily cleaned plant is difficult to quantify.

Safe work areas, inspection areas, and maintenance areas are essential for today's mill operations. While in the past it was not always easy to reconcile safety needs and initial costs, and while today rules and regulations which must be applied to milling plants are often onerous, it only makes good sense in both personnel relations and costs to make and keep a plant as safe as possible.

Excess noise is a problem which under OSHA regulations is coming under extremely close control. These controls are adding a new dimension to plant design, particularly in the grinding mill areas and in ore transfer chutes in crushing plants. Mill and motor noise will be difficult to quiet; but the adequate use of rock boxes, rubber chutes, and screens will aid substantially in reducing noise level.

PRIMARY CRUSHING PLANTS

The primary crusher is one of the most important units in the plant, since only in the largest operations is there more than one installed. The primary is also subject to more wear and abuse from large rock than any other piece of mechanical equipment. Good design of this installation is therefore of prime importance.

The size and shape of the structure housing the primary crusher is influenced by several factors:

1. Truck size
2. The possible need for a grizzly
3. Crusher size

4. Storage above and below the crusher
5. The need for adequate maintenance areas

While in the early days of minerals processing grizzlies ahead of primary crushers were standard design features, the larger gyratory crushers today do not, as a rule, need them to operate effecitvely unless the ore is very sticky or there is some special circumstance dictating their need. An example of the latter case is found at Butte, where there is wet ore and a need for removing fines to facilitate the picking of timber and tramp iron before crushing the ore. The absence of a grizzly will decrease the depth of the crusher enclosure by 20 to 40 ft (6.1 to 12.2 m), sharply decrease its cost, and decrease damage and wear on the crusher from ore falling so far.

While the grizzly or no-grizzly philosophy has been argued for many years, it would appear from the many crushers being installed without them that it is being realized by operators and designers that crushing rock-on-rock in a primary is better than crushing rock-on-steel in the fine crushing plant. Proof of this can be seen at Butte, where the grizzly opening was reduced from 7 to 4 in. (178 to 102 mm) and when hard ore is encountered the three secondaries are more nearly able to maintain their normal product size, thus permitting milling capacity to decrease substantially less than was usual under these conditions. Bougainville is another plant where benefits have been obtained by decreasing grizzly openings ahead of the primary crusher.

During the last few years both shovel and truck sizes have markedly increased. The former has meant more and larger oversize rocks which will not enter the crusher and thus need to be broken or removed from the dump pocket. Newer crushers are more and more being equipped with boulder breakers in their dump pockets as the preferred method for handling oversize instead of removing them with a hydraulic powered grapple.

Any primary crusher will produce at maximum capacity when it is choke fed, but between each load dumped it is virtually standard practice to allow the pocket to almost clear so that oversize pieces blocking the receiving opening can be detected. Therefore, it seems logical to design the dump pocket to contain only slightly more than the load of the largest truck to be used. A pocket larger than that wastes invested capital and makes crusher relining somewhat more difficult. A dump pocket over a gyratory of no more than about 9 to 14 ft (2.75 to 4.25 m) depth depending on the truck size is adequate and will generally preclude the use of a splitter to protect the crusher spider. Such a shallow pocket will also not require wear-resistant steel liners if it is made wide enough (the primary crusher dump pocket at Cities Service Company's Pinto Valley plant is of this design). It would seem, therefore, that the frequently stated position that a dump pocket should be capable of receiving ore from two trucks dumping simultaneously is just not tenable.

With only slightly over a truck load being in the dump pocket at one time, it therefore follows that the crushed ore pocket should be of about the same size, thus allowing further capital savings. In making these statements, it is assumed larger trucks will be introduced in the future and that the pocket will be designed accordingly.

Another large item of consideration in primary crushing plant design is the crushed ore feeder. While apron feeders have been standards of the industry for many years, belt feeders have now also proved their worth in this service. Aside from having lower height requirements and capital and installation costs than apron feeders, their operating and maintenance costs are minimal. One such 8 ft (2.44 m) wide unit at Similkameen Mining Company in British Columbia has now handled 34 million tons of hard, jagged copper ore while incurring no maintenance charges, a record which could not be equalled by an apron feeder. Operation of the feeder is best controlled by radioactive level indicators which control a stepless drive, preferably of the hydraulic type.

Gyratory crusher enclosures have traditionally been made up of square or rectangular structures of two or sometimes three compartments in plan view, with the three-compartment units having electrical and dust collection equipment in the third compartment. The two-compartment structures are less expensive and more functional because electrical gear is above ground near the maintenance area and the dust collector is similarly situated or is placed at the daylight end of the takeaway conveyor. The former has to draw dust upward but the latter draws it downward and out the tunnel. Dust generated in the dump pocket is best suppressed by water sprays located under the truck dumping position so the ore falls through a water fog curtain.

A few circular enclosures have been constructed; but, while they are economic in regard to concrete and reinforcing steel usage, support of the crusher on the resulting large-diameter diaphragm is structurally difficult and usually results in awkward support walls which limit maintenance access. The Figure-8 design of the new Palabora crusher, however, has combined low concrete and steel usage with good structural design.

FINE CRUSHING PLANTS

Since fine crushing plants using the stacked arrangement of a standard cone crusher located above two shortheads have proved to be unable to produce the operating availabilities and low capital and operating costs demanded of modern high-tonnage operations, secondary and tertiary crushing plants have generally followed the in-line arrangement developed for the Kaiser Eagle Mountain plant in California. The two variations of this design, screens either over or under the crushers (both of which are workable), have given rise to arguments as to which is best and which probably never will be resolved.

When screens are located under the crushers, vibrating loads are nearer ground level and the drop of ore from the feeders into the crushers is less, which permits simpler chute design. Since product has been removed from the circulating load, dust collection at the scissors conveyor transfer and tripper above the bins is simplified. Additionally, the bins above the shortheads can either be smaller, since they contain no final product, or for the same size they will provide a larger surge capacity.

Screens over the crushers require the lifting of final product by conveyor to the top of the crushing plant and more careful and heavier structural design to prevent the building from vibrating, a problem that has cursed a number of installations. On the other hand, screens above the crushers are more accessible; and with the 7 or 8 ft (2.13 or 2.44 m) wide belt feeders now available, they can be fed over their entire width, thus increasing screen efficiency. In addition, the product and circulating load conveyors are farther apart, which gives some more flexibility in design. All in all, it would appear that screens over the crushers may offer the best arrangement for today's high-tonnage fine crushing plant; but at the same time, it must be admitted that the Palabora fine crushing plant, with screens above and below the two standards and below the five shortheads, probably has a better availability and a higher capacity than any other crushing plant in the world. The crushing plant operates 24 hours a day, seven days a week, with one 8-hour and one 12-hour maintenance shutdown per week. Mill tonnage is 59,500 stpd or 54,000 mtpd. Each of the two standard crushers average 1,490 stph or 1,350 mtph.

There is a third crusher and screen arrangement which probably will be used more as larger concentrators are built. With the increasing use of very large single-stage ball mills, the feed to which must be held to minus 3/8 or 1/2 in. (9.5 or 12.3 mm), it becomes more efficient to use more screens than crushers. Consequently, the separate screen and crushing plants as used at Bougainville will probably be copied even though their capital costs are higher than conventional fine crushing systems. With this arrangement, adding screens or crushers is simplified and individual screens can be shut down for repairs without decreasing crushing capacities.

GRINDING PLANTS

In selecting the proper grinding circuit for a mineral processing plant, the designer must consider many alternatives such as:

1. Power - total power from primary crusher through to the final product for concentration
2. Crushing and grinding media cost
3. Operating and maintenance costs
4. Overall system availability
5. Ease of supervision, operation, and maintenance
6. Capital costs

While oftentimes a conventional grinding circuit can be chosen on the basis of laboratory testing, autogenous or semi-autogenous circuits must yet be sized only after testing in a pilot scale mill. In general, the harder the ore is the more there is to be gained in autogenous grinding, but some relatively soft ores can also be ground in this manner; and both Pima in Arizona and Palabora in South Africa have expanded their operations by adding semi-autogenous and totally autogenous milling respectively. Pima finishes the grinding by using second-stage ball mills whereas Palabora's autogenous grind is single-stage.

The selection of autogenous versus conventional grinding in most cases requires extensive studies, metallurgical engineering, and economic analyses. In certain cases, however, the decision for semi-autogenous milling is clear-cut from the start of the project. Two precious metals producers, Rosario's Pueblo Viejo gold operation in the Dominican Republic and Earth Resources' DeLamar Silver Mines in Idaho, are good examples of plants that would have had extreme difficulties in conventional crushing and milling if semi-autogenous grinding had not been selected. Both these mines produce a mixture of hard ore and soft, sticky material which would make fine crushing a virtual impossibility.

The classic rod mill-ball mill circuit, while far from disappearing from the milling scene, has in recent years almost entirely been displaced in the larger concentrators by single-stage ball mills. The reason for this change of course is the limitation on rod length and therefore rod mill horsepower and the reduction in availability from coupling two major pieces of equipment. The use of the large mills has also precluded the use of scoop feeders which were always a maintenance problem. Wheel-mounted spout feeders, adapted from autogenous mills, have also permitted relining these mills from the feed end with relining machines, which can enter the mill through a door in the rear wall of the mill building. The result of this design is a relatively short span crane and lower capital costs.

The change from scoops to spouts has resulted in some small increase in head on the cyclone feed pumps, but the greater flows from the large mills have permitted the use of large, slow-speed pumps; larger pipelines which have lower friction losses; and 30 and even 36 in (762 and 914 mm) cyclones which operate at low pressures. All these factors tend to counteract the slightly higher head pumping heads caused by the absence of mill feed scoops. Also, scoops are not without their need for power; but the use of large-diameter cyclones can cause reduced separating efficiencies which generate higher circulating loads. Double cycloning may become standard practice in the future in order to reduce circulating loads and increase mill capacities.

A materials-handling and design problem which has been caused by the use of these high-capacity single-stage mills is that of the large quantity of rock chips which can be rejected from the mills. The sources of these chips are, of course, worn or broken screens in the fine crushing plant, which must be better maintained than those furnishing feed to rod mills. A distinct aid in this problem is the separate screening plant mentioned earlier. With poorly maintained screens, the quantity of chips rejected from mills can be enormous, but their removal can be simplified by designing into the plant a truckway below the grinding mills. While a chip-handling conveyor would solve the problem more easily, better screen maintenance in the crushing plant is the only logical solution.

Ball-handling has always been a problem in grinding plants; and over the years plant designers have generally changed from the ball handling train (which adds to the width of the grinding bay) to handling and charging with the mill bay crane, although Phelps Dodge uses a unique building column-mounted travelling jib crane, and others such as Island Copper and Anaconda at Butte handle balls on a conveyor belt. While all such systems are workable and the travelling jib is particularly functional, it is thought best, at least in milder climates, to reduce congestion and movement inside the mill building by taking all grinding media handling outside where it will not interfere with operations. Adding balls to the mill feed belts outside the mill building is practiced in a number of cases and actually can be done for a lower capital cost.

At Anaconda's Butte concentrator, a unique ball handling system is in use. Grinding equipment in the original flowsheet consisted of six rod mills and twelve pebble mills with the latter drawing grinding media from the screens in the fine crushing plant. Pebbles were brought to the mill building by an inclined conveyor and tripper located over twelve pebble bins and rationed to the secondary mills by belt feeders. When the pebble mills were converted to ball mills, this system was no longer used. At that time, the inclined belt and tripper were each equipped with a second belt lying over the original belt and powered by it by friction. It was then possible to convey balls up 15-degree inclines when they were fed between the two belts to form a ball and belt "sandwich." Balls are now stored in the old pebble bins and fed to the ball mills by the belt feeders on a daily basis.

Modern autogenous and semi-autogenous milling plants have many of the same materials handling problems that do conventional plants, except the grinding units can handle far more tonnage. The enormous flows over conveyors and through sumps, however, can present clean-up problems of heretofore unencountered magnitude if the plant is not designed with this in mind.

In a paper published in 1971 (1), the advantages of open-air mills were described; and in recent years, three open-air autogenous plants (Cobar and Kanmantoo in Australia and the Palabora expansion in South Africa) and at least one rod mill-ball mill plant in the non-ferrous industry (the Anamax Twin Buttes Oxide Plant) have been built in the open air. The Palabora plant consists of two 32 ft by $15\frac{1}{2}$ ft (9.75 by 4.7 m) totally autogenous mills, each powered by two 4500 hp (3355 kw) wound rotor motors, which drive the mills through gearboxes. Each mill discharges to an 8 ft by 20 ft (2.44 m by 6.1 m) vibrating screen. Screen underflow is cycloned and both the screen oversize and cyclone underflow are returned to the mills. Cyclone overflow from each mill line is pumped by five stages of pumps 5277 ft (1640 m) to the concentrator for flotation. Mill spacing, 72 ft 2 in. (22 m) center to center, was not constrained by the need to minimize building volume and maintenance is considerably eased by the room provided. A truck and small crane ramp between the mill feed conveyors permits ready access to the relining floor behind the mills. Although such a design could not be installed in extremely cold areas, it is suggested that it would not be necessary to limit open-air mills to those areas where the climate is extremely temperate.

FLOTATION

With the long overdue adoption of very large flotation cells, distinct design advances have been made recently in this area. At the Bougainville plant, with its two spindle 600 cubic ft (17 m^3) units, walkways were installed on top of the cells and large launders were used to handle froth from adjacent cell rows. Concepts concerning the number of cells per row have also changed and no longer are there heard the dogmatic statements that the minimum number of cells should be 14, 24, or even 30. Rougher rows of as few as six cells are common and cleaners and recleaners can use as few as two or even one. Three hundred cubic ft (8.5 m^3) cells are now common in this latter service. Too short a row still leaves the possibility of short-circuiting with a loss of recovery. Flotation cell manufacturers have seemingly only scaled up to larger sizes, and it is not completely certain that the aeration efficiency is as good as with the smaller cell. Considerable work needs to be done in this area.

Plant designs using these concepts can result in substantially reduced capital costs through radically smaller flotation areas and can also favorably affect operating costs. It is also suggested that complete replacement of older and smaller flotation cells by 600 to 1,000 cubic ft (17 to 28 m^3) cells could, in high power cost areas, be justified on the basis of power and maintenance savings alone and should be seriously studied by the owners of older flotation plants.

While the use of large flotation cells permits radically smaller flotation buildings, they also permit shorter launder runs, fewer pumps, and cell floors built much closer to the pump floor, additional savings which cannot be ignored.

Flexibility in plant design, mentioned previously, is perhaps most important in flotation. The balance between recoveries and grades from the roughers and rougher scavengers can often be adjusted by the use of double launders.

THICKENING

Since the development of large diameter thickeners it has somehow become almost an axiom that a mineral processing plant could not operate without water reclaim by thickening. With the advent of these larger thickeners, however, their installed costs have become enormous and they have come to represent a possibly appreciable source of downtime when they become stalled and a potential large maintenance cost when they need repairing.

Thickeners should be equipped with large, high-pressure headers on the bridge that can be used with water or slimes to wash out the pulp if the thickener becomes stalled. Washing a thickener out after a stall will save many repair bills, and the design should provide for a means to rapidly dump the thickener.

The alternate to thickening, of course, is to either pump all the tailings to the pond or pump all the return water from the pond to the plant, depending on the relative elevations of the two. In some cases, both tailings and return water must be pumped. With no thickeners, all pumps and both tailings and return water lines must be larger than when thickeners are used. Pumping all tailings or water also requires additional power. While all these items can be easily summarized by comparing capital and operating costs of each system, there are additional factors to be considered. Topographic conditions in many areas dictate that tailings dams must be made of borrow materials rather than the coarser fractions of the tailings themselves, and increasingly stringent regulatory authority rulings often do the same. When this happens, the necessity of distributing tailings around the dam is often nullified.

Examples of mineral processing plants which were designed to operate without thickeners are Reserve and Similkameen; and one large plant, Lornex, at 38,000 tpd, used three thickeners originally and then changed to gravity flow of total tailings and pumping all return water.

On the average, the use of thickeners permits the reclaim of about 50% of the water contained in the tailings, and a well-managed tailings pond can permit the recovery of an additional 25%. On the other hand, it is reasonable to assume a well-managed tailings pond could act as a thickener as well as a pond and return virtually all of that 75% if no thickeners were used. In summary, during the design phase of the plant the careful consideration of all these factors and an economic study of them may result in a substantial savings in capital cost.

When thickeners are used, it is most important that their initial filling and operation be carefully programmed and controlled. Ignoring proper construction and start-up procedures is, unfortunately, commonly done and has resulted in many costly repairs and delays. While local topography often prohibits gravity discharge of thickeners, this method of operation is generally less troublesome than pumping the underflow. It must be remembered, however, that such thickeners are difficult to fill with straight tailings because of a lack of discharge head at the start. Because of this difficulty, such thickeners should be equipped with cyclones to provide slimes only at startup.

CONVEYOR BELTS

Conveyor belts and their associated equipment and structures probably cause more problems than most other equipment in a mineral processing plant. Since conveyors are almost never duplicated, they can easily shut down an entire plant when they fail; and it is difficult to completely redesign or rebuild a poorly designed conveyor system without serious interruptions to production.

Most importantly, conveyors should be designed for the highest availability with low maintenance and the ability to operate without causing housekeeping problems. Good housekeeping requires that belts be mounted high but good maintenance requires them to be low; thus, belt height from the floor must always be a compromise. In no case should it be impossible to clean under the return idlers and tail pulley.

Conveyor loading points deserve special consideration since belt wear is a direct result of loading point geometry. Crossover conveyors connecting the two legs of a scissors are subject to extreme wear because of their short length and should be wider and run slower than the scissors conveyors. A common error is to minimize the height loss at the loading point. With coarse ore, a grizzly at the loading point will allow the belt to be bedded with fines ahead of the coarse rock.

Every loading station on a conveyor belt presents a hazard to the belting from tramp iron and sharp rocks, and most belt tears occur at this point. Belt rips can best be avoided by installing hinged and spring-loaded impact idlers. When tramp iron penetrates the belt and is pushed against the idler, a microswitch on the idler brakcet stops the conveyor before extensive damage can be done.

Of particular concern is the construction of pulleys and trippers and as much standardization of belt widths, pulley sizes, shafts, and bearings as possible. The purchase of the heaviest duty equipment in all these categories will save the extra cost many times through less maintenance and higher availability. Lightweight trippers will commonly flex and skew as they travel and as a result will be subject to high maintenance charges. Purchase of conveying equipment on price alone can only result in problems. Heavy-duty conveying equipment is not a luxury.

FLOORS

One of the most important and yet often overlooked areas in plant design is floors and walkways. Elevated floors in crushing and screening plants and walkways along conveyors or in any areas where rock is being handled should be made of checkerplate which can be easily swept clean. Gratings or expanded metal in these areas will become choked with rocks which present precarious footing and thus become hazards. Grating should be used where fine ore is being processed such as grinding bays and flotation areas. Expanded metal with its sharp edges is particularly good in molybdenite processing areas. Concrete floors should in all cases be sloped at least 3/8 in. per foot (30.6 mm per meter). Floor trenches covered with grating are almost useless for drainage as they quickly become completely filled with settled solids.

SUMMARY

The purpose of this paper was not to explore all the plant design variables available but rather to discuss criteria and objectives. No one design can be deemed best; however, general criteria are applicable to all mineral processing plants. Ever decreasing ore grades and increasing capital and operating costs are strongly influencing plant designs and are continually applying pressure to the owner and the engineer to produce designs that will meet changing economic conditions. The direct results of these designs are capital and operating costs and at the outset of any project there should be a mutually agreed upon balance between these two all-important items. First cost, however, should not be an overriding consideration in equipment selection. Operating availability, cost of spare parts, ease of maintenance, and overall quality are all as important as original equipment cost. In addition to the traditional design problems a new plant presents, new environmental restraints previously unencountered now impose their own unique considerations on designs.

Air, water, dust, noise, and safety all pose special new problems for the designer.

Any plant design should provide for an orderly expansion with minimum interference to production. Even though there are no initial plans for such a move, it is almost inevitable that more production will be required. When it is, the new design problems presented should not be solvable at excessive cost.

REFERENCES

1. Shoemaker, R. S., and Taylor, A. D., "Mill Design for the Seventies," paper presented at the AIME Centennial Annual Meeting, New York, Feb. 26-March 4, 1971.

Chapter 41

The Future of Plant Design In Mineral Processing and Extractive Metallurgy

Walter Riethmeier, Vice President and Assistant to the General Manager
Arthur J. Lynch, Senior Project Engineer

Western Knapp Engineering
a Division of Arthur G. McKee & Company

ABSTRACT

The future of plant design in mineral processing and extractive metallurgy is discussed in relation to present trends and the future need to build plants that will treat larger tonnages and lower grades of ore. The future equipment and plant design of crushing, grinding, screening, classification, flotation, magnetic concentration, gravity concentration, hydrometallurgy, thickening, filtration, tailings disposal, health, safety, environment, and materials of construction are discussed. Future plants are expected to follow upon present evolutionary trends. The impact of escalation, energy costs, and increased regulation of the industry is foreseen as a major influence on plant design.

INTRODUCTION

The future of plant design in mineral processing and extractive metallurgy is expected to continue along present trends of evolutionary development.

The plants of the future are expected to continue to reflect the inherent differences found in mineral deposits, their locations and the experience and priorities of the engineers who plan to exploit those deposits.

The future plant that will receive most attention will be the high capacity facilities for treatment of low grade ores.

The discussion of future plant design which follows includes sections on primary crushing, fine crushing and screening, grinding, concentration flotation, magnetic separation, gravity concentration, hydrometallurgy, thickening, filtration, tailings disposal, materials of construction, and health, safety, and environment.

Primary crushing plants will be designed to accept feed from larger pieces of mining equipment and some open pit operations will use in-pit crushing. Designers of large volume secondary and tertiary crushing and screening plants will have the option of selecting the 3.3-m (10-foot) cone crusher which, because of its size, may dictate new plant arrangements.

Trends in milling practice towards larger "autogenous" or single-stage ball milling will continue, but the increasing pressure of energy costs may cause alternative solutions to be sought.

Flotation concentrators will use the next generation of flotation cells in the 28 cubic meter (1000 cu ft) range, and computer controlled automation of comminution and concentration will become common practice.

Magnetic concentration plants will benefit from current development work on wet drum low-intensity separators and on high-intensity magnetic separators. Gravity concentration techniques have recently enjoyed renewed attention with application to the treatment of iron ores, beach sands, and flotation tailings; this trend will continue.

Hydrometallurgy is expected to expand its important role in mineral processing.

Six hundred-foot thickeners are expected to be installed for heavy metallurgical duties in the future and the new, compact, high-capacity thickeners will attract additional interest.

Filtration duties that might attract most attention in the future are countercurrent washing of leach pulps and clarifying of leach liquors ahead of ion exchange.

There will be increased attention to tailings impoundment, as this can be the controlling factor on project development schedule, the largest item on the capital cost estimate, a major operating cost item, and the determining factor in plant location.

New materials of construction are expected to be developed in the future with perhaps the greatest need being in the handling of corrosive solutions and abrasive and corrosive slurries.

The impact of increasing regulation and control in the fields of health, safety, and enviroment is expected to grow and to require both an increase in the awareness and experience of designers and an increase in the designers' reliance on specialists.

The effects of escalating energy costs and increasing regulation of the industry are foreseen as major influences on plant design and on the time required to bring major projects to production.

PRIMARY CRUSHING

The primary crushing plants of the future will reflect requirements of the specific mining operations with which they are associated. The gyratory crusher will continue to be the preferred crusher for most large scale operations, with the possible exception of some mobile, underground or in-pit crushing plants where other types of crushers may be considered.

For mines in capacity ranges and operating depths similar to current operations, few changes to the general concept of primary crushing plants are anticipated. For open pit mines of higher than present-day operating capacities and greater than present depths, changes in design concepts to allow for these conditions can be expected.

For primary crushing plants servicing open pits in the current capacity range, few changes are expected to basic concepts. These plants are expected to be located as near as possible to the pit with haulage from the pit being by means of mine haul trucks, as at present. Crushers in these operations will be 1.37 m to 1.52 m (54 in. to 60 in.) gyratory crushers. Because of escalating capital and operating costs, additional attention will be given to design in order to minimize cost with little or no sacrifice in function. Crusher structures will remain massive, but careful design will reduce the installed cost of the concrete, steel, and other materials, to a minimum. To reduce capital cost, electrical equipment and switchgear will be located adjacent to the crusher structure, rather than within the main structure. Crusher structures of both conventional rectangular design and of cylindrical design will be built in the future. Belt feeders for discharge of the crushed rock beneath the crusher will gain wider acceptance. Improved maintenance access, particularly at the eccentric service level, will be incorporated in future plants.

For many future plants servicing pits of larger capacity and depth than present operations, the basic concept will be similar to present-day plants, incorporating as many crushers as required to provide necessary capacity. Other plants will be similar, but will have larger equipment to provide required capacity and ability to accept larger feed. At least one l.83 m (72 in.) crusher has been installed, and other manufacturers have started design on crushers of this size. Two meter or even larger equipment may be utilized in the long term.

Increased attention will be given to getting the best economic fit between blasting techniques, shovel and truck operations, and comminution equipment selection. Recent primary crushing plants have been designed to match 19 cubic meter (25 cu yd) shovels and 155 to 182 tonne (170-200 short ton) trucks. In the future, plants may be designed to match the 38 cubic meter (50 cu yd) shovel and the 320 tonne (350 short ton) truck.

It is expected that in-pit crushing will be used to a greater extent in the future, allowing conveyor belt transport of the ore and waste from the pit. The in-pit installations will be either truly mobile plants, allowing easy relocations, or will be fixed installations which will remain in one location for extended periods of time. In the latter case, a portion of the crusher structure may be fabricated steel, allowing for disassembly and relocation.

Dust suppression and control will be of increasing importance in the future. Because of this, it is anticipated that some future installations will have complete enclosures over the individual crushers, and will be provided with both water sprays and powerful dust removal systems; and maintenance within the enclosures will be provided by overhead travelling cranes. In other future installations we will see primary crushers without building enclosures and with maintenance provided by mobile cranes. Impact breakers will find increasing use in clearing the crusher cavities of oversize material. Hydraulic rock-handling equipment may also be used to aid in clearing of hangups in the cavity.

Because of the high capital costs involved, and other considerations, scalping grizzlies will only be installed in specific instances when there are demonstrated advantages.

Instrumentation concepts will remain similar to the present. Work will continue on improvement of methods of measuring some of the difficult variables, including crusher discharge opening and level of coarse crushed rock in the crusher discharge rock-box.

FINE CRUSHING AND SCREENING

Few changes are anticipated in fine crushing and screening plants for preparing feed to conventional grinding circuits. Cone crushers will continue to be the accepted equipment for secondary and tertiary crushing services. The various arrangements now in use in crushing plants will continue to be used. Selection of the most suitable arrangement will be made on the basis of the particular circumstances. The most popular arrangement for secondary and tertiary fine crushing systems will consist of a scalping screen ahead of the secondary crusher, followed by closed circuit tertiary crushing and screening, with screens located either before or after the tertiary crusher or housed in a separate screening building.

Three meter (ten-foot) cone crushers have been manufactured and successfully operated in the Republic of South Africa. This could be a trend in larger capacity operations of the future where we could expect conventional crushing circuits to consist of 3 meter (10 ft) secondary and 3 meter (10 ft) tertiary cone crushers. The 3 m by 7.3 m (10 ft by 24 ft) screen is becoming available, but it is thought by some that 2.4 m by 6.0 m (8 ft by 20 ft) mechanical vibrating screens are the reasonable maximum size, and, if so, the coupling of screens and crushers in the 3 m (10 ft) cone crushing plant may not be practical.

The impracticality of coupling screens and crushers will dictate that the system consist of secondary crusher feed scalping screens, with undersize going directly to fine ore products and oversize to a secondary crusher bin or surge pile, followed by belt feeders to the secondary crushers and secondary crushing discharge material, in turn, conveyed to the closed-circuit tertiary crushing and screening section. The screens in the tertiary crushing section will be remote from the crushers, either ahead of the crusher or at the crusher discharge and tertiary feed bins, or surge piles will be provided ahead of the tertiary crushers.

The fine crushing plant of the future, when preparing feed for single stage ball mill grinding or for direct feed to leaching circuits, may incorporate a quaternary crushing and screening stage with the quaternary crushers being either conventional shorthead cones or gyra-disc type crushers.

The preferred equipment will continue to be the cone crusher for secondary and tertiary crushing duty with either cone crushers or gyra-disc crushers being preferred for quaternary duty. Impact-type crushers and hammermills will be used for special applications, including some of the softer, clayey, or easily crushable ores.

The preferred screen will continue to be the incline-type, motor-driven, mechanical, vibrating screen of either two-bearing or four-bearing design. For fine screening applications, particularly those associated with quaternary circuits, high-frequency, high-intensity, vibrating screens may be increasingly used. Due to increasing concern regarding working environment, dust control and noise control will play a more important part in the design of the future fine crushing and screening plants. Improved designs, plastic materials for liners and screen cloths, and enclosures will be used to reduce noise levels. Dust control concepts will remain essentially as at present. Dust and noise in the working areas may need to be monitored by strategically placed probes with continuous strip chart recordings of the dust and noise levels. Outdoor installation of the complete fine crushing and screening plant, serviced by mobile equipment, will become more common where climate permits.

GRINDING

The future of grinding plant design is expected to follow several different paths. It is possible that, given similar circumstances, two different operations would solve comminution problems in radically different ways; one way might be to design a plant with large autogenous mills fed directly with run-of-mine ore, while another might use three or four crushing stages followed by small conventional mills.

Each of these and other approaches may be seen in the future, as different engineers interpret current experience and try to achieve optimum trade-offs with respect to capital and operating costs.

The trend toward larger mills and the use of autogenous or semi-autogenous mills has taken advantage of the cost benefit of scaling up plant equipment. While this trend may continue, there are those who feel that future plant designs will focus more on operating costs and energy conservation than on minimizing capital expenditure. A trend towards autogenous or semi-autogenous grinding is discernible and, at the same time, there is a trend in conventional grinding toward single-stage ball milling instead of the rod-mill/ball-mill combinations.

Opinions differ in the industry as to where these trends will lead. Some engineers expect a resurgence of rod milling. Most agree, however, that increasing study time and attention is being, and will be, paid in the future to the comminution circuit as a whole.

There has been, in times past, a tendency to compartmentalize thinking into, say crushing, grinding, and beneficiation or extraction.

In the future, the total communition system will be evaluated including crushing, screening, grinding and classification, all considered in conjunction with the final concentration process, and the system will be selected, based on capital and operating costs of the system as a whole.

We may expect grinding equipment in the future to increase in size, except for rod mills which are felt to be reaching their limit. We may expect to see 11 m (36 foot) and possibly 12.2 m (40 foot) diameter autogenous mills in lengths of 4.6 m to 6.1 m (15 to 20 feet). Ball mills will probably reach diameters in excess of 6.1 m (20 feet) and, for some applications, lengths of 12.2 m (40 feet).

The use of larger grinding equipment will result in a reevaluation of classification techniques. High frequency screening, utilizing abrasion-resistant elastomer decks, offers one possibility of positive classification based on particle size which could increase grinding effectiveness.

In treating the low-grade ores of the future, renewed consideration will be given to rejection of tailings at as coarse a grind as possible. This may be achieved by several stages of grinding and classification with concentration equipment (flotation cells, gravity separators, magnetic separators, or others) between stages to reject a low-grade tailings as early as possible, with a consequent reduction in grinding energy.

As the size of the grinding mills increases, foundation and support design will become more scientific in the interest of reducing capital cost. The present rules-of-thumb based on rotating mass will be replaced by rigorous technical analysis.

The cost of grinding and its importance to the concentration operations will intensify research aimed both at improving existing grinding methods and discovering radically new methods of achieving particle size reduction.

FLOTATION - CONCENTRATION

The trend toward larger flotation cells is continuing. The 14 cubic meter (500 cu ft) single spindle and the 17 cubic meter (600 cu ft) double spindle flotation cells are now widely used and are accepted by most operating personnel. Major equipment suppliers are now prepared to offer the next larger machine which will be in the 28 to 35 cubic meter (1000 to 1250 cu ft) size range.

One 28 cubic meter (1000 cu ft) single spindle flotation cell has been put on test at Kennecott's Chino mine in New Mexico.

While the major flotation equipment suppliers have increased the sizes of the cells, these suppliers have been relatively conservative in the size increases when compared to others in the industry. A copper mine in Canada has installed and operated 56 cubic meter (2000 cu ft) cells, and a U.S. operator processing saline salts has used 56 cubic meter (2000 cu ft) tank cells for 40 years, and many years ago, successfully converted an 11 meter (35-foot) thickener into a flotation machine.

The progress towards larger flotation cells has been a steady, rather than a spectacular, development. While some operators will not be ready, the 28 cubic meter (1000 cu ft) cell is expected to be adopted for new large mills within the next five years. Some engineers feel that the 28 cubic meter (1000 cu ft) or larger cell would not be considered except for mills of 36,000 tonne per day (40,000 stpd) or greater capacity.

The 28 cubic meter (1000 cu ft) cell is about the limit of size for offsite fabrication and transportation to the mill site in one piece. Larger cells will be fabricated on the jobsite from steel, concrete, or other materials.

Metallurgically, on a cell volume basis, larger cells may or may not prove equal to smaller ones, but the important comparison in the future, as now, will be return on investment, taking all factors into account.

Recent additions to the flotation cells available to operators include the the Davcra cell and the Outokumpu machine.

Future cell design changes will be aimed at further reduction in power consumption. More instrumentation and control is being used in flotation concentrators, and computerized automatic control will be used in some plants in the future. With all necessary analysis of process streams from instream (or onstream) analysers, mass flow meters, density recorders, particle size analysers, the flotation circuit will be controlled automatically by the computer.

MAGNETIC SEPARATION

The recent trends in magnetic separation equipment include the adoption of larger, low-intensity wet drum separators and the development and large-scale use of high-intensity wet magnetic separators.

Among the range of conventional, low-intensity, magnetic separation equipment, the wet drum separator has received most attention from engineers and suppliers. These machines are generally available in single to quadruple drum arrangements, with the flow of the pulp being cocurrent, semi-countercurrent, or countercurrent.

The permanent magnets used for these wet drum separators are now made of ceramic materials. The size range that the wet drums have been available in has changed from 0.61 m (two-feet) diameter to 0.91 m (three-foot diameter), and now to 1.22 m (four-foot) diameter. Manufacturers have settled on a 3 meter (ten-foot) length of drum so that the 1.22 m (four-foot) diameter by 3 m (ten-foot) long unit is thought to be the largest proven equipment available.

The increased diameter of the drum produces overall unit throughput cost savings and a higher grade concentrate per stage of concentration.

The wet drum separator is finding continued application in the concentration of magnetic iron ores, for the removal of magnetic components from various flotation products, and for the cleaning of ferrosilicon in sink and float plants.

Equipment suppliers are directing their development work toward improving the magnetic circuitry of existing hardware in order to achieve improved recovery.

The wet high-intensity magnetic separator has been applied to many duties and has been predicted to have a future in processing many feebly-magnetic ores and tailings, and for the cleaning up of concentrates, kaolin, coal, and waste water streams.

Examples of recent applications of the wet high-intensity magnetic separator have been seen in the iron ore industry in Brazil and Quebec. In Brazil, Cia. Vale do Rio Doce installed 28 Jones separators. Each of these machines weigh 100 tonnes and treats one million tonnes per year of ore. In Quebec, Sidbec Normines, Inc., at Port Cartier has purchased six separators that are used for upgrading spiral concentrates.

Much has been reported on the research and development of wet high-intensity, magnetic separators; and equipment suppliers are working on the improvement of magnet and matrix design.

GRAVITY CONCENTRATION

The technique of gravity separation for the concentration of ores is currently enjoying renewed attention and is predicted by some to have a bright future.

The development of new higher efficiency and higher capacity equipment is continuing. This work is intended to match the new developments in large-scale equipment now used in communition and flotation.

Gravity concentration will be given more consideration in the future as an alternative process step.

The spiral concentrator was recently selected for the preparation of iron ore concentrated at Quebec Cartier Mining Company's Mt. Wright Project. This concentrator, reported as the world's largest based on production capacity at startup, uses spirals to produce 45,000 tonnes (50,000 short tons) of concentrate per day. It has six mill lines using wet autogenous milling. In each mill line, the roughers total two hundred and eight 0.61 m (24-inch) double start spirals of five-turn with 0.44 m (17½-inch) pitch. The cleaner and the recleaner spirals in each line amount to two hundred and sixteen 0.61 m (24-inch) double start units of four-turn with 0.44 m (17½-inch) pitch. This gravity concentrator is designed so that the double stage heads pumps transfer slurry feed to the top of the concentrator building from where it flows by gravity through the rougher, cleaner, and recleaner spirals directly onto the product concentrate filters.

The Reichert cone developed in Australia is a gravity concentrator device that has gained widening acceptance in recent years. The Reichert cone, initially used in Australia's mineral sands industry, can be made of fiberglass coated with polyurethane elastomer. These cones are mounted in circular steel frames to provide a self-supporting unit weighing only 2.5 tonnes for 50 to 100 tonne per hour capacity.

A typical cone concentrator plant operating in low-grade mineral sands comprises Reichert cones on rougher, scavenger, cleaner, and recleaner duties with a Reichert spiral used to produce a final product. This circuit provides for ratio of concentration of 100:1 with a recovery of better than 90 percent of the heavy minerals, (rutile and zircon).

Other recent applications include an iron sands concentrator in New Zealand; preconcentrating of tin ore in Australia; recovery of heavy minerals from flotation tailings at Palabora, South Africa; and the Richards Bay beach sand processing project, also in South Africa.

The Palabora application, which upgrades values (uranium and zirconium) by 1000:1, demonstrates the possible future potential for recovery of minerals from flotation tailings.

HYDROMETALLURGY

The role of hydrometallurgy has come more clearly into focus during recent years. In the past there has been speculation that hydrometallurgy would be the panacea of minerals processing and that pyrometallurgy would be eclipsed by the new methods. The importance of hydrometallurgy in the future should not be understated but it should be recognized that its application and development will be parallel to and, in some instances, in combination with pyrometallurgy.

Copper and uranium are examples of metals where hydrometallurgy is developing. Copper values are being won by heap leaching, vat leaching, and agitation leaching.

The future use of in-situ leaching for copper ore bodies has been studied. Recent copper plants developed on a prototype or small commercial scale include the Arbiter plant, using an ammonia leach, and the Cymet and CLEAR plants, which employ a chloride leach system. These plants, designed to treat copper sulfide concentrates are in the early stages of their operations development. The next generation of these, and similar, plants will be of great interest to the mining community.

The Anamax oxide plant at Twin Buttes, Arizona, is a recently commissioned facilitity that uses crushing, grinding, agitated acid leaching, solvent extraction, and electrowinning to produce cathode copper from oxide ore. At the Chingola Division of Nchanga Consolidated Copper Mines in Zambia, similar processes are used to produce cathode copper from flotation tailings. At Lakeshore in Arizona, Hecla-El Paso Natural Gas uses a combination of hydro- and pyrometallurgical process including flotation and roast-leach-electrowin oxide ore vat leaching and cementation of copper with sponge iron and sulfide ore.

The trends that can be seen now suggest that future copper oxide ore plants will be built with leaching and with solvent extraction and electrowinning to produce a finished copper product that requires no further metallurgical treatment.

Designs for plants treating as much as 27,000 tonnes per day (30,000 short tons per day) of ore using a single process line for vat leaching and solvent extraction have been developed. Such a design requires very large vats, 62.5 m x 53.5 m (205 ft x 175 ft), and some development design work on the required high capacity vat loading and unloading equipment.

The future design of solvent extraction plants will benefit from the experience of recent installations. Priorities for designers will include avoidance of excessive crud formation problems, reduction of settler sizes, and selection of appropriate materials of construction and construction methods.

The design of electrowinning tankhouses has shown some progress without fundamental changes. The future electrowinning plant t will concentrate on improving removal of organic material in the pregnant electrolyte feed from solvent extraction; improving present systems for suppressing acid mist generation from the electrowinning cells; and practices that will allow the use of higher current densities in combination with control of impurities and copper quality.

While liquid ion exchange is now well established, the development of continuous ion exchange techniques using solid resins is progressing to a point where it can be anticipated that solid ion exchange will be used in copper processing.

Uranium processing has led the way in the development of solvent extraction and ion-exchange process plants. While the copper processing plants have now developed solvent extraction technology to the point where reference to uranium experience is no longer frequent, in ion exchange, uranium is taking the lead with the development of new continuous ion exchange systems.

In the long term, the future of hydrometallurgical plants is very good, especially if fuel costs for pyrometallurgical plants and the cost of meeting enviromental controls increase even more than present levels. Hydrometallurgical plants using acid leach processes are especially attractive where sulfur or sulfuric acid is in oversupply.

The future attractiveness of hydrometallurgical plants is enhanced by the fact that they can be designed from the smallest scale and with semi-moveable potential (say, 200 tpd) to plants of 50,000 tpd capacity, or perhaps even greater.

Hydrometallurgical plants can be designed so that the engineering/construction durations are significantly less than in pyrometallurgical plants.

The similarities of hydrometallurgical treatment of ores to the practices in the chemical processing industry will bring increasing influence and construction staffs to both high pressure process equipment and sophisticated corrosion-proof materials.

THICKENING

The trend to date and probably for the future in thickener installations is toward larger units.

Thickeners of 90 m to 120 m (three to four hundred-foot) diameter are becoming common, and some larger units, including 150 m (five-hundred-foot) diameter thickeners, are in use on phosphate slimes duties. We can expect to see some 180 m (600-foot) thickeners on heavy metallurgical duties in the future. These larger thickeners, using a center caisson with the underflow pumps installed at the bottom are being accepted for the thickening of tailings and for countercurrent decantion operations.

One challenge for future thickener circuits will come from the need to design large tonnage washing circuits where agitation leaching is selected to win values from low-grade copper, uranium, and other ores.

The uranium industries in the U.S. A. and South Africa have tended to follow separate routes, with thickeners being used for washing circuits in the U. S. A. and vacuum filters being used in South Africa. While these different approaches may continue, it is expected that future plants will show more diversity with more American users selecting filters and vice-versa.

Some interest is being developed in the mineral industry in the use of compact, high-capacity thickeners. The Lamella thickener and the Enviro-Clear thickener, together with other competing equipment, will be considered as alternatives to conventional equipment for many future plants. While these types of thickeners may have many attractive features, and will, no doubt, increase their share of the market, conventional thickeners are expected to dominate for some time to come.

Some changes in thickener rake mechanisms have been seen (e.g., cabletorq) and more are expected. Greater use of hydraulic driveheads is also expected on the large thickeners of the future.

There is a wide variety of materials available for thickener tank construction, with wood, stainless steel, and rubber-lined steel being popular for small and medium-sized thickeners on acid duties.

The future successful development of large thickeners, especially those required for acid leach plants that use a raffinate from solvent extraction to wash the leach pulp, will depend on the performance of plastic linings and membranes exposed to acid solutions containing organic solvent material.

FILTRATION

The two filtration duties that might attract most attention in the future are countercurrent washing of leach pulps and clarifying of leach liquors ahead of ion exchange.

Drum filters have been used for countercurrent washing of uranium leach pulps in U.S.A. and South Africa and for leached copper concentrates and calcines in Zambia. New interest is currently being shown in the horizontal belt filter. This type of filter, available from one supplier in sizes up to 6 m x 37 m (20 ft x 120 ft), allows the cake to be washed countercurrently in many stages as it is carried along on top of the filter belt. This avoids interstage repulping and transport of cake, when compared to a drum filter countercurrent filtration alternative, and the interstage pumping of underflows, when compared to a thickener countercurrent decantation alternative. This equipment will be considered for some future plants.

The need to clarify leach liquors ahead of ion exchange plants varies with the type of installation with liquid/solid fixed-bed ion exchange columns and liquid/liquid solvent extraction systems requiring a cleaner feed than the feed needed for moving bed ion exchange systems.

The application of high flow-rate sand filters has developed to replace precoat filters as the equipment selected for this duty. Work is continuing on the improvement of sand filter design and selection of bed media. Future improvements that would relieve cementing of the media in high acid-consuming operations that produce heavy gypsum deposits would be welcome, together with improved clarity of filtrate that would reduce crud formation in solvent extraction or in some solid/liquid ion exchange systems.

In general, we can expect clarification filtration problems to increase in the future because of enviromental controls on liquid effluent streams, which may, in turn, bring new solutions to in-process filtration duties.

TAILINGS DISPOSAL

The disposal of tailings for future projects will be a subject of primary importance both to the mine owner and the engineer/constructor.

The tailings dam of the future will be for some projects the determining factor in governmental decisions on whether to allow a project to proceed. There will be projects where the tailings dam is the controlling factor on the project development schedule, the largest item on the capital cost estimate, a major operating cost item, and the determining factor on plant location.

There is a trend towards regulations that would require zero liquid effluent discharge or leakage from process plants and tailings dams.

We have seen a recent example of a uranium project where more than 30 permits were required to build a tailings dam. This bears little resemblance to some tailings dams constructed in the copper mining industry where construction of the impoundment dike proceeds on an operating cost basis, using spiggoting or cycloning, following a minimum capital expense for a small starter dam and with expected significant leakage through the dam wall and seepage into the ground under the tailings dam area. Now, in uranium plant tailings dams, we are seeing designs that include a completely engineered dam constructed to full height sufficient to contain tailings for the estimated operating life of the mine. Dams like this with a clay core may be highly instrumented to detect any movement in the dam structure. Requirements may include a clay- or bentonite-type surfacing of the entire impoundment area, or perhaps in some cases, impervious plastic membrane systems. Progressive covering of the solidified portion of the dam with material suitable for reseeding is being required.

At Homestake Mining Company's Grizzley Gulch dam in South Dakota, the tailings system includes a clay core to retard seepage and a drainage system that will collect any seepage through the dam that does occur for recycle to the impoundment area.

Recycling of water from tailings thickeners and/or tailings dams will be required, and treatment of this water will, in many cases be required, especially in some hydrometallurgical applications where the recycle water stream contains impurities that would build up in the process circuit.

MATERIALS OF CONSTRUCTION

While no radical departure from present materials of construction are foreseen in the future beneficiation plant, evolution will take place, and many of today's materials will be replaced by different and better ones.

Ultra-hard ceramic materials may find increasing applications. Ceramic discharge spigots on cyclones are now common. It is possible that the future will see increased use of ceramic pump liners and impellers for extremely abrasive applications, such as rod mill discharge pumps.

Rubber, elastomers, and plastics will find increasing use for corrosion and abrasion resistant applications in the plant of the future. Various types of rubber are now commonly used in this service, as are plastics, including polyurethane and polyethylene. Use of these plastics, and similar materials not yet developed will be common . Plastic pipe, such as high-density ultrahigh molecular weight polyethylene will replace rubber-lined steel pipe now commonly used for abrasive, corrosive slurry handling. The future will see increasing use of all plastic vessels for many applications, such as acid solution storage tanks, solvent extraction storage tanks, thickener tanks, and similar vessels.

Rubber lining for abrasion and corrosion resistance in large vessels, such as leach tanks, slurry holding tanks and similar applications, is a laborious process requiring highly skilled labor and extensive equipment, including a steam generator. The future will see replacement of rubber linings by materials that are easier to apply, possibly by spraying or by cold cementing.

Plastic liners will also be increasingly used in solids handling bins and chutes where their low coefficient of friction can aid material flow and reduce noise levels. Polypropylene sheets have been successfully used to line damp concentrate surge bins, and similar plastics in the future could be used advantageously to line fine ore bins and chutes.

HEALTH, SAFETY, AND ENVIRONMENT

The impact on future plant design of increasing regulation and control in the fields of health, safety, and environment will grow.

The designer of mineral processing plants in the future will need to be experienced in the requirements of established regulations and have access to specialists who have the time and resources to interpret new and upcoming regulations.

In addition to regulations, there are other trends that could be assumed will lead to higher standards of health and safety. In relation to plant production, future process plants will have fewer operators with fewer routinely tiring and hazardous tasks, and the general level of education, training, and skill will tend to rise. Along with these trends, it can be expected that the operating work force will be increasingly aware of, and place more emphasis on, the importance of working conditions.

The design of plants with good working conditions, including adequate access for operation and maintenance, good cleanup facilities, proper ventilation and reasonable noise and lighting levels will be looked upon as a sound investment that can improve production levels and efficiency, as well as a necessity imposed by codes and regulations.

The obvious effects of increasingly stringent enviromental regulations will be seen in increased capital and operating costs.

Another major effect is the stretching of project development schedules. A grassroots mining and mineral processing project may be ten to fifteen years in the making and require in excess of a hundred permits from various regulatory bodies.

The greatest environmental concerns for the minerals industry (excluding smelters) are land use and tailings disposal and the containment and/or treatment of liquid effluents.

Chapter 42

DESIGN OF MINERAL PROCESSING FLOWSHEET-- CASE HISTORY

by Paul R. Smith, Jr. and Gary E. Butts

Project Managers
Colorado School of Mines Research Institute

INTRODUCTION

The design of a mineral processing flowsheet involves many factors. The importance of the sample can not be overemphasized. Competent and thorough mineralogical examinations are important as an aid in designing the test program. Designing and conducting the laboratory test program is basic to the design of the flowsheets.

In the project under consideration a flotation pilot plant was conducted to prove the process and to provide a realistic prediction of metallurgical results. Moreover, the pilot plant operation provides important pilot plant design data.

Finally, accurate and clear reporting is essential so that the information developed can be utilized advantageously by the client and the engineering firm.

The design of a minerals processing flowsheet is a team effort. Engineers from the mining company, the research organization, and the engineering firm all contribute to the effort.

SAMPLES AND SAMPLE PREPARATION

The type and amount of sample collected for flowsheet development investigations varies widely but all engineers recognize that gathering a representative sample of the ore body is the single most important aspect of any process development effort. The process developed for any given ore body is only as reliable as the sample on which the flowsheet was based. For this reason, a great deal of time and money must be allocated for sampling.

Industry wide generalizations cannot be offered relative to the amount of material which must be collected to insure the integrity of the sample. The type of deposit and the occurrence of the mineralization will dictate the extent of the sampling program. Obviously, a rich, spotty gold deposit will require more extensive sampling than will a uniform, porphyry copper deposit such as found throughout the southwestern United States.

For a typical sulfide deposit, samples are obtained as drilling progresses to delineate the limits of the deposit. The core recovered from the drilling program is carefully labeled and is often preserved when the possibility of oxidation could affect the metallurgical test results. The whole core is then split and one-half is logged and placed in storage for future reference. The remaining material is crushed to approximately 6 mesh and a portion of the crushed core is pulverized finer and analyzed to determine the grade of the particular sample. The results of the analytical determinations are evaluated and, based on projected economics, the difference between ore and waste is established.

The crushed core which is classified as ore is then prepared for preliminary laboratory testing and mineralogical examinations. Normally, the individual core segments are composited to represent ore types, ore grade, or material which would be mined at various time intervals throughout the life of the mine. Each composite is then blended well, the particle size of the material reduced further, and the ore is split into appropriate size test lots for the subsequent metallurgical investigation. Head samples from each composite are submitted for rigorous chemical, particle size, and mineralogical analyses to characterize the samples. Depending on the capabilities of the property owner, the sampling and subsequent metallurgical investigation may be performed internally or the services of contract research and development firms may be engaged. Of course, the actual laboratory investigation would not begin unless the results of the coring program suggested the existence of an economically minable reserve.

Assuming the existence of an ore body has been established, the property owner proceeds with mine planning and actual mine development work. For either open pit or underground mining operations, the preliminary mine development work often consists of driving declines, adits, shafts, and drifts into the deposit. The purpose of the expensive but necessary exercise is twofold:

1. To assess conditions which would be encountered at actual mining depths, and

2. To obtain bulk samples of the ore body for use in continuous pilot plant experiments.

Typically, a few hundred tons of representative sample are required to operate a pilot plant but many thousand tons will be removed from the exploratory underground operations. Obviously, it is not practical to manually sample thousands of tons of material to obtain a sample for pilot plant testing. Basically, the developer has two choices:

1. Prepare a sample of run of mine ore and ship to the testing facility for further processing, or

2. Process all run of mine ore through a sampling plant located at the mine site.

If the mineralization of the deposit is uniform, the first alternative is by far the least expensive and most simple. On the other hand, if the mineralization is somewhat variable, the second alternative is preferred.

When the sample of ore is received at the testing facility, all of the material is processed through the crushing plant to reduce the size of the particles to approximately 1/4 inch. The typical crushing plant consists of a jaw and cone crusher operating in closed circuit with vibrating screens. The crushed material is mixed well and placed in storage for subsequent portions of the investigations.

If the ore body appears variable, then all of the run of mine ore should be processed through a sampling plant to insure the reliability of the test sample. Several different types of sampling plants have been built and successfully operated but a typical "tower" type facility is illustrated in Figure 1. Of course, the percentage of sample collected at each stage and the number of stages incorporated in the plant will be dictated by the type of ore. The cuts shown in Figure 1 are sufficient to satisfy the most stringent sampling requirements for rich and spotty ores. R. H. Richards has presented recommended sample weights at various particle sizes for ore types ranging from very uniform to very spotty(1).

Material from the sampling plant is shipped to the test facility where it is blended and/or composited prior to testing.

FIGURE 1

Tower Sampling Facility

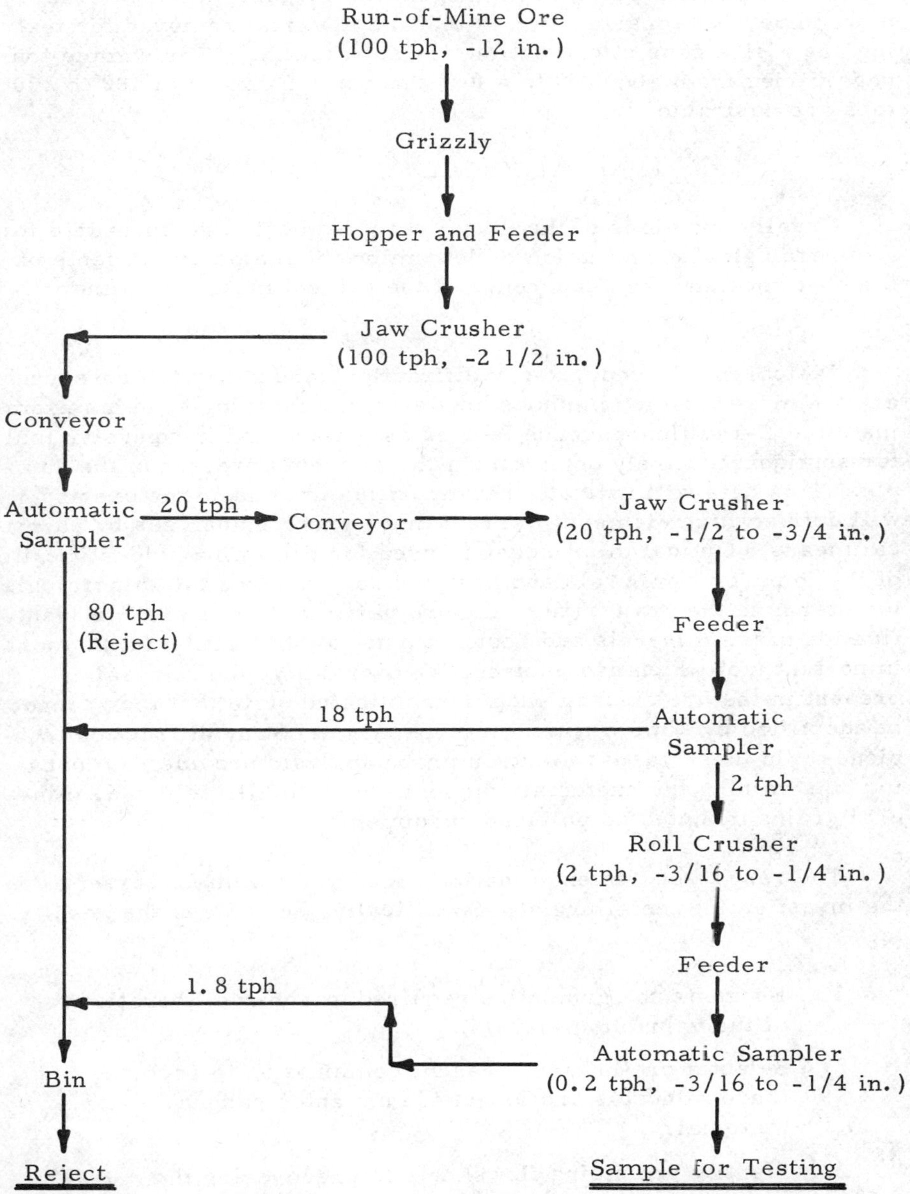

Samples for Autogenous Grinding

The principles described above generally are true for samples taken for autogenous, or semiautogenous, grinding testing. The most important consideration is that the material removed for testing has a size consist compatible for such testing. The sample tonnage varies from about 50 to a few hundred. Samples of 100 to 200 tons are desirable.

MINERALOGICAL EXAMINATION

Usually, portions of the initial core segments are submitted for a mineralogical examination to determine the major components of the host rock and the occurrence of the subject metals or mineralization.

Optical microscopy, x-ray diffraction, and x-ray fluoresence are the most common methods used by the mineralogist in his examination. X-ray fluorescence is used as a rapid and inexpensive tool for semiquantitatively determining the elements present in the sample. This data will indicate the impurities present in the ore and will define which elements warrant more precise analyses by chemical means. Optical microscopy is used for preliminary identification of the important minerals and host rock and enables the mineralogist to determine the grain size of the ore particles. For precise identification of the minerals and host rock, x-ray diffraction is the most important tool which can be used. Occasionally, the minerals present in the ore exist in such a complicated state that they cannot be identified by simple optical microscopy or x-ray diffraction techniques. In these cases, electron probe analysis provides a means for determining the chemical composition of small, selected, mineral grains mounted on polished specimens.

The results of this examination constitute a valuable asset to the investigating metallurgist. Specifically, he can use the results to:

1. Estimate the grind size required to achieve liberation of the valuable minerals.
2. Select a preliminary reagent combination to recover these minerals and reject the unwanted gangue material.
3. Identify promising flowsheets for recovering the minerals into marketable concentrates or pure metal.

The mineralogical examination is not limited to evaluating ore samples. Many times, examination of various process products can aid the metallurgist in solving a particular problem and delineating the direction of future efforts.

LABORATORY TEST PROGRAM

As previously mentioned, the preliminary process development investigations can proceed simultaneously with the drilling and mine development program. Usually, the preliminary studies will be conducted on drill core size-reduced to minus 6-mesh.

Equipped with the results of the mineralogical examination, the metallurgist will generally begin the program by conducting tests to determine the time required to grind the -6 mesh ore to various sizes in a laboratory rod or ball mill. Attempts are made to establish grinding times which result in particle size distributions coarser and finer than the liberation size indicated by the mineralogist. These size distributions are developed from sieve analysis performed on the ground pulp and can be plotted as a function of grinding time. This plot, in turn, can be used to predict the grinding time required to generate any desired particle size distribution.

Upon completion of the grinding tests, the engineer will select a flowsheet and reagent suite which has the greatest potential for successfully treating the ore. Selection of these process parameters is based on the experience of the metallurgist coupled with current industry practice. In other words, the test engineer will select the simplest conventional technology which has been used to treat similar ores successfully. He will then conduct preliminary tests to acquaint himself with the ore and determine the amenability of the deposit to the particular flowsheet chosen. If the results of the preliminary study are not encouraging, then the engineer will continue to "scope" other processes until favorable results are achieved. However, the ore may prove refractory to all known conventional technology at which point the R & D organization must use all of its available resources to solve the problem. In the past, such situations have rarely developed but as lower grade and more refractory deposits are exploited, the need for original thinking and new process techniques will increase.

Optimization of Process Variables

Nevertheless, a favorable processing flowsheet will eventually be identified and the operating parameters must be optimized.

Optimization of a flotation process to treat Cu and Mo, such as we are addressing today as an example, involves several steps. The unit operation of primary grinding, rougher flotation, regrinding, cleaner flotation, copper-molybdenum separation, and production of marketable Cu and Mo concentrates must be studied.

Optimization of the primary grinding and rougher flotation steps involve performing tests to evaluate the following variables:

1. The effect of particle size on recovery of Cu and Mo into a rougher flotation concentrate.
2. The effect of reagent types and reagent concentration on recovery of the two metal sulfides.
3. The effect of pH on recovery of the two metals, and on gangue rejection.
4. The effect of rougher flotation time on recovery of the valuable minerals.
5. The effect of flotation pulp density on recovery of Cu and Mo.

Although the list of variables seems relatively short, it may be necessary to perform in excess of 50 flotation tests to optimize the aforementioned process parameters. Ideally, plots can be developed which will indicate minimum process requirements and the point of diminishing returns (i. e., grinding finer than a certain nominal size does not improve the metallurgical performance).

In many cases the total number of tests required can be reduced by employing factorially designed test series. These test series provide an indication of optimum process conditions with a minimum number of tests. It is always necessary, however, to conduct additional tests to determine exactly the optimum process conditions.

Once the optimum primary grind and rougher flotation conditions have been established, efforts are directed towards delineating the regrinding and cleaner flotation requirements. Essentially, the same variables studied for the rougher flotation step are investigated for the regrinding and cleaner flotation unit operations. The objective of this phase of the investigation is to define the parameters which maximize overall recovery while at the same time, enable production of a high grade concentrate suitable for treatment in a Cu-Mo separation circuit.

Since most porphyry copper ores contain only minor amounts of molybdenum, optimization of the process conditions associated with Cu-Mo separation circuits is usually postponed until a sufficient quantity of bulk concentrate is available from the pilot plant operations.

Testing Various Ore Types and Composites

The aforementioned test work is usually conducted on a composite of drill core samples which are considered representative of the entire ore body. However, significant differences in ore type and mineralization may occur within any one particular deposit. For this reason, it is advantageous to determine the amenability of the "extreme" ore types to the process developed for the "average" ore. It may also prove beneficial to develop similar information on ore composites which represent the type of ore to be processed at various time intervals throughout the life of the mine. The additional data developed from testing various ore types not only reassures the lending institutions but also provides the mill operator with valuable guidelines. Approximately 300 lbs of representative sample is required to complete the initial bench scale investigation.

Hopefully, by the time the process development studies have been completed on drill core, the bulk sample is available for testing. Flowsheet simulation tests are then performed on the bulk sample to confirm the results of the previous investigation.

The next step in the process development program is to conduct locked cycle flotation tests to establish equilibrium recoveries, concentrate grades, and reagent consumptions. Locked cycle experiments consist of a series of bench scale tests performed in such a manner as to simulate continuous plant operations. For example, suppose the preliminary bench scale experiments indicate the ore is amenable to a conventional grinding, rougher flotation, regrinding, two stage cleaner flotation and scavenger flotation circuit. When the locked cycle test is conducted, the only products which are filtered, dried, and prepared for analyses are the rougher tailing, the scavenger tailing and the bulk Cu-Mo concentrate (2nd cleaner concentrate). The 2nd cleaner flotation tailing together with the scavenger concentrate are saved and recycled to the appropriate unit operation during the next batch flotation test. Water is recovered from filtration of the flotation tailings and decantation of the recirculating products. This water is used during the succeeding stage of the locked cycle test. The flowsheet shown in Figure 2

FIGURE 2

Locked-Cycle Flotation Test

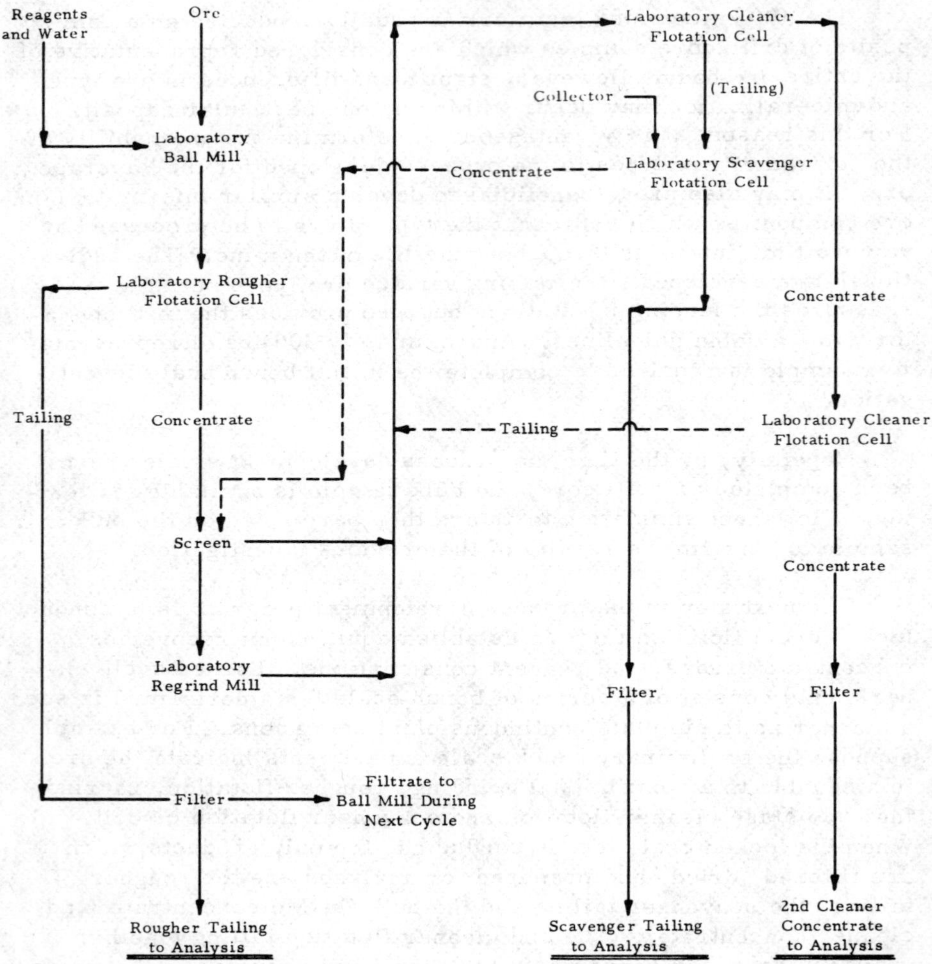

Note: Dashed dotted lines indicate that the products are recycled to next batch flotation step.

illustrates the manner in which a locked cycle experiment might be performed. The experiment continues until the concentrate grade and recoveries remain relatively constant for 2 cycles of operation. Normally, equilibrium is achieved within 5-6 cycles.

PILOT PLANT OPERATIONS

There are several reasons for conducting pilot plant operations:

1. To demonstrate the process on a continuous basis. This is of particular importance to investors.
2. To provide engineering data, such as flow rates, flotation time, etc.
3. To determine the long-term effect of recycled water and middling products on metallurgical performance.
4. To generate large quantities of products which can be submitted to potential customers for evaluation. In addition, a comprehensive study of pilot plant tailings is often performed to provide guidelines for construction of tailing pond systems.

Based on the results of the bench-scale test work, a flowsheet and material balance is prepared for the pilot plant. Quite often, the metal content of the ore and the desire to recover by-products will dictate the feed rate to the plant. For example, if the developer wishes to continuously demonstrate production of a marketable Mo concentrate from ore containing 0.02% Mo, then the feed rate to the plant must be rather substantial. On the other hand, it may be advantageous or necessary to store the concentrate from the copper circuit and campaign this material through the molybdenum circuit after enough concentrate has been accumulated. In these cases, a much lower feed rate is in order. Most contract R & D organizations maintain flotation pilot plants capable of processing 500-2,000 lbs of ore per hour. Occasionally, lesser throughputs are employed when only smaller samples are available. Larger pilot plants have been constructed at the mine site when conditions warrant greater tonnage throughput.

Pilot plants are assembled according to the flowsheet developed in the laboratory. No organization can possess the amount of grinding equipment required to always have exactly the correctly sized grinding mill(s). This is overcome by variations in the classification circuit, by varying the ball charge, and by varying the pulp density in the grinding mill. Most frequently the grinding circuit consists of a rodmill in open circuit followed by a ball mill in closed circuit.

Usually the retention time provided in each individual flotation circuit is double the flotation time required in the laboratory. For example, assume eight minutes rougher flotation time was found to be optimum in the laboratory. The flotation machines installed in the pilot plant would provide approximately 16 minutes retention time. This is due to more rapid concentrate removal in batch tests and to short-circuiting during continuous operation.

Probably the most critical equipment in a flotation pilot plant are the reagent feeders and the pumps. Reagent feeders must be employed which provide exact flow of minute streams regardless of the quantity of material in the reservoir. The slurry pumps must provide flow which is not subject to surges. This can usually be accomplished by employing variable speed peristoltic pumps.

Thickeners are usually provided for the tailings and filters for the concentrates. After break-in operations, the maximum amount of reclaimed water is recycled throughout the plant.

Break-In Operations

Once the pilot plant is assembled, water is run through the circuit to check for leaks and insure that all equipment is in good mechanical condition.

Next, the ore sample is fed to the plant. Reagent feeders are calibrated and the plant is operated for a few days to acquaint the crew with the circuit and characteristics of the ore. The most important aspect of this break-in period is establishing the primary grind which closely approximates the particle size distribution used in the laboratory experiments.

During the break-in period, operations are restricted to day shift and, usually, the grind is established and the "bugs" worked out within four to five days.

Operations for Optimization

After the break-in period, the pilot plant is operated to optimize the process parameters for each unit operation. The effect of particle size, flotation times, reagent consumption, etc. on product and grade are determined.

During this period, operations are restricted to day shift and on any one day, emphasis is placed on one particular unit operation. Optimization usually requires one week or more.

Continuous Operation

Once the optimum operating conditions have been established in the pilot plant, the circuit is operated around the clock to demonstrate the process on a continuous scale and determine the effect of recycled water and middling products on metallurgical performance. Quite often, the effects of recirculation do not appear for 5 days so at least this much time should be allotted for this phase of the program.

Three operators are usually assigned to each shift--one engineer and two technicians. The support staff consists of a technician for sample preparation, an analyst, another engineer on a part-time basis for data collection and evaluation, and a mechanic on call to perform routine maintenance. Laboratory flotation tests may also be performed in conjunction with all phases of pilot plant operations. The results of these laboratory tests provide guidelines for control in the pilot plant.

Sampling procedures are formulated and practiced during the break-in period. Initially, the sampling program tends to be very comprehensive, but as familiarity is gained with the response of the ore to the process, the intensity of the sampling program can frequently be reduced. Throughout the pilot plant campaign, special samples are collected to ascertain the effect of a specific variable on a particular unit operation and "profile" samples are often taken of the entire circuit to determine the response of the ore to a certain set of operating conditions. In addition to the special samples, shift composites are collected on a routine basis to guide daily operations of the pilot plant. Depending on the manpower and resources available, these composites may encompass three 8-hr shifts or one 24-hr composite. The most common compromise is two 12-hr shifts.

A typical sampling schedule for a pilot plant producing a bulk Cu-Mo concentrate is presented in the table below.

Sample	Screen Analysis	Data Required: Chemical Analysis %					
		Insol	Cu	Mo	Au	Ag	Fe
Feed (Rougher Flotation)	X		X	X	X	X	X
Rougher Concentrate			X	X			
Rougher Tailing			X	X			
Regrind Classifier O'Flow	X		X	X			
1st Cleaner Concentrate			X	X			
Scavenger Concentrate			X	X			
Scavenger Tailing			X	X			
2nd Cleaner Tailing			X	X			
2nd Cleaner Concentrate	X	X	X	X	X	X	X

Sampling frequency and the size of the sample cuts will depend on the feed rate but typically, samples will be taken no less than once each hour. Screen analysis of each of the above products are obtained at some point during the investigation, but only those listed are performed on a routine basis. Pulp density measurements are performed routinely for operating control and the specific gravity of each product is determined for design purposes. Furthermore, each of the aforementioned products may be analyzed extensively to determine the impurity content, but such analyses are generally restricted to the bulk Cu-Mo concentrate (2nd cleaner concentrate).

As previously mentioned, the importance of sampling cannot be over-emphasized and for this reason, a technician is usually assigned to the pilot plant whose total responsibility is preparation of the samples for chemical and sieve analysis.

In addition to the sample preparation individual, an analyst can also be assigned to the operation to insure the chemical determinations are performed quickly and accurately. With the aid of the sample preparation and analytical technicians, the metallurgist can receive results within 24 hours after submitting the sample and thus decide the direction of future efforts.

From sample weights and analysis, a metallurgical balance can be computed for any sampling period. Several methods can be used to calculate the balance, but past experience has shown that multiproduct formulas yield the most consistent and reliable results.

Cu-Mo Separation Testing

As bulk Cu-Mo concentrate becomes available from the plant,

a laboratory program is initiated to develop a process for recovering the Mo as a marketable by-product.

Basically, two flotation methods are used to effect the initial Cu-Mo separation: (1) depression of the molybdenum and flotation of the copper into a saleable concentrate, or (2) depression of the copper minerals and flotation of the Mo. Although this appears to be relatively simple, the initial separation is often quite difficult and requires a comprehensive effort. The correct reagent combination and pH control are critical and several experiments are required to optimize the parameters for the initial separation.

Once the initial separation is perfected, additional test work is performed to delineate the cleaner flotation, regrinding, etc. requirements necessary to produce a high-grade molybdenum product.

Miscellaneous Testing

Miscellaneous tests are conducted throughout the development program to provide additional data for design and scale-up of a commercial facility. The miscellaneous determinations include:

1. Bond crushing indicies to determine relative crushing requirements.
2. Bond rod mill and/or ball mill grindability indicies to determine the power required to grind the ore to the desired size.
3. Settling tests on various pilot plant streams to determine the thickening unit area and flocculation requirements.
4. Filtration tests on various pilot plant streams to determine the filter unit area requirements.
5. Specific gravity and bulk density measurements of pertinent mill products.

AUTOGENOUS GRINDING TESTS

Since autogenous, or semiautogenous, grinding has increasingly been incorporated into the design of minerals processing plants, this also is frequently subjected to testing. Such installations sometimes can reduce substantially the capital cost of a plant, since fewer pieces of size-reduction equipment, less bins, fewer conveyors, and less screens are required. Moreover this results in a smaller building, and a reduction in the electrical equipment requirements. Moreover, operating costs can be reduced since less grinding media is required and maintenance costs are reduced. Finally, by eliminating crushing, screening, and conveying equipment, the handling of wet, sticky, and frozen ores is facilitated.

For the test program the ore must be from minus 8 to a maximum of minus 12 inches. The sample is screened into three or four size fractions to enable feeding a constant size consist to the mill. The grinding mill is set-up in closed circuit with a classifier.

The ore is fed at a constant rate, and mill checked periodically to determine if the charge in the mill is increasing or decreasing. Adjustments in the feed rate are made until the charge becomes constant. The mill is then fed at a constant rate. The following conditions are determined: mill power, ore feed rate, ore feed size, final product size, circulating load, circulating load particle size, and pulp density. At the completion of the test the size distribution of the charge in the mill is also determined, to be certain that a critical size is not developing. If the load in the mill continues to build with a reduced feed rate, the ore cannot be ground autogenously.

Semiautogenous tests are conducted in a similar manner, obtaining data as described for autogenous tests. Tests are conducted with different ball loadings such as 2%, 4%, 6%, and 8% of mill volume. With each ball charge, the feed rate must be adjusted until a stable circulating load is established.

LIAISON

Several organizations and individuals may participate in the process development program. If the work is performed at a contract R & D facility, the responsibility for the project will be assigned to one engineer who in turn assembles a team of competent professionals and support personnel to insure the program objectives are attained on time and within budget. The sponsor may desire to assign one or more of his employees to the project in which case these individuals will work closely with the responsible engineer at the R & D facility. Many small and/or emerging mining companies do not have experienced professionals on their staff and may therefore engage the services of a consultant to act in their behalf.

The sponsor may also contract the services of an engineering company to begin feasibility studies of the total project. Quite often, representatives of the engineering firm will actively participate in the development effort to insure that all data needed for design of the commercial facility is obtained.

REPORTS

Upon completion of all phases of the investigation, a report must be written to record all of the data developed during the effort. Such reports are quite lengthy due to the voluminous amount of information collected from the study. All pertinent data and recommended design information are summarized in a few pages near the front of the report, to provide information for the decision makers. Discussion of results, and detailed data are provided for the close scrutiny required by metallurgists.

REFERENCES

1. Taggart, p. 19-22, 10th printing.

Chapter 43

PROBLEMS IN MILL PROCESS DESIGN

by Howard W. Jacky and John D. Vincent

Manager, Metallurgical Engineering
The Anaconda Company
Mineral Resources Group
Denver, Colorado

Consulting Metallurgical Engineer
Tucson, Arizona

ABSTRACT

This chapter covers the normal problems encountered in the building of an ore concentrating facility. To produce such a facility in today's environment requires correlation by the engineers involved of geologic data, mine planning, scheduling, future planning, water resources, power, land availability, testing, design, start-up and environmental considerations.

Of course, the program scheduling and character of the ore will determine the weight given to the above considerations in the final design and construction of the plant.

INTRODUCTION

The ultimate result that the engineer in a process mill design program seeks is a plant that gives him the optimum combination of the following factors:

Low capital cost

Low operating cost

Simple and easy operation

Optimum tonnage throughput, concentrate grade and recovery

Minimum maintenance

Quick start-up

To achieve this goal, planning should start on the design program during the early stages of mine exploration. Ore testing should be a part of the development drilling of the orebody. To facilitate the complete planning, a definite time-sequence program should be initiated and followed through the entire project. The program should define the parameters of the various steps and establish tentative dates of completion. Part of the program, of necessity, must be completely coordinated with mine exploration and planning. A simple listing of the various steps involved could be as follows:

Selection of process engineer or group

Sample procurement

Sample analysis and study

Bench testing

Pilot plant testing

Stockpiling system

Plant pre-engineering study

Operational crew study

Operational cost estimate

Project decision

Selection of design engineering and construction firm

Design and engineering

Purchasing

Construction and start-up

As the program advances, timing and sequence changes will require constant updating and modification of the steps. In many cases the phases will overlap, producing a condition of multiple steps proceeding simultaneously.

The problems encountered and the possible pitfalls that may be encountered by the personnel involved in the project is the subject of this paper. These can be many, intermeshed and complicated. We will attempt to point out some of these problems and in some cases suggest reasons for proceeding in certain ways through this sequence.

To confine this paper within limits, it is assumed that certain portions of the project have been determined. These will not be discussed. They are:

Land acquisition

Adequate water acquisition

Shipping facilities

Marketing

Power Acquisition

Fuel acquisition

Environmental considerations

Selection of Process Engineer or Group

The size and makeup of the process engineering group naturally depends on the size and complexity of the project. The group should be changed, enlarged or decreased as the program advances and the requirements change. It is most important, however, that a leader for this group (Chief Process Engineer) be selected very early in the program. Selection of the Chief Process Engineer should be based on the following qualifications:

Expert knowledge of the field

Operational experience in this type plant

Plant start-up experience

Ability to work with people

Continuity on program

It is not particularly true that the Chief Process Engineer should be the future mill superintendent. He may or may not be the best man for the job. It is more important that the Chief Process Engineer be selected for his knowledge and experience. One important qualification is a requirement for continuity on the job from start to plant start-up. It is very important that at least one person be available with the knowledge of why certain routes are followed during the various stages of the entire project. Frequently it is desirable to have the future mill superintendent on the job as an assistant to the Chief Process Engineer.

The ultimate decision for process design and equipment selection should rest with the Chief Process Engineer. Too often, these decisions are given to a plant operational manager who has inadequate experience for decisions in this field.

In summary, the importance of the selection of Chief Process Engineer can not be over-stressed. His influence in the operation will be felt for many years after the start-up, even if he is long gone. His input will have a lasting effect on the long-range operating costs and production results.

Sample Procurement

One of the first jobs in which the newly selected process engineer must become involved is representative sample procurement for testing. It is always a problem to get enough representative samples at the right time and particularly so if the property is new. To facilitate this, the process engineer should immediately start working with the geological group in charge of ore reserve development to effect a program for sample procurement. Arrangements should be made with the geologists for samples from drill cores, cuttings, surface cuts, underground exploration and any other type sampling being done. The process engineer should, at this time, arrange for preliminary bench testing of the samples as soon as they are available. He should quickly determine the effect of sample aging on his metallurgical results. If this is a program, correct sample procurement, handling and storage become more difficult. It may be necessary to continue this program throughout the entire property exploration program. Sample testing results will show the extent of the program required. The degree of uniformity of the orebody usually determines the volume of testing required.

The process engineer should be thoroughly cognizant of the geological sampling program. From this and from the results of his initial bench test work, he should ascertain the sample requirements for a detailed test program. He should have enough input in the program to secure the necessary samples for his detailed testing at the correct time.

Sample Analysis and Studies

It is obvious that correct sample assay results are basic for the success of the complete operation. Errors at this point may certainly have a lasting effect. To reduce the possibility of this problem, the process engineer should closely check the sample preparation and analysis procedures.

Sample preparation is extremely critical for the production of correct analyses. The major problem here is one of constant monitoring of the process. A correct procedure for handling samples should be established and retained in its entirety. Too many times, the sample preparation employee finds a short-cut that makes his job easier, to the detriment of the sample. A reliable employee in this dirty job is truly an asset.

Although it seems rather elemental, a hard look should be given to the type of assay procedure being used, particularly in the areas of oxide and by-product mineral analysis. Failure to initiate the correct analytical procedures has in the past created problems that have returned to haunt more than one metallurgist or plant manager in the form of invalid mineral recovery estimates. For example, particular care must be taken that the oxide copper analysis is the type needed to give the most information for the process. An analysis that is proper for a copper leaching plant may be of little use for a concentrator flotation operation.

Many times, molybdenum assay procedures have resulted in misleading information. Undetected oxide mineralization can upset an otherwise reasonable recovery estimate. Other assays that supply information used for determination of the mineral type are necessary for reliable estimates of recoveries and concentrate grades.

A knowledge of the orebody mineralization is also required in order to determine mineral type segregation in the mine. This information should be used in the mine planning as an important tool for programming ore feed to the mill. If the orebody rock type and mineralization vary widely, it may be necessary to test potential composites of various ore types. These may be vital to an efficient mine plan. The process engineer should work closely with the mine planning group towards supplying them with information indicating which types of mine ore can be combined for successful treatment.

Ore hardness determination is another study that must be made on a complete cross section of samples from the potential orebody. The range of hardness proportioned to the orebody area tonnages is a necessary estimate required for crushing and grinding plant design. Orebody hardness variations may also become a factor in the mine planning area important in the determination of proper mill feed required for planned tonnage throughput. The installed cost of today's crushing and grinding plants decrees that design must be kept to close tolerances. This further emphasizes the importance of the study.

Bench Testing

Bench testing is a key operation in the program. The test results not only define the preliminary flow sheet and reagent requirements but they also are required to determine the necessity of pilot plant testing. If the decision is made that pilot plant work is not necessary, the bench test work becomes the basis for the final refinement of the flow sheet. At this point, the degree of ability of the bench test technician becomes very important. Ideally, the test technician or engineer should have enough test experience along with operating experience to be able to relate the two together. In many cases, a good experienced technician will produce results that will closely match in both recovery and grade those produced by the concentrator. The credibility of this man's work may also be a factor in the decision for pilot plant need. The test technician's ability can also modify the length of the test program. If it is decided that piloting is necessary, the bench test work may help to decide to what degree and in what direction it should be pursued. In the final analysis, a good, knowledgeable, experienced bench test operator can save both time and money in the development of data required for design.

Pilot Plant

If a decision is to be made on type of grind circuit, i.e. conventional, autogenous or semi-autogenous, piloting is probably necessary. Detailed study of by-product metallurgy usually also commits the process engineer to a pilot plant program.

During this piloting operation, special care must be taken to ascertain that the plant feed is representative of the proposed mine feed to the plant. If it is not, the results of the tests may not be fully significant for plant design. It may then be necessary, if representative feed is not available, to design more broadly, e.g. larger size grinding or flotation, to compensate for this situation. At the conclusion of the pilot test work, the process engineer should have data that will determine the decisions to be made on the following areas of the plant:

Final balanced flow sheet

Type and size of crushing plant

Type and size of grinding and classification equipment

Flotation plant size and equipment selection for cells and pumps

Thickener area required and equipment selection

Reagent requirement

Water requirements

By-product plant requirements

Filter plant size requirement and equipment selection

Concentrate disposal

Tailings disposal

Stockpiling System

Metallurgical test results should also at this point help to make decisions regarding stockpiling as the mine is developed. The process engineer should work with the mine planning group to determine stockpile planning for anything the metallurgical plant may be involved with at a later date. This includes such things as:

Plant feed that is mined as part of the development program

Oxidized and partially oxidized ore

Dump leachable low grade ore

The process engineer should have input in decisions made on such things as:

Stockpile separation according to grade

Stockpile separation according to oxide content

Stockpile separation according to ore type

Leach dump design and location

These are areas often not fully thought out in the early days of mine development, usually because some extra effort and cost are involved. Neglect in proper planning, however, can result in untreatable leach dumps and stockpiles that mill poorly and are costly to reclaim.

Feasibility Study

At this point, most of the data is available for a project feasibility study. A decision must be made between in-house study, outside contract study, or both. The decision will usually depend on personnel availability, and sometimes the project financing will require an outside study. In any case, the process engineer must be set on all phases of the design and have data available to back his decisions. He should have a balanced flow sheet and a preliminary plant layout plot established. Tailings pond location should be tentatively set.

It will also be necessary to have made decisions in the crushing plants on such things as equipment size, number of sections and operating shifts per day. The operating shifts for the primary crusher usually must follow mine planning. Equipment size at this point normally must be selected to fit feed size. Sometimes this results in excess tonnage capacity for a crusher. When this occurs, a decision must be made whether to balance the rest of the crushing plant to the crusher output or to the crushing plant tonnage requirements. One method can lead to an oversized crushing plant. The other results in unbalanced equipment in the plant that can cause design problems.

Another problem that often arises at this point, again, is size selection of the primary crusher. Normally, an open pit mine wants to operate with the largest shovels and trucks available. When it comes to primary crusher selection, management suddenly becomes conservative and proposes installation of a crusher sized on required throughput. For the next 20 operating years, the shovels and trucks get larger, the feed ore size gets larger, but the primary crusher does not change. The operators continue to fight ever increasing oversized rock in the crusher. The economics of this type of decision is questionable. The process engineer should make this decision, not management.

Along with the crushing plant decisions, the process engineer should be set on such questions in the remainder of the plant as equipment size, type and number of sections. Each decision and its alternatives should be analyzed. In some cases, it may be desired that the feasibility study include alternate schemes. Available land and type of terrain may be deciding factors in other cases.

An estimate of the crew requirements must also be included in the study. This requires a detailed manpower analysis for the plant and includes such things as crew compositions and shift scheduling. It can be important to schedule plant crew changes with mine crew changes. Such things as car-pooling and crew housing availability can enter into these decisions.

Maintenance crews usually operate from either a central service, directly under the mill superintendent, or split between the two systems. Either way has its advantages and disadvantages. Most generally, maintenance is most efficient if the split system is used. Personalities of the supervisory personnel involved play a large role in the decision for this division and success of the maintenance program. A system should be chosen at this point, for it will affect the shop and equipment requirement analysis for the feasibility study.

Support facilities must be preliminarily established in this study. Such things as sample handling rooms and equipment, assay laboratories, change rooms, offices, water handling and tailing system must all be evaluated. Failure to completely include all required support facilities in this study has many times resulted in an incomplete and incorrect financial estimate.

The final result of this study, of course, is a complete estimated picture of the operation, including all capital costs, operating costs, production costs, equipment lists, crew, table of organization, balanced flow sheets and metallurgy. This, in turn, is combined with the mine operation estimates, overhead costs and sale costs to complete the full financial picture of the proposed project. On this report, the final decision of either to proceed or not to proceed is based.

Selection of Design Engineering and Construction Firm

Assuming that the project has been approved, the next step becomes selection of the design engineering and construction firm. A separate chapter or book could be written on this area alone, covering problems that evolve and decisions that must be made.

One of the first things to decide is whether to give the entire project to one firm or split it. Normally, each case has to be evaluated for this decision. A single firm has advantages in project completion speed since they can easily start construction before engineering completion. This system also normally does not experience the interfacing problems to the extent that multiple firms might. There can be other reasons why multiple firms are more advantageous in certain cases, such as (1) a very large contract or (2) special technical abilities. Another common type that might be selected is design engineering and supervisory construction. In this case, all construction may be contracted to a number of specialized firms.

Usually the contracts are let for bid, and this brings up the question of what type. Do we want a cost-plus, fixed fee, and what about incentive payments? Again, the answers to these questions depend largely on the timing, economic conditions and personnel involved.

When the bids are returned by the potential contractors, they must, of course, be analyzed and the contractor selected. Bids normally contain three major sections. The first is financial. The second contains some resumes of the principal proposed supervisors and the third is a sales pitch covering the firm's past history of projects. The first part may require some financial analyzing, averaging and rearrangement for equal comparison. Later, there may be some modifications and horse trading.

The personnel section can have a far-reaching effect on the project. The major people involved should be thoroughly checked out and interviewed. Design and construction people have a history of moving around, so their past should be well investigated for job stability as well as ability. Loss of principal supervisors in the middle of a job is usually costly to the client. The availability of good key personnel for the job should weigh heavily in the final selection of the contractor.

The sales portion of the bid normally may require checking with firms that have worked with the particular contractor and, possibly, visits to plants the contractor has designed and built.

Design and Engineering

Once the actual design engineering for the plant starts, man-hour cost savings and project speed can be greatly benefited if the client has frozen on the basic components of the plant. This includes such things as flow sheets, plant layout, equipment selection and support facilities. There is a tendency to want to continually change the basics as the project expands and additional avenues of possibilities reveal themselves. The further along the project has gone, the more costly it is to make basic changes.

Another problem that arises at this point concerns design plant capacity. Client management usually initially has set capacity. However, at this point there is suddenly a great deal of concern about getting a few thousand more tons per day through the plant without changing design specifications. This can be poor thinking because, if it is a well designed plant, there will be no single bottleneck point which, if expanded, would allow tonnage increase. A maximum tonnage plant throughput should be selected and to this the designers should adhere. Considerable care, however, should usually be taken to allow for possible future expansion. This can make design difficult, for it is very easy to inadvertently box in a design expansion area as the project advances.

Design direction and monitoring by the process engineer or engineering group should be continued throughout the design engineering period. Too often, not enough qualified people are available for this work or they are withdrawn from the job too soon. A good process engineer should contribute a great deal to the finalized layouts of such things as walkways, platforms, stairways and area access points. His contribution can improve access for maintenance to equipment as well as reduce required travel for operators. He can contribute to improving the ever present housekeeping problem. His input into the design of chutes, ore transfer points, bins, conveyor, feeders, slurry piping and pump sumps can cut dollars out of operating and maintenance costs after start-up.

The process engineer must today take in the environmental studies that would be required for the use of tailing dams, tailing lines and the implication of pollution to existing aquifers. He would also be required to study where the necessary water for the operation will come from and what its effect on the environment will be. If the plant uses processes for which the heat is produced by coal or natural gas, then the evolution of waste products must be considered. If the plant is a hydrometallurgical plant, it may be necessary to consider the evolution of toxic gases such as carbon dioxide, hydrogen sulfide, etc. and dispose of them so they are not harmful to the workers or to the general public. All of these questions, including the use of certain reagents, such as cyanide, acids or gases, become a matter of serious study in the environmental sense for the process engineer.

Even the appearance of the plant, or the esthetics, as one might say, are of importance today, and this is particularly true of tailing dams and waste piles where it will be necessary in most cases to attempt a reforestation program to satisfy environmental concerns.

Detailed monitoring of design is a tedious assignment. It pays off, however, not only in the major areas but also by correcting any minor mistakes before they reach the field and are built into the plant. Plant modeling can be a tremendous assist to the process engineer by greatly reducing the time required to review drawings. If modeled early enough, mistakes can be easily discovered and corrected before construction. Design improvements are much easier to see and many can be readily inserted without excess cost. In most cases, modeling will easily pay its own way.

Sometimes a design firm will subcontract out various areas of the plant. These can include sections of piping or conveying systems. The detailed design and, possibly, the partial shop assembly may be done at a remote area. If the process engineer is not alert to this system, he can find himself suddenly confronted with a bunch of half assembled process slurry lines that look like they were built for a bathroom plumbing job.

Probably during this period the engineering firm will assemble a final cost estimate for the plant. This always seems to be much higher than anyone expects, management is unhappy, so a cost cutting operation evolves. In order to pacify management, some reduction must be made. The process engineer should have a good preconceived knowledge of this probability and know where the cuts can be made with the least amount of damage to the plant design.

Purchasing

Purchasing may be handled completely by the design engineering firm, completely by the client or split between the two in any number of ways. There are a few areas that should be closely monitored by the process engineer in this facility. Equipment selection is most important. Equipment purchase is normally put out to bid. The process engineer must make certain that the purchased product will deliver the designated duty with a minimum of downtime and repair cost. Selection of equipment many times hinges on personal judgment, operating experience and knowledge of how the product has performed at other operations. These must be weighed against the equipment cost in order to make the proper decision.

Usually, some system of equipment inspection is established during the purchasing. Many times the inspection ends up in the hands of the design engineering firm's professional inspector. For many items this is adequate. However, usually the inspector may never have operated the equipment or even seen it in operation. The additional inspection by the process engineer of at least the major items can be important. It is much better to spot areas of deficiencies at the factory than in the half constructed plant.

Another area of concern at this point must be the spare parts listing and purchase. Usually the equipment manufacturer is requested to submit his recommended spare parts list. Normally the listing is quite good, but it should be checked out by qualified people using equipment detailed parts lists and prints. Personal judgment has the greatest impact in this selection. This is especially true under conditions of multiple units of operating equipment or in-place spares. In general, the old slogan of "better to have a few extra unused spares than be shut down from lack of spares" still holds.

Construction and Start-up

In most cases, by the time the plant is under construction it is too late for major design modifications. Many times it is less costly to build in a mistake and correct it after construction than it is to make the change during construction. The preferred policy is to do everything possible before construction to prevent this occurrence. Construction monitoring by the process engineer will, however, pick up small items that can be changed for plant improvement without excess engineering cost, as well as areas of sub-standard construction work.

During the construction period, operating and maintenance crews selection should be made and a training program initiated. Plant and maintenance superintendents, at least, should be on hand during the entire construction period. Assistants and foremen should be phased in as construction develops. Ideally, a full crew should be on hand for familiarization and training one month prior to start-up. Plant operational manuals can be helpful for crew training. They usually are best when written, or at least edited, by the process engineer or operating superintendent.

Personality conflicts can easily arise between operators and construction workers during this period. A close watch for this potential problem is very necessary. Conflicts are most likely to occur during the early stages of start-up when construction personnel are still engaged in their finishing work. Long working hours during this time tend to shorten tempers.

Summary

This is a brief description covering some of the problems and potential problem areas the authors have encountered in past construction projects. Each project presents its own new and particular problems that must be encountered. Many of them will appear as a sudden surprise to those involved. With good planning, sequential development of the project and constant monitoring, it should be possible to minimize those unforeseen occurrences that appear during construction and delay start-up as well as quick achievement of plant design tonnage, recovery and concentrate grade.